STRUCTURAL SEISMIC AND CIVIL ENGINEERING RESEARCH

Structural Seismic and Civil Engineering Research focuses on civil engineering research, anti-seismic technology and engineering structure. These proceedings gather the most cutting-edge research and achievements, aiming to provide scholars and engineers with preferable research directions and engineering solutions as reference. Subjects in these proceedings include:

- Engineering Structure
- Materials of Civil Engineering
- Structural Seismic Resistance
- Monitoring and Testing

The works in these proceedings aim to promote the development of civil engineering and earthquake engineering. Thereby, promoting scientific information interchange between scholars from top universities, research centers and high-tech enterprises working all around the world.

PROCEEDINGS OF THE 4TH INTERNATIONAL CONFERENCE ON STRUCTURAL AND SEISMIC AND CIVIL ENGINEERING RESEARCH (ICSSCER 2022), QINGDAO, CHINA, 21–23 OCTOBER 2022

Structural Seismic and Civil Engineering Research

Edited by

Ankit Garg
Department of Civil and Environmental Engineering, Shantou University, China

Bingxiang Yuan
School of Civil and Transportation Engineering, Guangdong University of Technology, China

Yu Zhang
College of Pipeline and Civil Engineering, China University of Petroleum, China

CRC Press
Taylor & Francis Group
Boca Raton London New York Leiden

CRC Press is an imprint of the
Taylor & Francis Group, an **informa** business

A BALKEMA BOOK

First published 2023
by CRC Press/Balkema
4 Park Square, Milton Park, Abingdon, Oxon, OX14 4RN
e-mail: enquiries@taylorandfrancis.com
www.routledge.com – www.taylorandfrancis.com

CRC Press/Balkema is an imprint of the Taylor & Francis Group, an informa business

© 2023 selection and editorial matter Ankit Garg, Bingxiang Yuan & Yu Zhang; individual chapters, the contributors

The right of Ankit Garg, Bingxiang Yuan & Yu Zhang to be identified as the authors of the editorial material, and of the authors for their individual chapters, has been asserted in accordance with sections 77 and 78 of the Copyright, Designs and Patents Act 1988.

All rights reserved. No part of this book may be reprinted or reproduced or utilised in any form or by any electronic, mechanical, or other means, now known or hereafter invented, including photocopying and recording, or in any information storage or retrieval system, without permission in writing from the publishers.

Although all care is taken to ensure integrity and the quality of this publication and the information herein, no responsibility is assumed by the publishers nor the author for any damage to the property or persons as a result of operation or use of this publication and/or the information contained herein.

Library of Congress Cataloging-in-Publication Data
A catalog record has been requested for this book

ISBN: 978-1-032-47040-5 (hbk)
ISBN: 978-1-032-47042-9 (pbk)
ISBN: 978-1-003-38434-2 (ebk)

DOI: 10.1201/9781003384342

Typeset in Times New Roman
by MPS Limited, Chennai, India

Table of contents

Preface	xiii
Committee Members	xv

Civil material analysis and seismic design

Research on TFM imaging for detecting internal defects in concrete using ultrasound array L.F. Rong & J.L. Hu	3
Experimental study of seismic performance of precast composite wall with different types of thermal insulation layers W. Huang, J.S. Li, Y.C. Guo & X.W. Miao	10
Study on strength characteristics and solidified mechanism of calcareous sand Z.M. Wang, J.J. Yang, Y.L. Wu & L. Wang	23
An analytical study based on 3D characteristics of finite element negative Poisson ratio composite structure mechanical properties X.M. Liu & X. Jiang	33
Seismic responses and damage analysis of pretensioned high-strength concrete pipe pile reinforced with non-prestressed steel bars H.Q. Hu, G. Gan, Y.J. Bao, X. Han & X.P. Guo	41
Experimental study on shear connection performance of high strength bolts with single side tapping Z.H. Zhang & G.X. Mao	47
Seismic analysis of continuous rigid frame bridge with double thin-wall piers based on SHDR K.Q. Wang & R.C. Ji	55
Study on seismic performance and fragility assessment of offshore jacket platform with fracture damage S. Zhang, H. Lin, A.M. Uzdin, Y.H. Guan, L. Yang, P.P. Han & H. Xu	64
Seismic fragility of continuous beam bridges based on two kinds of correlations X.Y. Gou, L. Yan, L.F. Cheng, S.R. Liu, T.S. Chen, X.Y. Wang, P. Zhang, W.J. Xie & H.Y. Chen	77
Strength level earthquake and ductility level earthquake seismic analysis of jack-up platform Y. Zhang & L.J. Qian	85
Shrinkage and creep effects for prestressed concrete beam bridge based on beam length shortening L. Cheng	90

Dynamic response analysis of metro station under bidirectional seismic load *Q. Wang, Y.C. Zhao, X.Z. Zhang & T. Shen*	99
Study on mechanical properties of foamed concrete with multiple factors *Y.W. Dai & H.Y. Xie*	105
Experimental study on mechanical performance of basalt-polypropylene hybrid fiber reinforce concrete *T.S. Li, L.H. Wang, C.F. Li, S.F. Sun, Z.G. Pang & Q.H. Shu*	111
Research on sulfate-freezing-thawing resistance of mineral admixture concrete in alpine-cold regions *L. Zhou*	120
Analysis of bearing characteristics on soft soil foundation reinforced with bamboo grid and flow-solidified silt *D.K. Zhao, S.L. Yue, J.H. Ding, H.C. Wang & Z.H. Yu*	129
Concrete proportion for assembled laminated slabs *J.X. Li, Q. Li, Y.Q. Qiu, Q. Wang & J.J. Wu*	137
Study on temperature monitoring and deformation characteristics of steam curing prestressed concrete box girder *F. Zeng, L. Zeng, Y.F. Sun, H.T. Tian & Y.X. Xu*	142
Analysis of the influence of springback rate of wet-sprayed concrete based on aggregate characteristic analysis *Y. Zhang & J. Zhang*	150
Thermal buckling analysis of submarine pipe-in-pipe systems with initial defect under seismic loading *H. Xu, H. Lin, H. Karampour, P.P. Han, S. Zhang, C. Han, H.C. Luan & L. Yang*	156
Study on the impact resistance of CFST column with an inner circular steel tube *Z.K. Jiao & S.P. Lei*	165
Application of concrete-filled steel tubular support in trackage roadway of Yangcheng mine *L. Qiu, K.-M. Liu, J.-W. Wang, G. Feng, C.-H. Zhang, Z.-Y. Zhou & S.-S. Liu*	171
Study on the effect of pre-treatment of recycled aggregate on the durability of concrete *B.B. Yan, X.N. Zhang, Z.Y. Wang & Y.C. Shi*	178
Research on RAP dispersion in recycled asphalt concrete with steel slag *P. Guo, F.Y. Liu, R.N. Pang, S.F. Fang & F. Shen*	186
Rational analysis of water intake and utilization in Pearl River basin irrigation area *G.L. Mu*	196
A comprehensive evaluation of water saving in Weifang, Shandong Province, China *H.J. Wang, S. Han, L.X. Lei, Z.H. Cai, X. Cong & Y.Y. Liu*	203

Comprehensive evaluation method of concrete beam bridge based on Bayesian network *W. Ji, S.J. Ge, M.J. Zhang, M. Ding & Z.L. Li*	209
Study on characteristics of soil and water loss based on different vegetation combinations in runoff plots *M.Q. Zhao & X.H. Wang*	216
Material selection and construction technology application of weak magnetic reference laboratory *H. Guang, H.C. Zhang, L. Fang, H.L. Li, W. Kong & H. Wang*	221
Temperature control measures for winter construction of super large bridge concrete on expressway *Y.H. Shen*	227

Engineering structure and building quality reinforcement

Validation of wind tunnel numerical simulation of super high-rise buildings *Y. Zhang, M.L. Wang, Z.H. Wen, F. Deng & T. Hu*	235
Experimental study on RC beams strengthened with prestressed and anchor steel plates *H.D. Lei, Y.H. Liu, A.J. Li & Z.X. Li*	243
The influence of the underground structure on surrounding soils under earthquake and its application *Z.-Y. Yu, J.-K. Zhang, H.-R. Zhang & Y.-J. Qiu*	253
Study on explosion resistance of ceramic/steel composite structure coated with polyurea *Q.-L. Liu, X. Jia, Z.-X. Huang, W. Xia, Y. Wang & T. Zhang*	264
Tunnel stability analysis of karst cave location distribution in karst area *X.Z. Zhang, Y.C. Zhao, Q. Wang & Z.Y. Dai*	274
Seismic analysis of structures with friction dampers *X.X. Wu, D.W. Zhou & Y. Li*	280
Seismic response analysis of shield tunnel considering the internal structure *Q.Z. Fu & X.W. Zhang*	287
Analysis of vibration effect of pile foundation parameters on superstructure under blasting in a tunnel *B.F. Duan, W.S. Xu, Z.W. Yu & Z.J. Sun*	294
Analysis of influencing factors of segment floating during the construction period of the shield tunnel *Y.L. Jiang, F.G. Lin, Y.C. Hu & X. Tang*	302
Technical analysis of lattice super-high pier in bridge construction *M. Guo & R.A. Jiang*	310
Research on adit rock mass mechanics test of Baihetan Hydropower Station *D. Liu & A. Liu*	317

Key construction technology of complex "Xumishan" spatial bending-torsion aluminum alloy structure X.Y. Sun, F. Yang & H. Zhao	323
The feasibility analysis of a new prefabricated steel structure system R. Xing, J.J. Zhang & H.G. Lei	330
Experimental method for verifying truck weight limits of highway bridges S.H. Li & D. Liu	337
Wind vibration response analysis of derrick steel structure based on time domain analysis method D.Y. Han, N. Liu, G.Q. Zhu, Y. Huang, X.J. Yang & L.M. Zheng	348
Study on overall and local mechanical properties of tower-beam longitudinal restraint of three tower cable-stayed bridge J.H. Tong, P. Wu & Y.Y. Zhu	355
Damage and permeability of surrounding rocks during blasting excavation of tunnels in karst areas W.H. Li, D.X. Qu, Q.Y. Zhang & B. Yang	362
Risk factors affecting tunnel collapse and treatment L.P. Yang & T. Wang	372
Collapse risk assessment of highway tunnel construction based on game theory and mutation progression method Y.W. Ren, B.X. Wang, R. Ma & H.Y. Wang	379
Study on construction technology optimization of post-cast strip in underground engineering Y.H. Shen	391
Study on the conception, static and dynamic performance of the ribbed flat grid L.C. Jiang, R.C. Nam & J.M. Nam	396
Study on the shape of flushing system of diversion pool in urban drainage network B.C. Liu, G.G. Li, Y.C. Han, Z.S. Chen & G.Q. Dou	408
Research on characteristics and quality defects of aluminum molds in high-rise buildings J.C. Xie	417
Application of green low-carbon prefabrication technology in specific buildings L. Xu, J. Yang, Z. Li, J.X. Wang, D. Li, C.H. Zhang, L.F. Mo, H.G. Xue, W. Yang, L.H. Peng & Y.W. Zeng	424
Construction engineering quality management based on BIM and Big Data S. Zong	432
Research on key technology of prefabricated building construction N.N. Liu, Y.J. Wang & T. Hu	439

Effect of clay on microstructure and mechanical properties of Guilin red clay *L.G. Wang, Q.Y. Shi, B. Yang, J.H. Chen & S.J. Li*	445
Comparative analysis of domestic and foreign tunnel maintenance codes *J.T. Yu, M. Pang, Z.X. Chen & M.K. Shi*	456

Intelligent monitoring and engineering technique optimization

Research on the optimization of joint dispatching between the quay crane and AGV in automated container terminal *J. Zhang*	465
Comprehensive evaluation method and application of foundation pit construction safety based on measured data *J.F. Liu, K.P. Wu, J.X. Luo, H.X. Zhou & Q.C. Qiu*	472
Fast building method of bridge OpenSees dynamic model based on Python language *G.P. Zhu & H.J. Lei*	480
Dynamic displacement calculation of constrained steel plate based on rigid plastic model under blast load *G.Q. Hu & T.C. Yang*	487
Analysis of measured data of upper span tunnel foundation pit engineering *T. Bao, X.B. Sun, J. Pang, L.C. Shou & L.F. Wang*	494
Analysis of the spatiotemporal pattern of the gathering and dispersal of evacuated people in large and medium-sized cities after strong earthquakes *W.Y. Shen, D.P. Li, J.F. Yin, Y.B. Jian, X.H. Miao & D. Yao*	501
Study on the regulation of pentamode lattice ring structure on impact stress wave *B.Y. Han & Z.H. Zhang*	512
Integrated launching technology of the beam-arch combination system *J. Li & H.J. Yu*	524
Failure probability assessment of Fujian earth buildings under multi-disaster coupling *S.B. Pan, Z.Y. Liang & Q. Liu*	530
Structural response prediction based on blind Kriging model *Z.X. Li, J.S. Liu, H.J. Gao, Y.W. Chen, C.Y. Yu, W.R. Zhang & Z. Yang*	537
Research on the digital twin intelligent building management platform in communication industry existing office buildings *W.W. Kou, S.H. Ye & S.G. Guo*	547
A simplified method for calculating fundamental frequency of concrete-filled double skin steel tubular structure for onshore wind turbine tower *S.-Z. Li, X.H. Zhou, Y.H. Wang, D. Gan, R.H. Zhu & L.-L. Ning*	557

Numerical study on GFRP-strengthened offshore T-joints under
earthquake cyclic loading 564
P.P. Han, H. Lin, A.M. Uzdin, L. Yang, H.C. Luan, C. Han,
H. Xu & S. Zhang

Harmonic response in deep sea truss spar platform 573
N. Liu, J.S. Liu & W. Liu

Design and performance simulation of hydraulic system of walking
launching equipment 579
M.X. Shi & Q. Hu

Analysis of Sag effect of auxiliary cable of cable-stayed bridge and
replacement of CFRP cable 586
F. Pan, P. Wu, Y.Y. Zhu & W.J. Sun

Numerical analysis on fully bolted beam-to-column joints in assembled
steel structure 593
H.S. Liang, W. Xie, X.X. Tang & X.Q. Luo

Estimation method of bulk material quantity of engineering cost based on
BP algorithm 600
S. Zong & B. Chen

Comparative study on prediction models of EPB shield tunnelling
parameters in the water-rich round-gravel formation 606
J. Wang, J.P. Zhao, Z.S. Tan & S.Y. Fu

Study on the current situation and solutions of aquatic and biological
channel connectivity in urban rivers – The Kunyu River and
Shuiya Ditch in Beijing as an example 614
L.J. Wang, G.N. Li, Y.C. Han, S.T. Liu, S.K. Sun & X.Y. Wang

Application of 3D design with multi-platform collaboratively in the
centralized control building of the Wudongde Hydropower Station 624
N. Wang, W. Wang, D.H. Chen & L.J. Li

Application of residual modified gray model in dam safety monitoring 631
Z.Q. Fu

Analysis and suggestions on the construction of a water-saving society in
Yunnan Province 640
Q.-J. Zhu & S. Wang

Study on the improvement of spatial effects of the old residential
life circle in the context of aging suitability 646
X.L. Xu, T. Liu, J. Liang, J.H. Huang & T.F. Wu

Numerical simulation of freeze-thaw damage of root-soil complex based on
discrete element method 655
Y. Sun & H. Li

Research on DEM construction in mountain areas based on airborne
LiDAR data 662
W. Chen

Simulation experimental study on aggregate filling and interception in mine water inrush channel *P.L. Su, C. Li & F. Liu*	667
A comprehensive dam safety monitoring information system for catchment/area-level hydropower station groups *G. Cui, Q. Ling & L. Zhang*	674
Human factor failure pathways in dam failures *D.D. Li & H.W. Wang*	681
Ecological park design and sustainability evaluation for metropolis based on emergy method *J.X. Zhang, L. Huang, Y. Zhang & D. Xu*	686
Study on the status and conservation of stone carvings in Lianyungang *Y.Y. Chen, D.H. Zhang & Y.Q. Wu*	692
Optimization design of aluminum alloy connections on PV roof against wind-uplift load *X.X. Tang, Y.F. Liu, H.S. Liang, D. Wu & X.Q. Luo*	698
Author index	707

Preface

The 2022 4th International Conference on Structural Seismic and Civil Engineering Research (ICSSCER 2022) was successfully held on October 21st–23rd, 2022, in Qingdao, China (virtual conference). With a display of the latest research outcomes in theoretical research, experimental study and engineering application, ICSSCER 2022 created a platform for scholars, researchers and technical personnel to exchange ideas and learn the latest research developments.

We had the honor of having invited Prof. Limin He from China University of Petroleum (East China), China to serve as our Conference Chair. The conference was composed of keynote speeches, oral presentations, and online Q&A discussion, bringing together over 150 delegates from all over the world. Firstly, keynote speakers were each allocated 30–45 minutes to hold their speeches. Then in the next part, oral presentations, the excellent papers we had selected were presented by their authors one by one.

During the conference, six distinguished professors were invited to address their keynote speeches. Among them, Prof. Zhiqiang Zhangi from Southeast University, China delivered a speech on High-performance Viscous Fluid Damper Series Development and Experimental Research. In this project, a series of high-performance viscous fluid dampers (VFD) were developed. Three studies were conducted respectively, the performance test study of high-performance VFD, the test and application study of a new type of variable coefficient VFD, and the seismic performance study of self-resetting viscous energy dissipation support-steel frame structure system. And then, Prof. Zehra Canan Girgin from Yildiz Techinal University, Turkiye made a report on the title Short and Long Term Flexural Characteristics of Basalt Fibre Cementitious Composites. In this study, the effects of pozzolanic ingredients such as ground granulated blast furnace slag (GGBS), nano-clay (NC), metakaolin (MK) were investigated in basalt fibre cementitious composites (BRC). And after the detailed experimental studies, it was revealed that especially MK usage promises to enhance the durability performance of BRC. Their brilliant keynote speeches had sparked heated discussion in the conference. And every participant praised this conference for disseminating useful and insightful knowledge.

The proceedings of ICSSCER 2022 are a compilation of the accepted papers and represent an interesting outcome of the conference. These papers feature but are not limited to the following topics: Engineering Structure, **Materials of Civil Engineering, Structural Seismic Resistance, Monitoring and Testing**, etc. All the papers have been checked through rigorous review and processes to meet the requirements of publication.

We would like to express our sincere gratitude to all the keynote speakers, peer reviewers, and all the participants who supported and contributed to ICSSCER 2022. Particularly, our special thanks go to the CRC Press, for the efforts of all its colleague in publishing this paper volume. We firmly believe that ICSSCER 2022 had turned out to be a forum for excellent discussions that enable new ideas to come about, promoting collaborative research.

The Committee of ICSSCER 2022

Committee Member

Conference Chair
Professor. Limin He, *China University of Petroleum (East China), China*

Academic Committee Chair
Professor. Dayong Li, *China University of Petroleum (East China), China*

Academic Committee Members
Professor. Bing Bai, *Beijing Jiaotong University, China*
Professor. Xiaohua Bao, *Shenzhen University, China*
Professor. Wengui Cao, *Hunan University, China*
Professor. Zhangnv Zeng, *Henan University of Technology, China*
Professor. Fusheng Zha, *Hefei University of Technology, China*
Professor. Jinjian Chen, *Shanghai Jiao Tong University, China*
Professor. Rui Chen, *Harbin Institute of Technology, Shenzhen, China*
Professor. Xiaoping Chen, *Jinan University, China*
Professor. Xingzhou Chen, *Xi'an University of Science and Technology, China*
Professor. Wenwu Chen, *Lanzhou University, China*
Professor. Xudong Cheng, *China University of Petroleum (East China), China*
Professor. Chunyi Cui, *Dalian Maritime University, China*
Professor. Yanjun Du, *Southeast University, China*
Professor. Tianhui Fan, *South China University of Technology, China*
Professor. Hongmei Gao, *Nanjing Tech University, China*
Professor. Weiming Gong, *Southeast University, China*
Professor. Chong Jiang, *Central South University, China*
Professor. Liang Kong, *Qingdao University of Technology, China*
Professor. Diyuan Li, *Central South University, China*
Professor. Jinhui Li, *Harbin Institute of Technology, Shenzhen, China*
Professor. Lianxiang Li, *Shandong University, China*
Professor. Xia Li, *Southeast University, China*
Professor. Yurun Li, *Hebei University of Technology, China*
Professor. Yuanhai Li, *China University of Mining and Technology (Xuzhou), China*
Professor. Fayun Liang, *Tongji University, China*
Professor. Shihua Liang, *Guangdong University of Technology, China*
Professor. Hang Lin, *Central South University, China*
Professor. Yuliang Lin, *Central South University, China*
Professor. Donghai Liu, *Tianjin University, China*
Professor. Gang Liu, *China University of Petroleum (East China), China*
Professor. Hai Liu, *Guangzhou University, China*
Professor. Junwei Liu, *Qingdao University of Technology, China*
Professor. Run Liu, *Tianjin University, China*
Professor. Xianshan Liu, *Chongqing University, China*
Professor. Yongjian Liu, *Guangdong University of Technology, China*
Professor. Mengmeng Lu, *China University of Mining and Technology (Xuzhou), China*
Professor. Xiaoming Luo, *China University of Petroleum (East China), China*
Professor. Jianbing Lv, *Guangdong University of Technology, China*
Professor. Guoxiong Mei, *Hohai University, China*
Professor. Huafu Pei, *Dalian University of Technology, China*

Professor. Xiusong Shi, *Hohai University, China*
Professor. Hongyan Shi, *Guangdong University of Technology, China*
Professor. Dong Su, *Shenzhen University, China*
Professor. Liang Tang, *Harbin Institute of Technology, China*
Professor. Liyuan Tong, *Southeast University, China*
Professor. Chenghua Wang, *Tianjin University, China*
Professor. Dongxing Wang, *Wuhan University, China*
Professor. Dong Wang, *Ocean University of China, China*
Professor. Wenbin Wu, *China University of Geosciences (Wuhan), China*
Professor. Yonghong Wu, *Kunming University of Science and Technology, China*
Professor. Yang Xiao, *Chongqing University, China*
Professor. Zhijun Xu, *Henan University of Technology, China*
Professor. Jiang Xu, *Chongqing University, China*
Professor. Zhenhao Xu, *Shandong University, China*
Professor. Yiguo Xue, *Shandong University, China*
Professor. Changbin Yan, *Zhengzhou University, China*
Professor. Rongtao Yan, *Guilin University of Technology, China*
Professorate Senior Engineer. Guanghua Yang, *Guangdong Research Institute of Water Resources and Hydropower, China*
Professor. Junjie Yang, *Ocean University of China, China*
Professor. Minghui Yang, *Hunan University, China*
Professor. Xueqiang Yang, *Guangdong University of Technology, China*
Professor. Zhongxuan Yang, *Zhejiang University, China*
Professor. Jiangtao Yi, *Chongqing University, China*
Professor. Rangang Yu, *China University of Petroleum (East China), China*
Professor. Bingxiang Yuan, *Guangdong University of Technology, China*
Professor. Jingke Zhang, *Lanzhou University, China*
Professor. Junwen Zhang, *China University of Mining and Technology (Beijing), China*
Professor. Liaojun Zhang, *Hohai University, China*
Professor. Mingyi Zhang, *Qingdao University of Technology, China*
Professor. Qianqing Zhang, *Shandong University, China*
Professor. Yongjun Zhang, *Qingdao University of Technology, China*
Professor. Wengang Zhang, *Chongqing University, China*
Professor. Gang Zheng, *Tianjin University, China*
Professor. Dequan Zhou, *Changsha University of Science & Technology, China*
Professor. Feng Zhou, *Nanjing Tech University, China*
Professor. Qizhi Zhu, *Hohai University, China*
Professor. Haiyang Zhuang, *Nanjing Tech University, China*
Professor. Yan Zhuang, *Southeast University, China*
Professor. Guoliang Dai, *Southeast University, China*
Professor. Xuanming Ding, *Chongqing University, China*
Associate Professor. Xiaoyu Bai, *Qingdao University of Technology, China*
Associate Professor. Yulei Bai, *Beijing University of Technology, China*
Associate Professor. Keping Chen, *Central South University, China*
Associate Professor. Beibing Dai, *Sun Yat-sen University, China*
Associate Professor. Youkou Dong, *China University of Geosciences (Wuhan), China*
Associate Professor. Qinwen Du, *Chang'an University, China*
Associate Researcher, Yumin Du, *Northwestern Polytechnical University, China*
Associate Professor. Hailei Kou, *Ocean University of China, China*
Associate Professor. Jiale Li, *Hebei University of Technology, China*
Associate Professor. Weichao Li, *Tongji University, China*
Associate Professor. Peiyuan Lin, *Sun Yat-sen University, China*
Associate Professor. Chao Liu, *Guangzhou University, China*

Associate Professor. Kaiwen Liu, *Southwest Jiaotong University, China*
Associate Researcher, Qingbing Liu, *China University of Geosciences (Wuhan), China*
Associate Professor. Congshuang Luo, *Henan University of Urban Construction, China*
Associate Professor. Linlong Mu, *Tongji University, China*
Associate Professor. Feifan Ren, *Tongji University, China*
Associate Professor. Xiang Ren, *Xi'an University of Science and Technology, China*
Associate Professor. Jiangwei Shi, *Hohai University, China*
Associate Professor. Lei Su, *Qingdao University of Technology, China*
Associate Professor. Liqiang Sun, *Tianjin University, China*
Associate Professor. Yixian Wang, *Hefei University of Technology, China*
Associate Professor. Lei Wang, *Anhui University of Science & Technology, China*
Associate Professor. Wanying Wang, *Guangdong University of Technology, China*
Associate Professor. Yonghong Wang, *Qingdao University of Technology, China*
Associate Professor. Yong Wen, *Zhongkai University of Agriculture and Engineering, China*
Associate Professor. Yang Wu, *Guangzhou University, China*
Associate Professor. Zhongnian Yang, *Qingdao University of Technology, China*
Associate Professor. Junhong Yuan, *Inner Mongolia University, China*
Doctor, Chunshun Zhang, *Monash University, China*
Associate Professor. Jianlong Zhang, *Guangdong University of Technology, China*
Associate Professor. Lijuan Zhang, *Guangdong University of Technology, China*
Associate Professor. Ling Zhang, *Hunan University, China*
Associate Professor. Sulei Zhang, *Qingdao University of Technology, China*
Associate Professor. Wei Zhang, *South China Agricultural University, China*
Associate Professor. Xiaoling Zhang, *Beijing University of Technology, China*
Associate Professor. Yaguo Zhang, *Chang'an University, China*
Associate Professor. Heng Zhao, *Hunan University, China*
Associate Professor. Haizuo Zhou, *Tianjin University, China*
Associate Professor. Xinjun Zou, *Hunan University, China*
Senior Engineer. Zhongwei Wang, *Shenzhen Modern City Architectural Design Co. Ltd., China*
Senior Engineer. Zongjie An, *Shenzhen Modern City Architectural Design Co. Ltd., China*
Researcher, Hang Zhou, *Chongqing University, China*
Doctor, Jun Bi, *Nanjing Tech University, China*
Lecturer. Binbin Yang, *Xuchang University, China*
Lecturer. Hao Zhang, *Shanghai Normal University, China*
Lecturer. Deluan Feng, *Guangdong University of Technology, China*
Lecturer. Xing Gong, *Guangdong University of Technology, China*
Lecturer. Mingjie Jiang, *Guangxi University, China*
Lecturer. Qingzi Luo, *Guangdong University of Technology, China*
Chief Engineer, Houbing Xing, *CSCEC4 Civil Engineering Co. Ltd., China*

Civil material analysis and seismic design

Research on TFM imaging for detecting internal defects in concrete using ultrasound array

Lifan Rong*
Collaborative Innovation Center for Performance and Security of Large-scale Infrastructure, Shijiazhuang Tiedao University, Shijiazhuang, China

Jinlin Hu*
School of Safety Engineering and Emergency Management, Shijiazhuang Tiedao University, Shijiazhuang, China

ABSTRACT: The total focusing method (TFM) based ultrasound array is considered the "gold standard" of the detection of internal defects and is the potential for high-precision imaging of internal void defects in concrete. However, the fact that the ultrasonic wave field of concrete structures is extremely complicated and brings about the reduction of internal void defects image quality. In order to research the influence of ultrasonic wave filed on the quality of TFM imaging, a three-dimensional finite element model was established for the concrete structure to simulate three types of ultrasonic wave, i.e. P-wave, SV-wave and SH-wave propagation in the concrete and interaction with the void defect. The ultrasonic images of concrete using the TFM method were obtained. And the influence of the ultrasonic wave field on the character of received signals and the TFM imaging for concrete is discussed. The result shows that the wave field excited by SH-wave is more regular and the echo wave of the SH-wave from the void detect in the received signals is more obvious than P-wave and SV-wave. The quality of the TFM image based on SH-wave is obviously higher than those based on P-wave and SV-wave, both of which exists artifact due to wave-type conversion. The normalized value in the TFM image is closely related to the void defect, and the regions surrounded by a higher-value normalized counter line have a larger probability consistent with the void defect.

1 INTRODUCTION

Ultrasonic testing is a common method for detecting internal defects in concrete structures, and the test results depend on the interpretation of test data. Traditional A-scan analysis methods (Cosmes-López 2017; Luo 2016; Xu 2017) used for defect detection can only qualitatively determine the defect existing in the concrete and cannot measure the position and size of the defects (Xu 2013; Zhu 2018). The Synthetic Aperture Focusing Technique (SAFT) is a mature industrial ultrasonic imaging method and has been widely used for defect detection in concrete (Freeseman 2016). However, the number of signals obtained during the SAFT ultrasonic test is small. The SAFT imaging quality is often suboptimal in fact, due to the non-homogeneity of concrete (Hao 2011).

Based on SAFT, a method taking advantage of more received signals to improve the imaging quality is developed and called as Total focusing method (TFM) (Weston 2012). It is considered the "gold standard" of the detection of internal defects. TFM has been used

---

*Corresponding Authors: rong-lifan@foxmail.com and chillworld@163.com

for the detection of the inner defect in concrete (Ge 2022; Wang 2021). However, the ultrasonic wave field of concrete during the tests is extremely complicated and has an obvious effect on the imaging quality.

In order to research the influence of ultrasonic wave filed on the quality of TFM imaging, a three-dimensional finite element model was established for the concrete structure to simulate three types of ultrasonic wave, i.e. P-wave, SV-wave and SH-wave propagation in the concrete and interaction with the void defect. The influence of the ultrasonic wave field on the character of received signals is discussed. The ultrasonic images of concrete using the TFM method were obtained and analyzed.

2 TOTAL FOCUSING METHODS FOR CONCRETE STRUCTURE

The principle of the Total Focusing method (TFM) is shown in Figure 1. In a set of ultrasonic transducer array, the transducers emit ultrasonic waves one by one, and all transducers receive ultrasonic waves. After the ultrasonic test, there are n×n ultrasonic waves received, which is considered full matrix capture (FMC).

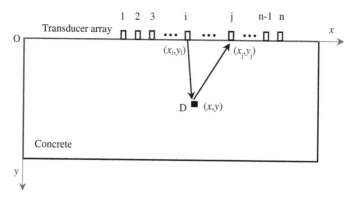

Figure 1. Principle diagram of TFM imaging.

It is assumed that the ultrasonic velocity of the concrete is c. The exciting transducer position coordinates are (x_i, y_i). The receiving transducer position coordinates are (x_j, y_j). And the position coordinate of any pixel D is (x, y). The time delay t_{ij} experienced by the ultrasonic wave is transmitted from the exciting transducer and then reaches the position of the receiving transducer via pixel D, which can be calculated by equation (1):

$$t_{ij}' = \frac{\sqrt{(x_i - x)^2 + (y_i - y)^2} + \sqrt{(x_j - x)^2 + (y_j - y)^2}}{c} \quad (1)$$

The value of any pixel can be obtained in the imaging area grid by equation (2):

$$I'(x,y) = \sum_{i=1}^{n} \sum_{j=1}^{n} S_{ij}(t_{ij}(x,y)) \quad (2)$$

Where $I'(x, y)$ represents the value of any pixel, S_{ij} represents the amplitude envelope of the received signal when the exciting and receiving transducer positions are (x_i, y_i) and (x_j, y_j).

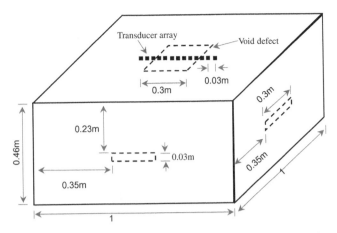

Figure 2. Three-dimensional finite element model of concrete with inner void defect.

3 FINITE ELEMENT SIMULATION

3.1 Finite element model

A three-dimensional finite element simulation model was established by using commercial finite element software Abaqus. As shown in Figure 2, both the length and width of the finite element model are 1 m, while the height of the finite element model is 0.46 m. A void defect of 0.3×0.3×0.03 m in size was placed in the model, which was simulated by the solid element with lower mechanical parameters. The material parameters are shown in Table 1.

Table 1. Material parameters of the concrete with inner void defect.

Property	Concrete	Void defect
Density (kg/m^3)	2400	1100
young' modulus (Gpa)	34.5	1×10^{-4}
Poisson's ratio	0.2	0.45

As shown in Figure 1, thirteen transducer positions were arranged on the concrete surface, with 0.03 m in distance between each other. A three-cycle sine wave modulated by the Hanning window with 50 kHz central frequency was used as the excitation signal in this paper. Meanwhile, the four side faces of the model were set as infinite boundaries to avoid wave reflection from the side faces of the model.

In order to analyse the influence of the wave types on the TFM imaging quality, three types of wave, such as P-wave, SV-wave and SH-wave, were adopted as the excitation sources, respectively. P-wave was excited by impact load, while S-wave was excited by dynamic shear loads.

3.2 Analysis of finite element simulation results

Figure 3 shows the ultrasonic wave propagation and interaction with void defect at each moment in the finite element simulation process when the 7th transducer is employed as the excitation source. When three groups of cloud images from different times are compared, the SH-wave is scattered and reflected the most regularly at the void defect.

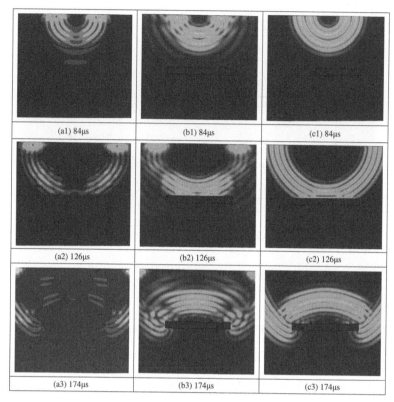

Figure 3. Cloud pictures of wavefield (a) P-wave (b) SV-wave (c) SH-wave.

Figure 4 shows the signals received from the 1st transducer when the P-wave, SV-wave and SH-wave were generated at the 7th transducer position respectively. The time-domain data of three signals had been normalized. As shown in Figure 4, the first arriving wave of the receiving signals, called the direct wave, is a surface wave that contains larger energy. And the scattered wave from the void defect is behind the direct wave, with small energy. Especially, the amplitude of the scattered wave for the P-wave is so small that hardly recognized, while the amplitude of the scattered wave for the SH-wave is almost equal to that of the direct wave.

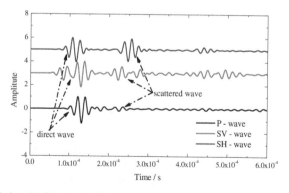

Figure 4. Received signals of P-wave, SV-wave and SH-wave.

From Figure 4, it can be seen that the direct wave and scattered wave are separated from each other in the time domain. Thereby, a window function method is used to suppress the surface wave in receiving signals to eliminate its effect on the TFM image quality. The window function is expressed as equation (3):

$$D_{ij}'(t) = \begin{cases} \lambda D_{ij}(t), & t_1 \leq t \leq t_2 \\ D_{ij}(t), & t < t_1, t > t_2 \end{cases} \quad (3)$$

In the equation, $D_{ij}'(t)$ is the signal after suppressing the surface wave, $D_{ij}(t)$ is the original signal, and λ is the suppressing coefficient, which is 0.01 in this paper. t1 is the surface wave delay and t2 is the time domain width of the excitation signal.

Figure 5(a) is a set of signals received from all 13 transducers when the SH-wave is excited at the 1st transducer position. It is shown that the direct wave propagates forward steadily, and the scattered wave is unclear in signals received from 1st-4th transducers due to low amplitude. After suppressing the surface wave by windowing the time domain, the processed signals are shown in Figure 5(b). The surface wave is almost removed, while the normalized amplitude of the scattered wave increases significantly.

There are 169 sets of received signals for each simulation with different excitation conditions P-wave, SV-wave and SH-wave. And the collected 169 sets of received signals are ultrasonically imaged using the total focus method. Figure 6 shows the three TFM images for the middle section of the concrete structure based on P-wave, SV-wave and SH-wave, respectively. And TFM image region width reduces to 0.5 m for ease of analysis, which also contains all the void defect middle profile. In Figure 6, the black rectangle represents the void defect, the red color represents more scattered energy, and the blue color represents low scattered energy.

As shown in Figure 6, the color of the void defect region is red and yellow, and the normalized value of the void defect region is approximately greater than 0.7, which is significantly larger than most other regions. However, in Figure 6(a) and (b) some other regions also have high normalized values, while in Figure 6(c) all the other regions have significantly low normalized values. Obviously, the TFM image based on SH-wave indicates the void defect region more accurately than the other two images.

According to the theory of stress wave, the excitation wave arriving at the interface between the concrete and void defect is scattered and then received by the transduce array. Hence, the normalized value of the void defect region is great in the TFM image. But in Figure 6(a) and (b), some other regions having great normalized values do not scatter waves in fact, so it is obvious that the images in these regions are artifacts.

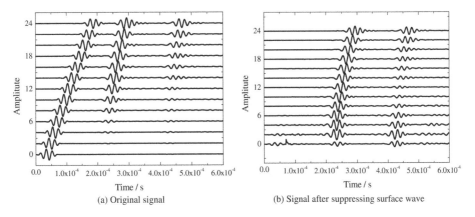

Figure 5. Signal before and after suppressing surface wave.

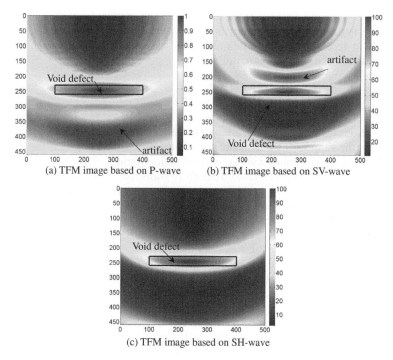

Figure 6. TFM images based on three wave types.

4 DISCUSSION

Wave-type conversion occurs when the P-wave or SV-wave arrives at the interface between concrete and void defect. A part of the P-wave is converted to SV-wave, called P-SV wave, while the other part of the P-wave is scattered as P-wave. Also, when the excitation source is SV-wave, a part of it is converted to P-wave, called SV-P wave, while the other part is scattered as SV-wave. The converted wave (P-SV or SV-P) contained in the received signals is superimposed at the region without void defect due to the difference in arrival time of P-wave and SV-wave and results in an artifact in the TFM images based on P-wave or SV-wave. It is worth noting that wave-type conversion does not occur for SH-wave, so there is no artifact existing in the TFM image based on SH-wave.

The normalized value in the TFM image is closely related to the void defect. As shown in Figure 7, the regions having higher normalized values are more close to the void defect.

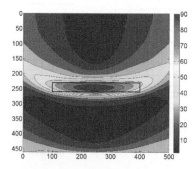

Figure 7. Contour map for TFM image based on SH wave.

The counter map shows that the region with a normalized value greater than 0.8 is contained in the void defect region, while the majority of the region with a normalized value greater than 0.6 coincides with the void defect region. Hence, the region surrounded by the normalized counter line of 0.8 can be certainly considered as the void defect, and the region surrounded by the normalized counter line of 0.6 can be considered as the void defect with a large probability in this paper.

5 CONCLUSIONS

In order to research the ultrasonic TFM imaging problem of detecting a void defect in concrete structure, a three-dimensional finite element model was established to simulate three types of ultrasonic wave, i.e. P-wave, SV-wave and SH-wave propagation in the concrete and interaction with the void defect. And ultrasonic images of concrete using the TFM method were obtained and analyzed. The specific results are as follows:

(1) The SH-wave is scattered and reflected more regularly from void defect than P-wave and SV-wave, and the echo wave of the SH-wave from the void detect in the received signals is more obvious than P-wave and SV-wave.
(2) The quality of the TFM image based on SH-wave is obviously higher than those based on P-wave and SV-wave, both of which exists artifact due to wave-type conversion.
(3) The TFM image's normalized value is closely related to the void defect, and regions surrounded by a higher-value normalized counter line are more likely to contain the void defect.

ACKNOWLEDGEMENTS

The authors thank the National Key Research and Development Program of China (2021 YFB2601000) and the National Natural Science Foundation of China (U2034207). The authors also thank the Science and Technology Project of Hebei Education Department (QN2020411).

REFERENCES

Cosmes-López M F et al. (2017). Ultrasound Frequency Analysis for Identification of Aggregates and Cement Paste in Concrete. *J. Ultrasonics*. 73: 88–95.
Freeseman K et al. (2016). Nondestructive Monitoring of Subsurface Damage Progression in Concrete Columns Damaged by Earthquake Loading. *J. Engineering Structures*. 114: 148–157.
Ge Lulu et al. (2022). High Resolution Ultrasonic Imaging Method for the Reinforced Concrete Based on Total Focusing. *Chinese Journal of Sensors and Actuators*. 35(3): 361–366.
Hao Z X et al. (2011). Design and Application of a Small Size SAFT Imaging System for Concrete Structure. *J. The Review of Scientific Instruments*. 82(7): 073708.
Luo M et al. (2016). Concrete Infill Monitoring in Concrete-filled FRP Tubes Using a PZT-based Ultrasonic Time-of-flight Method. *J. Sensors*. 16(12): 2083.
Wang Guan et al. (2021). A TFM Based Ultrasonic Array Detection and Imaging Method for Concretes. *J. Technical Acoustics*. 40(4): 482–489.
Weston M et al. (2012). Calibration of Ultrasonic Techniques Using Full Matrix Capture Data for Industrial Inspection. *J. Insight – Non-Destructive Testing and Condition Monitoring*. 54(11): 602–611.
Xu B et al. (2013). Active Debonding Detection for Large Rectangular CFSTs Based on Wavelet Packet Energy Spectrum with Piezoceramics. *Journal of Structural Engineering*. 139(9): 1435–1443.
Xu B et al. (2017). Numerical Study on the Mechanism of Active Interfacial Debonding Detection for Rectangular CFSTs Based on Wavelet Packet Analysis with Piezoceramics. *J. Mechanical Systems and Signal Processing*. 86: 108–121.
Zhu W et al. (2018). A Feasibility Study on Fatigue Damage Evaluation Using Nonlinear Lamb Waves With Group-velocity Mismatching. *J. Ultrasonics*. 90: 18–22.

Experimental study of seismic performance of precast composite wall with different types of thermal insulation layers

Wei Huang & Jiashen Li
Xi'an University of Architecture and Technology, Xi'an, Shaanxi, China

Yongchao Guo
Shaanxi Provincial Natural Gas Co. Ltd, Xi'an, Shaanxi, China

Xinwei Miao
Xi'an University of Architecture and Technology, Xi'an, Shaanxi, China

ABSTRACT: In this study, three kinds of thermal insulation layers (polystyrene particles insulation mortar, XPS, and foam cement) were proposed and applied to the conventional precast composite wall. To obtain the mechanical properties such as the failure process of the precast composite wall with different thermal insulation layers, three 1/2-scaled precast composite wall specimens with different thermal insulation layers were manufactured and then tested under a pseudo-static experiment; meanwhile, the same scaled specimen without thermal insulation layer was made for comparison. The results and analysis showed that the seismic performance of precast composite walls is influenced by the thermal insulation layer structure and thermal insulation material. Four specimens have similar experiment phenomena and show the bending-shear failure mode finally. The trend of the stiffness degradation curve of the four specimens is similar, and the three thermal insulation layers improve the initial stiffness and the energy consumption capacity of the precast composite wall obviously.

1 GENERAL INSTRUCTIONS

With the rapid development in society and economy, the development of precast fabricated concrete structures for residential buildings has been a hot topic in the construction industry across China. Meanwhile, the requirement for quantity and quality of buildings is higher. Thermal insulation performance is one of the focuses of residential buildings. According to statistics, walls account for 60%–70% of building envelopes in terms of heat transfer loss (Yan 2016). It is becoming increasingly important to look for a new type of wall structure that has better thermal insulation and load-bearing capacity while being relatively simple in technical construction.

In recent years, scholars and institutions around the world have carried out comprehensive research on different kinds of walls with thermal insulation layers. Shi et al. (2019) designed a new self-insulating structural wall with an inner skeleton. and investigated the seismic performance of the wall under a horizontal low-cyclic load test. The experiment results showed that this kind of wall had characteristics of seismic behavior, construction convenience and economic performance. Salmon et al. (1995) studied the mechanical and thermal performance of composite insulated concrete sandwich panels. Hao et al. (2015) proposed a new integrated structure of composite wall with inner thermal insulation. They studied the load-bearing capacity and the seismic performance of the wall through a comparative trial of low-frequency cyclic loading of the composite wall with a middle insulation layer and ordinary concrete wall under vertical load with small eccentricity. The results showed that the composite wall with middle insulation had a higher load-bearing capacity and better seismic capacity than the ordinary concrete wall. Boscato et al. (2018) studied the mechanical performance of a prefabricated composite wall made of a reinforced concrete slab and a glulam frame under a load-displacement test and evaluated the thermal performance of the wall by means of a hot-box apparatus. The results showed that the wall showed a good thermal performance, low environmental impact with respect to similar construction systems and promising structural behavior.

In China, the most common wall thermal insulation structure is exterior wall thermal insulation, which requires the construction of a thermal insulation layer after the completion of the building construction. A secondary construction is required and is technically complicated during the construction process. Meanwhile, residential industrialization is vigorously developing in China to improve the quality of residential buildings and to reduce environmental pollution, where prefabricated concrete structures play an important part. The combination of prefabricated concrete structure technology and energy conservation technology is a trend in the development of construction industrialization in China, which has a very good perspective of application (Wang 2019; Ye 2016).

The precast composite wall structure system, consisting of precast composite walls, cast-in-situ concrete restraint parts, assembled composite beams and composite slabs, has characteristics of rapid construction, energy dissipation, energy conservation and pollution reduction, and thermal insulation. Besides, due to its special material and structure, its bearing system has multiple anti-seismic lines (Figure 1) (Yang 2015). Based on the analysis in the foregoing introduction and to combine the precast fabricated concrete structure with thermal insulation technology, three kinds of thermal insulation layers were applied to conventional precast composite walls. Their seismic performance was obtained from a pseudo-static experiment in this study.

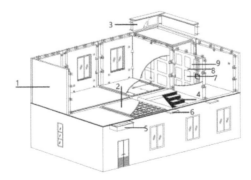

Figure 1. Precast composite wall structure system. 1-Precast composite wall; 2-Composite slab; 3-Prefabricated parapet; 4-Prefabricated stair; 5-Prefabricated canopy; 6-Prefabricated air-conditioning slab; 7-Aerated concrete block; 8-Rib beam; 9-Rib column.

2 EXPERIMENTAL PROGRAM

An experiment was conducted in the Lab of Structural Engineering and Earthquake Resistance at Xi'an University of Architecture and Technology to study the seismic performance of precast composite walls with different kinds of thermal insulation layers. In this study, three precast composite walls were designed with polystyrene particles insulation mortar thermal insulation layer, XPS thermal insulation layer, and foam cement thermal insulation layer, referred to as PCW2, PCW3, and PCW4, respectively. One conventional precast composite wall without a thermal insulation layer was made for comparison and numbered PCW1.

2.1 Details of specimens

According to the technical specification for precast composite wall structure, four 1/2-scaled precast composite walls were manufactured. The specimen consisted of a precast composite wall, foundation beam, and cast-in-situ boundary element (Huang 2007; Yao 2004). The horizontal connection between the precast composite wall and the cast-in-situ boundary element was a dovetail joint connection. The vertical connection between the precast composite wall and the foundation beam was slurry sitting; meanwhile, a weld plate was embedded in the foundation beam to connect the vertical steel rebars. Dimensions, steel rebars, and connection details of the precast composite wall, cast-in-situ boundary element, and foundation beam are illustrated in Figure 2. Based on PCW1, three different thermal insulation layers were applied, whose construction is given in Figures 3 and 4.

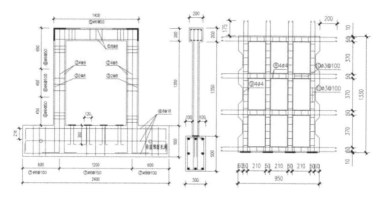

Figure 2. Dimensions, steel rebars, and connection details of the specimens. (a) Cast-in-situ boundary element and foundation beam. (b) Precast composite wall.

1–100 mm precast composite wall; 2–10 mm polymer binding mortar; 3–40 mm polystyrene particles insulation mortar or foam cement; 4-alkali-resistant mesh fabric.

1–100 mm precast composite wall; 2–10 mm polymer binding mortar; 3–40 mm XPS thermal insulation panel; 4-galvanized steel wire mesh; 5–10 mm fine aggregate concrete cover; 6-Thermomass MS/MC connecter.

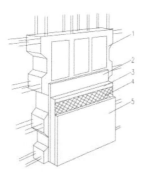

Figure 3. Thermal insulation layer construction of PCW2 and PCW4.

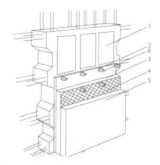

Figure 4. Thermal insulation layer construction of PCW3.

2.2 Material properties

Materials were tested to acquire accurate mechanical properties before the experiment. The concrete grade was designed as C30, and the actual mean concrete cubic compression strength of the precast composite wall and cast-in-situ boundary elements were 43.5 MPa and 42.1 MPa, respectively, after 28 days of standard curing. The mean compressive strength of the aerated concrete block was 4.1 MPa. The stirrup reinforcement was HPB300, while the longitudinal steel rebar was HRB400. The mechanical properties of steel rebar, polymer binding mortar, and thermal insulation materials are shown in Tables 1–3, respectively.

Table 1. Material propriety of steel rebar.

Type	D(mm)	f_y(MPa)	f_u(MPa)	Elasticity moduli (MPa)
HPB300	6	205	605	2.1×10^5
HRB400	6	376	665	2.3×10^5
HRB400	8	440	590	2.3×10^5

d, f_y, and f_u represent the diameter, yield strength, and ultimate strength of steel rebar, respectively.

Table 2. Material propriety of steel rebar.

Type	Compression strength (Mpa)	Bending strength (Mpa)	Elasticity moduli/ (Mpa)	Poisson ratio
Polymer binding mortar	31.9	8.7	1336	0.16

Table 3. Material propriety of thermal insulation materials.

Type	Dry bulk density (KN/m³)	Compression strength (MPa)	Elasticity moduli/ (MPa)	Poisson ratio
Polystyrene particles insulation mortar	305	0.78	324	0.12
XPS board	25	0.46	237	–
Foam cement	280	0.46	467	0.13

2.3 Experiment setup and loading method

The vertical force calculated by axial compression ratio (0.3) was 350 kN according to a shear wall on the first floor of an eight-storied residential building. The vertical force was generated by a 2000 kN hydraulic jack above a steel beam and then distributed to the specimen by two steel beams. After the specimen was stable vertically, the horizontal load was delivered to the top of the wall by the reaction wall and the MTS actuator. The experiment setup is illustrated in Figure 5.

First, a reversed cyclic quasi-static lateral loading was applied in force control. In this stage, each magnitude was 10 kN and cycled only once. After yielding, the method of loading changed into displacement control. The displacement control was changed in 2 mm increments, and the horizontal displacement of each stage was applied three times.

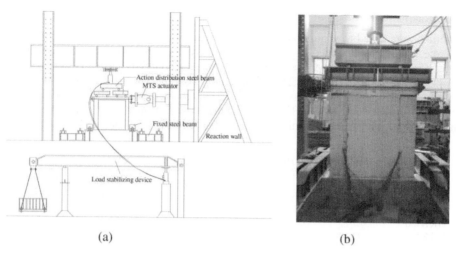

(a) (b)

Figure 5. Test setup.

The loading stage should stop when the horizontal load-bearing capacity of the specimen drops to 85% of its maximum carrying capacity or at the failure stage of the specimen.

3 EXPERIMENTAL RESULTS AND ANALYSIS

3.1 *Experimental phenomenon and failure mode*

The failure process of the four specimens could be classified into three stages according to the experiment: elastic stage, elastic-plastic stage, and failure stage. The pictures taken at the failure stage of the four specimens are shown in Figure 6; no cracks were visible when the vertical load was applied.

(a)-1 the north (a)-2 the south

(b)-1 the north (b)-2 the south

Figure 6. Failure phenomenon and mode of specimens. (a) PCW1; (b) PCW2; (c) PCW3; (d) PCW4.

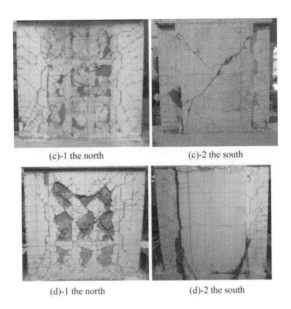

(c)-1 the north   (c)-2 the south

(d)-1 the north   (d)-2 the south

Figure 6. (Continued)

PCW1

During the elastic stage, some vertical micro-cracks occurred in the middle aerated concrete blocks first, then the original cracks expanded and developed into diagonal cracks with the increase of horizontal load. When it entered the elastic-plastic stage, diagonal cracks extended to rib beams and rib columns. Some longer diagonal cracks and horizontal cracks appeared at the bottom of the wall and at the bottom of the vertical cast-in-situ boundary elements. When it reached the failure stage, with the load increased, mounts of horizontal cracks appeared in the vertical cast-in-situ boundary elements and then penetrated through the wall gradually. Vertical micro-cracks appeared in the bottom of the cast-in-situ boundary column, original diagonal cracks expanded and then penetrated through rib beams and rib columns, and the most aerated concrete block fell off. Finally, the concrete of the cast-in-situ boundary column foot on the compression side was crushed; the longitudinal steel bars were bent under stress; the vertical cracks in the dovetail joint penetrated; the visible 'X' shape crossing cracks formed in each grid; the specimen was unstable, out-of-plane, and unable to resist load; and the experiment ended. The failure modes of the precast composite wall and the cast-in-situ boundary column were, respectively, the shear failure and bending failure.

PCW2

The failure process and mode of PCW2 were similar to PCW1. The obvious differences are summarized as follows:

Some micro-cracks occurred in the bottom aerated concrete blocks rather than in the middle part at the beginning of the horizontal load. With the increase in load, some horizontal cracks occurred in the bottom of the tension-side boundary element. When it entered the elastic-plastic stage, many micro-cracks appeared in the thermal insulation layer, With the increase in horizontal load, the micro-cracks developed into long diagonal cracks, finally forming crossing cracks. When it reached the failure stage, a gap appeared between the thermal insulation layer and the precast composite wall, the width of which reached 5 mm. Cracks in the thermal insulation layer even reached 6 mm. Finally, the longitudinal steel bars were broken, the thermal insulation layer had a tendency to wrap and fall, and the alkali-resistant mesh fabric in the diagonal cracks had a total tear.

PCW3

The failure process and mode of PCW3 were similar to PCW1 as well. The obvious differences are summarized as follows:

During the elastic stage, two transverse cracks occurred in the bottom of the precast composite wall on the tension side, and a few inclined cracks extended to rib beams. Besides, some vertical cracks appeared in the joint seam of the dovetail joint. During the elastic-plastic stage, a gap appeared between the XPS thermal insulation board and the precast composite wall. With the increase in load, a gap occurred between the XPS board and also on the fine aggregate concrete cover. When it reached the failure stage, visible crossing cracks formed in the XPS thermal insulation layer, and the maximum crack width reached 5 mm. The slurry sitting connection between the precast composite wall and the foundation beam was broken.

PCW4

The failure process and mode of PCW4 were similar to PCW1, too. The obvious differences are summarized as follows:

Some inclined micro-cracks occurred in the aerated concrete blocks at the beginning of the horizontal load. With the increase in load, some vertical cracks appeared in the joint seam of the dovetail joint. When it entered the elastic-plastic stage, two long horizontal cracks appeared in the upper and middle parts of the thermal insulation layer. When it reached the failure stage, the width of the gap between the thermal insulation and the precast composite wall reached 6 mm. Finally, the longitudinal steel bars were broken, and only a few micro-cracks appeared in the thermal insulation layer, but many cement mortars fell off at the bottom.

3.2 *Hysteretic curve*

Figure 7 shows the hysteresis curves of the four specimens obtained by the test. The hysteretic behavior of the four specimens is summarized as follows:.

In the initial stage, the hysteresis curves were nearly linear. With the constant increase in load, the hysteresis loops shaped like spindles, the area within the hysteresis loop increased, and the plastic deformation of specimens accumulated gradually with the accompanied pinch phenomenon. During the elastic-plastic stage, the hysteresis loops turned into a bow shape. With the increase of the controlled displacement, hysteretic curves of the four specimens turned the bow-shape into an s-shaped with the accompanied pinch phenomenon in varying degrees. The curves of PCW2 and PCW3 clearly reflected the response of the anchorage slip and resisting shear stresses.

After peak load, the bearing capacity of the four specimens reduced rapidly and then tended to stabilize; meanwhile, the displacement increased significantly, indicating that the four specimens showed good ductility. During the displacement loading stage, the stiffness of the four specimens degraded rapidly with more obvious pinch phenomenon and sliding phenomenon companies due to the fall of blocks and concrete and the yield of steel bars.

PCW4, PCW1, PCW2, and PCW3 had the greatest degree of fullness of hysteresis curves from maximum to minimum. The results showed that the foam cement thermal insulation layer improved the energy consumption capacity, shear-bearing capacity, and delayed cracking of the structure, compared with the conventional precast composite wall. The energy dissipation capability of the precast composite wall with XPS thermal insulation layer or polystyrene particles insulation mortar thermal insulation layer both decreased.

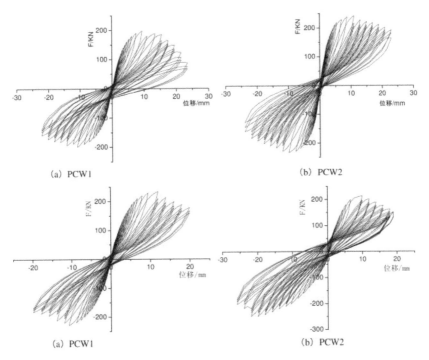

Figure 7. Hysteresis curves of specimens.

3.3 Skeleton curve and characteristics

The cracking load, yield load, peak load, and ultimate load of specimens are expressed as f_{cr}, f_y, f_p and f_u, respectively. The corresponding displacements are cracking displacement, yield displacement, peak displacement, and failure displacement and are expressed as Δ_{cr}, Δ_y, Δ_p, and Δ_u, respectively.

The defined characteristic points are as follows:

When some long cracks occurred in the aerated concrete blocks and inflection points appeared in the skeleton curve, the corresponding load was defined as the cracking load. The yield load is calculated according to the energy equivalence method. The average of the sum of absolute values of maximum pull force and maximum push force is defined as bearing capacity. When the bearing capacity decreases to 85% of the peak load, it is defined as the ultimate point.

The skeleton curves are shown in Figure 8, and the characteristic points of the four specimens are shown in Table 4.

Through comparative analysis, the conclusions that could be drawn are:

The skeleton curves of the four specimens are basically similar in shape. Before the load reached about 50% of the peak load, the curves were proximately linear. Later, with the increase in load, the curves were no longer linear, and there were obvious deviations between each curve. The specimens yielded when it reached about 75% of the peak load.

After peak load, compared with PCW1, the slopes of the descent part of the skeleton curves of PCW2~PCW4 were smaller, and the decrease of bearing capacities of PCW2~PCW4 was less, showing that the precast composite wall with thermal insulation layer had advantages over the conventional precast composite wall in deformation capacity.

Compared with PCW1, the crack load, yield load, peak load, and ultimate load of PCW2 increased by 48.05%, 21.52%, 17.13%, and 17.11%, respectively. The corresponding

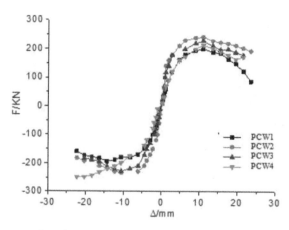

Figure 8. Skeleton curve of specimens.

Table 4. Loads and displacements of the specimens at characteristic points.

Specimen	Cracking point		Yield point		Peak point		Ultimate point	
	P_{cr}(kN)	Δ_{cr}(mm)	P_y(kN)	Δ_y(mm)	P_m(kN)	Δ_m(mm)	P_u(kN)	Δ_u(mm)
PCW1	74.50	1.89	160.42	4.68	199.52	10.99	169.59	17
PCW2	110.3	1.65	194.95	4.13	233.7	11.3	198.6	19
PCW3	85.7	1.34	178.51	3.45	236.1	9.35	200.7	15
PCW4	96.6	2.10	172.35	3.0	235.66	9.5	200.3	17.5

load of PCW3 increased by 15.03%, 11.28%, 18.33%, and 18.34%, respectively, whereas the corresponding load of PCW4 increased by 29.66%, 7.44%, 18.11%, and 18.11%, respectively. Experiment results showed that the thermal insulation layers proposed in this study could delay the cracking of precast composite walls.

3.4 Displacement ductility coefficient and relative deformation value

The displacement ductility coefficient reflects the component's ability to resist plastic deformation without collapse under a strong earthquake and is defined as the ratio between the ultimate displacement and the yield displacement.

To evaluate the deformation capability of specimens, a ratio of the horizontal displacement at the top of the specimen to the height of the specimen is defined, which can be expressed as $u = \Delta/H$ and named relative deformation value. When the value u approaches 1/60, the specimen is unstable and out-of-plane, and the experiment ends, although the specimen does not reach the failure stage.

The displacement ductility coefficients and relative deformation values of each specimen are shown in Table 5. It could be seen that the displacement ductility coefficients of the four specimens, from maximum to minimum, were PCW2, PCW4, PCW1, and PCW3. The ultimate relative deformation value of each specimen was bigger than 1/71, which was larger than the limit value of the elastic-plastic displacement angle between the layers given by *Technical Specification for Precast Composite Wall Structure*. The results indicated that the thermal insulation structure form had a great effect on the ductility of precast composite walls. All four specimens had well elastic-plastic deformation capacity and collapse-resistant capacity.

Table 5. Displacement ductility coefficient and relative deformation value of the specimens.

Specimen		PCW1	PCW2	PCW3	PCW4
Displacement ductility coefficient		3.63	4.60	3.12	3.78
Relative deformation value	Δ_{cr}/H	1/715	1/906	1/675	1/643
	Δ_y/H	1/288	1/327	1/391	1/450
	Δ_u/H	1/123	1/120	1/144	1/142
	Δ_{max}	1/79	1/71	1/90	1/77

3.5 *Stiffness degradation*

Stiffness is an essential property of precast composite wall structures. In the low-cycle reversed loading test, stiffness will decrease constantly. Secant stiffness is adopted to estimate the stiffness degeneration (Dang 2014) and calculated according to the formula:

$$K_i = \frac{|F_i| + |-F_i|}{|\Delta_i| + |-\Delta_i|} \qquad (1)$$

in which F_i and Δ_i stand for the peak load and corresponding displacement of each cyclic stage. To draw the stiffness degradation curves K_i is taken as the X-axis and Δ_i is taken as the Y-axis. The stiffness degeneration curves are illustrated in Figure 8.

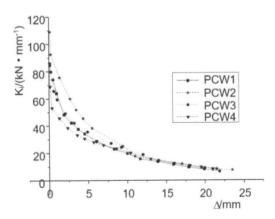

Figure 9. Stiffness degradation curves of specimens.

Some conclusions could be obtained from Figure 9:

The trend of stiffness degradation curves of the four specimens was similar. In the preliminary stage, stiffness degradation was rapid and then tended to be slow.

Compared with PCW1, the initial stiffness of PCW2~PCW4 increased by 106%, 60%, and 30%, respectively. It indicates that the initial stiffness of the precast composite wall was influenced by the thermal insulation layer structure and thermal insulation materials.

3.6 *Energy dissipation*

In a cyclic loading process, the energy dissipation due to inelastic deformation is generally represented by the area surrounded by the load-deformation hysteresis curve (Huang 2018). To evaluate the energy consumption capacity of the precast composite wall specimens, the equivalent viscous damping coefficient h_e is defined and calculated by the formula (Yao 2001).

$$h_e = \frac{1}{2\pi} \cdot \frac{S_{\widehat{ABC}+\widehat{AEC}}}{S_{\Delta OBD} + S_{\Delta OEF}} \quad (2)$$

Where $S_{\widehat{ABC}+\widehat{AEC}}$ is the area within the hysteresis loop; $S_{\Delta OBD}$ and $S_{\Delta OEF}$ stand for the area of triangle OBE and ODF, respectively. Figure 10 shows the calculation parameters in Equation (2).

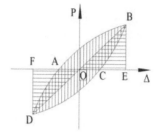

Figure 10. Calculation of energy dissipation coefficient.

The equivalent viscous damping coefficients of PCW1~PCW4 at four states were calculated and illustrated in Table 6, from which some conclusions could be obtained and summarized as follows:

Table 6. Equivalent viscous damping coefficient of specimens.

Specimen	PCW1	PCW2`	PCW3	PCW4
Creaking load	3.12	2.88	2.92	2.98
Yield load	4.38	4.41	4.13	4.08
Peak load	6.93	7.76	7.58	7.82
Ultimate load	7.38	5.97	4.79	6.12

At the cracking stage, the equivalent viscous damping coefficient of PCW1 was slightly greater than PCW2~PCW4 as the relationship between load and displacement was linear, and the area within the hysteresis loop was very small when the four specimens worked in the elastic stage. From the cracking state to the peak state, the equivalent viscous damping coefficients of the four specimens showed a steady upward tendency on different degrees, indicating cumulative energy dissipation. From the peak state to the ultimate state, the equivalent viscous damping coefficients of PCW1 continued to increase. However, by contrast, PCW2~PCW4 decreased; that is, the energy dissipation capacity showed a descent tendency after the thermal insulation layer was broken.

When PCW1~PCW4 reached the yield state, energy accounted for 63.2%, 16.8%, 54.5%, and 52.2% of energy dissipation at peak state. After specimens were yielded, with the damage of the precast composite wall accumulated and the cracks of the thermal insulation layer expanded and developed, the equivalent viscous damping coefficients of PCW2~PCW4 improved obviously compared to PCW1, showing that precast composite wall with a thermal insulation layer has better energy consumption capacity, especially the precast composite wall with foam cement thermal insulation layer.

4 CONCLUSIONS

A pseudo-static experiment was conducted to study the seismic performance of the precast composite wall with three different thermal insulation layers in terms of experiment phenomenon, failure modes, hysteresis curves, ductility, stiffness degradation, and energy dissipation compared to the precast composite wall without thermal insulation layer.

The main results obtained from the tests can be summarized as follows:

The experiment phenomenon of the precast composite wall with a thermal insulation layer is similar to that of the conventional precast composite wall. All of them underwent an elastic stage, elastic-plastic stage, and failure stage. The failure mode of them is a bending-shear failure.

Three thermal insulation layers increase the bearing capacity and inhibit the wall's cracking process, especially the polystyrene particles insulation mortar thermal insulation layer. Compared with the conventional precast composite wall, the bearing capacity of precast composite walls with polystyrene particles insulation mortar thermal insulation layer, XPS thermal insulation layer, and foam cement thermal insulation layer increased by 17.13%, 18.33%, and 18.11%, respectively. The crack load, yield load and ultimate load all increased to some extent.

The four specimens have well elastic-plastic deformation capacity and collapse-resistant capacity. The displacement ductility coefficient of PCW2 and PCW4 increased by 26.72% and 0.41%, respectively, while the displacement ductility coefficient of PCW3 decreased by 14.05% compared with the conventional precast composite wall. The relative deformation values of the four precast composite walls are bigger than the limit value of the elastic-plastic displacement angle between the layers of the precast composite wall structure given by the technical specification for precast composite wall structure.

The trend of stiffness degradation curves of the four precast wall specimens is similar, and the initial stiffness of precast composite walls with polystyrene particles insulation mortar thermal insulation layer increases most obviously. The thermal insulation layer structure and the thermal insulation materials enhance the energy consumption capacity of conventional precast composite walls.

ACKNOWLEDGMENT

The Scientific Research Planning Project of the Education Department of Shaanxi Province (No. 19JC024).

Beilin District Science and Technology Plan Project (GX2131).

REFERENCES

Dang Zheng, Liang Xingwen, et al. Experimental and Theoretical Studies on Seismic Behavior of Fiber Reinforced Concrete Shear Walls[J]. *Journal of Building Structures*, 2014,35(6):12–22.

Boscato G., Dalla Mora T., Peron F., Russo S., Romagnoni P. A New Concrete-glulam Prefabricated Composite Wall System: Thermal Behavior, Life Cycle Assessment and Structural Response[J]. *Journal of Building Engineering*, 2018,19.

Hao Yuhong, Li Jiaqi, Xu Huizi, Wang Yuqing, Shi Yong. Experimental Study on Seismic Performance of Composite Wall with a Thermal Insulation Layer in the Middle[J] *Journal of Building Structure*, 2015,3 6 (S2):237–243.

Huang W, Zhang M, Yang Z. A Comparative Study on Seismic Performance of Precast Shear Walls Designed with Different Variables[J]. *KSCE Journal of Civil Engineering*, 2018, 22.

Huang Wei. A Study on the Seismic Performance of the Composite Wall in the Proposed Dynamic Test[J]. *Xi'an Univ. Of Arch. & Tech.* (Natural Science Edition), 2007, 26(3): 49–56.

Kisa M H, Yuksel S B, Caglar N. Experimental Study on Hysteric Behavior of Composite Shear Walls with Steel Sheets[J]. *Journal of Building Engineering*, 2020, 33:101570.

Li Zhongxian. *Engineering Structure Test Theory And Technology*[M]. Tianjin: Tianjin University Press. 2003.

Light Frame Structure Study Group. Research on the Theory and Application of Light Frame Structure of Dense Rib Board[R]. *Xi'an: Xi'an Univ. of Arch. & Tech.*, 2000.

Miao X W, Huang W, Ling K, et al. Experimental Study and Numerical Simulation of Reinforced Concrete Walls with Lightened Blocks Connected with Bolts[J]. *Advances in Structural Engineering*, 2020, 23(4):136943321990068.

Salmon D C, Einea A. Partially Composite Sandwich Panel Deflections[J]. *Journal of Structural Engineering*, 1995, 121(4):778–783.

Several Opinions of the Central Committee of the Communist Party of China on Further Strengthening the Management of Urban Planning and Construction[J]. *Home*, 2016(08):1-4.

Shi Fengkai, Liu Fusheng, Wang Shaojie, Qiu Yibo, Liu Kang, Huang Xinghuai. Experimental Research and Analysis on Seismic Performance of Self-insulating Dark Frame Load-bearing Wall[J]. *Engineering Mechanics*, 2019,36(04):158–166.

Technical Specification for Precast Composite Wall Structure: DBJ61/T 94-2015.[S]. Xi'an: China Building Materials Press, 2015.

Wang Dongyan, Wang Lin, Liu Kang, Zhang Yangang, Hu Fengqin, Zhang Zongjun. Research on the Structural Performance of Load-bearing and Thermal Insulation Integrated External Wall Panel[J]. *Industrial Building*, 2019,49(12):84–87.

Yan Yiceng. *Study on the Combination Between Energy Conservation and Optimization of Public Building Envelope Under Green Building System.* [D] Chongqing University, 2016.

Yang Zengke, Huang Wei. Research on the Construction Process of Eco-composite Wall Structure[J]. *International Journal of Earth Sciences and Engineering*, 2015, 8(6):1209–1214.

Yao Qianfeng, Huang Wei, Tian Jie. The Study on the Stress Mechanism and Aseismic Performance of the Composite Wall of the Laminates[J]. *Journal of Architectural Structure*, 2004, 25(6): 67–74.

Yao Qianfeng, Chen Ping. *Civil Engineering Structure Test*[M]. Beijing: China Construction Industry Press, 2001.

Ye Zengping. The Development Status and Trend of Prefabricated Composite Wallboard of Industrialization of the Construction Industry[J]. *Fujian Construction Science & Technology Industrialization of the Construction Industry*, 2016, (01):28–30.

Study on strength characteristics and solidified mechanism of calcareous sand

Zimou Wang, Junjie Yang & Yalei Wu*
Ocean University Of China, Qingdao, China

Liang Wang*
South China Sea Institute of Planning and Environment, Ministry of Natural Resources, China

ABSTRACT: In order to achieve the strategic goal of safeguarding national maritime rights and building maritime power, large-scale reclamation projects have been carried out by China on islands and reefs in the South China Sea. These projects are usually far away from the land, and it is extremely expensive to transport a large amount of engineering-filling materials from the land. So the calcareous sand widely distributed near the islands and reefs is usually used as the filling material. However, the friability of calcareous sands will lead to the characteristics of high compressibility and low bearing capacity of hydraulic reclamation soil foundations accumulated by coral calcareous sands. In order to explore the strength characteristics and solidified mechanism of calcareous sand under different curing agents, the unconfined compressive strength test was carried out on the solidified calcareous soil specimens under different conditions, and the influencing factors of its strength were studied from a microscopic perspective combined with the SEM test. This study provides a theoretical basis for future strength properties of solidified calcareous soils.

1 INTRODUCTION

As the global population expands and land resources are increasingly scarce, marine resources have become an important way for countries to seek development. The South China Sea is the top priority of China's marine development. Most of the islands in the South China Sea are composed of coral reef rock masses. It is the best way to use local materials and calcareous sand widely distributed around the islands and reefs as fillers. However, due to the special physical properties of calcareous sand, it needs to be reinforced to improve the quality of the project and reduce the risk of the project.

The coral reef rock mass is a rock and soil mass formed by the remains of reef-building corals through complex and long-term geological processes, which is often called reef limestone. The loose sand-gravel layer on the surface of the coral reef rock mass is mainly composed of coral reef calcareous sand, and its main chemical component is calcium carbonate (Shen & Wang 2010).

The special properties of calcareous sand make the solidification of terrigenous sand very different from the thickness of terrigenous sand, mainly because: 1) It is difficult to completely fill the voids during solidification, which is one of the reasons that affect the solidification effect because the solidification solution is difficult to enter the closed inner voids. These voids often occupy a large proportion of the voids of calcareous sand; 2) The strength

---

*Corresponding Authors: wuyalei@ouc.edu.cn and 35585930@qq.com

and brittleness of calcareous sand particles make it easy to break the particles during solidification, and the closed inner voids become fully connected or semi-connected voids, changing the initial curing solution ratio reduces the curing effect. Therefore, in the study of solidified calcareous soil, in-depth research on its microstructural characteristics and the microscopic mechanism of strength growth is necessary. This research is of great significance to understanding the special engineering mechanical properties of calcareous sand and to serve the construction of the South China Sea.

The size of calcareous sand particles varies in size, and the shape is roughly divided into spherical, needle-like, flake-like, multi-legged or limb (Figure 1). The shape can also be defined by surface area and volume. Definitions such as sphericity and circularity can also be used in engineering (Chen & Wang 2005).

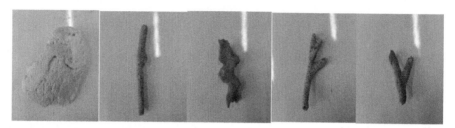

Figure 1. Physical map of calcareous sand.

Calcareous sand is called calcareous sand when it is divided into sands according to gradation. Calcareous sand particles have basic physical and mechanical properties such as porosity, low hardness, low strength and brittleness. Therefore, the calcareous sand foundation has the characteristics of high compressibility and low bearing capacity in engineering applications (Wang et al. 1997). Cement and other curing agents are a common method for strengthening soft soil foundations. When the foundation soil is mainly sandy soil, cement and other solidifying agents are used to reinforce the reclaimed foundation inside the cofferdam, and a high-strength calcareous sand solidified body with a width of tens to hundreds of meters is constructed to form an island reef construction with the cofferdam. It can prevent the loss of sand and soil and perform in-situ cement reinforcement for areas with high bearing capacity requirements (such as building foundations, airport runways, etc.). Therefore, the strength of the reinforced calcareous sand foundation meets the design requirements.

Unconfined compressive strength is an important mechanical index reflecting the ability of cement soil to resist axial pressure under unconfined pressure conditions. The maximum particle size of the soil sample proposed is not more than 5 mm, and the mold is a cylindrical test mold with a diameter of 50 mm.

Different from general terrigenous sand, calcareous sand is a special geotechnical medium with more than 50% calcium carbonate mineral content (Liu et al. 1995). Under the same test conditions, the strength of calcareous sand is much smaller than that of quartz sand. The main reason is the high calcium carbonate content, usually greater than 80%. The Mohs hardness of calcite is 3, while that of quartz is 7. So calcareous sand particles are easier to break than quartz sand when subjected to force (Zhou et al. 2013). Chen Zhaolin et al. took the calcareous sand of the South China Sea as the basic aggregate and carried out a comparative test of the mechanical properties of seawater-mixed calcareous sand aggregate concrete and seawater-mixed ordinary concrete. According to (Chen et al. 1991). Shen Jinlin studied the factors affecting the strength of calcareous sand concrete by uniaxial compression test by means of a parallel test and concluded that the strength of calcareous sand concrete increases with the growth of age. The water-cement ratio has a greater impact on calcareous sand concrete. The larger the water-cement ratio, the higher the strength; The strength of calcareous sand concrete is restricted by the type of cement (Shen 2016). Lu Bo and Liang Yuanbo pointed out that it is feasible to use seawater and calcareous sand to make cement concrete (plain concrete) in practical application. If there is a lack of fresh water and river sand, as long as the concrete grade is appropriately increased, the amount of cement is increased by a little more than that of ordinary concrete. By reducing the water-cement ratio, sufficient strength can be obtained for concrete to be used in plain concrete projects such as roads, floors and breakwaters. The effect of sulfate-resistant cement is better than that of ordinary Portland and slag cement, and pozzolan cement is not used (Lu & Liang 1993). In order to study the influence of different cement content, water-cement ratio and different sand content on the unconfined compressive strength and impermeability of calcareous sand cement soil. Xu Chao and Li Zhao, taking into account the mixing time, formulate a formula for cement reinforcement of calcareous soil. The proportioning scheme of sand (Xu et al. 2014), set 4 factors, namely cement content, water-cement ratio, mixing time, the quality ratio of silty sand, and calcareous sand. Each factor has three levels. In order to observe the changing law of various factors of cement-soil with the curing age, 14 d, 28 d and 60 d were selected for curing according to the compressive strength test regulations of cement-soil (Leng et al. 2010) Among the four factors, the cement content has a significant effect on the unconfined compressive strength and impermeability of cement-soil, and it is the biggest factor affecting the unconfined compressive strength and impermeability of cement-soil at each age. Qin Xiuyun proposed according to the experiment that the compressive strength of cement coral sand mortar decreases nonlinearly with the increase of the sand-ash ratio. When the sand-ash ratio is 5:1, the mortar strength decreases by about 29.31%, and the cement mortar with high sand-ash ratio decreases by about 29.31%. The strength increases greatly with age, and the compressive strength of cement coral mortar is relatively high in the early stage of curing. However, with the increase of curing time, the compressive strength of cement coral mortar is lower than that of cement mortar made of river sand after 3–7 days, and the difference in strength between the two decreases with the increase of sand-cement ratio (Qin et al. 2019).

To sum up, predecessors have made abundant achievements in the strength of solidified calcareous soil with the help of relevant tests, but they are limited to a relatively single solidifying agent as the test condition. In this paper, the design test uses ASW curing agents (Wu et al. 2021) and cement as a comparison. A single factor unconfined compressive strength test is designed and then combined with the SEM test (scanning electron microscope test). The microscopic analysis is carried out to obtain its curing mechanism, which is the follow-up calcareous sand. The research on strength characteristics has laid the foundation of theoretical research.

2 EXPERIMENTAL STUDIES

2.1 *Tested materials*

The calcareous sand used in the test was taken from the South China Sea. The outer contour of the calcareous sand particles is irregular, and the particle surface is rough and covered with fine pores. There are two kinds of test curing agents, among which the cement is No. 42.5 ordinary Portland cement produced by Weifang Luyuan Building Materials Co., Ltd., and the materials used for ASW (Wu et al. 2021) curing agent are carbide slag (CCR), slag (GBFS), fly ash (FA) and (Figure 1), of which CCR is an alkali activator, and GBFS, FA are pozzolanic materials. The test water is tap water.

Figure 2 shows that the maximum particle size Dmax of the soil used for the test is 8 mm. After the gradation is determined according to the actual sieving situation, the particle gradation curve is drawn.

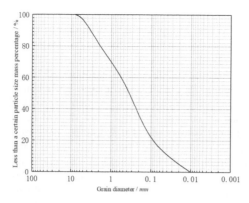

Figure 2. Particle size distribution curve of calcareous sand.

According to the "Geotechnical Test Method Standard"(Ministry of Construction of the People's Republic of China 2019), the specific gravity of calcareous sand is 2.71 and the water content is 0.02%.

2.2 Testing method

Influence factors of cement solidified soil mechanical properties and microscopic characteristics can be divided into to be solidified soil conditions (moisture content, grain and grain group content, composition and content of water-soluble salt, ph, pollution, types and content of organic matter, etc.), curing agent conditions (cement kinds and dosage of admixture type and dosage, etc.) and curing conditions (maintenance mode, temperature and ages, etc.). In this paper, cement and ASW are used as curing agents, and the curing conditions are standard curing. Table 1 shows the designed cement mixing ratio, water-cement ratio and age scheme.

Table 1. Specimen preparation plan for UCS.

Soil Used for test	Curing agent	The sample size (Diameter × high)/mm	Curing agent mixing ratio %	Grey water than	Curing time /d
Calcareous Sand	Cement ASW	50 × 100	5,7,10	1.5	7,14,28,90,180,360

The soil for testing and the curing agent were mixed and put into a mixer for stirring, and tap water was added and then stirred again. The test mold is evenly coated with release oil (neutral) and air-dried, and the cement soil is filled into the test mold in 5 layers. After each layer is filled, it is tamped 10 times with a metal rod. After filling, scrape the surface with a scraper, cover the surface with a plastic cover, put it into the curing box for standard curing for 48 hours, then take it out and demould, and then put it into the curing box for standard curing to the set age, as shown in Figure 3 ∼ Figure 4. The setting temperature of the maintenance equipment is 20±1°C, and the humidity is ≥ 95%. The unconfined compressive strength test was carried out on the cement-added solid samples reaching the curing age, and their stress-strain curves and unconfined compressive strength values were obtained.

Figure 3. The sample under curing.   Figure 4. Unconfined compressive strength test.

3 TEST RESULTS AND ANALYSIS

Figure 5 shows the results of the unconfined compressive strength test of the solidified calcareous soil. The stress increases with the strain, reaches a peak and then softens with the strain. This property does not change with the original soil type, cement mixing ratio, curing age and other factors. The stress corresponding to the peak is defined as the unconfined

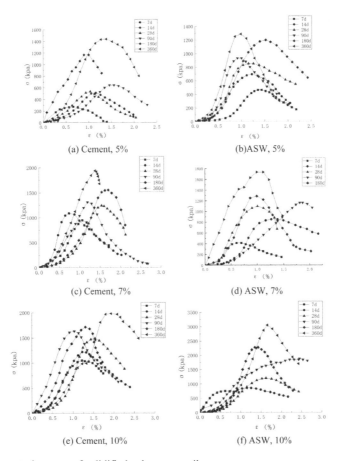

(a) Cement, 5%   (b) ASW, 5%

(c) Cement, 7%   (d) ASW, 7%

(e) Cement, 10%   (f) ASW, 10%

Figure 5. Stress-strain curve of solidified calcareous soil.

compressive strength. The unconfined compressive strength test results of the solidified calcareous soil are shown in Table 2.

Table 2. Unconfined compressive strength of cured calcareous soils.

Soil Used for test	Curing agent mixing ratio %	Unconfined compressive strength (KPa)					
		7d	14d	28d	90d	180d	360d
Cement	5	274.9	513.6	535.4	643.6	1168.2	1439.5
	7	947.51	1085.1	1244.1	1309.2	1550	1938.6
	10	1048.8	1215.2	1490.3	1630.2	1712.9	1982
ASW	5	464.1	694.7	853.5	933.3	1200.8	1287.6
	7	413.6	832.1	1097.1	1164.6	1251.4	1734.5
	10	752.2	853.5	1200.8	1876.4	2271.4	3052.5

From Table 2, the relationship between the strength of two kinds of solidified calcareous soils and the ratio of curing agent and age can be obtained. As shown in Figure 6, the strength of solidified calcareous soil increases with the increase of the mixing ratio. When ASW mixing ratio reached 10%, the strength of the solidified calcareous soil is obviously improved, and the strength at 360d is 1.76 times that of the solidified calcareous soil with the additional ratio of 7% ASW.

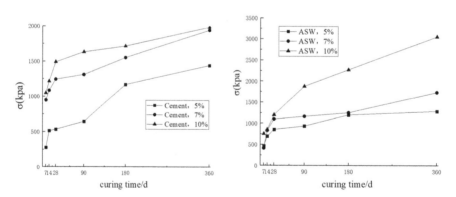

Figure 6. Strength-age relationship of solidified calcareous soil under different mixing ratios.

As shown in Figure 7, when the mixing ratio is 5%, the strength of cement-solidified calcareous soil is higher than that of ASW-solidified calcareous soil after a curing age of 180d. When the mixing ratio was 7%, the strength of cement-solidified calcareous soil was higher than that of ASW-solidified calcareous soil at all ages before age 360d. When the mixing ratio is 10%, the strength of ASW-solidified calcareous soil is higher than that of cement-solidified calcareous soil after 90 days of age. From this, we can explore the mechanism of its strength change through microscopic experiments.

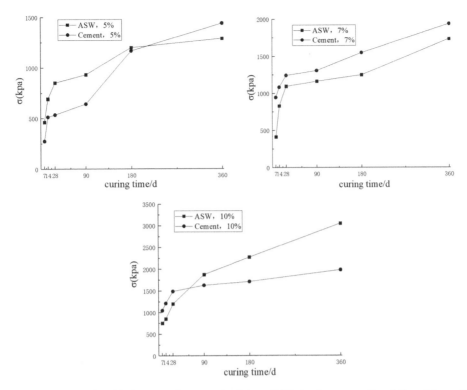

Figure 7. The relationship between the strength of the solidified calcareous soil and the mixing ratio under different types of curing agents.

4 MICROMECHANISM ANALYSIS OF CALCAREOUS SAND SOLIDIFIED SOIL

Figure 8~13 compares and analyzes the effects of two curing agents on the strength of the cured calcareous soil through the microscopic mechanism. Among them, when the mixing ratio of the curing agent is 7% and 10%, the strength of cement-solidified calcareous soil at 28d is higher than that of ASW-solidified calcareous soil. Under the condition of a 10% mixing ratio, ASW-solidified calcareous soil after 90 d. The strength is better than that of cement-solidified calcareous soil, so the mechanism of its strength change is explored through microscopic experiments.

500x   2500x   5000x

Figure 8. SEM image of solidified calcareous soil (ASW, 7%, 28 days).

Figure 8 shows the 28d-cured ASW-cured calcareous soil after crushing, in which the curing agent does not have an obvious hydration reaction with the original soil, resulting in lower strength. As can be seen from Figure 9, the cement-solidified calcareous soil cured for 28 days has begun to produce cemented substances and crystals, and the sand particles are cemented together to form the structural skeleton of the solidified calcareous soil. We compared it with the original soil to measure the solidified calcareous soil at this time. The integrity of the original calcareous sand is improved. The pore size of the original calcareous sand becomes smaller, and it is impossible to accurately distinguish the hydration products of the solidified calcareous soil from the original soil particles. The macroscopic performance is that the strength is higher than that of ASW-solidified calcareous soil.

Figure 9. SEM image of solidified calcareous soil (cement, 7%, 28 days).

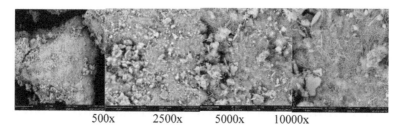

Figure 10. SEM image of solidified calcareous soil (cement, 10%, 28 days).

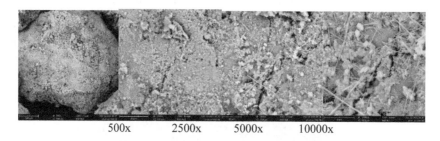

Figure 11. SEM image of solidified calcareous soil (ASW, 10%, 28 days).

As shown in Figures 10 and 11, the crushed solidified calcareous soil particles have produced a large number of fibrous and needle-like crystals interspersed in the structural voids, making the original void structure dense. Among them, the structure of cement-solidified calcareous soil is more compact, that is, the macroscopic strength is higher.

According to Figures 12 and 13, the solidified calcareous soil cured for 180 d has a more compact structure, and has no obvious voids compared with the original soil, revealing the

Figure 12. SEM image of solidified calcareous soil (cement, 10%, 180 days).

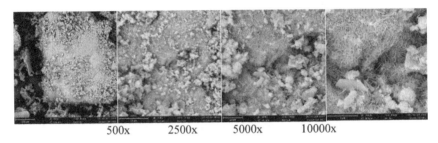

Figure 13. SEM image of solidified calcareous soil (ASW, 10%, 180 days).

characteristics of higher strength with age. In particular, the crystalline state of ASW-solidified calcareous soil is fuller, which proves that the hydration reaction has been fully carried out and a more compact microstructure is produced. The macroscopic performance is stronger than that of cement-solidified calcareous soil.

5 CONCLUSION

In this paper, different curing agents are used to solidify calcareous sand, and the strength characteristics of solidified calcareous soil are revealed through the strength of an unconfined compressive test. The main conclusions are as follows:

(1) The strength of the solidified calcareous soil increases with the increase of the incorporation ratio. When the ASW incorporation ratio reaches 10%, the strength of the solidified calcareous soil increases significantly. At 360d, the strength is the strength of the ASW solidified calcareous soil with an incorporation ratio of 7%. 1.76 times.
(2) When the mixing ratio is 5%, the strength of cement-solidified calcareous soil is higher than that of ASW-solidified calcareous soil after a curing age of 180 d.
(3) When the mixing ratio is 7%, the strength of cement-solidified calcareous soil is higher than that of ASW-solidified calcareous soil before the age of 360 d.
(4) When the mixing ratio is 10%, the strength of ASW-solidified calcareous soil is higher than that of cement-solidified calcareous soil after 180 days of age.

REFERENCES

Chen Hai-yang, Wang Ren, Li Jian-guo, Zhang Jia-ming: Grain Shape Analysis of Calcareous Soil[J]. *Rock and Soil Mechanics*, Vol. 26 No. 9(2005),1389–1392.
Chen Zhao-lin, Chen Yue-tian, Qu Ji-ming. A Feasibility Study of Application of Coral Reef Sand Concrete, *The Ocean Engineering*[J], Vol. 9, No. 3(1991), 67–80.

Leng Fa-guang Rong Jun-ming, Ding Wei, et al. Introduction of Revised Test Methods of Long-term Performance and Durability of Ordinary Concrete GB/T 50082-2009[J], Vol. 39 No. 2(2010), 6–9.

Liu Chong-quan, Yang Zhi-qiang, Wang Ren. *The Present Condition and Development in Studies of Mechanical Properties of Calcareous Soils*[J], Vol. 16, No. 1 (1995), 74–84.

Lu Bo, Liang Yu, *Experimental Study of Concrete Prepared with Coral Reef and Sea Water*[J], Vol. 12 No. 5 (1993), 69–74.

Ministry of Construction P.R. China, GB/T50123-2019, *Standard for Geotechnical Testing Method* [s]. Beijing: China Planning Press, 2019.

Qin Xiu-yun, Zhao Jun, Liu Mao-jun. *Compressive Strength and Microstructure of Coral Sand Cement Mortar* [J], Vol. 19 No. 21(2019), 239–244.

Shen Jian-hua, Wang Ren, Study on engineering properties of calcareous sand[A]. *Journal of Engineering Geology*, 1004-96645/2010/18(suppl.)-0026-07.

Shen Jin-lin, *Experimental Study on Compressive Strength of Coral Aggregate Concrete Mixed with Seawater* [J], Vol. 30 No. 4 (2016), 524–526.

Wu Ya-lei, Yang Junjie, Li Si-Chen, et al. Experimental Study on Mechanical Properties and Micro-Mechanism of All-Solid-Waste Alkali Activated Binders Solidified Marine Soft Soil[J]. *Materials Science Forum*, 2021(1036): 327–3.

Wang Ren, *Study on engineering Geological Properties of Coral reefs and Feasibility of Large Project Construction on Nansha Islands* [M], Beijing: Science Press, 1997.

Xu Chao, Li Zhao, Yang Ji-bao. *Research on Mixing Proportion Test of Cemented Coral Reef Sand*[J], Vol 41 No. 5(2014), 70–74+89.

Zhou Yang, Liu Xiao-yu, Li Shi-hai. Pile Foundation Engineering in Calcareous Soil[J]. *Port & Waterway Engineering*, No. 9 Serial No. 483(2013), 143–150.

An analytical study based on 3D characteristics of finite element negative Poisson ratio composite structure mechanical properties

Xiaomei Liu*
Xiamen Institute of Technology, Xiamen, Fujian Province, China

Xu Jiang
The Eighth Engineering Bureau of China City Investment Group Co., Ltd, Shanghai, China

ABSTRACT: In response to the development of modern engineering technology to build buildings with longer service life, it is necessary to study new materials with special mechanical properties. In this paper, the structural mechanical properties of a bridge when using the negative Poisson ratio material are studied. A finite element analysis of the structure is carried out, which can more intuitively sense the influence of the use of the material on the structural mechanical properties of the bridge.

1 PREFACE

China witnessed the emergence of beams at a very early stage. Ancient bridges were built with soil and stone. However, since the use of soil bridge structures led to instability during rain, people began to use wood to build bridges. In modern times the level of science and technology applied in building bridges is getting higher and higher. People have begun to use steel and concrete as the main construction materials for beams. With the development of science and technology, the use of steel structure as the main construction materials for beams has become prevalent. Many articles are now being worked on for the use of new materials to make the structure of beams more stable with longer service life. (Chaphalkar 2015) In this paper, the use of negative Poisson ratio materials as the main building materials of beams is mainly studied, and the structural mechanical properties analysis based on finite elements is performed on beams. Negative Poisson's ratio material is a new type of material with great development prospects. The ratio effect of negative Poisson ratio material makes the mechanical properties of this material greatly enhanced compared with other materials, and the shear stress that this material can withstand is much greater than other materials. Negative Poisson ratio materials also have good wear resistance, corrosion resistance, and heat resistance. Pyrite, arsenic, and cadmium are natural negative Poisson ratio materials. In the 80 s of the last century, Lakes first proposed these natural negative Poisson ratio materials as designable materials (Lakes 1987; Lakes). After decades of development, many polymerization and composite materials with negative Poisson ratio effects have been studied. These materials are already in use in many engineering fields.

2 COMPOSITION AND MECHANISM OF NEGATIVE POISSON'S RATIO MATERIALS

2.1 *Composition of negative Poisson's ratio material*

In general, materials with a negative Poisson ratio effect can be divided into porous negative Poisson ratio materials, negative Poisson ratio composite materials, and molecular negative Poisson ratio materials.

---

*Corresponding Author: sindyqd@163.com

Porous negative Poisson ratio materials: Rock and wood in nature are porous materials. One phase of this material is solid, and the other phase is composed of liquid or void. The structure of these materials has advantages that most of the traditional materials or structures do not have, such as high strength, lightweight, good temperature, sound insulation, and good energy absorption.

Molecular negative Poisson's materials: These are mainly some polymer and crystal materials with special microstructures, such as zeolite, silica crystals, and some metal elements.

2.2 Negative poisson ratio material structure

2.2.1 Concave structure

A common negative Poisson's type of material is made up of a truss structure consisting of thin ribs and connecting hinges, as shown in Figure 1(a) Structure diagram. Figure 1(b) Force diagram is a two-dimensional concave hexagonal honeycomb structure, which is applied to a uniaxial pull. Its h rod moves outward because when the l rod receives the pull force, the concave angle θ expands, the l rod rotates, and the overall mechanism expands to produce a negative Poisson's ratio effect, as shown in Figure 1(c) Renderings.

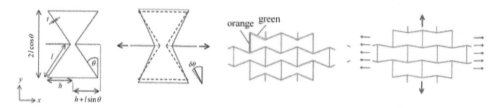

Figure 1. (a) Structure diagram, (b) Force diagram and (c) Renderings.

2.2.2 Metal negative poisson ratio material

In this article, a metallic material with a negative Poisson's ratio effect is defined as a metal negative Poisson's ratio material. The use of metal materials to make negative Poisson ratio porous materials can effectively improve the strength and hardness of the material, so now the research of negative Poisson ratio materials is mainly based on metal materials. A 3D metal negative Poisson has better mechanical properties than other materials – one of the most significant advantages of this material, which can be adjusted by modal scale factors, having almost identical negative Poisson ratio properties under the action of tensile forces and pressures. Another advantage of this material is that it has good geometric symmetry because the material still maintains negative Poisson's ratio properties over a wide range of strains, producing a large expansion effect.

2.2.3 Composite negative poisson ratio material

At present, the research of composite negative Poisson ratio materials is mainly manifested in four aspects: (1) Making negative Poisson ratio materials by superimposing specific angle laminates. (2) Combining materials with negative Poisson ratio characteristics with other materials to manufacture negative Poisson ratio materials. (3) Studying the properties of negative Poisson ratio materials from different aspects. (4) Studying more composite negative Poisson ratio materials. Materials with negative Poisson ratio characteristics in composite negative Poisson ratio materials can be used not only as the substrate of this material but also as a reinforcing fiber of the composite material. We choose the right material and the negative Poisson ratio characteristics of the material for combination according to the need. The study shows that the two Poisson ratio differences of the materials together form a new material. The performance of the new material can be improved a lot compared with the original one. Adding about 10% of the negative Poisson ratio concave structure of the material can make the new material in the case of the overall relative density unchanged, with the stiffness increasing by three times. This new material, compared with the performance of other raw materials, improves equivalent to 1:30. The application of this technology in engineering can significantly improve the stability and service life of buildings. The negative Poisson ratio material combined with the cement composite material can improve the strength and toughness of cantilever beams and greatly improve the stability of buildings.

3 CHARACTERISTICS OF NEGATIVE POISSON MATERIALS FOR BUILDING BEAMS

3.1 *What is a beam*

In a 3D space, when the scale of one structure in one direction is significantly greater than the scale of the other two directions, the stress effect in this direction being most important, it can be called a beam-column structure. Among them, the components that mainly bear the axial force are columns, and the components that mainly bear the bending moment are beams.

The deformation of a beam depends on its topological form, structural size, cross-sectional shape, material properties, load boundary conditions, and many other factors. When analyzing the deformation of the beam structure, reasonable assumptions and necessary simplifications of the deformation mode of the beam structure are made. Based on a material constitutive model, the mechanical equilibrium equation lists partial differential equations. According to the different assumptions of mechanical analysis simplification, the beam models commonly used in mechanics mainly include Euler-Bernoulli beam theory, Timoshenko beam theory, general beam theory (GBT), higher-order beam theory, etc.

3.2 *Advantages of negative poisson ratio material beams*

At present, most beam structures are reinforced concrete structures or steel structures. Compared with the bridge of these structures, the negative Poisson of the material of the beam has better explosion resistance and impact resistance. Compared with other materials, its shear modulus will be higher, and the negative Poisson is also better than the material's shock absorption ability. The beam made of negative Poisson material also has a strong notch fracture toughness, which can greatly improve the safety and stability of the beam. Cantilever beams occupy a great role in many fields (Chaphalkar 2015; Khalatkar 2012, 2017; Zeng 2017). However, cantilever beams are easily affected by external factors because of material and production problems, causing the structure of the beam to be unstable, which may finally lead to the weakening of the carrying capacity of the beam. If the negative Poisson ratio material is used to design the cantilever beam, the load-carrying capacity, and stability of the cantilever beam can be significantly enhanced.

3.3 *Using cross-section method to find shear forces and bending moments*

Figure 2 is a schematic diagram of a cantilever beam, which acts at the A-end-concentrating force F.

We draw the force of the beam to seek its binding F_Q, the binding moment M_B, and divide the beam into two segments with m-m and find any x interface internal force.

Shear force F_Q is an internal force parallel to the cross-section.

Bending moment M – the force couple moment perpendicular to the cross-section.

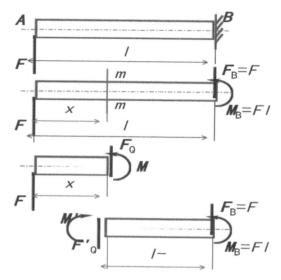

Figure 2. Force analysis diagram of a cantilever beam.

We take the left section beam of the cross-section as the equilibrium equation of the study object column.

$$\sum \phi_\psi = 0: \quad F - F_Q = 0 \quad F = F_Q \tag{1}$$

$$\sum M_x(\phi) = 0: \quad M - F_x = 0 \quad M = F_x \tag{2}$$

We take the right section beam of the cross-section as the equilibrium equation of the study object column.

$$\sum \phi_\psi = 0: \quad F'_Q - F_B = 0 \quad F'_Q = F_B = F \tag{3}$$

$$\sum M_x(\phi) = 0: \quad -M' - F_x*(l-x) + M_B = 0 = F*(l-x) - Fl = Fx \tag{4}$$

We take the left section beam of the cross-section as the equilibrium equation of the study object column.

$$\sum \phi_\psi = 0 \tag{5}$$

$$\sum M_x(\phi) = 0 \tag{6}$$

We take the right section beam of the cross-section as the equilibrium equation of the study object column.

$$\sum \phi_\psi = 0 \tag{7}$$

$$\sum M_x(\phi) = 0 \tag{8}$$

We take a section of the beam, as shown in Figure 3, and the positive and negative of F_Q and M specified as the shear force is positive left and right. The reverse is negative, and the bending moment is positive and negative.

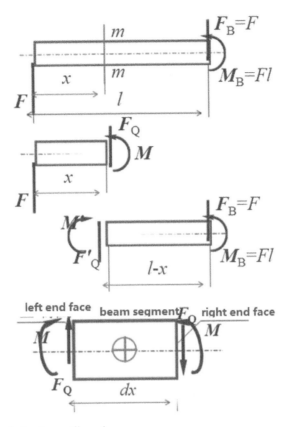

Figure 3. Force analysis of a cantilever beam.

4 MODELING AND FINITE ELEMENT ANALYSIS OF CANTILEVER BEAMS

4.1 Modeling of cantilever beams

At present, people's requirements for industrial development are becoming more and more stringent. The previous approach was based on the experience people gained in the course of their work to design, create entities, and then validate them. This kind of solution takes a long time, the work efficiency is greatly reduced, and the structural design cannot be optimized. It cannot catch up with the accelerated pace of the machinery industry. If you want to improve the efficiency of designing products, you need to reduce the cost of the product and reduce the error rate of the product. Around 1980, the first simulation of mechanical system dynamics appeared, which can provide users with assistance in the early stages of design through virtual prototypes, improve the authenticity of predicting mechanical structures, optimize system design, and gradually show its vigorous growth as an engineering analysis technology. The SOLIDWORKS platform provides a mechanical platform for simple simulation analysis and performing stress–strain simulation analysis of the resulting 3D model. By comparing the simulation data with the theoretical calculations, the user can reduce the time consumed and the error rate of the work.

As an excellent 3D modeling software, SOLIDWORKS is the world's first 3D CAD system based on Windows development. SOLIDWORKS has three characteristics: powerful, easy to learn and easy to use, and technological innovation. SOLIDWORKS can provide different design solutions to reduce errors in the design process and improve product quality. Not only does SOLIDWORKS offer such powerful features, but it also provides every engineer and designer with features that are easy to operate, learn, and use. The model built is shown in Figure 4.

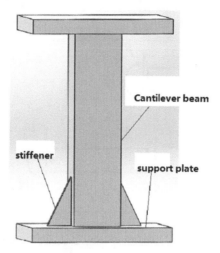

Figure 4. Structural model of a cantilever beam.

4.2 *Simplification of the cantilever beam model*

In all walks of life cantilever beams play a vital role. In the past cantilever beam materials led to problems when beam bearing capacity is not enough or corrosion or heat resistant. The study selects a new composite negative Poisson ratio material for static and dynamic analysis of cantilever beams to judge whether the composite negative Poisson ratio material design of the cantilever beam meets some important parameters under actual working conditions.

The simplified model is established using **SOLIDWORKS**. The finite element analysis has the function of removing the small details that do not affect the final result. The simulation analysis of the cantilever beam can be directly used in this study to simplify the model. The simplified model is shown in Figure 5.

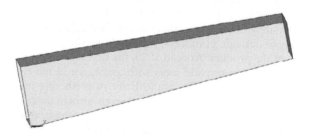

Figure 5. A simplified model diagram of a cantilever beam.

4.3 *Static and dynamic analysis of cantilever beams*

The model shown in the image above is composed of support plates, stiffeners, and cantilever beams. It is made of composite negative Poisson's ratio material, which is meshed in ANSYS and constrained after meshing. The processed model is shown in Figure 6.

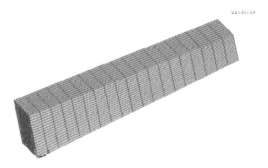

Figure 6. Force analysis graph.

When setting the boundary conditions of the cantilever beam during static analysis, it is not necessary to consider the influence of other environmental factors on the analysis results. It is only necessary to consider the constraints of the cantilever beam and the magnitude of the load in combination with the actual situation in the condition-setting process. In this study, the fixed end of the cantilever beam is set as fully constrained, and a load of 70000 N is applied to the upper surface of the cantilever beam.

The resulting load distribution result plot is shown in Figure 7.

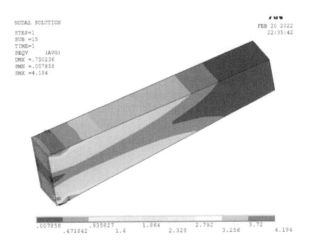

Figure 7. Load distribution.

As can be seen from the above figure, the cantilever beam designed with a negative Poisson's ratio mainly generates a bending moment at the fixed end after carrying a load of 70000 N. The static analysis results of cantilever beams designed using reinforced concrete are much better than those found in the data.

Figure 8 is the total deformation of the cantilever beam designed using the composite negative Poisson ratio material. From the figure, it can be seen that under the action of uniform load, the maximum displacement of the cantilever beam occurs at the free end. According to the degree of deformation in the figure, the cantilever beam designed with the negative Poisson ratio is much better than the static analysis results of the cantilever beam designed with reinforced concrete.

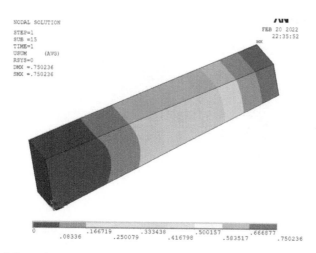

Figure 8. Total deformation.

5 CONCLUSION

This article mainly introduces the application of negative Poisson ratio materials and negative Poisson ratio materials in engineering and other aspects, modeling through SOLIDWORKS, and finally importing the model into ANSYS, setting the model material, constraints, and loads to be borne. After analyzing the model and comparing it with the cantilever beam built of ordinary reinforced concrete, the result is that the cantilever beam built with negative Poisson is superior to the material and can withstand more pressure. However, the results of this study should not be put into practice. This study only performs an ANSYS static analysis of the material. The material to be used requires LS/DYNA software for dynamic impact analysis. The energy absorption and shock absorption effect will be studied according to the software analysis results.

ACKNOWLEDGMENT

2022 Education and Scientific Research Project for Young and middle-aged Teachers in Fujian Province.

REFERENCES

Chaphalkar S. P., Khetre S. N., Meshram A. M. Modal Analysis of Cantilever Beam Structure Using Finite Element Analysis and Experimental Analysis[J]. *American Journal of Engineering Research*, 2015, 4(10): 178–185.

Deraman A. S., Niirmel R., Mohamad M. R. Analysis of Rectangular Flexible Horizontal Piezoelectric Cantilever Beam Base on ANSYS[C]//IOP Conference Series: *Materials Science and Engineering*. IOP Publishing, 2020, 917(1): 012076.

Khalatkar A. M., Haldkar R. H., Gupta V. K. Finite Element Analysis of Cantilever Beam for Optimal Placement of Piezoelectric Actuator[C]//*Applied Mechanics and Materials*. Trans Tech Publications Ltd, 2012, 110: 4212–4219.

Lakes R. Foam Structures with a Negative Poisson's Ratio[J]. *Science*, 1987, 235(4792): 1038–1040.

Lakes R. Advances in *Negative Poisson's Ratio Materials* [J]. 1993.

Zeng J., Ma H., Zhang W., et al. Dynamic Characteristic Analysis of Cracked Cantilever Beams Under Different Crack Types[J]. *Engineering Failure Analysis*, 2017, 74: 80–94.

Seismic responses and damage analysis of pretensioned high-strength concrete pipe pile reinforced with non-prestressed steel bars

Hongqiang Hu
The Architectural Design & Research Institute of Zhejiang University Corporation Limited, Hangzhou, China
College of Engineering, Zhejiang University City College, Hangzhou, China
Center for Balance Architecture, Zhejiang University, Hangzhou, China
College of Civil Engineering and Architecture, Zhejiang University, Hangzhou, China

Gang Gan*
The Architectural Design & Research Institute of Zhejiang University Corporation Limited, Hangzhou, China
Center for Balance Architecture, Zhejiang University, Hangzhou, China

Yangjuan Bao
School of Civil Engineering and Architecture, Zhejiang University of Science & Technology, Hangzhou, China

Xu Han
China Railway Siyuan Survey and Design Group Corporation Limited, Wuhan, China

Xiaopeng Guo
Xi'an Center of Geological Survey, China Geological Survey, Xi'an, China

ABSTRACT: Pretensioned high-strength concrete pipe pile reinforced with non-prestressed steel bars (PRHC piles) is a kind of pile that adds non-prestressed steel bars to ordinary pretensioned high-strength concrete pipe piles (PHC piles) to improve its horizontal bearing capacity and seismic performance. To investigate the seismic behavior of PRHC piles, a three-dimensional dynamic nonlinear finite element model of PHRC pile embedded in the soft soil is constructed. To analyze the seismic response and damage analysis, the nonlinear behavior of soil, high-strength concrete damage behavior, and the interaction between the soil and the pile are systematically considered. The seismic acceleration responses, shear forces, bedding moments, and concrete damage states of PRHC pile under one seismic ground motion with different seismic intensities are comprehensively presented and analyzed in this study. The results can provide the theoretical basis for the seismic design and seismic performance assessment of PRHC piles.

1 INTRODUCTION

A pretensioned high-strength concrete pile (PHC pile) is a hollow pipe pile produced in a prefabrication plant by tensioning prestress, centrifugal forming and high-temperature, and high-pressure steam curing. PHC piles have a lot of advantages, such as high bearing

*Corresponding Author: gang@zuadr.com

capacity, high concrete strength, good ability of crossing soil layers, fast construction speed, cheaper cost per unit bearing capacity than ordinary piles, and reliable pile quality. PHC piles are consequently being widely used in high-rise buildings, bridges, ports, wharves, and other engineering construction in recent years.

However, many post-earthquake investigation results have found that some PHC piles have experienced various damage states, which have demonstrated that PHC piles are vulnerable to seismic hazards and might have a relatively poor earthquake-resistant capacity. Its application in the earthquake area is thus limited (Hu et al. 2021, 2022; Karkee & Kishida 1997; Kishida & Karkee 1998; Koseki et al. 2012; Tokimatsu et al. 2012). As a consequence, a lot of scholars and engineers have devoted great efforts to investigating the seismic response and behavior of PHC piles through experiments and numerical models, as well as suggested measures to improve the seismic performance of PHC piles.

Adding non-prestressed steel bars in PHC piles is a reasonable way of improving the horizontal bearing capacity and seismic performance of PHC piles, which is called PRHC pile in some Chinese standards and codes. Liu et al. (2018) studied the seismic performance of PRHC piles with full-scale field tests. Wang et al. (2015) also performed experimental research on the aseismic behavior of pipe piles with hybrid reinforcement. The results of hysteresis curves, skeleton curves, ductility coefficient, and curvature distribution were analyzed. Although a lot of reciprocating load tests and monotonic load tests have been conducted to investigate the horizontal bearing capacity behavior of PRHC piles, few studies are found on the dynamic time history analysis of the seismic behavior of PHRC piles.

In the present study, a three-dimensional dynamic nonlinear finite element model of PHRC pile embedded in soft soil is constructed for seismic response and damage analysis of PRHC pile, in which the nonlinear behavior of soil, high-strength concrete damage behavior, and the interaction between the soil and pile are systematically considered. The results presented in this study can provide the theoretical basis for the seismic design and seismic performance assessment of PRHC piles.

2 THREE-DIMENSIONAL DYNAMIC FINITE ELEMENT MODEL OF PRHC PILES

In this study, a three-dimensional dynamic finite element model of PRHC piles embedded in soft soil is constructed by the Abaqus software, as illustrated in Figure 1. The PRHC pile is a pipe pile with a diameter of 0.6 m and a height of 30 m. The 16ϕ10.7 prestressed steel bars and 16ϕ14 non-prestressed steel bars are arranged inside the PHRC pile. The effective concrete precompressed stress of the PRHC pile is set as 6 Mpa. The nonlinear behavior and damage characteristic of the concrete pile body is modeled by the concrete damaged plasticity (CDP) model (Hibbit et al. 2010). Some parameters of high-strength C80 Concrete are as follows: Young's modulus E_c = 37968.7 MPa, Poisson's ratio v_c = 0.2, compressive strength f_c = 53.6 MPa and tensile strength f_t = 3.5 MPa.

The parameters of prestressed steel bars are as follows:

Young's modulus E_1 = 200 GPa, yield strength f_1 = 1280 Mpa.

The parameters of non-prestressed steel bars are:

Young's modulus E_2 = 200 GPa, yield strength f_2 = 400 Mpa.

The corresponding reinforcement ratio with prestressed steel bars and reinforcement ratio with no-prestressed steel bars is 0.75% and 1.28%, respectively.

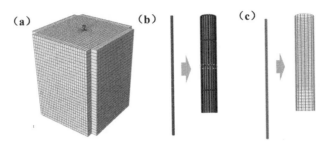

Figure 1. The constructed finite element model of the pile foundation in the present study: (a) the whole finite element model; (b) the finite element model of the pile; (c) the finite element model of steel bars.

In the numerical simulation, the Mohr-Coulomb model is employed to characterize the nonlinear behavior of soil. The relevant parameters are as follows:

Elastic modulus = 50 Mpa, Poisson's ratio = 0.35, Cohesion = 25 kPa and Friction angle = 20 degrees.

The dynamic interaction between the PRHC pile and soil is considered by the shear-strength criterion of the Mohr-Coulomb type with a friction coefficient of 0.4. To model the vertical load on the PRHC pile and the inertial effect from the superstructure, a mass block with a height of 3 m, a volume of 3.7052 m^3, and a density of 74393.3 kg/m^3 is assumed and tied at the pile head to simulate the 2756.421 kN vertical load.

The seismic ground motion recorded at the station of Christchurch Resthaven in the 2011 New Zealand earthquake is input in the present study. The corresponding acceleration time history and Fourier amplitude are illustrated in Figure 2. During the numerical simulation, the peak acceleration value of this ground motion is scaled to 0.1 g, 0.2 g, 0.3 g, and 0.4 g, respectively, to investigate the effect of seismic intensity on the seismic response and damage characteristic of PHRC pile under earthquake.

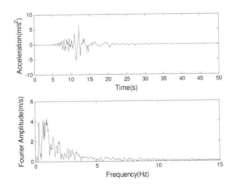

Figure 2. The input seismic acceleration time history and corresponding Fourier amplitude.

3 RESULTS

To reveal the seismic responses and damage state of the PHRC pile under the seismic ground motion with various seismic intensities from the view of dynamic time history analysis, the acceleration, shear force, and bending moment (BM) time histories at the pile head, as well as the compression and tension damage states are comprehensively presented in this section.

Figure 3 is the acceleration time history responses on the top of the pile under the ground motion with various input intensities (0.1 g, 0.2 g, 0.3 g, and 0.4 g). From the figure, it can be found that the seismic responses at the pile head increase with the rise of the intensity of input ground motion, and the corresponding maximum acceleration responses reach about 0.5021 m/s^2, 1.618 m/s^2, 2.853 m/s^2, and 4.475 m/s^2 for the input seismic intensities of 0.1 g, 0.2 g, 0.3 g, and 0.4 g, respectively. In addition, the amplification effect of the soil foundation is also revealed from the results of Figure 3 based on this three-dimensional finite element model of the PHRC pile foundation.

Figure 4(a) is the shear force time histories on the top of the pile underground motion with various input intensities (0.1 g, 0.2 g, 0.3 g, and 0.4 g). The sheer force of the pile is also increased with the rise of input seismic intensities and the corresponding maximum shear force values reach about 30.83 kN, 111.7 kN, 115.9 kN, and 149.4 kN for the input seismic

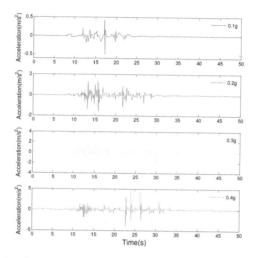

Figure 3. The acceleration time responses on top of the pile under the ground motion with various input intensities (0.1 g, 0.2 g, 0.3 g, and 0.4 g).

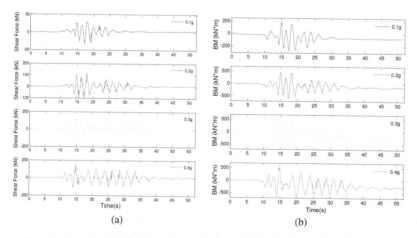

Figure 4. The (a) shear force and (b) bending moment responses on the top of the pile underground motion with various input intensities (0.1 g, 0.2 g, 0.3 g, and 0.4 g).

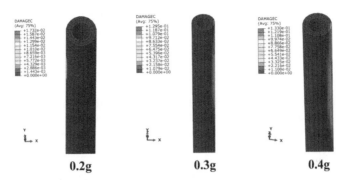

Figure 5. The compression damage states of pile head underground motion with various input intensities (0.2 g, 0.3 g, and 0.4 g).

intensities of 0.1 g, 0.2 g, 0.3 g, and 0.4 g, respectively. The time histories of bending moment (BM) on the top of the pile underground motion with various input intensities (0.1 g, 0.2 g, 0.3 g, and 0.4 g) are illustrated in Figure 4(b), where it can also be found that the BM of the pile is also increased with the rise of input seismic intensities and the corresponding maximum BM value reach about 270.7 kN*m, 602.6 kN*m, 627.7 kN*m, and 656.1 kN*m for the input seismic intensities of 0.1 g, 0.2 g, 0.3 g, and 0.4 g, respectively.

In the study, the CDP model is also adopted in the present dynamic finite element time history to investigate the concrete damage characteristic of PHRC piles under various seismic intensities of ground motion. According to the dynamic time history analysis, there is no damage to pile underground motion with an input intensity of 0.1 g. However, there are different levels of comprehension and tension damage with input intensities of 0.2 g, 0.3 g, and 0.4 g. Figure 5 shows the compression damage states of PRHC pile underground motion with various input intensities (0.2 g, 0.3 g, and 0.4 g). The corresponding maximum comprehension damage factor is 0.01732, 0.1295 and 0.133 for the seismic intensity of 0.2 g, 0.3 g, and 0.4 g, respectively. Figure 6 shows the tension damage states of pile head underground motion with various input intensities (0.2 g, 0.3 g, and 0.4 g). The corresponding maximum tension damage factor is 0.6993, 0.8257, and 0.8141 for the seismic intensity of 0.2 g, 0.3 g, and 0.4 g, respectively. It is obvious that the tension damage is more serious than the comprehension damage because the concrete has larger comprehension strength than the tensile strength. Furthermore, from these two figures, it is possible to see not only that the damage to the PHRC pile becomes more severe as input seismic intensities increase but also that the

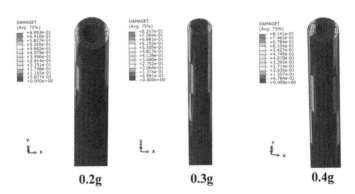

Figure 6. The tension damage states of pile head underground motion with various input intensities (0.2 g, 0.3 g, and 0.4 g).

damaged area moves down from the top of the pile and expands as input seismic intensities increase.

4 CONCLUSIONS

To investigate the seismic behavior of PRHC pile, a three-dimensional dynamic nonlinear finite element model of PHRC pile embedded in soft soil is constructed for seismic response and damage analysis, in which the nonlinear behavior of soil, high-strength concrete damage behavior, and the interaction between the soil and pile are systematically considered. The seismic acceleration responses, shear forces, bedding moments, and concrete damage states of PRHC pile under one seismic ground motion with different seismic intensities are comprehensively presented and analyzed in this study. It has been found that the acceleration, shear force, and bending moment of the PHRC pile increase with the rise of the input intensity of ground motion, and thus increase the risk of the pile foundation under earthquakes. The tension damage of the PHRC pile is more serious than the comprehension damage because the concrete has larger comprehension strength than the tensile strength. The seismic damage of the PHRC pile becomes more serious, and the damaged area enlarges with the increase of input seismic intensities.

ACKNOWLEDGMENTS

This work was funded by the fellowship of the China Postdoctoral Science Foundation (Grant No. 2021M702791), the projects of the Center for Balance Architecture in Zhejiang University (Grant Nos. K20212159 and K20203330C), and the National Natural Science Foundation of China (Grants Nos. 42272323 and 41902274).

REFERENCES

Hibbit, H. D., Karlsson, B. I., & Sorensen, P.: *Abaqus Theory and User's Manual.* (2010).
Hu, H., Huang, Y., Xiong, M., & Zhao, L.: Investigation of Seismic Behavior of Slope Reinforced by Anchored Pile Structures Using Shaking Table Tests. *Soil Dynamics and Earthquake Engineering*, 150, 106900 (2021).
Hu, H., Huang, Y., Zhao, L., & Xiong, M.: Shaking Table Tests on Slope Reinforced by Anchored Piles Under Random Earthquake Ground Motions. *Acta Geotechnica*, 1–18 (2022).
Karkee, M. B., & Kishida, H.: Investigations on a New Building with Pile Foundation Damaged by the Hyogoken-Nambu (Kobe) Earthquake. *The Structural Design of Tall Buildings*, 6(4), 311–332 (1997).
Kishida, H., & Karkee, M.: Behavior of Friction Piles During the Hyogoken-Nambu Earthquake and its Implications to the Design Practice. *Geotechnical Engineering Bulletin*, 7(2), 79–89 (1998).
Koseki, J., Koda, M., Matsuo, S., Takasaki, H., & Fujiwara, T.: Damage to Railway Earth Structures and Foundations Caused by the 2011 off the Pacific Coast of Tohoku Earthquake. *Soils and Foundations*, 52(5), 872–889 (2012).
Liu, C., Sun, Y., Zheng, G., Liu, Y., Hu, Q., & Liu, Y.: Experimental Study on the Seismic Performance of the Prestressed High-strength Concrete Pipe Pile Reinforced with Non-prestressed Steel Bars. *China Civil Engineering Journal*.54(4): 77–86 (2018).
Tokimatsu, K., Tamura, S., Suzuki, H., & Katsumata, K.: Building Damage Associated with Geotechnical Problems in the 2011 Tohoku Pacific Earthquake. *Soils and Foundations*, 52(5), 956–974 (2012).
Wang, T., Du, Z., Zhao, H., Liu, XS., & Wang, M.: Experimental Research on Aseismic Behavior of Pipe Piles with Hybrid Reinforcement. *Journal of Civil Engineering and Management.* 32(3): 27–32 (2015).

Experimental study on shear connection performance of high-strength bolts with single side tapping

Zaihua Zhang* & Guangxiang Mao*
College of Civil Engineering, Hunan City University, Yiyang, China

ABSTRACT: The load-bearing capacity of high-strength bolts with single-side tapping shear connection was experimentally studied. The ultimate shear capacity of a single bolt in the friction stage and the ultimate shear capacity in the pressure stage were measured. The test results show that when the tapping high-strength bolt connection is designed according to the pressure type high-strength bolt, the average value of the ultimate shear capacity of the bolt measured by the test has a larger safety reserve than the calculated ultimate shear capacity of the bolt, which can meet the design requirements of the pressure type high-strength bolt. However, when it is designed according to the friction type high-strength bolt, the average value of the shear bearing capacity of the bolt obtained by the test is less than that calculated by the standard, and the relative slip between plates obtained by the test is relatively large, which cannot meet the normal use limit state requirements of the friction type high-strength bolt in shear.

1 INTRODUCTION

The traditional fastening connection of steel structure high-strength bolts is shown in Figure 1(a). As a fastener, the high-strength bolt connection pair includes a bolt, nut, and two washers. The joint anti-slip bearing capacity can be calculated in accordance with the relevant provisions of the *Design Code for Steel Structures (GB50017-2003)* and the *Technical Code for High-strength Bolt Connection for Steel Structures (JGJ82-2011)*. Considering the difficulties in the assembly of traditional high-strength bolts in the case of closed sections, one-side tightening of high-strength bolts has received extensive attention in engineering (Chen et al. 2016; Manuela et al. 2021; Xu et al. 2015). Currently, the widely used one-side bolts include Hollo-bolt (Lindapter 2019), modified Hollo-bolt (Cabrera et al. 2021; Jeddi et al. 2022), Ajax One-side Bolt (Ajax 2005), T-Head bolt (Wang et al. 2021; Wan et al. 2020), Flow drill (Flow drill et al. 2020), Huck Bolt (Huck 2020), and Blind Bolt (Javora & Skejić 2017), etc. (as shown in Figures 1b–1h). There are also numerous reports on the performance tests and finite element analysis results of these bolts, especially the Hollo-bolt system developed by Lindapter International (Wang et al. 2009; Wang et al. 2013). Although these bolts can effectively solve the difficult problem of assembly and construction of the closed section, most bolts are patented products, and the price is more expensive than ordinary high-strength bolts. At the same time, most of them need different professional tools for the construction of ordinary high-strength bolts, which is not conducive to the mass promotion of products. Therefore, a tapping-type high-strength bolt assembly and connection method have received attention in engineering.

*Corresponding Authors: zaihua_zhang@163.com and 33384222@qq.com

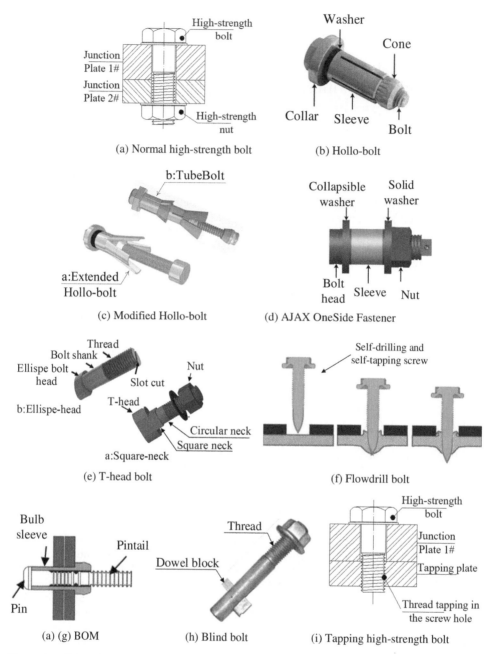

Figure 1. Comparison of two kinds of high-strength bolt connections.

The tapping high-strength bolt adopts the connection construction shown in Figure 1(a). It uses the screw rod of an ordinary high-strength bolt, and at the same time in the inner wall of the screw hole of the connecting plate tapping to form threads; the plate with tapping thread is used instead of high-strength nuts. Such a closed-section assembly connection can be very useful for achieving unilateral construction.

However, due to the great difference between the material properties of the steel plate and the nut, the failure mode, stress mechanism, and bearing capacity of this new connection mode under the shear state is not clear. The present code of steel structure does not contain this kind of connection type, so the study on the performance evaluation of the new bolt joint shear research work is very necessary. As for the research on unilateral fastening connections, most of the current studies focus on the application of such connections (Bin et al. 2021; Xu et al. 2019), and few studies have investigated the mechanical behavior of a single bolt. This paper discusses the shear-bearing capacity of tapping-type connections with high-strength bolts.

2 TEST OVERVIEW

2.1 *Specimens design*

A single bolt was taken as the research object, and three groups of specimens were designed. M20, M24, and M30 commonly used in practical engineering were selected as high-strength bolts in the test, and the thickness of the connecting plate with internal thread was 20 mm and 24 mm. The specific situation of each specimen is shown in Table 1. Figure 2 shows the design and fabrication size of each specimen, and the specimens after processing and forming are shown in Figure 3.

Table 1. Grouping of specimens.

Specimen grouping	Performance classes of bolts	Nominal diameter of bolts	Thickness of plate with tapping hole	Specimen number
Group 1	Grade 10.9	M20	20 mm	M20-1 M20-2 M20-3
Group 2	Grade 10.9	M24	20 mm	M24-1 M24-2 M24-3
Group 3	Grade 10.9	M30	24 mm	M30-1 M30-2 M30-3

2.2 *Loading and measuring scheme*

The test was loaded using a displacement loading method on a 1000 kN class electro-hydraulic servo universal testing machine (Figure 4). Before the formal loading, the specimen is preloaded to ensure that the bolt connection smoothly enters the normal shear state. At the same time it is ensured the test equipment is working properly. After preloading to 5% of the predetermined load, the load was unloaded to zero, and then the monostatic load was applied, and the loading rate was controlled at 2 mm/min until the specimen was significantly damaged or the bearing capacity was significantly decreased. The force and displacement applied by the actuator during the test were recorded by the MTS loading system.

During the test, the main measurement content includes: (1) the shear force value when the relative slip starts to occur between two friction surfaces; (2) the ultimate shear force value; (3) the load-deformation curve.

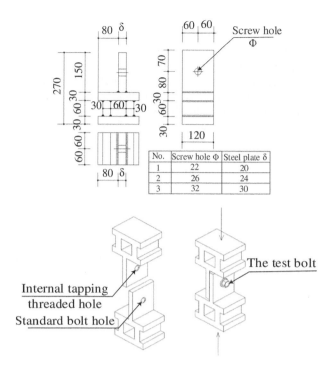

Figure 2. Specimen size and assembly method.

Figure 3. Machining condition of the specimen connection hole.

Figure 4. Test loading device.

3 TEST RESULTS AND ANALYSIS

3.1 *Experimental phenomenon*

Through preloading, the support and constraint part of the specimen is in good contact with the loading part, and the formal loading stage is entered. Formal loading is the first stage, the bolt connection at the elastic stress deformation stage, the growth of the shear force, and displacement relationship is close to the linear change, plate, and bolt deformation are not obvious. When the load reaches a certain value, the connecting plate and screw thread tapping steel appear obvious relative slip, slip after a certain distance, the growth of the shear force and displacement relationship and close to the linear change, continue to load. When approaching the ultimate bearing capacity of the bolt connection, plastic deformation of the screw occurs, and finally, the screw is cut off and destroyed. Small extrusion deformation occurs at the thread of the bolt hole wall of the steel plate, and the shear bearing capacity of the bolt decreases sharply. The failure state is shown in Figure 5.

Figure 5. Specimen failure mode.

3.2 *Test results*

1) The relation curve between the shear force on each bolt and the relative displacement on both sides of the friction surface of the connecting plate is shown in Figure 6. It represents the shear capacity of a friction-type connection high-strength bolt and the shear capacity of a pressure-bearing connection high-strength bolt, and are respectively calculated according to the relevant provisions of GB50017-2017.

   It can be seen from the curve in the figure that the AB section is the friction force transmission stage of the tapping high-strength bolt connection. The BC section is the relative slip stage, and the CD section is the pressure force transmission stage after the jacking contact between the two connecting steel plates and the bolt. Both AB and CD sections exhibit a linear force state. The test results of all the three groups of specimens show that the tapping high-strength bolt connection has a long slip phase, which affects the normal use of the connection for friction type connection that does not allow slip.

2) The slip shear values at the beginning of the relative slip between two friction surfaces of each specimen and the ultimate shear values are summarized in Table 2. The test values are compared with the shear bearing capacity design values calculated according to the friction bolt and bearing bolt, respectively.

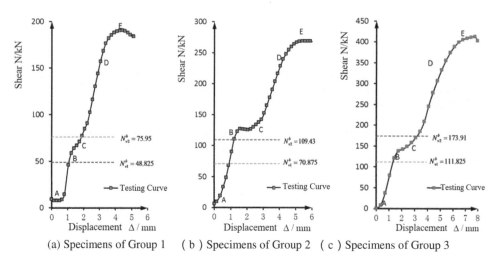

(a) Specimens of Group 1　(b) Specimens of Group 2　(c) Specimens of Group 3

Figure 6. Shear-displacement curve for tapping connection of high-strength bolts.

Table 2. Comparison between shear test results and standard shear capacity of a single high-strength bolt.

Specimen grouping	Specimen number	Shear force test value for slip Q_1/kN	Ultimate shear test value Q_{max}/kN	Design value of shear capacity of friction bolt N_{v1}^b/kN	Design value of shear capacity of bearing bolt N_{v2}^b/kN	$\dfrac{Q_1}{N_{v1}^b}$	$\dfrac{Q_{max}}{N_{v2}^b}$
Group 1	M20-1	67.060	194.685	48.825	75.950	1.373	2.563
	M20-2	42.365	191.025	48.825	75.950	0.868	2.515
	M20-3	54.770	183.070	48.825	75.950	1.122	2.410
Group 2	M24-1	138.450	266.080	70.875	109.430	1.953	2.432
	M24-2	105.610	266.830	70.875	109.430	1.490	2.438
	M24-3	108.385	272.165	70.875	109.430	1.529	2.487
Group 3	M30-1	124.435	410.495	111.825	173.910	1.113	2.360
	M30-2	162.235	415.475	111.825	173.910	1.451	2.389
	M30-3	129.075	412.805	111.825	173.910	1.154	2.374

The data in the table shows that during the frictional force transfer stage, the average friction shear bearing capacity of each specimen test value of the tapping high-strength bolt connection was greater than the corresponding computation according to the specification of ordinary high-strength bolts friction shear bearing capacity, but the data is discrete. The main reason is the torque method imposed on the bolt pre-tightening force itself has large discreteness. The friction coefficient between plates in the test is also different from the standard value. At the same time, the test value (M20-2) is lower than the standard calculation value, and the maximum displacement value of the friction phase of the test is relatively large, which does not meet the normal use limit state requirements of friction type high-strength bolts, and is not suitable for engineering design according to the traditional design criteria of friction type high-strength bolts.

For each specimen in the confined force stage, tapping type high-strength bolt connection was greater than the limit of the shear bearing capacity of a single pressure type high-strength bolt for the standard calculated value. The surplus amount is larger, and the visible tapping type high-strength bolt connection according to the traditional pressure type high-strength bolt design criteria for engineering design is relatively appropriate.

4 CONCLUSION

1) The test results show that the mechanism and failure mode of the tapping high-strength bolt connection are similar to those of the traditional high-strength bolt connection in various stages of shear. There are also elastic stages of friction force transmission, plate slip stage, elastic stages of screw force transmission, and elastic-plastic stages.
2) During the stage of shear resistance by friction, the experimental values of shear strength of all specimens are relatively discrete, and the relative slip between plates is relatively large, which cannot meet the normal use limit state requirements of shear resistance of friction-type high-strength bolts. It is not suitable for engineering design according to the design criteria of traditional friction-type high-strength bolts.
3) In the pressure shear stage of the tapping high-strength bolt connection, the ultimate shear capacity of the connection is much higher than the design value of the ordinary high-strength bolt according to the standard. It is appropriate to carry out the engineering design of this type of connection according to the design criteria of the traditional pressure-type high-strength bolt connection.

REFERENCES

Ajax Fasteners Innovations. (2005). *Joint Design us in Oneside Structural Fastener.*
Arconic Fastening Systems and Rings. (2020). Huck BOM brochure, https://www.arconic.com.
Bin Jia, Juan Ding, Zeyu Ding, et al. (2021). Experimental Research on the Mechanical Behavior of Steel Frame Joints Using High Strength Bolt Connection with Tapping Steel Plate[J]. *Progress in Steel Building Structures*, 23(1): 48–58 (in Chinese).
Cabrera M., Tizani W., Ninic J., et al. (2021). Experimental and Numerical Analysis of preload in Extended Hollo-Bolt blind bolts[J]. *Journal of Constructional Steel Research*, 186: 106885.
Flowdrill Ltd, Flowdrill Brochure. (2020). https://www.flowdrill.com.
Javora, A. & Skejić, D. (2017). Resistance Assessment of Beam-to-column Joints with Different Blind Bolt Systems. Technical Gazette, 24(4): 1103–1112.
Jeddi M. Z., Sulong N. H. R., Ghanbari-Ghazijahani T. (2022). Behaviour of double-sleeve TubeBolt moment connections in CFT columns under cyclic loading[J]. *Journal Of Constructional Steel Research*, 194: 107302.
Kefan Chen, Yuhan Li, Jinyu Lui. (2016). Research Progress on the Application of Unilateral Fastening Bolts in Structural Engineering[J]. *Jiangsu Construction*, 174(1): 27–30 (in Chinese).
Lindapter T. H. B. Hollo-Bolt®[J]. *Cavity Fixings*, 2: 41–43.
Lindapter. (2019). Hollo-bolt product brochure, uk.
Manuela Cabrera, Walid Tizania, Jelena Ninic. (2021). A Review and Analysis of Testing and Modeling Practice of Extended Hollo-Bolt Blind Bolt Connections[J]. *Journal of Constructional Steel Research*, 183, 106763.
Ministry of Housing and Urban-Rural Development of the People's Republic of China. (2011). *Technical Specification for High Strength Bolted Connection of Steel Structures*, China building industry press, Beijing.
Ting Xu, Wei Wang, Yiyi Chen. (2015). A review on foreign research status of one-side bolt[J], *Steel Construction*, 30(200): 27–33 (in Chinese).
Wan C., Bai Y., Ding C., et al. (2020). Mechanical performance of novel steel one-sided bolted joints in shear [J]. *Journal of Constructional Steel Research*, 165: 105815.
Wang J.-F., Han L.-H., Uy B. (2009). Behaviour of Flush End Plate Joints to Concrete-filled Steel Tubular Columns, *Journal of Constructional Steel Research*, 65: 925–939.

Wang J.-F., Han L.-H., Uy B. (2009). Hysteretic Behaviour of Flush End Plate Joints to Concrete-filled Steel Tubular Columns, *Journal of Constructional Steel Research*, 65: 1644–1663. 740

Wang J., Zhang L., Spencer B. (2013). Seismic Response of Extended end Plate Joints to Concrete-filled Steel Tubular Columns, *Engineering Structures*, 49: 876–892.

Wang P., Sun L., Zhang B., et al. (2021). Experimental Studies on T-stub to Hollow Section Column Connection Bolted by T-head Square-neck One-side Bolts Under Tension[J]. *Journal of Constructional Steel Research*, 178: 106493.

Zhenhua Xu, Zhong Liu, Xiongping Shu. (2019). Analysis on the Mechanical Properties of Diagonal Bracing Joints on Prefabricated Steel Frame Structure Connected with Steel Tapping High Strength Bolts[J]. *Progress in Steel Building Structures*, 21(6): 114–128 (in Chinese).

Seismic analysis of continuous rigid frame bridge with double thin-wall piers based on SHDR

Keqin Wang* & Richen Ji*
College of Civil Engineering, Lanzhou Jiaotong University, Lanzhou, China

ABSTRACT: In this paper, the seismic performance of a continuous rigid frame bridge with double thin-wall piers based on super high-damping seismic isolation rubber bearing has been studied and established in a dynamic mode by using nonlinear time history analysis. The study simulates four groups of seismic wave acceleration time history curves manually and takes the international land port municipal infrastructure project Weihe No. 5 bridge as the engineering background. The bridge site is located in the Weihe River valley section, which belongs to the temperate semi-humid and semi-arid climates. The seismic fortification intensity of the bridge is 8 degrees, and the peak acceleration is 0.3 g. The main beam of the bridge is a prestressed concrete continuous box girder of variable height, and the bridge tower and the main beam are in the form of consolidation. The results show that the super high-damping seismic isolation rubber bearing can not only reduce the seismic response but also avoid the excessive displacement of the structure. It is more suitable for continuous rigid frame bridges with double thin-wall piers in high-intensity areas, which is compared with ordinary plate rubber bearings. The isolation effect of super high-damping seismic isolation rubber bearing is influenced by the geometric parameters of the double-wall piers. The increase of wall thickness and center-to-center distance of double piers can give rise to the increase of seismic response of the structure.

1 INTRODUCTION

The continuous rigid frame bridge combines the characteristics of the continuous beam bridge and the T-shaped rigid frame bridge. The reasonable choice of pier stiffness can make the distribution of internal forces of the bridge more reasonable (Wang 2014). It has been widely used in bridge construction because of its continuous and consolidated structure form of main girder and pier girder, which is convenient for construction and can meet the stress requirements of a large span (Zhang 2006). Compared with other types of piers and columns, the double thin-wall piers have smaller horizontal compressive stiffness, which is an ideal flexible pier for continuous rigid frame bridges. It is effective to reduce the bending moment value generated by horizontal displacement on piers (Zhang 2017).

Based on the structural form of the continuous girder bridge with double thin-wall piers, many experts and scholars have conducted in-depth research on its mechanical and dynamic characteristics in the earthquake effect. Zheng et al. (2007) discussed the shock absorption scheme of setting elastic-plastic connection beams in continuous rigid frame bridges with double thin-wall high piers and compared the internal forces and displacements of piers under the action of earthquake effects; Zhou et al. (2009, 2007, 2003) studied the influence of pier section form and single-limb pier arrangement form on the dynamic characteristics of

---

*Corresponding Authors: 1367063292@qq.com and 1165251533@qq.com

continuous rigid frame bridge structure; Chen (2020), Jiang (2020), and Zhang et al. (2020) obtained the influence of basic structural parameters on the seismic response of bridges, discussed the seismic isolation system applicable to the double thin-wall pier system; Li Jie et al. (2012) determined the natural vibration characteristics and the most unfavorable seismic excitation direction of the curved continuous rigid frame bridge with double thin-wall and high pier; Wang et al. (2021) paid attention to the influence of the free length of pile foundation on the seismic response of continuous rigid frame bridges with double thin-wall piers.

In the previous seismic analysis of continuous rigid frame bridges with double thin-wall piers, there are few studies on the isolation effect of super high-damping seismic isolation rubber bearing (SHDR) on such bridges. The seismic isolation performance of the SHDR system in high-intensity areas and the influence of geometric parameters of the thin-wall pier on the seismic isolation effect of the SHDR system are systematically discussed, which adopts the nonlinear time history analysis method and takes the international land port municipal infrastructure project Weihe No. 5 bridge as the engineering background. The above provides a reference for the application of SHDR in the seismic isolation of such bridges.

2 ENGINEERING BACKGROUND AND FINITE ELEMENT SIMULATION

2.1 Project profile

Weihe Bridge No. 5 spans 75 m+126 m+75 m and is a continuous rigid frame bridge with double thin-wall piers. The main beam is a continuous prestressed concrete box beam of variable height, with a single box and a five-chamber section with a straight web. The beam height at the fulcrum is 510 cm, and the beam height at the span and side pier is 220 cm. The bridge tower and main beam are in the form of consolidation. The tower is 20 m high, and the distance between the cable beam and the tower is 4 m and 1 m, respectively. The cast-in-place layer of the box girder and bridge deck adopts C50 concrete, and C40 concrete is used for the piers and columns. The layout of key parts of the main bridge and support plane is shown in Figures 1 and 2.

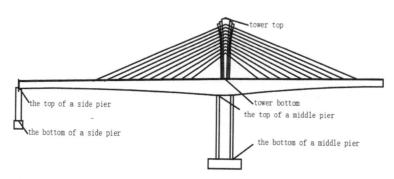

Figure 1. Layout of key parts of the main bridge.

2.2 Finite element analytical model

MIDAS CIVIL is used to establish the full bridge dynamic analysis model, which is divided into 364 nodes and 443 units. The main bridge, cable tower, and pier all adopt beam elements; the stay cables adopt truss elements only under tension. The equivalent elastic modulus of the stay cable was calculated according to Ernst's sag effect formula (Yi 2018).

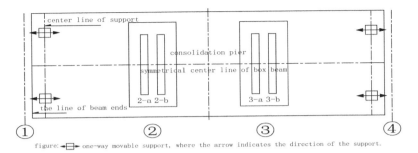

Figure 2. Supporting floor plan.

Ernst's formula is as follows:

$$E_{eq} = E \bigg/ \left(1 + \frac{\gamma^2 L^2 E}{12\sigma^3}\right) \quad (1)$$

In the formula: E_{eq} is the equivalent elastic modulus of cable (kPa); E is the elastic modulus of steel strand (kPa); L is the horizontal projected length of the cable (m); γ is the unit weight of the stay cable (kN/m³); σ is tensile stress of the stay cable (kPa).

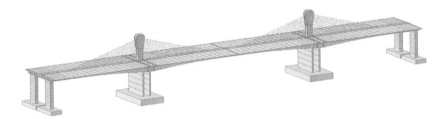

Figure 3. Finite element analytical model.

2.3 Seismic wave selection

Four groups of artificially input seismic waves are used to analyze the isolation effect of ultra-high damping rubber bearing under high-intensity earthquakes. Four sets of peak ground motion accelerations are: $A = 0.1$ g, $A = 0.2$ g, $A = 0.3$ g, and $A = 0.4$ g. Figure 4 shows four groups of seismic wave acceleration response spectrum curves. Figure 5 shows one group of the time history curves of seismic wave acceleration.

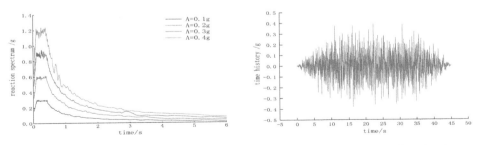

Figure 4. Reaction spectrum curves of acceleration.

Figure 5. Time history curves of acceleration.

3 RESTORING FORCE MODEL OF SHDR

SHDR is composed of upper and lower connecting plates, ultra-high damping rubber, and internal stiffening plates. Figure 6 shows the bearing diagram.

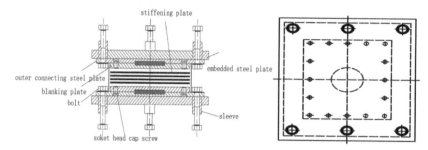

Figure 6. Schematic diagram of super high-damping seismic isolation rubber bearings.

The connecting steel plate acts to transfer the supporting force between the components, which is the connecting foundation; the internal stiffener can effectively improve the vertical stiffness of the support; the super high-damping rubber not only absorbs the seismic wave energy but also realizes the reset function in the elastic stage. As a result, the good bearing characteristics and high damping performance make the ultra-high damping bearing have excellent ability of earthquake resistance and impact damage.

The SHDR is usually defined as a bilinear restoring mechanical model in finite element simulation, as shown in Figure 7. In the diagram, K_1 is pre-yield stiffness, K_2 is post-yield stiffness, S_y is yield displacement quantity, Sd is designed damping displacement, F_y is yield force, and F_d is the design damping force. The hysteresis curve of the actual restoring force of one set of SHDR is shown in Figure 8.

The basic parameters of the support system are shown in Table 1.

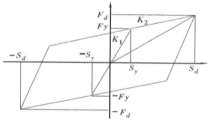
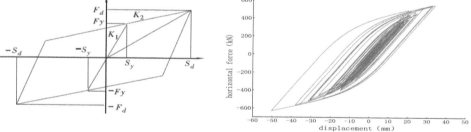

Figure 7. Theoretical hysteretic curve model.

Figure 8. Actual hysteretic curve model.

Table 1. The basic parameters of super high-damping seismic isolation rubber bearings.

Bearing plane dimension (mm)	Bearing capacity (kN)	Displacement (mm)	Bearing height (mm)	Shearing modulus of elasticity (Mpa)
1020×1020	10000	±100	240	1.0

Horizontal stiffness before yield (kN/mm)	Horizontal stiffness after yield (kN/mm)	Equivalent horizontal stiffness (kN/mm)	Yield force (kN)
24.7	4.5	7.3	497

4 ANALYSIS OF ISOLATION OF SHDR IN THE HIGH-INTENSITY REGION

4.1 *Analysis of dynamic characteristics of bridges*

In this modeling, multiple Ritz vector methods were used to compare and analyze the natural vibration characteristics of laminated rubber bearings and super high-damping seismic isolation rubber bearings. The following results can be obtained from the first five modes in Tables 2 and 3: The super high-damping seismic isolation rubber bearing compared with the plate rubber bearing not only can reduce the natural vibration frequency and the overall stiffness of the structure, but also reduce the seismic response and make the structure tend to be safe in the earthquake.

Table 2. Natural vibration frequency and mode of laminated rubber bearings.

Modal number	Natural frequency of vibration (Hz)	Mode of vibration
1	1.26308	First vertical bend of the main beam
2	1.67813	First side bend of the main tower
3	3.50405	Lengthways wave
4	4.24618	Transverse bending of main beam + secondary bending of main tower
5	5.06335	Second vertical bend of the main beam

Table 3. Natural vibration frequency and mode of super high-damping seismic isolation rubber bearings.

Modal number	Natural frequency of vibration (Hz)	Mode of vibration
1	0.94085	First vertical bend of the main beam
2	1.28157	Lengthways wave + the second vertical bend of the main beam
3	3.15184	Main tower has a symmetrical lateral bend
4	3.55721	Transverse bending of girder
5	4.82425	Third vertical bend of the main beam

4.2 *Seismic response analysis of the isolation system of SHDR*

The analysis of the bending moment and displacement deformation at the bottom of the double thin-wall pier under different peak seismic accelerations, as illustrated in Tables 4–6.

From Table 4 it can be concluded: The super high damping bearing is better than the ordinary plate rubber bearing in reducing the bending moment at the bottom of the pier, the seismic isolation rate of 2# pier bottom is between 4% and 31%, and the seismic isolation rate of the 3# pier bottom is between 6% and 31%. The isolation rate of SHDR increases with the increase of earthquake acceleration, which explains that SHDR can play a better isolation effect in areas with high seismic intensity.

Tables 5 and 6 show the super high-damping rubber bearing isolation system, which, compared with the plate rubber bearing, can reduce the 2# pier top displacement by 16%~23%, and decrease the tower top displacement by 13%~24%. The displacement of pier 3# is reduced by 15%–23%. The displacement of the tower top is reduced by 9%–24%. The beam end displacement is reduced by 16%~24%. The maximum displacement produced is 53 mm,

Table 4. Comparison of bending moment at the bottom of piers 2# and 3#.

Earthquake acceleration (g)	2-a# bending moment at the bottom of the pier (kN·m)			2-b# bending moment at the bottom of the pier (kN·m)		
	Plate rubber bearing	SHDR	Isolation rate	Plate rubber bearing	SHDR	Isolation rate
0.1	22230	21212	−5	28089	27058	−4
0.2	24517	21225	−13	30892	27080	−12
0.3	32674	24208	−26	43211	32340	−25
0.4	39091	26945	−31	51355	37061	−28

Earthquake acceleration (g)	3-a# bending moment at the bottom of pier (kN·m)			3-b# bending moment at the bottom of pier (kN·m)		
	Plate rubber bearing	SHDR	Isolation rate	Plate rubber bearing	SHDR	Isolation rate
0.1	17221	16196	−6	14990	14001	−7
0.2	25837	21583	−16	23931	18338	−23
0.3	43154	32246	−25	32745	24321	−26
0.4	51954	37031	−29	39094	26989	−31

Table 5. Displacement and deformation comparison of pier 2#, pier 3#, and tower top.

Earthquake acceleration (g)	2# the pier top displacement (mm)			2# displacement of the top of the tower (mm)		
	Plate rubber bearing	SHDR	Isolation rate	Plate rubber bearing	SHDR	Isolation rate
0.1	21.5	18	−16	36.7	32	−13
0.2	41.3	32	−23	45.9	35	−24
0.3	57.1	48	−16	63.7	53	−17
0.4	65.2	53	−19	74.5	57	−23

Earthquake acceleration (g)	3# pier top displacement (mm)			3# displacement of the top of the tower (mm)		
	Plate rubber bearing	SHDR	Isolation rate	Plate rubber bearing	SHDR	Isolation rate
0.1	21.2	18	−15	60.4	55	−9
0.2	41.3	32	−23	65.3	55	−16
0.3	57	48	−16	68.8	56	−19
0.4	65.2	53	−19	74.6	57	−24

which is within the allowable range of the specification. The SHDR relies on good damping performance to avoid excessive displacement and deformation of the structure and has a good protective effect on the structure as a whole.

As a whole, SHDR has a good isolation effect and greatly improves the structural stress, which can prolong the natural vibration period of the continuous rigid frame bridge with double thin-wall piers during earthquakes. In addition, the analysis data shows that a seismic intensity of 8 degrees and above can best support the seismic isolation effect and reduce the overall cost of the bridge.

Table 6. Comparison of beam end displacement and deformation.

Earthquake acceleration (g)	Small lane beam end displacement (mm)			Large distance beam end displacement (mm)		
	plate rubber bearing	SHDR	isolation rate	plate rubber bearing	SHDR	isolation rate
0.1	21.7	18	−17	21.5	18	−16
0.2	41.9	32	−24	41.8	32	−23
0.3	57.9	48	−17	57.8	48	−17
0.4	66.1	53	−20	66.1	53	−20

5 INFLUENCE OF PARAMETERS OF DOUBLE THIN-WALL PIERS ON SHDR

5.1 *Influence of geometric parameters on dynamic performance*

The natural vibration frequencies of structures with a wall thickness of 1.0, 1.1, and 1.2 m and center-to-center distance of double piers of 5, 6, and 7 m were compared in this paper, to study the influence of wall thickness and center-to-center distance of double piers on the dynamic performance of the SHDR support system. The effects of wall thickness and center-to-center distance of double piers on the SHDR system are shown in Tables 7 and 8.

Calculation results show that the natural vibration frequency increases gradually with the increase of wall thickness. The wall thickness has a significant effect on the second-order frequency. When the thickness increases from 1 m to 1.2 m, the natural vibration frequency increases by 29%, the other frequencies also showed an increasing trend, but the influence range was between 4% and 8%. The natural vibration frequency of the structure also increases gradually with the increase of the center-to-center distance of double piers, but there is no obvious change. Its range is between 3%–10%.

Table 7. Influence of wall thickness on natural vibration frequency of SHDR system.

Wall thickness/m	Natural frequency of vibration/Hz				
	First stage	Second stage	Third stage	Fourth stage	Fifth stage
1.0	0.940	1.282	3.152	3.557	4.824
1.1	1.061	1.458	3.283	3.676	4.934
1.2	1.084	1.805	3.438	3.792	5.133

Table 8. Influence of center-to-center distance of double piers on natural vibration frequency of SHDR system.

Center-to-center distance of double piers/m	Natural frequency of vibration/Hz				
	First stage	Second stage	Third stage	Fourth stage	Fifth stage
5.0	0.940	1.282	3.152	3.557	4.824
6.0	1.062	1.378	3.328	3.596	4.854
7.0	1.084	1.432	3.408	3.682	5.013

5.2 Influence of geometric parameters on internal force response

The shear moment of pier 2-b# is taken as the object of nonlinear time history analysis so as to study the influence of the center distance and wall thickness on the seismic response of the SHDR system. The time history curve simulation parameters are set according to engineering background requirements. The seismic intensity of this area is 8 degrees, the peak acceleration of ground motion is 0.30 g, the site category is Class II, and the site characteristic period is 0.40 s. The influence of wall thickness on the internal force of pier 2-b# is shown in Figure 9.

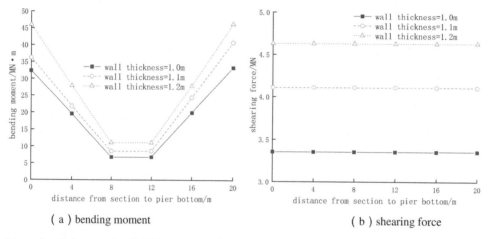

Figure 9. Influence of wall thickness on the internal force of 2-b# pier

Calculation results show that the bending moment shear force of pier 2-b# increases significantly with the increase in wall thickness. When the wall thickness increases from 1.0 m to 1.2 m, the bending moment at the bottom of the pier increases by 30%, and the shear force increases by 28%. The internal force of the 2-b# pier almost does not change along with the increase of the center-to-center distance of double piers. These results indicate that the geometric parameters of the thin-wall double-limb pier can change the structural stiffness and also have a certain impact on the isolation effect of the SHDR system.

6 CONCLUSION

This paper systematically analyzes the seismic performance of the continuous rigid frame bridge with double thin-wall piers based on SHDR and obtains the following conclusions:

(1) Compared with the plate rubber bearing, the super high-damping rubber bearing is more suitable for the isolation system of the double thin-wall continuous rigid frame bridge in the high-intensity area.
(2) The natural vibration frequencies of the isolation system gradually increase along with the increase of the distance between the center-to-center distance of double piers and wall thickness, whereas the vibration modes almost do not change.
(3) The internal force response of the isolation system will increase significantly with the increase of the wall thickness and the center-to-center distance of double piers, so it is necessary to consider the influence of geometric parameters in the isolation study of such bridges.

FOUNDATION ITEM

Study on Seismic Performance of New Bearing Bridges (52068041).

REFERENCES

Chen Aijun, Peng Rongxin, Wang Jiejun, et al. (2020). Study on Seismic Behavior of Twin-limb Thin-walled Piers of Long span Continuous rigid frame bridge. *Vibration and Shoc.* 39(1):1–7.
Jiang Jianjun, Zheng Wanshan, Tang Guangwu, et al. (2020). Seismic Analysis and Seismic Mitigation Measures for Box Girder of Long-span Continuous Rigid Frame Bridge. *Road Works.* 45(2):28–33.
Li Jie, Chen Huai, Wang Yan, et al. (2012). Seismic Response Analysis of Curved Continuous Rigid Frame Bridge With Double Thin Wall and High Pier. *J. World Earthquake Engineering.* (04).
Maneetes H., Linzell D. G. (2003). Cross Frame and Lateral Bracing Influence on Curved Steel Bridge Free Vibration Response. *Journal of Constructional Steel Research.* 59(9):1101–1117.
Wang Shuai, Song Shuai, et al. (2021). Influence of Free Pile Length on Seismic Response of Continuous Rigid Frame Bridge with Double Thin-walled Pier Under the World Seismic Roof. *World Earthquake Engineering.* 37(01).
Wang Kehai (2014). *Research on Seismic Resistance of Bridges* (2nd Edition). M. China Railway Publishing House.
Yi Jiang, Mo Jinsheng, Li Jianzhong (2018). Study on Cable Relaxation of Single-pylon Cable-stayed Bridge Under Strong Earthquake. *Engineering Mechanics.* 35(6):97–104.
Zhang Hui (2006). *Bridge Engineering Technology.* M. Northeastern University Press.
Zhang Yongliang, Lu Xiaosu, Chen Xingchong, et al. (2017). Seismic Design of the Continuous Rigid Frame Bridge with Low Double Thin-Walled Piers in High Intensity Earthquake Area. *Journal of Railway Engineering Society.* 34, (11):45–50.
Zhang Yue, Xue Lei, Chen Shuai, et al. (2020). Parametric Sensitivity Analysis of Continuous Rigid Frame Bridge in High Seismic Intensity region. *Journal of Earthquake Engineering.* 42(2):311–317.
Zheng Kaifeng, Wen Shudong, LI Huaiguang, et al. (2007). Application of Inelastic Beams on Double Thin-wall High piers of Continuous Rigid Frame Bridge. *Northwestern Seismological Journal.* 29(4):303–306.
Zhou Yongjun, He Shuanhai, Zhang Gang, et al (2009). Effect of Pier Section Type on Seismic Response of Curved Continuous Rigid Frame Bridge. *Journal of Highway and Transportation Research and Development.* 26(2):68–72.
Zhou Yongjun, He Shuanhai, Song Yifan, et al. (2007). Contrast Analysis for Seismic Response of Continuous Rigid Frame Bridge with Single or Double-thin-wall. *Journal of China and Foreign Highway.* 27(3):114–117.

Study on seismic performance and fragility assessment of offshore jacket platform with fracture damage

Shuo Zhang
College of Pipeline and Civil Engineering, China University of Petroleum (East China), Qingdao, China

Hong Lin*
College of Pipeline and Civil Engineering, China University of Petroleum (East China), Qingdao, China
Center for Offshore Engineering and Safety Technology (COEST), China University of Petroleum (East China) Qingdao, China

Alexander Moiseevish Uzdin
Department of Mechanics and Strength of Materials and Structures, Emperor Alexander I St. Petersburg State Transport University, Russia

Youhai Guan
College of Pipeline and Civil Engineering, China University of Petroleum (East China), Qingdao, China

Lei Yang
College of Science, China University of Petroleum (East China), Qingdao, China

Pingping Han & Hao Xu
College of Pipeline and Civil Engineering, China University of Petroleum (East China), Qingdao, China

ABSTRACT: The offshore jacket platform, which is in long-term operation in harsh conditions at the ocean, will be confronted with damages such as fractures. As an extremely external strong load, marine earthquake often causes great harm to offshore structures. Therefore, it is of great significance to study the dynamic response and seismic fragility of the damaged offshore platform under seismic loading and to clarify the corresponding seismic limit criteria. In this study, the offshore jacket platform with fracture damage is taken as the research object, and the method is based on nonlinear dynamic analysis results of the platform subjected to possible ground motions. By using the finite element (FE) software ANSYS, the FE model of the damaged offshore jacket platform is established. Based on the incremental dynamic analysis (IDA) method, the amplitude modulation of seismic waves was carried out, and the elastoplastic dynamic time history analysis of the offshore jacket platform under seismic loadings of different intensities is carried out. According to the results of IDA analysis, quantile curves were drawn to study the damage degree of the platform under different seismic intensities; besides, the seismic vulnerability curve of the platform was drawn by using linear regression fitting methods, and the failure probability under different states is calculated to reveal its limit capacity to withstand seismic hazard. The study provides theoretical support for ensuring the safe operation of offshore jacket platform structures under seismic loading.

---

*Corresponding Author: linhong@upc.edu.cn

1 INTRODUCTION

In recent years, the exploitation of offshore oil and gas resources has entered a rapid development stage. Offshore jacket platform is widely used due to its advantages such as high safety, simple configuration and low manufacturing cost. As an extremely external strong load, marine earthquake often causes irreversible local damage to the structure and even leads to the collapse of the structure beyond the limit state.

Limited research has been done on the dynamic response of offshore platforms under earthquakes. Bea et al. (1979) analyzed the dynamic response of jacket platform under earthquake and confirmed that the design of the offshore platform in accordance with API RP2A standard issued by the American Petroleum Institute can ensure its seismic performance in the same environment is sufficient. Naggar et al. (2005) established a simplified Winkel dynamic beam model, which considered the nonlinear response of the offshore platform under earthquake.

At present, there are no specific provisions on the seismic design of offshore platform structures in China. Wang (Yamada et al. 1989) compared and analyzed the current seismic design standards of offshore platforms, and explored the classification of offshore platforms, determination of seismic loading, and determination of platform toughness and strength. Xu et al. (2016) adopted the incremental dynamic analysis method and determined the failure modes of offshore jacket platforms under different seismic loadings. Shao et al. (2016) proposed a set of seismic toughness analysis methods for offshore jacket platforms based on structural nonlinearity.

As a parameter analysis method, incremental dynamic analysis (IDA) can determine the response of structures under different seismic intensities (Vamvatsikos & Cornell 2002). Lee et al. (2014) proposed an IDA-based seismic vulnerability analysis program for steel moment-resistant frames, which shows that the IDA method can fully express the nonlinear behaviour beyond yielding. Asgarian et al. (2010) used the IDA method to evaluate the nonlinear dynamic response of frame structures, and the performance-based assessment data showed that the maximum exceeding probability under different limit states can reveal the superiority of ductile structures.

The aim of this research is to comprehensively evaluate the seismic performance of the offshore jacket platform by simulating the dynamic time-history response of the structure from elasticity, elastic-plastic to overall collapse, and to further analyzes the seismic vulnerability of the structure using statistical methods, so as to provide a reference for the seismic design of the offshore jacket platform. In this study, based on the incremental dynamic analysis method (IDA) and the ANSYS finite element analysis software, 10 ground motion records are selected to study the seismic performance of the offshore jacket platform and analyze its seismic vulnerability.

2 FE MODELLING AND NON-LINEAR DYNAMIC RESPONSE ANALYSIS

2.1 *Geometry model and FE model*

As shown in Figure 1, the offshore jacket platform studied in this paper is a four-legged structure, with a total height of 26.5 m and maximum horizontal width of 12.7 m. The platform consists of fourth levels with X-braces. The dimensions of the upper deck are 15 m × 14 m. The four main pipes have a diameter of 800 mm and a wall thickness of 75 mm, while the horizontal brace pipe and the diagonal brace have a diameter of 500 mm and a wall thickness of 75 mm. The dimensions of the jacket are shown in Figure 1.

Moreover, a fracture in the leg was simulated to represent the potential damage occurring due to the long-term service in the harsh ocean. The position of the fracture could be seen in Figure 1.

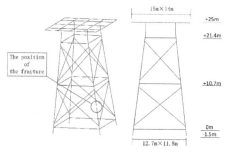

Figure 1. Geometry model of offshore jacket platform.

Figure 2. FE model of the offshore jacket platform.

The ANSYS software was used to establish the FE model of the damaged jacket platform. The jacket was modelled using shell 181 elements, while the I-beam constructing the upper deck was modelled using beam188 elements. To consider the total mass of 32000 kg of the topside, mass21 element was used. For the established FE model, there are 86906 elements and 86219 nodes in total.

The equivalent pile method is adopted to consider the soil-structure interaction, specifically, the pile legs of the jacket platform are fixed at six times the diameter of the pile below the mud line.

2.2 *Material model*

The commonly used Q235 steel was selected for the offshore jacket platform. The density is 7850 kg/m^3. The elastic modulus is 206 GPa. The Poisson's ratio is 0.3. The yield strength is 235 MPa, and the tangent modulus is 12.36 GPa. Here, the Bilinear kinematic (BKIN) model was adopted as the option for plasticity analysis.

2.3 *Modal analysis*

ANSYS provides several modal analysis methods, among which, the subspace method has advantages combining subspace iterative technology and the generalized Jacobian iterative algorithm. The first 4 natural frequencies are 4.8943 Hz, 5.6243 Hz, 5.7786 Hz and 6.6900 Hz.

2.4 *Non-linear dynamic response analysis*

The commonly used methods for seismic dynamic response analysis are spectral analysis and time history analysis. Compared with other analysis methods, the time history analysis method has a characteristic of nonlinear dynamics, which allows the development of plastic deformations that can be examined in detail. Thus, here the time history analysis was used to carry out the dynamic analysis of the jacket platform under seismic loading.

It has been proven that horizontal vibration caused by the earthquake is the main factor leading to the failure of the offshore jacket platform. Therefore, in this study, only the shear seismic wave was applied to the offshore jacket platform. The EL-CENTRO records (as shown in Figure 3) are selected to calculate the dynamic response of the platform.

Figure 4 shows the displacement curve of node 419 which is located in the corner of the deck. It could be seen that the maximum displacement is 0.092425 m, which occurs at t = 2.52 s.

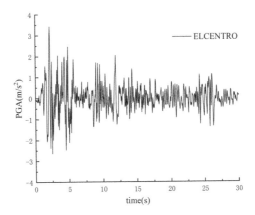

Figure 3. The EL-CENTRO records.

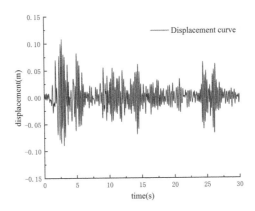

Figure 4. Displacement curve of node 419.

Figure 5 shows the Von-mises stress contour of the offshore jacket platform at this moment before and after t = 2.52 s. The maximum mises stress appears at the bottom of the pile leg. The contour of equivalent displacement is shown in Figure 6. The maximum displacement appears on the upper I-beam.

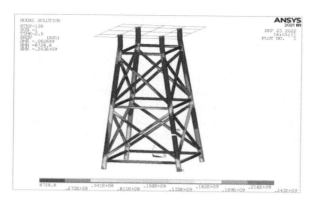

(a) t=2.5s

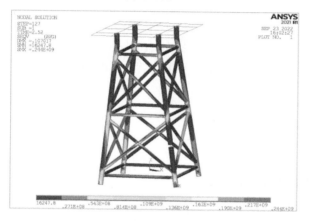

(b) t=2.52s

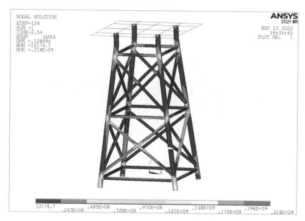

(c) t=2.54s

Figure 5.　Equivalent stress contour of the platform.

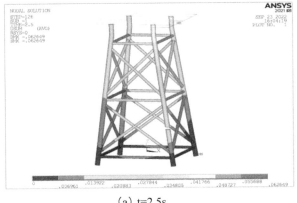
(a) t=2.5s

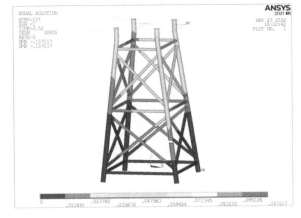
(b) t=2.52s

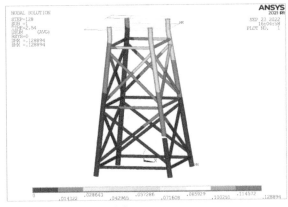
(c) t=2.54s

Figure 6. Equivalent displacement contour of the platform.

3 INCREMENTAL DYNAMIC ANALYSIS (IDA)

By using the IDA method, one or more ground motion records could be modulated to different intensities by a series of Scale Factor (SF), and then the modulated records could be inputted into the structural, to conduct structural elastoplastic analysis. After the relationship curves (i.e. the IDA curves) between intensity measures (IM) of earthquake records and damage measures (DM) of a structure got, the seismic performance of the structure could be evaluated through the behaviour points on these IDA curves.

3.1 *The ground motion intensity measure and the structural damage measure*

Currently, the ground motion intensity measures IM include peak ground acceleration (PGA), peak ground velocity (PGV) response spectrum acceleration Sa (T_1,5%) corresponding to the fundamental period of the structure with a damping ratio of 5%, etc. On the other hand, the structural damage measures DM is an important index to measure the demand of the structure under seismic loading. The commonly used DM includes the maximum Inter-story Drift Ratio of the structure, the maximum base shear force and the maximum displacement of the structure. In this study, the peak ground acceleration (PGA) is selected as the intensity measure of ground motion, and the maximum Inter-story Drift Ratio θ_{max} is adopted as the damage measure of the structure.

3.2 *Ground motion selection*

In IDA analysis, it is necessary to select multiple different ground motion records so as to reduce the uncertainty of seismic loading. However, with the increase in the number of seismic records, the calculation costs in the analysis process will increase sharply. According to the existing research (Luco 2000), more than 10 ground motion records should be adopted in order to achieve better results. Thus, in this study, 10 ground motion records were selected for IDA analysis, including (a)TRI_TREASURE ISLAND_90 record, (b) TRI_TREASURE ISLAND_00 record, (c)El-CENTRO record, (d)Shanghai artificial wave 1 record, (e)Shanghai artificial wave 2 record, (f)Shanghai artificial wave 3 record, (g) Shanghai artificial wave 4 record, (h)Tianjin wave 1 record, (i)Tianjin wave 2 records and (j) Tianjin wave 3 record, respectively, as shown in Figure 7.

3.3 *Modulation of seismic records*

Currently, there are four commonly used modulation methods (Cao 2008; Lv et al. 2009; Vamvatsikos 2002), among which, the hunt & fill tracing algorithm is adopted considering its advantage of both accuracy and efficiency.

The amplitude modulation of peak ground acceleration (PGA) could be written as follow,

$$a'(t) = \frac{A'_{max}}{A_{max}} a(t) \qquad (1)$$

Where, $a(t)$, A_{max}, $a'(t)$ and A'_{max} are acceleration time history and peak ground acceleration before amplitude modulation and acceleration time history and peak ground acceleration after amplitude modulation, respectively. Where $\frac{A'_{max}}{A_{max}}$ is defined as the amplitude modulation coefficient λ.

3.4 *Structural limit state*

In order to evaluate the seismic performance of the structure, it is necessary to define the damage degree and the limit state point. Referring to the seismic code (GB50011 2010). The damage degree of the offshore jacket platform could be divided into five stages, namely, intact, slightly damaged, moderately damaged, severely damaged and collapsed, which are separated by four limit state points of DS1, DS2, DS3 and DS4, as shown in Table 1.

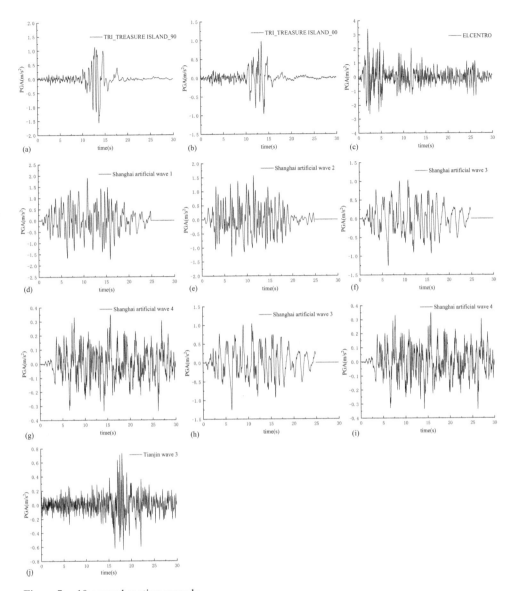

Figure 7. 10 ground motion records.

Table 1. Damage degree of offshore jacket platform under earthquake.

Damage degree	Intact	Slightly damaged	Moderately damaged	Severely damaged	Collapsed
Limit state points	<DS1	DS1-DS2	DS2-DS3	DS3-DS4	>DS4

4 SEISMIC VULNERABILITY ANALYSIS

4.1 Statistical analysis based on IDA results

After IDA analysis, it is necessary to carry out statistical analysis on the IDA curve cluster, so as to evaluate the seismic performance of the structure from the perspective of probability. Therefore, the quantile statistical method was used to summarize IDA curves and calculate the quantile curves of 84%, 50% and 16% of the IDA curve cluster, so as to reduce the dispersion of the IDA curve cluster. The main steps of the statistical method are as follows:

(1) Calculate the median η_D and standard deviation of the natural logarithm β_D of different maximum Inter-story Drift Ratios of different ground motion records under the same PGA, and draw all points (η_D, PGA) into a curve, which is the 50% quantile curve.
(2) Calculate $\eta_D e^{\pm \beta_D}$ and draw all points $\eta_D e^{\pm \beta_D}$ into dot plots respectively, which are 16% and 84% quantile curves.

These three quantile curves represent the exceedance probability of the maximum Inter-story Drift Ratio θ_{max} under a certain PGA in 10 ground motion records, which can be used to evaluate the seismic performance of the offshore jacket platform based on probability.

4.2 Seismic vulnerability

The seismic vulnerability of a structure refers to the conditional probability that the structural response reaches or exceeds a certain limit state under different earthquake intensities. The purpose of analyzing seismic vulnerability is to describe the relationship between different earthquake intensities and structural damage degree from the perspective of probability analysis. In this study, the linear regression fitting method is used to draw the seismic vulnerability curve.

For the linear regression fitting method, the relationship between structural damage measures (DM) and ground motion intensity measures (IM) satisfies the following formula:

$$\widehat{D} = a(IM)^\beta \tag{2}$$

Take the logarithm of both sides of formula (2):

$$\ln \widehat{D} = \ln a + b \ln(IM) = A + B \ln(PGA) \tag{3}$$

Where A and B are regression coefficients obtained by linear regression analysis on all data points on the IDA curve.

The vulnerability probability analysis function is:

$$P_f = P(C/D \leq 1) = (C - D < 0) \tag{4}$$

Assuming that both C and D obey lognormal distribution, the failure probability of each damage state can be expressed as:

$$P_f = \Phi\left(\frac{\ln \widehat{D} - \ln \widehat{C}}{\sqrt{\beta_C^2 + \beta_D^2}}\right) \tag{5}$$

Where D is the structural damage measures, C is the limit state point mentioned above, $\ln \widehat{D}$ and $\ln \widehat{C}$ are the log average. β_C and β_D are the log standard deviation.

5 RESULTS

5.1 Identification of structural limit state based on IDA curve

This section takes the TRI_TREASURE ISLAND_90 record as an example, and adopts the hunt & fill tracing algorithm, with amplitude modulation step $a=0.6$ g, incremental step $b=0.2$ g and initial value $A_1=0.2$ g. A total of 19 amplitude modulations are performed on the TRI_TREASURE ISLAND_90 record and obtain the corresponding maximum Inter-story Drift Ratio value.

According to the literature (Huang 2018; Wang 2021), combined with the results of IDA analysis presented above, four limit state points DS1, DS2, DS3 and DS4 are defined as $\theta_{max1} = 1/770$, $\theta_{max2} = 1/330$, $\theta_{max3} = 1/250$, $\theta_{max4} = 1/100$.

Taking the peak ground acceleration (PGA) as the ordinate and the maximum Inter-story Drift Ratio θ_{max} as the horizontal ordinate, the IDA curve of the offshore jacket platform under the action of this ground motion record is drawn, as shown in Figure 8. It can be seen from IDA curve that the maximum Inter-story Drift Ratio of the structure gradually reaches DS1, DS2 and DS3 until it exceeds 0.01, which means that the offshore jacket platform will collapse, with the increase of the PGA value.

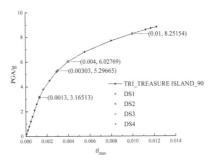

Figure 8. IDA curve.

5.2 Results of IDA curve cluster

According to the IDA analysis process of a single ground motion record, a series of IDA curves can be obtained by performing incremental dynamic analysis on the remaining 9 seismic ground motion records. By drawing 10 IDA curves in the same coordinate system, the IDA curve cluster of the offshore jacket platform under seismic loading can be obtained, as shown in Figure 9.

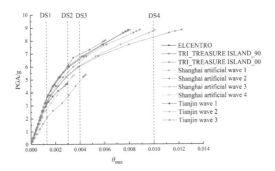

Figure 9. IDA curve clusters of 10 ground motion records.

From the IDA curve clusters shown in Figure 9, it can be seen that when the peak ground acceleration PGA value is small, the IDA curve clusters are relatively concentrated, and the offshore jacket platform structure is in the linear elastic stage at this time; With the increase of PGA, the discreteness of IDA curve cluster gradually increases, and the difference of curve characteristics is more and more obvious, which indicates that there are great differences in the structural response of jacket platform under the action of different ground motion records. Therefore, selecting multiple ground motion records to perform IDA analysis on the offshore jacket platform can more comprehensively reflect the possible dynamic response of the offshore jacket platform structure under seismic loading.

5.3 Results of quantile curves

Mathematical statistical analysis was conducted on the IDA analysis results obtained, and 16%, 50% and 84% quantile curves were drawn, as shown in Figure 10. If we combined with the quantile curve and according to the four limit state points defined in Table 3, the PGA values of the limit state points of the IDA curve with the exceedance probability of 16%, 50% and 84% can be statistically calculated, as shown in Table 2.

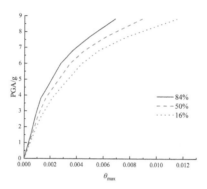

Figure 10. 16%, 50%, 84% quantile curves.

Table 2. The PGA value of the limit state points with the probability of exceeding 16%, 50%, and 84%.

Exceedance probability	DS1 $\theta_{max1} = 1/770$	DS2 $\theta_{max2} = 1/330$	DS3 $\theta_{max3} = 1/250$	DS4 $\theta_{max4} = 1/100$
84%	3.852 g	6.203 g	7.034 g	10.655 g
50%	3.142 g	5.424 g	6.392 g	9.288 g
16%	2.648 g	4.670 g	5.644 g	8.374 g

5.4 Results of seismic vulnerability analysis

Based on the linear regression fitting method and formula (2), carry out linear regression analysis on all PGA and θ_{max} values obtained in the IDA analysis, and obtain the linear regression equation (6).

$$\ln \hat{D} = 1.2157 \ln(PGA) - 7.7866 \tag{6}$$

Substituting the obtained linear regression equation (6) into formula (5), the failure probability is obtained as follows:

$$P_f = \Phi\left(\frac{\ln \widehat{D} - \ln \widehat{C}}{\sqrt{\beta_C^2 + \beta_D^2}}\right) = \Phi\left(\frac{1.2157 \ln(PGA) - 7.7866 - \ln \widehat{C}}{0.5}\right) \tag{7}$$

According to reference (Wang 2021), when the structural damage index is the maximum Inter-story Drift Ratio, $\sqrt{\beta_C^2 + \beta_D^2} = 0.5$. The value takes the four limit state points defined in Section 5.1.

According to the failure probability, the seismic vulnerability curves of the offshore jacket platform are drawn, as shown in Figure 11.

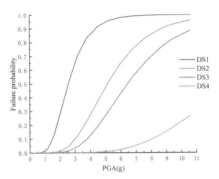

Figure 11. Seismic vulnerability curves.

From the seismic vulnerability curves of the offshore jacket platform, it can be seen that it is easy for the offshore jacket platform structure to reach the degree of slightly damaged under seismic loading, but the failure probability of jacket platform structure collapse is very low, indicating that the platform structure shows good seismic performance after entering the elastoplastic stage. From this, it can be seen that the probability of damage to offshore jacket platform structure under different seismic intensities can be accurately analyzed by vulnerability analysis of offshore jacket platform structure.

6 CONCLUSION

Based on the IDA method, this study analyzes the seismic performance of offshore jacket platforms, and draws the following conclusions:

(1) By drawing the IDA curve of the offshore jacket platform under seismic loading, we can see the whole change process of the platform structure from elasticity to elastoplasticity and finally to collapse state under seismic loading.
(2) By drawing the IDA curve cluster and quantile curves, it can be seen that the dynamic response of the jacket platform is also different due to different ground motion records. The IDA quantile curves drawn in the sense of probability can more accurately evaluate the seismic performance of the structure.
(3) By drawing the seismic vulnerability curves, it can be seen that the platform structure is easy to produce plastic strain and enter a slightly damaged state under the seismic loading; However, when the structural response enters the elastoplastic stage, it will show better seismic performance.

FUNDING

This research was funded by the National Natural Science Foundation of China, China, grant No. 51879272, No. 52111530036; and the Fundamental Research Funds for the Central Universities, China, grant No. 22CX03022A.

REFERENCES

Ao Huang. *Typhoon Vulnerability Assessment of Offshore Jacket Platform*[D]. Qingdao. China University of Petroleum, 2018.

Asgarian B., Sadrinezhad A., Alanjari P., Seismic Performance Evaluation of Steel Moment Resisting Frames Through Incremental Dynamic Analysis, *J. Constr. Steel Res.* (2010), 66(2): 178–190.

Bea R. G., Audibert J., Akky M. R. Earthquake Response of Offshore Platforms[J]. *J Struct Div. Am Soc Civ Eng;* (United States), 1979, 105(2): 377–400.

Che Xu, Yang Yang, Zheng He, et al. Study on Seismic Failure Characteristics of Jacket Platform [J]. *Journal of Harbin Engineering University*, 2016, 37, 232(02): 34–41.

Dagang Lv, Xiaohui Yu, Guangyuan Wang. Structural Collapse Analysis Based on Single Ground Motion Recording IDA Method[J]. *Earthquake Engineering and Engineering Vibration.* 2009, 29(6): 33–9.

GB50011-2010. *Code for Seismic Design of Buildings* [S]. Beijing: China Building Industry Press, 2010.

Jinyue Wang. *Seismic Performance Analysis of GFRP Reinforced Concrete Frame Structures Based on IDA Method*[D]. Jinnan: Jinan University, 2021.

Luco N., Effects of Connection Fractures on SMRF Seismic Drift Demands. *ASCE Journal of Structural Engineering* 2000, 126: 127–136.

Naggar M.H.E., Shayanfar M.A., Kimiael M., et al. Simplified BNWF Model for Nonlinear Seismic Response Analysis of Offshore Piles with Nonlinear Input Ground Motion Analysis[J]. *Canadian Geotechnical Journal*, 2005, 42(2): 365–80.

Lee S.W., Yi W.H., Kim H.J., Seismic Fragility Functions for Steel Moment Resisting Frames Using Incremental Dynamic Analyse, *J. Comput. Struct. Eng. Inst. Korea* 27 (6) (2014) 509–516.

Vamvatsikos D., Cornell C.A. Incremental Dynamic Analysis[J]. *Earthquake Engineering and Structural Dynamics*, 2002, 31(3): 491–514.

Vamvatsikos D. *Seismic Performance Capacity And Reliability Of Structures As Seen Through Incremental Dynamic Analysis:* [D]. Stanford: Stanford University, 2002.

Weidong Shao, Jinlin Hou, Liqin Wang, et al. Seismic Toughness Analysis of Jacket Platform Based on Structural Nonlinearity[J]. *China Offshore Oil and Gas*, 2016, 28(4): 125–131.

Xiujuan Cao. *Research and Application of Elastic-Plastic Analysis Method for High-Rise Building Structure* [D]. Changsha: Hunan University, 2008: 29–30.

Yamada Y., Iemura H., Kawano K., et al. Seismic Response of Offshore Structures in Random Seas[J]. *Earthquake Engineering & Structural Dynamics*, 1989, 18(7): 965–81.

Seismic fragility of continuous beam bridges based on two kinds of correlations

Xiaoying Gou
School of Civil Engineering, Chongqing Three Gorges University, Chongqing, China

Lei Yan*
School of Civil Engineering, Chongqing Three Gorges University, Chongqing, China
School of Civil Engineering, Chongqing Jiaotong University, Chongqing, China

Longfei Cheng, Songrui Liu, Tiansen Chen & Xinyong Wang
School of Civil Engineering, Chongqing Three Gorges University, Chongqing, China

Ping Zhang
Chongqing Wanzhou District Transportation Comprehensive Administrative Law Enforcement Detachment, Chongqing, China

Weijia Xie & Hongyu Chen
School of Civil Engineering, Chongqing Three Gorges University, Chongqing, China

ABSTRACT: The study aims to carry out in-depth research on seismic fragility and realize the performance-based seismic design (PBSD) of continuous girder bridges. For this, the influence of the correlation between uncertain factors on a single component of a bridge and the correlation of failure modes of each component on the vulnerability of the bridge system have been studied. First, the application situation, correlation research method, and damage quantification index of three kinds of uncertainties in the fragility study of a continuous beam bridge are proved. Second, based on the comparison of various research methods, an applicable method for quantifying the two types of correlations is proposed. Finally, the problems to be solved in the study of bridge vulnerability and the development trend are pointed out.

The results show that the seismic fragility analysis method of continuous beam bridge proposed in this paper can not only consider the influence of various uncertainties on the vulnerability of the components but also reasonably quantify the two kinds of correlation problems to carry out the research on the fragility of the system. The study is of great significance for the development of a performance-based seismic design and practical engineering application on continuous beam bridges.

1 INTRODUCTION

The total number of bridges in China is enormous, and earthquakes occur frequently. Bridge damage induces road traffic blocks, and smooth roads turn into a moat. Earthquake resistance and disaster relief rely on footing and handling, which leads to large damage to people's lives and property. Therefore, continuous beam bridges, the most

---

*Corresponding Author: yanlei1988413@163.com

commonly used and the most easily fragile type, have been stubborn in the hearts of countless builders for many years (Zhuang et al. 2009). The seismic fragility analysis is critical for improving bridge seismic performance and carrying out performance-based seismic design (PBSD).

With the development of theories on structural design, the study of bridge seismic fragility developed from the fragility concept based on overall resistant force to local fragility on bridge components (Lu & Wang 2006). But, the key to this research is quantizing the impact of various uncertainty factors on bridge components' capacity, thereby carrying out PBSD to improve the seismic performance of the whole bridge system. A wide range of uncertainty studies (Cornell & Krawinkler 2000; Fema 1999; Hwang & Liu 2004; Lv et al. 2022; Li et al. 2012; Lu et al. 2022; Nielson & Pang 2011; Padgett et al. 2007; Park et al. 1985; Pan et al. 2010; Yu & Lu 2016) found the obvious relation between pier diameter and span and height, and proposed that structural parameters cannot be used as independent variables for component fragility analysis. Li et al. (2012) discovered the failure probability of the mid-span continuous beam bridge system to further exceed its component. Evidently, the two types of correlations are not ignored – various uncertainty factors to a single bridge member and the failure mode of components to the bridge system. Although a great number of researchers emphasized the significance of considering both the factors in previous research, the quantitative analysis method is difficult to propose. Therefore, it is tough to meet the requirement for considering kinds of factors and their correlation in actual engineering and guide the construction of the project. To sum up, previous research shows that it is of great significance for PBSD to quantify two kinds of correlations. Therefore, establishing a systematic seismic fragility analysis method for continuous beam bridges considering both types of correlations under the guidance of PBSD is an urgent subject for future study.

2 UNCERTAINTY FACTORS AND DAMAGE QUANTIFICATION INDEX

2.1 *Uncertainty factors and damage quantification index*

It is crucial for PBSD that quantitative analysis methods about uncertainty factors and correlations are emphasized in the research. However, the judgment for the key factor from three types of uncertainties (including modeling parameters and structural and ground motion uncertainties) induced by differences in bridge characteristics is extremely complicated. Therefore, the paper will explore this problem to ascertain the application, relation analysis method, and damage index about various uncertainty factors and lay the foundation for the next research of correlations.

First, for the scene of three kinds of uncertainties, Cornell et al. (2000) pointed out the general formula for quantitative research for uncertainty factor; Padgett et al. (2007) studied and selected suitable modeling parameters by specific formula to reduce the fragility analysis work, but its impact was regularly ignored than structural and ground motion uncertainty factors. For this, Hwang et al. (2004) researched the influence of both on the fragility of bridges lacking seismic-damaged data. Moreover, Yu et al. (2016) explored the correlation of both and proposed the conclusion: If seismic response characteristics of structures are presented linearly, it merely considered structural factors. Otherwise, it must simultaneously consider both due to the effects of ground motion uncertainty being amplified. It can be seen that modeling parameters have less effect than other parameters for the overall continuous beam bridge. Therefore, the paper regards them as deterministic variables and only considers structural and ground motion uncertainties.

Second, scholars at home and abroad have conducted in-depth research on how to judge the key factor and quantify the two correlations by adopting reasonable methods to face the problem of numerous impact factors and their complex relationships in two types of uncertainties. Padgett. (2007) and Nielson. (2011) found that the method of using multiple ground motion time-history records and each bridge sample to form an independent sample could effectively decrease the effect of ground motion spectrum characteristics and inputting direction, thereby only considering the impact of structural uncertainties. Pan Y et al. (2010) explored the effects of fixed pier structural parameters for the fragility with the above method and comparatively analyzed pier curvature ductility under various conditions; the result showed that the sensitivity of the following parameters was highest, such as superstructure volume weight (W), reinforcement yield strength (f_y), compressive strength of concrete (f_c) and thus they are the key factor of structural uncertainties.

Finally, it is necessary to select suitable damage indexes to quantify the failure mode of components caused by key factors to bridge system fragility. However, previous research treated damage-controlling indexes of fragile components as quantifying indexes for fragility analysis. This paper chooses the fragile pier as the study object and carries out the comparison in three kinds of damage indexes being widely used to lay the foundation for research for corresponding various damage indexes of the two correlations.

The displacement ductility ratio is the most widely used index with the highest relative quantification accuracy among all the indexes presented in Table 1. Furthermore, Li et al. (2012), Song (2017), and Feng (2020) also proved the above conclusion. Therefore, this index can be the most effective tool for describing pier damage statistics in this study.

Table 1. Comparison of three types of damage indexes.

Type	Damage state	Quantitative formula	Application
Park–Ang index	Referring to HAZUS 99 (1999)[11]: five type of damage states	① Park–Ang[12-13] ② Improved Park–Ang[14]	Single degree of freedom systems Bending deformation damage criterion
Capability requirement index[15]	Referring to ATCl996: three Type of damage states	$ID_1 \geq 0.5$, $0.5 \geq ID_2 \geq 0.33$ and $ID_3 > 0.33$	Capability requirement ratio satisfying lognormal distribution
Displacement ductility ratio index	Referring to HAZUS 99 (1999)[11]: five type of damage states	$\mu'_{cy} \geq ID_1$, $\mu_{cy} \geq ID_2 > \mu'_{cy}$, $\mu_{d,0.002(d,0.004)} \geq ID_3 \geq \mu_{cy}$, $\mu_m \geq ID_4 > \mu_{d,0.002(d,0.004)}$ And $ID_5 > \mu_m$	Studying various uncertainties

2.2 *Various uncertainties corresponding to damage indexes*

Based on the above research, the paper proposes quantitative indexes corresponding to structural and ground motion uncertainties for continuous beam bridges. In terms of ground motion uncertainties, SA (spectral acceleration) is widely used as the quantitative index of ground motion intensity in the Code for Seismic Design of the United States and Japan. However, Nielson. (2011) found PGA (peak ground acceleration) more suitable for the index of ground motion intensity than SA by comparative analysis of PSDM (pre-stack depth migration) of PGA, Sa-gm, Sa-0.2, and Sa-1. Besides, the Seismic Ground Motion Parameters Zonation Map of China (GB18306-2015) and a great number of previous research also approve this. Also, the curvature and displacement ductility ratio of cross sections are used as damage indexes of bridge components. For the damage degree of each component under various states by crossing moment-curvature analysis and its specific calculation method the analysis of P-M-Φ for the RC beam crossing section proposed by Zhang (2006) have been used.

3 SEISMIC FRAGILITY ANALYSIS METHOD BASED ON QUANTIFYING TWO TYPES OF CORRELATIONS

The research proposes the damage indexes for correlations and simplifies the analytical methods for ground motion uncertainty. Therefore, the chapter will mainly focus on two types of correlations caused by structural uncertainties. On the one hand, as for the correlations between the factors of the first type of structural uncertainty, the impact of the two correlations can be accurately quantified only by selecting a suitable method and function, thereby drawing a fragility curve based on various factors and displacement ductility ratio. On the other hand, the key factors for continuous beam bridge and the second type of correlation can be resolved by furtherly selecting a suitable method to study component failure relation caused by this factor to the overall bridge fragility. The core of quantifying this correlation is choosing a suitable fragility analysis method and function. For this, the paper carries out a deep research. First, the commonly used bridge seismic vulnerability analysis methods are divided into traditional fragility analysis and experimental-analytical methods based on their data resource and function. Second, the characteristics of the two methods are compared in Table 2. Third, the capture will propose the appropriate method.

In previous research (Du 2007), the traditional method shows characteristics such as simple calculation, strong subjection, and single influence factors, as shown in Table 3. Therefore, its vulnerability assessment results usually lack reliability. Contrarily, the experimental-simulated method has great controllability and overcomes this disadvantage, which will be used to quantify the first type of relation and judge the key factor (in Sections 3.1 and 3.2). The second type of relation based on it will be carried out in Section 3.1.

Table 2. Comparison between two types of seismic fragility analysis method.

Type	Fragility function	Application
Traditional methods	Referring to Cornell. (2000) from PEER proposed the first-order structural reliability limit probability calculation formula	Bridges with more full historical seismic damage data
Experimental-analytical methods	Referring to the general fragility formula based on critical variables (CV) proposed by Lv. (2009) (2019)	Continuous beam fragility analysis

Table 3. Five types of fragility formula.

Type	Critical variable CV	Limit state LS_i	Fragility formula	Fragility mode
$F_{R_i,demand}(x)$	$CV=IM$	$EDP \geq EDP_i$	$P_{f,demand} = P(EDP \geq EDP_i|IM = x)$	Traditional fragility model
$F_{R_i,damage}(x)$		$DM \geq DM_i$	$P_{f,damage} = P(DM \geq DM_i|IM = x)$	
$F_{R_i,decesion}(x)$		$DV \geq DV_i$	$P_{f,decision} = P(DV \geq DV_i|IM = x)$	
$F_{R_i,capacity}(x)$	$CV=EDP$	$DM \geq DM_i$	$P_{f,capacity} = P(DM \geq DM_i|EDP = x)$	Capability fragility model
$F_{R_i,loss}(x)$	$CV=DM$	$DV \geq DV_i$	$P_{f,loss} = P(DV \geq DV_i|DM = x)$	Loss fragility model

3.1 Vulnerability function classification considering different conditional variable values

The core of the experimental-simulated method is selecting the appropriate fragility function to calculate structural failure probability. The two fragility models based on research at home and abroad: (1) Seismic fragility based on traditional risk theory and (2) PBEE (performance-based earthquake engineering)-2 based on reliability and risk performance objectives. The former was defined as the conditional probability of bridge component response exceeding its limit state under specific seismic intensities ($IM=x$), which not only quantitatively characterized the seismic capacity of the bridge but also described the relation between the seismic intensity and the structural damage degree. However, the PBEE-2 treated the engineering demand parameters (EDP) of a structure as the conditional variable to establish the seismic fragility model (Hwang & Liu 2004); therefore it is a narrow capacity fragility mode. From the analysis, it can be seen that the fundamental reason for the difference between the two is different condition variables. The unified form can be realized by utilizing a generalized variable to replace both condition variables. For this, Lv. (2019) established the general fragility formula based on critical variables (CV), as given in Formula 1 (Lv et al. 2019).

$$P_{f,\ integrate} = F_{R_i}(x) = P(LS_i|CV = x) \quad (1)$$

Besides, under CV and LS_i, the existing structure seismic fragility conditional probability formulas are summarized in the following five types: demand fragility, damage fragility, decision fragility, capacity fragility, and loss fragility, as shown in Table 3.

The traditional fragility model adopts IM (intensity measure) as the input variable. It not only directly quantifies the structural and ground motion uncertainty but also further studies the influence of the correlation. Therefore, it is more suitable to analyze the first relation and judge the key factors than others.

3.2 Component theoretical fragility function based on traditional fragility model

Based on the difference between the mathematical model and the calculating method, the traditional fragility model can be attributed to four methods: maximum likelihood estimation, cloud method based on PSDM, quadratic regression analysis method, and increasing dynamic analysis method (IDA). Wu et al. (2017) introduced the function establishment method. The paper merely carries out a comparative analysis of characters and application in Table 4, which provides a reference for choosing a suitable method in quantifying the first

Table 4. Four fragility function establishment methods of bridge components.

Type	Character	Application
Maximum likelihood estimation	Only needing to know the binary damage state of the bridge under earthquake	Empirical seismic fragility analysis
Cloud method based on PSDM	Only needing a small number of seismic waves to predict bridge response accurately	Studying ground motion uncertainties
Quadratic regression analysis method	No need for PSDM analysis	It can accurately quantify the impact of type first relation on component
Increasing dynamic analysis method	NLTHA led to a huge amount of data	It can study structural and ground motion uncertainties

relation by judging the key factors. In conclusion, the quadratic regression analysis method and IDA method are more appropriate than others.

3.3 *Fragility analysis method of continuous beam bridge system based on the second correlation*

The research presents the significance of the second correlation. For this, scholars at home and abroad have conducted many studies. Pan Y et al. (2010) proved the approximated joint probability distribution function (JPDF) can well reflect the relationship between various components; Li et al. (2012) explored the boundary estimation method of the first and second order to establish continuous beam bridge system fragility. The result shows that the second order, by introducing relative coefficient ρ, can effectively consider the correlation between the failure probabilities of each component than the first order. Therefore, the second order method is more suitable for carrying out this study. Besides, in the comparative analysis of the boundary estimation method, Monte-Carlo simulation, product of conditional marginal method (PCM), and improved product of conditional marginal method (IPCM) (Feng 2020; Wu et al. 2017), the applicable scene of four types of methods are: (1) Boundary estimation method and Monte-Carlo method must be based on the result of component fragility analysis; (2) Monte-Carlo method can not only directly establish the fragility curve of bridge system but also independently consider the impact of various uncertainties than the boundary estimation method; (3) PCM method can efficiently establish the fragility curve of the bridge system than both; therefore, it is suitable for investigating bridge systems considering multiple failure modes; (4) above all, the IPCM method can better consider the influence of the interaction between different components with external force and reflect the actual condition of the bridge.

In the study using the correlation function to establish the fragility function of the continuous beam bridge system, Shen et al. (2014) established a Nataf function that can consider uncertain structural factors to analyze the system fragility of the three-span continuous beam bridge. Song et al. (2017) also studied system fragility by building a Copula function with the characteristics of reflecting the complex nonlinear correlation between components and found that it effectively reduced the difficulty of system fragility analysis. Zhang et al. (2021) combined BOX-COX with the Monte-Carlo method to explore the three-span prestressed concrete continuous beam bridge. They proposed a method that not only avoids the three assumptions of the cloud map method but also does not increase the number of NLTHA calculations. Therefore, it can effectively close the gap between simulation and actual engineering.

To sum up, the above analysis selects a suitable method to quantify the second correlation by proposing the scene in various ways. Although the fragility analysis methods based on the correlation functions can greatly improve the efficiency of calculation and accuracy of quantization, it is mainly used to solve specific engineering problems. Therefore, such methods often do not possess universal applicability. As such, the commonly used methods are adopted to carry out fragility analysis of continuous beam bridges. The methods also explore a highly applicable correlation function that aims at the complex actual condition of the Chinese continuous beam bridges with a huge difference in site condition, construction, and so on, establishing a new system fragility analysis method.

4 PROSPECT

This paper reviewed the problems of judging key uncertainty factors and quantifying correlations for continuous beam bridges. However, the above conclusions only provide a reference method for this problem. How to provide a systematic method that aims at the great challenges with difficult quantitative analysis under the complex actual condition of

Chinese continuous beam bridges is for further study. In recent years, bridge seismic fragility research in China has sprung up with the rapidly developing numerical simulation software and fragility theory. Therefore, the methods for seismic fragility of bridge systems also have made great progress and gradually achieved conversion from fragility theory to guiding engineering application. However, to achieve phased progress in the practical engineering application of bridges, it is necessary to consider the characteristics of the actual bridge. This would require further systematic research aimed at a specific type of bridge.

5 CONCLUSION

The paper points out the urgent problem to be further studied based on the analysis of three types of uncertainty: How to suitably quantify two types of correlations for the overall bridge. However, for the challenge still existing in the research of bridge seismic fragility, the author puts forward the following work to be aimed at future research:

(1) The complex actual problems in continuous beam bridges, which will be helpful in choosing an effective analysis method for guiding PBSD on bridge construction.
(2) The limitation involving the previous fragility analysis methods, which mainly resolved specific engineering problems. It would be necessary to establish a set of bridge fragility analysis methods applicable to actual bridge engineering.
(3) The problem that most fragility analysis mainly considered structural uncertainty, meanwhile regarding ground motion as a deterministic parameter, which was against the reality of bridges affected by multi-factors. Future research should be carried out by selecting a suitable method considering such factors.

ACKNOWLEDGMENTS

This project was supported by the General Program of China Postdoctoral Science Foundation (2019M663442), the Scientific and Technological Research Program of Chongqing Municipal Education Commission (KJQN202001216), Wanzhou District Science and Technology Innovation Project (20210306, Study on bearing mechanism and buried depth optimization of corrugated guardrail column of low-grade highway in mountainous area); Chongqing Three Gorges Reservoir Bank Slope and Engineering Structure Disaster Prevention and Control Center Civil and Hydraulic Master's Degree Open Fund (TMSL20YB06); Postgraduate Research and Innovation Project of Chongqing Three Gorges University (YJSKY22066); Cooperation projects between the universities of Chongqing and institutes affiliated with the Chinese Academy of Sciences (HZ2021012).

REFERENCES

Ang A. H., Kim W. J. and Kim S. B. "*Damage Estimation of Existing Bridge Structures,*" ASCE, (1992).
Cornell, C. A. and Krawinkler, H. "Progress and Challenges in Seismic Performance Assessment". (2000).
Du, P., "Seismic Vulnerability Analysis of the Reinforced Concrete Bridge," *Earthquake Resistant Engineering and Retrofitting*, 89–93 (2007).
FEMA. HAZUS99: Technical Manual. 1999.
Feng Y. "A seismic Vulnerability Analysis Method for Bridge System Based on the Improved Product of Conditional Marginal Method," *Journal of Highway and Transportation Research and Development*, 66–72 (2020).
Hwang, H. and J. B. Liu, "Seismic fragility Analysis of reinforced Concrete Bridges," *China Civil Engineering Journal*, 47–51 (2004).

Hwang H., Jernigan J. B. and Lin Y. W. "Evaluation of Seismic Damage to Memphis Bridges and Highway Systems," *Journal of Bridge Engineering*, 322–330 (2000).

Li, L. F., Wu, W. P., Huang, J. M. and Wang, L. H., "Study on system Vulnerability of Medium Span Reinforced Concrete Continuous Girder Bridge Under Earthquake Excitation". *China Civil Engineering Journal*, 152–160+2 (2012).

Lu, D. G. and Wang, G. Y., "Local seismic Fragility Analysis of Structures Based on Reliability and Sensitivity," *Journal of Natural Disasters*, 157–162 (2006).

Lv, G. Y., Wang, K. H., Qiu, W. H. and Zhang, B. Z., "Seismic fragility Models of Double column Piers of Small and Medium-span Bridges in highway Based on random Forest," *Earthquake Engineering and Engineering Dynamics*, 94–103 (2022).

Lv, D. G., Yu, X. H., Song, P. Y., Zhang, P. and Wang, G. Y., "Simplified fragility Analysis Methods for optimal Protection Level Decision Making And Minimum Life-cycle Cost Design of Seismic Structures," *Earthquake Engineering and Engineering Dynamics*, 24–29 (2009).

Lv, D. G., Liu, Y. and Yu, X. H., "Seismic Fragility Models and Forward-backward Probabilistic Risk Analysis in Second-generation Performance-based Earthquake Engineering," *Engineering Mechanics*, 1–11 +24 (2019).

Nielson B. G. and Pang W. C. "Effect of Ground Motion Suite Size on Uncertainty Estimation in Seismic Bridge Fragility Modeling," *Structures Congress*, 23–34 (2011).

Padgett, Jamie., Asce, M. and Desroches, Reginald. "Sensitivity of Seismic Response and Fragility to Parameter Uncertainty," *Journal of Structural Engineering* 133:12(2007).

Pan, Y., Agrawal, Anil., Ghosn, Michel and Alampalli, S. "Seismic Fragility of Multispan Simply Supported Steel Highway Bridges in New York State. II: Fragility Analysis, Fragility Curves, and Fragility Surfaces," *Journal of Bridge Engineering*, (2010).

Park, Y., Ang, Alfredo, H.-S. "Mechanistic Seismic Damage Model for Reinforced Concrete," *Journal of Structural Engineering*, 722–739 (1985).

Shen, G. Y., Yuan, W. C. and Pang, Y. T., "Bridge Seismic Fragility Analysis Based on Nataf Transformation," *Engineering Mechanics*, 93–100 (2014).

Song, S., Qian, Y. J. and Wu, G. "Seismic Fragility Analysis of a Bridge System Based on Multivariate Copula Function," *Journal of Vibration and Shock*, 122–129+208+130 (2017).

Stone W. C. and Taylor A. W. "Seismic Performance of Circular Bridge Columns Designed in Accordance with AASHTO/CALTRANS Standards," *Concrete Bridges*, (1993).

The compilation of seismic zoning map of China, GB 18306–2015 (2015).

Wu, W. P., Li, L. F., Hu, S. C., and Xu, Z. J., "Research Review and Future Prospect of the Seismic Fragility Analysis for the Highway Bridges," *Earthquake Engineering and Engineering Dynamic*, 85–96 (2017).

Yu, X. H., Lu, D. G. and L, B., "Estimating Uncertainty in Limit State Capacities for Reinforced Concrete Frame Structures Through Pushover Analysis," *Earthquakes and Structures*, 141–161 (2016).

Zhang, J. H. "Study on Seismic Vulnerability Analysis of Normal Beam Bridge Piers Based on nuMerical Simulation," *Tongji University*, (2006).

Zhang, P. H., Guo, J. J., Zhou, L. X. and Yuan, W. C. "Seismic Vulnerability Analysis of Bridge Structure Based on BOX-COX Transformation," *Journal of Vibration and Shock*, 192–198 (2021).

Zhuang, W. L., Liu, Z. Y. and Jiang, J. S., "Earthquake-induced Damage Analysis of Highway Bridges in Wenchuan Earthquake and Countermeasures," *Chinese Journal of Rock Mechanics and Engineering*, 1377–1387 (2009).

Strength level earthquake and ductility level earthquake seismic analysis of jack-up platform

Yong Zhang* & Lijun Qian
Marine Design and Research Institute of China, Shanghai, China

ABSTRACT: To ensure that jack-up platforms can survive after earthquakes, seismic analysis for strength level earthquake (SLE) and ductility level earthquake (DLE) should be conducted. SLE seismic analysis assures that the platform structures can still work after earthquakes. In DLE seismic analysis some members of the platform structures are destroyed, but the whole structure does not fall. This paper first introduces the different steps for SLE and DLE seismic analysis. The loads considered for SLE and DLE seismic analysis are studied. Lastly, the criteria for the structure under SLE and DLE are discussed. The structure natural period, code check, and collapse analysis is checked for the three-leg jack-up platform. This research provides a basic design scheme for the SLE and DLE seismic analysis of the jack-up platform.

1 INTRODUCTION OF SLE AND DLE STRUCTURE DESIGN APPROACH

An analysis is carried out using the structural software package SACS.

1.1 *Seismic analysis of SLE*

The strength level earthquake (SLE) seismic analysis is performed using the response spectrum procedure in accordance with API RP 2A (Yang 2019). The SLE analysis is used to design all primary members, joints, and single piles of the jack-up platform (Ghazi 2022).

The analysis is carried out based on the following steps (Wang 2019):

a) A SACS (structural analysis computer system) model is developed such that the inertial properties, including self-weight, buoyancy, variable weight, and entrapped fluid, are accounted for.
b) Dynamic characteristics are calculated by the DYNPAC module of SACS.
c) An earthquake response spectrum analysis is performed using the SACS dynamic response program.
d) The gravitational and dynamic stresses from the analysis are combined to obtain the total member stresses.

1.2 *Seismic analysis of DLE*

The ductility level earthquake (DLE) seismic analysis is performed using the equivalent static load procedure (Konstandakopoulou 2020). These equivalent static loads are generated for 24 directions in 15-degree increments. The number of directions is chosen to ensure

---

*Corresponding Author: wurui986@163.com

maximum structural response for the eight primary directions and those in-between (Tondini 2018).

The analysis is carried out based on the following steps (Abdel Haifiez 2019):

a) A SACS model is developed such that inertial properties, including self-weight, buoyancy, variable weight, and entrapped fluid, are accounted for.
b) Dynamic characteristics are calculated by the DYNPAC module of SACS.
c) An earthquake response spectrum analysis is performed using the SACS dynamic response program to generate the equivalent static loads in 24 directions.
d) Collapse/pushover analysis is performed for the DLE analysis. Equivalent static loads generated from DLE analysis are converted into the structural model. All gravity, buoyancy, and hydrostatic loads are combined, converted, and applied to the converted structural model. The analysis is carried out by incrementing the 0.1×DLE loads to 2.0×DLE loads at a step size of 0.05×DLE loads. Therefore, a total of 48 equivalent static load cases in 24 directions are analyzed.

2 MODEL AND ASSUMPTIONS

2.1 *Main data*

The platform is a three-leg truss jack-up platform. Its main data and environmental conditions are shown in Table 1.

Table 1. Main data and environmental conditions.

Length	72 m
Width	70 m
Molded depth	10 m
Operating water depth	50 m
Penetration depth	15 m
Air gap	15 m
Light weight	20000 t
Variable weight	5000 t

The whole model of the jack-up platform is shown in Figure 1.

Figure 1. The whole model of the jack-up platform.

2.2 Peak ground acceleration

Peak Ground Acceleration (PGA) of 200 years' return period is used for SLE. PGA of 1000 years' return period is used for DLE. A uniform modal damping ratio of 5% is applied for all modes in computing the dynamic characteristics of the structure (Deng 2021).

Response seismic spectrum is listed as follows:

$$\beta(T) = \begin{cases} 1 & T \leq T_0 \\ 1 + (\beta_{max} - 1)\dfrac{T - T_0}{T_1 - T_0} & T_0 < T \leq T_1 \\ \beta_{max} & T_1 < T \leq T_g \\ \beta_{max}\left(\dfrac{T_g}{T}\right)^c & T_g < T \end{cases} \quad (1)$$

where, T is the period of the response spectrum, β_{max} is the maximum value of the response spectrum, T_g is the characteristic period of the response spectrum, and c is the attenuation coefficient.

Parameters of standard design spectra for the site are listed in Table 2.

Table 2. Parameters of standard design spectra.

Probabilities	β_m	T_0	T_1	T_g	C
200a	2.25	0.04	0.10	0.55	0.9
1000a	2.25	0.04	0.10	0.80	0.9

2.3 Splash zone and corrosion allowance

The corrosion allowance for all steel members in the splash zone is assumed to be 0.3 mm/year for primary members. Therefore, a total reduction in thickness of 9.0 mm should be considered for analysis in this paper.

2.4 Ice contact zone and abrasion allowance

The abrasion allowance for all steel members in the ice contact zone is assumed to be 3.0 mm for in-place analysis.

2.5 Seismic loads

The response of the first 30 natural modes is combined using the complete quadratic combination (CQC) method for the three individual directions, as shown in Table 3. The responses of the three directions are then combined using the square root of the sum of the squares (SRSS) method.

Table 3. Load factors.

Direction	Spectrum Direction Factor	Direction Combination Factor
X	1.00	1.00
Y	1.00	1.00
Z	0.50	1.00

2.6 Gravity-seismic load combination

For SLE analysis, the results of gravity load and dynamic load are combined. No sign (direction) is associated with the dynamic load. For DLE analysis, after combining the modal response to obtain the global structural response, equivalent static loads are generated from the structural displacements at all joints. Static loads were generated for all 24 earthquake directions. These loads were then combined with gravity loads and hydrostatic loads to obtain an overall seismic response.

3 RESULTS AND CONCLUSIONS

3.1 Stress criteria

All structural members and joints are checked for compliance with the requirements of API RP 2A-WSD and AISC 335 codes. An allowance stress increase of 70% is used for member code checks. For joint punching shear checks, 1.6 factor of allowance stress increase is used.

3.2 Structure natural period

The first three natural periods and the corresponding deformed modes for SLE and DLE are summarized in Table 4.

Table 4. Natural periods.

Mode	Period (sec) SLE	Period (sec) DLE	Deformed Modes
1	5.785	5.841	Global swaying in Y-dir.
2	5.747	5.821	Global swaying in X-dir.
3	4.985	5.126	Global twisting about Z-axis

3.3 Code check results

The joint and member code check results are shown in Table 5.

Table 5. Code check results.

Items	UC
Members	0.86
Joints	0.88

3.4 Collapse/Pushover for DLE

For SLE analysis, the results of gravity load and dynamic load are combined. No sign (direction) is associated with the dynamic load. Therefore, when the results of gravity load and dynamic load are combined, there were a total of two load combinations for each seismic load case.

For DLE, collapse/pushover analysis with equivalent static earthquake load is performed. The results of collapse/pushover analysis indicate that the reserve strength ratio (the coefficient of collapse DLE load) is not less than 1.0. The platform can withstand earthquake loads in 1000-year ductility level earthquake conditions.

3.5 *Conclusions*

These calculation results show that the structural members are of adequate size to withstand all forces that may occur during its in-place condition. Deflections are found to meet the correlative requirement.

REFERENCES

Abdel Hafiez, H. E. (2019). Strength and Ductility Level Earthquake Design for Gulf of Suez Oil Platforms. Journal of Seismology, 23(2), 199–215.

Deng, W., Tian, X., Han, X., Liu, G., Xie, Y., & Li, Z. (2021). Topology Optimization of Jack-up Offshore Platform Leg Structure. Proceedings of the Institution of Mechanical Engineers, Part M: Journal of Engineering for the Maritime Environment, 235(1), 165–175.

Ghazi, Z. M., Abbood, I. S., & Hejazi, F. (2022). Dynamic Evaluation of Jack-up Platform Structure Under Wave, Wind, Earthquake and Tsunami Loads. Journal of Ocean Engineering and Science, 7(1), 41–57.

Konstandakopoulou, F. D., Evangelinos, K. I., Nikolaou, I. E., Papagiannopoulos, G. A., & Pnevmatikos, N. G. (2020). Seismic Analysis of Offshore Platforms Subjected to Pulse-type Ground Motions Compatible with European Standards. Soil Dynamics and Earthquake Engineering, 129, 105713.

Tondini, N., Zanon, G., Pucinotti, R., Di Filippo, R., & Bursi, O. S. (2018). Seismic Performance and Fragility Functions of a 3D Steel-concrete Composite Structure Made of High-strength Steel. *Engineering Structures*, 174, 373–383.

Yang, Y., Wu, Q., He, Z., Jia, Z., & Zhang, X. (2019). Seismic Collapse Performance of Jacket Offshore Platforms with Time-variant Zonal Corrosion Model. *Applied Ocean Research*, 84, 268–278.

Yunlong, Wang, Kai, L., Guan, G., Yanyun, Y., & Fei, L. (2019). Evaluation Method for Green Jack-up Drilling Platform Design Scheme Based on Improved Grey Correlation Analysis. Applied Ocean Research, 85, 119–127.

Shrinkage and creep effects for prestressed concrete beam bridge based on beam length shortening

Lei Cheng*

Research Institute of Highway, Ministry of Transport, Beijing, China

ABSTRACT: To obtain the time-dependent regularity of shrinkage and creep in long-span prestressed concrete beam bridges accurately, enhance the prediction accuracy of long-term deformation and prestress losses, the longitudinal shortening of beam length of a prestressed concrete bridge with five-span continuous rigid frame-continuous beam structure is analyzed. The analysis adopts four general and two modified calculation models according to the theory of uniformity distribution on the cross-section of shrinkage and creep and the calculation method of increment superposition in stages. The calculation model that can reflect the long-term shrinkage and creep regularity of the structure is obtained through the comparison and analysis of calculated and measured data. The data is verified through the test of the longitudinal prestress losses caused by shrinkage and creep. The results show that the modified model B3 has high prediction accuracy, which can be used to analyze the shrinkage and creep effects in long-span prestressed concrete beam bridges.

1 INTRODUCTION

Over the past 30 years, with the improvement of bridge design level and construction technology, long-span prestressed concrete beam bridges have developed rapidly in China. However, in recent years, the problem of mid-span deflection has become increasingly prominent, which is a relatively common phenomenon (Xie 2010). At present, most studies suggest that shrinkage and creep of concrete are the main causes of mid-span deflection (Feng 2009; Xie 2010; Hu 2014).

Creep and shrinkage are inherent properties of concrete. The phenomenon that the deformation of concrete increases with time under the continuous action of the constant load is called creep. All concrete under load will undergo creep, and the creep effect changes with time. Shrinkage refers to the phenomenon of volume reduction of concrete in the initial stage of condensation in the air or during the hardening process without stress. It is generally believed that the shrinkage and creep of concrete have basically the same influencing factors.

The prestressed concrete main beam is subjected to bending moment, shear force, and axial force. As such the influence of shrinkage and creep on its deformation is not only reflected in the continuous deflection of the main beam, but also in the shortening of beam length. In this paper, a comparative study on the shrinkage and creep effect of the structure is carried out based on the analysis of the length shortening of the main beam of a large-span variable-area prestressed concrete beam bridge.

2 ANALYSIS ON BEAM LENGTH SHORTENING

2.1 Research background

The analysis object is a five-span one-connection variable-area prestressed concrete continuous rigid frame-continuous beam bridge. Its span arrangement is (65+160+210+160+65)

---

*Corresponding Author: L.cheng@rioh.cn

m = 660 m. The overall arrangement is shown in Figure 1. The main beam is in a three-way prestressed concrete structure with a single box and single chamber box section. There is a diaphragm on the top of the pier. The top plate of the box beam is 17.15 m wide and has a 2 % one-way cross slope, while the bottom plate is 8.35 m wide. The cantilever length is 4.4 m. The beam heights of the rigid frame pier top, continuous pier top, and common pier top are 10.5 m, 5.8 m, and 3.5 m, respectively. The beam height of the closure section is 3.5 m, which changes according to the quadratic parabola along the span direction. The thickness of the top plate of the box beam on the top of the rigid frame pier is 60 cm, and the thickness of the remaining positions is 28 cm. The thickness of the bottom plate of the rigid frame pier top, continuous pier top, and common pier top box beam is 120 cm, 60 cm, and 32 cm, respectively. The thickness of the bottom plate of the closure section is 32 cm, which varies along the span direction according to a quadratic parabola. The web thickness of the box beam is divided into 120 cm, 60 cm, 50 cm, and 40 cm. In the tapered beam section, the web thickness varies in a diagonal line along the inner side. The main beam was designed according to the bridge design specifications 85 (JTJ 023-85 1985) with fully prestressed concrete components, using hanging baskets for symmetrical overhanging construction. The bridge was completed and opened to traffic in 1999.

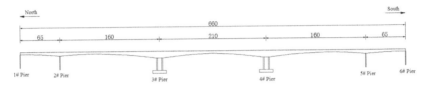

Figure 1. General layout of Bridge (Unit: m).

Recent inspections have found that the gap between the 320 mm modulus expansion joints at both ends of the main beam was 4 × 53 mm in summer, and the rubber waterstop is overstretched and cracked due to the inability to provide effective shrinkage at low temperatures in winter. Due to the large longitudinal slope of the bridge deck (3%), it was initially believed that the excessive gap of the expansion joint was caused by the slip of the T beam of the approach bridge to the abutment. However, after detailed investigation, it was found that the shear deformation of the approach plate bearing at the common pier was biased toward the mid-span of the main bridge. The center of the upper seat plate of the bearing at the common pier and the continuous pier of the main bridge was also offset to the mid-span of the main span relative to the center of the bottom basin. The anchor bolts of the upper seat plate of some continuous piers were cut off due to their contact with the bottom basin. By measuring the longitudinal offset of the center of the upper seat plate at the shared pier relative to the center of the bottom basin, and considering the temperature effect correction, the beam length of the bridge when it had been in operation for 20 years was shortened by about 410 mm compared with the beam length of the bridge when the support of the common pier was installed (the side span was closed). Therefore, the excessively large expansion of the joint gap of the main bridge was caused by the beam length shortening of the main bridge.

2.2 *Reasons for beam length shortening*

The main reasons for the change in beam length are the shrinkage and creep of concrete, prestressed load, dead load, and vehicle load. To accurately analyze the influence of various factors, according to the construction steps, duration and load changes provided by the design and construction documents, a plane rod system model was established for structural finite element calculation and analysis.

Table 1 lists the changes in beam length caused by prestressed load, dead load, and vehicle load during the stage from the side span closure to the completion of the bridge (because the bridge has a large traffic flow that could not be closed, the impact of the vehicle load was included). From the data in Table 1, it can be concluded that the prestress effect caused the beam length to shorten, while the dead load and vehicle load caused the beam length to elongate. The total beam length change under the combined action of the three was 25.9 mm, and the main beam was in a shortened state.

Table 1. Values of beam length caused by different loads (Unit: mm).

Effect	Prestressed load	Dead load	Car load
Beam length change	46.4	−10.0	−10.5

Note: Positive and negative values indicate beam length shortening and elongation.

The creep deformation is related to the load on the concrete. For a fully prestressed concrete beam, there is no tensile stress in the normal section of the concrete along the longitudinal prestressed reinforcement direction. That is, the normal stress of each section of the main beam is the compressive stress. Therefore, qualitatively, the creep will cause the length of the main beam to shorten.

From the above analysis, it can be concluded that the reasons for the shortening of the beam length are shrinkage and creep of the concrete and the effect of the prestressed load, while the effect of the dead load and the vehicle load leads to the elongation of the beam length.

3 COMPARATIVE ANALYSIS ON THE SHRINKAGE AND CREEP EFFECT BASED ON THE ANALYSIS ON BEAM LENGTH SHORTENING

3.1 Calculation method

3.1.1 Shrinkage

The concrete structure is not caused by an external force, but is determined by the characteristics of the structural material itself; the shrinkage of concrete changes with time, and its growth rate is affected by conditions such as air temperature and humidity. At present, a large number of researches have been carried out on the causes and mechanisms of various shrinkages, and a relatively clear conclusion has been reached (Lu 2012). The shrinkage has nothing to do with the stress state of the structure, but is only related to the constraint state of the structure itself. Reinforcement will affect the development of concrete shrinkage and deformation, and produce structural internal force redistribution and section stress redistribution. To simplify the calculation, it is generally assumed that the variation law of shrinkage is similar to the variation law of creep (Fan 1993).

3.1.2 Creep

Studies have shown that when concrete is stressed $\sigma_c < 0.4 f_{ck}$, the creep behavior is basically linear. The creep caused by the load increment at a certain time is basically independent of the creep caused by the previous load, and the superposition principle is effective (ACI Committee 209 2008). Since the working stress of the concrete structure during service usually meets the above requirements, the creep effect analyses of the concrete structure basically adopt the linear creep theory (Hu 2014).

Creep is related to stress level. For sections with different stress distributions along the section height, the creep at different positions of the section height affected by the stress is also different. However, it is generally assumed that the shrinkage and creep of each part of the same section of the concrete structure are the same. The regular change is called the coherent assumption of shrinkage and creep.

According to the above theory, a method for calculating the creep of a statically indeterminate prestressed concrete beam is obtained (Lu 2012) on the assumption that the creep of the same section of the main beam is equal to the creep produced by the concrete at the center of gravity of the prestressed steel bar, the concrete creep is calculated separately. The internal force redistribution of the statically indeterminate structure and the cross-sectional stress redistribution caused by the creep of the prestressed reinforced concrete are accumulated into the final calculation result by the incremental superposition algorithm in stages. The displacement, internal force, and stress increment caused by concrete creep at each stage were calculated according to the concrete creep time function from the cumulative value of concrete element stress at the beginning of each stage. The age and period of each concrete element was taken into consideration in the calculation stage to calculate the displacement of each stage. The internal force and the stress increment are linearly superimposed, and the calculation result of the whole process of concrete creep is obtained.

Many scholars have confirmed that the assumption of shrinkage and creep consistency is inaccurate to a certain extent, and the distribution of shrinkage and creep on the cross-section is not consistent. The distribution property will have a great influence on the internal force and deflection of the structure (Xie 2007, 2017). But for the prestressed concrete beam as a flexural member, the shortening of the beam length caused by the increase of deflection can be ignored. (For this bridge, the ratio of the mid-span deflection increment to the shortened value of the beam length is 1:0.037.)

3.2 *Calculation model of shrinkage and creep*

3.2.1 *Basic idea*

Over the years, scholars at home and abroad have carried out a large number of theoretical and experimental studies on the shrinkage and creep of concrete, and have proposed a variety of theories and calculation methods. Many countries have also adopted different models to introduce norms and recommend their application according to their national conditions.

However, shrinkage and creep are affected by many factors, and some factors are coupled with each other. Different models deal with the coupling effects of each factor differently. As such it is difficult to describe them with a unified expression. At the normative level, the calculation models of shrinkage and creep of national codes are not uniform, even if the codes of different regions or different industries in the same country have different calculation models. At the regional level, regional differences lead to differences in ground materials and environments in various regions. Since the properties and influencing factors of concrete are random variables, it is difficult to obtain parameter values that are consistent with the actual situation during the design and construction process. The results calculated by the same model are also different. Therefore, many researchers suggest that after a short-term test of concrete shrinkage and creep should is carried out for a specific project, and different models are fitted and revised according to the short-term test data that a suitable model should be selected as the applicable model for the project (Wang 2010; Xu 2010; Zeng 2014).

Since a short-term test of concrete shrinkage and creep was not carried out during the construction of the bridge, it is difficult to simulate the mix ratio and environment during construction. As such the shrinkage and creep coefficient of the main beam concrete could not be fitted through the test. Although the calculation models recommended by various specifications have limitations, they are also a summary of many theories and experiments, so they have certain representativeness and applicability and can simulate the changing laws of concrete shrinkage and creep to a large extent. Similarly, various correction formulas obtained by relevant researchers through short-term shrinkage and creep tests of laboratory standard specimens or real bridge concrete also have high reference values. Therefore, in the absence of reliable real bridge test data, it is effective to refer to the existing research results for the analysis of the shrinkage and creep of the bridge.

This paper assumed that the variation law of bridge concrete shrinkage and creep was the same as the standard formula or revised formula, used the standard formula or revised formula to analyze and calculated the shrinkage and creep law of the bridge and the structural deformation caused by it, and compared the theoretical calculation results with the measured data to find a suitable model for the bridge and use it to predict the long-term deformation of the bridge under shrinkage and creep.

3.2.2 *Model selection*

At present, the creep models adopted by various countries can basically be divided into three categories (Hu 2014). The first category is to describe the creep law as a whole according to the performance characteristics of concrete creep in the load-bearing stage, not subdivided into basic creep and dry creep or recoverable and irrecoverable creep; the representative model is CEB-FIP MC1990-2010, ACI 209R-92, AASHTO LRFD 2007, GL2000, etc. The second category is based on the different properties of creep under the condition of water exchange between concrete and the environment, and the creep behavior is represented by basic creep and dry creep; representative models include BP-KX, B3, B4 and so on. The third category is based on the deformation recovery characteristics of concrete during unloading, and the creep behavior is represented by recoverable creep and non-recoverable creep; representative models include CEB-FIP MC1978 and so on.

In the three editions of the *Code for Design of Highway Reinforced Concrete and Prestressed Concrete Bridges and Culverts* (namely JTJ 023-85, JTG D62-2004 and JTG 3362-2018), the shrinkage and creep models refer to the CEB-FIP model specifications, however, the difference is that JTJ 023-85 adopts the MC1978 model, JTG D62-2004 and JTG The 3362-2018 uses the MC 1990 model.

Four calculation models were selected corresponding to the classification of creep models, namely: CEB-FIP MC1990 (CEB-FIP Model Code 1990 1993) model of the first category (applicable to JTG D62-2004 and JTG 3362-2018) with ACI 209R-92 (Hu 2014) model (the model recommended by the American Concrete Institute), B3 (Bazant 1995) model of the second category (International Federation of Testing and Research in Materials [RILEM] recommended model) and Three types of CEB- FIP MC1978 model (also applicable to JTJ 023-85).

Given that the coarse and fine aggregates, cement varieties and mix ratios used in the concrete of the main beam of the bridge are similar to those of a prestressed concrete box beam bridge in the south measured in the article of Zeng (2014), only the environment is different, so the shrinkage rate is selected. Variable correction formula (based on JTG D62-2004 specification) for comparative calculation. In addition, many pieces of literature (Ding 2004; Wang 2010, 2017) pointed out that the revised B3 model has higher prediction accuracy, so the B3 model revised formula of the article of Wang (2017) is selected to calculate together.

Based on the above six shrinkage and creep models, the time-dependent curves of shrinkage strain and creep coefficient of the bridge are drawn as shown in Figures 2 and 3, respectively.

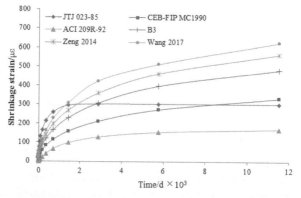

Figure 2. Comparison of shrinkage strain.

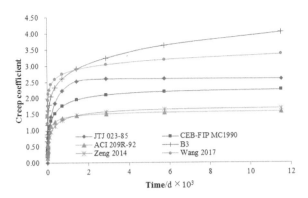

Figure 3. Comparison of creep coefficient.

3.3 *Results comparison and analysis*

The time functions of the shrinkage strain and creep coefficient of the six models are substituted into the analysis program, and the structural deformation caused by shrinkage and creep is calculated by the step-by-step incremental superposition algorithm according to the consistency theory of shrinkage and creep.

The comparison results of the beam length shortening value and the measured value caused by different shrinkage and creep models are listed in Table 2. The measured value deducted the beam length change value caused by the prestressed load, dead load and vehicle load. The year value started when the side span is closed. It can be concluded from the data in Table 2: Compared with the measured values, the calculated values of the B3 model and the B3 modified model (Wang 2017) are larger, and the calculated values of the other models are smaller; among the existing general models, CEB-FIP MC The calculated values of the 1990 model and the B3 model are relatively close to the measured values, with errors of −34.4% and 33.3%, respectively, while the difference is 51.8% calculated according to the specification JTJ 023-85, which is a reference for the design of the bridge; the revised model based on the test results-the B3 correction model (Wang 2017) and the JTG D62-2004 correction model (Zeng 2014) have the closest calculation values to the measured values, and the B3 correction model has the highest accuracy, with an error of only 5.7%.

Table 3 lists the ratio of the beam length shortening value caused by creep to the total beam length shortening value under the combined action of shrinkage and creep in the four models with relatively small errors. increases, the creep effect value also gradually increases; the creep effect in the B3 model is significantly greater than the shrinkage effect, and the

Table 2. Comparison of shortening values of beam length caused by shrinkage and creep unit: mm.

Model category	1 year	5 years	10 years	20 years	50 years
JTJ 023-85	137	181	185	185	185
CEB-FIP MC1990	90	162	204	252	306
ACI 209 R - 92	72	132	160	188	210
B3	166	304	400	512	650
Zeng 2014	142	222	266	308	356
Wang 2017	134	236	308	406	566
Measured value	/	/	/	384	/

Note: Positive and negative values indicate beam length shortening and elongation respectively; "/" indicates no measured data.

Table 3. Ratios of shortening values of beam length caused by creep and creep & shrinkage.

Model category	1 year	5 years	10 years	20 years	50 years
CEB-FIP MC1990	0.47	0.54	0.59	0.63	0.69
B3	0.73	0.77	0.78	0.78	0.78
Zeng 2014	0.28	0.34	0.38	0.43	0.48
Wang 2017	0.46	0.54	0.59	0.63	0.67

creep and shrinkage effect values of the JTG D62-2004 revised model (Zeng 2014) are basically the same after 20 years of loading. Equivalent; the proportion of the creep effect in the CEB-FIP MC1990 model and the B3 modified model (Wang 2017) is basically the same, and both are slightly greater than the shrinkage effect after 5 years of loading. In general, except for the JTG D62-2004 revised model, the creep effect of the other three models is greater than the shrinkage effect.

4 LOSS OF LONGITUDINAL PRESTRESS CAUSED BY SHRINKAGE AND CREEP

The prestress losses of the prestressed tendons of the prestressed concrete structure after the tension and anchorage due to the shrinkage and creep of the concrete and the relaxation of the prestressed tendons are called long-term prestress losses, which increase with time. Due to the coupling effect of concrete shrinkage and creep and the relaxation of prestressed tendons, the calculation of long-term prestress losses is complicated. To simplify the calculation, the longitudinal prestress losses caused by shrinkage and creep discussed in this section did not consider the coupling effect of the two.

Twenty bundles of prestressed tendons on the bottom of the box beam near the 2nd to 5th spans were selected to test the permanent stress, and the parameters were corrected through the full-scale test to obtain the final result. The test site is shown in Figure 4. It was assumed that the actual prestress losses caused by other factors except for shrinkage and creep is the same as the specified value, after deducting the existing effective stress and the prestress losses caused by other factors except for shrinkage and creep from the construction tension control stress. The measured value of the prestress losses caused by shrinkage and creep was obtained, and the average value was 126 MPa.

The results based on JTJ 023-85 model, CEB - FIP MC1990 model, B3 model, JTG D62-2004 revised model (Zeng 2014) and B3 revised model (Wang 2017) are listed in Table 4,

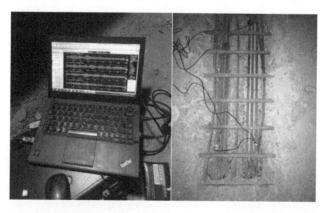

Figure 4. Stress test of prestressing tendon.

Table 4. Theoretical mean values of prestress losses caused by shrinkage and creep.

Model category	The average value of prestress loss / Mpa 5 years	Error between the theoretical mean value and the measured value / % 20 years
JTJ 023-85	82.7	−34.4
CEB-FIP MC1990	78.0	−38.1
B3	117.5	−6.8
Zeng 2014	101.7	−19.3
Wang 2017	110.8	−12.1

including the theoretical mean values of the prestress losses caused by shrinkage and creep and the errors between the theoretical mean values and the measured values. It can be concluded from the data in the table that the calculated values of the B3 model and the B3 correction model (Wang 2017) are the closest to the measured values, with errors of −6.8% and −12.1%, respectively. Therefore, the B3 correction model was used to analyze the shrinkage and creep effect of the bridge with high precision.

It is worth mentioning that, given the difference between the theoretical value of the tension control stress and the losses of various prestresses under the anchor and the actual situation, the error of the permanent stress data collection, the single and the insufficient number of prestressed tendon sample locations, and the shrinkage of the section. For the effects of the non-uniform distribution of creep and the compactness of prestressed pipeline grouting, the measured values and model calculated values are different from the actual situation. Further research is needed to obtain more accurate results. In addition, the theoretical average values of prestress losses obtained based on the five shrinkages and creep calculation models listed in Table 4 are all smaller than the measured average values. It indicates that the calculated values of each model underestimate the actual shrinkage and creep of concrete, the actual stress level is far lower than the existing effective stress value calculated according to the original design, and the compressive stress reserve of the main beam is greatly reduced. If it is allowed to develop, it will cause excessive deflection of the box beam, excessive width of cracks in the web and bottom plate, and even occur defects that affect the safety of bridge structures.

5 CONCLUSION

(1) Six kinds of shrinkage and creep models are used to analyze the beam length change of a large-span variable-section prestressed concrete beam bridge. By comparing the theoretical calculation value with the measured value, the B3 correction model, a model that can reflect the long-term shrinkage and creep change law of the bridge, is obtained. At the same time, the consistency between the theoretical value and the measured value of the longitudinal prestress losses caused by shrinkage and creep also verifies the good applicability of the B3 correction model to the bridge.
(2) The commonly used shrinkage and creep models all take into account, the environmental humidity, concrete compressive strength, theoretical thickness or body surface ratio of components, cement type, mix ratio, water-cement ratio, concrete age and load holding time during loading. Several or all of the main parameters are introduced into mathematical expressions, so various shrinkage and creep models have certain applicable conditions, but each model can basically cover most concrete types and can be used for the analysis of shrinkage and creep of conventional concrete structures.

(3) The influence of concrete shrinkage and creep on the structure is also related to the structure type, construction method, stress state, etc. The influence of shrinkage and creep on the internal force and deformation of the structure should be specifically analyzed according to the characteristics of different bridge structures.
(4) Given that the calculation results of different shrinkage and creep models are quite different, data and calculation models that conform to local actual conditions should be selected for analysis. For new bridges, it is recommended to revise the selected model through short-term tests of actual bridge specimens, and then conduct subsequent internal force and deformation analysis. For existing bridges without measured shrinkage and creep data, it is recommended to use various calculation models to analyze the law of shrinkage and creep of the bridge and the structural internal force and deformation caused by it. Then the theoretical calculation results should be compared with the test data of the actual bridge to find the applicable model to predict the internal force and long-term deformation of the bridge under the action of shrinkage and creep.

REFERENCES

ACI Committee 209. (2008). *Guide for Modeling and Calculating Shrinkage and Creep in Hardened Concrete (209.2R-08)*. Farmington Hills: America Concrete Institute.
Bazant Z. P., Murphy W. P. (1995). Creep and Shrinkage Prediction Model for Analysis and Design of Concrete Structures-Model B3. *Materials and Structure* 28 (1): 357–365.
Comite Euro-International Du Beton. (1993). *CEB-FIP Model Code 1990*. London: Thomas Telford Services Ltd.
Ding W. S., Lu Z. T., Meng S. P., et al. (2004). Analysis and Comparison of Prediction Models for Concrete Shrinkage and Creep. *Bridge Construction* 2004(06): 13–16.
Fan L. C. (1993). *Bridge Engineering*. Beijing: China Communications Press.
Feng P. C. (2009). Study of Key Technical Issues on Design of Continuous Rigid-Frame Bridge. *Bridge Construction* 2009 (04): 46–49.
Hu D. (2014). *Theory of Creep Effects in Concrete Structures*. Beijing: Science Press.
Lu J. M. (2012). *Research on Prestress Losses and Calculation Method of Shrinkage and Creep in Concrete*. Beijing: Research Institute of Highway, Ministry of Transport.
Ministry of Transport of the People's Republic of China. (1985). *Specifications for Design of Reinforced Concrete and Prestressed Concrete Highway Bridges and Culverts (JTJ 023-85)*. Beijing: China Communications Press.
Wang H., Qian C. X. (2010). Modification of Creep Prediction Models Based on Short-term Concrete Test Data of Sutong Bridge. *Bridge Construction* 2010 (02): 32–36.
Wang Y. B., Jia Y., Liao P., et al. (2017). Comparison and Analysis of Prediction Model of Concrete Shrinkage and Creep. *Railway Engineering* 57 (8): 146–150.
Xie J., Wang G. L., Zheng X. H. (2007). State of Art of Long term Deflection for Long Span Prestressed Concrete Box beam Bridge. *Journal of Highway and Transportation Research and Development* 24 (01): 47–50.
Xie J., Zheng X. H. (2010). *Research of Treatment for Deflection of Long-span Prestressed Concrete Bridge*. Beijing: Research Institute of Highway, Ministry of Transport.
Xie J., Pan B. L., Zhou Y. S. (2017). Time-dependent Correlation Between Non-uniformity Distribution of Cross-section Shrinkage and Creep and Vertical Cracking in Web of T-shaped Beam Bridge. *Journal of Highway and Transportation Research and Development* 34 (11): 79–83, 109.
Xu H., Peng X. M., Zou L. Q. (2010). Prediction Research of Long-term Creep for Bridge Concrete Based on Short-term Test. *Journal of Chongqing Jiaotong University: Natural Science Edition* 29 (4): 518–520.
Zeng D., Xie J., Zheng X. H., et al. (2014). Prediction of Long-term Deflection and Correction of Shrinkage and Creep Model of High-strength Concrete of Bridge Based on Measured Short-term Data. *Journal of Highway and Transportation Research and Development* 31 (11): 72–77.

Dynamic response analysis of metro station under bidirectional seismic load

Quan Wang, Yucheng Zhao*, Xuzhe Zhang & Ting Shen
School of Civil Engineering, Shijiazhuang Railway Shijiazhuang, China

ABSTRACT: In recent years, the safety of the structure of subway stations, as the transit center of urban rail transportation, has become a key research direction. Most urban areas in China are in seismic zones, and earthquakes are one of the main factors causing the destabilization of underground structures. In order to investigate the dynamic response of conventional subway stations under the action of seismic waves in different directions, a box-type subway station is used as the research object, and finite element dynamic time analysis is used to investigate the seismic response of subway stations under horizontal, vertical and horizontal-vertical coupled action of seismic waves. The results show that the horizontal seismic wave input has a large effect on the seismic response of the structure and the structure has good vertical seismic performance.

1 INTRODUCTION

The 1995 Hanshin earthquake in Japan led to the complete collapse of the Ookai station, which directly aroused the attention of the world for the seismic resistance of underground structures (Du et al. 2018, 2016). Li Xia et al. (Li & Lu 2022), Mao Nianhua (Mao 2020), Hu, Shuangping (Hu & Gao 2020), Chen et al. (2020) analyzed the dynamic response and damage mechanism of subway stations by numerical simulation. The current traditional view of seismic resistance is that it is mainly horizontal seismic waves that cause the structure to undergo vibration damage, and vertical seismic waves have almost no effect, but the Wenchuan earthquake damage data showed that vertical earthquakes also have a large effect on underground structures (Li et al. 2020).

Therefore, scholars have gradually started to pay attention to vertical seismic waves and carry out research. Zhang et al. (2019) studied the specific effects of vertical seismic motion on subway stations through model tests. Liu Xu et al. (2015) evaluated and analyzed the force and deformation of underground structure floor under vertical seismic action. Deng Zehan, Song Lin, Zhang Han, et al. (Deng 2018; Song et al. 2010; Zhang & Zhang 2021) analyzed the seismic response law of the subway station with seismic waves in different directions as input conditions, respectively.

In this paper, based on finite element analysis software, dynamic time analysis is used to analyze the seismic response laws and force states of subway structures under horizontal, vertical and horizontal-vertical coupled action seismic waves, to provide a reference for the design of similar underground subway projects.

2 MODEL OVERVIEW

In this paper, a box-type subway station in Beijing is taken as the research background. The subway station is a two-story, three-span, double-column station. The station structure is

*Corresponding Author: zyc36306@163.com

22.6 m wide, and 16.2 m high, and the overburden is 4 m thick. The thickness of the top plate is 700 mm, the thickness of the bottom plate is 800 mm, the thickness of the middle plate is 400 mm, and the thickness of the side wall is 600 mm. The spacing between two longitudinal columns is 6.2 m, and the diameter is 0.8 m. According to the Code for Investigation of Geotechnical Engineering, the seismic fortification intensity of the proposed site is VII, the peak design basic acceleration is 0.1 g, and the design characteristic period is 0.4 s. The impact of soft soil seismic subsidence and foundation soil liquefaction is not considered.

In the numerical calculation scheme, the three-dimensional problem is transformed into the two-dimensional plane problem, and the plane strain analysis of the structure is carried out. Grid division and monitoring point layout are shown in Figures 1 and 2 respectively.

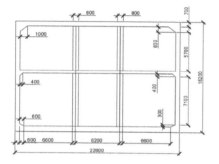

Figure 1. Grid model.   Figure 2. Structural grid model.

3 PARAMETER SELECTION

Mohr-Coulomb model is adopted for the constitutive relation of soil mass. Soil mass material parameters are shown in Table 1. The structure adopts C30 concrete and beam element simulation, and the boundary is undrained. In this paper, the fixed constraint at the bottom and the free field boundary around can eliminate the influence of the boundary effect. Time history analysis is adopted for dynamic analysis. The characteristic period of the model is obtained through eigenvalue analysis, and modal damping is set. In order to ensure the calculation accuracy and convergence, the dynamic calculation step and seismic frequency are 0.02 s.

Table 1. Basic parameters of the soil mass.

Layer	Thickness	Bulk density/ (kg·m^3)	Elastic modulus/ (mpa)	Poisson's ratio	Cohesion/ (kpa)	Internal friction angle/(°)
Miscellaneous fill	2	19.4	8.4	0.25	30	24
Silt 1	5	17.4	9.7	0.43	26	25
Silt 2	9	16.6	36	0.35	13	24
Sandy silt	7	18.8	55	0.32	6	22
Silty clay 1	11	18.6	43	0.30	22	15
Silty clay 2	26	22	20	0.27	40	13

Kobe wave with obvious near-field seismic wave characteristics is selected as the ground motion input. Since the model may generate residual displacement after the movement is completed after the time history is completed, baseline correction and filtering adjustment are conducted through Seismosignal software to clear the integral of the time history. From the observation records of strong earthquakes, the peak value of vertical seismic waves is generally less than that of horizontal seismic waves, which is generally 2/3 of that of horizontal seismic waves. The anti-seismic analysis is adopted for the Project according to the fortification intensity of a 7-degree earthquake.

4 CALCULATION RESULTS AND ANALYSIS

4.1 Stress response analysis

The structural column plays an important role in the stability of the structure. Table 2 shows the stress values of the structural column at different measuring points under the earthquake action. The principal stress at the lower end of the upper structural column is less than that at the upper end, while the principal stress at the lower end is the opposite. The principal stress of the roof is smaller than that of the floor, which may be due to the energy loss caused by the seismic wave input from the bottom of the structure, and the overall stress is smaller under the vertical seismic wave input.

Table 2. Maximum principal stress of structural column measuring point (Company: kN/m^2).

Monitoring point	Horizontal seismic wave	Vertical seismic wave	Coupled seismic wave
Z1	3265.06	1846.48	3121.68
Z2	2016.92	773.13	1927.26
Z3	1441.05	388.82	1393.80
Z4	3934.96	1933.48	3892.40

In order to understand the spatial effect of the structure, due to the limited space, the laws of the maximum axial force, shear force and a bending moment of the sidewall along the height change under the horizontal seismic wave input are analyzed, as can be seen in Figure 3. For the axial force, the bottom of the side wall is subjected to the largest axial force, because most of the weight of the whole structure and the overlying soil layer is basically borne by the bottom of the side wall and the central column. For shear, it can be found that the unit shear stress near the middle plate is small, and the maximum shear stress occurs at the connection between the side wall and the bottom plate. When the shear stress is large, cracks will appear when it exceeds the structural shear strength. For the bending moment, the bending moment at the upper and lower edges of the middle plate is the smallest. The bending moment along the axis of the middle plate is basically symmetrically distributed. The bending moment values at both ends are large. At the same time, it also bears the upper load. The connection between the side wall and the middle column and the top and bottom plate is easy to reach the ultimate compressive stress, thus causing serious compression bending damage.

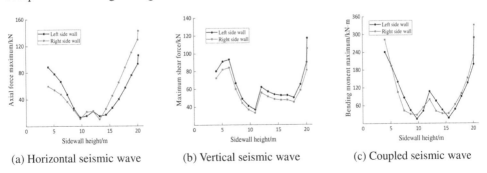

(a) Horizontal seismic wave (b) Vertical seismic wave (c) Coupled seismic wave

Figure 3. Variation curve of internal force of side wall along longitudinal height.

It can be seen from the above analysis that the axial force distribution of the structure has a strong spatial effect, and the maximum axial force of the left and right side walls are 32.7% and 34.6% different at the top and bottom plates, respectively. The internal forces at the connection between the side wall and the floor slab of the underground station structure, and at the connection between the central column and the floor slab, are relatively large. These locations are prone to be damaged under the earthquake action, resulting in the overall collapse of the structure, which should be taken into account in the seismic design.

4.2 Acceleration response analysis

The peak acceleration response values are summarized by extracting the acceleration time history curve of the monitoring point. As shown in Table 3, there is no significant difference in the acceleration response of the same floor. Under the horizontal and coupled seismic wave input, the peak acceleration decreases with the increase of the structure depth. The vertical seismic wave has the most obvious amplification effect on the structural acceleration, which does not change with the change of the structure height. The peak value of acceleration under the coupled seismic wave input is greater than that under the horizontal seismic wave input, which indicates that the vertical seismic wave has a greater impact on the structural acceleration. Therefore, the vertical seismic wave input should not be used when considering the structural acceleration response. The peak acceleration occurs at the top plate of the structure, where the acceleration response is the largest, which may be the vulnerable position of the structure, and should be considered emphatically.

Table 3. Peak Acceleration of Monitoring Points (m/s^2).

Earthquake direction	A	B	C	D	E	F	G	H	I
Horizontal seismic wave	0.7520	0.7419	0.7419	0.5152	0.5022	0.5022	0.4386	0.4202	0.4202
Vertical seismic wave	1.9646	1.9476	1.9476	1.9642	1.9469	1.9469	1.9596	1.9460	1.9460
Coupled seismic wave	0.8653	0.8423	0.8602	0.6667	0.6409	0.6630	0.6148	0.5768	0.6020

4.3 Deformation displacement response analysis

From the numerical point of view 4, the maximum displacement vector of the structure under horizontal, vertical and coupled seismic wave input is 6.04 cm, 1.25 cm and 6.87 cm

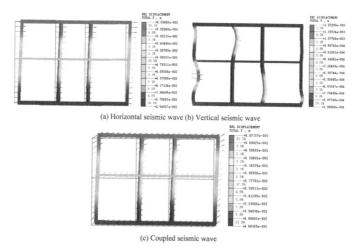

(a) Horizontal seismic wave (b) Vertical seismic wave

(c) Coupled seismic wave

Figure 4. Structure displacement vector diagram.

respectively. From the perspective of the overall displacement direction of the structure, the structure will move horizontally under the input of a horizontal seismic wave, and the structure will float slightly under the input of coupled seismic wave, but the movement direction is mainly horizontal. Under the input of vertical seismic waves, the structure hardly moves, but the side walls and structural columns will deform significantly.

4.4 Horizontal deformation analysis

Through the analysis of the horizontal relative displacement of the structure, the time history curve of the relative horizontal displacement between the layers of the structure is obtained, as shown in Figure 5. It can be seen from the figure that the time history diagram of relative horizontal displacement is almost the same under the input of horizontal seismic wave and coupling seismic wave, while the horizontal relative displacement caused by vertical seismic wave input is small. The inter-story displacement angle is introduced to judge the overall horizontal displacement of the structure. The inter-story displacement angle is respectively under the horizontal, vertical and coupled seismic waves, 7.11×10^{-4}, 3.34×10^{-5} and 7.32×10^{-4}.

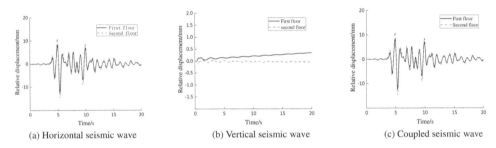

(a) Horizontal seismic wave   (b) Vertical seismic wave   (c) Coupled seismic wave

Figure 5. Time course diagram of relative horizontal displacement between layers.

4.5 Vertical deformation analysis

Figure 6 shows the time history curve of the relative vertical displacement between the middle of the structural bottom plate and the side wall. The overall trend of seismic wave input in different directions is similar, and the relative vertical displacement under coupled seismic wave input is slightly greater than that of horizontal seismic wave. In order to further understand the impact of structural overturning under the earthquake, Figure 7 shows the variation diagram of the vertical displacement peak along the width direction of the bottom plate. It can be seen that the middle part is subject to the largest vertical displacement,

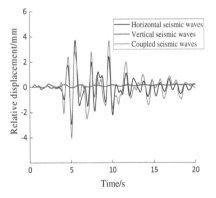 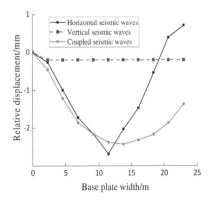

Figure 6. Time course of vertical displacement of the bottom plate.

Figure 7. Peak vertical displacement of the bottom plate along the width direction.

mainly bending deformation. The vertical displacement values of the left and right side walls are different, which indicates that the structural floor will produce tilt deformation under the earthquake action, and the deformation is inconsistent. Under the input of vertical seismic wave, the vertical displacement of the structural floor has no obvious change along the width direction.

5 CONCLUSION

In this paper, the time history analysis method is used to analyze its dynamic response under the action of seismic waves in different directions, and the conclusions are as follows:

(1) Under the action of different seismic wave directions, the change rule of the principal stress of the central column does not change with the earthquake input direction, but only changes in numerical value. The maximum internal force response occurs at the structural connection, which should be considered in the seismic design.
(2) The peak acceleration of the same floor has no obvious change, and the acceleration response increases with the decrease of the structure depth. The acceleration amplification effect of the structure under vertical seismic wave input is the most obvious.
(3) From the displacement vector, the relative horizontal displacement increases with the increase of the side wall height. The fluctuation trend of the maximum relative horizontal displacement time history curve is basically the same.
(4) The vertical deformation of the structural floor is U-shaped, mainly bending deformation.

REFERENCES

Chen Juntao et al. Seismic Response and Damage Mechanism of the Subway Station in Rock and Soil Strata [J]. *IOP Conference Series: Materials Science and Engineering*, 2020, 741: 012086–012086.

Deng Zehan. Seismic Design and Analysis of Subway Underground Stations Based on Time Course Analysis Method [J]. *Engineering Technology Research*, 2018(16): 1–4.

Du Xiuli, Li Yang, Xu Chengshun, et al. Progress of Research on the Causes of Earthquake Damage and Mechanism Analysis of Disaster Formation in Okaikai Subway Station of 1995 Hanshin Earthquake in Japan [J]. *Journal of Geotechnical Engineering*, 2018, 40(02): 223–236.

Du Xiuli, Wang Gang, Lu Dechun. Analysis of Seismic Damage Mechanism of Daikai Subway Station in the Hanshin Earthquake in Japan [J]. *Journal of Disaster Prevention and Mitigation Engineering*, 2016, 36(02): 165–171.

Hu Shuangping, Gao Zhihong. Seismic Response Analysis of Soil-column-free Large-span Subway Station Structure [J]. *Tunnel Construction (in English and Chinese)*, 2020, 40(03): 352–363.

Li Jiangle, Zhang Shirong, Wang Sheliang. Shaking Table Tests on the Seismic Response of a Columnless Subway Station with Asymmetric Load under Bidirectional Seismic Action [J]. *International Journal of Safety and Security Engineering*, 2020, 10(4).

Li X., Lu W. D. Structural Deformation Analysis of T-shape Interchange Subway Stations Under Horizontal Earthquake [J]. *Urban Express Transportation*, 2022, 35(01): 87–95+106.

Liu X, Qi C, Ye F, et al. Dynamic Response of Shallowly Buried Underground Structural Footings Under Vertical Seismic Action [J]. *Journal of Underground Space and Engineering*, 2015, 11(04): 969–974.

Mao Nianhua. Simulation Analysis of Seismic Response of Underground Subway Stations Using Split Columns [J]. *Urban Rail Transit Research*, 2020, 23(10): 58–61+66.

Song Lin, Meng Zhaobo, Wu Minzhe, et al. Seismic Response Analysis of Double-level Island Subway Station Structure [J]. *World Earthquake Engineering*, 2010, 26(02): 187–192.

Zhang Han, Zhang Xiwen. Dynamic Response of Subway Stations Under Coupled Ground Shaking [J]. *Journal of Earthquake Engineering*, 2021, 43(01): 245–250.

Zhang Zhiming, Emilio Bilotta, Yong Yuan, Yu Haitao, Zhao Huiling. Experimental Assessment of the Effect of Vertical Earthquake Motion on Underground Metro Station [J]. *Applied Sciences*, 2019, 9(23).

Study on mechanical properties of foamed concrete with multiple factors

Yiwen Dai & Hongyang Xie*

School of Civil Engineering and Architecture, Nanchang Hangkong University, Nanchang, Jiangxi, China

ABSTRACT: In order to explore the optimal mixing ratio of foamed concrete under the dosage of a bubble group, brick powder, water-material ratio and HPMC (hydroxypropyl methylcellulose), orthogonal experiments with 4 factors and 5 levels were carried out to determine the optimal mixing ratio of foamed concrete through the efficiency coefficient method. The results showed that the compressive strength of foamed concrete decreased with the increase of bubble group content, gradually decreased with the increase of brick powder content, increased first and then decreased with the increase of water-material ratio, and gradually increased first and then decreased with the increase of HPMC content. When the mixing ratio of foamed concrete is 3% bubble group content, 30% brick powder content, 0.55 water-material ratio and 0.05% HPMC content, the efficiency coefficient value is the maximum, that is, the optimal mixing ratio of foamed concrete with brick powder is A1B3C3D3. According to the range analysis table, the factors affecting the performance of foamed concrete from high to low are foam content > water-material ratio > HPMC content > brick powder content.

1 INTRODUCTION

With the continuous speeding up of the Chinese industrialization process, the problem of energy consumption and pollution also becomes more and more acute, and adopting new materials with more environmental protection and lower energy consumption becomes an urgent matter in all walks of life (Li et al. 2011; Zhang et al. 2020), in the process of engineering construction, foam concrete is often used as thermal insulation material and roadbed filling material (Chen et al. 2021; Guo & Yan 2019; Sun et al. 2021; Wang et al. 2018; Wei et al. 2019; Yu 2018; Zhang et al. 2018). This field has aroused more and more attention around the world (Benazzouk et al. 2008; Chandni & Anand 2018; Kirubajiny et al. 2021; Soon-Ching & Kaw-Sai 2010; Zaher et al. 2018). The performance of the foam concrete is decided by many factors and mutual influence. Based on this situation, the author conducted the orthogonal experiment with the four factors of five levels on multiple sets of foam concrete to explore the optimal mixture ratio, and join in the project common waste brick powder to achieve the goal of recycling. In order to provide a theoretical basis for the application of brick powder foamed concrete in practical engineering in the future.

2 EXPERIMENTAL MATERIALS

2.1 Cement

The cement used in the experiment is Conch P.O.42.5 ordinary Portland cement. The chemical composition and basic properties of cement are shown in Table 1.

---

*Corresponding Author: dyw2458751868@163.com

Table 1. The chemical composition of cement.

Composition	SiO$_2$	Fe$_2$O$_3$	Al$_2$O$_3$	CaO	MgO	SO$_3$
Content (%)	22	4.2	5.2	63	1.5	2.2

2.2 Foaming agent

The physical properties of the HT compound foaming agent produced by Henan Huatai Building Materials Company in line with market specifications are shown in Table 2.

Table 2. Physical properties of HT composite foaming agent.

Measure results	Color	Density (Kg/m^3)	Solid content (%)	PH	Foam multiple	1 h Settlement from (mm)	1 h Exudation rate (%)
	Dark brown	1.05	23.8	9.2	15–30	≤50	≤70

According to the Technical Specification of Bubble Mixed Lightweight Soil Filling Engineering, the performance test of the blowing agent shows that the HT compound blowing agent diluted 30 times has the least leakage water and the least settlement distance.

2.3 Stabilizing agent

HPMC produced by Fuqiang Fine Company is used, and its physical performance indicators are shown in Table 3.

Table 3. Physical performance indicators of HPMC.

Measure results	Appearance shape	Viscosity (myriad)	PH	Water content (%)
	White powder	15	6.5	2.1

3 EXPERIMENTAL PROCEDURES AND METHODS

3.1 Block preparation

The experimental material accords with the way designed by the experimental scheme. We accurately weigh the quasi amount of each component and then mix it with a blender for 150 s, so that the component material is fully mixed, and then we add a quasi amount of water, and stir for 150 s.

The foaming agent aqueous solution will be added to the cement foaming machine. Through air pressure foaming, the foam will be quantified into the mixed cement slurry. With uniform mixing for 120 s, the slurry surface does not emerge bubbles. Running a 160–200 mm mobility test, it meets the requirements after pouring into the mold.

After 36 h of static preparation, the surface of the mold is scraped and the mold is removed with an air gun, and then stored in a standard curing room with a temperature of 20±20 C and humidity above 95% for 28d (Li et al. 2015).

3.2 Design orthogonal experiment mix ratio

In the process of the experiment, there are various factors affecting the experimental results, and the influence among them will also change with the change in the amount. Therefore, the orthogonal experiment (Gao 1988; Ma et al. 2021; Yang et al. 2019) is designed, and four groups of factors including the amount of foam, the amount of brick powder, the ratio of water to material, and the amount of HMPC are set, as well as five levels of change. Study the specific strength of different components of brick powder foamed concrete and conduct data processing. The details are shown in Table 4.

Table 4. Factor level of orthogonal experiment.

Level	Factors			
	Bubble content	Brick powder content	Water material ratio (c)	HMPC dosage
1	3%	10%	0.45	0.03%
2	4%	20%	0.5	0.04%
3	5%	30%	0.55	0.05%
4	6%	40%	0.60	0.06%
5	7%	50%	0.65	0.07%

4 EXPERIMENTAL RESULTS AND ANALYSIS

4.1 Orthogonal experimental results

According to the orthogonal experiment results, the range analysis of the compressive strength is carried out, and the results are shown in Table 5.

Table 5. Range analysis table of compressive strength of brick powder foamed concrete.

	Bubble content	Brick powder content	Water material ratio	HMPC dosage
K1	5.584	3.030	2.338	2.803
K2	2.318	2.944	2.992	2.863
K3	2.182	2.911	3.715	2.955
K4	2.176	2.419	2.678	2.824
K5	1.340	2.296	1.877	2.154
Poor value R	4.552	0.734	1.838	0.801
Row rank	1	4	2	3

In Table 5, each level factor is calculated by the orthogonal experiment of poor value. The size of the poor value, to a certain extent, reflects the influence of various factors on compressive strength. It can be concluded that the bubble content > water material ratio > HMPC dosage > brick powder content, and dosage of HMPC bubble content can be obtained for mixed brick powder's influence on the compressive strength of foam concrete is most significant.

With the increase of foam content, the compressive strength of foamed concrete with brick powder decreases gradually. The reason is that with the increase of foam content, the pores of foamed concrete increase, and the proportion of cementing materials as the main source of strength of foamed concrete with brick powder decreases naturally, and the compressive strength also decreases (Hu et al. 2019).

With the increase of brick powder content, the compressive strength of brick powder mixed foamed concrete gradually decreases. The reasons are analyzed as follows: with the increase of brick powder content, the effective bond between aggregates providing strength for concrete decreases, which leads to the decrease of compressive strength. Moreover, effective foams in concrete will burst, harmful pores will increase, and compressive strength will decrease (Li et al. 2018).

With the increase in the water-material ratio, the compressive strength of brick powder foamed concrete rises first and then decreases. The reasons are analyzed as follows: with the increase in the water-material ratio, the hydration reaction is carried out more and more thoroughly. However, as the ratio of water to material continues to increase, the effective foam in concrete is destroyed, the harmful holes increase, and the strength decreases.

4.2 Select the optimal mix ratio

The requirements of dry density and compressive strength, two important indexes of foamed concrete, are contradictory. In this paper, the efficiency coefficient method is adopted to calculate the efficiency coefficient and total efficiency coefficient of these two indexes, as shown in Table 6.

Table 6. Efficiency coefficient table of brick powder foamed concrete.

Set no	Investigation target		Efficacy coefficient		Total efficiency factor $d = \sqrt[2]{d_1 d_2}$
	Dry density Kg/m^3	28d compressive strength/ Mpa	Dry density d_1	28d compressive strength d_2	
1	1312.7	5.7292	0.65	0.74	0.69
2	1284.9	6.4432	0.66	0.83	0.74
3	1182.1	7.7256	0.72	1.0	0.85
4	1186.9	5.4355	0.72	0.7	0.71
5	1295.5	2.5885	0.66	0.33	0.47
6	1139.2	2.5821	0.75	0.33	0.25
7	966.4	2.9913	0.88	0.39	0.59
8	1034.2	1.9719	0.82	0.26	0.46
9	1213.5	1.8555	0.70	0.24	0.41
10	1165.3	2.1913	0.73	0.28	0.45

It can be seen from Table 6 that in the experimental group mixed with brick powder, the third group has the largest total efficiency coefficient D = 0.85, the dry density is 1182.1 kg/m^3, and the compressive strength is 7.726 Mpa. The mix level of this group is A1B3C3D3, that is, 3% foam content, 30% brick powder content, 0.55 water-material ratio, and 0.05% HPMC content.

5 THE SEM EXPERIMENT

The specimens in the third group in Table 6 were taken as 10 mm × 10 mm × 10 mm for Sem microscopic scanning and analysis experiment, and the experimental results were shown in Figure 1.

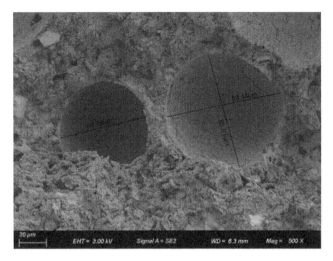

Figure 1. Microscopic images of the experimental group and control group after hydration for 28 days at 500X.

According to the results of the SEM test, it can be concluded that 1) the improvement of compressive strength of foamed concrete mixed with brick powder is due to the addition of brick powder to refine the pores and increase the density of the concrete matrix; 2) HPMC, as a foam stabilizer, reduces the destruction of bubble groups in the preparation process of the specimen, reduces the generation of connected holes and other harmful holes in the specimen, and makes the distribution of pores more uniform.

6 CONCLUSION

Various factors affect the performance of foamed concrete, and the effect of different matching ratios on the performance of foamed concrete is different. The following conclusions can be drawn from the experimental results:

1) According to the range analysis in Table 6, the factors influencing the performance of brick powder foamed concrete from high to low are as follows: foam content > water-material ratio >HPMC content > brick powder content
2) The compressive strength of brick powder foamed concrete showed a downward trend with the increase of foam content and brick powder content, and a downward trend with the increase of water-material ratio and HPMC
3) According to the efficiency coefficient method, the optimal mixing ratio of brick powder foams that meets the requirements of light and high strength is 3% foam content, 30% brick powder content, 0.55 water-material ratio, and 0.05% HPMC content.

FUND PROJECT

University-level project of Nanchang Hangkong University (YC2022-130)

REFERENCES

Benazzouk A., Douzane O., Mezreb K., Laidoudi B., Que'neudec M. Thermal Conductivity of Cement Composites Containing Rubber Waste Particles: Experimental Study and Modelling[J]. *Construction and Building Materials*, 2008, 22(4): 573–579.

Chen Liyan, Yang An, Hong Fen, Ma Yongjiong, Wang Jun, Qiao Hongxia. Effect of different Fly Ash Content on the Performance and Pore Size Of Foams Concrete [J]. *Concrete*, 2021(08): 137–140.

Gao Yunyan. *Orthogonal and Regression Test Design Method* [M]. Beijing: Metallurgical Industry Press, 1988. (In Chinese).

Guo J.Y., Yan Y.J., Experimental Study on Microbeads Foamed Concrete with Large Fly Ash Content [J]. *New Building Materials*, 2019, 46(01): 79–81.

Hu Yanli, Hao Jingao, Zhao Xiangmin, Peng Hailong, Yang Weibo, Gao Peiwei. Study on the Relationship Between Performance and Pore Structure of Foamed Lightweight Concrete [J]. *Journal of Nanjing University of Science and Technology*, 2019, 43(03): 363–366.

Kirubajiny Pasupathy, Sayanthan Ramakrishnan, Jay Sanjayan. Influence of Recycled Concrete Aggregate on the Foam Stability of Aerated Geopolymer Concrete [J]. *Construction and Building Materials*, 2021, 271.

Li Fangxian, Yu Qijun, Luo Yunfeng, Wei Jiangxiong. Mathematical Characterization and Analysis of Pore Structure of Foamed Concrete [J]. *Journal of Southwest Jiaotong University*, 2018, 53(06): 1205–1210.

Li Qiuyi, Quan Hongzhu, Qin Yuan. *Concrete Recycled Aggregate* [M]. Beijing: China Architecture and Building Press, 2011.

Li Yingquan, Zhu Lide, Li Juli, Hu Shikai, Duan Ce, Wang Xiaofan. Design of Mixture Ratio of Foamed Concrete, 2011, 26(02).

Ma Yongjiong, Yang An, Hong Fen, Chen Liyan, Wang Jun, Qiao Hongxia, Mix Design of Foamed Concrete Based on Orthogonal Test [J]. *Concrete*, 2021 (07): 147–150.

Sun Jing, Wang Hong, Chen Jing, An Yanling, Lan Jianwei, Liu Hongbo. Study on the Performance of Undisturbed Fly Ash and Iron Tailings Foamed Concrete [J]. *Concrete and Cement Products*, 2021(05): 100–104.

Soon-Ching Ng, Kaw-Sai Low. Thermal Conductivity of Newspaper Sandwiched Aerated Lightweight Concrete Panel [J]. *Energy and Buildings*, 2010, 42(12): 2452–2456.

Chandni T.J., Anand K.B. Utilization of Recycled Waste as Filler in Foam Concrete [J]. *Journal of Building Engineering*, 2018, 19: 154–160.

Wang Xuanfu, An Zhushi, Xie Tao, Li Fan, Cui Kai, Liu Zili. Research on Foam Concrete Recycled From Construction Waste [J]. *Construction Technology*, 2018,47 (S2): 45–48.

Wei Xiangming, Dong Chao, Feng Jingjing, Yang Jinbo. Experimental study on the performance of slag foamed concrete with double fly ash [J]. *Concrete and Cement Products*, 2019 (07): 63–66.

Yang Fanxuan, Shi Minghui, Niu Yangyang, Wang Junyi, Zhang Jiaxuan, Yin Guansheng. *Study on Mixture Ratio of Foamed Concrete Based on Orthogonal Test* [C]. Proceedings of 28th National Structural Engineering Academic Conference (Vol. iii). 2019.

Yu Junyan. *Study on the Roadbed of Foam Lightweight Soil for the Reconstruction and Expansion of Binlai Expressway* [D]. Shandong University, 2018.

Zhang Xiaoming, Huang Peizheng, Cui Qingyi. Research on the Performance of Micro-Powder Foam Concrete Recycled From Construction Waste [J]. *Concrete and Cement Products*, 2020(05): 96–98.

Zhang Song, Li Ruyan, Dong Xiang, Zhang Lan, Li Zhixiong. Preparation of Foamed Concrete with Recycled Micro Powder Instead of Cement [J]. *Bulletin of the Chinese Ceramic Society*, 2018, 37(09): 2948–2953.

Zaher Mundher Yaseen, Ravinesh C. Deo, Ameer Hilal, et al. Predicting Compressive Strength of Lightweight Foamed Concrete Using Extreme Learning Machine Model [J]. *Advances in Engineering Software*, 2018, 115: 112–125.

Experimental study on mechanical performance of basalt-polypropylene hybrid fiber reinforce concrete

Tongshuai Li, Lihua Wang*, Chunfeng Li, Shifu Sun, Zengguang Pang & Qinghua Shu
College of Civil Engineering and Architecture, Shandong University of Science and Technology, Qingdao, Shandong, China
Shandong Key Laboratory of Civil Engineering Disaster Prevent and Mitigation, Qingdao, Shandong, China

ABSTRACT: Adding fibers to concrete can improve the shortcomings of concrete and enhance the mechanical properties in practical engineering. Hybrid fibers can enhance concrete from multiple scales compared with single fibers. In order to study the hybrid effect of fibers, the paper prepares concrete materials with two kinds of fibers, and adds the basalt fibers and polypropylene fibers with different aspect ratios in the concrete with the different fiber content by volume of 0.05%, 0.10%, 0.15% and 0.20%. The performance test compares the performance of ordinary concrete, single-fiber concrete and hybrid-fiber concrete. The results show that the addition of basalt fiber and polypropylene fiber can significantly improve the mechanical properties of concrete. Among them, hybrid fiber concrete combines the advantages of the two fibers, and the basic mechanical properties of concrete are improved more significantly, and the comprehensive performance is more excellent. The cube compressive strength, tensile, and flexural strength were increased by 16.62%, 21.33% and 25.40% respectively compared with ordinary concrete. At the same time, the addition of fibers improves the brittle failure of concrete, and the overall failure characteristics show ductile failure.

1 INSTRUCTIONS

As one of the most widely used materials in the civil engineering industry, concrete also has disadvantages such as high brittleness and poor crack resistance (Zeng 2020). Through the research of some scholars, it is found that adding fibers into the concrete matrix can effectively improve the mechanical properties of concrete. A single fiber can only enhance the performance of concrete on one scale (Yoo 2017). A large number of scholars have conducted research on how to enhance the performance of concrete on multiple levels and scales (Afroughsabet 2015; Aslani 2013; Malgorzata 2016; Wu 2003; Yin 2014), and hybrid fiber reinforced concrete emerges as the times require. Hybrid fiber-reinforced concrete refers to fiber-reinforced concrete prepared by mixing the same type of fibers with different aspect ratios or two or more different types of fibers into the matrix (Zhao 2018). With the help of the hybrid effect formed between the fibers, the strengths and weaknesses of each other are complemented, and the mechanical properties of concrete are enhanced from different scales. A large number of experimental studies at home and abroad have shown that compared with no-fiber or single-fiber, hybrid fiber shows a better reinforcement effect on the mechanical properties of concrete.

In this paper, the hybrid fiber reinforced concrete is prepared by mixing basalt fiber and polypropylene fiber. The performance of fiber-reinforced concrete is compared with ordinary concrete and single fiber-reinforced concrete through the compressive test, split tensile test, and microscopic test of the flexural test.

---

*Corresponding Author: wlh@sdust.edu.cn

2 MATERIALS AND EXPERIMENT

2.1 *Materials*

Ordinary Portland cement is used as cement, and medium sand with a fineness modulus of 2.51 is used as fine aggregate. Coarse aggregate is crushed stone with a particle size of 5 ~ 25 mm. The additive is polycarboxylate superplasticizer and 3-Aminopropyltriethoxysilane. Hybrid-reinforced materials were produced using chopped basalt fiber and polypropylene, the physical properties as provided by the suppliers are presented in Table 1.

Table 1. Basalt and polypropylene fiber properties.

Fiber type	Length (mm)	Specific gravity (kg/m^3)	Diameter (μm)	Tensile strength (MPa)	Elastic modulus (GPa)	Ultimate elongation (%)
Basalt	6, 12, 18	2650	17	4100	100	2.8
Polypropylene		910	18~48	486	>4.8	30

2.2 *Mix proportions*

According to the Chinese code (JGJ 55-2011), this test is based on the mix ratio of C30 concrete. The specific mix ratio is shown in Table 2.

Table 2. Concrete mix proportions (kg/m^3).

Concrete	Water	Sand	Stone	S.P.
328	114	668	1217	0.66

Note: S.P.=Superplasticizer.

2.3 *Testing methods*

The mechanical properties of concrete in this test were tested according to the test method in GB/T 50081-200. Under the premise that the benchmark mix ratio remains unchanged, the fiber incorporation method is carried out in the following two ways: (1) Preparation of single-doped basalt fiber concrete, single-doped polypropylene fiber concrete, and mixed fiber concrete according to 0.05%, 0.10%, 0.15%, and 0.20% fiber content; (2) Considering the different aspect ratio of fiber, three kinds of fiber concrete were prepared by controlling the total content of 0.20% fiber. The influence of fiber on concrete performance was studied from three aspects: fiber type, volume fraction and length. Specific test groups are shown in Tables 3 and 4.

Table 3. Hybrid fiber concrete test groups of the same length.

Mix No.	Length (mm) BF	Length (mm) PVA	Fiber content (%)
C0	–	–	0
HF6-0.05	6	6	0.05
HF6-0.10	6	6	0.10
HF6-0.15	6	6	0.15

(*continued*)

Table 3. Continued

Mix No.	Length (mm) BF	Length (mm) PVA	Fiber content (%)
HF6-0.20	6	6	0.20
HF12-0.05	12	12	0.05
HF12-0.10	12	12	0.10
HF12-0.15	12	12	0.15
HF12-0.20	12	12	0.20
HF18-0.05	18	18	0.05
HF18-0.10	18	18	0.10
HF18-0.15	18	18	0.15
HF18-0.20	18	18	0.20

Note: HF means for hybrid fibers; 6, 12, and 18 stand for fiber length; 0.05, 0.10, 0.15, and 0.20 are fiber content, and the above-mentioned fibers are mixed in 1:1.

Table 4. Different fiber length hybrid fiber concrete test grouping.

Mix No.	Length (mm) BF	Length (mm) PVA	Fiber content (%)
B6P6	6	6	0.20
B6P12	6	12	0.20
B6P18	6	18	0.20
B12P6	12	6	0.20
B12P12	12	12	0.20
B12P18	12	18	0.20
B18P6	18	6	0.20
B18P12	18	12	0.20
B18P18	18	18	0.20

3 TEST RESULTS AND ANALYSIS

3.1 *Failure analysis of concrete*

Figure 1 shows the failure image of the concrete, there is a big difference between ordinary concrete and fiber-reinforced concrete in compression failure. With the increase of the load,

(a) Ordinary concrete   (b) Fiber concrete

Figure 1. Concrete compressive failure.

the ordinary concrete around the specimen begins to peel off, and a large number of vertical cracks begin to appear on the surface and gradually develop. As the loading process progresses, the concrete spalls more seriously, and it suddenly fails near the ultimate load. After adding fibers, with the increase of the load, the surface of the specimen fell off less, cracks were less, and there was a tearing sound during loading. When the ultimate load was reached, the shape of the specimen remained good.

Figure 2(a) shows the splitting tensile image of the concrete, The ordinary concrete specimen did not change significantly in the first half of loading. With the increase of the load, a number of vertical cracks appeared near the upper and lower battens, and the width gradually increased with the further increase of the load, which finally caused the specimen to split and separate, and the time was damaged. There are no signs before failure, and the sound of failure is severe, showing the characteristics of brittle failure.

As shown in Figure 2(b), the failure modes of basalt fiber-reinforced concrete, polypropylene fiber-reinforced concrete and hybrid fiber-reinforced concrete are similar in splitting and tensile strength. With the progress of loading, cracks appear near the edge of the bead on the upper and lower surfaces of the loaded specimen and show ductile failure characteristics.

(a) Ordinary concrete　　　(b) Hybrid fiber

Figure 2.　Splitting tensile failure mode of concrete.

Figure 3 shows the flexural failure image of concrete. During the flexural test of ordinary concrete, before approaching the peak load, small cracks began to appear at the bottom of the specimen, and as the loading progressed, the cracks developed upward, and the concrete immediately broke when the peak load was loaded, showing typical brittle failure characteristics. Compared with ordinary concrete, fiber-reinforced concrete has a relatively long loading process, small crack width, and good toughness in failure characteristics.

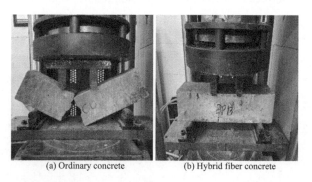

(a) Ordinary concrete　　　(b) Hybrid fiber concrete

Figure 3.　Concrete flexural failure.

3.2 Mechanical property

Table 5 shows the compressive strength, tensile strength and flexural strength of hybrid fiber concrete under the first incorporation of two kinds of fibers. Figure 4 shows the trend of compressive strength of single-fiber concrete and the trend of compressive strength of concrete with fiber content under the first incorporation of mixed fibers. For single-fiber concrete, controlling 0.10 % fiber content and 18 mm fiber length has the best enhancement effect on the compressive strength of concrete. The hybrid fiber with a length of 18 mm and a content of 0.10 % shows higher compressive performance, which is 16.62 % higher than that of ordinary concrete. The hybrid fiber with a length of 6 mm and a content of 0.20 % shows a negative hybrid effect on the compressive performance of concrete, and the compressive strength is 1.78 % lower than that of ordinary concrete.

At the same time, it can be seen in Figure 5 that under the second mixing method, when the fiber content is constant, the compressive strength of concrete is positively correlated with the fiber length, and the change in polypropylene fiber length has a significant effect on

Table 5. Test results of hybrid fiber reinforced concrete of the same length.

Mix No.	Compressive strength (MPa)	Tensile strength (MPa)	Flexural strength (MPa)
C0	33.7	2.86	4.33
HF6-0.05	34.5	3.00	4.73
HF6-0.10	35.3	3.11	5.08
HF6-0.15	35.1	3.14	5.23
HF6-0.20	33.1	3.01	5.04
HF12-0.05	35.0	3.10	4.98
HF12-0.10	37.5	3.24	5.23
HF12-0.15	35.8	3.36	5.43
HF12-0.20	34.9	3.12	5.20
HF18-0.05	36.2	3.25	4.88
HF18-0.10	39.3	3.43	4.96
HF18-0.15	38.5	3.57	5.35
HF18-0.20	36.1	3.33	5.19

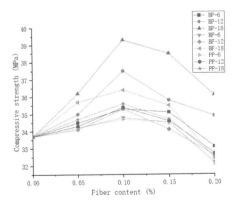

Figure 4. Effect of fiber content of single and mixed fiber on the compressive strength of concrete.

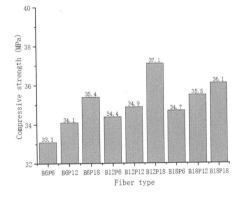

Figure 5. Effects of different lengths of hybrid fibers on the compressive strength of concrete.

the compressive strength of concrete. The intensity effect is more pronounced. The compressive strength of the B6P6 group is 33.1 MPa, compared with the B18P6 group, the length of the polypropylene fiber is fixed, and the length of the basalt fiber is changed, the compressive strength is increased to 34.7 MPa, and the strength is increased by 2.97% compared with the ordinary concrete; compared with the B6P18 group, the fixed the length of basalt fiber, changing the length of polypropylene fiber, the compressive strength is increased to 35.4 MPa, the strength is increased by 5.04% compared with ordinary concrete, for the strength improvement of hybrid fiber concrete, increasing the length of polypropylene fiber can play a more significant effect. By comparison, it is found that the compressive strength of the hybrid fiber reinforced concrete of the B12P18 group is 10.09% higher than ordinary concrete, and the enhancement effect is the best.

It can be seen from the trend in Figure 4 that in terms of the enhancement effect of concrete compressive strength, the hybrid fiber-reinforced concrete is better than the single fiber-reinforced concrete. The possible reasons are: (1) Compared with the basalt fiber, the polypropylene fiber has a better performance in matrix mixing. Especially in single-fiber concrete, with the increase of fiber content, basalt fiber will produce an agglomeration phenomenon, which will increase the weak area inside the concrete matrix and have a negative reinforcement effect on the concrete. And polypropylene fibers, by virtue of their superior dispersing properties, overlap in the concrete matrix to form a three-dimensional network structure, which has a good constraining effect on the micro-cracks in the concrete due to self-shrinking and other reasons, and it is beneficial to the improvement of concrete strength (He 2013). (2) The treatment of the coupling agent reduces the smoothness of the fiber surface, thereby enhancing the bond between the fiber and the concrete matrix; at the same time, the two fibers have greatly different elastic moduli and elongations, which effectively inhibits the generation and development of microcracks in the concrete matrix (Cao 2017).

Figure 6 shows the variation trend of concrete tensile strength with fiber content in single-fiber concrete and hybrid fiber in the first fiber mixing method. When the length of the hybrid fiber is 18 mm and the content is 0.15%, the tensile strength is the largest, which is 21.33% higher than ordinary concrete. With the further increase of the fiber content, the dispersibility of the fibers in the concrete matrix becomes poor, and the tensile strength of the hybrid fiber concrete will drop significantly.

It can be seen from Figure 7 that under the second mixing method, when the fiber content is constant, the tensile strength of concrete is positively correlated with the fiber length, and the change of the basalt fiber length has a greater impact on the tensile strength of concrete.

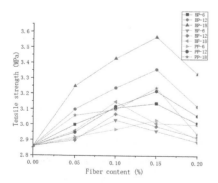

Figure 6. Effect of the fiber content of single and mixed fiber on the tensile strength of concrete.

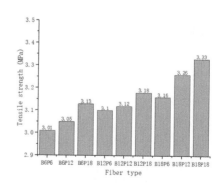

Figure 7. Effects of hybrid fibers of different lengths on the tensile strength of concrete.

As shown in Figure 5, the tensile strength of the B6P6 group is 3.01 MPa, compared with the B18P6 group. The length of the polypropylene fiber is fixed, and the length of the basalt fiber is changed, the tensile strength is increased to 3.16 MPa, and the strength is increased by 10.48% compared with ordinary concrete. For the B6P18 group, we fix the length of basalt fiber, and change the length of polypropylene fiber. The tensile strength is increased to 3.13 MPa, and the strength is increased by 9.44% compared with ordinary concrete. It can be seen that for the improvement of the tensile strength of hybrid fiber concrete, the length of basalt fiber increases has a more pronounced effect. In comparison, it is found that the splitting tensile performance of the hybrid fiber concrete is the best in the B18P18 group. Compared with ordinary concrete, the tensile strength is increased by 16.43%.

Figure 8 shows the variation trend of hybrid fiber-reinforced concrete and single fiber-reinforced concrete. The flexural strength of the hybrid fiber-reinforced concrete with a length of 12 mm is the highest. When the content is 0.15%, the flexural strength reaches the maximum value. Compared with ordinary concrete, the flexural strength increased by 25.40%.

It can be seen from Figure 9 that after the 18 mm basalt fiber and the 12 mm polypropylene fiber are mixed, the flexural strength of the concrete is 5.25 MPa, which is 21.25% higher than ordinary concrete, and greater than hybrid fiber with the same length. The hybrid effect of hybrid fiber-reinforced concrete with a suitable length is better than that of fiber-reinforced concrete with the same length.

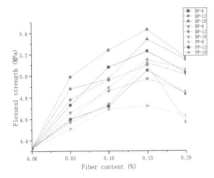

Figure 8. Effect of fiber content on flexural strength of concrete.

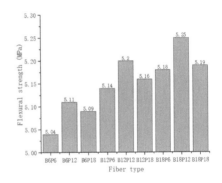

Figure 9. Effect of hybrid fibers of different lengths on the flexural strength of concrete.

The reason for the above situation is that the fibers are mixed into the concrete matrix to form a three-dimensional network structure, which inhibits the generation and expansion of cracks in the concrete, and enhances the strength and toughness of the concrete. When hybrid fibers of different lengths are mixed into concrete, fibers with smaller lengths form bridging inside the concrete, preventing microscopic cracks from further penetrating into macroscopic main cracks. Fibers with different lengths play their respective roles and complement each other, thereby enhancing the flexural properties of concrete (Kong 2017).

3.3 *Microstructures test*

It can be seen from Figure 10 that the surface of basalt fibers is relatively smooth and has a cylindrical solid structure. Because of the low elongation of basalt fibers and poor ductility, when cracks occur in the matrix, the basalt fibers are pulled out or broken. It can also be seen from the microscopic morphology that the fracture surface is flat and has no necking phenomenon.

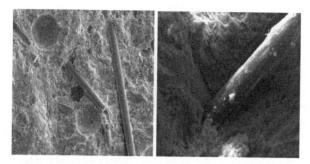

Figure 10. Micromorphology of basalt fibers.

Figure 11 shows an SEM image that there are more granular hydration products attached to the surface of polypropylene fibers, which increases the bonding between the fibers and the matrix. At the same time, due to the high elastic modulus and ultimate elongation of polypropylene fibers, they play a bridging role when cracks occur. However, when the fiber roots are stretched, a necking phenomenon occurs, and pores are formed with the matrix, which makes the overall performance lower than basalt fibers, which weakens the reinforcement effect to a certain extent.

Figure 11. Micromorphology of polypropylene fibers.

Figure 12 shows that the two kinds of fibers are randomly distributed in the concrete matrix. When the fiber concrete is configured, a coupling agent containing a large amount of Si element is added to pretreat the fibers to reduce the smoothness of the fiber surface. The increased production of hydration products enhances the bonding between fibers and cement paste, optimizes the microstructure of concrete, and thus improves the performance of concrete.

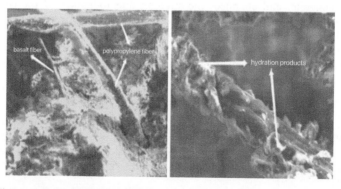

Figure 12. Micromorphology of hybrid fibers.

4 CONCLUSION

Using the fiber content and length as variables, the reinforcement effect of basalt fiber and polypropylene fiber on concrete under two mixing methods was studied, and a series of mechanical tests were carried out. According to the statements studied, the conclusions were drawn as follows:

(1) For single-fiber concrete and hybrid fiber-reinforced concrete, the appropriate amount of fiber helps to improve the compressive strength. The test results show that the compressive strength of the specimen can be increased by up to 15.96% in the first mixing method; in the second mixing method, the change in the length of polypropylene fibers has a greater impact on the compressive strength of concrete. Significantly, the compressive strength of the hybrid fiber-reinforced concrete in the B12P18 group was 10.21% higher than ordinary concrete.

(2) In terms of splitting tensile strength, the addition of fibers makes the tensile properties of concrete better than ordinary concrete. In the first mixing method of hybrid fibers, HF18-0.15% has the highest tensile strength, which is 24.66% higher than that of ordinary concrete. In the second mixing method, the change of basalt fiber length has a more significant effect on the tensile strength of concrete, and the tensile strength of the B18P18 group is 14.95% higher than ordinary concrete.

(3) The addition of hybrid fibers is more effective in enhancing the early flexural properties of concrete, and the flexural strength of fiber-reinforced concrete is better than ordinary concrete in the entire range of content. Under the optimal content of 0.15%, the flexural strength of the first blending method of the hybrid fiber can be increased by 23.95%. In the second blending method, the flexural strength of the B18P12 group is increased by 20.05% compared with ordinary concrete.

REFERENCES

Afroughsabet V., Ozbakkaloglu T. (2015) Mechanical and Durability Properties of High-Strength Concrete Containing Steel and Polypropylene Fibers [J]. *Construction and Building Materials*, 94: 73–82.

Aslani F., Nejadi S. (2013) Self-compacting Concrete Incorporating Steel and Polypropylene Fibers: Compressive and Tensile Strengths, Moduli of Elasticity and Rupture, Compressive Stress-Strain Curve, and Energy Dissipated Under Compression [J]. *Composites Part B*, 53(5): 121–133.

Cao H.L., Lang H.J., Meng S.H. (2017) Experimental Research on the Basic Structure and Properties of the Continuous Basalt Fiber [J]. *Hi-Tech Fiber and Application*, (05):8–13.

Doo-Yeol Yoo, Sung-Wook Kim, Jung-Jun Park. (2017) Comparative Flexural Behavior of Ultra-High-Performance Concrete Reinforced with Hybrid Straight Steel Fibers [J]. *Construction and Building Materials*, 132.

GB/T 50081-2002, Standard for Test Method of Mechanical Properties on Ordinary Concrete, Beijing, China, 2002. (in Chinese)

He R., Li Y.P., Chen S.F., Ji S.H. (2013) Influence of Fiber Combination on Flexural Properties of Hybrid Fiber Reinforced Concrete [J]. *Journal of Guangxi University(Natural Science Edition)*, (06): 1306–1312.

JGJ 55-2011, Specification for Mix Proportion Design of Ordinary Concrete, Beijing, China, 2011. (in Chinese).

Kong X.Q., Bao C.C., Gao H.D., Zhang W.J., Qu Y.D. (2017) Experimental Study on Mechanical Properties of Hybrid Fiber Reinforced Recycled Concrete [J]. *Concrete*, (11): 105–109.

Malgorzata Pajak. (2016) Investigation on Flexural Properties of Hybrid Fibre Reinforced Self-compacting Concrete [J]. *Procedia Engineering*, 161.

Wu Y., Li J., Wu K.R. (2003) Mechanical Properties of Hybrid Fiber-Reinforced Concrete at Low Fiber Volume Fraction [J]. *Cement & Concrete Research*, 33(1): 27–30.

Yin C., Xu L.H., Zhang Y.Y. (2014) Experimental Study on Hybrid Fiber-Reinforced Concrete Subjected to Uniaxial Compression [J]. *Journal of Materials in Civil Engineering*, 26(2): 211–218.

Zhao K.Y., Wang Y., Zhang J.T., Xu S.C. (2018) Research Status of Hybrid Fiber Reinforced Concrete [J]. *Concrete*, (03):132–137+140.

Zeng Z.H., Li C.X., Chen Z.Y., Ke L. (2020) Study on Mechanical Properties and Optimum Fiber Content for Basalt/Polyacrylonitrile Hybrid Fiber Reinforced Concrete [J]. *Journal of Railway Science and Engineering*, 17(10): 2549–2557.

Research on sulfate-freezing-thawing resistance of mineral admixture concrete in alpine-cold regions

Li Zhou*
CCCC Fourth Highway Bureau, Fifth Engineering Ltd., Xi'an, China

ABSTRACT: Concrete structures in alpine-cold regions of western China are under the combined effects of sulfate attack and freezing-thawing cycles all year round, with significant durability problems. Based on a highway project and the quadratic general rotary design method, this paper studies the influence pattern of four factors, i.e., fly ash, slag, coarse aggregate and water-cementitious ratio, on the relatively dynamic elastic modulus and mass loss rate of concrete under the coupled sulfate-freeze-thaw environment. The results show that fly ash and slag could help to improve the denseness of microstructure and the sulfate-freeze-thaw resistance of concrete and that the degree and pattern of influence of these four factors on durability are subject to the circumstances. The optimum sulfate-freeze-thaw resistance can be obtained when the fly ash admixture is 32%, the slag admixture is 19%, the coarse aggregate volume ratio is 0.34, and the water-cementitious ratio is 0.4.

1 INTRODUCTION

The western areas of China (such as Xinjiang, Tibet, and Qinghai, etc.) are located in a cold, high-altitude region characterized by a considerable temperature difference, heavy snowfall, and intense evaporation throughout the year, as well as sulfate-rich water and soil. In this region, concrete structures such as bridges and culverts are subject to the superimposed effects of sulfate attack and the freeze-thaw cycle. The sulfate attack accelerates the deterioration of concrete. The presence of sulfate, on the other hand, generates more osmotic pressure in concrete pores due to an increase in the concentration difference between the interior and exterior, thereby decreasing its freeze-thaw resistance (Wang & Ma 2017). This coupling effect causes more severe damage to concrete structures than a single factor (Wan 2013), accelerating the durability degradation of concrete. Therefore, the research related to the durability pattern of concrete under the coupling effect of sulfate-freeze-thaw cycles has also become popular in recent years.

Fang (Fang et al. 2019) reviewed the experimental method and deterioration mechanism under the combined action of the freeze-thaw cycle and sulfate attack, finding that more research is required to unify the distinction between laboratory tests and simulation with practical projects. Wang (2019) found that freeze-thaw cycles can accelerate sulfate attack, and the mass loss rate and structural failure degree of hydraulic concrete increase under coupling conditions. Yuan (2013) et al. tested the impact of salt solution types and concentrations on the freeze-thaw resistance of concrete as determined by the fast freeze-thaw test method. By a vast quantity of durability test data, Yu (2011) et al. determined the relationship between the initial damage velocity (IDV), damage acceleration (DA) of concrete and the environmental condition and stress state of the structure.

On the other hand, fly ash and slag are beneficial to improve the durability of concrete under sulfate-freeze-thaw cycles coupling environment. For example, Qin (2017) et al. studied

---

*Corresponding Author: 1570365944@qq.com

the chloride penetration resistance and freeze-thaw cycle resistance of C50 concrete prepared with different fly ash and slag content, and obtained the optimal content of slag and fly ash (50%) and its ratio (4:1). Li (2015) studied the freeze-thaw resistance of different mineral admixtures in different salt solutions, and discovered the influence of sulfate freeze-thaw resistance: FA < SG < OPC < FA + SG. Jin (2021) studied the durability of concrete in saline soil areas such as Qinghai under triple factors of freeze-thaw, carbonation and brine corrosion.

However, these studies mainly focus on the salt attack and freeze-thaw resistance of OPC (Wang 2021), with water-cement ratio, salt concentration, salt type and admixture type and concentration as the main factors. Moreover, the selected evaluation indexes are mostly concrete appearance quality, compressive strength and relative dynamic elastic modulus, etc. It is well known that mineral admixtures (such as slag and fly ash) improve the sulfate resistance of concrete (Ge et al. 2008; Qin 2017; Wang & Du 2011) and have relatively high economic value. In the current research, however, the types of mineral admixtures in concrete, the specific number of mineral admixtures, and the influence pattern on the freeze-thaw resistance of concrete when they are used together remain unclear, resulting in the absence of rules for mix design in practical project applications.

In this study, twenty groups of mineral admixture concrete with different mix ratios are designed using the quadratic general rotary design approach. The design is based on the performance of raw materials in high and cold regions. Through the test data, a coupling model is established to analyze and compare the sulfate-freeze-thaw cycle performance of concrete with different proportions of mineral admixture (fly ash, slag), water-cement ratio and coarse aggregate volume ratio, to determine the effect laws of various factors and obtain the optimal mix ratio, which serves as a guideline for achieving good economy and durability of concrete.

2 MATERIALS AND METHODS

2.1 Materials

Based on the project, all the materials in this study are consistent with the materials used in the construction site. The cement was produced by Xinjiang Tianshan Cement Co., Ltd, with the type of P.O 42.5 OPC. The apparent density of crushed natural aggregates from the local stone factory was 2650 kg·cm^{-3}, with 5~20 mm continuous gradation. The sand was local natural river sand with an apparent density of 2700 kg·cm^{-3} and a fineness modulus of 2.3. The superplasticizer was from Qinghai Zhongsui New Materials Co., Ltd., with the type of CT-1 and content of 0.6~1.2% (mass of the cementitious material). In this study, fly ash and slag were used to replace the cement partly, and the parameters are shown in Table 1 below.

Table 1. Performance index of mineral admixtures for experiment.

Admixture	density/kg·cm^{-3}	Loss on ignition/%	Specific surface area/cm kg^{-1}	Fineness 45um/%	Activity index/%	
					7d	28d
Fly ash	2300	1.96	275.8	18	–	–
Slag	2890	0.6	425	–	78	100

2.2 Mix design

Considering that the bridge project is located in the alpine region at the junction of Xinjiang and Qinghai provinces, the use of sulfate and deicing salt in the soil will lead to salt corrosion. Furthermore, the freeze-thaw environment brought by the rain and snow season for half a year also puts forward higher requirements for the durability performance of concrete structures. To determine the coupling effect of water cement ratio, coarse aggregate content (volume), and mineral admixtures content (fly ash and slag) on the freeze-thaw resistance of concrete under sulfate attack, the primary and secondary relationship, and the optimum mix design, twenty groups of concrete with different mixture were designed and tested for their mass loss ratio and dynamic elastic modulus.

In this paper, the quadratic general rotary design method (Li et al. 2012; Li 2013; Lu et al. 2002) was used to design the mixture. Compared with orthogonal design, this method can not only consider the mineral admixture content, water cement ratio and coarse aggregate content, which have a great influence on the mixture of concrete but also can test the influence degree of independent variables on dependent variables, significantly reducing the number of mixture groups and workload (Lu et al. 2002). According to the 4-factor (1/2 implementation) quadratic general rotary combination design method, the necessary parameters of this study are the experimental factor $p = 4$, asterisk wall value $\gamma = 1.682$, and the corresponding mixture can be designed according to the structure matrix (Lu et al. 2022). Based on the research results of Qin (2017), Wang (Wang & Du 2011), and Cui (Cui et al. 2017), to reduce the amount of cement and obtain the same level of strength. The replacement rate (mass) of fly ash in this paper is 0 ~ 40%, the replacement rate of slag is 0~30%, and the volume rate of coarse aggregate is 0.3~0.4. The total amount of cement and slag is fixed at 360 kg·m^{-3}, the water-cement ratio (the mass ratio of water to cement plus slag) ranges from 0.3 to 0.5, and the gas content is controlled at about 5.5%. The combination design code can be found in Table 2, while the corresponding mixture design is shown in Table 3.

Table 2. Quadratic rotation-orthogonal combination design codes.

Code of the design method	Z_1/kg·m^{-3}	Z_2/kg·m^{-3}	Z_3	Z_4
$-\gamma$ (−1.682)	0	0	0.300	0.30
−1	29	21	0.320	0.34
0	72	54	0.350	0.40
+1	114	86	0.380	0.46
$+\gamma$ (+1.682)	144	108	0.400	0.50

Table 3. Mixture design and related basic properties of the specimen.

Specimens' Codes	Fly ash/ kg·m^{-3}	Slag/ kg·m^{-3}	Coarse aggregates/ kg·m^{-3}	Water/ kg·m^{-3}	Cement/ kg·m^{-3}	Sand/ kg·m^{-3}	Air content/ %	Compressive strength (28d)/ MPa
1	114	86	1007	162	274	698	5.5	44.7
2	114	86	848	162	274	746	5.7	46.9
3	114	21	1007	126	339	817	6.0	34.6
4	114	21	848	126	339	864	6.0	39.0
5	29	86	1007	126	274	914	6.2	38.2
6	29	86	848	126	274	962	6.1	41.0
7	29	21	1007	162	339	804	5.7	30.0
8	29	21	848	162	339	851	5.9	32.0
9	0	54	928	144	306	917	5.8	32.5
10	144	54	928	144	306	747	5.8	42.8
11	72	0	928	144	360	835	5.4	34.1
12	72	108	928	144	252	829	5.6	36.0
13	72	54	795	144	306	872	5.7	39.4
14	72	54	1060	144	306	792	5.7	37.2
15	72	54	928	108	306	928	5.6	46.7
16	72	54	928	180	306	736	5.	44.5
17	72	54	928	144	306	832	5.8	45.6
18	72	54	928	144	306	832	5.9	45.7
19	72	54	928	144	306	832	6.2	46.2
20	72	54	928	144	306	832	6.0	44.4

According to the design code in Table 2 and the quadratic general rotation combination design rules, the following equation exists:

$$Z_1 = 42.806X_1 + 72, Z_2 = 32.105X_2 + 54, Z_3 = 0.030X_3 + 0.35, Z_4 = 0.060X_4 + 0.40 \quad (1)$$

where Z_1, Z_2, Z_3, Z_4 represent the fly ash content, slag content, coarse aggregate volume ratio and water cement ratio, respectively.

2.3 Testing processes

Three 100×100×400 specimens were prepared in each group according to the above mixture. According to the "fast freezing method" of the standard (Ministry of Construction of the People's Republic of China, GB/T 50081–2002), the specimens cured for 24 days were immersed in 5% sodium sulfate solution by mass for 4 days, and then moved to a rubber box with 5% sodium sulfate solution for fast freeze-thaw cycle test. After every 50 cycles, the relative dynamic elastic modulus and mass loss rate of concrete are measured and calculated according to Equation (2) and Equation (3), respectively.

$$P_r = \frac{E_n}{E_0} \times 100\% \quad (2)$$

where P_r is the dynamic elastic modulus of the specimen after freeze-thaw cycles of n. E_n is the dynamic modulus of elasticity (MPa) after freeze-thaw cycles of n. E_0 is the initial dynamic modulus of elasticity (MPa) of the specimen before freeze-thaw cycle;

$$W_r = \frac{G_0 - G_n}{G_0} \times 100\% \quad (3)$$

Where W_r is the mass loss rate of the specimen after freeze-thaw cycles of n; G_n is the mass of specimen (kg) after freeze-thaw cycles of n; G_0 is the initial mass (kg) of the specimen before the freeze-thaw cycle.

3 ANALYSIS OF THE EFFECT OF EACH FACTOR ON THE SALT-FREEZING PERFORMANCE OF CONCRETE

The results of the relative dynamic elastic modulus and mass loss rate of concrete after sulfate attack and 300 freeze-thaw cycles are shown in Figure 1. It can be seen that the

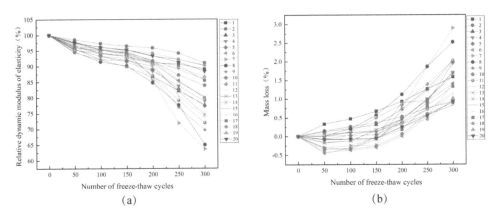

Figure 1. The curve between the number of freeze-thaw cycles and: (a) the relative dynamic elastic modulus; and (b) the mass loss rate.

relative dynamic elastic modulus of concrete decreases gradually with the increase of freeze-thaw cycles, and the decreasing rate of some experimental groups accelerates gradually in the later period. Generally, the relative dynamic elastic of all groups is not less than 60% of the initial value after 300 freeze-thaw cycles, which indicates that all groups can withstand 300 freeze-thaw cycles and meet the engineering requirements. The relative dynamic modulus of elasticity of groups 2, 15, and 18 is higher than 90%, with group 2 having the maximum value of 91.01%, whereas the value of groups 7, 8, and 9 is lower than 70%, with the minimum value of 63.61% for group 7.

In addition, the mass loss rate of 70% of the specimens exhibited a decreasing trend at first, followed by a progressive increase that accelerated with increasing cycle periods. After 300 freeze-thaw cycles, the rate of mass loss for all groups was below 3%. The mass loss rate of groups 3, 7, and 11 exceeded 2 percent, while the mass loss rate of group 7 reached 2.89 percent. The mass loss rate for groups 2, 10, 15, 18, and 20 was less than 1%, with the lowest rate occurring in group 15.

These test findings indicate that, under the coupling conditions of sulfate attack and freeze-thaw cycle, the rate of concrete deterioration changes little when the cycle periods (< 150 times) are relatively short. As cycle periods rise, the pace of degradation gradually accelerates. This is comparable to the findings in reference (Wang & Ma 2017). To further investigate the effect of various parameters on the interior damage of concrete specimens after a freeze-thaw cycle, the following quaternary quadratic regression models of relative dynamic elastic modulus (Y_1) and mass loss rate (Y_2) under sulfate and freezing-thawing cycles are established as follows:

$$Y_1 = 86.986 + 5.471X_1 + 4.056X_2 - 1.195X_3 - 0.511X_4 \\ -3.205X_1^2 - 4.568X_2^2 - 3.004X_3^2 + 0.76X_4^2 \\ -0.313X_1X_2 + 0.037X_1X_3 - 0.114X_1X_4 \tag{4}$$

$$Y_2 = 0.990 - 0.250X_1 - 0.252X_2 + 0.128X_3 + 0.049X_4 \\ +0.095X_1^2 + 0.387X_2^2 + 0.279X_3^2 + 0.037X_4^2 \\ +0.053X_1X_2 - 0.003X_1X_3 + 0.033X_1X_4 \tag{5}$$

Table 4 displays the results of the variance analysis of influencing factors. Results indicate that the relative influence of each factor on the relative elastic modulus is as follows: fly ash

Table 4. Analysis of variance of the relative dynamic elastic modulus (Y_1) test results.

Source of variation	Sum square	Degree of freedom	Mean square	F value	p value
X_1	317.012	1	317.012	93.4939	0.0002
X_2	174.289	1	174.289	51.4017	0.0008
X_3	151.145	1	151.145	4.4576	0.0885
X_4	27.608	1	27.608	0.8142	0.4082
X_{12}	113.457	1	113.457	33.461	0.0022
X_{22}	230.451	1	230.451	67.9652	0.0004
X_{32}	99.640	1	99.640	29.3859	0.0029
X_{42}	63.714	1	63.714	1.8791	0.2288
X_1X_2	3.0419	1	3.0419	0.0897	0.7766
X_1X_3	0.0422	1	0.0422	0.0012	0.9732
X_1X_4	0.4059	1	0.4059	0.012	0.9171
Regression	841.8944	14	84.213	F_2=5.51076	0.0824
Residuals	117.8983	5	33.9072		
Lack-of-fit	104.7835	2	78.2106	F_1=8.58872	0.0053
Errors	13.1148	3	4.3716		
Sum	959.7927	19			

content > slag content > coarse aggregate volume ratio > water cement ratio. The test of misfit and significance ($F_1 < F_{0.05}(2, 3) = 9.55$, $F_2 > F_{0.05}(14, 5) = 4.524$) shows that there are no other factors that cannot be ignored, and the regression equation fits well with the testing data and can predict the relative dynamic modulus of elasticity. After removing the unimportant factors, the following simplified regression model was obtained:

$$Y_1 = 86.986 + 5.471X_1 + 4.056X_2 - 1.195X_3 + 3.205X_1^2 - 4.569X_2^2 - 3.004X_3^2 \quad (6)$$

In the same way, the relative impact of each factor on the mass loss rate of concrete is as follows: slag> fly ash> aggregate volume ratio> water cement ratio. The regression model obtained after eliminating the insignificant items is as follows:

$$Y_2 = 0.990 - 0.250X_1 - 0.252X_2 + 0.128X_3 + 0.049X_4 + 0.095X_1^2 + 0.387X_2^2 + 0.279X_3^2 \quad (7)$$

3.1 Analysis of the effect of the single factor

Regarding the relative dynamic modulus of elasticity, the effect curve depicted in Figure 2 can be produced if all other parameters are set to zero and just one element is examined. When the test level is within [−1.682, 1.682], the trend of other factors is a downward-opening parabola; that is, with the steady growth of each factor, the relative dynamic modulus of elasticity first grows and then decreases, and each reaches a maximum value.

As shown in Figure 2 (a), the relative dynamic elastic modulus of concrete reaches the maximum when the fly ash content is: $X_1 = 0.5$ (i.e., the content is 25%), and a relatively high fly ash content can provide a higher relative dynamic elastic modulus. When the content of slag is 17.3% ($X_2 = 0.45$), the relative dynamic elastic modulus of concrete reaches the maximum value. Similarly, the relative dynamic elastic modulus of concrete reaches the maximum value when the volume ratio of coarse aggregate is 0.335 ($X_3 = 0$). In comparison, an increase in the water-to-cement ratio will result in a drop in the relative dynamic elastic modulus of concrete, although the magnitude of the change is minimal.

Similarly (as shown in Figure 2 (b)), within the test range, the mass loss rate of concrete will initially reduce and then increase with the steady increase in fly ash content, slag content, coarse aggregate volume ratio and water cement ratio. The mass loss rate is minimal when the content of fly ash is 34% ($X_1 = 1.2$). When the slag content is 15% ($X_2 = 0$), the mass loss rate has a minimal value, and the influence of the slag on its mass loss rate is greater than other factors. The mass loss rate is at its minimum when the volume ratio of coarse aggregate is 0.335% ($X_3 = 0$) while the rate of mass loss is minimal when the ratio of water to cement is 0.37 ($X_4 = -0.5$).

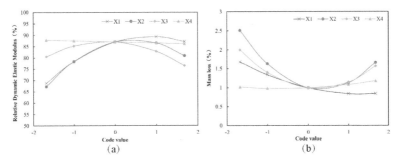

Figure 2. Influence curve of factors on (a) the relative dynamic elastic modulus (Y_1) and (b) the mass loss rate (Y_2).

In general, the application of mineral admixture enhances the durability of concrete, and its effect is greater than that of coarse aggregate and water cement ratio, which is compatible with the action principle of fly ash, slag, and aggregate in concrete. The tiny particles of fly ash have a filling effect and pozzolanic ash properties, which, through a series of physical and chemical processes (Ge et al. 2008), can minimize the pores between concrete and increase the compactness of concrete. This also helps improve the concrete's durability and resistance to freezing and thawing. However, excessive addition harms the concrete's strength and resistance to freezing and thawing since it reduces the cement content. The addition of identical slag can form more C-S-H gels to fill the microscopic pores, so enhancing the freeze-thaw resistance of concrete; however, an excess of slag will decrease its strength and durability. Increasing the volume fraction of coarse aggregate to a certain degree can improve the strength of concrete and reduce the porosity, but beyond a certain range, it will also reduce the cement paste, thereby weakening the strength and cohesion between aggregate and cement paste, thereby diminishing the freeze-thaw resistance.

3.2 Analysis of the joint effect of multi-factor

When the relative dynamic modulus of elasticity is taken as the evaluation index and the other two factors are set to zero, the joint influence surface diagram is shown in Figure 3. It can be seen that the relative dynamic elastic modulus of concrete initially increases and subsequently declines, and there is a maximum point: $P_{(0.78, 0.45)} = 90.11\%$, with the increasing of fly ash and slag content as shown in Figure 3 (a). Similarly, as Figure 3 (b) shows, as the ratio of fly ash to coarse aggregate content and volume increases, the relative dynamic elastic modulus increases firstly and then decreases, but within the value range, the effect of fly ash on the relative dynamic elastic modulus is greater and has a maximum value: $P_{(0.75, 0.25)} = 89.67\%$. Figure 3 (c) demonstrates that the effect of fly ash content on relative dynamic elastic modulus is greater than that of water-cement ratio, and the change of water cement ratio has little influence on relative dynamic elastic modulus, with the maximum value of $P_{P_{(0.85, -1.682)}} = 90.76\%$.

When the mass loss rate is taken as the evaluation index (as shown in Figure 4), the effect and the best dosage of different factors can be obtained by the same token: the influence of slag dosage is greater than that of fly ash mass loss rate, and there is a minimum value c. The

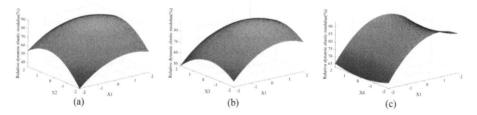

Figure 3. Surface diagram of the combined effect of multiple factors on the relative dynamic elastic modulus: (a) $X_1-X_2-Y_1$; (b) $X_1-X_3-Y_1$; (c) $X_1-X_4-Y_1$.

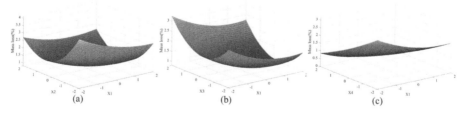

Figure 4. Surface diagram of the combined effect of multiple factors on the mass loss: (a) $X_1-X_2-Y_2$; (b) $X_1-X_3-Y_2$; (c) $X_1-X_4-Y_2$.

influence of coarse aggregate volume ratio on mass loss rate is greater than the minimum value of fly ash: $P_{(1.04, -0.28)} = 0.85\%$. The smaller the water-cement ratio, the larger the fly ash content, and the smaller the mass loss rate and the minimum value: $P_{(1.66, 0.23)} = 0.87\%$.

3.3 *Design optimal*

To explore the effect of four factors on the relative dynamic elastic modulus of concrete and find out the best mixture design of sulfate freeze-thaw resistance, the frequency analysis method was used in this study based on the above data. The results are shown in Table 5, after 300 freeze-thaw cycles, the optimum mixture within a 95% confidence interval is

Table 5. Frequency analysis of relative dynamic elastic modulus (Y_1).

Code value	X_1	Frequency	X_2	Frequency	X_3	Frequency	X_4	Frequency
−1.6818	0	0	0	0	20	0.1481	27	0.2
−1	0	0	5	0.037	35	0.2593	27	0.2
0	40	0.2963	60	0.4444	50	0.3704	27	0.2
+1	55	0.4074	50	0.3704	30	0.2222	27	0.2
+1.6818	40	0.2963	20	0.1481	0	0	27	0.2
Weighted average		0.906		0.582		−0.286		0
standard error		0.056		0.06		0.078		0.107
Distribution range (95%)		0.796~1.016		0.465~0.700		−0.439~−0.134		−0.209~0.209
Corresponding dosage (Z)		0.295~0.321		0.191~0.212		0.337~0.346		0.388~0.412

obtained so that the relative dynamic elastic modulus of concrete specimens is greater than 88.15%. According to Equation (1), the optimum content of each factor can be calculated as follows: the content of fly ash is 106.074~115.491 kg·m^{-3} (29.5~32.1%), the content of slag is 68.929~76.474 kg·m^{-3} (19.1~21.2%), the volume ratio of coarse aggregate is 0.337~0.346, and the water-cement ratio is 0.388~0.412. The results are comparable to those predicted by the mass loss ratio. The ratio of mass loss is less than 1.53% in a 95% confidence interval: fly ash content is 116.090 kg·m^{-3} (32.25%) slag content is 68.447 kg·m^{-3} (19.01%), while coarse aggregate volume ratio is 0.341 and water cement ratio is 0.413.

To sum up, by testing the relative dynamic elastic modulus and mass loss rate of concrete specimens after 300 freeze-thaw cycles respectively, the mathematical models of corresponding indexes and effect factors are obtained from the analysis of testing data, and then the optimal mixture is obtained as follows: fly ash content 32%, slag content 19%, coarse aggregate volume ratio 0.34, and water cement ratio 0.41, which can make the concrete reach more than 88.15% relative dynamic elastic modulus and less than 1.53% mass loss rate.

4 CONCLUSION

By using the quadratic general rotary design method [and taking the relative dynamic modulus of elasticity and mass loss rate as criteria, this paper quantitatively designs and analyzes the effect of fly ash content, slag content, coarse aggregate volume ratio, and water cement ratio on the freeze-thaw cycle resistance of concrete under sulfate attack condition, and obtains the relevant mathematical model and the optimal mixture:

(1) The addition of fly ash and slag is helpful to the durability of concrete under salt-freeze-thaw conditions and there is an optimum amount of fly ash and slag when only one

factor is considered. Coarse aggregate volume ratio and water-cement ratio have relatively little influence on the durability of concrete under the same condition.
(2) The regression model for the dynamic modulus of elasticity and mass of concrete under the combined influence of multiple factors is established. In addition, the primary and secondary relationships between each element and concrete durability vary slightly based on evaluation criteria, with fly ash and slag having a stronger impact than coarse aggregate and water cement ratio.
(3) The optimal mixture is designed by numerical analysis and microstructure analysis, taking into account the combined effect of the four parameters, with the ash content of fly of 32%, slag content of 19%, coarse aggregate volume ratio of 0.34, and water cement ratio of 0.41. This mixture could provide an improved freeze-thaw resistance under sulfate attack circumstances.

REFERENCES

Cui C., Peng H., Liu Y., et al. Influence of GGBFS Content and Activator Modulus on Curing of Metakaolin Based Geopolymer at Ambient Temperature. *Journal of Building Materials*, 2017, 20(4): 535–542.
Fang X.W., Lou Z.K., Gao Y.L. Research Progress on Frost Resistance Durability of Concrete Under Sulfate Attack. *Concrete*, 2019, 362(12): 6–11.
Ge Y., Yuan J., Yang W.C., et al. Property of Fly-ash Concrete in Sodium Sulfate Solution Under Freeze-thaw and Dry-wet Cycling. *Journal of Wuhan University of Technology*, 2008, 185(6): 33–36.
Jin W.F. *Experimental Study on Anti-frozen Durability of Concrete in Saline Soil Region*. XI'AN, Chang'an University, 2010.
Li C., Yang S.H., Li Y.Q., et al. The Quadratic General Rotary Design Method to Optimize the Process Conditions of Tea Oil Extraction by Water Substitution. *Food and Nutrition in China*, 2012, 18(10): 42–45.
Li G.F. *Study on the Cotton Yarn Process of Quadratic General Rotary Unitized Design*. Xinjiang, Xinjiang University, 2013.
Li Q. Effect of Mineral Admixtures on Salt-freezing Resistance of Concrete. *Bulletin of the Chinese Ceramic Society*, 2015, 34(3): 239–242.
Lu E.S., Song S.D., Guo M.C. Several Questions of Quadratic General Rotary Design Method. *Journal of Northwest A & F University (Natural Science Edition)*, 2002(5): 110–113, 120.
Ministry of Construction of the People's Republic of China, Standard for Long-term Performance and Durability Test Methods for Ordinary Concrete: GB/T 50081–2002. Beijing, 2009.
Qin L., Ding J.N., Zhu J.S. Experiment on Anti-permeability and Frost Resistance of High Strength Concrete with High-ratio of Fly Ash and Slag. *Transactions of the Chinese Society of Agricultural Engineering (Transactions of the CSAE)*, 2017, 33(6): 133–139.
Wan L.D. *Experimental Study on the Durability of Concrete Under the Combined Action of Sulfate Erosion and Freeze-thaw Cycle*. Xi'an, Xi'an University of Architecture and Technology, 2013.
Wang J. Experimental Study on the Durability of Hydraulic Concrete Based on the Combined Effect of Freeze-thaw Cycles and Sulfate Attack. *Water Resources Planning and Design*, 2019, 190(8): 82–85.
Wang P., Du Y.J. Research on Frost-resistance and Anti-permeability Durability of High-volume fly ash concrete. *Concrete*, 2011, 266(12): 76–78.
Wang W. Research on Optimization of Mixture Ratio of Salt-freezing Performance of Concrete for Large Bridges in Alpine Region. *Concrete*, 2021, 379(5): 119–122.
Wang Z. W., Ma F. Study on the Damage of Concrete Under the Couple Action of Sulfate Corrosion and Freeze-thaw. *New Building Materials*, 2017, 44(11): 40–43.
Yu H.F., Sun W., Ma H.Y., et al. Analysis of Damage Degradation Parameters of Concrete Subjected to Freezing-thawing Cycles and Chemical Attack. *Journal of Architecture and Civil Engineering*, 2011, 28(4): 1–8.
Yuan L.D., Niu D.T., Jiang L., et al. Study on Damage of Concrete Under the Combined Action of Sulfate Attack and Freeze-thaw Cycle. *Bulletin of the Chinese Ceramic Society*, 2013, 32(6): 1171–1176.

Analysis of bearing characteristics on soft soil foundation reinforced with bamboo grid and flow-solidified silt

DongKui Zhao, Shaolin Yue & Jihui Ding*
CCCC Road & Bridge Special Engineering Co., Ltd., Wuhan, Hubei, China

Haochen Wang
China Construction Third Bureau First Engineering Co., Ltd., Wuhan, China

Zenghui Yu
China Construction Second Bureau Eighth Engineering Co., Ltd., Jinan, China

ABSTRACT: When the surface layer of a deep soft soil foundation is super soft soil, the surface layer of silt or super-soft soil foundation reinforced with a bamboo grid can meet the requirements of construction equipment. The existing calculation methods and calculation parameters of bearing capacity that soft soil foundation reinforced by bamboo grid mainly references to geotextile reinforced foundation. In this paper, the bearing characteristics of soft soil foundations reinforced by the bamboo grid and flow-solidified silt are studied by model test. The test results show that the addition of basalt fiber improves the failure mode of the flow-solidified silt composite foundation, and avoids brittle cracking when the impact shear failure occurs; when the curing age is 3d, 5d and 7d, the ultimate bearing capacity of bamboo grid and solidified silt foundation increases by 25.4%, 32.9% and 40.1% respectively; with the increase of age, the pressure diffusion angle of the bamboo grid and solidified silt layer increases slightly, and the addition of fiber increases the pressure diffusion angle; the pressure diffusion angle of cement flow-solidified silt layer is $35.3° \sim 42.5°$, and that of fiber flow-solidified silt layer is $37.2° \sim 47.2°$.

1 INSTRUCTION

The bamboo grid has high tensile strength, bending rigidity, and certain shear resistance (Deng & Cheng 2018; Wang et al. 2022). After the surface layer of silt or super-soft foundation is reinforced with a bamboo grid, the bearing capacity of the composite foundation is obviously improved (Aazokhi et al. 2018; Yuan et al. 2014), which can meet the requirements of construction equipment. Premixed flow-solidified silt is a very good construction material with high strength, controllable quality, wide application range, and a friendly environment. The flow-solidified silt with cement alone often suffers from brittle failure. The addition of fiber can improve the elastic modulus of the sludge-solidified silt and is conducive to the development of soil ductility (Carruth & Howard 2013). Namdar et al. (2012) analyzed the deformation characteristics of cement-solidified soft soil before and after polymer fiber reinforcement. The fiber improved the strength of solidified soil and changed its brittle failure mode to ductile failure mode, which increased the water stability of solidified soil. Khattak et al. (2006) showed that the addition of fiber can improve the tensile strength of

---

*Corresponding Author: dingjihui@126.com

silt-solidified soil. Carruth et al. (2013) pointed out that the incorporation of fiber improves the elastic modulus of solidified silt and is conducive to the development of soil ductility. The research of Yetimoglu T (Yetimoglu et al. 2005) and Tang Chaosheng (Tang et al. 2007) showed that with the addition of fiber, the strength of the sample can be effectively improved, the strain of the sample during failure can be increased, and the ductility of the sample can be improved. He Qingju (He 2018) found that under the same reinforcement rate when the fiber length is 12 mm, the strength of fiber-solidified silt is the largest. Chen Wei, Gao Wenbo, et al. (2020) took the marine dredged sludge deposited in Dalian Bay as the object, carried out pilot test research on the filling of mobile solidified soil, and analyzed the change characteristics of the strength index of the sludge-solidified soil with the amount of cement and the curing age. Hu Zhenhua, Wang Ying, et al. (2020) pointed out through model tests that the artificially formed solidified soil hard shell can greatly improve the bearing capacity of the foundation; based on the punching shear failure theory of double-layer foundation, the bearing capacity formula of square foundation in the model test is derived.

Although scholars have conducted a lot of research on the mechanical properties of the flow-solidified soil, the research results are mostly limited to the indoor mix ratio, the mechanical properties of the flow-solidified soil and the analysis of the influencing factors and so on, and the model and field tests and engineering practice are less involved, especially the research on the bearing characteristics of the flow-solidified silt and bamboo grid reinforced soft soil or super-soft soil is almost not involved. Therefore, based on the soft soil subgrade project in a city, this paper carries out the research on the bearing capacity and deformation characteristics of the soft foundation strengthened by the flow-solidified silt and bamboo grid through the indoor model test, analyzes the influence factors of the pressure diffusion angle, and provides the basis for the similar soft soil foundation treatment design.

2 INDOOR MODEL TEST

2.1 *Engineering background*

The upper surface layer of a soft soil foundation is soft soil. The thickness of the soft soil layer is 15 m $\sim$ 20 m, the density is 1.41 g/cm^3 $\sim$ 1.69 g/cm^3, the water content is 50.0% $\sim$ 94.0%, the void ratio is 0.952 $\sim$ 2.714, the plasticity index is 25 $\sim$ 70, and the undrained strength is 3.9 kPa $\sim$ 14.6 kPa. In the early stage of the foundation, there is no mechanical inserting plate, so two inserting plates are used. First, manual inserting plates are used. The depth of silt treatment is not more than 5 m, and the spacing is 0.8 m. After the shallow silt has a certain strength, the mechanical inserting plate shall be carried out. The spacing of plastic drainage plates is 1.2 m and the depth is 15 m $\sim$ 20 m. The stacking height is 5 m (0.5 m sand, 4.5 m soil), and the treatment width is 34 m. Unloading shall be carried out after the consolidation degree of the soft soil layer reaches 90%. The characteristic value of foundation-bearing capacity after foundation treatment is more than 80 kPa. Considering the low bearing capacity of the surface layer of the sludge, the inability of mechanical equipment to construct, and the slow consolidation speed of the sludge, it is proposed to adopt the bamboo grid and flow-solidified silt composite foundation.

2.2 *Experimental materials*

(1) Soil material
   The silt used was taken from the Economic and Technological Development Zone of a city. The density is 1.65 g/cm^3, the water content is 55.2% and 52.1%, the plastic limit is 20.0, the plasticity index is 32.1, the liquid index is 1.10, and the pore ratio is 1.55. The

Figure 1. Model box and layout of bamboo grid.

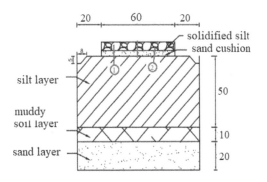

Figure 2. Soil layer profile of model box (unit: cm).

size of the model box is 100 cm×100 cm×100 cm (Figure 1). According to the actual soil layer of the project, add the appropriate amount of water to the test soil, mix it evenly, and put it into the model box layers. The soil layer in the model box is divided into three layers (Figure 2). The thickness of the silt layer is 50 cm, the water content is 72%, and the density is 1.68 g/cm^3; The thickness of the muddy soil layer is 10 cm, the water content is 46%, and the density is 1.78 g/cm^3; The bottom sand layer has a thickness of 20 cm, the water content of 30%, and a density of 1.54 g/cm^3.

(2) Bamboo and geotextile

The bamboo grid adopts bamboo from Jiangsu Province, with a diameter of 1.0 cm~1.2 cm. The moisture content is 11.0%, and the length of the bamboo joint is 17.5 cm~23.3 cm. The mass per unit area of Geotextile is 150 g/m^2, the tensile strength is 28 kN/m ~ 250 kN/m, and the elongation is 35%.

(3) Mix the proportion and strength of flow-solidified silt

The cement is ordinary portland cement P.O 42.5. Basalt fiber length is 12 mm, and diameter is 16 μm~22 μm. The density is 2.62 g/cm^3 ~ 2.65 g/cm^3, the tensile strength is equal and more than 3200 MPa, the elastic modulus is 90 GPa ~ 105 GPa, and the elongation at break is 2.7% ~ 3.6%. The solidified silt layer is divided into two test zones ① and ②. The mix ratio and unconfined compressive strength are shown in Table 1. Zone ① is cement-solidified silt, and zone ② is fiber-flow-solidified silt. Compared with the strength of zone ①, the 3d, 5d and 7d strength in the ② zone increased by 128%, 115% and 66%, respectively.

Table 1. Parameters of flow-solidified sludge.

zone	Water solid ratio	Cement content %	Flow value/cm	Wet density/ g/cm^2	Fiber content/%	Unconfined compressive strength /kPa		
						3d	5d	7d
①	1.2	11	17	1.46	0.0	43.1	66.0	104.6
②	1.2	11	15	1.51	0.8	98.1	142.2	173.8

2.3 Analysis of experimental results

(1) After the self-weight consolidation of the soil in the model box is completed, the drainage plate shall be inserted (the drainage plate is 3 cm wide, 15 cm long, space is 15 cm, and arranged in a regular triangle). Laying 15 cm × 15 cm bamboo grid and 4 cm sand cushion. Load tests are carried out on soft soil foundation, bamboo grid + soft soil foundations, and sand cushion + bamboo grid + soft soil foundation.

(2) Flow-solidified silt is filled on the sand cushion, the schematic diagram of the subgrade model is shown in Figure 2. Zone ① is cement flow-solidified soil (fiber content is 0) and zone ② is basalt fiber cement flow-solidified silt (fiber content is 0.8%). The load tests are carried out after the flow-solidified silt is cured for 3d, 5d and 7d.

The load plate size is 15 cm × 15 cm × 2 cm steel plate. Figures 3 and 4 are load settlement curves under different working conditions. Strictly speaking, the bearing capacity of the foundation is not a fixed value. In addition to the physical and mechanical properties of the soil, it is related to the size, form and burial depth of the foundation (GB50007 2011). The load settlement curve of soft soil in Figure 3 shows that the bearing capacity characteristics of soft soil foundation correspond to $s/b = 0.06$ and the limit value $s/b = 0.10$; The load settlement curves of bamboo grid + soft soil foundation and sand cushion + bamboo grid + soft soil foundation are basically linear. Figure 4 is the load settlement curve of different curing ages. When the curing age is 3d, the curve is an S-shape (nonlinear section + linear

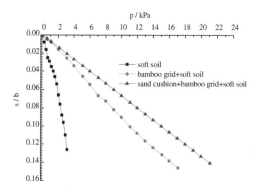

Figure 3. Load settlement curve before solidified silt is filling.

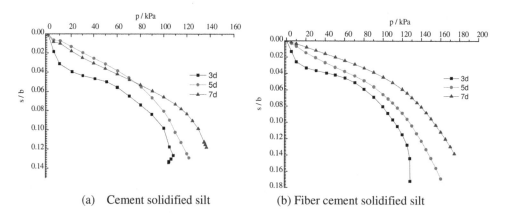

(a) Cement solidified silt (b) Fiber cement solidified silt

Figure 4. Load settlement curve at different curing ages.

section + nonlinear section). The curing age is short, the strength of the solidified soil is insufficient, and the load is small when the deformation begins. When s/b=0.03, the load increment increases until it is destroyed. When the curing age is 5d and 7d, the strength of the flow-solidified silt increases with the increase of the curing age, and there is an obvious straight line at the beginning of the curve, followed by a nonlinear curve. Therefore, the bearing capacity corresponding to different settlement ratios is mainly analyzed (as shown in Table 2).

Table 2. Bearing capacity under different working conditions.

Working conditions	Before filling of solidified soil		Age	Cement solidified silt		Fiber cement solidified silt	
	$P_{s/b=0.06}$/ kPa	$P_{s/b=0.1}$/ kPa		$P_{s/b=0.06}$/ kPa	$P_{s/b=0.1}$/ kPa	$P_{s/b=0.06}$/ kPa	$P_{s/b=0.1}$/ kPa
Soft soil	2.0	2.8	3d	63.3	101.2	82.9	110.7
grid + soft soil	6.4	11.0	5d	86.1	109.5	95.1	128.0
Sand+grid + soft soil	9.6	15.0	7d	93.1	130.4	122.2	153.0

Figure 3 and Table 3 show that the bamboo grid has a high elastic modulus, which can improve the deformation resistance when combined with soft soil. Compared with soft soil foundation, when s/b = 0.06 and 0.10, the bearing capacity of bamboo grid + soft soil foundation increases by 220% and 293% respectively, and that of sand cushion + bamboo grid + soft soil foundation increases by 380% and 436% respectively. Figure 4 and Table 3 show that with the increase of age, the bearing capacity of the solidified silt + bamboo +grid soft soil foundation increases significantly. When the curing age is 3d, 5d and 7d, the bearing capacity of bamboo grid + solidified silt foundation increases by 31.0%, 10.5% and 31.4% respectively when s/b = 0.06, and 16.9% and 17.3% respectively when s/b = 0.10.

2.4 Foundation failure mode

As shown in Figure 5, the failure mode of solidified silt hard shell layer is punching shear failure. For the hard shell layer without basalt fiber, there are cracks around the failure area.

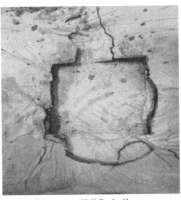
(a) Cement solidified silt

(b) Fiber cement solidified silt

Figure 5. Failure mode of composite foundation.

When basalt fiber is added, there are no cracks in the shear failure area. The basalt fiber enhances the ductility of the solidified silt hard shell layer, avoids brittle cracking in the event of punching shear failure, and enhances stability.

Figure 6 shows the microstructure of cement-solidified silt obtained by SEM scanning, and the surface is an uneven porous structure. The fiber will be embedded in the solidified silt (Figure 6 (c) and (d)). The fiber can form a fiber grid in the solidified silt. The fiber reinforcement has a gripping effect, and there is friction between the solidified silt particles and the fiber. The fiber is embedded in the solidified soil particles to form a whole. When the fiber is damaged by force, the friction force will be generated between the soil and the fiber to prevent brittle failure and achieve the effect of enhancing the shear strength.

(a) Cement solidified silt (100 times) (b) Cement solidified silt (2000 times)
(c) Fiber cement solidified silt (100 times) (d) Fiber cement solidified silt (2000 times)

Figure 6. Microstructure of flowing solidified silt.

3 PRESSURE DIFFUSION ANGLE OF SOLIDIFIED SILT LAYER

3.1 *Calculation of bearing capacity of bamboo grid + solidified silt composite foundation*

When the bamboo grid is filled with a certain thickness of flow-solidified silt layer, the solidified silt layer has a stress diffusion effect. When the soft soil foundation of the bamboo grid reinforced reaches the limit state, the tension of the bamboo grid is the tensile strength (T_u) of the bamboo grid, and the vertical load is the ultimate bearing capacity of the composite foundation f_u (Bai 1994):

$$f_u = [C_u N_c + 2KT_u \sin\theta_{hu}/(b + 2H\tan\alpha)](b + 2H\tan\alpha)(l + 2H\tan\alpha)/(bl) \quad (1)$$

Where b is the width of the load working surface (m); l is the length of the load working surface (m); θ_{hu} is the angle between T_u and the horizontal plane of the angle; α is the pressure diffusion angle; K is the coefficient related to the foundation shape, K = 1~2. When $l/b \geq 10$, K = 1; if $l/b = 1$, K = 2.

When the thickness of solidified silt layer H = 0, (1) is converted to:

$$f_u = C_u N_c + 2KT_u \sin\theta_{hu}/b = f_{us} + f_{uN} \quad (2)$$

Where f_{us} is the ultimate bearing capacity of soft soil foundation(kPa); f_{uN} is the ultimate bearing capacity of bamboo grid + soft soil foundation When H = 0 (kPa).

When substituting formula (2) with formula (1), the formula becomes:

$$f_u = [f_{us} + bf_{uN}/(b + 2H\tan\alpha)](b + 2H\tan\alpha)(l + 2H\tan\alpha)/(bl) \quad (3)$$

3.2 Pressure diffusion angle of the bamboo grid and solidified silt layer

When substituting the results in Table 1 into formula (2) to determine f_{uN}, and substituting f_{us} and f_{uN} into formula (3), the pressure diffusion angle of the solidified silt layer can be calculated by trial calculation method, as shown in Table 4.

Table 4. Pressure diffusion angle (unit: °).

Age	Cement solidified silt layer		Fiber cement solidified silt layer	
	$\alpha_{s/b=0.06}$	$\alpha_{s/b=0.10}$	$\alpha_{s/b=0.06}$	$\alpha_{s/b=0.10}$
3d	35.3	35.5	40.3	37.2
5d	41.1	37.1	42.9	40.0
7d	42.5	40.3	47.2	43.2

The pressure diffusion angle of the solidified silt layer can be calculated by the trial calculation method, as shown in Table 4. Table 4 shows that the pressure diffusion angle of cement solidified silt layer is 35.3° ~ 42.5°, and that of fiber solidified silt layer is 37.2° ~ 47.2°. Only when the buried depth ratio is greater than 0.5 and the compression modulus ratio of the upper and lower layers is greater than 10, the stress diffusion angle is taken as 30° in the current code for the design of the foundation (GB50007 2011), and other cases are less than 30°. The pressure diffusion angle of bamboo grid + solidified soil increases slightly with the increase of age, and the addition of fiber increases the pressure diffusion angle. Compared with the case of s/b = 0.06, the pressure diffusion angle corresponding to s/b = 0.10 is slightly smaller; the pressure diffusion angle of the cement solidified sludge layer is close to that of cement solidified sludge layer at 3d, and increases by 9.7% and 5.2% at the 5d and 7d respectively; at 3d, 5d and 7d, the fiber cement solidified silt increased by 7.7%, 6.8% and 8.5% respectively. Compared with cement solidified silt layer, when s/b = 0.06, the fiber cement solidified silt increased by 14.5%, 6.8% and 8.5% respectively at 3d, 5d and 7d; when s/b = 0.06, it increased by 4.8%, 7.8% and 7.2% respectively.

4 CONCLUSION

(1) The model test shows that the bearing capacity of the composite foundation of bamboo grid + solidified silt increases with the increase of curing age. The addition of fiber forms a fiber net and the fiber has the function of gripping and wrapping, which enhances the ductility of the solidified silt hard shell layer, avoids brittle cracking in the event of punching shear failure, and enhances stability. The ultimate bearing capacity of the

composite foundation is increased by 25.4%, 32.9% and 40.1% respectively when the curing age is 3d, 5d and 7d.
(2) Based on the pressure diffusion theory, the pressure diffusion angle is calculated according to the bearing capacity formula of a composite foundation reinforced with bamboo grid + solidified silt. With the increase of age, the pressure diffusion angle of bamboo grid + solidified silt increased slightly, and the addition of fiber increased the pressure diffusion angle. The pressure diffusion angle of cement solidified silt layer is $35.3° \sim 42.5°$, and that of fiber flow solidified soil layer is $37.2° \sim 47.2°$.

REFERENCES

Aazokhi Waruwu, Husni Halim, et al. Bamboo Grid Reinforcement on Peat Soil under Repeated Loading [J]. *Journal of Engineering and Applied Sciences*. 2018, 13(8): 2190–2196.

Bai Bing, Calculation of bearing capacity of reinforced soft soil foundation under three-dimensional stress of Geotextile [J]. *East China Highway*, 1994, 91(6): 54–56.

Carruth W.D., Howard I.L. Use of portland cement and polymer fibers to stabilize very high moisture content fine-grained soils [J]. *Advances in Civil Engineering Materials*, 2013, 2(1): 124.

Chen Wei, Gao Wenbo, et al. Pilot study on flow solidification of marine dredged sludge [J]. *Geotechnical Mechanics*, 2020, 41(S2): 2–11.

Deng Yousheng, Cheng Zhihe. Study on compression test of bamboo structure[J]. *Science Technology and Engineering*, 2018, 18(27): 223–227.

GB50007 (2011), *Code for Design of Building Foundation* [C]. China Construction Industry Press.

He Qingju, Study on the Influence of Fiber Reinforced Solidified Sludge On Mechanical [J]. *Energy and Environmental Protection*. 2018, 40: 20–28.

Hu Zhenhua, Wang Ying, et al. Experimental and Theoretical Study on Bearing Capacity of Double-Layer Foundation with Artificial Hard Shell Solidified By Geotechnical Mechanics [J]. *People's Yangtze River*, 2020, 51(02): 0147–06.

Khattak M.J., Alrashidi M. Durability and mechanistic characteristics of fiber reinforced soil-cement mixtures [J]. *International Journal of Pavement Engineering*, 2006,7(1): 53–62.

Tang Mingchao, Shi Bin, et al. Experimental Research on Polypropylene Fiber Reinforced Soft Soil. *Rock And Soil Mechanics*. 2007, 28(9): 1796–1800.

Namdar P., Estabragh A.R., Javadi A.A. The Behavior of Cement-stabilized Clay Reinforced with Nylon Fiber [J]. *Geosynthetics International*. 2012, 19(1): 85–92.

Wang Guixiang, Yan Shengkang, et al. Stress and Deformation Analysis of Bamboo Net in Soft Soil Foundation Reinforcement [J]. *Journal of Hebei University (Natural Science Edition)*. 2022, 42(4): 443–448.

Yetimoglu T., Inanir M., Inanir O.E. A Study on Bearing Capacity of Randomly Distributed Fiber-Reinforced Sand Fills Overlying Soft Clay [J]. *Geotextiles & Geomembranes*. 2005,23(2): 174183.

Yuan Man, Ding Jihui, Cao Yanliang. The Application of Bamboo Network Reinforcement Technology on Hydraulic Fill Soft-Soil Foundation Treatment [J]. *World Journal of Engineering & Technology*, 2014, 02(2): 68–72.

Concrete proportion for assembled laminated slabs

Jiaxing Li*, Qiang Li*, Yiqin Qiu*, Qian Wang* & Junjie Wu*
Department of Civil Engineering, Zhejiang Ocean University, China

ABSTRACT: The research was carried out to study the concrete compound ratio that is impermeable and suitable for high-temperature steaming, and to obtain a new laminated slab compound ratio through indoor testing. For the production of laminated slabs with thin slabs, with high impermeability requirements, early strength requirements, and rapid production through steam curing, we have developed a new concrete formulation with different superfine compound admixtures and additives through experimental research methods. The new concrete formulation for laminated slabs meets the impermeability requirements, has good compatibility, and is suitable for high-temperature steam curing.

1 GENERAL INTRODUCTION

The increasing use of prefabricated composite floor slabs requires improving product quality and production efficiency for the factory production of laminates in building industrialization. It is important to conduct an in-depth study of stacked panels with good mechanical properties, high precision, high production efficiency, easy storage and transfer, and environmental protection, which are the key tasks of building industrialization in recent years. To realize the industrialization of the new laminated panel, the first necessary part is controling the precision from the concrete formulation and the production process.

Stacked floor slabs are not suitable for ordinary concrete formulation because of their thin shape, which is generally slightly more than 60 mm. Since the maximum particle size of coarse aggregate is limited, the requirement for compatibility is high, and the impermeability of the thin slab is also high, fine stone concrete with added admixtures is often used.

Laminated floor slabs have high requirements for production efficiency, such as the use of early strength agents and steam-raising technology to shorten the production time of laminated floor. The concrete formulation of stacked slabs requires a special design. Shen et al. (2018) studied the effect of incorporating metakaolin on the performance of steamed concrete and showed that the incorporation of metakaolin significantly improved the demolding strength and chloride ion penetration resistance of steamed concrete and the sulfate attack resistance of steamed concrete. Zeng et al. (2014) studied the early hydration characteristics and pore structure of high-temperature steamed high-strength mortar by incorporating metakaolin and mineral powder and the result showed that the introduction of the admixture substantially reduced the porosity of the material. Li et al. (2020) studied the effect of the use of admixtures and their compound on the strength of steamed concrete products and found that under the same water-cement ratio, the use of naphthalene high-efficiency water-reducing agent is better. The addition of a retarder and air-entraining agent in the process of concrete steaming will reduce the strength of concrete steaming, while the appropriate

---

*Corresponding Authors: 1010710485@qq.com, qiangli1972@163.com, qiuyiqin8845@163.com, wq2547557136@163.com and 1739537491@qq.com

amount of expansion agent or early strength agent can increase the strength of concrete steaming. Julie et al. (2002), and Picandet et al. (2009) studied the effects of steel fibers on the permeability of cracked concrete. The study of the permeability of sand concrete is mostly limited to the influence of a single material factor on permeability (Cheng & Huang 2016). However, the current research on the concrete formulation of thin steamed laminated slabs suitable for factory production is not deep enough, and further research is needed.

This paper conducts a study of the mix ratio of concrete that is impermeable and suitable for high-temperature steaming and obtains a new type of stacked plate mix ratio through indoor tests. Through the improvement of the new additive compound, a new formula of fine stone concrete superposition, which meets the requirements of anti-seepage, has good and easy properties, and is suitable for high-temperature evaporation, has been proposed.

2 KEY ISSUES OF CONCRETE RATIO DESIGN FOR LAMINATED SLAB

The key issues concerning concrete ratio design for laminated slabs are summarized as follows:

(1) Focusing on the problem of insufficient compactness in the concrete formulation of laminated slabs prone to water seepage to resolve the purpose of high impermeability requirement of laminated slabs.
(2) Adding admixtures for high requirement of early strength of laminated slabs to achieve the purpose of rapid production.

3 LAMINATED SLAB CONCRETE RATIO DESIGN CHARACTERISTICS

The study conducts a comprehensive analysis of the characteristics of laminated slabs:

1. The ordinary concrete mix design method
2. The proposed superfine composite admixture concrete mix design method
3. The introduction of the concept of the surplus coefficient and the coefficient of cementitious materials
4. The reference water-cement ratio
5. The preparation of the concrete mix in line with the high impermeability of laminated slabs
6. High early strength of the concrete mix

4 PRINCIPLES AND WAYS TO IMPROVE THE IMPERMEABILITY OF LAMINATED SLAB CONCRETE

4.1 *Reducing water-cement ratio to control the slump*

The water-cement ratio of the concrete mix plays a decisive role in determining the size and number of concrete porosity after hardening, which directly affects the compactness of the concrete. Theoretically, in the premise of meeting the water required for complete hydration of cement and wetting of sand and gravel, the smaller the water-cement ratio, the better the concrete compactness, strength, and seepage resistance. If the water-cement ratio exceeds the water retention limit of cement, the seepage resistance of concrete will be significantly reduced. In the hydration process of cement, the evaporation of free water in the concrete will leave a large number of pores inside the concrete. These pores penetrate each other to form open capillary secretion channels so that the concrete structure's impermeability performance is reduced and permeability is increased. Therefore, the water-cement ratio is the main factor affecting the seepage resistance of concrete. Tests show that when the water-cement ratio is greater than 0.65, the seepage resistance of concrete decreases sharply. The specification for quality acceptance of underground waterproofing works" (GB50208) stipulates that the water-cement ratio should not be greater than 0.55 and the slump should not be greater than 50 mm.

4.2 *Select the best sand rate*

In the case of the same amount of cement, the size of the sand rate affects the permeability of the concrete. The value of sand rate is related to many factors: the gradation of aggregate, void ratio, particle shape, maximum particle size, soil content, fineness modulus of sand, cement dosage, etc. Because the particle size of sand is much smaller than that of stone, the change in sand rate will make a big change in the total area of the aggregate. The change in sand rate will also make a big change in the compatibility of the concrete mix. In the case of a certain cement paste, when the sand rate is too large, the total surface area of the aggregate is large, the void rate increases and the concrete mix liquidity is small. When the sand rate is very small, due to insufficient sand, it cannot form enough pores around the coarse aggregate and play a lubricating role in the mortar layer. If the proportion of cement and water consumption is relatively increased, the concrete is prone to uneven phenomena – the concrete mix liquidity is reduced, coarse aggregate segregation, cement paste loss, and even collapse. To make waterproof concrete, which has good compatibility, adhesion, and water retention, the cement mortar layer between the aggregates has to maintain a certain thickness. The stones should be around a sufficient thickness of cement mortar wrapping layer to prevent water penetration, and the sand rate should be larger than ordinary concrete. The underground waterproofing project quality acceptance specification specified a sand rate of 35% to 45% to be appropriate so that the concrete waterproofing performance between the stones and sand particles is good as it maintains a certain distance between the cement mortar or cement stone. In addition, this not only ensures preventing the infiltration of certain pressure water and avoiding cracks in the cement stone but also that the gap between the stones and sand particles is not too thick or too thin for the construction of the concrete with ease.

4.3 *Appropriate increase in the amount of cement*

Concrete coarse aggregates are available as crushed stones or pebbles. Since the surface of a crushed stone is rough and angular, cement-stone adhesion is better than pebbles. The strength of crushed concrete is good and it has good seepage resistance. The concrete settlement gap is caused by the different degrees of settlement caused by the different densities and particle sizes of aggregates and cement when its structure is formed. The greater the cement secretion, the larger the coarse aggregate particle size, the worse the water retention of the concrete mix, and the greater the relative settlement. After the concrete is poured, the coarse aggregate settles faster and is fixed earlier. The cement mortar continues to sink between the coarse aggregate. Part of the water precipitates on the surface, and part of the water secretion is blocked by the concrete in the coarse aggregate or reinforced and accumulated in the coarse aggregate, and cement stone contact with the lower part of the coarse aggregate, the formation of water accumulation layer The free water evaporates to form a tree-root-like connected pore. When the gel is blocked, permeable channels are created inside the concrete.

In the concrete hardening process, the stone does not shrink. The surrounding cement mortar produces shrinkage, but its deformation is different. The larger the stone, the greater the shrinkage difference, resulting in cracks on the surface of mortar and stone. To reduce delamination, increase the resistance to pressure water penetration and reduce mortar and stone cracks, waterproof concrete stone particle diameter should be $5 \sim 40$ mm.

4.4 *Reasonable choice of cement varieties*

Waterproof concrete is widely used in ordinary silicate cement, silicate cement, volcanic ash silicate cement, and slag silicate cement. As the silicate cement slag admixture contains more vitreous, resulting in large concrete secretion, it is not conducive to concrete impermeability. Therefore, priority should be given to ordinary silicate cement.

4.5 *Reasonable choice of admixtures*

The purpose of using admixtures is (1) to inhibit and reduce the generation of pores in concrete from waterproof materials and (2) to change the characteristics of pores to block and cut off permeable channels. Complete hydration of cement is required to combine water for about 20% to 25% of the weight of cement. If the consumption of construction water is far more than that of hydration water and the evaporation of excess water to form capillary seepage channels, the concrete seepage resistance is reduced, for which water-reducing agents should be used. The water-reducing agent has a strong dispersion effect on the cement and can reduce the mixing water, thereby reducing the porosity and increasing the compactness and impermeability of the concrete. At the same time, the concrete mixed with the water-reducing agent has good compatibility. Increasing ultra-fine silica fumes, fly ash plugging concrete pores, and reducing the rate of water secretion can improve the permeability of the concrete. However, strict control should be exercised with the amount of admixture; otherwise, it will significantly reduce the strength of the concrete. Adding an early-strength waterproofing agent can speed up cement hydration. With the increase in the number of hydration products cementite crystallizes into a fine, dense structure.

5 CONCRETE IMPERMEABILITY TEST

The seepage height method is applicable in determining the average seepage height of hardened concrete under constant water pressure to indicate the concrete's resistance to water penetration performance. The test procedure is as follows.

(1) After the test piece is prepared, start the impermeability meter and open the valves under the six test positions so that water seeps out from the six holes. The water should fill the test pit, and the sealed test piece should be installed on the impermeability meter after closing the valves under the six test positions.
(2) After the test piece is installed, the valve under the six test positions should be opened immediately so that the water pressure is controlled at a constant (1.2 Shi 0.05) MPa within 24 h. The pressurization process should not be greater than 5 min, and the time to reach a stable pressure should be used as the starting time of the test record (accurate to 1 min). In the process of pressure stabilization, at any time, observe the seepage of the specimen end. When there is a specimen end seepage, the test should be stopped, and the time and the height of the specimen as the height of the seepage of the specimen should be recorded. For the test specimen that does not have water seepage, the test should be stopped after 24 h, and the test specimen should be promptly removed. During the test, when water is found to seep out from the perimeter of the specimen, it should be sealed again according to the method specified in the previous section.
(3) The test piece taken out from the seepage resistance instrument should be placed on the press, and a steel spacer of 6 mm diameter should be placed at the center of the upper and lower ends of the test piece along the diameter direction. It should be ensured that they are in the same vertical plane. Then the press should be started, and the test piece along the longitudinal section should be split into two halves. After the test piece is split, the watermark should be traced with a waterproof pen.

(4) The plate on the split surface of the test piece should be trapezoidal. A steel ruler along the watermarks at equal intervals is used to measure the water seepage height value of 10 measurement points; readings should be accurate to 1 mm. When encountered with a blocked measurement point by the aggregate, the arithmetic average of the seepage height near the two ends of the aggregate should be taken as the seepage height of the measurement point, and the sample state is recorded. The test data is shown in Table 1.

Table 1. Impermeability test data.

Sample numbering	Sample state	Devise grade	Engineering parts	Date of production Detection date Age (days)	Conservation methods	Test results		Design strength
						Maximum water pressure (MPa)	Water infiltration	
Concrete osmosis2022-007128-1	effective	C30	PC component stacking board	2022-05-27 2022-06-24 28	Standard maintenance	0.6	Seepage water	Meet the P6 impermeability grade requirements
						0.6	Seepage water	
						0.6	Seepage water	
						0.6	Seepage water	
						0.6	Seepage water	
						0.6	Seepage water	

6 CONCLUSION

(1) To improve the anti-seepage performance of the concrete, it is recommended to improve the anti-seepage performance of the laminated slab by improving the concrete ratio of the laminated slab, focusing on solving the problem of insufficient compactness in the concrete formulation of the laminated slab that is prone to water seepage, and studying the addition of ultrafine silica fume, fly ash, and appropriate admixtures.
(2) The test for improving the impermeability performance – the impermeability test block production test – was conducted after the optimization of the ready-mixed concrete performance ratio, and the impermeability grade reached P6 grade.

REFERENCES

Cheng Y.H. & Huang F. (2016). Test Research on Effects of Waste Ceramic Polishing Powder on the Permeability Resistance of Concrete. *Mater Struct*, 49(3): 729–738.
Li D.-L. & Lu Z.-T. (2020). Effect of Additive and its Composite use on the Strength of Steamed Concrete Products. *J. Gre. build. mat*. 6: 13–14.
Picandet V. & Khelidj A. (2009). Crack Effects on Gas and Water Permeability of Concretes *Cem Concr Compos*. 39(6): 537–547.
Rapoport J. & Aldea C. M., (2002). Permeability of Cracked Steel Fiber-reinforced Concrete. *J Mater Civil Eng*, 14(4): 355–358.
Shen H.B. & He Z.M. (2018). Study on the Effect of Doping Partial Kaolin on the Performance of Steamed Concrete. *J. Ningbo Univ. (Sci. and. Tec. Ed.)*. 31(4): 74–80.
Zeng, J., Shui, & Wang, S.N. (2014). Study on Early Hydration Characteristics and Pore Structure of Evaporated High-strength Mortar Doped with Metakaolin and Mineral Powder. *J. Cent. South. Univ. (Nat. Sci. Ed.)*. 45(8): 2857–2863.

Study on temperature monitoring and deformation characteristics of steam curing prestressed concrete box girder

Feng Zeng
Guangzhou Highway Engineering Group Co., Ltd., Guangdong, Guangzhou, China

Lei Zeng
Guangzhou Expressway Co., Ltd., Guangdong, Guangzhou, China

Yifang Sun
Guangzhou Highway Engineering Group Co., Ltd., Guangdong, Guangzhou, China

Haitao Tian & Yingxuan Xu
CCCC Highway Long Bridge Construction National Engineering Research Center Co., Ltd., Beijing, China

ABSTRACT: Through the method of experimental monitoring, the temperature inside the concrete of the steam curing concrete box girder is monitored to evaluate the cracking risk during the steam curing process, the shrinkage change of the concrete under the steam curing condition is studied, and the prestress loss of the concrete beam do research. The results show that the maximum temperature difference between the inside and outside of the concrete box girder during the steam curing process is less than 20°C, the risk of cracking is small, and the shrinkage of the concrete develops rapidly during the steam curing process, and the shrinkage after the steam curing is smaller than that of the standard curing concrete, and after tensioning, the prestress loss of the cured specimen is less than that of the standard curing. Compared with the precast and curing process of traditional precast beams, the steam curing process can rapidly increase the strength of concrete, the compressive strength of concrete can reach the level of standard curing for 28 days within 24 hours under proper curing conditions.

1 INTRODUCTION

The steam curing process can improve the early strength of concrete, speed up the turnover of molds, and improve production efficiency (Shao 2019). It has been widely used in the production of reinforced concrete components for prefabricated buildings and large-scale key infrastructure, and it has achieved good economic benefits. Steam curing can accelerate the hydration of cement inside the concrete and contribute to the development of concrete properties (Bentur et al. 2001; Li & Zhao 2003; Martinez-Ramirez & Frias 2009; Rojas & Cabrera 2002). However, the large temperature difference between the inside and outside of the concrete box girder during the steam curing process may cause cracking, and there is no unified conclusion on the effect of steam on the volume stability of concrete (Hanehara et al. 2001; Nagataki et al. 1988).

The purpose of this study is to explore the distribution law of the internal temperature field of concrete under steam curing, to clarify the effect of curing temperature on the shrinkage and creep of concrete, and finally to realize the rapid production of concrete precast beams, greatly shorten the precast period, and improve the turnover efficiency of formwork.

2 TEST PLAN

2.1 Raw materials

The cement adopts P·II 52.5 R grade cement produced by Sinoma Hengda Cement Co., Ltd. The performance indicators are shown in Table 1. F-class I fly ash from Shenhua Guangdong and S95 slag powder from Tangshan Caofei are used. The fineness modulus of the medium sand is 2.81, and the mud content is 1.0%, with 5~20 mm continuous graded crushed stone. The needle flake content is 3.7%, the crushing value is 8%, and the mud content is 0.4%. The water-reducing agent is a polycarboxylate high-performance water-reducing agent produced by a company in Guangdong, with a solid content of 35%.

Table 1. Cement performance index.

Project	Fineness	Density (g/cm^3)	Stability (Rey's method /mm)	Initial setting time (min)	Final setting time (min)	Standard consistency water consumption (%)
index	364	3.14	1.5	168	223	27.0

2.2 Mix ratio

The concrete mix ratios are shown in Table 2. The standard curing 28d strength of concrete is 75.4 MPa, and the elastic modulus is 46.5 GPa.

Table 2. Concrete mix ratio.

Cement kg/m^3	Fly ash kg/m^3	Mineral powder kg/m^3	Sand kg/m^3	Stone kg/m^3	Water kg/m^3	Admixture kg/m^3	Water-to-glue ratio
362	48	67	639	1186	148	5.250	0.31

2.3 Maintenance system

2.3.1 Steam curing

After the concrete is poured, let it rest for 12 hours at 20°C and 95% RH, and then release the mold. From 20°C, the temperature is increased at a rate of 10°C/h until 50°C. After maintaining the constant temperature for 24 hours, the temperature is lowered to 20°C at 5°C/h.

2.3.2 Standard curing

The formed specimens were rested for 16 hours at 20°C and humidity of 95%, and the standard curing was continued after demoulding.

2.4 Shrinkage test

The shrinkage test was tested by JMZX-215HAT high-performance intelligent digital string-embedded strain gauge produced by Changsha Jinma Measurement and Control Technology Co., Ltd. The specific process is as follows: brushing a layer of butter on both ends of the tester, and fixing it horizontally at the middle height of the test mold with a fishing line. The data is read every 3 hours for the first three days, and every 24 hours after three days (Liu 2018). The test process is shown in the Figure 1:

Figure 1. Concrete shrinkage test procedure.

2.5 *Prestress loss test*

The cross-sectional form of the test beam is b×h = 200 mm×300 mm, the beam length is 3450 mm, the steel bars are HRB400, the stirrup diameter is 8 mm, the bellows diameter is 30 mm, and the hoisting point is 1 m away from both ends. The tensile stress is 0.6 fpk. One is stretched after standard curing for 7d, and the other is stretched after steam curing. The prestress loss adopts the RH-600 vibrating wire anchor cable meter produced by Jiangsu Ruihe Measurement and Control Co., Ltd., and the instrument specification is 20 T. The schematic diagram of the test beam is shown in Figure 2:

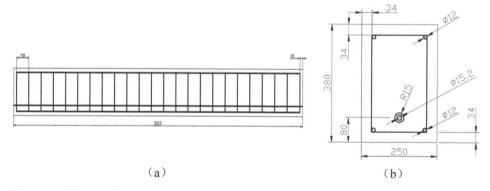

(a)　　　　　　　　　　　　　　(b)

Figure 2. Schematic diagram of the test beam.

2.6 *Temperature monitoring test*

The temperature monitoring test was carried out with a 3-meter-long segment beam, and the layout of the measurement was based on the principle of highlighting key points and taking into account the overall situation. The inner edge beam segment beam layout temperature measuring point is selected. Due to the smaller wall thickness of the beam concrete, the internal temperature gradient is smaller compared to the mass concrete. Therefore, the number of each positioning measurement point is appropriately reduced. The temperature

Figure 3. Test procedure.

monitoring area refers to the location layout requirements of "Technical Specifications for Temperature Measurement and Control of Mass Concrete (China Construction Industry Press 2016), a total of 6 locations are designed, and each location has one measurement point. The layout of the measuring points is as follows. The temperature was detected with a JMT-36A intelligent thermometer produced by Changsha Jinma Company.

The temperature detection system is set up after the reinforcement cage is tied and hoisted to the prefabricated formwork. The sensor for monitoring the internal temperature of the beam body is tied to the corresponding position of the reinforcement cage, and the temperature is measured since the concrete is poured. The time step is set to 1 h. When the difference between the concrete center temperature and the minimum ambient temperature is less than 25°C continuously, the monitoring is stopped. (Luo 2014)

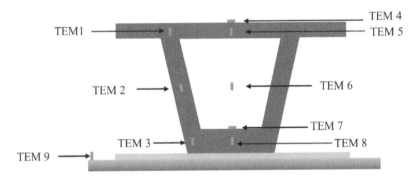

Figure 4. Schematic diagram of prefabricated beam section measuring point layout.

3 TEST RESULTS AND ANALYSIS

3.1 Temperature monitoring test

The temperature monitoring test results of each measuring point are shown in Figure 5. It can be seen from the results that the ambient temperature change in the steaming shed is basically consistent with the temperature determined by the curing system. When the temperature is constant, the temperature of each measuring point fluctuates around 50°C. In the steam curing condition, the top plate of the box girder is directly exposed to the high-temperature steam environment, and the early heating is slightly faster than that of the bottom plate; however, due to the large thickness of the bottom plate, more heat of hydration is generated during the hydration of concrete, and due to the heat preservation effect of the bottom mold, the temperature of the bottom plate is higher (People's Communications Press 1993).

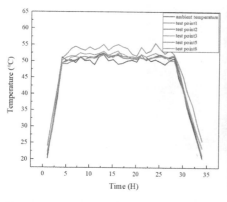

Figure 5. Concrete temperature monitoring test results.

Under the condition of steam curing, the temperature change trend of each measuring point is basically consistent with the ambient temperature, and the highest temperature is slightly higher than the ambient temperature, which is caused by the heat release of cement hydration. The temperature difference at each point is not over 20°C, which is in line with the requirements of "Technical Specification for Highway Bridge and Culvert Construction" (Neville et al. 1983), and the risk of cracking is small. After the steam curing of the stage beam specimen, the surface is smooth, the color is uniform, and there are no appearance quality problems such as cracks and pits.

3.2 Concrete shrinkage test

The results of the concrete shrinkage test are shown in Figure 6. It can be seen from the test results that under different curing conditions, the shrinkage of concrete increases gradually with the extension of the adjournment period. The shrinkage of concrete develops rapidly in the first 15 days, thereafter the shrinkage rate of concrete gradually slows down. Before 7 days, the shrinkage of steam-cured concrete was greater than that of standard-cured concrete, and after 7 days, the shrinkage of standard-cured concrete was greater than that of steam-cured concrete. This is because early steam curing significantly increases the hydration rate of concrete, resulting in faster development of early shrinkage and a full release of

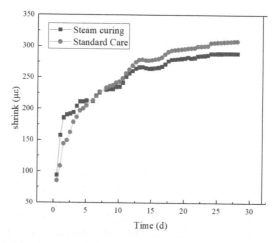

Figure 6. Concrete shrinkage test results.

shrinkage. In the later stage, the shrinkage of concrete continued to develop, but at this time, the elastic modulus of steam-curing concrete was higher, and the creep coefficient was lower, so the overall shrinkage of steam-curing concrete was smaller than that of standard curing concrete (Acker & Ulm 1997; Feldman 1972; JTG 3362 2018; Yu 1995; Wittmann 1971).

3.3 Prestress loss test

3.3.1 Prestress loss calculation

According to the "2018 Edition Highway Reinforced Concrete and Prestressed Concrete Bridges and Culverts Design Specification" (Neville et al. 1983), the prestress loss caused by concrete shrinkage is calculated, which can be calculated as follows (1):

$$\sigma_{l6}(t) = \frac{0.9 E_p \varepsilon_{cs}(t, t_0)}{1 + 15 \rho \rho_s} \quad (1)$$

where $\rho = \frac{A_p + A_s}{A}$, $\rho_{ps} = 1 + \frac{e_{ps}^2}{i^2}$.

σ_{l6} – The loss of prestressing caused by the shrinkage of concrete at the center of gravity of the longitudinal prestressed steel bar section of the member;

E_p – Modulus of elasticity of prestressed reinforcement;

ρ – The reinforcement ratio of longitudinal reinforcement of members;

A – the cross-sectional area of the member, for the pre-tensioned member, $A = A_0$; for the post-tensioned member, $A = A_n$;

i – the radius of gyration of the section, $i^2 = I/A$, the pre-tensioned member takes $I = I_0$, $A = A_0$; the post-tensioned member takes $I = I_n$, $A = A_n$;

e_p – The distance from the center of gravity of the prestressed steel bar section in the tension area and the compression area of the member to the center of gravity of the member section;

e_s – The distance from the center of gravity of the longitudinal ordinary steel bar section to the center of gravity of the member's section in the tension zone and compression zone of the member;

e_{ps} – The distance from the center of gravity of the section of prestressed steel bars and ordinary steel bars in the tension zone and compression zone of the member to the center of gravity of the member section;

$\varepsilon_{cs}(t, t_0)$ – The force-transmission anchorage age of the prestressed steel bar is t_0, and the shrinkage strain of concrete when the age is t is considered in the calculation.

The calculation results are shown in Figure 7:

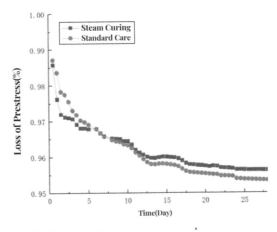

Figure 7. Calculation result of prestress loss.

3.3.2 Prestress loss test

The results of the prestress loss test are shown in Figure 8. The prestress loss of the early steam curing specimen is slightly larger than that of the outdoor spray curing specimen, which may be because the humidity of the specimen is relatively high under the condition of steam curing, and the humidity inside the concrete decreases after the steam curing is stopped. This results in a larger drying shrinkage; the prestress loss of the steam curing specimens for about 10 days is less than that of the outdoor spray curing specimens, which may be due to the accelerated hydration rate of concrete during the steam curing process, and the release of concrete shrinkage is completely related, and the steam curing process. It can accelerate the development of the elastic modulus of concrete, and the better mechanical properties can resist the shrinkage and creep caused by an external force (Dhir et al. 1986; JTG 3362-2018, Soroka & Bentur 1978).

Comparing the calculation results of prestress loss and the test results of prestress loss, it can be seen that the development trend of prestress loss based on shrinkage calculation is basically the same as the development trend of prestress loss measured by actual test, which shows that the prestressing of prestressed concrete in the early stage is basically the same. The loss has a greater correlation with the shrinkage of the concrete itself, and steam curing can reduce the shrinkage of the concrete, thereby reducing the prestress loss after prestressing.

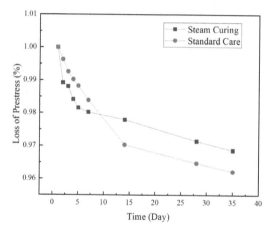

Figure 8. Prestress loss test results.

4 CONCLUSION

(1) In the early stage of steam curing, the hydration rate of concrete is accelerated, which promotes the release of concrete shrinkage. Therefore, steam curing can reduce the shrinkage of concrete in the later stage.
(2) Due to the drying shrinkage of early-rise concrete, the prestressed loss rate of prestressed concrete in the early stage is faster, and in the later stage, the prestressed loss of steam-cured concrete is lower than that of standard curing concrete.
(3) Steam curing according to the steam curing system determined above, the internal temperature of the concrete box girder is stable, and the appearance quality of the concrete is good after the steam curing.

ACKNOWLEDGMENT

Financial support provided by CCCC (China Communications Construction Co., Ltd) to conduct this research under grant number 2016-ZJKJ-PTSJ05 is greatly acknowledged.

REFERENCES

Acker P., Ulm F.J. Creep and Shrinkage of Concrete: Physical Origins and Practical Measurements – ScienceDirect [J]. *Nuclear Engineering & Design*, 1997, 203(2):143–158.

Bentur A., Igarashi S.I., Kovler K. Prevention of Autogenous Shrinkage in High-Strength Concrete by Internal Curing Using Wet Lightweight Aggregates [J]. *Cement & Concrete Research*, 2001, 31(11):1587–1591.

B M H V A, A J S, B J B L. Effect of the Curing Temperature on the Creep of a Hardened Cement Paste [J]. *Cement and Concrete Research*, 2012, 42(9):1233–1241.

Dhir R.K., Munday J., Ong L.T. *Investigations of the Engineering Properties of OPC/pulverized-fuel Ash Concrete: Deformation Properties*. 1986.

Feldman R.F. Mechanism of Creep of Hydrated Portland Cement Paste [J]. *Cement and Concrete Research*, 1972, 2(5): 521–540.

Hanehara S., Tomosawa F., Kobayakawa M., et al. Effects of Water/powder Ratio, Mixing Ratio of Fly Ash, and Curing Temperature on Pozzolanic Reaction of Fly Ash in Cement Paste [J]. *Cement & Concrete Research*, 2001, 31(1): 31–39.

JTG 3362-2018, Design Code for Highway Reinforced Concrete and Prestressed Concrete Bridges And Culverts [S].

Li G., Zhao X. Properties of Concrete Incorporating Fly Ash and Ground Granulated Blast-Furnace Slag [J]. *Cement and Concrete Composites*, 2003, 25(3): 293–299.

Luo. Junli. *Experimental Research on Shrinkage Creep and Temperature Time-varying Effects of Prestressed Box Girder in High-speed Railway* [D]. Central South University, 2014.

Martinez-Ramirez S., Frias M. The Effect of Curing Temperature on White Cement Hydration [J]. *Construction & Building Materials*, 2009, 23(3): 1344–1348.

Ministry of Housing and Urban-Rural Development of the People's Republic of China. *Technical Specification for Temperature Measurement and Control of Mass Concrete* [M]. China Construction Industry Press, 2016.

Nagataki S., Ohga H., Sakai E. Mechanical Properties of Concrete With Fly Ash Under High Temperature Curing [J]. *Doboku Gakkai Rombun—Hokoknshu/Proceedings of the Japan Society of Civil Engineers*, 1988, 8(390):189–197.

Neville A. M., Dilger W. H., Brooks J. J. Creep of Plain and Structural Concrete. *Construction Pr*, 1983.

Quanwei. Liu. *Research on the Seismic Performance of Prefabricated Concrete Shear Walls with Different Connection Methods* [D]. Chang'an University, 2018.

Rojas M F., Cabrera J. *The Effect of Temperature on the Hydration Rate and Stability of the Hydration Phases of Metakaolin–Lime–Water Systems* [J]. 2002, 32(1):133–138.

Shuangshuang. Shao. *Analysis on Performance and Cracking of Steam Curing Concrete at Super Early Age* [D]. Qingdao University of Technology, 2019.

Soroka, I., and C. Bentur. "Short-term Steam-Curing and Concrete Later-Age Strength." *Materials & Structures* (1978).

The First Highway Engineering Corporation of the Ministry of Communications. *Technical Specifications for Construction of Highway Bridges and Culverts* [M]. People's Communications Press, 1993.

Wittmann F.H. Discussion of Some Factors Influencing Creep of Concrete. 1971.

Yu, Khristova, K., Aniskevich. Prediction of Creep of Polymer Concrete [J]. *Mechanics of Composite Materials*, 1995.

Analysis of the influence of springback rate of wet-sprayed concrete based on aggregate characteristic analysis

Yuan Zhang* & Jun Zhang
CCCC First Highway Engineering Group Co. Ltd., Beijing, China

ABSTRACT: The rebound rate of wet shotcrete is directly related to the construction cost and quality. In practice, the technical index of the rebound rate of wet-sprayed concrete cannot be effectively controlled, and the rebound rate is relatively high. Therefore, the influence analysis of the rebound rate of wet-sprayed concrete based on the analysis of aggregate characteristics is put forward. In this paper, three projects, Nantian No.6 bid, Renzun No.4 bid and Chongzun No.8 bid, are taken as the research objects. From the two characteristics of aggregate particle size and aggregate stone powder content, ten working conditions are designed, namely, the coarse aggregate particle size of 13.2 mm, 9.5 mm, 4.75 mm, 2.36 mm and 0.6 mm, and fine aggregate stone powder content of 47.5%, 50%, 52%, 56% and 60%. At the same time, the laser section scanner is used to measure the rebound data, calculate the rebound rate, and analyze the influence of aggregate particle size and aggregate stone powder content on the rebound rate of wet-sprayed concrete. The results show that aggregate particle size and aggregate stone powder content have a great influence on the rebound rate of concrete. There is a positive correlation between aggregate particle size and the rebound rate of wet-sprayed concrete, and the rebound rate can be reduced by an appropriate amount of aggregate stone powder. When the aggregate stone powder content is 52%, the rebound rate is the lowest.

1 INTRODUCTION

Wet concrete spraying is a common way of concrete sprinkler irrigation, which means that the mixed concrete mixture is sprayed to the sprayed surface by a special spraying device through spraying pressure and power. Compared with dry-sprayed concrete, wet-sprayed concrete has many advantages, such as simple operation, low construction cost, fast sprinkler irrigation speed, fast concrete strength growth, etc., and has been widely used in the construction field. Wet-sprayed concrete also has some disadvantages. In the wet-sprayed process, the concrete aggregate will rebound when it comes into contact with the sprayed surface, which makes some aggregates unable to reach the sprayed point. The aggregate that bounces back to the ground can no longer be used as concrete raw materials, but can only be recycled for other aspects, thus causing the waste of concrete materials and the pollution of the construction environment. Therefore, the rebound rate is an essential technical index in wet-sprayed concrete construction and needs to be strictly controlled (Xie 2020). To control it effectively, it is necessary to know the influence mechanism of the rebound rate of wet-sprayed concrete. There are many factors influencing wet-sprayed concrete's rebound rate. The domestic research on the rebound rate of wet-sprayed concrete started relatively late, and the existing theories are not enough. So far, there is no systematic analysis theory of

*Corresponding Author: 276106823@qq.com

rebound rate, which can't provide a solid basis for the optimization and innovation of wet-sprayed concrete technology, resulting in a high rebound rate in the actual operation process, and the expected wet-sprayed concrete effect can't be achieved. This seriously increases the construction cost. Therefore, the influence analysis of the rebound rate of wet-sprayed concrete based on the analysis of aggregate characteristics is put forward.

2 TEST METHODS AND MATERIALS

2.1 Raw material

The raw materials used in the test mainly include cement, accelerator, water reducer, coarse aggregate and fine aggregate, among which the cement is PO.624.15 Portland cement produced by Shanghai FAWE Cement Co., Ltd. The accelerator is YD-5D alkali-free liquid accelerator produced by Tianjin OFOU Chemical Reagent Co., Ltd. (Zhao et al. 2020). Ti-25 KG early-strength high-efficiency water-reducing agent is adopted, and the water-reducing rate of this water-reducing agent is 15.48% (Zeng et al. 2020). The medium-coarse aggregate is limestone crushed stone provided by OUH crushed stone yard, with a silt content of 0.56% and a lump content of 0.16%. In contrast, the medium-fine aggregate is marble artificial sand provided by OUH crushed stone yard, with a fineness modulus of 2.46 and a dry absorption rate of the saturated surface of 1.49% (Zou 2020). After sampling inspection of raw materials, the properties of cement and aggregate meet the current national standards.

2.2 Experimental method

The test is based on the influence of wet-sprayed concrete's rebound rate, so from the angle of aggregate particle size and aggregate stone powder content, the influence of these two factors on concrete's rebound rate is analyzed (Guo 2021). Six kinds of wet-sprayed concrete were prepared. Wet-sprayed materials are made of limestone with relatively homogeneous stones. Three items, namely, Nantian No.6 Bid, Renzun No.4 Bid and Chongzun No.8 Bid, are taken as the research objects, and their aggregate and sand grading compositions are shown in the following Table 1.

Table 1. Composition of aggregate and sand gradation.

Project name	Aggregate gradation composition 9.5:4.75:2.36:0.6	Composition of sand grain gradation 4.75:2.36:0.6: 0.075:<0.075
Nantian No.6 Bid	1.8:93.2: 3.7:1.3	0.6:28:32:25.2:14.2
Renzun No.4 Bid	2.6:92.9:3:1.5	0.6:24.4:38.5:22.8:13.7
Chongzun No.8 Bid	19.7:78.6:0.8:0.9	2.5:30:25.9:22.1:19.5

The internal mortar composition refers to 2.36 mm:0.075 mm: < 0.075 mm: < 0.075 mm: cement: water: admixture. The external aggregate mortar composition refers to 2.36 mm: 0.6 mm: 0.075 mm: < 0.075 mm: cement: water: admixture. The mix design of wet shotcrete for three projects is as follows.

In addition, the internal mortar composition of wet shotcrete in three projects is 17.3:1.1:0.1:27.3:85.6:38.8:1.1 (w/c = 0.45), 20:15.2:1.3:0.1:26.19:98.11:37 respectively. 29:0.98 (w/c = 0.38), 15.1:1:0.1:22.2:76.8:33.4:0.88 (w/c = 0.43), and the external mortar composition of aggregate is 283.4:318.6:243.2:109.8 respectively. To analyze the influence of aggregate particle size on it, five working conditions of coarse aggregate particle size of

Table 2. Mix proportion design table of wet shotcrete.

Project	Nantian			Chongzun	Renzun
Sand ratio (%)	56	52	47.5	50	60
And slump expansion (mm)	220	200	185	210	220
	565	530	380	515	560
The internal volume of mortar (m³)	0.127	/	/	0.112	0.11
The internal void volume of aggregate (m³)	0.251	0.273	0.30	0.275	0.220
The volume of external aggregate mortar (m³)	0.604	0.56	0.523	0.537	0.641
The true volume of aggregate (m³)	0.292	0.317	0.348	0.319	0.255
The thickness of the peripheral mortar layer of aggregate (mm)	2.7	2.2	1.9	2.5	2.7

13.2 mm, 9.5 mm, 4.75 mm, 2.36 mm and 0.6 mm are designed. To analyze the influence of aggregate stone powder content on it, five working conditions with fine aggregate stone powder content of 47.5%, 50%, 52%, 56% and 60% were designed, and the rebound rate of wet-sprayed concrete under different aggregate characteristics was tested.

3 REBOUND RATE TEST

In the process of wet-sprayed concrete, due to the influence of pump pressure, concrete mix proportion, wind pressure, and other factors, some wet-sprayed concrete can't wholly adhere to the sprayed surface, and the coarse aggregate and some fine aggregate in the concrete will rebound and fall to the ground (Li et al. 2021). According to the rebound mechanism of wet-sprayed concrete, the wet-sprayed process can be divided into two parts: the initial wet-sprayed stage and the viscous-sprayed stage (Yang 2022). In the initial wet spraying stage, the concrete moves forward discretely after leaving the spray nozzle, and the surfaces of fine aggregate and coarse aggregate are wrapped by cement slurry. However, in the initial stage, there is less cement slurry attached to the aggregate, and the slurry on the surface of the aggregate can't offset the kinetic energy generated by wet spraying. After the aggregate touches the sprayed surface, it collides, bounces due to inertia, and falls to the ground to form rebound aggregates (Huang 2022). As the volume of wet-sprayed concrete increases, the cement slurry attached to the aggregate increases. After a large amount of concrete with certain elasticity is accumulated on the sprayed surface, its original rigid body is transformed into an elastic body. At this time, the dense spraying stage is entered, and the collision between the aggregate and the sprayed surface belongs to an inelastic collision. Combined with this test scheme, the rebound process is defined as the process from dynamic to static when the concrete aggregate arrives in the mortar layer of the sprayed surface. According to the quantitative analysis of energy conservation, the rebound of concrete aggregate to the mortar layer is expressed by the formula:

$$\frac{1}{2}ne^2 - \frac{1}{2}ne_1^2 - q = 0 \qquad (1)$$

where, n represents wet shotcrete coarse aggregate quality; e indicates the spraying speed before wet shotcrete rebounds to the mortar layer; e_1 represents the movement speed of wet shotcrete after it bounces back to the mortar layer; q represents the energy absorbed by an aggregate rebound to mortar layer, and its calculation formula is:

$$q = kh \qquad (2)$$

where, k represents the bonding resistance of aggregate in the mortar layer; h represents the rebound movement length of an aggregate rebound into the mortar layer (Hu et al., 2022). According to the above rebound mechanism of wet-sprayed concrete, in the experiment, an OFAKI-4544 laser cross-section scanner was used to measure the rebound data, and the designed gross cross-section area before wet-sprayed concrete and the mortar layer area after wet-sprayed concrete were scanned and measured by this instrument (Zhang et al. 2022). To obtain the effective rebound area, the rebound rate of wet-sprayed concrete is calculated according to the measured data, and its calculation formula is as follows:

$$\eta = \frac{b \cdot \sum(s - s_1)}{mv} \times 100\% \qquad (3)$$

where, η represents the rebound rate of wet-sprayed concrete; b represents the road length of the test section of the rebound rate of wet shotcrete; s indicates laser cross-section scanner in a single scan; s_1 represents the cross-sectional area of highway mortar layer after concrete wet spraying; m indicates the number of measured sections in the test section of the rebound rate of wet shotcrete; v indicates the total volume of wet-sprayed concrete in the test section of the rebound rate of wet-sprayed concrete.

4 TEST RESULTS AND DISCUSSION

4.1 *Influence of aggregate particle size on rebound rate of wet-sprayed concrete*

Formula (3) is used to calculate the rebound rate of wet-sprayed concrete under different aggregate particle size characteristics and analyze the influence of aggregate particle size on the rebound rate of wet-sprayed concrete according to the test data shown in Figure 1.

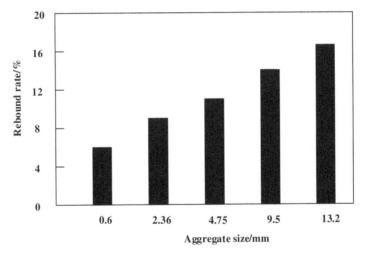

Figure 1. Influence of aggregate particle size on rebound rate of wet-sprayed concrete.

As seen from the above figure, with the increase of aggregate particle size, the rebound rate of wet-sprayed concrete gradually increases, and relevant specifications require that the rebound rate of wet-sprayed concrete should be controlled below 10%. When the aggregate size is more significant than 15 mm, the rebound rate of wet-sprayed concrete will exceed 10%. When the aggregate size is larger than 4.75 mm, the rebound rate of wet-sprayed concrete exceeds 10%. When the aggregate particle size is 13.2 mm, the rebound rate reaches 17.62%, far exceeding the specification requirements. This is because there is a negative correlation between aggregate particle size characteristics and the workability of wet-sprayed concrete. There is also a negative correlation between the workability of concrete and the rebound rate. When the aggregate particle size gradually increases, the workability and shotcrete of concrete are weakened, and the cohesiveness of aggregate is worse. The rebound rate of wet shotcrete rises slowly. Therefore, in practical engineering, the aggregate size of wet-sprayed concrete can be relaxed to 13 mm at most.

4.2 Influence of aggregate stone powder content on rebound rate of wet-sprayed concrete

Formula (3) is used to calculate the rebound rate of the wet-sprayed concrete side wall, and the rebound rate of sprayed surface and the rebound rate of the side wall are added to obtain the total rebound rate. According to the calculated data, the test results are shown in the following Figure 2.

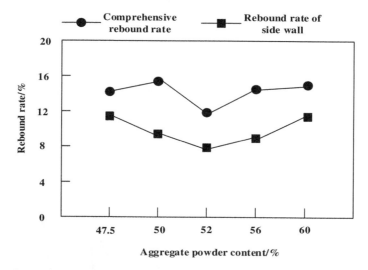

Figure 2. Influence of aggregate stone powder content on rebound rate of wet-sprayed concrete.

It can be seen from the above figure that when the aggregate stone powder content is in the range of 47.5%–50%, the rebound rate and comprehensive rebound rate of the side wall will increase with the increase of stone powder content. When the content of aggregate stone powder is 52%, the two springback rates reach the lowest value. This is because the content of aggregate stone powder has a certain influence on the cohesiveness and wrapping of wet-sprayed concrete. With the increase of stone powder content, the wrapping and cohesiveness of concrete will be improved, making it difficult to rebound. However, the increase of stone powder content will also increase the weight of wet-sprayed concrete and the overall particle size of coarse aggregate, so when the lime content reaches a certain value, the rebound rate of concrete will increase instead. To sum up, in the actual project, the content of stone powder in aggregate will be controlled at about 52%, to ensure the quality of wet-sprayed concrete.

5 CONCLUSION

To further control the construction cost and improve the construction quality, this paper analyzes the influence of aggregate characteristics on the rebound rate of wet-sprayed concrete by experiment. The test results show that with the increase in aggregate particle size, the rebound rate of wet-sprayed concrete gradually increases. When the aggregate size is larger than 4.75 mm, the rebound rate of shotcrete exceeds 10%, so the aggregate size can be relaxed to 13 mm in practical engineering. The content of aggregate powder also has an important influence on the rebound rate of wet-sprayed concrete. When the content of aggregate stone powder is 52%, the two springback rates reach the lowest value. This research has important practical significance for controlling the technical index of the rebound rate of wet-sprayed concrete, improving the construction quality of wet-sprayed concrete and controlling the construction cost, and can promote the further refinement of building construction standards.

REFERENCES

Guo Yongzhong. (2021) Research on the Influencing Factors and Construction Technology of Tunnel Wet Shotcrete Resilience. *Railway Construction Technology*, 3, 18–21+178.

Hu Shi, Cai Haibing, Ma Zuqiao, et al. (2022) Uniaxial Compression Test of Water-Saturated High Ductility Shotcrete Under Different Loading Rates. *Materials Reports*, 36, 106–115.

Huang Yunhai. (2022) Experimental Study on Performance Indicators of Sprayed Concrete Accelerators for Water Conveyance Tunnels in Northern Cold Regions. *Heilongjiang Hydraulic Science and Technology*, 50, 5–7.

Li Deming, Wen Shuyi, Zhang Jianwei, et al. (2021) Performance Test and Engineering Application of Single-Layer Lined Polyolefin Coarse Fiber Shotcrete. *Modern Tunnelling Technology*, 58, 195–203.

Xie Zhiwei. (2020) Research on the Influencing Factors of the Resilience Rate of Wet Shotcrete in Hydraulic Engineering. *Water Resources Planning and Design*, 4, 64–67.

Yang Weihua. (2022) Discuss the Influence of the Thickness of Shotcrete of Wet Spraying Manipulator on the Rebound Rate. *Scientific and Technological Innovation*, 3, 119–122.

Zeng Luping, Zhao Shuang, Wang Wei, et al. (2020) Characteristics of Bubble Structure, Water Penetration Resistance and Frost Resistance of Hardened Shotcrete. *Journal of the Chinese Ceramic Society*, 48, 1781–1790.

Zhao Shuang, Hong Jinxiang, Qiao Min, et al. (2020) Construction Test of Early-strength Shotcrete in Wushan Tunnel of Zhengwan High-speed Railway. *Tunnel Construction*, 40, 369–373.

Zhang Xiaodong, Chen Wenyuan, Ji Yafeng. (2022) Technical Measures for Controlling the Rebound Rate of Wet Shotcrete in Tunnels. *Journal of Shijiazhuang Institute of Railway Technology*, 21, 39–43.

Zou Qian. (2020) Research on the Effect of Polymer Tackifier on the Rebound Rate of Shotcrete and its Indoor Evaluation Method. *China High and New Technology*, 14, 100–103.

Thermal buckling analysis of submarine pipe-in-pipe systems with initial defect under seismic loading

Hao Xu
College of Pipeline and Civil Engineering, China University of Petroleum (East China), Qingdao, China

Hong Lin*
College of Pipeline and Civil Engineering, China University of Petroleum (East China), Qingdao, China
Center for Offshore Engineering and Safety Technology (COEST), China University of Petroleum (East China) Qingdao, China

Hassan Karampour*
School of Engineering and Built Environment, Griffith University, Australia

Pingping Han, Shuo Zhang, Chang Han & Haochen Luan
College of Pipeline and Civil Engineering, China University of Petroleum (East China), Qingdao, China

Lei Yang
College of Science, China University of Petroleum (East China), Qingdao, China

ABSTRACT: Submarine pipe-in-pipe systems (PIP) are often subjected to high temperature, high pressure, or other external loads such as seismic loads. When an earthquake occurs, the simultaneous action of multiple loads will cause serious damage to the pipeline. Moreover, local initial defects often occur in pipes during manufacture, installation, and operation. The existence of initial defects will greatly increase the possibility of pipeline damage under load. The purpose of this paper is to analyze the buckling behavior of PIP under high temperatures and pressure and consider the seismic load caused by earthquakes. Based on the finite element software ANSYS, the buckling failure of the pipeline with initial defect under combination loads is simulated by using the time-history dynamic analysis methods. It was found that the seismic load will increase the pipeline buckling response and the maximum stress of the pipeline. Moreover, pipeline buckling failure is greatly affected by different insulation materials and initial defects.

1 INTRODUCTION

The submarine pipeline is an important offshore infrastructure. Due to the harsh field environment, pipeline damage and failure accidents occur frequently. It will damage the pipeline transportation system and cause serious economic losses and environmental pollution. In recent years, a lot of research has been carried out on the numerical simulation and damage analysis of pipeline buckling failure, and great progress has been made (Psyrras et al. 2018; Pan et al. 2021) (Saeedzadeh & Hataf 2011; Wang et al. 2020). At present, the buckling failure research of pipelines is mainly limited to the single-layer pipeline, simple loading

---

*Corresponding Authors: linhong@upc.edu.cn and h.karampour@griffith.edu.au

conditions, or simple environments, without considering the buckling failure of double-layer pipelines under multiple loads in a complex environment. Pipeline damage is often the result of multiple factors.

As is known, marine seismic accidents occur frequently, which will seriously affect the stability of pipelines under high temperatures and high-pressure loads. At present, most of the studies are limited to analyzing the buckling failure of a pipe, a few loads, or a simple environment. There are few studies on buckling failure analysis of double-layer suspended span pipelines under various loads. Gong et al. (2012) conducted buckling propagation experiments and numerical simulations of pipes with initial geometric defects under external pressure. Zhang et al. (2019) studied the buckling of the PIP under high temperatures and high pressure. Mina et al. (2020) numerically studied the vulnerability of submarine high-pressure or high-temperature pipelines to seismic action. Demirci et al. (2018) conducted tests and numerical simulations on fault rupture of continuously buried pipelines. However, there are few studies on the interaction of multiple loads including seismic loads for suspended span pipelines.

This research aims to study the failure response of suspended span pipelines under high temperature, high pressure, and seismic load, and to find out the influence of insulation layer and initial defects on pipeline load failure, which provides a reference for future research on pipeline load failure accidents.

2 THEORY AND METHODS

2.1 *Framework*

In this paper, the buckling failure of PIP under high temperature, high pressure, and earthquake loads is studied. With the increase in temperature, the buckling deformation of the pipeline will enter the plastic stage, which will have a significant impact on the failure response of the pipeline under seismic load.

Figure 1 outlines the procedures of the proposed method for the simulation of thermal buckling and influence analysis of PIP under seismic load, which includes four steps listed as follows:

Step 1: Steady-state structural analysis. The critical buckling temperature is determined by considering the temperature load and the inner pressure load.

Step 2: Modal analysis. The first and third modes of the PIP system are calculated to determine the Rayleigh damping coefficient of the system.

Step 3: Seismic analysis. The pipeline seismic response under two conditions including pre-buckle and post-buckle is studied based on the time-history seismic analysis method.

Step 4: Influencing factor analysis and failure analysis. The influences of insulation layer materials and initial defects on the failure of pipelines are studied.

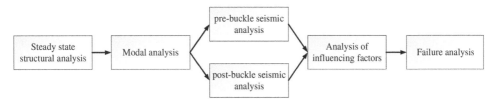

Figure 1. Flow chart of pipeline impact analysis by multiple loads.

2.2 Modal analysis

The damping ratio is added to the model properties in the form of the Rayleigh damping coefficient:

$$[C] = \alpha_d[M] + \beta_d[K] \qquad (1)$$

where [C] is the damping matrix, [M] is the mass matrix, [K] is the stiffness matrix, and α_d and β_d are the mass-proportional damping coefficient and stiffness-proportional damping coefficient respectively. In this paper, Kalinontzis (1998) proposed the method, which is widely used in submarine pipeline calibration. It is selected to calculate the damping coefficient:

$$\alpha_d = \frac{2\xi\omega_1\omega_3}{\omega_1 + \omega_3}, \quad \beta_d = \frac{2\xi}{\omega_1 + \omega_3} \qquad (2)$$

Where ξ is taken as 0.05, and ω_1 and ω_3 are the first and third mode frequency of the pipeline soil system.

2.3 Earthquake loading

The time history analysis method was adopted for seismic dynamic response analysis of the PIP in this paper, which is an effective method. Its advantage is that the detailed evolution process of the displacement, velocity, and acceleration of the pipeline structure is subjected to the dynamic action of an earthquake to track the stress and failure state of the structure of the elastic stage and the elastoplastic stage. The seismic record was selected as an El-Centro record and then applied to the pipeline along the transverse direction (y direction as shown in Figure 3). The time history curve of acceleration of the seismic record is shown in Figure 2:

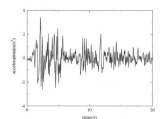

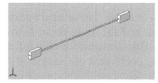

Figure 2. El-Centro Seismic record.

Figure 3. Schematic diagram of the suspended span pipeline model.

Figure 4. Schematic diagram of PIP.

3 FE MODEL AND MATERIAL PROPERTIES

3.1 Geometry model and FE model of the PIP

The research object is a submarine-suspended span pipeline (Figure 3), with both ends of the pipeline buried in the soil. Since the focus is the buckling analysis of the pipeline, the soil is simplified as a rigid body. The PIP system is composed of the inner pipe, the outer pipe, and the insulation layers between the inner and outer pipes (Figure 4). Three different types of commonly used materials were selected to be filled as the thermal insulation layer including Polyurethane foam, polyethylene, and phenolic resin.

Simple support constraint is adopted at both ends of the pipeline, that is, the displacement in three directions at both ends of the pipeline is constrained. The outer pressure caused by the seawater and the inner pressure caused by the oil transportation on the pipe wall was applied to the inner and outer walls of the pipeline model, respectively.

The initial temperature of the pipeline is 22°C and an increasing temperature change of ΔT is uniformly applied to the wall thickness. Due to the thermal loading, the pipeline tends to expand. However, this expansion is resisted by external pressure. Thus, it induces axial forces in the wall thickness and causes a global buckling in the pipeline (Karampour et al. 2013; Karampour 2018). To control the lateral buckling response, a transverse sinusoidal initial defect with a defect mode is applied at the midpoint of the pipe. The value of defect amplitude v_0 is 200 mm and the value of wavelength l_0 is 10000 mm. The detailed parameters of the model were listed in Tables 1 and 3.

In this paper, the Solid186 element was used for modeling the PIP system. The total number of the established FE model is 370228.

Table 1. Model geometric parameters.

Parameter	Value
Outer pipe diameter /mm	300
Inner pipe diameter /mm	240
wall thickness /mm	10
External pressure /MPa	10
Internal pressure /MPa	10
Length of suspended section /m	40
Length of buried section /m	5

3.2 Material models of PIP

Severe plastic deformation often occurs when submarine pipelines are subjected to complex loads. To characterize the plastic strain process of pipeline more accurately, the Ramberg-Osgood model (Ramberg 1943) was used to describe the constitutive relationship of pipeline materials:

$$\varepsilon = \frac{\sigma}{E}\left\{1 + \frac{n}{1+r}\left(\frac{\sigma}{\sigma_y}\right)^r\right\} \quad (3)$$

Where σ and ε are the stress and strain of the pipe respectively, E is the elastic modulus, σ_y is the yield strength, and n and r are model parameters. In this paper, n is 10 and r is 12 (O'Rourke & Liu 2012). The stress-strain curve of the pipeline is shown in Figure 5. At the

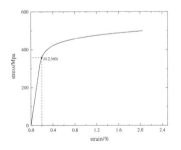

Figure 5. Stress-strain relationship of the pipeline.

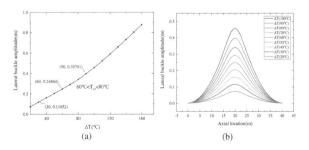

Figure 6. Lateral buckling response of the pipeline.

Table 2. Detailed parameters of the insulation layer.

Material	Density kg*m^{-3}	Thermal expansion coefficient °C^{-1}	Young's modulus MPa	Poisson's ratio
Polyurethane foam	192	0.000114	66.1	0.316
polyethylene	958	0.000145	1080	0.418
phenolic resin	1400	0.0000469	7440	0.345

Table 3. Model Material Properties.

Parameter	Value
density /kg·m^{-3}	7850
Thermal expansion coefficient/°C^{-1}	1.17×10^{-5}
Young's modulus /MPa	206000
Poisson's ratio	0.3

initial stage, the stress-strain relationship of the pipeline is linear. When the strain reaches about 0.2%, the stress-strain relationship enters the plastic stage and the pipeline strain starts to increase rapidly with the increase of stress until it reaches failure.

4 RESULTS

4.1 Thermal buckling results

To study the effect of high temperature on the buckling of compression spanning pipe with initial defects, a uniformly increased temperature variation ΔT is applied to the pipes. The maximum lateral buckling amplitude in the pipeline occurs at the defect in the middle of the pipeline. Figure 6 (a) shows the thermal buckling response of the pipe at the maximum buckling amplitude. When the temperature rises to the critical value, the buckling amplitude increases nonlinearly with the corresponding increase in temperature. The lateral buckling curves at different temperature variations are shown in Figure 6 (b). The comparison of the results in Figure 6 (a) and (b) shows that the critical buckling temperature (T_{cr}) of the pipe is between 60–80°C.

4.2 Post-buckle seismic results

After applying gravity and internal and external pressure to the pipeline, a temperature variation of ΔT=100°C (greater than the critical buckling temperature) was applied to the pipeline. At this temperature variation, the pipe has buckled (Figure 6a). Then, the seismic load was further applied to the pipeline.

The lateral buckling profile of the pipeline before and after the application of seismic load is shown in Figure 7. The curve of the maximum stress time history in the middle of the pipeline was shown in Figure 8. It could be seen that the maximum stress in the middle of the pipeline rises to a level slightly higher than the yield stress, from 429 MPa to 440 MPa. The maximum stress of the pipe is located in the middle of the outer pipe of the PIP, and the maximum stress in the middle of the inner pipe is 430 Mpa, which is slightly lower than the maximum stress of the outer pipe. The buried section of the pipeline is averagely distributed with stresses higher than the two ends of the suspended span section of the pipeline, and the stresses of the buried section of the inner and outer pipes are 228 Mpa and 358 Mpa respectively.

It can be seen that the application of seismic load has significantly increased the buckling amplitude to 0.79 m, while it was only 0.46 m when the earthquake did not occur.

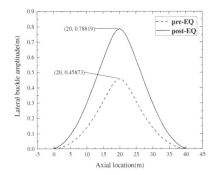

Figure 7. Lateral buckling profile of pipeline before and after the earthquake.

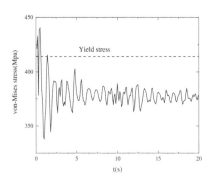

Figure 8. Von-Mises's stress -time history in the middle of the pipeline.

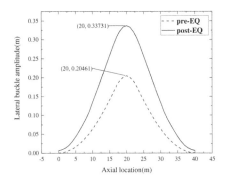

Figure 9. Comparison of lateral buckling profiles before and after earthquakes.

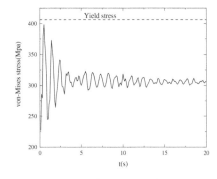

Figure 10. Von-Mises's stress-time history is in the middle of the pipeline.

4.3 Pre-buckle seismic results

In this case, the temperature variation is maintained at $\Delta T = 50°C$ (slightly below the critical temperature) after the application of self-weight and internal and external pressure. Then, the seismic input is applied. As a result of the earthquake, the pipeline bending amplitude increases, and the maximum buckling amplitude is 0.34 m, as shown in Figure 9 (less than the amplitude of the scenario shown in Figure 7). Due to the combination of thermal and seismic actions, the maximum stress at the crown increased from 297 MPa to almost 398 MPa, as shown in Figure 10. However, the maximum stress is still lower than the yield strength of the material (414 MPa). The maximum stress of the inner pipe is 367 MPa, which is still lower than that of the outer pipe. The maximum stress of the inner and outer pipes of the buried section is 137 MPa and 189 MPa respectively.

5 ANALYSIS OF INFLUENCING FACTORS

5.1 Influence of insulation layer

To study the influence of the insulation layer on pipeline seismic buckling behavior, three types of commonly used insulation layer materials were selected as shown in Table 2. The same constraint conditions were applied to the model, moreover, the same temperature variation (lower than the critical buckling temperature), and the same external load were considered, which means that only the influence of the material of the insulation layer was studied.

The calculation results of pipeline models using three types of insulation materials were extracted respectively. The maximum stress time history in the middle of the pipeline after the earthquake was shown in Figure 11. The profiles of the pipeline buckling amplitude was shown in Figure 12. Through comparison, it can be seen that the maximum stress of pipes with polyethylene as an insulation layer in the earthquake is smaller, while the maximum stress of pipes with phenolic resin as an insulation layer in an earthquake is larger. The post-earthquake buckling amplitude of the pipeline using polyurethane foam as an insulation layer is 0.34 m, which is the smallest of the three materials. The post-earthquake buckling amplitude of pipes with phenolic resin as an insulation layer is 0.39 m, which is the largest among the three materials. The failure behavior of pipelines is determined by a variety of material parameters, and different failure behaviors have different sensitivity to different material parameters.

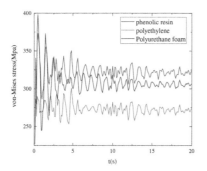
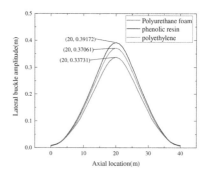

Figure 11. Curves of Mises stress- time of the pipeline with different materials.

Figure 12. Lateral buckling profiles after the earthquake.

5.2 Influence of initial defects

Since the defect mode of the initial defect is a sine function $l = v_0 \sin\left(\frac{\pi x}{l_0}\right)$, the initial defect is changed by changing the amplitude v_0 and wavelength l_0 of the initial defect function, and then the influence of the initial defect on the pipeline failure under load is studied. The wavelength l_0 is fixed at 10000 mm, and the amplitude v_0 is varied to 200 mm, 300 mm, and 400 mm, respectively, to study the influence of initial defect amplitude on pipeline failure mode under load. Moreover, the amplitude v_0 is fixed at 200 mm, and the wavelength l_0 is varied to 10000 mm, 15000 mm, and 20000 mm, respectively, to study the effect of wavelength of initial defect on pipeline failure under load.

The same constraints and external loads were applied to each model. The results are shown in Figures 13–16. According to the comparison, the larger the initial flaw amplitude

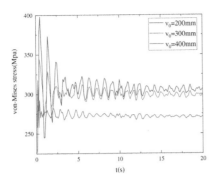

Figure 13. Influence of amplitude factors on maximum Mises's stress of the middle of the pipeline.

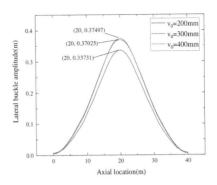

Figure 14. Influence of amplitude on pipeline buckling amplitude after the earthquake.

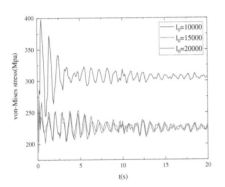

Figure 15. Influence of wavelength on maximum stress of the middle of the pipeline.

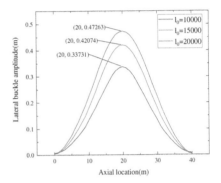

Figure 16. Influence of wavelength on pipeline buckling amplitude after an earthquake.

is, the smaller the maximum stress in the middle of the pipeline is, and the greater the post-earthquake buckling amplitude is. With the increase of the initial defect amplitude, the maximum stress in the middle of the pipeline decreases sharply, however, the post-earthquake buckling amplitude increases slightly. The greater the initial defect wavelength is, the greater the post-earthquake buckling amplitude of the pipeline is. With the increase of the initial defect wavelength, the slower the post-earthquake buckling amplitude increases, the greater the initial defect wavelength is, and the greater the post-earthquake buckling amplitude of the pipeline. With the increase of the initial defect wavelength, the slower the post-earthquake buckling amplitude increases.

6 CONCLUSIONS

The numerical simulation model of the suspended span PIP structure with the initial defect was established for global buckling analysis. The interaction between seismic load and high-temperature and the high-pressure load was studied through two different loading scenarios. The results show that the maximum stress of the outer pipe of the PIP structure is greater than that of the inner pipe. There is a large stress uniformly distributed in the buried section of the pipe, and the input of seismic load will significantly increase the buckling amplitude of the pipe.

The influence factors including the amplitude and wavelength of initial defects and the insulation material on the buckling behavior of the PIP structure were studied by using the control variable method. Using polyethylene as a thermal insulation layer can significantly reduce the maximum stress that the pipeline bears in an earthquake. While using polyurethane foam as a thermal insulation layer can significantly reduce the pipeline's bending amplitude after the earthquake. The stress of the PIP structure in the earthquake is sensitive to the amplitude of the initial defect, and the wavelength of the initial defect has a greater impact on the post-earthquake buckling amplitude.

The simulation results of suspended span pipelines obtained in this paper can provide a theoretical basis for failure analysis of submarine pipelines in the future. Some optimization measures can be taken according to the conclusions drawn from the simulation results to improve the resistance of the submarine pipeline to sudden disasters to fully ensure the integrity and reliability of the submarine pipeline transportation system.

FUNDING

This research was funded by the National Natural Science Foundation of China, China, grant No. 51879272, No. 52111530036; and the Fundamental Research Funds for the Central Universities, China, grant No. 22CX03022A.

REFERENCES

Demirci H.E., Bhattacharya S., Karamitros D., et al. (2018). Experimental and Numerical Modeling of Buried Pipelines Crossing Reverse Faults. *Soil Dynamics and Earthquake Engineering*, 114: 198–214.
Gong S., Sun B., Bao S., et al. (2012). Buckle Propagation of Offshore Pipelines Under External Pressure. *Marine Structures*, 29(1): 115–130.
Kalliontzis C. (1998). Numerical Simulation of Submarine Pipelines in Dynamic Contact with a Moving Seabed. *Earthq Eng Struct Dynam*; 27(5): 465–86.
Karampour H. (2018). Effect of the proximity of Imperfections on Buckle Interaction in Deep Subsea Pipelines. *Mar Struct*; 59: 444–57.
Karampour H., Albermani F., Gross J. (2013). On lateral and Upheaval Buckling of Subsea Pipelines. *Engineering Structures*, 52(Jul.): 317–330.
Mina D., Forcellini D., Karampour H. (2020). Analytical Fragility Curves for Assessment of the Seismic Vulnerability of HP/HT Unburied Subsea Pipelines. *Soil Dynamics and Earthquake Engineering*, 137.
O'Rourke M.J. & Liu X. (2012). Seismic Design of Buried and Offshore Pipelines.
Pan H., Li H.N., Li C. (2021). Seismic Behaviors of Free-spanning Submarine Pipelines Subjected to Multi-support Earthquake Motions Within Offshore Sites. *Ocean Engineering*, 237.
Psyrras, Nikolaos, K., et al. (2018). Safety of Buried Steel Natural Gas Pipelines Under Earthquake-induced Ground Shaking: A Review. *Soil Dynamics and Earthquake Engineering*, 106(Mar.): 254–277.
Ramberg W. (1943). Description of Stress-Strain Curves by Three Parameters. *National Advisory Committee for Aeronautics*, Technical Note.
Saeedzadeh R. & Hataf N. (2011). Uplift Response of Buried Pipelines in Saturated Sand Deposit Under Earthquake Loading, *Soil Dynamics and Earthquake Engineering*, 31: 1378–1384.
Wang Y., Zhang P., Hou X.Q., et al. (2020). Failure Probability Assessment and Prediction of Corroded Pipeline Under Earthquake by Introducing in-line inspection data. *Engineering Failure Analysis*, 115:104607-.
Zhang Z., Yu J., Liu H., et al. (2019). Experimental and Finite Element Study on Lateral Global Buckling of the Pipe-in-pipe Structure by Active Control Method. *Applied Ocean Research*, 92:101917–101917.

Study on the impact resistance of CFST column with an inner circular steel tube

Zike Jiao* & Shaopeng Lei

School of Civil Engineering, Xi'an University of Architecture and Technology, Xi'an, Shanxi, China

ABSTRACT: To study the impact performance of concrete-filled steel tubular columns with inner steel tubes, a numerical model of drop hammer impact was established using the finite element software LS-DYNA. The structural response under different impact forces was investigated by changing the steel strength and the concrete strength. It is found that the impact resistance of concrete-filled plain steel tubular columns can be significantly improved by reasonably adding an inner steel tube. To verify the correctness of the numerical simulation method and strategy, the existing impact test of CFST is simulated, and the simulation results are in good contrast with the test results.

1 INTRODUCTION

In recent years, concrete-filled steel tubular structures have been widely used in industrial plants, large-span high-rise buildings, and bridge piers because of their good bearing capacity, ductility, and seismic resistance. As a load-bearing member, concrete-filled steel tubular (CFST) will inevitably receive the effect of accidental impact. Given this, the research and development of new structures to improve impact resistance are increasingly urgent.

To further improve the strength and ductility of concrete-filled steel tubular columns, scholars have proposed concrete-filled steel tubular columns with different internal structures, such as vertical stiffener (Tao et al. 2005, 2008), circumferential stiffeners (Lai & Ho 2014) welded on steel tubes, and steel reinforcement cages (Ding et al. 2016, 2020) arranged in steel tubes. A series of studies have been carried out on the seismic performance of concrete-filled steel tubular columns with internal structures at home and abroad, but research on the impact resistance of this type of column is rarely reported. Therefore, the research on its dynamic response needs to be deepened.

Therefore, based on the comparison and verification of the existing impact tests, the numerical model of a concrete-filled steel tube column with an inner steel tube is established by using the finite element analysis software LS-DYNA. The impact quality, impact speed, steel strength, and concrete strength are analyzed.

2 FINITE-ELEMENT MODELING AND VERIFICATION

2.1 *Numerical model*

The finite element analysis software LS-DYNA is used to establish a circular concrete-filled steel tube column with an inner round steel tube to analyze the impact quality, impact speed, steel strength, and concrete strength on its impact resistance. The numerical model is divided into five parts: outer steel pipe, inner steel pipe, concrete, impact body, and fixed support. Each part of the model is modeled by the Solid164 solid element. During the contact setting,

---

*Corresponding Author: 846766117@qq.com

the steel pipe and concrete, the steel pipe concrete and the support, and the impact body and the steel pipe all adopt Automatic Surface to Surface contact. According to the test results, the relative slip between the steel tube and the concrete is small and negligible, so the common node coupling treatment is adopted. The fixed ends at both ends restrict the displacement of the end face in all directions. When dividing the mesh, based on the careful consideration of improving the calculation accuracy and saving the calculation cost, the mesh is locally densified at the impact site. The established numerical model is shown in Figure 1.

Figure 1. Schematic diagram of a numerical model.

2.2 Material models

In this paper, a bilinear elastic-plastic constitutive model MAT_PLASTIC_KINEMATIC, which can reflect the isotropic and dynamic hardening plastic characteristics of steel, is selected as the steel pipe material and the strain rate effect of steel is reflected by Cowper-Symonds model. The yield condition is calculated according to Formula (1):

$$\sigma_y = \left[1 + \left(\frac{\dot{\varepsilon}}{C}\right)^{\frac{1}{p}}\right]\left(\sigma_0 + \beta E_P \varepsilon_p^{eff}\right) \quad (1)$$

where $\dot{\varepsilon}$ is the strain rate; C and p are the material strain rate parameters, and their values are $40s^{-1}$ and 5; σ_0 is the initial yield stress; ε_p^{eff} is equivalent plastic strain; β is the dynamic hardening coefficient; E_p is the plastic hardening modulus.

A plastic damage constitutive model MAT_CONCRETE_DAMAGE_REL3(72R3) is used for concrete materials. A dynamic increasing coefficient (DIF) is introduced to quantify the influence of strain rate on the strength of concrete materials. In this study, DIF curves of concrete compressive strength and tensile strength given by Hao et al. (Hao & Hao 2014) were used.

The compressive DIF (CDIF) and tensile DIF (TDIF) at the strain rate $\dot{\varepsilon}$ are determined by the following equations:

$$CDIF = \frac{f_{cd}}{f_{cs}} = \begin{cases} 0.0419(\log \dot{\varepsilon}) + 1.2165 & for(\dot{\varepsilon} \leq 30s^{-1}) \\ 0.8988(\log \dot{\varepsilon})^2 - 2.8255(\log \dot{\varepsilon}) + 3.4807 & for(\dot{\varepsilon} > 30s^{-1}) \end{cases} \quad (2)$$

$$TDIF = \frac{f_{td}}{f_{ts}} = \begin{cases} 0.26(\log \dot{\varepsilon}) + 2.06 & for(\dot{\varepsilon} \leq 1s^{-1}) \\ 2(\log \dot{\varepsilon}) + 2.06 & for(1s^{-1} < \dot{\varepsilon} \leq 2s^{-1}) \\ 1.44331(\log \dot{\varepsilon}) + 2.2276 & for(2s^{-1} < \dot{\varepsilon} \leq 150s^{-1}) \end{cases} \quad (3)$$

where f_{cd} and f_{td} are the dynamic compressive and tensile strengths of concrete, respectively; f_{cs} and f_{ts} are the static compressive and tensile strengths of concrete, respectively.

2.3 Model validation

To verify the effectiveness of concrete-filled steel tubular columns under vehicle impact, the drop weight test of concrete-filled steel tubular columns made by Hou Chuanchuan was simulated and verified in this paper. The test adopts a circular section concrete-filled steel tube column with outer diameter d = 180mm, a wall thickness of 3.65mm, a concrete strength grade of C60, a steel strength grade of Q235, and an end plate of 30mm thick square steel plate with a strength grade of Q345. In this paper, the two working conditions of a drop hammer height of 5.5m, drop hammer mass of 465kg, drop hammer height of 2.5m, and drop hammer mass of 920kg are simulated and analyzed.

Figure 2 shows the comparison between the simulation results and the test results of the impact force time history curve under the two working conditions. Figure 3 shows the comparison between the simulation results and the test results of the mid-span deflection curve of the concrete-filled steel tube column. The results show that the numerical model established by using the selected modeling method, material model, and input parameters can accurately predict the impact load response, impact force time history curve, and mid-span displacement time history curve of CFST structure.

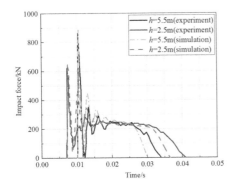
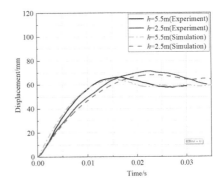

Figure 2. Impact force history curves.

Figure 3. Mid-span displacement history.

2.4 Numerical simulation results

Based on verifying the model, the impact model of a concrete-filled steel tube column with an inner circular steel tube is established. Model parameters and results can be seen in Table 1. Figure 4 is the time history curve pair of impact force of concrete-filled steel tube column with inner circular steel tube and plain concrete-filled steel tube column. Figure 5 is the comparison of the mid-span deflection of the two. It can be seen from Figures 4 and 5 that when the built-in round steel pipe is included, the peak value of impact force is significantly increased, and the deflection is significantly reduced, which indicates that when the built-in steel pipe is included, the restraint effect of concrete is significantly enhanced, so the overall rigidity of the member is improved and the impact resistance is significantly improved.

Table 1. Parameters of model and results of impact resistance.

Specimen	D/mm	L/mm	t_0/mm	t_1/mm	m/kg	v/(m/s)	F_{max}/kN	Δ_{max}/mm	Δ_{stab}/mm
CC2	90	1940	3.65	—	920	6.4	654.24	68.01	64.28
RCFST	90	1940	3.65	2.4	920	6.4	700.01	58.45	54.86

Note: D is the diameter of the CFST column; L is the effective length of the specimen; t_0 is the thickness of the steel tube; t_1 is the thickness of the inner steel pipe; M is the impact body mass; V is impact velocity; F_{max} is impact peak; Δ_{max} is the mid-span deflection; Δ_{stab} is the mid-span residual deflection.

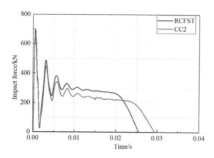
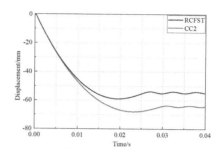

Figure 4.　Impact force history curves.

Figure 5.　Mid-span displacement history.

3 PARAMETER ANALYSIS

To better study, the impact performance of concrete-filled steel tubular columns with inner circular steel tubes, nine-member models with different parameters are established. The values of model parameters and calculation results can be seen in Table 2.

Table 2.　Parameters of model and results of impact resistance.

Specimen	D/ mm	L/ mm	t_0/ mm	t_1/ mm	m/ kg	v/ (m/s)	F_y/ Mpa	F_c/ Mpa	F_{max}/ kN	Δ_{max}/ mm	Δ_{stab}/ mm
RCFST-1	90	1940	3.65	2.4	1000	7	247	40.7	715.44	84.34	79.88
RCFST-2	90	1940	3.65	2.4	1000	6	247	40.7	587.12	61.05	57.18
RCFST-3	90	1940	3.65	2.4	1000	5	247	40.7	503.60	43.56	38.40
RCFST-4	90	1940	3.65	2.4	1000	5	247	40.7	501.55	33.17	27.36
RCFST-5	90	1940	3.65	2.4	750	5	247	40.7	497.47	23.08	17.03
RCFST-6	90	1940	3.65	2.4	500	5	340	40.7	527.84	35.55	28.65
RCFST-7	90	1940	3.65	2.4	1000	5	423	40.7	565.44	31.50	24.22
RCFST-8	90	1940	3.65	2.4	1000	5	247	50.2	541.94	41.91	36.42
RCFST-9	90	1940	3.65	2.4	1000	5	247	68.3	605.77	39.28	33.67

Note: F_y is the steel yield strength; F_c is the compressive strength of concrete.

3.1 *Impact speed*

To study the dynamic response of concrete-filled steel tubular columns with inner circular steel tubes under impact velocity, other parameters are kept unchanged, and only the velocity of the impact block is changed. It can be seen from Figures 6 and 7 that with the increase

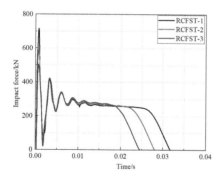
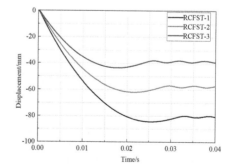

Figure 6.　Impact force history curves.

Figure 7.　Mid-span displacement history.

in impact speed, the peak value of the impact force of the member increases, and the mid-span deflection increases significantly. This is because the increase in impact energy leads to the increase of energy absorbed by the component and the increase in plastic deformation.

3.2 *Impact mass*

To keep the impact speed constant, and change the mass of the impact body to study the influence of different energy, the impact mass varies from 500 to 1000kg. It can be seen from Figures 8 and 9 that with the increase of the mass of the impactor, the impact force changes little, the platform value increases slightly, the impact time increases significantly, the energy absorbed by the component increases and the mid-span deflection increases.

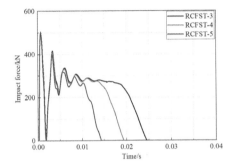 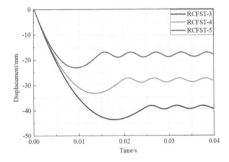

Figure 8. Impact force history curves.   Figure 9. Mid-span displacement history.

3.3 *Steel strength*

In the common scope of engineering, the influence of the yield strength of steel is analyzed. As can be seen from Figures 10 and 11, with the increase of the yield strength of the steel, the peak value of the impact force time history curve of the member does not change much; the impact force platform value is significantly increased, the impact duration is decreased, and the mid-span deflection of the member is decreased. It shows that with the increase of the yield strength of the steel, the flexural strength of the member section increases, and the impact resistance increases.

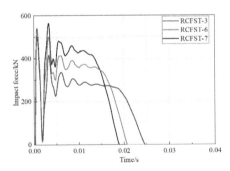 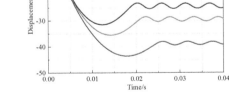

Figure 10. Impact force history curves.   Figure 11. Mid-span displacement history.

3.4 *Concrete strength*

The influence of concrete strength is analyzed in the common scope of engineering. It can be seen from Figures 12 and 13 that with the increase of concrete strength, the peak value of impact force increases, the impact force platform value slightly increases, the impact duration decreases, and the mid-span deflection of the member decreases. It shows that with the increase of concrete strength, the flexural strength and impact resistance of the member section increase.

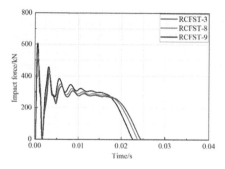

Figure 12. Impact force history curves.

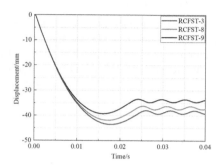

Figure 13. Mid-span displacement history.

4 CONCLUSIONS

The dynamic response of a concrete-filled steel tube (CFST) with an inner steel tube under lateral impact is analyzed by the finite element method. According to different factors such as impact speed, impact quality, steel strength, and concrete strength, the impact force time history curve and mid-span deflection time history curve of concrete-filled steel tubular members with inner steel tubes are obtained by using the finite element software LS-DYNA. Through the above research, the following conclusions are obtained:

(1) Setting an internal circular steel tube in circular CFST can significantly improve the impact resistance of components.
(2) The impact mass and impact velocity have significant effects on the impact force time history curve and mid-span deflection time history curve of the member. With the increase of impact mass and impact velocity, the peak value of the impact force of the member increases, and the mid-span deflection increases obviously.
(3) With the improvement of steel strength and concrete strength, the peak value of the impact force of the member is slightly increased, the mid-span deflection is significantly reduced, and the overall rigidity and section bending strength of the column are increased.

REFERENCES

Ding, F. X., Lu, D. R., Bai, Y., Zhou, Q. S., Ni, M., & Yu, Z. W., et al. (2016). Comparative Study of Square Stirrup-Confined Concrete-Filled Steel Tubular Stub Columns Under Axial Loading. *Thin-Walled Structures*, 98(JAN.PT.B), 443–453.

Ding, F., Liu, Y., Fei, L., Lu, D., & Chen, J. (2020). Cyclic Loading Tests of Stirrup Cage Confined Concrete-Filled Steel Tube Columns Under High Axial Pressure. *Engineering Structures*, 221, 111048.

Hao, Y., & Hao, H. (2014). Influence of the Concrete DIF Model on the Numerical Predictions of RC Wall Responses to Blast Loadings. *Engineering Structures*, 73 (Aug. 15), 24–38.

Lai, M. H., & Ho, J. (2014). Confinement Effect of Ring-confined Concrete-filled-steel-tube Columns Under Uni-axial Load. *Engineering Structures*, 67(67), 123–141.

Tao, Z., Han, L. H., & Wang, D. Y. (2008). Strength and Ductility of Stiffened Thin-walled Hollow Steel Structural Stub Columns Filled with Concrete. *Thin-Walled Structures*, 46(10), pp. 1113–1128.

Tao, Z., Han, L. H., & Wang, Z. B. (2005). Experimental Behavior of Stiffened Concrete-filled Thin-walled Hollow Steel Structural (hss) Stub Columns. *Journal of Constructional Steel Research*, 61(7), 962–983.

Application of concrete-filled steel tubular support in trackage roadway of Yangcheng mine

Lu Qiu, Ke-Ming Liu*, Jia-Wei Wang, Gang Feng, Chang-Hao Zhang,
Ze-Yuan Zhou & Shao-Shuai Liu
School of Civil Engineering and Architecture, Linyi University, Linyi, China

ABSTRACT: To solve the problems such as large rock deformation around the roadway and long deformation time during the support of the −650 m horizontal trackage roadway of Yangcheng mine, the research methods of field geological survey combined with theoretical calculation and field monitoring were used to analyze the deformation characteristics of surrounding rock and to investigate causes of roadway instability and failure. On this basis, a high-strength composite supporting scheme based on Concrete-filled Steel Tubular Support was proposed according to the strengthening support theory of pressure rings in deep-mine roadways. The bracket section was designed as an arc wall semicircle arch, and the main steel tube was $\Phi 194$ mm $\times 8$ mm seamless steel tube, filled with the C40 strength class concrete and enforced with shotcrete and rock bolt supporting and grouting reinforcement in surrounding rock. The research results showed that this supporting system could provide more than 1.57 MPa supporting reaction force to the surrounding rock, which could effectively inhibit the moving of surrounding rock into the roadway space. Besides, the accumulative deformation of the surrounding rock of the roadway was found to be less than 100 mm after the supporting was stabilized, satisfying the roadway deformation requirements.

1 INTRODUCTION

After the coal mine enters deep mining, it will be affected by high crustal stress, high ground temperature, and strong mining disturbance. Meanwhile, the complexity of the geological conditions and stress field where the deep rock mass is located causes dramatic ore pressure on the roadway, and the surrounding rock of the roadway exhibits significant characteristics of large plastic deformation and continuous rheology. Controlling the rock that surrounds the road is significantly more problematic than that of the shallow strata (Huang et al. 2021; Kang et al. 2022; Kong et al. 2020; Li et al. 2022; Li 2012); hence, it is challenging to effectively control the convergence deformation of the surrounding rock of the deep mine roadway by simply relying on an anchor bolt (cable), U-shaped steel, and other support forms.

Concrete-filled steel pipe support (CFSTS) is a kind of high-strength support whose bearing capacity can reach 3 times that of U-shaped steel support with the same amount of steel consumption (Gao et al. 2010); therefore, it is especially suitable for supporting the deep mine roadway and the soft rock roadway. The performance research of CFST structures in deep underground engineering mainly focuses on two aspects: axial compression bearing capacity and flexural capacity. Li X. B. (Li 2012) and Wang Jun (Wang 2014) studied the

*Corresponding Author: 717525855@qq.com

mechanical properties of steel-concrete short columns with different wall thicknesses and inner diameters of steel tubes. Gao et al. (2010) and Liu Guolei (Liu 2013) studied the mechanical properties of CFST supports in different specifications under concentrated load. At present, this support has been successfully applied in more than 20 coal mines in China (Gao et al. 2015; He et al. 2015; Liu 2017; Wang 2019; Xia 2020).

The current study, taking the −650 m south wing trackage roadway of Yangcheng Mine as the engineering background, introduces the application of composite support technology with CFSTS as the principal part in the trackage roadway in detail, which provides a reference for controlling the stability of surrounding rock of other deep mine roadways with similar geological conditions.

2 ENGINEERING BACKGROUND

The −650 m horizontal south wing trackage roadway of Yangcheng Mine has a burial depth of 690 m, and the bulk density of surrounding rock was 22 kN/m^3. The surrounding rock of the roadway was mainly composed of argillaceous cemented sandstone, with well-developed rock stratification and joint, which was soft and easy to be weathered. The dip angle of coal-series strata was found to be 25°–30°. The coal-series strata in this area had stable occurrence and simple structure. Most of the trackage roadway traverses strata were composed of siltstone, middle sandstone, and mudstone, and the unidirectional compressive strength of the rock was 30–40 MPa. The vertical crustal stress in this area was about 15.2 MPa, and there was higher horizontal tectonic stress, with a lateral pressure coefficient of 1.7–2.0.

The original section of the −650 m horizontal trackage roadway was horseshoe-shaped, with the clear width × clear height = 4800 mm×4200 mm for the roadway. After excavating it to form the roadway, a 50 mm-thick C20 concrete layer was sprayed as the temporary support, and then shotcrete and rock bolt supporting +U29 steel support was implemented for the permanent support. The spray layer was 150 mm thick, and the bolt was Φ20 mm×2400 mm high-strength spiral steel resin cartridge bolt, with a row spacing of 1000 mm×1000 mm. The metal mesh was welded using Φ5 mm steel mesh, with a mesh size of 100 mm×100 mm.

After the roadway was supported, the surrounding rock deformation was monitored for 180 days. The results showed that the two sides of the main roadway were subjected to severe deformation, with the moving distance of the two sides over 1000 mm. The phenomenon of floor heave and roof carving was also apparent. The moving distance between the roof and floor was detected as more than 280 mm, the roof support was exposed, the floor heave was more than 800 mm, and the concrete spraying layer was cracked at many locations. Even after multiple repairs, the roadway's stability could not be achieved. Figure 1 shows the deformation and failure of the roadway.

Figure 1. Deformation and failure diagram of the roadway.

3 DESIGN OF THE COMPOSITE SUPPORT SCHEME FOR CFSTS

3.1 *Physical and mechanical property tests on the surrounding rock of the roadway*

According to the test results on the mechanical and hydraulic properties of surrounding rock by field sampling, the rock's natural uniaxial compressive strength was noted as 5.1–12.9 MPa, and the saturated uniaxial compressive strength was 3.6–5.3 MPa. The total amount of clay minerals in the surrounding rock samples was determined as 53.9 wt%; in the relative content of clay minerals, the illite/smectite mixed layer accounted for 39 wt%, illite 9 wt%, and kaolinite 52 wt%. The rock has a stronger water swell ability, and its water-saturated absorptivity was measured as 33.3 wt%. After the water absorption, it had a high expansion rate, and its strength decreased significantly, with the rock subjected to disintegration and uralitization.

3.2 *Deformation and failure analysis of the trackage roadway*

The main causes of surrounding rock failure of the −650 m horizontal trackage roadway can be summarized as follows:

1) The burial depth of the roadway was 690 m. According to the measured results on crustal stress, the crustal stress field of the trackage roadway belongs to the horizontal tectonic stress field on a macroscopic level, and the angle between the direction of the maximum horizontal principal stress and the roadway was 55.4°. Under the influence of horizontal tectonic stress, the stability of the roadway's surrounding rock gradually deteriorated, which is manifested mainly by the rheological behavior of soft rock and the difficulty with roadway maintenance.
2) The lithology of surrounding rock altered significantly due to roadway cross-measure, aggravating the deformation of different surrounding rock segments, and parts of the roadway.
3) The supporting design was unreasonable, and the supporting strength was insufficient. The bearing capacity of U29-type steel support was lower, failing to provide a more significant support reaction force to control the deformation of surrounding rock; the floor support was weak, aggravating the floor heave and reducing the stability of the two sides of the support system; the coordination between different supporting forms was weak.

4 SUPPORT SCHEME DESIGN FOR THE CFSTS

4.1 *Support scheme design for the −650 m horizontal trackage roadway*

In support of the deep mine roadway, the load of the surrounding rock is more significant than its strength; therefore, the surrounding rock fails to achieve self-stabilization. Therefore, it is necessary to strengthen the rock mass within a certain width range around the roadway to enable it to have a higher bearing capacity, thereby forming a pressure-bearing ring to control the stability of surrounding rock outside the bearing ring and realize the purpose of roadway stability. The following support strengthening technologies are required to form the pressure-bearing ring: (i) shotcrete and rock bolt supporting; (ii) CFSTS supporting; (iii) grouting reinforcement in surrounding rock.

The support scheme design is as follows:

1) The roadway section was expanded and repaired, and the temporary support was performed: the bottom plate was re-excavated to facilitate the formation of closed support, and after the expansion and repair, the shotcrete was applied and rock bolt supporting was constructed. The reinforced concrete is poured at the bottom plate, with the concrete strength grade of C40;

2) Secondary support: CFSTS is employed for the secondary support, and the gangue bags are filled between the concrete spray layer and CFSTS, with a thickness of 100 mm, as a flexible deformation space;
3) Grouting reinforcement: After the completion of the CFSTS support, the shotcrete process is applied again to monitor the bearing capacity of the support. When the bearing capacity exceeds 80% of the designed bearing capacity of the support, grouting reinforcement was implemented and performed on the surrounding rock.

In the scheme, the CFSTS was the main support method, and the shotcrete and rock bolt supporting and grouting reinforcement in surrounding rock were the auxiliary support methods.

4.2 Design of the CFSTS

1) Support structure design and parameter selection
Based on the engineering analogy method, the support section was designed as a semi-circular arch of the arc wall. φ194 mm×8 mm seamless steel tube was selected as the main steel tube; the joint coupling had the size of φ219 mm×10 mm. The main structure of the support included the top arch section, the two sides, and the reverse bottom arch section. The supports were connected by concrete-filled steel tubular connecting rods, with a spacing of 0.8 m.
2) CORE CONCRETE RATIO
The strength grade of the core concrete was C40, with slumps $\geq$ 160 mm.

4.3 Design of the shotcrete and rock bolt supporting

1) After expansion and repair of the roadway, a 30–50 mm-thick shotcrete layer was applied immediately on the surrounding rock to seal the surrounding rock, prevent rock weathering and falling, and isolate the moisture exchange between the surrounding rock and air.
2) The φ22 mm×2400 mm high-strength spiral steel resin anchor bolt was used. The upper spacing was 1000 mm, and the spacing between the bottom angles was 800 mm. For the principal parts, the number and length of anchor bolts can be appropriately increased, with a row spacing of 800 mm.
3) A 50 mm-thick shotcrete layer was applied again. This step could be temporarily delayed. CFSTS was set up; first, the powerful anti-pull mesh was laid behind the support, and gangue bags were arranged between the powerful anti-pull mesh and the anchored mesh as a flexible deformation layer. When the flexible deformation layer reached the deformation limit, shotcrete was applied again.

Figure 2 demonstrates the design of shotcrete and rock bolt supporting+ CFSTS supporting.

The ultimate support resistance provided by the above composite support structure was calculated by the formula proposed by LIU et al. (LIU, GAO, ZHANG, 2017); also, the meaning of specific parameters was obtained by Liu et al. (2017). Finally, the ultimate bearing capacity of the composite support body was obtained to be 1.57 MPa.

$$p_i = \left\{p_s + p_c + \frac{\sigma_s \pi d^2}{4D_s D_d}\left[1 - \left(1 + \frac{t}{R_s + h}\right)^{-\xi}\right] + c \cot \varphi \left[1 - \left(1 + \frac{t}{R_s + h}\right)^{1-\xi}\right]\right\} \cdot \left(1 + \frac{t}{R_s + h}\right)^{\xi-1} \tag{1}$$

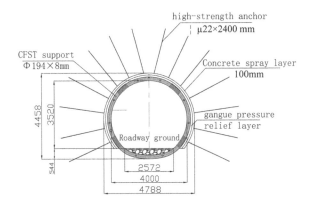

Figure 2. The support section of CFST support and bolt mesh spray.

4.4 *Design of grouting reinforcement in the surrounding rock*

1) Control mechanism of the surrounding rock grouting on the roadway deformation
   The control mechanism of the surrounding rock grouting on the roadway deformation is mainly explained in:
   a) The grouting slurry plays a role in filling and strengthening the fissure surface in the surrounding rock. It improves the physical and mechanical properties of fractured rock mass, increases the static and dynamic elastic modulus of rock mass, and weakens the anisotropy of rock mass, thus guaranteeing the integrity of surrounding rock and subsequently improving the overall bearing capacity of surrounding rock, and providing better surrounding rock conditions for further stabilizing roadway.
   b) Through the surrounding rock grouting, the surrounding rock is re-cemented, and the cohesion C and internal friction angle φ of the rock mass improve, thereby improving the strength and overall strength of the rock mass surrounding the fracture and increasing the bearing capacity of the rock mass in the pressure-bearing ring.
2) Highly efficient surrounding rock grouting technology
   a) Timing for grouting
   The bearing capacity of CFSTS was monitored. When the bearing capacity reached 80% of the designed bearing capacity, the grouting reinforcement was performed on the surrounding rock.
   b) Grouting material
   The cement grout was selected for grouting. Ordinary Portland cement Grade 42.5 was used as the cement, with a water-cement ratio of 0.6–0.7. The $\varphi 6$ mm holes arranged in a twist shape in the middle of $\varphi 22$ mm × 3 mm ordinary steel tube were used as grouting pipe, which was 3 m long, with a row spacing of 1600 mm × 2400 mm.
   c) Grouting control parameters
   The grouting pressure was 2.0–2.5 MPa, and the grouting time was controlled within 120 min. The full-section porous simultaneous grouting technology was used to complete grouting in one or more sections at a time to achieve the effects of fast and highly efficient grouting. The amount of grouting was less than 1 t cement per section.

5 CONSTRUCTION TECHNOLOGY AND DEFORMATION MONITORING FOR THE CFSTS

5.1 *Construction processes of the roadway support*

Roadway support processes are listed as follows: ① Performing roadway wall-caving and roof brushing; ② Applying Initially shotcrete 50 mm thick concrete layer; ③ Performing anchor mesh support; ④ Installing empty steel tubular supports and laying the powerful anti-pull net behind the supports; ⑤ Pouring concrete in steel tubular supports; ⑥ Applying shotcrete layer again; ⑦ Monitoring the bearing capacity of CFSTS to reach 80% of the designed value; ⑧ Performing grouting reinforcement on surrounding rock.

5.2 *Monitoring of the roadway surrounding rock deformation*

After the construction was completed, the convergence deformation of the roof and floor of the roadway and the surrounding rock on both sides were continuously monitored for 90 days, the result of which was shown in Figure 3 below. The monitoring results showed that the deformation of surrounding rock tended to be stable after nearly 50 days of support; in more than three years of roadway support up to now, the accumulated deformation of the support was less than 100 mm, which fully demonstrated that CFSTS had good bearing performance and could meet the requirements of roadway deformation and the long-term stability of roadway; in addition, the overall structure of the support was intact, with no significant damage.

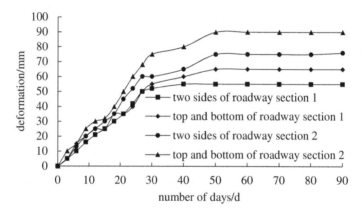

Figure 3. Deformation monitoring figure of the surrounding rock.

6 CONCLUSIONS

1) Under the action of high deep stress on the −650 m horizontal south swing trackage roadway, the surrounding rock deformation of the roadway was largely manifested as characteristics of soft rock, and the original shotcrete and rock bolt supporting +U29 type steel supporting could not meet the requirements for the stability of the roadway.
2) A composite support scheme with CFSTS as the principal part was designed. The cross-sectional shape of the bracket adopted an arc wall semicircle arch. $\varphi 194$ mm×8 mm seamless steel pipe was used for the principal part of the support; and C40 concrete was poured, which was assisted by shotcrete and rock bolt supporting and reinforcement grouting on surrounding rock.
3) The engineering application indicated that the composite support scheme based on CFSTS has a stronger bearing capacity, which can effectively maintain the long-term stability of the surrounding rock of the trackage roadway and the safety and reliability of the support structure.

FOUNDATION ITEM

Innovation and entrepreneurship training program for college students of Linyi University "Study on preparation and mechanical property test of mine pump concrete" (X202210452090).

REFERENCES

Gao Y. F., Wang B., Wang J., et al. Test on Structural Property and Application of Concrete-filled Steel Tube Support of Deep Mine and Soft Rock Roadway [J]. *Chin J. Rock Mech. Eng.*, 2010: 29 (S1) 2604–2609.

Gao Yanfa, Liu Keming, Feng Shaowei, et al. Early Strength Concrete Experiment and Applied Research of Early Strength Concrete-filled Steel Tubular Supports in Extremely Soft Rock Roadways [J]. *Journal of Mining & Safety Engineering*, 2015, 32(4): 537–543.

He Xiaosheng, Liu Keming, Zhang Lei, et al. Structural Design and Application of Concrete-Filled Steel Tube Support at an Extremely Soft Rock Roadway Intersection. *Journal of China Coal Society*, 2015,40(09), 2040–2048.

Huang Wanpeng, Sun Yuanxiang, Chen Shaojie. Theory of Creep Disturbance Effect of Rock and its Application in Support of Deep Dynamic Engineering [J]. *Chinese Journal of Geotechnical Engineering*. 2021, 43(9): 1621–1630.

Kang Yong-Shui, Geng Zhi, Liu Quan-sheng, et al. Research Progress on Support Technology and Methods for Soft Rock with Large Deformation Hazards In China [J]. *Rock and Soil Mechanics*, 2022, 43(8): 1–25.

Kong X. S., Shan R. L., Yuan H. H., et al. Study on the Sustaining Effect of Concrete-Filled Steel Tubular Supports in Deep Mining Roadways[J]. *Arabian Journal of Geosciences*, 2020, 13: 1–14.

Li Guichen, Yang Sen, Sun Yuantian, et al. Research Progress of Roadway Surrounding Strata Rock Control Technologies Under Complex Conditions[J]. *Coal Science and Technology*, 2022, 50(6): 29–45.

Li Nan. Mechanism of Floor Heave in High-stress Soft Rock Roadway and Technology of Synergistic Control of Wall and Floor[J]. *Rock and Soil Mechanics*, 2022, 41(4): 53–58.

Li X. B. Steel Tube Confined Concrete Strength and the Roadway Compression Ring Enhanced Support Theory. Beijing: China University of Mining and Technology-Beijing, 2012.

Liu G. L. *Research on Steel Tube Confined Concrete Supports Capability and Soft Rock Roadway Compression Ring Strengthening Supporting Theory*. Beijing: China University of Mining And Technology-Beijing, 2013.

Liu Keming Gao Yanfa Zhang Fengyin. Composite Supporting Technology of Concrete-filled Steel Tubular Support in Extremely Soft Rock Roadway with Large Sections [J]. *Journal of Mining & Safety Engineering*, 2017, 34(2): 243–250.

Wang J. *Flexural Mechanical Properties Experiment and Application Research of Concrete-Filled Steel Tube Beam and Arch*. Beijing: China University of Mining and Technology-Beijing, 2014.

Wang Jun, Hu Cunchuan, Zuo Jianping, et al. Mechanism of Roadway Floor Heave and Control Technology in the Fault Fracture Zone. *Journal of China Coal Society*, 2019, 44(2): 397–408.

Xia Fang Qian, Wang Jun. Supporting Technology of Flexural Strengthened Concrete-Filled Steel Tube[J]. *Journal of Mining & Safety Engineering*, 2020, 37(3): 490–497.

Study on the effect of pre-treatment of recycled aggregate on the durability of concrete

Baobao Yan*
CCCC Wuhan Harbour Engineering Design and Research Co., Ltd., Wuhan, Hubei, China
CCCC SHEC Wuhan Harbour New Materials Co., Ltd., Macheng, Hubei, China
Hubei Key Laboratory of Advanced Materials & Reinforcement Technology Research for Marine Environment Structures, Wuhan, Hubei, China

Xianan Zhang
Hubei Traffic Engineering Testing Center Co., Ltd., Wuhan, Hubei

Zhouyuan Wang
CCCC Wuhan Harbour Engineering Design and Research Co., Ltd., Wuhan, Hubei, China
CCCC SHEC Wuhan Harbour New Materials Co., Ltd., Macheng, Hubei, China
Hubei Key Laboratory of Advanced Materials & Reinforcement Technology Research for Marine Environment Structures, Wuhan, Hubei, China

Yichun Shi
Ningbo Fubang Highway Engineering Construction Co., Ltd., Ningbo, Zhejiang

ABSTRACT: The effects of pre-saturated water treatment, cement silica fume coating treatment, and polymer styrene-acrylic emulsion treatment of recycled aggregates with three different pre-treatments on the working properties, mechanical properties, and durability of concrete were compared and analyzed. The microhardness of the interface structure of recycled concrete with different pre-treatment methods was tested by a microhardness tester. The results show that the three aggregate pre-treatment methods can improve the performance of concrete, and the grouting treatment and pre-saturated water treatment can improve the compressive strength of concrete; The coating treatment can improve the microhardness of the interface between the old aggregate and the new slurry and reduce the width of the interface transition zone. The improvement effect on the transition zone between the old aggregate and the old slurry is not obvious; Coating treatment can reduce its concrete shrinkage to a certain extent and improve its frost resistance.

1 INTRODUCTION

My country is in a period of rapid development of transportation infrastructure. At the same time, the amount of waste concrete generated by the demolition of old buildings has repeatedly hit new highs (Wang & Wang 2011). Waste concrete consumes a large amount of non-renewable natural sand and gravel resources in my country, and even causes the dilemma of sand and gravel depletion in some areas. The efficient recycling of waste concrete has become an urgent problem for scholars today (Gan et al. 2018).

The waste concrete can be prepared into recycled aggregate. Compared with the natural aggregate, a layer of old mortar is attached to the surface of the recycled aggregate, which leads to a larger number of cracks, porosity, water absorption, and other indicators on the surface of the recycled aggregate. The reason why the old mortar is attached makes a

*Corresponding Author: bby1124@163.com

complex interface structure inside the recycled concrete, resulting in poor performance of the recycled aggregate (Zheng 2018). For this reason, scholars have carried out related research on recycled aggregate. Tsujino et al. (2007) and Spaeth et al. (2014) used polymers such as siloxane and silane to treat recycled aggregate and found that a hydrophobic film was attached to the surface of the aggregate, which reduced the water absorption of the recycled aggregate and improved the compressive strength of concrete. Li Ying et al. (Li et al. 2016) used pozzolan slurry to strengthen recycled aggregate and found that the slurry can directly fill the cracks and pores of recycled aggregate or the pozzolan slurry and the un-hydrated cement particles in the old mortar attached to the aggregate undergo secondary hydration reaction. The product is filled and recycled aggregate pores, thereby reducing its water absorption and porosity; Zhu Yaguang et al. (Zhu et al. 2018) used fungi microbial spray to strengthen recycled aggregate and found that calcium carbonate generated by the microbial precipitation method can fill the pores of recycled aggregate and improve its quality; R. V. Silva et al. (Silva et al. 2015) and Dong et al. (Dong and Chi 2017) used CO_2 to react with calcium hydroxide and calcium silicate hydrate in old mortar to generate carbonate to fill the pores of recycled aggregate and improve its quality. This method can also solidify and store CO_2 gas, which has the advantages of significant environmental benefits. However, there are few related studies on the durability of polymer-treated recycled aggregate concrete.

Given this, this paper comparatively studies the effects of three different pre-treatments, including pre-saturated water, pre-saturated polymer emulsion, and cement silica mortar, on the working properties, strength, and interface properties of concrete. The shrinkage performance and frost resistance performance of recycled concrete were evaluated according to its implementation.

2 RAW MATERIALS AND METHODS FOR TESTING

2.1 *Raw materials*

The cement is Conch P·O42.5 ordinary Portland cement; Silica fume is an off-white powder; The measured density is 2.204 g/cm^3, the SiO_2 content is above 90%, the average particle size is 0.1–0.2 μm, and the specific surface area is 18500 cm^2/g. The fly ash is Class II ash, with a specific surface area of 380 cm^2/g. The main chemical components of the above three materials are shown in Table 1. Coarse aggregate (NS) is limestone with a continuous gradation of 5–20mm. The fine aggregate is river sand, and the fineness modulus is 2.60. The styrene-acrylate copolymer emulsion of the solid content is 50%, and the pH value is 8–9. The superplasticizer is a polycarboxylate superplasticizer with a water reduction rate of 27.9%.

Table 1. The main chemical components of cement silica are fume and fly ash.

Type	Chemical composition (%)							
	SiO_2	Al_2O_3	Fe_2O_3	MgO	CaO	SO_3	Na_2Oeq	Loss
Cement	22.45	5.69	3.33	2.59	60.49	2.83	0.54	2.08
Fly ash	46.20	35.15	6.54	3.78	4.34	1.10	0.69	2.20
Silica fume	94.80	0.92	0.98	0.54	0.68	0.18	0.76	1.14

2.2 *Coarse aggregate treatment method*

There are three main methods of pre-treatment of recycled coarse aggregate:

(1) Pre-saturated water treatment of coarse aggregate (WS): Under normal temperature and pressure, the regenerated coarse aggregate is soaked in water for 24 hours, taken out and wiped with a rag until it is dry.

(2) Treatment of coarse aggregate with polymer styrene-acrylic emulsion (ES): The chemical slurry with styrene-acrylic emulsion should be prepared and watered at a ratio of 1:1 while stirring the aggregate, the chemical slurry evenly should be sprayed on the surface of the aggregate, and then drained. The 3D in indoor natural conditions is conserved for use.

(3) Coarse aggregate (PS) treated with slurry: the recycled aggregate should be soaked in a slurry of cement + 10% silica fume, where the water-binder ratio of the slurry is 1:1, and the aggregate is turned every 2 hours until it is initially Set aside and dry for 24 hours.

2.3 Mixing ratio and test method

Studies have shown that when recycled aggregate replaces natural aggregate by 30%, the mechanical properties are not significantly reduced (Guo et al. 2016). To study the effect of pre-treated recycled aggregate on the performance of concrete, the fixed replacement rate was 30%, and the untreated recycled aggregate (RS) control group was set as the control group. The concrete mix is shown in Table 2.

Table 2. Design of concrete mix ratio (kg/m^3).

Serial number	Cement	Fly ash	Sand	NS	RS	WS	ES	PS	Water	Admixture
1	325	57	824	735	314				145	3.82
2						314				
3							314			
4								314		

Testing Methods of Cement and Concrete for Highway Engineering (JTG3420-2020) were referred to for concrete working performance and mechanical performance tests. The shrinkage and frost resistance of recycled concrete was evaluated by a non-contact shrinkage test and rapid freeze-thaw cycle test.

The microhardness test specimens are concrete slices. First, the pre-treated recycled coarse aggregates with a particle size of 10–20 mm are selected. The coarse aggregate in the concrete should be removed to obtain a mortar with the same proportion, be filled into a 100mm×100mm×100mm cube mould, and with a filling height of 50mm, and then the saturated surface-dried coarse aggregate should be placed on the mortar, and be filled up. The mortar was vibrated together with the coarse aggregate to form, cured for 28 days, and sliced, as shown in Figure 1(a). The surface should be polished and ground to be tested to meet the microhardness test requirements and the ITZ should be found under a

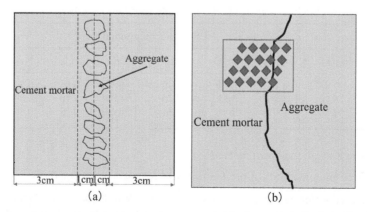

Figure 1. (a) Schematic diagram of microhardness test piece forming; (b) Dotting method of a microhardness test.

microscope of × 100 times to prevent the mutual influence between the indentations and select the inclined matrix dots. The distance between two adjacent points in the horizontal direction is 10μm, the load is 0.05kg, the action time is 5s, and the dotting method is shown in Figure 1(b).

3 RESULTS AND DISCUSSION

3.1 *Work performance*

Compared with the untreated recycled aggregate concrete, the slump of the strengthened recycled aggregate concrete showed an increasing trend, in which the PS and WS slumps reached 190 mm and 180 mm, respectively, while the ES slump was 175 mm. The results show that the three strengthening treatments can improve the slump of concrete, among which the slurry-coated treatment is the most effective, the pre-saturated water treatment is the second, and the polymer modification treatment has the worst effect on improving the slump of concrete. After the recycled coarse aggregate is treated with slurry, the cement slurry closes the pores and cracks of the recycled aggregate, which greatly reduces the water absorption rate of the recycled aggregate. The slurry treatment greatly improved the slump of the concrete, so PS showed the maximum slump. The pre-saturated water treatment can remove the old mortar attached to the surface of the recycled aggregate, and reduce the water absorption and roughness of the recycled aggregate, so the slump of the concrete prepared by the pre-saturated water treatment of the recycled aggregate is also improved accordingly. After the recycled aggregate is modified by polymer, a hydrophobic film is formed on the surface of the aggregate, which plays the role of sealing and blocking the pores on the surface of the recycled aggregate, reducing the water absorption rate of the recycled aggregate, and improving the slump of the mixture.

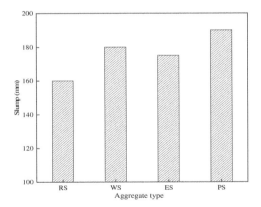

Figure 2. Influence of aggregate pre-treatment methods on a concrete slump.

3.2 *Mechanical properties*

Figure 3 shows the test results of the effect of recycled fine aggregate reinforcement on the compressive strength of concrete. It can be seen from Figure 3 that, compared with RS, the compressive strength of WS and PS is improved, while the compressive strength of ES shows a slight decrease. It is shown that the compressive strength of mortar can be improved by slurry coating treatment and pre-saturated water treatment, and the effect of pre-saturated water treatment is better than that of slurry coating treatment, while polymer modification treatment hurts the strength of mortar. After the recycled aggregate is treated with slurry, a layer of cement slurry is pre-coated on the surface of the aggregate to reduce the water absorption and porosity of the recycled aggregate. In addition, the activation effect of mineral admixtures, the pozzolanic effect, and the filling effect of micro-aggregates in the cement slurry all reduce the porosity of the mortar and improve the interface structure of the mortar. Therefore, the compressive strength of the mortar prepared by wrapping the recycled aggregate with the recycled aggregate is higher than that of the benchmark mortar. Pre-saturated water treatment can remove the old mortar attached to the recycled aggregate so that the quality of the recycled aggregate is the same as that of the untreated recycled aggregate. The compressive strength of WS is close to that of the RS control group and higher than that of ES. After the regenerated aggregate was modified by polymer, a layer of polymer film adhered to the surface, which weakened the bond between the aggregate and the slurry, so the compressive strength of the ES group was slightly lower than that of the benchmark mortar.

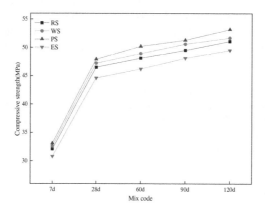

Figure 3. Influence of aggregate pre-treatment methods on compressive strength.

3.3 Interface features

A layer of old mortar is attached to the surface of the recycled aggregate because the existence of this layer of mortar makes the recycled concrete have multiple interfaces. Compared with conventional concrete, recycled concrete has three different interface transition zones surrounding the recycled aggregate, namely the old aggregate-new slurry interface (LG-XJ), the old aggregate-old slurry interface (LG-LJ), and old slurry new slurry interface (LJ-XJ). Figure 4 shows the microhardness test results of three interface transition zones of recycled aggregate concrete with different pre-treatments.

It can be seen from Figure 4 that, taking the LG-XJ interface formed by RS and slurry as an example, the microhardness of the LG-XJ interface formed by untreated RS and slurry is $88.4 kgf/mm^2 \sim 134.6 kgf/mm^2$. The microhardness of the LG-XJ interface formed by the slurry-coated recycled aggregate and slurry is $97.5 kgf/mm^2 \sim 143.5 kgf/mm^2$. The microhardness of the LG-XJ interface formed by pre-saturated water treatment of recycled aggregate and slurry is $89.2 kgf/mm^2 \sim 136.7 kgf/mm^2$. The microhardness of the LG-XJ interface formed by the polymer-modified recycled aggregate and the slurry was $94.8 kgf/mm^2 \sim 141.5 kgf/mm^2$. The width of the interfacial transition zone of the pulping treatment is the smallest, the width of the untreated RS is the largest, and the width of the pre-saturated water treatment is similar to that of the untreated RS. The results show that the reinforcement treatment of recycled aggregate can improve the interface structure. The improvement effect of the slurry treatment is the best, the effect of the polymer modification treatment is the second, and the pre-saturated water treatment is the worst. It can be seen from Figure 4(b) that after the pre-treatment of recycled aggregates, the difference in microhardness of the LG-LJ interface transition zone is significantly smaller than that of LG-XJ, and the width of the LG-LJ interface transition zone

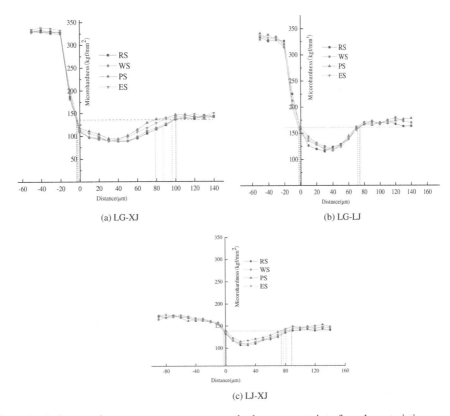

Figure 4. Influence of aggregate pre-treatment method on concrete interface characteristics.

is similar. It shows that the strengthening treatment of recycled aggregate only acts on the surface of the aggregate, the depth of the strengthening treatment is shallow, and the effect on the strengthening effect of the original interface transition zone of the recycled aggregate is weak. It can be seen from Figure 4 (c) that the microhardness of recycled concrete is PS > ES > WS > RS and the width of the interface transition zone is PS < ES < WS < RS. The test results are similar to those of the LG-XJ interface. The cement-silica slurry can be attached to the surface of the old recycled aggregate mortar, filling the surface pores to improve the interface structure with the new slurry. Comparing Figure 4 (a) to Figure 4 (c), it can be seen that the microhardness of the three types of interface transition zones formed by the same type of recycled aggregate and cement paste is as follows: LG-LJ interface is the best, LJ-XJ interface is the second; On the other hand, the LG-XJ interface is the worst, the LG-LJ interface has a high degree of hydration over time, and the pre-treatment has a better effect on the LJ-XJ interface.

3.4 *Shrinkage performance*

Figure 5 shows the test results of the effect of recycled fine aggregate reinforcement on the shrinkage of recycled concrete. It can be seen from Figure 5 that the shrinkage rate of WS, PS and ES specimens cured for 120d was 1.2%, 6.4%, and 4.5%, which is lower than that of RS, respectively. The results show that the strengthening treatment of recycled aggregate can reduce the shrinkage rate of mortar specimens. After pulping and polymer strengthening treatment, the water absorption rate of recycled aggregate and the water content of recycled concrete decrease, so the shrinkage rate of PS and ES decreases. The water absorption rate of recycled aggregate treated with pulping is lower than that of polymer treatment, so the

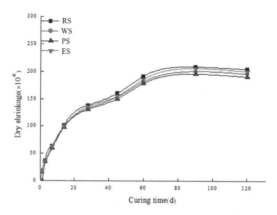

Figure 5. Effect of bone pre-treatment on shrinkage properties of concrete.

shrinkage rate of the pulping treatment is small. After the recycled aggregate is coated with slurry and polymer modified, a dense film is formed on the surface of the recycled aggregate, which reduces the water absorption rate. After the recycled aggregate is treated with pre-saturated water, the surface laitance falls off, and the old slurry absorbs sufficient water. The slow release, to a certain extent, supplements the loss of moisture from the new slurry, slows down the shrinkage of the concrete and reduces the shrinkage rate of the concrete.

3.5 *Antifreeze performance*

Figures 6 (a) and (b) show the experimental results of the effect of recycled aggregate treatment on concrete mass loss rate and compressive strength loss rate, respectively. It can be seen from Figure 6 that the mass loss rate of the strengthened recycled aggregate concrete is ranked from small to large: PS, WS, ES. After 150 freeze-thaw cycles, the compressive strength loss rate of RS was 13.22%, and the compressive strength loss rates of PS, WS, and ES were 8.23%, 9.21%, and 10.12%, respectively. Since the replacement rate of recycled coarse aggregate in this test group is 30%, the effect of recycled aggregate on the loss of concrete quality and compressive strength is not very significant. The results show that the pre-saturated water treatment has the best effect on improving the frost resistance of concrete, followed by the grouting treatment, and the polymer modification treatment is the worst. Slurry treatment greatly reduces the porosity of recycled aggregate and thus reduces

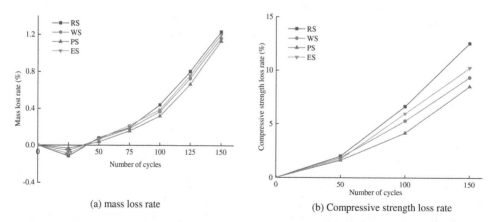

Figure 6. The influence of the treatment method of recycled coarse aggregate.

its water absorption. The amount of water immersed in the concrete is less, and the replacement rate of recycled coarse aggregate is 30%. Therefore, after the first 25 freeze-thaw cycles, the mass loss rate is slightly reduced. The pre-saturated water treatment method can remove the laitance attached to the aggregate, reduce the water absorption rate of the aggregate, and reduce the frost heave effect of the concrete, etc. Therefore, the loss rate of mass loss and compressive strength of PS specimens are the lowest. After the polymer modification treatment, the surface of the recycled aggregate is wrapped with a layer of polymer film, which reduces the water absorption rate of the recycled aggregate, thereby reducing the water absorption quality of the specimen and improving the frost resistance of the specimen.

4 CONCLUSION

(1) All three recycled aggregate pre-treatment methods can improve the concrete work performance, among which the recycled aggregate slurry treatment improves the concrete work performance the best; Both grouting treatment and pre-saturated water treatment can improve the compressive strength of recycled concrete, and polymer treatment has a certain deterioration effect on its concrete strength.
(2) Both recycled aggregate treatment and polymer treatment can improve the microhardness and reduce the interface width of the interface between the old aggregate and the new slurry and the interface between the old slurry and the new slurry. Pre-saturated water treatment has a weak effect on improving the interface transition zone of recycled concrete; recycled aggregate pre-treatment has no obvious effect on improving the interface between old aggregate and old slurry of concrete.
(3) Pre-treatment of recycled aggregate can reduce its concrete shrinkage, and compared with untreated recycled aggregate, slurry-coated aggregate has the greatest reduction in shrinkage; After 150 rapid freeze-thaw cycles, the mass loss and compressive strength loss of the recycled aggregate concrete treated with slurry were the lowest.

REFERENCES

Dong X.X., B J.Z., Chi S.P. The Durability of recycled Aggregate Concrete Prepared with Carbonated Recycled Concrete Aggregates. *Cement and Concrete Composites* 84 (2017) 214–221.

Gan F., Zhou M.X., Pu Q. Research Progress on the Properties of Recycled Coarse Aggregate and Recycled Concrete [J]. *Concrete and Cement Products*, 2018(09): 102–107.

Guo Z.G., Chen C., Fan B.J., et al. Experimental Study on Basic Mechanical Properties of Recycled Coarse and Fine Aggregate Concrete [J]. *Journal of Building Structures*, 2016, 37(S2): 94–102.

Li Y, Dai D.H., Yu H.F. Research on the Effect of Recycled Aggregate Strengthening Technology on the Performance of Recycled Concrete [J]. *Journal of Qinghai University (Natural Science Edition)*, 2016, 34(02): 1–4.

Silva R.V., Neves R., de Brito J., Dhir R.K. Carbonation Behaviour of Recycled Aggregate Concrete. *Cement & Concrete Composites* 62 (2015) 22–32.

Spaeth, V., Djerba Tegguer, A. Improvement of recycled Concrete Aggregate Properties by polymer Treatments. *Int. J. Sustain. Built Environ.* 2 (2014), 143–152.

Tsujino, M., Noguchi, T., Tamura, M. Application of Conventionally Recycled Coarse Aggregate to the Concrete Structure by Surface Modification Treatment. *J. Adv. Concr. Technol.* 5 (2007), 13–25.

Wang R.M., Wang LX. Analysis and Development Prospect of Construction Waste in China [J]. *China Urban Economy*, 2011, (5): 178–179.

Zheng J. Effects of Mortar Adhesion Rate and Mortar Adhesion Strength on the Properties of Recycled Aggregate Concrete [J]. *Commercial Concrete*, 2018(09): 1–2.

Zhu Y.G., Wu C.R., Wu Y.K., et al. Research Progress on Improving the Properties of Recycled Aggregate by Microbial Mineralization Deposition [J]. *Concrete*, 2018(07): 88–92.

Research on RAP dispersion in recycled asphalt concrete with steel slag

Peng Guo
Wuhan Municipal Road & Bridge Co. Ltd., Wuhan, China

Feiyi Liu
Wuhan Hanyang Municipal Construction Group Co. Ltd., Wuhan, China

Ruonan Pang & Sifa Fang
Wuhan Municipal Road & Bridge Co. Ltd., Wuhan, China

Fan Shen*
Wuhan Institute of Technology Materials Science and Engineering, Wuhan, China

ABSTRACT: The recycled asphalt pavement material RAP is affected by the aging asphalt wrapped on the aggregate surface during the hot mixing process, and cannot be completely dispersed during the mixing process, resulting in the difference between the actual gradation and the design gradation, which affects the accuracy of mineral aggregate gradation. To solve this problem, this paper studies the influence of the direct screening method and combustion furnace method on the gradation difference of RAP with different particle sizes. And through the orthogonal test, the influence law of different combinations of dispersion coefficients on the voids of recycled asphalt mixture is studied. On this basis, the best dispersion coefficient combination is used to modify RAP gradation to prepare the asphalt mixture, then the road performance of the asphalt mixture is measured and evaluated. The research results show that the gradation results obtained by the direct screening method and combustion furnace method are quite different, and RAP with different particle sizes will produce gradation variation in different situations. The orthogonal test results show that the Type III square of RAP-2 (0 ~ 5mm) is smaller, which is 1.327, indicating that the gradation composition of RAP in this particle size range is a more important factor affecting the gradation design of the mixture. The best-modified dispersion coefficients of RAP-1 (5 ~ 10mm) and RAP-2 are 0.4 and 0.8 respectively. The porosity, Marshall stability, dynamic stability, and low-temperature bending strain of recycled asphalt mixture prepared by the combination of these dispersion coefficients are 3.8%, 14.81kN, 5650 times/mm, and 2911.8 με respectively.

1 INTRODUCTION

By the end of 2020, the total highway mileage in China has reached 5.1981 million kilometers, and the highway maintenance mileage accounts for 99.0% of the total mileage (Finance 2020). In the maintenance project of asphalt pavement, a lot of RAP (recycled asphalt pavement) is produced. To avoid resource waste and RAP pollution, RAP is widely used in asphalt pavement recycling technology (Wang et al. 2019). Asphalt pavement recycling technology is to remix RAP that is crushed, screened, and mixed with regenerant,

---

*Corresponding Author: shenf@wit.edu.cn

new asphalt, and new aggregate into recycled asphalt mixture (Baskandi Deepak 2017; Kengo Akatsu et al. 2018). Now the asphalt pavement regeneration technology mainly adopts plant mixed hot regeneration (Zhou 2020). In the production of recycled asphalt mixture, the gradation design generally adopts two methods. The first method is the direct screening method, which regards RAP as "black aggregate", ignores the impact of RAP on the structure and performance of recycled asphalt mixture, and adopts *Test Methods of Aggregate for Highway Engineering* to screen RAP to design the gradation (Jamshidi et al. 2019; Liu 2016; Wei et al. 2017). The second method is the combustion furnace method, which burns the asphalt on the RAP surface completely before screening and gradation design (Liu 2020; Wang et al. 2019). However, due to the problems of binding agglomeration, uneven thickness of asphalt film, and different aging degrees of surface asphalt in RAP, it is difficult to determine the dispersion degree of RAP in the process of thermal regeneration. In addition, there is a mixing effect between new asphalt and old asphalt in the recycled asphalt mixture. Therefore, the gradation of recycled asphalt mixture designed by the above two methods is quite different from the actual gradation, which has a serious impact on the structure and performance of recycled asphalt concrete (Zhang & Tao 2018).

In conclusion, the research on the dispersion degree of RAP in recycled asphalt mixture is conducive to optimizing the mixture gradation and improving the road performance and durability of asphalt pavement. Therefore, based on the gradation difference analysis results of the direct screening method and the combustion furnace method, this paper designs different discrete coefficients, and uses the orthogonal test and variance analysis method to study the change law of the porosity of the recycled asphalt mixture mixed with steel slag under different discrete coefficient combinations to obtain the optimal discrete coefficient combination and the modified RAP gradation which get access to the real gradation, Then the performance of recycled asphalt mixture mixed with steel slag under different dispersion conditions is also studied, and the gradation correction method of RAP is evaluated.

2 RAW MATERIALS AND TEST PLAN

2.1 *Raw materials*

In this paper, I-D modified asphalt is used as asphalt, hot disintegration steel slag (with low activity of hot disintegration steel slag) and natural limestone are used as aggregate, limestone mineral powder is used as mineral powder, polyester fiber is used as fiber, and RAP adopts pavement milling materials of an old road section in Jiangxia District, Wuhan City, Hubei Province. Among them, the technical indicators of steel slag meet the technical requirements of JT/T 1086-2016 *Steel Slag Used in Asphalt Mixture* (JT/T1086-2016), the technical indicators of limestone aggregate and limestone mineral powder meet the technical requirements of JTG F40-2004 *Technical Specifications for Construction of Highway Asphalt Pavements* (JTG F40-2004), and the technical indicators of polyester fiber meet the technical requirements of JTT 533-2020 *Fiber for Asphalt Pavements* (JTT 533-2020). According to JTG E20-2011 *Standard Test Methods of Bitumen and Bituminous Mixtures for Highway Engineering* (JTG E20-2011), the effective asphalt content in RAP is 4.9% measured by the combustion furnace method. The technical indexes of RAP, aged asphalt in RAP, and steel slag are shown in Tables 1 and 2.

Table 1. Technical indexes of aged asphalt in RAP.

	Usage time/a	Softening point/°C	Penetration degree/ 0.1 mm	Penetration ratio/%	15°C extensibility/ cm	135°C viscosity/ Pa·s
Aged Asphalt	15	76	31	74	38	1349

Table 2. Technical index of steel slag.

	Needle and plate particle content/%	Adhesion	Immersion expansion rate/%	Abrasion value/%/	Crush value/%	Water absorption/%	Apparent relative density/ g/cm^3
Specification requirement	(Aggregate ≥ 9.5 mm) ≤ 12	≥ 5	≤ 1.8	≤ 33	≤ 22	≤ 3	–
Experimental result	8.3	5	1.5	21.5	15.4	1.5	3.189

2.2 Test plans

2.2.1 Analysis of RAP gradation difference

RAP-1 (5-10 mm) and RAP-2 (0–5 mm) are screened by direct screening method and combustion furnace method, and then the influence of different screening methods on rap gradation is analyzed by comparing the screening residue.

2.2.2 RAP gradation correction and dispersion analysis

Different dispersion coefficients (0.2, 0.4, 0.6, 0.8) are used to modify RAP gradation, and the modified gradation curve is analyzed.

RAP-1 and RAP-2 are taken as test factors. Different dispersion coefficients are taken as factor levels to design and carry out orthogonal tests, the difference between the voids of the experimental group and the benchmark group is taken as the basis of variance analysis, and the variance analysis is carried out to study the influence of different dispersion coefficients on the dispersion degree of RAP.

2.2.3 Performance evaluation of recycled asphalt mixture with steel slag

Steel slag and RAP are used to prepare recycled asphalt mixture with steel slag. In the mixture, the content of RAP-1 is 23.0% of the mass of mineral aggregate, the content of RAP-2 is 7.0% of the mass of mineral aggregate, and the oil-stone ratio is 5.9%. Marshall specimens are prepared to study the voids, strength, water stability, and high and low-temperature performance of the specimens.

Steel slag and natural aggregate are used to prepare the benchmark group of asphalt mixture. In this group of the mixture, the amount of 5~10mm aggregate is 23.0% of the mass of mineral aggregate, the amount of 0~5mm aggregate is 7.0% of the mass of mineral aggregate, and the oil stone ratio is 5.9%. Marshall specimens are prepared, and the voids of the specimens are measured to be 3.9%. The aggregate gradation of the benchmark group and recycled asphalt mixture with steel slag is shown in Table 3.

Table 3. Gradation of the asphalt mixtures.

	Volume passing percentage/%							
Sieve mesh/mm	13.2	9.5	4.75	2.36	1.18	0.6	0.3	0.15	0.075
Synthetic gradation	100.0	99.7	43.9	26.1	20.1	15.7	13.2	11.6	10.3
Median gradation	100.0	95.0	44.0	26.0	20.0	17.0	14.0	12.5	10.5

2.3 Test methods

2.3.1 Determination method of gradation

According to JTG E42-2005 *Test Methods of Aggregate for Highway Engineering* (JTG E42-2005), the gradation of natural aggregate, steel slag, and RAP is determined by direct screening method; According to JTG E20-2011 test code for asphalt and asphalt mixture of highway engineering (JTG E20-2011), RAP gradation is determined by combustion furnace method.

2.3.2 *Mix design*

In this paper, the discrete equation is introduced to modify two gradations of RAP. The discrete equation is shown in Equation 1.

$$\gamma_{ni} = ni - \varepsilon \times (ni - Ni) \qquad (1)$$

In the formula, γ_{ni} is the passing rate of each sieve of RAP-n modified by the corresponding dispersion coefficient, ni is the passing rate of RAP-n sieve obtained by combustion furnace method, Ni is the passing rate of RAP-n sieves obtained by direct screening method; ε is discrete coefficient with values of 0.2, 0.4, 0.6 and 0.8 respectively. Referring to the treatment method of Niu Zhe in Southeast University for steel slag in gradation design, the mass percentage passing rate of 0–5 mm and 5–10 mm steel slag is replaced by the volume percentage passing rate, then the RAP-1 dosage is fixed as 23.0% of the mineral mass, and the RAP-2 dosage is 7.0% of the mineral mass. The mix design of steel slag recycled asphalt mixture shall be carried out by using the modified steel slag gradation and the modified RAP gradation by JTG F40-2004 *Technical Specifications for Construction of Highway Asphalt Pavements* (JTG F40-2004).

The content of 0–5 mm natural aggregate in the mineral aggregate is fixed at 7%, and the content of 5–10 mm natural aggregate is fixed at 23%. The mix design of the benchmark group asphalt mixture is carried out by using the natural aggregate gradation and the modified steel slag gradation by JTG F40-2004 *Technical Specifications for Construction of Highway Asphalt Pavements* (JTG F40-2004).

2.3.3 *Preparation of asphalt mixture*

According to JTG F41-2008 *Technical Specifications for Highway Asphalt Pavement Recycling*, the benchmark group asphalt mixture and recycled asphalt mixture with steel slag are prepared by using the standard of hot recycling in the plant (JTG F41-2008).

2.3.4 *Performance evaluation*

According to JTG E20-2011 *Standard Test Methods of Bitumen and Bituminous Mixtures for Highway Engineering*, the voids ratio, Marshall stability, immersion residual stability, dynamic stability, and low temperature bending strain of the benchmark group asphalt mixture and recycled asphalt mixture with steel slag are determined (JTG E20-2011).

3 RESULTS ANALYSIS

3.1 *RAP gradation difference analysis*

The influence of the direct screening method and combustion furnace method on RAP gradation is tested. The research results are shown in Figures 1 and 2.

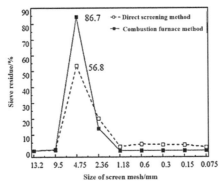

Figure 1. Screening results of RAP-1 by combustion furnace method and direct screening method.

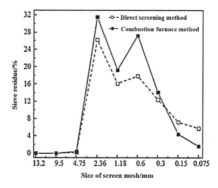

Figure 2. Screening results of RAP-2 by combustion furnace method and direct screening method.

The results show that RAP gradation results obtained by using the direct screening method and combustion furnace method are quite different. Compared with the combustion furnace method, the direct screening method has a higher proportion of coarse aggregate and less proportion of fine aggregate. For RAP with different particle sizes, the residue of RAP-1 with 4.75 mm sieve is 29.9% higher than that of the combustion furnace method, while the residue of RAP-2 with 2.36 mm, 1.18 mm, and 0.6 mm sieve is higher than that of the combustion furnace method, and the residue of RAP-1 with 0.15 mm and 0.075 mm sieve is lower than that of the combustion furnace method. During the milling and crushing process of RAP, the coating of asphalt film and the milling and crushing process jointly affect the gradation of RAP, and some independent block RAP is composed of several stones, which makes the coarse aggregate account for a high proportion of the screening results of the direct screening method. While the combustion furnace method burns all the asphalt to make the aggregate completely dispersed, so the screening results will be more "refined". In addition, RAP with different particle size ranges will produce gradation variation in different situations. The gradation difference of RAP-1 is concentrated in the large-size sieve (such as 4.75 mm), while the gradation difference of RAP-2 is reflected in both the large-size sieve and the small-size sieve. Therefore, when studying the dispersion of RAP in the hot mixing process, we must study the RAP with different particle size ranges.

3.2 RAP gradation correction and gradation dispersion analysis

Different dispersion coefficients are used to correct the gradation. The research results are shown in Figures 3 and 4, Tables 4–6.

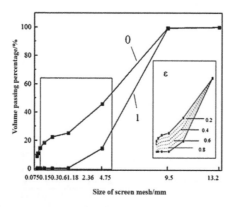

Figure 3. Modified grading curve of RAP-1.

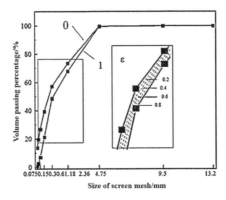

Figure 4. Modified grading curve of RAP-2.

Table 4. Orthogonal experiment results.

No.	RAP-1	RAP-2	Blank column	Voids difference/%		
				Experiment group a	Experiment group b	Experiment group c
1	1	1	1	1.10	1.15	1.20
2	1	2	2	2.00	1.80	1.60
3	1	3	3	0.40	0.70	1.00
4	1	4	4	1.00	0.80	0.70
5	2	1	4	0.80	0.70	0.60
6	2	2	3	0.40	0.41	0.40
7	2	3	2	0.80	0.75	0.70
8	2	4	1	0.30	0.10	0.00
9	3	1	2	0.60	0.62	0.60
10	3	2	1	0.90	0.93	0.90
11	3	3	4	1.50	1.30	1.20
12	3	4	3	0.60	0.90	0.70
13	4	1	3	0.60	0.40	0.10
14	4	2	4	1.20	1.10	0.90
15	4	3	1	0.60	0.40	0.10
16	4	4	2	0.60	0.72	0.70

Table 5. Between-subjects effect test (Dependent variable: VV).

	The square sum of type III	Degree of freedom	Mean square	F	Significance
Modified model	4.159[a]	6	0.693	6.671	0.000
RAP-1	2.832	3	0.944	9.084	0.000
RAP-2	1.327	3	0.442	4.257	0.010
Error	4.260	41	0.104	–	–
Total	37.841	48	–	–	–
Total after correction	8.419	47	–	–	–

Notes: [a]$R^2 = 0.494$ (adjusted $R^2 = 0.420$).

Table 6. RAP estimated marginal mean (Dependent variable: VV).

	Dispersion coefficient	Average value	Standard error	95% Confidence interval	
				Lower limit	Upper limit
RAP-1	0.20	1.121	0.093	0.933	1.309
	0.40	0.497	0.093	0.309	0.685
	0.60	0.896	0.093	0.708	1.084
	0.80	0.618	0.093	0.430	0.806
RAP-2	0.20	0.706	0.093	0.518	0.894
	0.40	1.045	0.093	0.857	1.233
	0.60	0.788	0.093	0.600	0.975
	0.80	0.593	0.093	0.405	0.781

The results show that different discrete coefficients have different correction results for RAP-1 and RAP-2. The gradation difference of RAP-1 is relatively large, and the larger difference is shown in the 4.75 mm sieve, with a difference of more than 20%; However, the gradation difference of RAP-2 is relatively small. The larger difference is shown on the 0.3 mm, 0.15 mm, and 0.075 mm sieves. Compared with RAP-1, the difference is relatively small, which is 10% ~ 20%. It shows that the dispersion coefficient has different degrees of correction for the gradation of aggregates with different particle sizes. After the aggregates with different particle sizes are corrected by different dispersion coefficients, the aggregate dispersion represented by the gradation curve is also different.

The results show that the significance of RAP-1 and RAP-2 is less than 0.050, which indicates that there is a difference between RAP-1 and RAP-2, and the dispersion degree of RAP-1 and RAP-2 is different at the same mixing temperature. It can indicate that aggregates with different particle size ranges cannot be corrected directly with the same dispersion coefficient. In this paper, the difference between the void ratio of the specimen and the void ratio of the reference group is taken as the evaluation standard of variance analysis, which represents the accuracy of the gradation design. And the smaller the void ratio difference is, the more accurate the gradation is. The square sum of Type III of RAP-2 is 1.327, and the square sum of Type III of RAP-1 is 2.382. The square sum of Type III of RAP-2 is small, which can indicate that RAP-2 has a more significant impact on the results, and RAP-2 gradation composition has a greater impact on the mixture gradation composition. It can be known from 2.1 that the particle size of RAP-2 is 0 ~ 5 mm, and its gradation difference is shown on the 2.36 mm and 0.075 sieves. The amount of fine aggregate controlled by these two sieve holes is the key to the formation of the void ratio of the mixture. If the gradation design is carried out by using the screening results obtained by the direct screening method, the results of the direct screening method are "too coarse", so excessive steel slag will be added to the mix design of the mixture. In addition, because RAP is not completely non-dispersible, the aggregates will be dispersed during the hot mixing process, resulting in more fine aggregates and smaller voids in the mixture. Similarly, if the screening results obtained by the combustion furnace method are directly used for gradation design, the results of the combustion furnace method are "too fine", which makes the amount of fine aggregate of steel slag too small. RAP cannot be completely dispersed, resulting in less fine aggregate and a larger void ratio of the mixture.

The smaller the voids difference is, the more accurate the gradation design is. As can be seen from Table 6, when the modified dispersion coefficient of RAP-1 is 0.4, the estimated marginal value is 0.497, and when the modified dispersion coefficient of RAP-2 is 0.8, the estimated marginal value is 0.593. The experimental results using the combination of these discrete coefficients show that the void difference reaches the minimum, indicating that using these two discrete coefficients to modify the gradation of RAP-1 and RAP-2 respectively can minimize the gradation difference of RAP. The modified RAP gradation by using the combination of dispersion coefficients is closest to the real dispersion state of the mixture under this hot mixing condition.

3.3 *Performance evaluation of recycled asphalt mixture with steel slag*

The optimum dispersion coefficient combination is used to modify RAP gradation, prepare steel slag mixed recycled asphalt mixture, and measure its Marshall stability, immersion residual stability, dynamic stability, and low temperature bending strain. The research results are shown in Table 7.

Table 7. Main performance parameters of recycled asphalt mixture with steel slag.

	Marshall stability/kN	Voids ratio/%	Dynamic stability/(times/mm)	Low temperature bending strain/με
Direct screening method	12.26	2.9	3233	3300.9
Modification method	14.81	3.8	5650	2911.8
Combustion furnace method	15.79	4.9	6531	2144.7

The research results show that among the three groups of asphalt mixtures, the asphalt mixture prepared with RAP gradation by direct screening has the lowest Marshall stability, voids ratio, and dynamic stability, which are 12.26kN, 2.9% and 3233 times /mm respectively, and it has the highest low temperature bending strain, which is 3300.9 με. However, the asphalt mixture prepared with RAP gradation by combustion furnace method has the highest Marshall stability, voids ratio, and dynamic stability, which are 15.79kN, 4.9%, and 6531 times/mm respectively, and it has the lowest low temperature bending strain, which is 2144.7 με. In the case of the same composite gradation and asphalt aggregate ratio, the voids difference between the asphalt mixture prepared by the modified method and the benchmark group is only 0.1%, indicating that the RAP gradation obtained by the modified method is the closest to the real dispersion state of RAP, so the gradation design is the most accurate. In the case of the same composite gradation and asphalt aggregate ratio, the voids ratio of the experimental group adopting the direct screening method is lower, 0.9% lower than that of the control group, and 1.1% higher than that of the combustion method, which is consistent with the analysis in 2.2 that the dispersion degree of RAP affects the voids of the mixture. The high-temperature performance and low-temperature performance of the three groups of mixtures are different. It is because there are too many fine aggregates in this group of mixtures when the direct screening method is used for gradation design. Since the high-temperature rutting resistance of asphalt mixture largely depends on the intercalation between aggregates, the reduction of coarse aggregates weakens the intercalation between coarse aggregates, and the strength and high-temperature performance of the mixture decline. In addition, due to a large amount of fine aggregate, the specific surface area of mineral aggregate is increased, and the adhesion between mineral aggregate and asphalt is improved, so the shrinkage of the mixture is small, and the low-temperature bending strain is higher than the other two groups. On the contrary, in the mixture of the combustion furnace test group, the fine aggregate is less, the voids ratio is higher, and the proportion of coarse aggregate in the mineral aggregate is increased, thus strengthening the impaction effect of coarse aggregate. Therefore, its Marshall stability and dynamic stability are higher than the other two groups, and the low-temperature bending strain is lower than the other two groups.

4 CONCLUSIONS

The conclusions are as follows:

1) The RAP gradation results obtained by the direct screening method and the combustion furnace method are quite different. Compared with the combustion furnace method, the coarse aggregate accounts for a higher proportion, and the fine aggregate accounts for a smaller proportion of the aggregates obtained by the direct screening method.
2) RAP with different particle sizes will produce gradation variation in different situations. The gradation difference of RAP-1 (5–10mm) is concentrated on the large-size sieve, while that of RAP-2 (0–5mm) is concentrated on both the large-size sieve and small-size sieve.
3) The orthogonal test results show that the square sum of Type III of RAP-2 is 1.327. Compared with RAP-1, the gradation composition of RAP-2 has a greater impact on the gradation composition of the mixture, and its gradation accuracy affects the voids of the asphalt mixture.
4) The modified method can improve the accuracy of the mixture gradation design. The asphalt mixture prepared by the modified method in this paper has balanced strength, water stability, high-temperature performance, and low-temperature performance. Its Marshall stability, dynamic stability, and low temperature bending strain are 14.81kn, 5650 times/mm, and 2911.8 μ respectively ε, and the difference between the void's ratio of the asphalt mixture and the voids of the benchmark group is only 0.1%.

REFERENCES

Baskandi Deepak. Bituminous Pavement Recycling – Effective Utilization of Depleting Non-Renewable Resources[J]. *The International Journal of Engineering and Science*, 2017, 6(03).

Jamshidi A., White G., Hosseinpour M, et al. Characterization of effects of reclaimed asphalt pavement (RAP) source and content on the dynamic modulus of hot mix asphalt concrete[J]. *Construction and Building Materials*, 2019, 217(AUG.30):487–497.

Kengo Akatsu, Yousuke Kanou, Shouichi Akiba. Separation Recycling Technology for Restoring Reclaimed Asphalt Pavement[J]. *Journal of JSCE*, 2018, 6(1).

Liu Liang. *Surface Characteristics and Performance Improvement of RAP* [D]. Chang'an University, 2016.

Liu Long. Comparative Analysis of Asphalt Content Determination by Centrifugal Separation Method and Combustion Furnace Method [J]. *Anhui Architecture*, 2020, 27(11).

Liu Yanqiang, Zhang Jie. The Experimental Study on the Determination of the Asphalt Content Correction Coefficient by the Combustion Furnace Method [J]. *Shanxi Science & Technology of Communications*, 2019(01): 28–30.

Research institute of highway ministry of transport. JT/T1086-2016, Steel Slag used in Asphalt Mixture [M]. China communication press, 2017.

Research institute of highway ministry of transport. JTG F40-2004 Technical Specifications for Construction of Highway Asphalt Pavements [M]. China communication press, 2009.

Research institute of highway ministry of transport. JTT 533-2020 Fiber for Asphalt Pavements [M]. China communication press, 2020.

Research institute of highway ministry of transport. JTG E20-2011 Standard Test Methods of Bitumen and Bituminous Mixtures for Highway Engineering [M]. China communication press, 2011.

Research institute of highway ministry of transport. JTG E42-2005 Test Methods of Aggregate for Highway Engineering [M]. China communication press, 2005.

Research institute of highway ministry of transport. JTG F41-2008 Technical Specifications for Highway Asphalt Pavement Recycling [M]. China communication press, 2008.

Statistics Bulletin of Transportation Industry Development in 2020 [J]. *Finance & Accounting for Communications*.

Wang Xuelian, Hu Lin, Huang Xiaoming. Research on Key Technology of Asphalt Pavement Plant Mixing Hot Recycling Process[J]. *Journal of China & Foreign Highway*, 2019, 039(001): 210–214.

Wei Wanfeng, Guo Peng, Tang Boming. Review of the Research on Diffusion Efficiency of Virgin-Aged Asphalt in Recycled Asphalt Mixture [J]. *Materials Reports*, 2017, 31(11): 109–114.

Wang D., Cannone Falchetto Augusto, Moon Ki Hoon, et al. Artificially prepared Reclaimed Asphalt Pavement (RAP)-an experimental investigation on re-recycling [J]. *Environmental Science and Pollution Research International*, 2019, 26(35): 35620–35628.

Zhou Zhou. *Research on Cracking Behavior and Fracture Mechanism of Plant Produced Reclaimed Asphalt Pavement Mixtures* [D]. Southeast University, 2020.

Zhang Yongli, Tao Qingrong. Improvement of Design Method of Marshall Mix Ratio of Hot-recycled Asphalt Mixture in Plant[J]. *Municipal Engineering Technology*, 2018, 36(1): 203–205.

Rational analysis of water intake and utilization in Pearl River basin irrigation area

Guiling Mu*

Pearl River Water Resources Research Institute, Pearl River Water Resources Commission, Guangzhou, China

ABSTRACT: Irrigation water consumption accounts for about 60% of the total water consumption in the Pearl River Basin. The unit water consumption for farmland irrigation is 10170 m^3/hm^2, 91% higher than the national level of 5325 m^3/hm^2. It is an important basis for confirming the initial water right of water intake in the irrigation area, ensuring reasonable agricultural water demand, strengthening agricultural water management, and implementing the strictest water resources management system to analyze and demonstrate the rationality of the Pearl River Basin irrigation water intake. Taking the Doumen irrigation area of the Pearl River Basin as an example, this paper analyzes and demonstrates the water intake and consumption, water source, and influence of water withdrawal, so as to provide the basis for the approval of a water intake permit in this irrigation area.

1 INTRODUCTION

Agriculture is a major water user in China, but the current agricultural water management is relatively extensive, especially in South China's water-rich areas. The irrigation water consumption in Pearl River Basin accounts for about 60% of the total water consumption in the basin. The unit water consumption for farmland irrigation is 10170 m^3/hm^2, which is the largest in China and 91% higher than the national level of 5325 m^3/hm^2. Strengthening the management of water intake permission in irrigation areas is an important measure to safeguard the rights and interests of rational water use in irrigation areas, and ensure agricultural production and food security. It is also an inevitable requirement for comprehensively implementing the strictest water resource management system and continuously deepening water reform, and is a prerequisite for ensuring the comprehensive reform of agricultural water prices.

The analysis and demonstration of irrigation water intake and hydration rationality is an important basis for the approval of water intake permits in irrigation areas (Liu 2018). It is also the basis for ensuring reasonable agricultural water demand, strengthening agricultural water management (Chen 2021; Huang 2022), and implementing the strictest water resource management system (Dai 2020). At present, there are few studies on rational analysis of agricultural water intake and consumption and they mainly focus on the "water-saving and grain increasing" project (He 2015; Li 2016; Liu 2014; Qin 2017; Sun 2015; Shi 2016).

This study takes the Doumen Irrigation Area in the Pearl River Basin as an example. Through the collection and investigation of information and data on water volume, water quality, water supply projects, water conveyance projects, and water withdrawal methods in the irrigation area, it analyzes the rationality and reliability of water intake in the irrigation

---

*Corresponding Author: mumudull2012@163.com

area, and the impact of water withdrawal methods in the irrigation area on the surface water environment. Besides, it puts forward reasonable and feasible water-saving measures and water resources protection measures in the irrigation area, providing a basis for the approval of a water intake permit in the irrigation area.

2 RESEARCH METHODS AND DATA SOURCES

2.1 Research methods

(1) On the basis of sorting out and analyzing the regional water resources and their development and utilization, the basic conditions of the climate, area, planting structure, supporting projects, and water intake of the irrigation area are investigated and analyzed.
(2) Analyzes the rationality of water intake and hydration, the reliability of water intake, and the impact of water recession in the irrigation area in detail. And the management measures of water resources in the irrigation area are proposed.

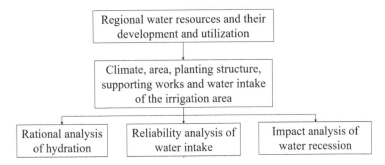

Figure 1. Research method framework.

2.2 Data sources

The data required are mainly from the water resources bulletin of the study area in recent five years, water function zoning, water resources comprehensive planning, water conservation planning, calculation report of effective utilization coefficient of farmland irrigation water, survey report of irrigation area, water supply project planning, sewage project planning, water quality monitoring data, etc.

3 CASE STUDY

3.1 Basic information on irrigation area

The Doumen Irrigation Area is located in Doumen Town, between Chifen Waterway and Hutiao Waterway. According to the latest agricultural survey data, the area of cultivated land in Doumen Irrigation Area is about 2000 hm^2, including 733 hm^2 of agricultural planting area and 1267 hm^2 of the fish pond area. The area of rice in the current irrigation area is shrinking, and the proportion of fish pond area accounts for more than 60%. There are few rice fields in the whole area, which are separated by fish ponds and fruit trees. According to a remote sensing image map and field survey, the cultivated area of farmland in the Doumen Irrigation Area has been severely reduced, and only 3 irrigation areas have been preserved completely. Area 1 mainly consists of DA Chikan Village and XIAO Chikan Village. The irrigation water source is mainly Chifen Waterway, and the water intake project site is the DA Chikan sluice. Area 2 is dominated by SHANG Zhou Village, XIA Zhou Village, and XIN Xiang Village. The irrigation water source is Hutiaomen Waterway, and the water intake project site is the LENG Chong sluice. Area 3 is mainly composed of NAN Men Village, DOU Men Village, BA Jia Village, DA Haochong Village, and XIAO Haochong Village. The irrigation water source is Hutiaomen Waterway, and the water intake project points are Nan Menchong sluice, HAO Chong sluice, and DA Miaokeng sluice.

According to the field survey, the crops with large irrigation water consumption are mainly rice and vegetables. And rice is double cropping rice, and vegetables are mainly leaf vegetables. Other fields have been divided by fish ponds, residential buildings, and factories. The natural precipitation of dry crops and forest fruit fields in the irrigation area can meet the growing demands, and no irrigation is carried out. At present, agricultural water metering facilities have not been installed in the irrigation area, and there is no statistical data.

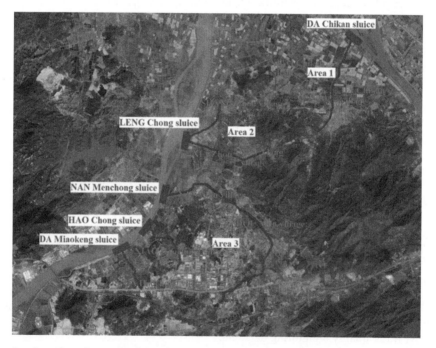

Figure 2. Location of water intake project in Doumen storage diversion irrigation area.

3.2 *Rationality analysis of irrigation district hydration*

3.2.1 *Implementation of "Three Red Lines"*

According to the "Notice of Zhuhai People's Government Office on Printing, the Assessment Method for the Implementation of the Strictest Water Resources Management System in Zhuhai", the total water consumption control index of Doumen District where Doumen Irrigation District is located from 2016 to 2030 is 193 million m^3, in which the thermal power is based on water consumption statistics. The water consumption per million yuan GDP control index of Doumen District in 2020 is 4500 m^3, and per million yuan of industrial added value control index is 2200 m^3. And the water quality standard rate of Doumen District is 85%.

According to the "Zhuhai Water Resources Bulletin in 2020", the actual total water consumption in Doumen District in 2020 is 180 million m³, in which the thermal power is based on water consumption. The actual value of water consumption per million yuan GDP is 4140 m³, and the actual value of water consumption per million yuan industrial added value is 2160 m³. According to "Zhuhai Environmental Quality Status in 2020", in the provincial examination section, the water quality standard rate of the Doumen District Water Function Zone is 100%. Therefore, the water use indicators of Doumen District in 2020 meet the most stringent water resource management requirements of Zhuhai.

3.2.2 *Water intake and consumption in the irrigation area*

There are many water intakes and outlets in the tidal discharge irrigation area, and it is difficult for metering facilities to accurately measure agricultural water intake. Therefore, the water intake in the irrigation area is calculated according to the water quota of Guangdong Province.

According to "Guangdong Water Quota Part 1: Agriculture", the irrigation water quota is specified at the bucket mouth of large and medium-sized irrigation areas. So the gross irrigation water consumption is the net irrigation water consumption divided by the effective utilization coefficient of irrigation water. The calculation formula is:

$$W = W_j/\eta = \left(\sum A_i m_i\right)/\eta \tag{1}$$

Where W—gross irrigation water consumption, m³; W_j—net irrigation water consumption, m³; η—effective utilization coefficient of irrigation water; A_i—planting area of the i crop, hm²; m_i—Water quota of the i crop, m³/hm².

The water quota is 90% guarantee rate quota, the rice quota is 19950 m³/hm², and the leaf vegetable quota is 5625 m³/hm². The irrigation area is taken from the latest census data of the Doumen District. According to "the Calculation Report on Effective Utilization Coefficient of Farmland Irrigation Water in Zhuhai", the effective utilization coefficient of farmland irrigation water in medium-sized irrigation areas in Zhuhai is 0.595.

According to formula (1), the water consumption for irrigation in the Doumen Irrigation Area is 16.5155 million m³, including 8.1810 million m³ in Area 1, 2.9339 million m³ in Area 2, and 5.4005 million m³ in Area 3.

3.2.3 *Rationality analysis of water intake and consumption*

Since the water consumption in the irrigation area has not been measured, the water consumption is calculated by the quota method. The water quota meets the requirements of the specification. And the crop irrigation area is calculated by the survey data of the Agriculture and Rural Bureau. Therefore, the calculation data of the water consumption is reliable, the method is appropriate, and the calculation results are reasonable.

According to the results of the first national water conservancy census, there are 4 medium-sized irrigation districts in Zhuhai, all located in Doumen District, respectively, the Baijiaochao Irrigation District, the Doumen Irrigation District, the Qianwu Water Storage, the Diversion Irrigation District, the Wushan Water Storage, and Diversion Irrigation District. The current total irrigation area of farmland is 936.55 hm², of which the Doumen Irrigation District accounts for 43.2%. The total irrigation water demand of the Doumen Storage and Diversion Irrigation Area is 16.5155 million m³, accounting for 26.7% of the total farmland irrigation water volume of 61.73 million m³ in Doumen District in 2020. There are many leafy vegetables planted in the Doumen Irrigation Area, and the water quota of leafy vegetables is smaller than that of rice, melon, and fruit vegetables, so the proportion of irrigation water consumption is less than that of the irrigation area. This is consistent with the actual situation.

Table 1. Calculation of irrigation water consumption in the irrigation area.

Area	Village	Rice area/ hm²	Leaf vegetable area/ hm²	Water quota/ m³·hm⁻²		Net irrigation water consumption/10⁴m³			Gross irrigation water consumption/ 10⁴m³
				Rice	Vegetable	Rice	Vegetable	subtotal	
Area 1	XIAO Chikan	1326.93	56.67	19950	5625	264.72	3.19	267.91	450.27
	DA Chikan	946.67	533.33			188.86	30.00	218.86	367.83
	subtotal	2273.60	590.00			453.58	33.19	486.77	818.10
Area 2	SHANG Zhou	186.67	186.67	19950	5625	37.24	10.50	47.74	80.24
	XIA Zhou	71.33	71.33			14.23	4.01	18.24	30.66
	XIN Xiang	531.13	46.67			105.96	2.63	108.59	182.50
	subtotal	789.13	304.67			157.43	17.14	174.57	293.39
Area 3	NAN Men	555.07	133.33	19950	5625	110.74	7.50	118.24	198.71
	DOU Men	240.00	26.67			47.88	1.50	49.38	82.99
	BA Jia	33.33	466.67			6.65	26.25	32.90	55.29
	DA Haochong	292.00	70.00			58.25	3.94	62.19	104.52
	XIAO Haochong	292.00	6.67			58.25	0.38	58.63	98.54
	subtotal	1412.40	703.33			281.77	39.56	321.34	540.05
	total	4475.13	1598.0	19950	5625	1598.0	89.89	982.68	1651.55

3.3 *Reliability analysis of water intake in the irrigation area*

3.3.1 *Analysis of water intake in the irrigation area*

The Doumen Irrigation Area is located between Chifen Waterway and Hutiao Waterway. The irrigation water is taken from Chifen Waterway through the CHONG Kou sluice, and from Hutiao Waterway through the NAN Menchong sluice and LENG Chong sluice. The agricultural irrigation water consumption in the Doumen Irrigation Area is 16.5155 million m³, and the water intake flow is 0.61 m³/s. Of which the water intake from Chifen Waterway is 8.1810 million m³, and the water intake from Hutiaomen Waterway is 8.3346 million m³.

3.3.2 *Analysis of water intake guarantee in the irrigation area*

The water intake flow from Hutiaomen Waterway in the Doumen Irrigation Area is 0.31 m³/s. After the implementation of the Pearl River Delta Water Resources Allocation Project, when the water supply guarantee rate of Tiger Leaping Gate is 95%, the average flow in the driest three days is 53.4 m³/s. The water intake of the Doumen Irrigation Area accounts for 0.58% of the flow of the Hutiaomen Waterway, accounting for a small proportion. Therefore, the amount of water taken from the Hutiaomen Waterway in the Doumen Irrigation Area can be guaranteed.

The water intake flow from Chifen Waterway in Doumen Irrigation Area is 0.56 m³/s. The proportion of water intake to the average flow of the ebb tide and the average flow of flood tide is 0.39% and 0.38% respectively under the condition of a 97% guarantee rate of the driest month flow in the year. Therefore, the amount of water taken from the Chifen Waterway in the Doumen Irrigation Area can be guaranteed.

Doumen Irrigation Area involves two primary water function areas, namely Hutiaomen Waterway Development and Utilization Area and Chifen Waterway Development and Utilization Area. According to "the Environmental Quality Status of Zhuhai in 2020", the water quality of the two water function areas involved is up to the standard, and the water intake quality meets the requirements.

3.3.3 *Water intake impact analysis*

The water intake flow from Hutiaomen Waterway in the Doumen Irrigation Area accounts for 0.58% proportion of Hutiaomen Waterway flow. The water intake flow from Chifen Waterway in Doumen Irrigation Area accounts for 0.39% proportion of water intake to the average flow of ebb tide under the condition of 97% guarantee rate of the driest month flow in the year. Therefore, the water intake in the Doumen Irrigation Area will not have a significant adverse impact on other water intakes.

3.4 *Impact analysis of water recession in the irrigation area*

The Doumen Irrigation Area is used for tidal drainage irrigation, and the water withdrawal point is also the water intake point. The water withdrawal method is natural water withdrawal through the drainage channel near the farmland. The main crops in the irrigation area are vegetables, rice, and fruit trees. The planting method is reasonable and the pollutants generated are less.

According to "the Zhuhai Water Resources Bulletin in 2020", the agricultural water consumption rate is about 60%. According to this, the water withdrawal is estimated to be about 40% of the water intake, so the water withdrawal in the Doumen Irrigation Area is 6.606 million m^3.

According to "the Environmental Quality of Zhuhai in 2020", the water withdrawal has not had a significant adverse impact on the water function area, and the way of water withdrawal in the Doumen Irrigation Area is reasonable.

4 CONCLUSIONS AND SUGGESTIONS

(1) With the urban development and agricultural transformation, the farmland in the Doumen Irrigation Area gradually shrinks, and the actual irrigation area is less than the designed irrigation area. According to the quota method, the irrigation water intake in the irrigation area is 16.515 million m^3, which is reasonable. The water intake points in the irrigation area are scattered. According to the distribution of water intake points, it is reasonable to calculate the irrigation area by zoning.
(2) The main water sources of the Doumen Irrigation Area are Hutiaomen Waterway and Chifen Waterway. As the water source of the irrigation area, the water from Hutiaomen Waterway and Chifen Waterway can meet the irrigation water demand of the irrigation area, and the irrigation water intake in the irrigation area will not cause adverse effects on other water users. The water intake in the irrigation area is reliable without major changes in the water inflow and the timely maintenance of the water intake project by the management department of the irrigation area.
(3) The farmland irrigation backwater in the irrigation area is a natural backwater, which eventually flows back to Hutiaomen Waterway and Chifen Waterway without causing serious pollution to the water environment of the backwater reaches. Therefore, the current water withdrawal mode in the Doumen Irrigation Area is reasonable and feasible.

(4) The irrigation mode in the irrigation area is extensive, and there is a lack of metering facilities. It is recommended to carry out a water-saving transformation and improve the water intake metering device.

ACKNOWLEDGMENTS

This research was supported by the National Key R&D Program of China (2021YFC3001000), the Special Foundation for National Science and Technology Basic Research Program of China (2019FY101900), and the Guangxi Key R&D Plan (902229136010).

REFERENCES

Chen Kaifeng, Wang Kaifeng. Analysis of the Water Intake Management in the Irrigation Area of the Second Management Office of Renmin Canal in Dujiangyan, Sichuan [J]. *Sichuan Water Resources*, 2021 (S1): 62–64.

Dai Qianqian, Zhou Fei, Liu Xiao, Ma Jun. Thoughts on Levying Water Resources Tax on Agricultural Production Water Consumption [J]. *Research on Water Resources Development*, 2020, 20 (04): 15–17.

He Feifei. Rational Analysis of Water Saving and Grain Increasing Action Water Consumption and Hydration in Haicheng [J]. *Research on Water Resources Development*, 2015, 15 (03): 65–67+70.

Huang Wenpu, Zheng Xiaoqing. Measurement Calibration Results and Experience Analysis of Water Intake Monitoring Station – Taking the Guangxi Water Resources Monitoring Capacity Construction Project as an Example [J]. *Guangxi Water Resources and Hydropower*, 2022 (03): 31–34.

Li Genglei, Sun Guixi. Demonstration and Analysis of Water Resources for Water Saving and Grain Increasing Project in Nanpiao District [J] *Groundwater*, 2016, 38 (001): 92–93.

Li Pengjun. Rationality Analysis on Water Consumption and Hydration of Yixian Water Saving and Grain Increasing Project [J] *Water Science and Engineering Technology*, 2016, No.195 (01): 73–75.

Liu Xu. *Evaluation Study on Water Intake and Drainage of Malan Irrigation District Project in Ning'an City* [D]. Northeast Agricultural University, 2018.

Liu Zhongyi, Liu Yang. Rational Analysis of "Water Saving and Grain Increasing Action" Water Consumption and Hydration in Heishan County, Liaoning Province [J]. *Heilongjiang Science and Technology of Water Conservancy*, 2014, 42 (02): 203–206.

Qin Xiaolei. Rational Analysis of Water Consumption and Hydration Based on "Water Saving and Grain Increasing Action" and Suggestions on Compensation Measures [J]. *Water Resources Development and Management*, 2017 (05): 63–65.

Shi Ning. Discussion on Water Resources Demonstration of "Water Saving and Grain Increasing Action" [J]. *Water Resources Development and Management*, 2016 (05): 13–15.

Sun Jihui. Rational Analysis of Water Consumption and Hydration in Fumeng County Water Saving and Grain Increasing Project Area [J]. *Water Conservancy Technical Supervision*, 2015, 23 (06): 57–59.

A comprehensive evaluation of water saving in Weifang, Shandong Province, China

Haijun Wang
Hydrological Center of Shandon Province, Jinan, China

Song Han & Linxi Lei
School of Water Conservancy and Environment, Jinan University, Jinan, Shandong, China

Zhenhua Cai
Hydrological Center of Shandon Province, Jinan, China

Xin Cong & Yuyu Liu*
School of Water Conservancy and Environment, Jinan University, Jinan, Shandong, China

ABSTRACT: Based on the data of the Water Resources Bulletin in Weifang City, referring to the "regional water saving evaluation method", nine indicators were selected to research the water saving of Weifang City. Analytic Hierarchy Process (AHP) and Fuzzy Comprehensive Evaluation Model were used to evaluate the water-saving level of Weifang City in 2020. The result showed that: on the whole, water saving in Weifang is at a relatively advanced level; Weicheng District, Kuiwen District, and Linqu County are at an advanced level, Zhucheng District is at a relatively advanced level, Qingzhou City, Shouguang City, and Anqiu City are at a general level, and Hanting District, Fangzi District, Gaomi City, Changyi City, and Changle County are at a backward level. The evaluation results are more in line with reality. Then, some targeted water-saving countermeasures are put forward to guide the sustainable utilization of water resources in Weifang City.

1 INTRODUCTION

With the rapid development of the economy and society, water conservation has become an important part of achieving sustainable economic and social development and ensuring the construction of ecological civilization. Water saving has important strategic significance for realizing the rejuvenation of the Chinese nation and promoting the construction of ecological civilization. Besides, water conservation supports sustained and healthy economic and social development. It is an urgent requirement to make water saving a prerequisite for the development, utilization, protection, allocation, and regulation of water resources. It is an effective way to ensure scientific and reasonable access to water for planning and construction projects and to promote the formation of a spatial layout and industrial structure suitable for the conditions of water resources. It is a powerful starting point to force the economical and intensive use of water resources and improve the efficiency of water use in the whole society.

At present, the research on water-saving evaluation mainly focuses on the construction of a water-saving evaluation index system and evaluation methods (Chen et al. 2004; Qin et al. 2021; Zhang & Wang 2015). Wang et al. (2018) proposed a water-saving evaluation model based on multi-level uncertain comprehensive evaluation with six water-saving standards,

*Corresponding Author: 152838200@qq.com

including comprehensive water-saving, domestic water-saving, ecological water-saving, industrial water-saving, agricultural water-saving, and social economy. Zhang and Liang (2021) applied the analytic hierarchy process and similar ranking technology to build a water-saving evaluation model and analyzed the water-saving situation in Henan Province. Ding and Liu (2018) used the fuzzy comprehensive evaluation method to evaluate the development level of water-saving irrigation in Sichuan Province.

Located in the west of the Shandong Peninsula, Weifang is an important transportation hub connecting the Shandong Peninsula urban agglomeration. Weifang is short of water resources. The average annual precipitation is only 664.8 mm, and the total amount of water resources is 2.728 billion m^3, while the per capita water resources are only 298 m^3 lower than the provincial average. With the rapid development of the economy and society, the contradiction between the distribution of water resources and the economic layout is becoming increasingly prominent. The coexistence of resource-based water shortage, engineering water shortage, and polluting water shortage has seriously affected the sustainable development of Weifang's economy and society. Therefore, it is important to assess the water saving for the regional sustainable utilization of water resources.

2 METHOD

2.1 Evaluation index system

Referring to the documents of the water-saving society and the actual situation of Weifang City, nine indicators were selected in four categories: comprehensive indicators, agricultural water indicators, industrial water indicators, and domestic water indicators. The water-saving evaluation indicators were divided into four levels: V1, V2, V3, and V4, which correspond to advanced, more advanced, commonly, and backward water-saving levels respectively (Table 1). This paper used the Analytic Hierarchy Process (AHP) to determine the weight of the index.

Table 1. Water saving index weight.

Target layer	Criterion layer	Index layer	V1	V2	V3	V4	Weight
Water-saving level evaluation	Comprehensive (0.17)	Water consumption per 10000 yuan GDP (m^3/10000 yuan)	≤13	13~24.5	24.5~36	>36	0.1417
		Utilization level of unconventional water source (%)	≥26.6	15.8~26.6	5.0~15.8	<5.0	0.0283
	Agriculture (0.35)	Effective utilization coefficient of farmland irrigation	≥0.732	0.64~0.732	0.548~0.64	<0.548	0.1323
		The proportion of water-saving irrigation area (%)	≥95.8	74.8~95.8	53.8~74.8	<53.8	0.1016
		Average water consumption per mu for farmland irrigation (m^3/Mu)	≤175	175~276	276~377	>377	0.1160
	Industry (0.24)	Water consumption of 10000 yuan industrial added value (m^3/10000 yuan)	≤6.3	6.3~10.9	10.9~15.5	>15.5	0.1309
		Reuse rate of Industrial Enterprises above Designated Size (%)	≥97.1	93.3~97.1	89.5~93.3	<89.5	0.1091
	Life (0.24)	Leakage rate of water supply network (%)	≤7.5	7.5~11.1	11.1~14.7	>14.7	0.1309
		Popularization rate of domestic water-saving appliances (%)	≥100	88.1~100	76.2~88.1	<76.2	0.1091

2.2 Fuzzy comprehensive evaluation

The Fuzzy Comprehensive Evaluation Method is a comprehensive evaluation method based on fuzzy mathematics. According to the membership theory of fuzzy mathematics, this method transforms qualitative evaluation into quantitative evaluation, that is, fuzzy mathematics is used to make an overall evaluation of things or objects restricted by many factors. It has the characteristics of clear results and strong systematicness. It can better solve fuzzy and difficult-to-quantify problems and is suitable for solving all kinds of uncertain problems (Li et al. 2015).

In this paper, the Cauchy distribution function is used as the membership function to calculate the membership matrix, which core formula is:

$$r(x) = \frac{1}{\left[1 + a_2(x - a_1)^2\right]} \quad (1)$$

Where, $r(x)$ is the membership function; x is the current value of index x of the year to be evaluated; a is the function parameter.

1) When x is class V_1:

$$a_1 = x_u \quad (2)$$

$$a_2 = \frac{4}{(x_u - x_i)^2} \quad (3)$$

2) When x is class V_2-V_3:

$$a_1 = \frac{x_u + x_i}{2} \quad (4)$$

$$a_2 = \frac{4}{(x_u - x_i)^2} \quad (5)$$

3) When x is class V_4:

$$a_1 = x_i \quad (6)$$

$$a_2 = \frac{4}{(x_u - x_i)^2} \quad (7)$$

Where x_u and x_i are the upper and lower boundary values of x corresponding to different levels of standard values respectively.

The membership matrix R is obtained from the membership function, and the comprehensive evaluation vector D is obtained by fuzzy multiplication with the weight vector W.

$$D = W \cdot R \quad (8)$$

According to the principle of maximum membership, determine the level of the target layer.

3 RESULT

Analytic Hierarchy Process (AHP) and Fuzzy Comprehensive Evaluation Model were used to evaluate the water-saving level of Weifang City in 2020 (Table 2 and Figure 1).

Table 2. Water saving evaluation results of counties and urban areas in Weifang City in 2020.

County and urban area	V1	V2	V3	V4	Water saving level
Weicheng District	0.2963	0.2519	0.186	0.2643	Advanced
Hanting District	0.1806	0.1752	0.2713	0.3727	Backward
Fangzi District	0.1839	0.1784	0.2565	0.3807	Backward
Kuiwen District	0.3181	0.2661	0.1626	0.2547	Advanced
Qingzhou City	0.2111	0.2536	0.2913	0.2429	Commonly
Zhucheng District	0.2123	0.4154	0.1223	0.2488	More advanced
Shouguang City	0.264	0.2149	0.3366	0.1811	Commonly
Anqiu City	0.1464	0.2605	0.3906	0.201	Commonly
Gaomi City	0.2121	0.2201	0.2254	0.3461	Backward
Changyi City	0.1786	0.2173	0.259	0.3415	Backward
Linqu County	0.3186	0.2656	0.155	0.2611	Advanced
Changle County	0.2836	0.2711	0.1155	0.3299	Backward
Weifang City	0.234	0.2874	0.2837	0.1967	More advanced

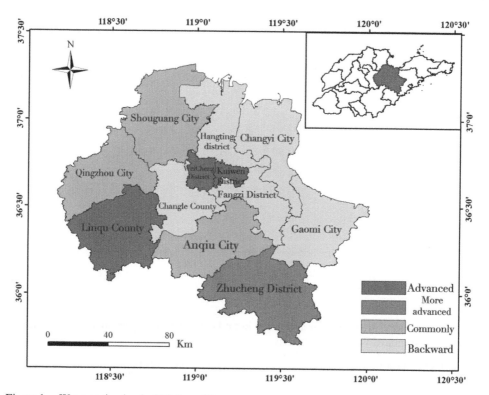

Figure 1. Water saving level of Weifang City.

(1) Water saving in Weicheng District, Kuiwen District, and Linqu County is at an advanced level. The reasons are nothing more than the following: less population, less arable land, and less high-water consumption industries, which fundamentally avoid water waste.

(2) Water saving in Zhucheng District is at a relatively advanced level. The popularization rate of water-saving appliances is at an advanced level, the reuse rate of industrial enterprises is above the designated size, the utilization level of unconventional water

sources is at a backward level, and other indicators are at a relatively advanced level. There are many water-consuming industries and water consumption, but the technical level of water utilization is not high, resulting in water waste.

(3) Water saving in Qingzhou, Shouguang, and Anqiu is at the general level. Most of the indicators in these areas are at the general level, due to the following reasons: large population; unreasonable industrial structure, and more water-consuming industries; the level of water-saving technology is not high.

(4) Water saving in Hanting District, Fangzi District, Gaomi City, Changyi city, and Changle County is at a backward level. The reasons lie in the following points: there are many industries with low output value but more water consumption, that is, the industrial structure is unreasonable; the pipe network is seriously aged, and the leakage rate is high; less water-saving irrigation area; the level of water-saving technology is not high, resulting in water waste.

4 WATER SAVING MEASURES

4.1 *Agricultural water saving*

(1) It is attempted to control the planting scale, optimize and adjust the planting structure, and develop agricultural water-saving irrigation; (2) improve agricultural management and planning; (3) improve infrastructure and increase capital investment in technology and equipment.

4.2 *Industrial water saving*

(1) Efforts are made to improve the recycling level of water resources; (2) make efficient use of process water; (3) restrict high water consumption industries in water-scarce areas, adjust the industrial structure, and eliminate high water consumption and high pollution industries.

4.3 *Water saving in urban life*

(1) Transform the urban water supply network to reduce the leakage rate of the network; use of water-saving appliances should be supported by policies; (2) take advantage of the role of price regulation; (3) vigorously carry out water-saving publicity and improve water-saving awareness; strengthen quota management and water use management to promote the construction of water-saving cities.

5 CONCLUSION

By collecting the data from Weifang Water Resources Bulletin, this paper selected nine representative indicators and compared them with similar regions to preliminarily evaluate the water-saving level of Weifang; AHP fuzzy comprehensive evaluation model was used to evaluate the water-saving level of Weifang City in 2020, and the results were realistic. Counties and cities can analyze the weak links of water-saving work from the subordinate levels of each index, and put forward targeted opinions, so as to provide e reference for further water-saving work.

ACKNOWLEDGMENTS

This study was supported by the Fund of Major Subject of Jinan University (Research on the Coordinated and Efficient Utilization Technology of Water Resources in Mi River Basin under the Background of High-Quality Development), and the Fund of Evaluation and Optimization of Hydrological Network for Real-time Flood Forecasting.

REFERENCES

Chen Y., Zhao, Y. & Liu, C. M. (2004). Evaluation Indication System of Water Conservation Society. *Resources Science*, (06): 83–89.

Ding, L. J. & Liu, Y. H. (2018). Assessment of Water-saving Irrigation Development Level of a Regional Area Based on Fuzzy Comprehensive Assessment Method. *Fresenius Environmental Bulletin*, 27(5): 2891–2899.

Li, J. X., Li, C. K, Luo, S. H., et al. (2015). Evaluation of Sustainable Utilization of Water Resources in Quanzhou City Based on AHP Fuzzy Comprehensive Evaluation Method. *Bulletin of Soil and Water Conservation*, (01): 210–214+286.

Qin, C. H., Zhao, Y., Li, H. H., et al. (2021). Assessment of Regional Water Saving Potential. *South-to-North Water Transfers and Water Science & Technology*, 19(1): 36–42.

Wang, X. S., Li, X. N. & Wang, J. W. (2018). Urban Water Conservation Evaluation Based on Multi-grade Uncertain Comprehensive Evaluation Method. *Water Resources Management*, 32(2): 417–431.

Zhang, X. Q. & Liang, T. Y. (2021). A Comprehensive Evaluation of Urban Water-saving Based on AHP-TOPSIS. *Desalination and Water Treatment*, 213: 202–213.

Zhang, Y. & Wang, X. J. (2015). Research on the Index System of Water-saving Society Construction in China. *China Rural Water and Hydropower*, (08):118–120+125.

Comprehensive evaluation method of concrete beam bridge based on Bayesian network

Wei Ji*
CCCC Infrastructure Maintenance Group Co., LTD, Beijing, China

Sijia Ge*
School of Transportation Science and Engineering, Harbin Institute of Technology, Harbin, China

Mingjun Zhang*
Heilongjiang Communications Investment Group Co. LTD, Heilongjiang, China

Meng Ding*
Nantong Central Innovation District Construction & Investment Co., L LTD, Nantong, China

Zhonglong Li*
School of Transportation Science and Engineering, Harbin Institute of Technology, Harbin, China

ABSTRACT: This research expounds the basic theory and construction process of the Bayesian network model to construct the probability and importance of risk analysis assessment system for predicting the risk of concrete beam bridge structural disease factors. Finally, combined with engineering examples, the concrete beam bridge with many disease factors and complex influence factors is taken as the research object to establish a model based on Bayesian network for comprehensive evaluation of concrete beam bridge disease risk. It provides a reference for risk prediction and prevention of concrete beam bridge structures.

1 INTRODUCTION AND BACKGROUND

The development of bridge engineering has gradually advanced from the stage of large-scale new construction to the stage of maintenance, appraisal and evaluation, and reinforcement (Shao 2016). The evaluation of the bearing capacity and durability of old bridges has become a research hotspot for experts and scholars. The process of existing bridge assessment methods can be divided into three aspects: information collection, analysis and assessment, and decision-making. The influential factors of bridge engineering evaluation are complex, and it is not reasonable and accurate to only rely on fixed mathematical and mechanical formula models for evaluation. Therefore, experienced knowledge of experts is needed for analysis. With the development of artificial intelligence, especially the development of the artificial neural network, the combination of the expert system evaluation method and artificial intelligence technology has become a research hotspot in the field of bridge evaluation in recent years.

This research is proposed to solve the existing traditional evaluation method of a single bridge, low efficiency, one-sided, and polymorphism in probability based on Bayesian

---

*Corresponding Authors: 105901088@qq.com, gesijia990128@163.com, zmj-1965@163.com, hitdingmeng@163.com and lizhonglong@hit.edu.cn

network. It is a kind of disease risk comprehensive evaluation method of concrete beam bridge based on the computer-aided control software. The objectives of the study are predicting the risk of disease in concrete beam bridge probability level, analyzing the impact of disease on the whole structure safety level, and diagnostic reasoning of accidents. The research method for the advanced computer applications in the field of artificial intelligence technology in the bridge comprehensive evaluation provides a feasible method based on the Bayesian network. It combines the theory of the analytic hierarchy process (AHP), the fault tree theory, and the fuzzy set theory. The complementary advantages of the existing bridge assessment methods are considered for a bridge safety evaluation of intelligent prediction and real-time updates in providing solutions.

2 BAYESIAN NETWORK

2.1 *Bayesian network description*

Bayesian network is a probabilistic graph model proposed by Judea Pearl in 1985 (Zhang 2006). Although the introduction of the Bayesian network does not further reduce the complexity (Song 2011), it provides a powerful model tool for people to intuitively understand the causal relationship of the system. Bayesian network is not only a mathematical language suitable for computer processing but also a model for the combination of a graph and probability. There are two ways to construct the Bayesian network structure, one is to construct it by consulting experts in related fields, and the other is to acquire the network structure through training and learning by analyzing prior data.

2.2 *The advantages of Bayesian network*

(1) Advantages of system state
   The node state of the Bayesian network can be polymorphic and describe the causal uncertainty of the event fault state, with the ability to describe probability. As a representation of joint probability distribution decomposition, the Bayesian network can reduce the complexity of reasoning.
(2) Advantages of quantitative reasoning evaluation
   Bayesian network not only has reasoning functions, but can also describe the probability of reasoning elements, calculate the conditional probability, and use variable elimination algorithm, cluster number propagation algorithm, and other algorithms to carry out forward reasoning to evaluate the risk factors affecting system security and reverse reasoning to diagnose system fault causes.
(3) Model update and learning advantages
   Bayesian network can update the network in real-time according to the measured data and obtain a more scientific and reasonable prediction. Bayesian network can combine prior experience, deeply analyze data, and determine network structure and network parameters by means of a computer algorithm for network learning (Zhou 2006). According to the chain rules, the complex joint distribution can be decomposed to reduce the computational complexity of probabilistic reasoning, which is one of the core advantages of the Bayesian network model. By using triangular fuzzy quantitative expert evaluation information, the limitations of traditional Bayesian network in the probability of root nodes can be overcome, determining Bayesian network parameters more scientific and reasonable. Reasoning and learning are the two core functions of Bayesian network (Fan 2012).

3 COMPREHENSIVE EVALUATION METHOD OF CONCRETE BEAM BRIDGE BASED ON BAYESIAN NETWORK

3.1 *Constructing Bayesian network structure*

Based on the analytic hierarchy process (AHP), the framework of the concrete beam bridge disease risk assessment system was established with the concrete beam bridge as the research target. The disease of the main components in the main parts of the beam bridge to be evaluated was set as nodes in the network, and a Bayesian network model was established. In accordance with the highway bridge standard regarding maintenance, highway bridge technique condition evaluation standards, and city bridge maintenance technical standard of disease classification, the guidance of concrete beam bridge components was extracted from the main disease factors in each part of the concrete beam bridge. Preliminary classification evaluation system was established based on the AHP and combined with the transfer path of a concrete beam bridge. The overall technical status of the Bayesian network structure is shown in Figure 1.

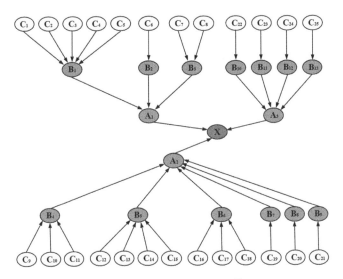

Figure 1. Overall Bayesian network structure technical status diagram.

3.2 *Determining node probability*

The starting point of the Bayesian network structure is to obtain the a priori probability of root nodes. The evaluation results of bridge diseases are obtained according to the experience and knowledge of experts. Through the design of an expert questionnaire, several experts (Z_i) were invited, and the main disease factors (C_i) of the actual bridge were investigated and rated considering their weightage by experts. Bridge disease factors are complex. It is difficult to get an accurate root node quantitative evaluation. Thus expert evaluation of the fuzzy number of evaluation set was established, seven natural language classification variables {is very small, small, small, medium, large, large, very large} were introduced to describe disease occurrence factors, corresponded with triangular fuzzy Numbers, and the fuzzy linguistic assessment was included into grade indices for appraisal accurate data.

The probability of the root node obtained according to the above steps is still the triangular fuzzy number. To determine the probability of disease occurrence between nodes more intuitively, the triangle fuzzy number is solved according to the mean area method and converted into the accurate probability value of the root node. The specific calculation is shown in Formula 1.

$$P = \frac{\tilde{l} + 2\tilde{m} + \tilde{u}}{4} \quad (1)$$

Where, P is the determined node probability value; $\tilde{l}$, $\tilde{m}$, and $\tilde{u}$ are the triangular fuzzy numbers after averaging.

After a root node probability, it is still necessary to determine the probability of intermediate nodes and leaf nodes, i.e., to determine the root node of the conditional probability table (Wang 2020). In this paper, the fault tree of the conditional probability of a Bayesian network logic gates mapping table was established and the relative weight of the root node of conditional probability tables was determined combined with standard and evaluating disease experts.

3.3 Results analysis

The prior probability and conditional probability tables of each determined root node are assigned to the Bayesian network structure to obtain the occurrence probability of each node and the overall structure technical condition probability (X) by positive reasoning. To score and grade, and comprehensively evaluate the actual bridge technical condition grade, the five kinds of scoring standards in *Highway Bridge and Culvert Maintenance Specification* or *Highway Bridge Technical Condition Assessment Standard*, and the five kinds of scoring standards in *Urban Bridge Maintenance Technical Standard*, were referred.

The probability of the overall technical conditions can be divided into I, II, III, IV, and V five categories, with the corresponding probability range of 0% ∼ 20%, 20% ∼ 40%, 40% ∼ 70%, 70% ∼ 90%, and 90% ∼ 100%, respectively. The worse condition of integral technology is, the greater the probability is.

4 EVALUATING THE APPLICATION OF THE MODEL

4.1 Project summary

Ningan Bridge (right) is a bridge on the G11 (Heda) line in Heilongjiang Province. The bridge is a PC-reinforced concrete supported T-beam bridge spanning the Mudan River.

According to a preliminary on-site survey, the main diseases of the Ningan Bridge include:

1. Watermarks on the bottom surface of the outer flange of the beam on both sides of each span, with local water seepage, whitening, and concrete aging.
2. Concrete spalling and reinforcing bars found at the lower edge of the main beam floor
3. Ruts found in the driveway
4. Pits, grooves, and damage are present in the pavement of the bridge deck
5. Pavement plate transverse cracking
6. Metal railing corrosion
7. Drain hole is seriously blocked.
8. Rubber strip of the expansion joint is damaged

4.2 Establishment of Bayesian network evaluation model

According to Sections 3.1–3.3, for the overall technical conditions of the fault tree and Bayesian network topology structure, the node names and numbers are entered into Python and Netica software to construct a Bayesian network structure model.

Five experts evaluated the root node of the Bayesian network for major disease factors, evaluating the sets {very small (VL), small (L), small (ML), medium (M) and large (MH), big (H), very large (VH)}, to identify root node diseases and assess concrete girder bridges.

Expert judgment is converted into triangular fuzzy numbers. After taking the experts' opinions into account, a weighted average of the individual experts is applied to the C1 node. The probability values of the 25 root nodes are calculated according to the above steps, as shown in Table 1.

Table 1. Root node probability table.

Root node	State1	State0	Root node	State1	State0
$P(C_1)$	0.86	0.14	$P(C_{14})$	0.39	0.61
$P(C_2)$	0.64	0.36	$P(C_{15})$	0.07	0.93
$P(C_3)$	0.58	0.42	$P(C_{16})$	0.08	0.92
$P(C_4)$	0.44	0.56	$P(C_{17})$	0.47	0.53
$P(C_5)$	0.09	0.91	$P(C_{18})$	0.07	0.93
$P(C_6)$	0.62	0.38	$P(C_{19})$	0.18	0.82
$P(C_7)$	0.23	0.77	$P(C_{20})$	0.23	0.77
$P(C_8)$	0.09	0.91	$P(C_{21})$	0.51	0.49
$P(C_9)$	0.63	0.37	$P(C_{22})$	0.63	0.37
$P(C_{10})$	0.80	0.20	$P(C_{23})$	0.49	0.51
$P(C_{11})$	0.54	0.46	$P(C_{24})$	0.83	0.17
$P(C_{12})$	0.87	0.13	$P(C_{25})$	0.93	0.07
$P(C_{13})$	0.51	0.49			

4.3 *Analysis of calculation results*

(1) Forward inference analysis

Forward reasoning analysis refers to the top-down reasoning from the root node to the leaf node. In the model, the overall technical status of the structure (leaf node) is deduced based on the main disease factors (root node). The probability grade of the overall technical status of the bridge structure is calculated by forward reasoning (Chen 2019). Known of the root node prior probability and the conditional probability of intermediate nodes, the input to the Bayesian networks, calculated in the absence of any evidence, leaf nodes in the State of X State 0 and State1 probability were 0.429, 0.571, according to the I ~ V class hierarchies of probability interval technology condition, to determine the overall structure of the whole bridge technique condition in tier III.

(2) Reverse diagnostic analysis

Reverse diagnostic analysis refers to the reasoning from bottom to top, from leaf nodes to root nodes. The model is based on the overall structure of technical conditions (leaf nodes) and the reasoning of main disease factors (root nodes). The backward reasoning root node is calculated in the different risk states of a posteriori probability based on the root node size of posterior probability to determine the most likely cause of risk factors (Zhou 2019). Taking the probability distribution of each root node, when the probability of leaf node X being State 1 is 0.571, for example, the root node C25 has the highest probability of occurrence, as shown in Figure 2.

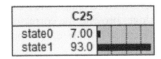

Figure 2. Node probability of C25.

C1, C12, and C24 are more likely to be in a serious state, as shown in Figure 3.

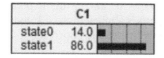 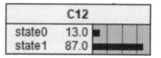 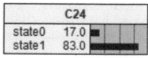

Figure 3. Node probability of C1, C12, and C24.

When the technology situation of the whole structure is in serious condition, the most possible disease risk factors, such as poor drainage, blocked drain pipes, flume defects (C25), pier concrete permeability (C12), main girder concrete cracks, water seepage, etc. (C1), sidewalks, railings, and damage to the lamps and lanterns marks (C24), should follow the order of the C25, C12, C1, and C24 for testing. Based on diagnostic reasoning, the staff should immediately strengthen the inspection of the drainage system condition (C25). When the drainage system is found to be in an extremely serious state, they can confirm P (C25 = State1) = 1, and update the established Bayesian network model for further analysis. When the drainage system damage (C25) is found to be normal, they should detect in turn on the pier concrete seepage (C12), the main beam concrete cracks, seepage (C1), sidewalks, railings, and lamp mark damage (C24).

According to the above analysis, the combination of disease factors is most likely to cause risks, and relevant measures are to be taken until the disease risk factors are controlled (Wang 2005).

5 CONCLUSIONS

This paper studies Bayesian network-based security risk assessment technology for concrete beam bridges. First, the analysis is based on the shortcomings of traditional methods in polymorphism and risk analysis of complex systems where accurate probability is difficult to obtain. Second, the comprehensive evaluation model of concrete beam bridge disease risk is established. Finally, taking the project of Ningan Bridge as an example, the practicability and scientificity of the evaluation model are verified. The relevant research contents and conclusions are summarized as follows:

(1) The study proposes a bridge disease analysis based on the Bayesian network comprehensive evaluation model of risk analysis, to introduce the AHP for the overall framework of the fault tree theory, through triangular fuzzy quantification expert evaluation information. Overcoming the traditional Bayesian network in the root node analysis is difficult to accurately access limitations in terms of probability.
(2) The evaluation model can carry out two-way inference of the Bayesian network, update the network in real-time according to the actual detection data, and obtain the real-time state of bridge safety. This method improves the automation, intelligence, and accuracy of the comprehensive evaluation of bridges, and can be used as a decision-making tool for risk management of the comprehensive evaluation of bridges. It provides a solution for the safety evaluation of bridge structures and a quantitative basis for the maintenance of bridges.
(3) The Bayesian network method is used to realize the real-time dynamic assessment of the disease risk of the Ningan Bridge, proving the correctness and feasibility of the evaluation method. It provides references for the daily operation and maintenance of the Ningan Bridge.

REFERENCES

Chen Lin, Xie Xuefei. Risk Analysis of Tank Car Transportation based on Bayes Network [J]. *Automotive Engineer*, 2016(02):44–46+58.

Chen Lu. *Research on the Dynamic Evaluation of Construction Safety Risk of Bridge Hanging Basket Based on Bayes Network* [D]. Huazhong University of Science and Technology, 2019.

Fan Xue-ping, LV Dagang. Bayes Prediction of bridge Structure Bearing Capacity based on DLM [J]. *Journal of Harbin Institute of Technology*, 2012, 44(12):13–17.

Shao Penglei. *Field test and Technical Status Assessment of Highway and Bridge* [D]. Zhengzhou University, 2016.

Song Naichao, Chen Suxia. Power grid Fault diagnosis Method based on Rough Set-Bayes [J]. *Chemical Automation and Instrumentation*, 2011, 38(07):816–819.

Wang Chengtang, Wang Hao, Tan Weimin, Zhong Guoqiang, Chen Wu. Possibility Evaluation of Deep foundation Pit collapse in Subway station based on Polymorbid Fuzzy Bayes Network [J]. *Rock and Soil Mechanics*, 2020(05):1–11.

Wang Guangyan, Ma Zhijun, HU Qiwei. Fault Tree Analysis based on Bayes Network [J]. *Systems Engineering Theory and Practice*, 2004(06):78–83.

Zhang Lianwen, Guo Haipeng. *Introduction to Bayes Networks* [M]. Science Press, 2006.

Zhou Jianguo, Xu Huiyun, Liu Fei. Application of Bayes Network in Risk Assessment of Operation Tunnel structure [J]. *Industrial Engineering*, 2019, 22(06):103–109.

Zhou Zhongbao, Dong Doudou, Zhou Jinglun. Application of Bayes Network in Reliability Analysis [J]. *Systems Engineering Theory and Practice*, 2006(06):95–100.

Study on characteristics of soil and water loss based on different vegetation combinations in runoff plots

Meng Qi Zhao*
Hangzhou Binjiang District Comprehensive Administrative Law Enforcement Bureau, Hangzhou, China
Hangzhou High-tech Human Resources Center, Hangzhou, China

Xiao Hong Wang*
Zhejiang Guang Chuan Engineering Consulting Co., Ltd., Hangzhou, China

ABSTRACT: In this paper, five runoff plots with different vegetation types were tested. The results show that: (1) The annual rainfall is mainly concentrated from April to September, accounting for 60.7% to 66.6% of the total annual rainfall. (2) Water and soil loss is mainly concentrated from May to September. (3) The amount of water and soil loss flows with the fluctuation of flow, indicating that water and soil loss is closely related to runoff and rainfall.

1 INTRODUCTION

Overland flow characteristic parameters are common overland flow dynamics parameters, such as slope runoff depth, velocity and other parameters, which are calculated flow shear stress and the premise of stream power. Therefore, the accurate determination of the slope flow characteristic parameters is very important. However, under natural conditions in the field, because the influence of overland flow deep is shallow and the factors like real-time rainfall and surface condition, etc. change strenuously (Chen 1992; Horton 1934), the overland flow characteristic parameters in field conditions are not easy to direct to determination while under the laboratory conditions, especially erosion static bed conditions, most of the factors can be artificially controlled so as to make the overland flow measurement possible (Zhang 2002). According to the literature review, there are 10 methods to measure the water flow on the surface of the slope at present.

In these measurements, the physical and mechanical method is currently the most widely used measurement method, but the probe method and multi-point, multi-section measurement are time-consuming and affect the test schedule because of the artificial judging of whether a stylus contact surface causing measurement errors (Hu 2015); Staining method to obtain the velocity is surface runoff maximum velocity which must be carried out to get the average flow velocity correction (Li 1996); Correction coefficient with sediment concentration changes (Bi 2009), when the sediment concentration is high, the water color changes, because the color reagent is not easy to observe and produce measurement error.

2 GENERAL SITUATION OF THE STUDY AREA

This paper mainly conducts tests on runoff plots in Anji County, Yongkang City, etc. The basic information is as follows:

*Corresponding Authors: iamfromtju@126.com and 656375320@qq.com

Anji Lake Pond Comprehensive Observation Field is located the upstream of Anji Lake Pond watershed, with a total area of 57.88 hm^2. The annual average rainfall near the runoff plot is 1153 to 1864 mm, and the erosive rainfall of more than 12 mm accounts for 73.9% of the rainfall. The soil in the watershed is mainly red soil and yellow soil.

There is a runoff plot of soil and water loss in Changshan County in the Tianma Small Watershed of the Qiantang River Basin. According to the rainfall data of Changshan Fangcun County Meteorological Station from 1980 to 2009, the average annual rainfall is 1319 to 2724 mm, and the rainfall causing water and soil loss is more than 12 mm, accounting for 78.3% of the rainfall.

3 SOIL AND WATER LOSS IN MAO ZHULIN COMPOSITE OPERATION

In the runoff plot of Anji County, three vegetation types including the pure bamboo forest and mixed bamboo forest were studied. From June to October, soil erosion monitoring and situation analysis were carried out 11 times, field rainfall was collected once, and soil loss, total nitrogen nutrient diversion loss and total phosphorus nutrient diversion loss were statistically analyzed. The specific analysis results are shown in Table 1. According to the results in Table 1, the greater the rainfall intensity is, the greater the runoff will be, and the easier the runoff will be. Therefore, the main driving force of water and soil loss is rainfall. According to the research results of the three vegetation types, the runoff coefficient of the runoff plot where the pure bamboo forest is located is large, while the runoff plot where the poplar, tung and bamboo mixed forest and the poplar, tung and bamboo mixed forest Hemerocallis fulva mixed forest are located is relatively small.

Table 1. Statistic of soil loss in the bamboo forest in different configuration modes.

Rainfall level	rainfall (mm)	Runoff coefficient (%)			Soil loss (kg/ha)		
		Pure bamboo	Bamboo – Yang Tong	Bamboo – Yang Tong – Lily	Pure bamboo	Bamboo – Yang Tong	Bamboo – Yang Tong – Lily
Moderate	30.0	1.18	1.02	1.05	9.67	10.00	7.00
Heavy	30.0	6.30	5.38	6.00	83.78	61.15	70.00
Heavy	70.0	10.00	6.00	8.00	270.00	196.00	119.00
Heavy	14.0	4.42	3.00	2.43	19.59	20.00	10.84
Heavy	24.0	5.71	5.00	2.76	55.00	32.00	20.30
Heavy	106.0	0.95	1.10	0.78	9.92	6.50	8.00
Rainstorm	17.0	1.15	1.11	0.73	9.60	10.00	5.66
Rainstorm	37.0	6.29	5.66	5.03	134.00	99.00	30.00
Rainstorm	90.0	4.12	5.00	3.00	15.76	16.00	12.00
Rainstorm	20.0	5.59	4.08	3.60	49.52	54.00	20.00
Rainstorm	35.0	1.79	1.67	1.19	4.95	5.00	2.65

The surface runoff of three runoff plots with different vegetation types was tested, and the results are shown in Figure 1. It can be seen from the research results in Figure 1 that the trend of the runoff coefficient of three runoff plots with different vegetation types is the same, with the maximum value reaching 10%, appearing on June 19, and the corresponding minimum rainstorm value is only 0.7%, appearing on July 13. The relatively heavy rainfall from June to August was collected. Among the 11 raindrops, 5 of which were distributed in June and July. Therefore, the runoff coefficient of runoff plots in June and August is relatively large.

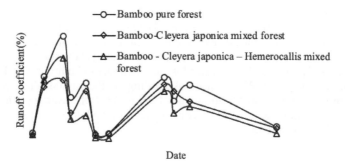

Figure 1. Variation of runoff coefficient.

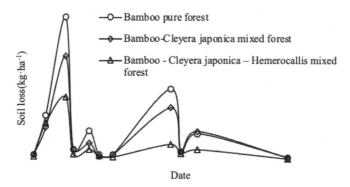

Figure 2. Changes in soil loss.

Table 2. Relationship model of surface runoff and soil loss.

Sample type	regression equation	R^2
Pure bamboo	y = 0.066x + 0.021	0.872
Bamboo – Yang Tong	y = 0.043x + 0.016	0.703
Bamboo- Yang Tong – Lily	y = 0.062x + 0.014	0.919

4 RAINFALL CHARACTERISTICS

In 2015, the statistics and analysis were carried out on the regional rainfall of five runoff plots in Anji County and Yongkang City. Figure 3 shows the details of the research results.

In 2015, five regions in Zhejiang Province were relatively rich in general, and the total rainfall after statistics was 1716.0 to 2834.5 mm. It can be seen from the research results (see Figure 1) that the rainfall fluctuation trend of the five runoff plots is obvious. Most of the rainfall is concentrated from April to September, accounting for 60.7% to 66.6% of the annual rainfall. The maximum rainfall occurs from June to July. The rainfall difference between the five runoff plots is obvious, among which Yuyao has the largest total rainfall, Changshan runoff plot is the second largest, and Yongkang and Anji are relatively small.

5 CHARACTERISTICS OF SOIL AND WATER LOSS

From the monthly runoff of the five runoff plots, the trend of runoff fluctuation is obvious. From May to September, the runoff is relatively concentrated, accounting for 67.4% to

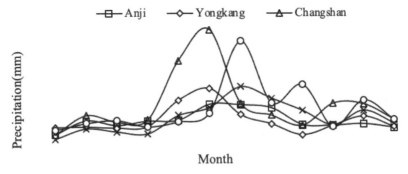

Figure 3. Monthly variation characteristics of rainfall in different regional.

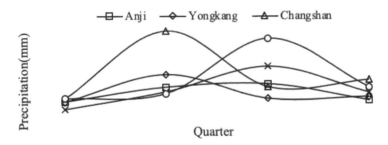

Figure 4. Seasonal variation characteristics of rainfall in different regions.

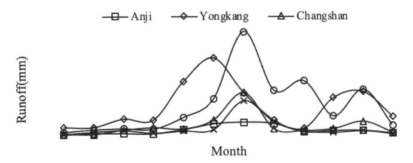

Figure 5. Monthly runoff depth in different regions.

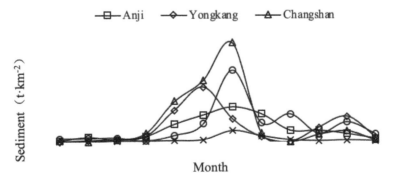

Figure 6. Monthly sediment yield characteristics in different regions.

76.2% of the total annual runoff. The communities with the largest monthly runoff are Yuyao and Yongkang, especially from May to September. The annual runoff of these two regions is concentrated in these months, and the minimum monthly runoff is Anji and Tiantai.

6 CONCLUSION

(1) Most of the rainfall is concentrated from April to September, accounting for 60.7% to 66.6% of the annual rainfall. The maximum rainfall occurs from June to July. The rainfall difference between the five runoff plots is obvious, among which Yuyao has the largest total rainfall, Changshan runoff plot is the second largest, and Yongkang and Anji are relatively small.
(2) From the monthly runoff of the five runoff plots, the trend of runoff fluctuation is obvious. From May to September, the runoff is relatively concentrated, accounting for 67.4% to 76.2% of the total annual runoff.
(3) Statistical analysis of rainfall runoff and water and soil loss in five runoff plots shows that water and soil loss fluctuates with runoff in runoff plots.

ACKNOWLEDGMENTS

This scientific research was supported by the major project of water conservancy science and technology of the Provincial Department of Science and Technology (Approval No. 2014F50015) and the Provincial Natural Science Foundation (Approval No. LY14E090005).

REFERENCES

Chen Guoxiang, Yao. *The Overland Flow Mechanics*. Hehai technology progress. 1992. 12(2): 7–13.
Horton R.E., Leach H.R., Van Vliet R. *Laminar Sheet-flow*. Transactions, American Geophysical Union. 1934. (15): 393–404.
Hu Guofang, Zhang Guanghui, Zhu Liangjun. Comparison of 3 Methods for Measuring Water Depth on the Sloping Surface. *Bulletin of Soil and Water Conservation*. 2015. 35(3): 152–155.
Li Gang, Abrahams A.D., Atkinson JF. Correction Factor in the Determination of Mean Velocity of Overland Flow. *Earth Surface Processes and Landforms*. 1996. 21(6): 509–515.
Sha Jide, Jiang Yongjing. The basic dynamic properties of nascent erosion slopes delamination water. *Journal of Soil And Water Conservation*. 1995. 9(4): 29–35.
Study on Mechanism of Slope Runoff and Sediment Transport Dynamics. Shaanxi: Northwest Agriculture and Forestry University, Yangling. 2011.
Wang Su, Sun Sanxiang, Hu Qinghua. The Domestic Research Status of Overland Flow. *Gansu Water Conservancy and Hydropower Technology*. 2004. 40(4): 332–333
Xie Chengdie. Slope flow types and calculation method of slope erosion. *Soil and Water Conservation Bulletin*. 1999. 19(4): 1–6.
Yao Wenyi, Tang Liqun. *Water Erosion and Sediment Yield Process and Simulation*. Zhengzhou: the Yellow River Water Conservancy Press, 2001.
Zhang Guanghui. Experimental Study on the Dynamic Characteristics of the Thin Layer Flow on the Slope Surface. *Advances in Water Science*. 2002. 13(2): 159–165.

Material selection and construction technology application of weak magnetic reference laboratory

Hao Guang*, Huichao Zhang, Lei Fang, Huilai Li, Wei Kong & Han Wang
China Institute of Metrology, Beijing, China

ABSTRACT: Taking the construction project of the national weak magnetic reference laboratory of China Academy of Metrology as an example, the material selection and construction technology application in the construction of buildings, power distribution, and heating are systematically expounded. The technical problems of iron, cobalt, and nickel, or building materials with magnetic strength greater than 5nT (nanoteslas), which are not allowed to be used in the weak magnetic reference laboratory, are solved, and a feasible scheme for the construction of the weak magnetic reference laboratory is proposed.

1 INTRODUCTION

Since the 18th National Congress of the Communist Party of China, under the strong leadership of the Party Central Committee with Comrade President Xi Jinping as the core, China's measurement industry has developed rapidly (Measurement Development Plan, 2021–2035). As a discipline with high requirements for experimental conditions and strong professionalism in metrology (Liu & Li 1973), magnetic metrology not only occupies a special position in the measurement technology system but also plays an important role in the rapidly developing quantum metrology reference system (Wang et al. 2020; Yao 2019; Zhang & He 2001). The detection, calibration, and research of magnetism need a standard magnetic field with high stability and a corresponding range to meet its requirements (Xu 2014). The classification into strong magnetism (ferromagnetism and ferrimagnetism), weak magnetism (paramagnetism and antiferromagnetism), and antimagnetism is based on the magnetic properties exhibited by materials (Zou et al. 2021). Taking the National Key Laboratory – Weak Magnetic Reference Laboratory as an example, this paper expounds on the technical problems of meeting the experimental conditions.

The weak magnetic reference laboratory is a laboratory for micro-magnetic detection, calibration, and research. With the rapid development of the chemical industry, shipbuilding, and other industrial fields in China, the application of the weak magnetic reference laboratory is becoming more and more wide, and the construction threshold is also getting higher and higher. As a project with special construction technology, the construction materials are required to strictly ensure to be non-magnetic or weakly magnetic, and any magnetic interference unrelated to the experiment will affect the

---

*Corresponding Author: guanghaowow@126.com

experimental results (Liu & Liu 2005; Mei et al. 2016; Wu & Zhang 2017; Zhong 2012; Zhang et al. 2012).

Therefore, a good non-magnetic experimental environment is the basic index of the construction of the weak magnetic reference laboratory.

As a national key laboratory belonging to the initial construction in China, its construction is of great significance. The evolution process, management system, and operation mechanism of its construction have been studied and explored by scholars (Wu 2022). However, the technical problems such as construction technology, selection of decoration materials, and other technical problems faced by infrastructure projects based on the building itself are ignored. This paper puts forward the construction scheme to meet the high stability and high precision environmental conditions of the weak magnetic reference laboratory from the aspects of construction site selection, equipment materials, and construction technology to innovate and achieve a breakthrough in technical problems, which would provide a reference experience for laboratory construction projects such as the weak magnetic reference laboratory.

2 ENVIRONMENTAL REQUIREMENTS AND ENGINEERING OVERVIEW

2.1 Environmental requirements

The construction of the non-magnetic experimental environment is mainly divided into site selection and construction. The interference of the electromagnetic field should be prevented in the site selection to ensure that there is no magnetic anomaly near the laboratory. The construction near large factories, such as power plants, iron and steel plants, heavy industrial manufacturing, and traffic facilities, and places that can produce electromagnetic interference, such as airports, subways, railways, and highways, should be avoided. At the same time, the magnetometer should also be used to measure the site selection. It is forbidden to use building materials and equipment containing iron, cobalt, nickel, or magnetic strength greater than 5nT in the construction of laboratories (Zhu 1986).

2.2 Engineering overview

In the early 1960s, with the consent and approval of the National Science and Technology Commission and the National Bureau of Metrology, it was decided by the National Institute of Metrology to establish a weak magnetic reference laboratory (Gan & Zhai 2005).

In 1965, with the help of the Beijing Municipal Government, the weak magnetic reference laboratory of the National Institute of Metrology, China, was constructed at the junction of Cherry Valley and XiShan Forest Farm in Beijing Botanical Garden, covering an area of about 1650 m^2. Located in the scenic forest areas, the magnetic field intensity is stable, and there would be no change in land property in the future, which is the best location (see Figure 1). Since its completion in 1966 and being put into use, the No. 6 Building in the courtyard is the weak magnetic reference laboratory (see Figure 2).

After years of use, the weak magnetic benchmark laboratory infrastructure is aging, equipment and materials are extremely damaged, and environmental conditions are deteriorating. The laboratory is unable to meet the needs of daily experiments. The 2017 Central Financial Renovation Fund of China decided to finance the project for the reconstruction of the weak magnetic reference laboratory area.

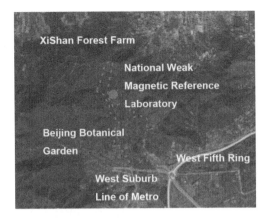

Figure 1. Construction site selection.

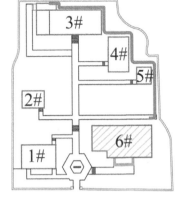

Figure 2. Campus plan.

3 CONSTRUCTION SCHEME

3.1 *Magnetic testing of building materials*

The magnetic testing of materials and equipment is the key to the construction of the weak magnetic reference laboratory. The first part is material testing in the design phase for the same category of different material products, such as roofing cement tiles and resin tiles selection, after comparing the magnetic strength preferred selection. The second step is on-site inspection. In the construction stage, the actual on-site materials are tested. It is worth noting that even in the first part of the detection of qualified products of the same brand and the same manufacturer, there will be magnetic differences due to different production batches of materials, such as white cement and limestone. Therefore, all construction materials must be tested after they arrive.

The materials and equipment of this project adhere to the principle of one inspection for each object and are subject to the reports issued by the Institute of Electromagnetics of China Academy of Metrology. The detection results are divided into three categories: (1) No magnetic, i.e., the detection magnetic is almost 0nT, which can be used in the project; (2) Smaller magnetism, i.e., magnetic strength less than 5nT, can be avoided in projects that can be used in small quantities; (3) Magnetic, i.e., magnetic strength greater than 5nT, cannot be used in projects. The test results are shown in Table 1.

3.2 *Selection of construction materials and process flow*

(1) Construction engineering

The selection of building structure forms is particularly important for the construction of the weak magnetic reference laboratory. For the common steel-concrete structure and masonry structure, the high magnetic content of steel bars and concrete in steel-concrete structures cannot meet the requirements of the weak magnetic reference laboratory conditions. In a masonry structure, limestone with magnetic content of less than 5nT is the best building material, while common red brick with magnetic content of more than 20nT is not recommended. Therefore, this project adopts the method of limestone masonry. The internal timber structure is used as the framework support, which is divided into two layers (see Figure 3).

The roof from abandoned corrugated cement asbestos tiles with a magnetic content of about 10 nT and resin tiles with lower and stable magnetism were used in the laboratory during the initial construction period.

Table 1. Comparison of magnetic materials.

Attribute	Testing materials
Magnetic-free (magnetic strength ≈ 0nT)	Copper rod, copper row, copper tube, nylon mesh, structural rubber, organic glass fiber reinforced plastic, polyethylene polypropylene cloth, PE tube, brass screw, PVC insulated electrical casing for construction, white sand, nylon bolt, inorganic aluminum salt waterproofing agent (liquid), rubber and plastic insulation, resin tile, white cement, putty powder, gypsum powder, atomic ash, wall (white rubber), self-adhesive asphalt waterproofing membrane, plastering white cement mortar
Magnetic (smaller) (Magnetic strength ≤ 5nT)	Portland cement, brass, aluminum film, aluminum sheet, aluminum sheet, cast aluminum, white sand brick, white stone Magnetic
Magnetic (Magnetic strength > 5nT)	Brass screws, red bricks, grass bricks, polymer waterproof mortar, high strength polymer mortar, cement bricks, mortar bricks, marble

Figure 3. Rubble masonry practice.

Figure 4. Distribution cabinet.

For decoration, white cement (Portland cement with calcium silicate as the main component) is more in line with the magnetic requirements of the weak magnetic reference laboratory than black cement (ordinary Portland cement) with high magnetic content. By comparing the white cement of several cement factories around Beijing, Hebei, and Anhui, six kinds of magnetic content standard products were selected for testing, and the white cement produced in An Qing City with the lowest magnetic content was selected. Laboratory insulation materials use flame retardant polyurethane foam insulation layer and epoxy resin-coated glass fiber reinforced plastic moisture-proof layer.

(2) Distribution engineering

The engineering power supply is introduced where the courtyard is located by the box transformer of Beijing Botanical Garden. The distribution system is composed of a low-voltage distribution cabinet and its outlet part. Common electrical equipment, such as copper wires, LED lamps, and socket switches in the market, are all tested to have met the magnetic requirements. Therefore, the focus of the construction of the distribution system in the weak magnetic reference laboratory is the selection of the material for the distribution cabinet box and the cable piping.

To meet the requirements of laboratory conditions, standardize the practice of laboratory power distribution system, and minimize the interference of materials in the experiment, the glass fiber reinforced plastic with magnetic content of about 0nT was selected as the material of power distribution cabinet box after comparing various materials (see Figure 4).

The customized finished glass fiber reinforced plastic distribution cabinet meets the magnetic requirements of the laboratory and conforms to the current national electrical related specifications.

The common copper and PVC pipes in cable piping materials have excellent non-magnetic properties. Compared with the two, the copper tube has good shielding characteristics, which can show a better-shielding effect after connecting to the grounding system. The disadvantage is the lack of mature accessories that need to be customized. PVC pipe has a set of standardized products, supporting pipe fittings, and mature construction technology, which can meet the specification requirements of design, construction, and acceptance and is low in cost and convenient in construction. The disadvantage is that it does not have electromagnetic shielding characteristics.

Comprehensive comparison of the characteristics of two materials, taking into account the actual use in the laboratory, to ensure that the material has minimal impact on the experimental results; the final choice is for copper tube as electrical piping.

(3) HVAC engineering

The weak magnetic reference laboratory conditions require temperature control at $23°C \pm 2°C$ and humidity control at 40% to 60%. To meet the requirements, the purified combined air conditioning unit is adopted with an air volume of $8000m^3/h$. The equipment includes direct expansion outdoor machines, fans, and ducts, which are laboratory non-magnetic customization. The main body of the unit box, ducts, water tanks, diffusers, rainproof shutters, closed control valves, fan shells, and blades are made of glass fiber-reinforced plastics. The internal surface cooler, fins, fire valves, insect nets, fan rotor, and stator of the unit are made of pure copper. The indoor temperature and humidity sensors are made of corundum materials.

Since the air conditioning unit still contains a small number of magnetic materials, it needs to be placed away from the laboratory. Therefore, the direct expansion of the outdoor unit is placed outside the west side of the air conditioning room of building 3#, and the purified combined air conditioning unit and fan are installed in the air conditioning room of building 3#. The air duct is connected to the weak magnetic reference laboratory (see Figure 2) to minimize the impact.

In the laboratory HVAC system, the selection of windpipe hoisting materials is also particularly important. The aluminum profiles and brass plates commonly used in the market cannot meet the magnetic requirements, and the strength of FRP (fiber reinforced polymers) and pure copper cannot be used. After testing and comparison, the glass fiber was finally selected, and the sling installation method was adopted. A single 3×50 mm sling is installed on the wooden keel inside the ceiling, which can withstand the 750 kg load (see Figures 5 and 6).

Figure 5. Air conditioning pipe hoisting.

Figure 6. Air conditioning pipe hoisting.

4 CONCLUSIONS

The weak magnetic reference laboratory was successfully accepted and put into use in May 2018. At present, the laboratory runs well, and the facilities and supporting systems meet the magnetic requirements of the laboratory.

The weak magnetic laboratory is very strict with the environment, and the selection of the laboratory location has a great impact on future experiments. The current situation and future actual development should be fully considered in the construction, and the local government should assist in completing the site selection if necessary.

Through a large number of tests and comparisons, the use of white cement, glass fiber reinforced plastic, copper, and other materials makes up for the blank of non-magnetic materials required for the construction of the weak magnetic reference laboratory environment and can be used as the main building materials for similar laboratories in the future.

The non-magnetic requirement of the laboratory environment makes many conventional methods in engineering unable to be implemented, which requires a lot of time and energy to study and solve. This paper gives several solutions that have certain guiding significance for the construction of weak magnetic reference laboratories in the future.

REFERENCES

Gan Chengde, Zhai Qingchang (2005). Records of the Preparation and Construction of the Wolong Buddha Temple Weak Magnetic Laboratory. *China Metrology* (11),46–48.

Liu Haiyang, Liu Yingjun (2005). *Exploration of China Ship Degaussing Laboratory: Any Magnetic Field on Earth can be Simulated*. China News Network, Liberation Army Newspaper. 2005.4.13, https://www.chinanews.com.cn/news/2005/2005-04-13/26/562332.shtml.

Liu Xingmin, Li Rong (1973). Magnetic Measurement Basis. *Electric Measurement and Instrument* (12),1–3. *Measurement Development Plan* (2021–2035).

Mei Xiangyang, Liu Zhifeng, Wu Huailiang, et al. (2016). Research on the Construction Technology of Demagnetization Laboratory. *Scientific and Technological Achievements*, 12,2,001–003.

Wang Qinping, Xuan Xiang, Zhao Ruojiang (2020). *China Metrology Science and Technology Development Research in 2020*. China Science and Technology Development Research and Scientist Seminar in 2020, Beijing, China

Wu Dandan. Prince Morning (2022). Evolution, Management System and Operation Mechanism of National Laboratory in China. *Laboratory Research and Exploration*, 41(2, 130–135).

Wu Huailiang, Zhang Lucheng (2017). On the Construction Technology Points of the Demagnetization Laboratory. *The Development Orientation of Building Materials*, (9),1672–1675.

Xu Yu (2014). *A High Accuracy Standard Magnetic Field Device*. *Shanghai Metrology Test*, (04),11–13.

Yao Hejun (2019). 90th Anniversary, Beijing Metrology Future Duration. *Metering Science and Technology*. 10, 001.

Zhang Zhonghua, He Qing. (2001) The prospect of electromagnetic measurement in the 21st century. *Modern Measurement Test*, (03),3–6.

Zhang Zongjian, Zhu Xuefeng, Tong Fei (2012). Laboratory Demagnetization Construction Technology. *Construction Technology*, (s2),351–353.

Zhong Hua (2012). Supervision Practice of Degaussing Laboratory Construction. *Construction Supervision*, (4), 154.

Zhu Yaoming (1986). Effects of iron, cobalt and nickel on the magnetic susceptibility of aluminum. *Light Metals*, (1), 55–58.

Zou Qin, Xiang Gangqiang, Luo Wenqi, Wang Mingzhi (2021). Research Progress of New Magnetic Materials. *Journal of Yanshan University*, 45 (1), 1–10.

Temperature control measures for winter construction of super large bridge concrete on expressway

Yaohui Shen*
Department of Civil Engineering, Xiamen University Tan Kah Kee College, Zhangzhou, Fujian Province, China

ABSTRACT: Based on the construction example of the bearing platform and pier abutment of a super large Expressway Bridge in Shanxi Province in winter, the temperature control measures for the reinforcement engineering and concrete mixing, transportation, pouring, curing, formwork removal and other links are formulated on the basis of thermal calculation, which can be used for reference for the quality control of concrete construction in winter.

1 INSTRUCTION

In winter engineering construction, due to the exothermic heat of cement cementitious materials in the hydration process, the internal temperature rises, while the external is frozen and contracted. If the temperature difference between the internal and external is too large, cracking will occur. In addition, the low internal temperature of concrete will also lead to the slow development of strength. When the internal temperature of concrete drops below 0°C, the ice expansion stress generated by water icing will lead to concrete cracking and strength reduction (Dong 2010). Therefore, it is of great practical significance to explore temperature control technology in winter construction.

2 PROJECT OVERVIEW

The project is located between Qixian County and Pingyao County, Shanxi Province. It is located in a cold area and generally lasts about 3 months in winter. The extreme maximum temperature over the years in the region is 37.4°C, the extreme minimum temperature over the years is −22.7°C, the average precipitation over the years is 431.2 mm, the average evaporation over the years is 1702.1mm, the average temperature in the coldest month over the years is −5.4°C, the average temperature over the years is 10.0°C, the average maximum wind speed over the years is 22 m/s, the maximum snow thickness over the years is 16 cm, and the maximum freezing depth of the soil is 77 cm. The project department is responsible for the construction of the passenger-dedicated line of the Daxi railway. The construction mileage is DK365+805-DK377+805, with a total length of 12 km, including the bored cast-in-place pile, bearing platform, pier body of 368 piers and one abutment. The total concrete volume is about 270000m^3. The construction period is 18 months. Due to the delay of demolition and other reasons, the construction progress lags behind a lot. To ensure the timely completion of the construction organization plan, the project department arranges the bored pile, bearing platform, pier and abutment for winter construction. According to

---

*Corresponding Author: yhshen@xujc.com

the local climatic conditions and the requirements of construction specifications, the winter construction period lasts from the end of November to the beginning of March of the next year, lasting for more than 3 months. Therefore, on the basis of thermal calculation, winter construction measures such as heating and thermal insulation are formulated for each link of raw materials, mixing, transportation, pouring and curing.

3 THERMAL CALCULATION

(1) Temperature of the concrete mixture $T_0(°C)$ (JGJ/t104-2011):

$$T_\circ = [0.92 \times (m_{ce}T_{ce} + m_f T_f + m_k T_k + m_{sa}T_{sa} + m_{g1}T_{g1} + m_{g2}T_{g2} + m_{g3}T_{g3})$$
$$+ 4.2T_W \times (m_W - W_{sa}m_{sa} - W_{g1}m_{g1} - W_{g2}m_{g2} - W_{g3}m_{g3})$$
$$+ c_w(W_{sa}m_{sa}T_{sa} + W_{g1}m_{g1}T_{g1} + W_{g2}m_{g2}T_{g2} + W_{g3}m_{g3}T_{g3})$$
$$- c_i(W_{sa}m_{sa} + W_{g1}m_{g1} + W_{g2}m_{g2} + W_{g3}m_{g3})]/$$
$$(4.2m_w + 0.92(m_{ce} + m_f + m_k + m_{sa} + m_{g1} + m_{g2} + m_{g3}))$$

In the formula: m_W, m_{ce}, m_f, m_k, m_{sa} and m_g are the amount of water, cement, fly ash, mineral powder, sand and stone (kg). T_W, T_{ce}, T_f, T_k, T_{sa} and T_g are the temperature of water, cement, fly ash, mineral powder, sand and stone (°C). W_{sa} and W_g are water content of sand and stone (%), c_w and c_i are the specific heat capacity of water (kJ/kg • K) and the heat of dissolution (kJ/kg). When the aggregate temperature is greater than 0°C: c_w = 4.2, c_i = 0; When the aggregate temperature is less than or equal to 0°C: c_w = 2.1, c_i = 335.

(2) Outgoing temperature of the concrete mixture is $T_1(°C)$ (JGJ / t104-2011):

$$T_1 = T_0 - 0.16 \times (T_0 - T_P), \text{ Where } T_P \text{ is the temperature in the mixer shed (°C)}$$

(3) The on-site mixed concrete shall be transported by loading and unloading vehicles, and the concrete molding temperature is $T_2(°C)$ (JGJ / t104-2011):

$$T_2 = T_1 - (A_1 \times n + A_2 \times t_2 + A_3 \times t_1) \times (T_1 - T_a)$$

Table 1. Selection of calculation parameters for the temperature of incoming formwork.

A_1	Temperature loss coefficient of loading, unloading and transfer, taken as 0.032
A_2	Temperature loss coefficient of tank car transportation, taken as 0.0042
A_3	Temperature loss coefficient of pouring and tamping, taken as 0.003
t_1	Pouring and vibrating time (min), taken as 15
t_2	Transportation time (min), taken as 50
n	Loading, unloading and transfer times, take 1
T_a	Ambient temperature

During thermal calculation, the ambient temperature is taken as 10°C below zero, the water temperature is heated to 60°C, the cement and fly ash are taken as 5°C, the stone is taken as 40°C, the temperature in the mixer shed is taken as 10°C, and the calculation is carried out in combination with the concrete mix ratio of winter construction.

According to the site construction conditions, the calculated results are: T_0 = 18.82°C, T_1 = 17.41°C, T_2 = 9.54°C.

On the basis of theoretical calculation, the temperature of the incoming formwork is ensured to meet the requirements. During the construction, the actual measurement results are combined, and the parameters are adjusted to reduce the error between theoretical calculation and actual measurement.

4 CONSTRUCTION ASSURANCE MEASURES

4.1 *Temperature assurance measures for winter construction of mixing station*

The mixing station is arranged at the station closest to the construction site to avoid heat loss due to long transportation time (Zhao 2010). In order to ensure the construction quality of concrete in winter, corresponding measures shall be taken according to the requirements of raw material temperature and ambient temperature. A 1t steam boiler is arranged in the mixing station, and four main pipes are laid respectively corresponding to the West bin, Sand and gravel upper hopper and reservoir, Laboratory, standard maintenance room and office building; The admixture storage bin and the mixing plant supply steam. (1) The first main pipeline is mainly used to heat the west side silos. The pipeline is connected from the steam boiler room to the underground heating pipeline embedded in the 7# heating silo on the west side. At the same time, a branch is laid to connect the remaining 6 silos. The pipeline is uniformly laid against the wall at the back of the silo and erected on the partition wall of each silo. At the same time, the front of each bin is set with colorful striped cloth for sealing to ensure the thermal insulation in the bin. The color-striped cloth can be opened when the loader is loading. (2) The second main pipeline is mainly used to heat the upper hopper of sand and gravel. At the same time, two branches are laid, one to the reservoir and the other to the two standard curing rooms in the north of the mixing station, to ensure that the aggregate is heated to 50°C, the mixing water is heated to 50°C to 60°C, and the temperature of the standard curing room of the test block reaches 20°C. (3) The third main pipeline is mainly used for the steam heating of the laboratory and comprehensive office building. (4) The fourth main pipeline is mainly used for the steam heating of the mixing building and the additive feeding tank. Pipes are laid around the mixing plant to supply heating, and colored steel plates are set around the mixing equipment in the mixing plant to prevent wind. The equipment is wrapped with a cotton quilt for thermal insulation. Ensuring that the temperature of the admixture reaching to 50°C, the temperature around the equipment in the mixing plant shall not be lower than 10°C. The above measures shall ensure that the concrete outlet temperature shall not be lower than 15°C, and the mixing time shall be about 50% longer than the normal temperature. The test personnel shall check the ex-factory temperature of the mixture at any time after opening and timely adjust the temperature of sand and gravel and mixing water according to the change of ex-factory temperature, pouring place and molding temperature. The pier column of the bearing platform shall be provided with a shed heater for heating or wrapped with a quilt for thermal insulation. See Figure 1 for the layout of the steam pipeline in the mixing station.

4.2 *Winter construction measures for reinforcement works*

All reinforcement processing sheds in the reinforcement processing workshop shall be closed with color strips to prevent wind and snow, so that the operating environment temperature can meet the requirements and reduce the temperature difference of welded parts. Heat preservation and antifreeze measures shall be taken for the gas source equipment. During welding, temperature control or current increase shall be adopted for each weld to reduce the welding speed and ensure that the welding quality meets the specification requirements.

The raw materials of the reinforcement shall be transported into the processing shed in advance. The welded joints shall not touch the ice and snow immediately. They can be

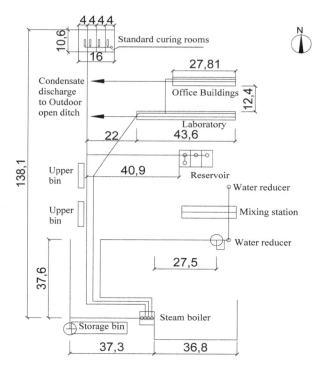

Figure 1. Steam boiler pipeline layout of mixing station (unit: m).

transported outdoors after being completely cooled. The processed reinforcement shall be stacked, covered and marked according to the use position.

During arc welding in winter, snow, wind and heat preservation measures shall be taken, and welding rods with good toughness shall be selected.

Five-point tack welding shall be used to fix the bead and the main reinforcement during the bead welding, and two points shall be used for the lap welding. The tack weld shall be more than 20 mm away from the bead or lap end. The thickness of the bead and lap weld shall not be less than 0.3d, and the weld width shall not be less than 0.8d.

4.3 Winter construction measures for concrete works

During concrete transportation, each mixer truck shall be insulated and wrapped with a rainproof quilt. The transportation time of concrete must be strictly controlled to shorten the transportation time as much as possible, and the construction shall be reasonably organized to shorten the stagnation time in the construction process. The concrete shall be cleaned timely after pouring to prevent freezing.

Concrete temperature measurement. The temperature change shall be measured at the pouring site at any time, and the molding temperature shall be timely detected to ensure that the molding temperature is not lower than 5°C. The change in concrete pouring temperature shall be fed back to the mixing station in time, and the ex-factory temperature shall be adjusted in time. When the ambient temperature is stable, it shall be monitored every two hours, and when the ambient temperature changes, each vehicle shall be monitored.

Concrete pouring. ① Before concrete pouring, the ice, snow and dirt on the formwork and reinforcement shall be removed. The pouring temperature of concrete shall not be lower than 5°C in any case (Chai 2020). The concrete shall be poured continuously in layers without interruption. The pouring thickness of each layer shall not be less than 20cm. ② When the old concrete surface and exposed reinforcement (embedded parts) are exposed to cold air, the old concrete and exposed reinforcement (embedded parts) within 1.5m and 1.0m away from the construction joint of the old and new concrete surface shall be protected from cold. ③ When the temperature is too low (lower than - 3°C), the formwork and reinforcement shall be preheated by means of a warm shed during the pouring of the bearing platform, and the temperature during the construction period of bearing platform concrete shall be ensured. The thermal insulation shed can be set up with plastic film, felt or color strip cloth. Heat sources such as the furnace and high-temperature lamp tubes are set in the shed to heat the whole pouring process and keep the temperature in the shed at no less than 5°C. The temperature at the completion of concrete pouring of the bearing platform and the beginning of curing shall be determined by thermal calculation, but shall not be lower than 5°C. During curing, the temperature at the bottom of the shed shall not be lower than 5°C, and the concrete surface shall be kept wet to ensure the construction quality of the bearing platform. ④ Before concrete pouring of the pier and abutment, tarpaulin or color strip cloth shall be used to cover and wrap the formwork. After the formwork is installed, the erected double-row scaffolds shall be used as the framework support, and a warm shed shall be set up to ensure the temperature during concrete pouring and the temperature difference between the inside and outside of the concrete before formwork removal. ⑤ During concrete mixing and pouring, when they are loaded into the mixer, the temperature of the water is measured, and coarse and fine aggregated the mixing temperature of concrete, pouring temperature and ambient temperature.

The specific measures for setting up the warm shed are as follows: (1) the warm shed uses double-row scaffolds as the support, and the tarpaulin is set up outside. The height is 2m higher than the height of the pier body. The top can be closed after the concrete is poured. (2) A coal-fired furnace is placed in the warm shed and heated with an open fire to ensure that the temperature in the shed is about 15°C. The exhaust pipe of the coal-fired boiler must be led out of the shed to discharge the flue gas outside the shed, so as to prevent gas poisoning and to prevent the high concentration of oxidized carbon from accelerating the carbonization of concrete. During concrete curing, special personnel shall be assigned to inspect the coal furnace, monitor the temperature in the shed for 24 hours and make records to ensure that the temperature at the bottom of the warm shed is not lower than 5°C. At the same time, the humidity in the warm shed shall be maintained. When the humidity is insufficient, the concrete surface and formwork shall be watered or covered with wet embankment straw bags. Concrete test blocks cured under the same conditions shall be set in the insulation shed. The formwork can be removed only after the strength of the test blocks under the same conditions reaches the design strength.

4.4 *Concrete curing for winter construction*

The bearing platform and pier body shall be cured by thermal storage method, i.e. spraying curing liquid, and then the surface shall be wrapped with plastic film, then two layers of geotextile and rainproof cloth. The moisture evaporated by the concrete itself is used to make it reach a humid environment. This method can not only keep moisture but also keep heat. Before the final setting of concrete, it shall be measured every 2 hours for the first three days and at least twice every day and night thereafter.

4.5 *Concrete formwork removal*

The formwork can be removed only when the concrete strength reaches the formwork removal strength under normal temperatures in winter, and the frost resistance requirements shall be met at the same time. The strength shall be subject to the test results of 28 days of curing under the same conditions. For the concrete cured by the warm shed method, when the ambient temperature is still below 0°C after curing, the formwork can be removed after the concrete is cooled to below 5°C.

5 CONCLUDING REMARKS

In winter engineering construction in the north, construction is difficult and there are many hidden dangers. During concrete construction, special attention shall be paid to temperature control measures. On the basis of thermal calculation, the concrete mixing, molding temperature and concrete curing temperature shall be strictly controlled. At the same time, attention shall be paid to the thermal insulation measures for reinforcement processing to ensure the construction quality of concrete in winter and finally achieve good results.

REFERENCES

Chai Maolin. Analysis and Research on Winter Construction Technology of Concrete in Plateau and Cold Regions [J]. *Scientific and Technological Innovation and Application*, 2020, 10(7): 154–156.
Dong Yanqiu. Analysis and Exploration of Concrete Winter Construction [J]. *Science and Technology Innovation Guide*, 2010 (26): 121.
JGJ / t104-2011, Winter Construction Specification of Building Engineering: JGJ/t104-2011 [S].
Zhao Xiaobin. Analysis on Matters Needing Attention in Winter Construction of Concrete Engineering [J]. *Value Engineering*, 2010, 29(21): 113.

*Engineering structure and building
quality reinforcement*

Validation of wind tunnel numerical simulation of super high-rise buildings

Yan Zhang, Mingli Wang, Zhihao Wen, Fang Deng & Ting Hu*
Chongqing College of Architecture and Technology, Chongqing, China

ABSTRACT: Wind load is not evenly distributed on the surface structure. As such, the average wind load shape coefficient based on load specification will not be able to reflect the actual wind pressure distribution. Wind tunnel tests or wind tunnel numerical simulations are required to determine the actual wind pressure distribution, which provides the basis for structural design. Wind tunnel tests can be omitted when the shapes of super high-rise buildings are relatively regular and simple. For the wind resistance design of structures, the wind pressure distribution can be determined through numerical wind tunnel analysis. It is verified that the simulated wind pressure distribution on the surface structure is similar to the test results, and the error between the numerical analysis and the test is small. The numerical simulation reflects the distribution of wind pressure on the surface of high-rise buildings. On the windward side, it is in good agreement with the test results, with an error of less than 10%. On the crosswind side and leeward side, the numerical simulation results are between the NPL (National Physical Laboratory) and TJ-2 test results. It can meet the accuracy requirements of engineering applications.

1 INTRODUCTION

Numerical wind tunnel technology is a method that uses computational fluid dynamics (CFD) to simulate the change of wind field around the structure on the computer and solve the wind load on the surface of the structure. CFD technology can not only reduce test costs and shorten test cycles, but it can also provide more detailed information than testing, which is useful for the simulation and analysis of problems in many aspects, such as studying the nature and mechanism of things. With the rapid development of computer hardware and software technology, the technology of using CFD to simulate and analyze the flow around a bluff body to obtain the wind pressure and wind speed streamline on the building surface has become more mature.

In recent years, a large number of super high-rise buildings have emerged. In structural design, wind load is often the main control load of such structures. The wind tunnel test is the most important means to obtain the wind-induced response of buildings. This paper systematically studies the wind load distribution of a super high-rise building based on the wind tunnel test and uses the finite element method to analyze and calculate the top floor displacement response to obtain meaningful conclusions (Liu 2016). Several axisymmetric 3D mountain models, such as the bell, Gaussian, and cosine, are established, and the average wind acceleration ratio at the top of the mountain under different mountain slopes is obtained through numerical calculation (Li 2011). Super high-rise buildings are flexible structures, and the wind-induced response is relatively large. Therefore, the wind effect of the

---
*Corresponding Author: 492049510@qq.com

structure has gradually become one of the main factors controlling the safety, comfort, and economy of super high-rise buildings. The research on wind effect and comfort of super high-rise buildings is an effective way to solve this problem (Liu 2012). Chen et al. (2010) simulated the average wind resultant force and resultant moment at the bottom of super high-rise buildings in different wind directions, which is similar to the wind tunnel test results, and generally, the difference between the two is no more than 15%. At the same time, the empirical expressions of fluctuating wind pressure auto spectral density and coherence function on the building surface are fitted, and the wind-induced dynamic response of the tower is analyzed using the CQC method of spatial random wind vibration. The comparison and analysis of the peak acceleration response of the top floor and the resultant force and moment of static equivalent wind load at the bottom of the tower show that the application of the special wind vibration analysis method for high-rise buildings in practical projects is feasible (Chen et al. 2010). Through the numerical simulation of wind pressure by using CFD software Fluent on the surface of an actual building project in Chongqing, the shape coefficient without the interference of surrounding buildings is calculated. Considering the influence of terrain interference on the height of the wind starting point and the climbing effect of wind on the slope, the most unfavorable shape coefficient is obtained. Based on the wind pressure analysis law of the building, a suggestion for the value of the wind load shape factor design of the project is proposed (Xu 2021). The numerical simulation analysis is carried out for a single-slab high-rise building. First, the impact of building shape change on the wind pressure distribution on the building surface is analyzed and discussed by changing the length – height ratio of the building. The results show that the length of the building has an effect on the distribution of wind pressure on the building surface. The taller the building is, the greater the maximum negative pressure on its negative pressure surface and the larger the positive pressure area on its positive pressure surface (Zhang 2022). Based on the wind tunnel test and CFD numerical simulation method, first, the deflection wind field is simulated, and second, the wind load characteristics of a square super tall building under deflection wind are studied. The influence mechanism of wind deflection is discussed based on the CFD numerical simulation results. Finally, the wind load characteristics of super-tall buildings with different aspect ratios in wind deflection fields are studied (Wei 2021). RNG k is adopted based on the project background of high-rise buildings in the commercial community of Zhonghua City. The turbulence model simulates the average wind pressure distribution on the surface of super high-rise buildings in the building group and the surrounding wind environment. It calculates two types of Reynolds number working conditions, such as Reynolds number and wind tunnel test. After increasing the incoming wind speed and model size to increase Reynolds number, the model compares them with the wind tunnel test results (Zhang 2011). Compared with the wind tunnel test, the numerical wind tunnel has many advantages, such as low cost, short cycle, wide application, etc. However, due to a lack of sufficient research and verification work, its accuracy and reliability have not been widely accepted. The numerical wind tunnel models of two typical high-rise buildings are established by using the CFD method to simulate and verify the wind pressure on the building surface. The calculation results show that the numerical wind tunnel can well reflect the distribution law of building wind pressure. Except for a few areas, such as the leeward side, the wind pressure simulation results have good accuracy and good reference value (Lv 2008).

This paper mainly establishes the verification model in Fluent, determines the verification parameters, and verifies the similarity between the wind pressure distribution on the simulated surface structure and the test results. Simultaneously, the accuracy of Fluent numerical simulation is validated by numerical wind tunnel simulation of a single CAARC standard model high-rise building in the atmospheric boundary layer, as well as analysis and comparison of numerical simulation results and wind tunnel test results. Based on this, the wind pressure distribution on the surface of the building (single and group) is determined. The CFD wind tunnel numerical simulation technology is used to simulate the wind pressure distribution on the surface of a super high-rise building under different wind directions. The validity of the wind tunnel numerical simulation analysis of the super high-rise building is verified, which provides a basis for structural design.

2 INTRODUCTION OF FLUENT SOFTWARE

The large-scale CFD commercial software Fluid16.0 is used for calculation and simulation. It is produced by Fluent, a famous CFD software company, and can be used to simulate the complex flow of fluid from incompressible fluids to highly compressible fluids. Fluent has been widely used in oil and gas production, heat exchange and ventilation, turbomachinery, material processing, electronics and HVAC industry, fire research, automotive industry, aerospace, and architectural design (Wang 2008).

Gambit is a supporting pre-processing software provided by Fluent, which can be used to generate geometric structures and computational domain meshes required for research or engineering problems or generate triangles, a tetrahedron or hybrid mesh with T grids under known boundary grid conditions. Refer to Table 1 for the block diagram of wind tunnel numerical simulation analysis of super high-rise buildings using this software.

Table 1. Analysis of block diagram.

	Method	Objectives
Numerical wind tunnel	Wind field fluid simulation Establishment of engineering model Numerical calculation analysis	Wind pressure distribution shape coefficient

3 VALIDATION OF ANALYSIS EFFECTIVENESS

3.1 *Verification I*

Establishing the model shown in Figure 1 in Fluent. Among them,

$L = 3660$ m, $B = 1260$ m, $H = 900$ m, $l = 60$ m, $b = 60$ m, $h = 300$ m.

The flow field is uniform, incoming velocity is $V = 20$ m/s and air density is $\rho = 1.225$ kg/m^3.

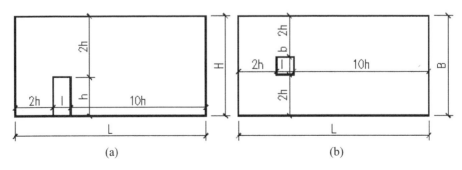

Figure 1. Verification of the geometric dimensions of the model.

Fluent calculation results are shown in Figure 2.

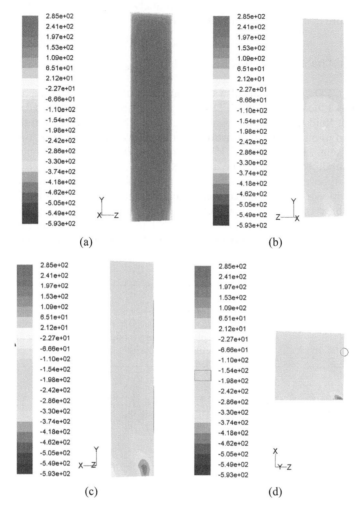

Figure 2. Verification of the wind pressure distribution on the model surface. (a) Front pressure distribution (b) Back pressure distribution (c) Side pressure distribution (d) Top pressure distribution.

According to the formula:

$$C_{pi} = \frac{\omega_i}{\frac{1}{2}\rho V^2} \tag{1}$$

The pressure coefficient of the building surface can be obtained C_{pi}, as shown in Table 2 and Figure 3.

Table 2. Surface pressure coefficient of validation model.

Pressure coefficient	Front	Top surface	Back	Side
C_{pi}	1.14	−1.14	−0.44	−0.79
	0.96	−0.79	−0.62	−0.62
	0.79		−0.79	−0.79
	0.61		−0.62	−0.97
	0.44			

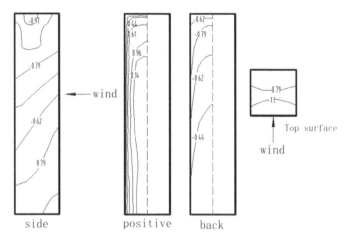

Figure 3. Pressure distribution on the surface of the validation model in a uniform flow field.

Figure 4 shows the measured average pressure coefficient with the free flow velocity as the reference speed. The dotted line in the figure represents the pressure, and the solid line represents the suction (Liu 1992).

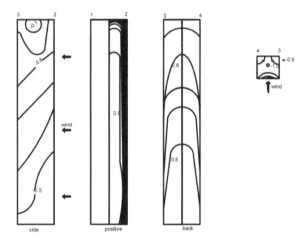

Figure 4. Pressure distribution on the surface of a high-rise building model in a uniform flow field.

3.2 Verification II

CAARC high-rise building standard model is an international standard wind engineering model proposed in 1969. It is mainly used to test the wind tunnel test simulation technology of high-rise building models to ensure the reliability of wind tunnel test measurement quality data.

This section verifies the accuracy of Fluent numerical simulation by conducting numerical wind tunnel simulation on a single high-rise building of the CAARC standard model in the atmospheric boundary layer and by analyzing and comparing the numerical simulation results with the wind tunnel test results.

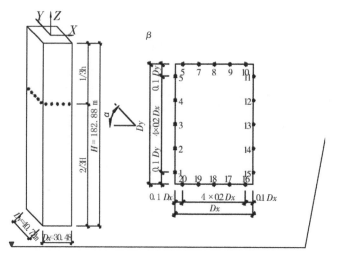

Figure 5. The layout of CAARC geometric dimensions and pressure measuring points.

The geometric dimension of the CAARC standard model is 30.48 m × 45.72 m × 182.88 m (100 ft × 150 ft × 600 ft). Full-scale modeling is conducted in Fluent, and the calculated watershed is 1800 m × 600 m × 1000 m, and the building is located 1/3 ahead of the river basin. See Figure 5 for the geometric dimensions of the model and the layout of pressure measuring points.

- Inflow surface: Average wind speed profile is $V(z) = V_0(z/z_0)^\alpha$, where z_0 and V_0 are the wind speeds at the two reference heights, respectively. According to the TJ-2 wind tunnel test of Tongji University (Luo 2004), the roof height (183 m) and top wind speed of the building are considered (this section only analyzes the wind field of category D; the test wind speed is 12.7 m/s); α is the ground roughness index; the value of Class D wind field is 0.3.
- Outflow surface: Since the outflow is close to full development, the fully developed outflow boundary condition is adopted.
- Top and both sides of the basin: Symmetry boundary condition is adopted, which is equivalent to the free sliding wall.
  ○ Building surface and ground: Wall condition without slip is adopted.

The turbulence model refers to the turbulence intensity value in Japanese specification: $I(z) = 0.1(z/H)^{-0.05-\alpha}$; Turbulent kinetic energy k and turbulent dissipation rate: $\varepsilon = 0.09^{0.75} k^{1.5}/L_x$. The L_x formula suggested by Japan is adopted $L_x = 100(z/30)^{0.5}$.

The average wind speed profile, turbulence intensity, k and ε above are all implemented using Fluent's UDF programming as the interface with Fluent.

When the wind direction angle is 0 degrees, the pressure coefficient of the building surface C_{pi} (the flow pressure at the height H of the model top is taken as the reference wind pressure) can be obtained according to the formula $C_{pi} = 2\omega_i/\rho V^2$, Where ρ is the air density, taken as 1.225 kg/m³, V is the wind speed at the top of the building, that is, the wind speed at height.

The international standard measuring point is located at a height of 2/3 H (122 m). The comparison between the calculation results in this section and the TJ-2 wind tunnel test data of Tongji University and the test data of NPL (National Physical Laboratory) at the standard measuring point is shown in Figure 6.

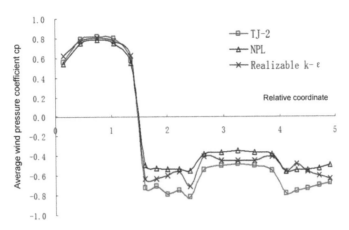

Figure 6. Comparison of shape coefficient distribution along the surface of buildings at 2/3H height under category D wind field at 0-degrees wind direction angle.

4 CONCLUSIONS

The actual wind pressure distribution cannot be obtained from the average wind load shape coefficient based only on load specification. Wind tunnel tests or wind tunnel numerical simulations are required to determine the actual wind pressure distribution, which also provides a basis for structural design. Wind tunnel tests can be excluded if the shape of super high-rise buildings is relatively regular and simple. The wind pressure distribution for wind resistance design of structures can be determined through numerical wind tunnel analysis. The large-scale CFD commercial software Fluid16.0 is used for calculation and simulation.

After verification, the conclusions are as follows:

(1) Verification 1 shows that the simulated wind pressure distribution on the surface structure is similar to the test results, and the error between the numerical analysis and the test is small.
(2) Verification 2 shows that the numerical simulation can better reflect the distribution of wind pressure on the surface of high-rise buildings; On the windward side, it is in good agreement with the test results, with an error of less than 10%; On the crosswind side and leeward side, the numerical simulation results are between the NPL and TJ-2 test results. It can meet the accuracy requirements of engineering applications.
(3) Instead of a wind tunnel test, the wind pressure distribution for the wind resistance design of structures can be determined through numerical wind tunnel analysis.

REFERENCES

Chen Wei, Huang Bencai, Zhao Jinsong, Bao Zuo. Numerical Simulation of Wind Load and Spatial Wind Vibration Analysis of Super High-rise Buildings. *Journal of Zhengzhou University* (Engineering Edition). 2010,31 (05): 60–64.

Liu Runfu. Wind Load Analysis of a Super High-rise Building [J]. *Low Carbon World*. 2016, (30): 149–150.

Li Zhengliang, Wei Qike, Huang Hanjie, Sun Yi. Research on Wind Induced Response of Mountain Super High Rise Buildings [J]. *Vibration and Shock*. 2011,30 (05): 43–48.

Liu Runfu. *Research on Wind Induced Vibration and Comfort of Super High Rise Buildings* [D]. Southwest Jiaotong University, 2012.

Liu Shangpei, Xiang Haifan and Xie Jiming. *Effect of Wind on Structure* [M]. Shanghai: Tongji University Press, 1992.

Luo Pan *Wind Tunnel Test Research Based on Standard Model* [D]. Tongji University, 2004.

Lv Fu, Yang Shichao, Zhan Jiemin, Li Qingxiang, Xu Wei, Tang Jian, Gao Ruofan. Numerical Wind Tunnel Simulation of Wind Pressure on High-rise Buildings. *Journal of Sun Yat-sen University* (Natural Science Edition). 2008,47 (S2): 117–121.

Wei Min. *Research on Wind Effect of Super Tall Buildings Under Deflection Wind Based on Wind Tunnel Test and Numerical Simulation* [D]. Chongqing University. 2021.

Wang Bin, Yang Qingshan *CFD Software and its Application in Building Wind Engineering*[J]. Industrial buildings, 2008, 28.

Xu Shilin. *Numerical Wind Tunnel Simulation of a High-rise Building in Chongqing Sichuan Architecture* [J]. 2021,41 (05): 157–159.

Zhang Dongbing, Liang Shuguo, Chen Yin, Zou Yao. Comparison of Numerical Simulation and Wind Tunnel Test Results of High-rise Building Wind Field [J]. *Journal of Wuhan University of Technology*. 2011,33 (04): 104–108.

Zhang Shiqi. *Numerical Wind Tunnel Simulation of Slab High-rise Buildings Based on CFD* [D]. Jilin University of Architecture and Architecture, 2022.

Experimental study on RC beams strengthened with prestressed and anchor steel plates

HaiDong Lei*, YanHai Liu* & AiJun Li*
Lanzhou Jiaotong University, China

ZhaoXiong Li*
Gansu New Urban Development & Construction Operation Group Co., Ltd., Gansu, China

ABSTRACT: In this paper, three T-shaped ordinary reinforced concrete beams are taken as the test objects. Through the four-point bending test, the experimental study of the test beam is carried out only in the tension side of the tensioned external prestressing reinforcement method and the tension side of the tensioned external prestressing combined with the compression side of the anchor plate. The stiffness, failure mode, strain change, crack development and deflection deformation of the beam under the two reinforcement methods are obtained, and the ultimate flexural bearing capacity of the beam under the two reinforcement methods is deduced. The comparison between the theoretical calculation results and the test results of the control beam and the reinforced test beam shows that both reinforcement methods can improve the ultimate bearing capacity of the beam. However, compared with the tensile side reinforcement method, the compression side is bonded with the steel plate and the tensile side is tensioned. External prestressing can significantly improve the flexural bearing capacity of the beam, and the experimental phenomena show that the combined reinforcement method can effectively suppress the development of cracks. The test results have certain reference significance for the reinforcement of ordinary concrete T beams.

1 INTRODUCTION

With the increase in bridge service life and the improvement of traffic load level, concrete bridges begin to show different diseases such as cracks, damage, ageing and so on. Direct demolition will bring a heavy economic burden, so the reinforcement of existing old bridges has become an important measure for the sustainable development of bridge construction.

In 1967, Hermite L.R. (1976) et al. carried out an experiment on strengthening concrete beams with glued steel plates. In 1975, Solomon S.K. (1975) et al. obtained the flexural capacity and failure mode of this component by means of an indoor failure test. In 1988, Jones R. (1988) et al. analyzed the end anchorage problem of steel plates on the tensile side of reinforced concrete beams. The test results show that there is stress concentration and peeling at both ends of the steel plate. According to the test data of steel plate reinforced beam in compression zone and theoretical analysis, Wang Linge 2007) put forward the calculation method of short-term deflection and stiffness of reinforced beam in serviceability limit state. Yang Suhang (2016) deduced the calculation equation of the local buckling strength of steel plates in a compression zone and established the calculation formula to prevent local buckling failure of steel plates in the compression zone. Sandile D. Ngidi et al.

---

*Corresponding Authors: lzjdlhd@163.com, 1657808489@qq.com, 1371381398@qq.com and 18336926637@163.com

(2018) used bonded steel plates with different width-to-thickness ratios to repair pre-cracked reinforced concrete beams and evaluated their performance. In order to study the bending behaviour of reinforced concrete beams, Rakgate, S.M et al. (2018) used externally bonded steel plates (EBSP) with different width-to-thickness ratios to reinforce the beams in bending. Rawan Al-Shamayleh et al. (2022) proposed experimental and analytical studies on the shear and flexural properties of reinforced concrete (RC) beams strengthened with two externally bonded carbon fibre reinforced polymer (CFRP) composites. Jamal A. Abdalla et al. (2022) explored the structural integrity of RC beams strengthened in bending using externally bonded aluminium alloy plates, taking advantage of the superior properties of aluminium alloy plates. Thamrin R et al. (2012) conducted a reinforcement test on reinforced concrete beams through bonded steel plates with webs to verify the stiffness and bending resistance of the reinforced beams. The work of the above scholars shows that the bonding steel plate has a significant effect on the improvement of the mechanical properties of the beam. But the disadvantage is that the discontinuity of the plate will cause stress concentration so the steel plate from the end of the plate began to debond, causing loss of reinforcement effect finally.

External prestressing reinforcement technology can effectively improve the bearing capacity of highway bridges, achieve the purpose of bridge reinforcement by improving the stress state of the bridge structure, and enhance the safety of bridges. In this paper, the prestressed reinforcement of the compression side of the T-section concrete beam is proposed, and the reinforcement method is studied.

2 TEST OVERVIEW

2.1 Test beam design

According to the *Code for Design of Highway Reinforced Concrete and Prestressed Concrete Bridges and Culverts (JTG 3362-2018)*, the section size and the design of the internal reinforcement were carried out. Considering the specific conditions of the school laboratory, the test object was the reinforced concrete simply supported T-beam. The section height of the T-beam was 250 mm, the width of the wing plate was 320 mm, the thickness of the wing plate was 50 mm, the thickness of the web was 80 mm, and the calculated span was 2400 mm. In order to facilitate loading, the total length of the beam was 2600 mm. In the test, the concrete strength is C30, and the thickness of the concrete net protective layer is 20 mm. Four HRB400 ribbed steel bars with a diameter of 12 mm were selected for the tensile longitudinal reinforcement at the bottom of the T-beam, and the reinforcement ratio was 1.56 %. Four HPB300 plain round steel bars with a diameter of 6 mm were selected for the top vertical reinforcement. The HPB300 plain steel bar with a diameter of 6 mm is selected as the stirrup, the spacing of the dense area is 100 mm, the spacing of the non-dense area is 125 mm, and the length of the dense area is 400 mm. The cross-sectional dimensions and reinforcements are shown in Figure 1.

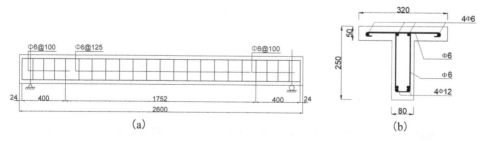

Figure 1. Section size and reinforcement diagram of test beam (unit: mm). (a) side view (b) front view.

2.2 Test reinforcement scheme

The test was loaded by four-point bending. Three concrete T beams are numbered A1, A2, and A3. A1 was the reference beam without any treatment, as shown in Figure 1. The A2 test beam was not drilled but reinforced by external prestress. When the concrete beam was strengthened by external prestressing, due to the existence of prestressing, the beam would arch upward and delay cracking. Therefore, in this test, the external prestressed steel bar adopted 1 × 7 standard prestressed steel strand with high strength and low relaxation. It had a nominal diameter of 15.2 mm, a cross-sectional area of 139 mm^2, and a standard tensile strength of $f_{pk,e}$ = 1860 MPa. The longitudinal arrangement was linear according to the specification (JTG/T J22-2008), as shown in Figure 2.

The A3 test beam was drilled, and the tension side was reinforced by external prestress while the compression side was reinforced by an anchor plate. In order to prevent the phenomenon of the steel plate and concrete stratification, the hole anchoring treatment was carried out at the four-point, fulcrum position and mid-span position, as shown in Figure 3. Combined with the section size of the concrete T beam, the diameter of the hole in the test was determined to be 8 mm. In order to ensure the reinforcement effect of the compression side anchor plate reinforcement, in this experiment, the steel plate was Q235 steel, and the design thickness was determined to be 4 mm. In order to ensure the maximum synergistic deformation of the reinforcing steel plate and the original bridge deck, the steel plate was bonded to the concrete beam by the building structural adhesive, and the steel plate was fixed by the anchor bolt, so that the concrete beam, the steel plate and the anchor bolt were integrated, as shown in Figure 4.

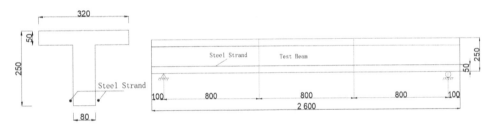

Figure 2. Prestressing steel strand arrangement diagram (unit: mm).

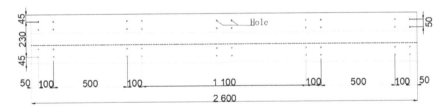

Figure 3. Pivot point, quarter point and span opening location diagram (unit: mm).

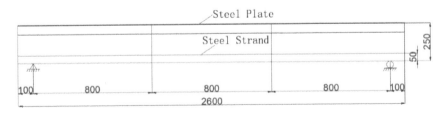

Figure 4. Anchored steel plate and external prestressing reinforcement diagram.

2.3 Test loading scheme

Two-point symmetrical loading was used to form a pure bending section area in the middle of the span to eliminate the influence of shear force on the bending performance of the normal section. The support was set at 100 mm from the beam end, and the pure bending area was 800 mm.

The loading process is carried out according to the *'Concrete Structure Test Method Standard' (GB / T50152-2012)*.

1) The specific loading process of reference beam A1 is: ① Preloading: loading to 6 kN with 2 kN/level, checking whether the support is stable, and unloading the load after the instrument and loading equipment are normal; ② Formal loading: 2 kN/stage until the first crack occurs, 5 kN per stage until the rebar yields, slow loading until the beam is finally destroyed.

2) The specific loading process of reinforced beam A2 is: ① preloading; ② The first loading: according to the benchmark beam A1 test results, set the crack width 0.1mm as the damage value, monotonic static loading to the damage value; ③ External prestressing: prestressing the steel strand on both sides of the web under unloading; ④ The second loading: 5 kN/grade loading until the steel yield, according to the mid-span deflection control loading until the beam finally destroyed.

3) The specific loading process of reinforced beam A3 is ① preloading; ② The first loading: According to the test results of the reference beam A1, the crack width of 0.1 mm is set as the damage value, and the damage value is monotonically loaded. ③ External prestressing after anchoring the steel plate: bolt anchoring the steel plate under unloading and prestressing the steel strand on both sides of the web; ④ The second loading: 5 kN/grade loading until the steel yield, according to the mid-span deflection control loading until the beam finally destroyed.

Five strain gauges were arranged on the mid-span side of the beam to measure the concrete strain. One dial indicator was arranged at both ends of the beam, midspan loading point and 1/4 and 3/4 positions of the beam to measure the deflection of the beam. Eight strain gauges were placed on the tensile steel to measure the strain. Three strain gauges were arranged on the steel plate to measure the strain. Four strain gauges were arranged in the middle of the steel strand to measure the strain of the steel strand, as shown in Figure 5.

The test focuses on measuring the following: The strain of concrete, the strain of steel bar and steel plate, the deflection under different loads, the development of cracks and so on. The loading method was shown in Figure 6.

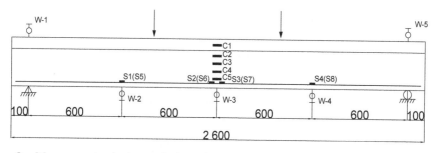

Figure 5. Measurement point layout (unit: mm).

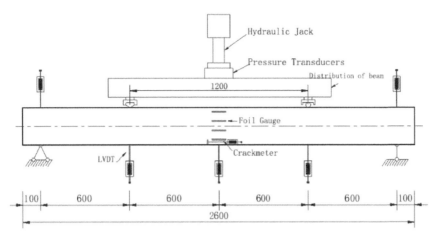

Figure 6. Panoramic view of test beam field loading.

3 TEST PHENOMENA AND ANALYSIS

3.1 *Test phenomena and failure modes*

The loading failure mode of each test beam is shown in Figure 7.

Figure 7. Destruction pattern diagram.

Cracks began to appear in the pure bending section of the A1 beam. As the load increases, the cracks continue to extend to the top of the beam, the width continues to expand, some cracks converge at the bottom of the beam, and finally, the test is stopped due to the crushing of the concrete in the compression zone. Compared with the A1 beam, the number of cracks in the pure bending section of the A2 and A3 beams is less and the width is smaller. Finally, the test was stopped because the concrete in the compression zone was crushed. The difference is that the cracks in the pure bending section of the A2 beam are evenly distributed. The inclined cracks of the A3 beam increased, and finally, shear failure occurred and the loading stopped.

3.2 *Load-strain analysis*

The load-strain curve of the tensile steel bar is shown in Figure 8. It can be seen from the figure: The load-steel strain curves of the A2 beam and A3 beam are the same. However, the externally prestressed reinforced beam enters the yielding stage earlier than the composite reinforced beam. It shows that the combined reinforcement can delay the yield of tensile reinforcement of the test beam. When the steel strain is the same, the load of the A2 beam is 66.03% higher than that of the A1 beam. The load of the A3 beam is 77.09% higher than that of the A1 beam and 6.66% higher than that of the A2 beam. That is, the combined reinforcement improves the yield load of the concrete test beam. Under the same external load, the strain of the A2 beam is 39% of that of the A1 beam. The strain of the A3 beam is 25% of that of the A1 beam and 64% of that of the A2 beam. That is, the composite reinforced concrete test beam can reduce the tensile strain of the longitudinal reinforcement, thereby improving the stiffness of the component.

3.3 Load-deflection analysis

According to the measured data, the load-midspan deflection curves of three beams under the four-point bending test are drawn, as shown in Figure 9.

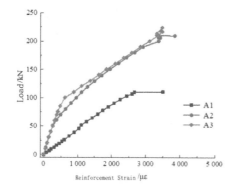

Figure 8. Load-strain curve comparison chart.

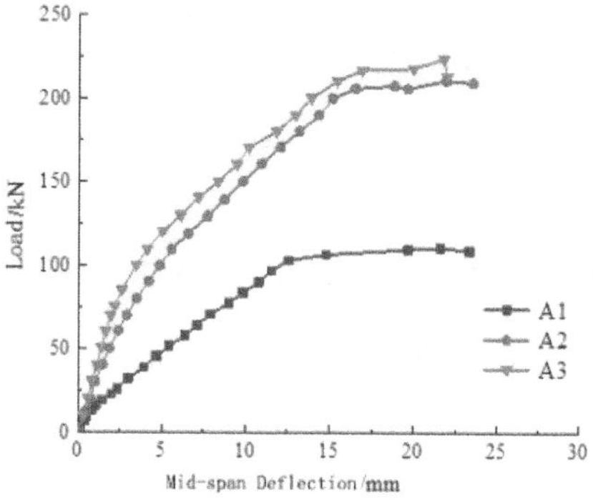

Figure 9. Load-deflection curves for beams A1, A2, A3.

From Figure 9, it can be concluded that: During the loading process, the load-midspan deflection curves of test beams A2 and A3 have almost the same trend. The slope of the load-mid-span deflection curve of the A3 beam is greater than that of A2, which is significantly greater than that of A1. Before the concrete cracks again, the load-midspan deflection curve of the A1 beam is linear. When the mid-span deflection is the same, the load that the A3 beam can withstand is higher than that of A2, which is 1.2 times that of A2, and significantly higher than that of A1, which is 2.05 times that of A1. It shows that compared with unilateral reinforcement, the combined reinforcement can improve the performance of the beam, further reduce the deflection of the beam, increase the stiffness and improve the deformation capacity of the concrete beam.

3.4 Analysis of bending capacity

This section mainly studies the bending capacity of three test beams under external load. By comparing the yield load and ultimate load of each test beam, the reinforcement effect of composite reinforcement on concrete beams is analyzed.

An equivalent rectangular compressive stress distribution diagram is used to simplify the processing. In the diagram, β denotes the ratio of the height of compressive stress zone x to the height of neutral axis xc assumed by plane section, $\beta = x/x_c$. γ represents the ratio of the stress σ of the compressive stress diagram to the maximum stress $\sigma 0$ of the concrete in the compression zone.

Figure 10 is the concrete compressive stress distribution map of the prestressed reinforced concrete T-beam under the ultimate state. Calculate the flexural bearing capacity of the normal section according to the diagram.

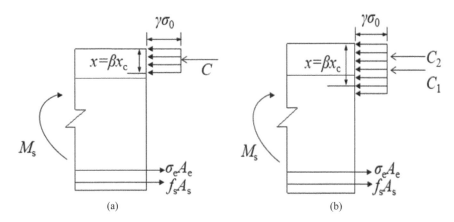

Figure 10. The compressive stress distribution map of concrete T-beam strengthened by external prestressing. (a) Sectional compressive stress distribution of the first kind (b) Sectional compressive stress distribution of the second kind.

For the first kind of T-sections ($x \leq h_f$):

$$\Sigma x = 0, f_s A_s + \sigma_e A_e = \gamma \sigma_0 x b_f = \gamma \sigma_0 \beta \xi_c h_0 b_f \tag{1}$$

$$\Sigma M_s = 0, \gamma \sigma_0 b_f x \left(h_{01} - \frac{x}{2} \right) = (f_s A_s + \sigma_e A_e)\left(h_{01} - \frac{x}{2} \right) \tag{2}$$

For the second kind of T-sections ($x > h_f$):

$$\Sigma x = 0, f_s A_s + \sigma_e A_e = \gamma \sigma_0 x b + \gamma \sigma_0 h_f (b_f - b) \tag{3}$$

$$\Sigma M_s = 0, \gamma \sigma_0 b x \left(h_{01} - \frac{x}{2} \right) + \gamma \sigma_0 (b_f - b) h_f \left(h_{01} - \frac{h_f}{2} \right) = (f_s A_s + \sigma_e A_e)\left(h_{01} - \frac{x}{2} \right) \tag{4}$$

In the formula:

σ_e-when the component reaches the ultimate bending bearing capacity, the stress value of the external steel strand;
A_e-cross-sectional area of external steel strand;
h_{01}-Distance from resultant point of internal tensile steel bar and external steel strand to top surface of the beam; C-Compressive stress resultant force of concrete in compression zone;
M_s-calculate the bending capacity of the normal section;
f_s-Design value of tensile strength of tensile steel bar;
A_s-cross-sectional area of tensile reinforcement;
b_f-calculating the width of the section flange plate;
h_f-calculate the height of the section flange;
b-calculating the width of the section web;
h_0-Calculate the effective height of the section.

After the external prestressing reinforcement, the height of the compression zone may be too high, and there is a risk of over-reinforcement ($\xi > \xi_b$), which cannot continue to improve the flexural bearing capacity, resulting in an unsatisfactory reinforcement effect. In order to improve the flexural bearing capacity to the expected reinforcement effect, the compressive side is strengthened by anchoring the steel plate. By reducing ξ to $\xi < \xi_b$, the concrete in the compression zone will not be crushed prematurely while making full use of the external steel strand.

We assume that the distance between the point of action of the steel plate's resultant force and the section's compression edge is a_s. Although the steel plate can be pressed together with the concrete, it does not mean that it can be as whole as the cast-in-place concrete. The stress value of the steel plate should be corrected, taking $\sigma_p = \kappa \sigma_{p1}$, where $\kappa = 0.2$.

Figure 11 is the compressive stress distribution of the normal section of the reinforced concrete flexural member under the limit state. The flexural bearing capacity of the normal section is calculated according to the diagram.

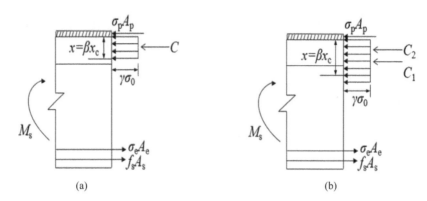

Figure 11. Compressive stress distribution of composite reinforced concrete T beam. (a) Sectional compressive stress distribution of the first kind (b) Sectional compressive stress distribution of the second kind.

The first kind of T-shaped section ($x \leq h_f$):

$$\Sigma x = 0, f_s A_s + \sigma_e A_e = \gamma \sigma_0 x b_f + \sigma_p A_p = \gamma \sigma_0 \beta \xi_c h_0 b_f + \sigma_p A_p \quad (5)$$

$$\Sigma M_s = 0, \gamma \sigma_0 b_f x \left(h_{01} - \frac{x}{2}\right) + \sigma_p A_p (h_{01} + a_p)$$
$$= (f_s A_s + \sigma_e A_e)(h_{01} + a_p) - \gamma \sigma_0 b_f x \left(\frac{x}{2} + a_p\right) \quad (6)$$

Type II T - section ($x > h_f$):

$$\Sigma x = 0, f_s A_s + \sigma_e A_e = \gamma \sigma_0 x b + \gamma \sigma_0 h_f (b_f - b) + \kappa \sigma_p A_p \quad (7)$$

$$\Sigma M_s = 0, \gamma \sigma_0 b x \left(h_{01} - \frac{x}{2}\right) + \gamma \sigma_0 (b_f - b) h_f \left(h_{01} - \frac{h_f}{2}\right) + \sigma_p A_p (h_{01} + a_p)$$
$$= (f_s A_s + \sigma_e A_e)(h_{01} + a_p) - \gamma \sigma_0 b x \left(\frac{x}{2} + a_p\right) - \gamma \sigma_0 (b_f - b) h_f \left(\frac{h_f}{2} + a_P\right) \quad (8)$$

In the formula:

σ_p-Steel plate stress design value;
A_p-Section area of steel plate;
a_p-Distance from the resultant force point of the steel plate to the top surface of the beam.

By comparing the yield load and ultimate load of each test beam, as shown in Table 1, analysis of the reinforcement effect.

Table 1. Test beam load capacity.

Test beam number	Yield load/kN	Ultimate load/kN	Bearing capacity improvement/%
A1	106	114	–
A2	206	210	84.2
A3	217	224	96.5

Compared with the reference beam A1, it is found that the cracking load of the externally prestressed reinforced beam A2 is 4 times higher than that of the A1 beam, the yield load is increased by 94.3%, and the ultimate bearing capacity is increased by 84.2%. External prestressing reinforcement is a very effective reinforcement method. It is found that the cracking load of A3 is 5 times higher than that of A1, the yield load of A3 is 1.04 times higher than that of A1, and the ultimate bearing capacity of A3 is 96.5 % higher than that of A1. It can be seen that the combined reinforcement is better than the single external prestressing reinforcement. It not only improves the cracking load of the concrete T beam, delays the cracking of the concrete in the tensile zone, but also improves the yield load and ultimate bearing capacity of the concrete beam, and the increase is considerable.

3.5 *Crack analysis*

The development and distribution of cracks in the three test beams are shown in Figure 12. From the crack distribution map and test results, it can be seen that: The cracks of the A1 benchmark beam are more and more densely distributed in the mid-span pure bending section. Most of them are vertical cracks and run through the bottom of the beam. The diagonal cracks of the shear span are mostly 45 degrees, extending from the beam end to the loading point. The cracks in the pure bending section of the A2 externally prestressed reinforced beam are mainly vertical cracks, and the number of cracks is small. Some cracks extend to the flange plate of the beam, and the crack width is smaller than that of the reference beam. Shear span diagonal cracks increased significantly and the spacing is more average; The crack distribution of the A3 composite reinforced beam is slightly different from that of other test beams. The mid-span pure bending section is basically vertical cracks, with a small number of cracks and uneven spacing. The oblique cracks in the shear span section are obviously increased. Finally, the test beam was stopped because the concrete beam was damaged by too many inclined cracks on the left side.

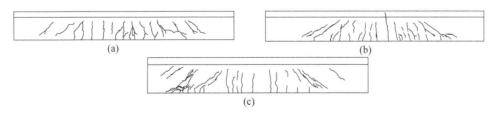

Figure 12. Distribution of cracks during the damage of the test beam. (a) A1 beam crack distribution map (b) A2 beam crack distribution map (c) A1 beam crack distribution map.

4 CONCLUSION

In this paper, the following conclusions are obtained by studying the mechanical properties and flexural bearing capacity of three test beams:

1) Under the same external load, the strain of the A3 beam is 25 % of that of the A1 beam and 64 % of that of the A2 beam. That is, the combined reinforcement effectively improves the yield load of the concrete test beam; Under the same external load, the composite reinforced concrete test beam can reduce the tensile strain of the longitudinal reinforcement, thereby improving the stiffness of the component.
2) When the mid-span deflection is the same, the load that the A3 beam can withstand is higher than that of A2, which is 1.2 times that of A2, significantly higher than that of A1, which is 2.05 times that of A1. It shows that compared with unilateral reinforcement, the combined reinforcement can improve the performance of the beam, further reduce the deflection of the beam and improve the deformation capacity of the concrete beam.
3) The cracking load of A3 is 5 times higher than that of A1. The yield load of A3 is 1.04 times higher than that of A1. The ultimate bearing capacity of A3 is 96.5 % higher than that of A1. It is proved that the combined reinforcement is better than the single external prestressing reinforcement, which can further delay the generation and expansion of new cracks. It not only improves the cracking load of the concrete T beam, delays the cracking of concrete in the tension zone, but also improves the yield load and ultimate bearing capacity of the concrete beam, and the increased range is considerable.

REFERENCES

Al-Shamayleh Rawan, Al-Saoud Huda, Abdel-Jaber Mu'tasim, Alqam Maha. (2022). Shear And Flexural Strengthening of Reinforced Concrete Beams with Variable Compressive Strength Values Using Externally Bonded Carbon Fiber Plates. *J. Results in Engineering*. 14.
GB/T 50152-2012. (2012). Standard for Test Method of Concrete Structures. S.
Hermite L. R. (1967). Concrete Reinforced with Glued Steel Plates[C] Synthetic Resins in Building Construction. *J. Paris*: RILEM International Symposium. 175.
Jones R., Swamy R. N., Charif A. (1988). Plate Separation and Anchorage of Reinforced Concrete Beams Strengthened by Epoxy-bonded Steel Plates. *J. Structural Engineer*. 66(5), 85–94.
Jamal A. A., Rami A. H., Hayder A. R. (2022). Behavior of Reinforced Concrete Beams Strengthened in Flexure using Externally Bonded Aluminum Alloy Plates. *J. Procedia Structural Integrity*.
JTG 3362-2018. (2018). *Specifications for Design of Highway Reinforced Concrete and Prestressed Concrete Bridges and Culverts*. S.
JTG/T J22-2008. (2008). Specincations for Strengthening Design of Highway Bridges. S.
Rakgate, S. M., Dundu. (2018). Strength and Ductility of Simple Supported R/C Beams Retrofitted with Steel Plates of Different Width-to-thickness Ratios. *J. Engineering Structures*.
Rendy Thamrin, Ricka Puspita Sari. (2017). Flexural Capacity of Strengthened Reinforced Concrete Beams with Web Bonded Steel Plates. *J. Procedia Engineering*. 171, 1129–1136.
Sandile D. Ngidi, Morgan Dundu. (2018). Composite Action of Pre-Cracked Reinforced Concrete Beams Repaired With Adhesive Bonded Steel Plates. *J. Structures*. 04, 005.
Wang L. G., Zhang Y. T. (2007). Short Period Rigidity and Deflection of Beams with Bonding Steel Plate in Compression Zone. *Journal of Huazhong University of Science & Technology*. 02, 81–85.
Yang S. H. (2016). *Experimental and Theoretical Analysis Study on Concrete Beams Strengthened with Composite Bonded Steel Plates*. D. Southeast University.

The influence of the underground structure on surrounding soils under earthquake and its application

Zhong-Yang Yu*
The First Construction Engineering Company Ltd. of China Construction Second Engineering Bureau, Beijing, China

Jing-Kun Zhang* & Hong-Ru Zhang*
School of Civil Engineering Beijing Jiaotong University, Beijing, China

Yan-Jia Qiu*
Changjiang Institute of Survey, Planning, Design and Research, Wuhan, China

ABSTRACT: To explore the dynamic interaction between the soil and structure and improve the seismic design method for the underground structure, numerical simulation was applied to discuss the influence of the underground structure on the surrounding soils. The seismic influence range of the underground structure on surrounding soils was explored with different influence factors, including structural form, site class, input ground motion and burial depth. The results showed that the more complex the structural form, the worse the soil condition, the lower the earthquake frequency, and the deeper the buried depth, the more obvious the influence effect on its seismic influence range. Nevertheless, the seismic influence range of the underground structure on the surrounding soils was basically within four times the structural width under different conditions. Then, an improved seismic design method was proposed. The improved method drew on the seismic influence range discussed above. Validation models with different boundary ranges were established and the results from two improved seismic design methods and the dynamic time-history method were compared. The results showed that it was necessary to determine the value range of the model boundary. The calculation accuracy of the two seismic design methods after boundary correction was higher, which can be applied to seismic design for shallow buried underground structures.

1 INTRODUCTION

The earthquake-resistance of underground structures has been an important research subject in the field of seismic engineering in recent years (Huo et al. 2005). Scholars have carried out a large number of investigations and research on this subject, and there is basically a consistent understanding of the seismic response characteristics of simple frame-type underground structures (Chen & Liu 2018; Dong et al. 2020; Lu & Hwang 2019; Xu et al. 2019; Yu & Zhang 2019; Zhang & Liu 2018). The seismic response of underground structures typically depends on the deformation of the surrounding soil, whereas its inertia is negligible. For a rectangular underground structure containing interior columns, interior columns are

---

*Corresponding Authors: 15115294@bjtu.edu.cn, 16115301@bjtu.edu.cn, hrzhang@bjtu.edu.cn and 17115316@bjtu.edu.cn

the weak points of the structure, and the seismic damage typically starts with their failure (Yu et al. 2021a, 2021b).

Based on these seismic characteristics of underground structures, some pseudo-static analysis methods have been proposed (Qiu et al. 2021). The response displacement method (RDM) is the most widely used seismic design method at present (Hamada et al. 1984; Tateishi 2005). The modelling process of the method is simple and easy to realize. The structural responses of the method can be expressed as simple equations. However, there are still some defects in the application of the method. For example, it is difficult to determine the parameter of the soil-spring accurately, and it is also difficult to be used for some underground structures with complex sections. Improved design methods, such as the static finite element method and integral RDM (Liu et al. 2013, 2014) have already solved the above problems and have been applied in the Chinese standards (GB50011 2010; GB50909 2014).

However, the value range of the model boundary is still not given during the application process of the improved method. When the static finite element method is applied, the finite element model without the station structure is established at first, and then the side and bottom boundaries of the model are fixed, as shown in Figure 1. The parameters of the soil spring are converted from the load applied by the hole in all directions (GB50909 2014). Different value ranges of the model boundary will obviously result in different parameters of the soil spring. This kind of calculation method easily leads to an accumulation of errors. Similar problems exist in the application process of integral RDM. The addition of the underground structure has an obvious influence on the seismic response characteristics of the surrounding soils. When the value range of the model boundary is larger than its influence range, the error caused by the value range can be ignored.

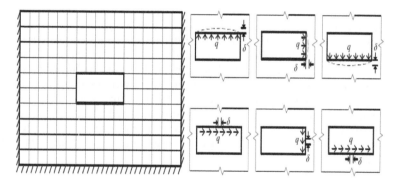

Figure 1. Static finite element method.

Therefore, it is of great significance to understand and explore the influence of underground structures on surrounding soils under earthquakes. This paper will discuss these seismic influence ranges by numerical simulation method. On this basis, an improved seismic design method is proposed and verified.

2 NUMERICAL MODEL

The interaction between the soil and structure is affected by many factors under earthquakes, such as structural form, site class, flexibility ratio between soil and structure, input ground motion, and buried depth. Therefore, it is very difficult to theoretically deduce the accurate seismic process of an underground structure, and this problem is explored by the finite element method.

2.1 Model parameter

There are three forms of underground structures, namely, single-story-two-span, two-story-three-span and three-story-three-span subway stations. The sectional sizes are shown in Figure 2. An elastic constitutive model is applied to the underground structure. The material properties are shown in Table 1.

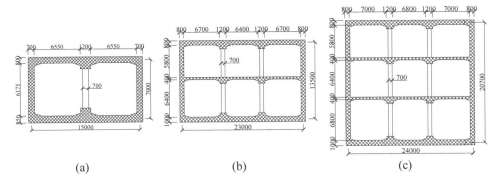

Figure 2. Sectional size of the underground structure /mm. (a) Single-story-two-span (b) Two-story-three-span (c) Three-story-three-span.

Table 1. Concrete parameter.

Type	Concrete grade	Density/ kg·m^{-3}	Elastic modulus/GPa	Poisson's ratio	Damping ratio
Interior column	C50	2500	34.5	0.2	0.05
Other structure	C35	2500	31.5	0.2	0.05

Three classes of site soils are selected as the model soil, referring to the Chinese standards (Bardet et al. 2000; Kuhlemeyer & Lysmer 1973). The soil parameters are shown in Table 2. Considering the modeling efficiency, the equivalent linear viscoelastic model is applied for the model soil (Bardet et al. 2000; Kramer & Paulsen 2004), as shown in Figure 3.

Table 2. Soil parameter.

Site class	Shear wave velocity/ m·s^{-1}	Density/ kg·m^{-3}	Poisson's ratio
Type II	300	1950	0.45
Type III	200	1950	0.45
Type IV	125	1950	0.45

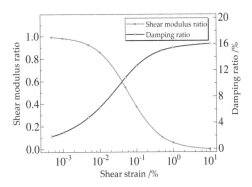

Figure 3. Shear strain with shear modulus and damping ratio.

2.2 Finite element modelling

Solid elements are used to simulate the model soil, and beam elements are used to simulate the station structure. The maximum mesh of the model is 0.5 m, to ensure efficient reproduction under all waveforms over the entire frequency range (i.e., frequencies between 0.2 Hz and 25 Hz) (Kuhlemeyer & Lysmer 1973). The "Master-Slave Surface" in ABAQUS is used to simulate the contact effect between the soil and structure. The "hard" contact is adopted in the normal direction, and the friction contact is adopted in the tangential direction. The friction coefficient is 0.4. These commands are quite common in engineering practices and have been widely adopted in previous papers (Chen & Liu 2018; Dong et al. 2020; Lu & Hwang 2019; Xu et al. 2019; Yu & Zhang 2019; Yu et al. 2021a, 2021b; Zhang & Liu 2018).

The Chinese standards (GB50011. 2010; GB50909 2014) require that the distance between the underground structure and the bottom boundary should be more than three times the structural height during time-history analysis. Based on this, the height of the model is 110 m. For the research subject explored in the paper, the value range of the model boundary has a great influence on the accuracy of the calculation result. Therefore, a remote artificial boundary used to provide accurate solutions is selected in the paper, namely, the width of the model on both sides is 30 times the structural width, as shown in Figure 4. In addition, the width and height of the model are slightly adjusted during analysis to ensure that the distance between the structure and each boundary is consistent under different structural forms and buried depths.

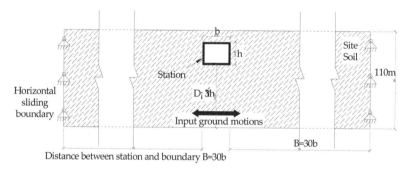

Figure 4. Numerical analysis model.

Considering the sensitivity of structural responses to input ground motions, three seismic records with obvious differences in frequency domain distribution are carefully selected for the time-history analysis. The input ground motions are vertically propagated from the bedrock surface as shear waves. Three acceleration time histories and Fourier spectra are shown in Figure 5. Meanwhile, five single sine waves with different frequencies are selected for comparative analysis.

3 SEISMIC INFLUENCE RANGE ANALYSIS

The addition of the underground structure will disturb the deformation and stress of site soils near the structure compared with the free field model of site soil. The change stabilizes with the increase in the distance from the structure. When the distance is far enough, the result will conceivably agree with that from the free field model. Therefore, this section will explore the influence range by changing the distance between the target position and structure. The peak relative displacement of the site soil at the horizontal height of the top or bottom of the

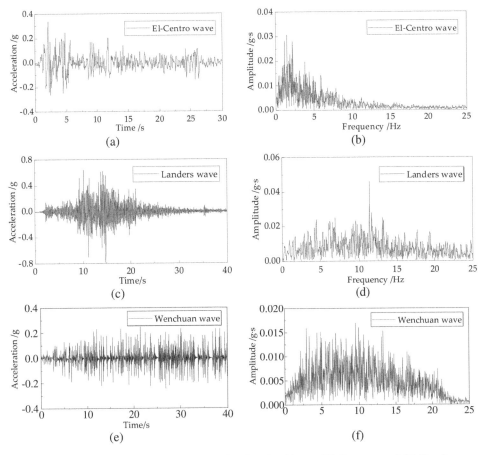

Figure 5. Three original seismic records. (a) Acceleration history (El-Centro wave) (b) Fourier spectrum (El-Centro wave) (c) Acceleration history (Lander's wave) (d) Fourier spectrum (Lander's wave) (e) Acceleration history (Wenchuan wave) (f) Fourier spectrum (Wenchuan wave).

underground structure can be used as the criterion to describe the influence range criteria of the underground structure (Wang et al. 2011), as shown in Figure 6. When the relative displacement of the target position is consistent with that of the free field model under the same condition, it is considered as the seismic influence range of the underground structure.

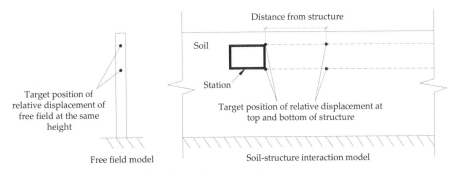

Figure 6. The target position of the finite element model.

3.1 Structural form and site class

Figure 7 shows the peak relative displacement of the target position with distance under different structural forms and site classes. The dotted line in the figure represents the peak of the horizontal relative displacement at the corresponding position in the free field model under the same conditions.

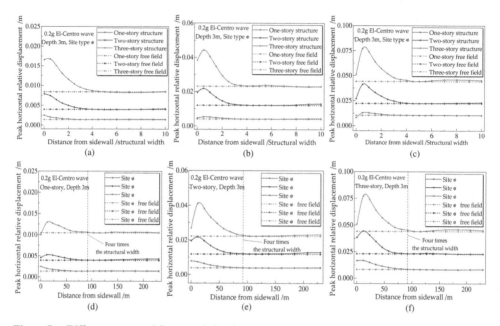

Figure 7. Different structural forms and site classes.

As seen from the graph, the more layers of the underground structure, the more complex the structural form, and the more obvious the influence effect on its seismic influence range. As seen from the attenuation slope of the curves, the worse the site soil condition, the faster the attenuation of the curve. Meanwhile, its seismic influence range tends to increase because of the larger deformation caused by the decrease in the soil strength. Using the structural width as the unit of measurement, the seismic influence range of the underground structure on the surrounding soils is basically within four times the structural width under different structural forms and site classes.

3.2 Input ground motion

Figure 8 shows the peak relative displacement of the target position with distance under different sinusoidal waves, seismic waves and earthquake intensities.

As seen from the graph, the low-frequency component of each seismic wave is the main factor affecting the deformation characteristic of the site soil. The deformation and seismic influence range gradually decreases as the frequency increases. In general, the low-frequency component has a more obvious influence on the seismic influence range of the underground structure than the high-frequency component. However, this influence range of the underground structure on the surrounding soils changes slightly under different seismic waves. As the intensity of the earthquake increases, the deformation of the site soil gradually increases, while the influence range basically has no significant change. This phenomenon is caused by

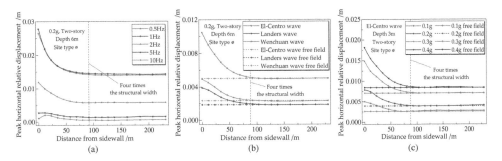

Figure 8. Different input ground motions.

the fact that the increase in earthquake intensity will increase the damping of the soil and weaken its influence range. Similarly, the seismic influence range discussed in the paper is still within four times the structural width under different input ground motions.

3.3 *Buried depth*

Based on the above influence range criteria, Figure 9 visually presents the curves of the seismic influence range with the buried depth of the underground structure. As seen from the graph, when the buried depth of the underground structure is very shallow, the seismic influence range will be increased due to the free boundary at the top of the model. When the buried depth exceeds 3 m, with the increase in the buried depth, the seismic influence range slowly increases, and similarly within four times the structural width.

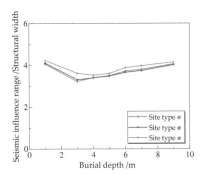

Figure 9. Different buried depths.

Therefore, based on the above three sections, it can be seen that there is a threshold value of the seismic influence range of the underground structure on surrounding soils under different structural forms, site classes, input ground motions and buried depths. The influence range discussed in the paper is four times the structural width. The above conclusions show that it is feasible to propose an improved seismic design method based on the seismic influence range.

4 VERIFICATION OF THE IMPROVED SEISMIC DESIGN METHOD

The response displacement method (RDM) uses a soil-spring element to quantitatively describe the interaction between the soil and structure (Liu et al. 2013). The static finite element method is suggested to solve the soil-spring parameter in the Chinese standards (GB50011 2010; GB50909 2014). The problem that different value ranges of model boundary will lead to different soil-spring parameters has been described in the Introduction. The integral RDM abandons the soil-spring element and describes the interaction between the soil and structure by establishing finite element models (Liu et al. 2014). The different value ranges of the model boundary will also lead to the different ground reaction forces provided by the surrounding soil during the application process of this method, thus affecting the calculation accuracy.

Therefore, it is necessary to determine the value range of the model boundary in the application of seismic design methods, which can improve the calculation accuracy. In this section, the seismic influence range obtained from the above analysis is taken as the value range of the model boundary in the improved seismic design method, and the calculation accuracy of the improved design method modified by the boundary range is verified.

The single-story-two-span and three-story-three-span subway stations in Section 1.1 are selected as two validation models. The model size, material parameter, modelling setting, etc., are consistent with those mentioned above. The buried depth is 3 m. According to the analysis in Section 2.2, the low-frequency component has a more obvious influence on the seismic influence range of the underground structure. Therefore, a 0.2 g El-Centro wave with rich low-frequency characteristics is selected as the input ground motion. The detailed modelling process of the RDM and integral RDM shall refer to the Chinese standards (Bardet et al. 2000; Kuhlemeyer & Lysmer 1973) and literature (Liu et al. 2013, 2014).

For further comparative analysis, the boundary conditions of one, four, seven and ten times the structural widths are selected as the value range of the model boundaries. The target sections A, B, C and D of the comparative analysis are shown in Figure 10. The calculated results of the two seismic design methods under different conditions are shown in Tables 4 and 5, including the peak bending moment of the structure and the interlayer displacement angles.

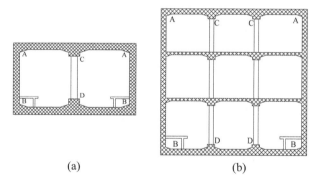

Figure 10. Target section. (a) Single-story-two-span station (b) three-story-three-span station.

According to the data in Tables 3 and 4, when the value range of the model boundary is different, the internal forces and deformations of the underground structure calculated by the two seismic design methods are quite different. So it is necessary to determine the value range of the model boundary. When the value range of the model boundary is one, four, seven and ten times the structural widths, the average error of the peak bending moments between the RDM and the time-history method is 20.1%, 7.1%, 10.2% and 17.2 respectively. The corresponding average errors calculated by integral RDM are 12.1%, 5.2%, 5.7% and 8.2%, respectively. Thus, it can be seen that when four times the width of the structure is selected as the value range of the model boundary, the average errors calculated by the two

Table 4. Results from the single-story-two-span station.

Calculation method		Target A	Relative error	Target B	Relative error	Target C	Relative error	Target D	Relative error	Interlayer displacement angle
Dynamic time history method		101.2		171.2		42.3		48.2		1/596
RDM	one time boundary	60.5	40.2	144.6	15.5	32.1	24.1	32.9	31.7	1/749
	four times boundary	88.3	12.7	167.7	2.0	44.8	5.9	45.8	4.9	1/576
	seven times boundary	108.6	7.3	182.6	6.7	48.9	15.6	49.6	2.9	1/535
	ten times boundary	117.7	16.3	186.8	9.1	52.7	24.6	53.8	11.6	1/486
Integral RDM	one time boundary	78.3	22.6	149	12.9	36.6	13.5	37.3	22.6	1/683
	four times boundary	91.2	9.8	174.5	1.9	43.6	3.1	44.6	7.5	1/564
	seven times boundary	98.6	2.6	176.1	2.8	44.7	5.7	45.7	5.2	1/565
	ten times boundary	101.5	0.3	179.8	5.0	46.8	10.6	47.9	0.6	1/559

Bending moment unit: kN·m·m^{-1} & relative error unit: %

Table 5. Results from the three-story-three-span station.

Calculation method		Target A	Relative error	Target B	Relative error	Target C	Relative error	Target D	Relative error	Interlayer displacement angle
Dynamic time history method		223.2		793.2		63.6		156.5		1/542
RDM	one time boundary	162.4	27.2	772.8	2.5	50.9	19.9	145.6	7.0	1/682
	four times boundary	201.3	9.8	813.6	2.5	68.6	7.8	174.4	11.4	1/535
	seven times boundary	233.9	4.7	834.1	5.1	75.4	18.5	188.6	20.5	1/496
	ten times boundary	249.3	11.7	843.9	6.4	83.6	31.4	197.6	26.3	1/445
Integral RDM	one time boundary	199.5	10.6	790.5	0.3	55.9	12.1	159.5	1.9	1/626
	four times boundary	218.6	2.1	815.5	2.8	66.9	5.1	170.6	9.0	1/549
	seven times boundary	232.6	4.2	820.3	3.4	70.2	10.3	174.4	11.4	1/521
	ten times boundary	240.5	7.7	826.5	4.1	76.6	20.4	183.2	17.1	1/514

Bending moment unit: kN·m·m^{-1} & relative error unit: %

methods are minimal, which truly reflects the interaction between the soil and structure. In addition, the average errors of the sidewalls and interior columns calculated by RDM are 11.2% and 16.5%, respectively. However, the corresponding errors of the integral RDM are 5.9% and 9.7%, respectively. Thus, it can be seen that the integral RDM is superior to the RDM in calculation accuracy.

These two validation models show that the calculation accuracy of the RDM and integral RDM after the modification of the boundary range is obviously improved. Meanwhile, it also shows that the improved seismic design method based on the seismic influence range proposed in the paper is feasible and effective, and can be applied to the actual seismic design for shallow buried underground structures.

5 CONCLUSION

1) The seismic influence range of the underground structure on surrounding soils is explored with different influence factors by the numerical simulation method, including structural form, site class, input ground motion and l burial depth. The results show that the more complex the structural form, the worse the soil condition, the lower the earthquake frequency, and the deeper the buried depth, the more obvious the influence effect on its seismic influence range. Nevertheless, the seismic influence range of the underground structure on the surrounding soils is basically within four times the structural width under different conditions.
2) The influence range obtained from the analysis is taken as the value range of the model boundary in the improved seismic design method. The calculation results from the response deformation method and the integral response deformation method with different boundary ranges are compared and discussed. The results show that the value range of the model boundary must be determined. The calculation accuracy of two improved seismic design methods is higher and can be applied to seismic design for shallow buried underground structures.

ACKNOWLEDGMENTS

This project is supported by the National Natural Science Foundation of China No.52078033 and the First Construction Engineering Company Ltd. of China Construction Second Engineering Bureau.

REFERENCES

Bardet, J.P.; Ichii, K.; Lin, C.H. *EERA-A Computer Program for Equivalent-Linear Earthquake Site Response Analyses of Layered Soil Deposits*. Los Angeles, CA, USA: Department of Civil Engineering, University of Southern California, 2000.

Chen, Z.Y.; Liu, Z.Q. Effects of Central Column Aspect Ratio on Seismic Performances Of Subway Station Structures. *Adv. Struct. Eng.*, 2018, 21, 14–29.

Dong, R.; Jing, K.; Li, Y.; Yin, Z.; Xu, K. Seismic Deformation Mode Transformation of Rectangular Underground Structure Caused by Component Failure. Tunn. Undergr. *Space Technol.*, 2020, 98:103298.

GB50011–2010 *Code for Seismic Design of Buildings*, in, China Architecture and Building Press, Beijing, China; 2010. (in Chinese)

GB50909–2014 *Code for Seismic Design of Urban Rail Transit Structures*, in, China Planning Press, Beijing, China; 2014. (in Chinese)

Hamada, M.; Izumi, H.; Iwano, M.; Shiba Y. Analysis of Dynamic Strain Around Rock Cavern and Earthquake Resistant Design. *Proc. Jpn. Soc. Civil Eng.*, 1984:197–205. Japan Society of Civil Engineers.

Huo, H.; Bobet, A.; Fernández, G.; Ramírez, J. Load Transfer Mechanisms Between Underground Structure and Surrounding Ground: Evaluation of the Failure of the Daikai Station. *J. Geotech. Geoenviron.* 2005, 131, 1522–1533.

Kramer, S.L.; Paulsen, S.B. Practical Use of Geotechnical Site Response Models. *Proceedings of International Workshop on Uncertainties in Nonlinear Soil Properties and their Impact on Modeling Dynamic Soil Response Richmond*, California, USA, 2004: 1–10.

Kuhlemeyer, R.L.; Lysmer, J. Finite Element Method Accuracy for Wave Propagation Problems. *J. Soil Mech. Found. Div.* 1973, 99, 421–427.

Lu, C.C.; Hwang, J.H. Nonlinear collapse simulation of Daikai Subway in the 1995 Kobe Earthquake: Necessity of Dynamic Analysis for a Shallow Tunnel. *Tunn. Undergr. Space Technol.*, 2019, 87: 78–90.

Liu, J.B.; Wang, W.H.; Zhang, X.B.; Zhao D.D. Research on Response Deformation Method in Seismic Analysis of Underground Structure. *Chin. J. Rock Mech. Eng.*, 2013, 32(1): 161–167. (in Chinese)

Liu, J.B.; Wang, W.H.; Zhao D.D.; Zhang, X.B. Integral Response Deformation Method in Seismic Analysis of Complex Section Underground Structures. *China Civil Engineering Journal*, 2014, 47(1): 134–142. (in Chinese)

Qiu, Y.J.; Zhang, H.R.; Yu, Z.Y. A Seismic Design Method of Subway Station Affected by Adjacent Surface Building. *Rock Soil Mech.* 2021;42(5):1443–52 [in Chinese)].

Tateishi, A. A study on Seismic Analysis Methods in the Cross Section of Underground Structures Using Static Finite Element Method. *Struct. Eng./Earthq. Eng.* 2005;22(1):41s–54s.

Wang, G.B.; Xie, W.P.; Sun, M.; Liu, W.G. Evaluation Method for Seismic Behaviors of Underground Frame Structures. *Chin J Geotech Eng*, 2011, 33(4):593–598. (in Chinese)

Xu, Z.G.; Du, X.L.; Xu, C.S.; Hao, H. Numerical Research on Seismic Response Characteristics of Shallow Buried Rectangular Underground Structure. *Soil Dyn. Earthq. Eng.*, 2019, 116, 242–252.

Yu, Z.Y.; Zhang, H.R. Seismic Characteristics and Analysis of Cross Transfer Station. *J. Southeast U: Nat. Sci. Ed.* 2019, 49, 1011–1018. (in Chinese)

Yu, Z.Y.; Zhang, H.R.; Qiu, Y.J.; Zhang, R.; Li, H. Shaking Table Tests for Cross Subway Station Structure. *J. Vib. Shock.*, 2021a, 40(9):142–151. (in Chinese)

Yu, Z.Y.; Zhang, H.R.; Qiu, Y.J.; Li, H. Study on Seismic Response Characteristics of A Seamless exchange subway station. *J. Hunan Univ.: Nat. Sci. Ed.* 2021b, 48(11): 166–176. (in Chinese)

Zhang, L.; Liu, Y. Seismic Responses of Rectangular Subway Tunnels in a Clayey Ground. *PLoS ONE*. 2018, 13(10): e0204672.

Study on explosion resistance of ceramic/steel composite structure coated with polyurea

Qing-Le Liu, Xin Jia*, Zheng-Xiang Huang, Wei Xia, Yu Wang & Tao Zhang
School of Mechanical Engineering, Nanjing University of Technology, Nanjing, Jiangsu, China

ABSTRACT: Polyurea elastomers have good physical and chemical properties that can improve the anti-detonation performance when coated on the surface structure. The dynamic response process of polyurea-coated ceramic/steel composite structure under explosive impact is simulated and discussed using ANSYS/LS-DYNA finite element software. The deformation damage, energy absorption, and shock attenuation of single steel plate, steel plate/polyurea, and ceramic/steel plate/polyurea under explosion response were analyzed. The correctness of the simulation is verified by experiments. The results show that the hole size of polyurea coating is smaller than that of a single steel plate structure when polyurea coating is used as a back blasting surface under the same explosion load. The ceramic/steel plate-coated polyurea structure has no obvious cracks and holes, the anti-explosion ability is improved, the energy absorption effect is good, and the shock wave attenuation ability is better than the other two kinds of structures.

1 INTRODUCTION

With the improvement of weapon damage performance, the anti-explosion protection capability of vehicles and ships has become the focus of attention. In the past, the improvement of structural antiknock ability was based on increasing the thickness of a single medium, adopting composite structures, and developing new materials (Wang 2006). However, for occasions with high mobility requirements, simply increasing the thickness of the medium increases the weight of the protective structure, which has played an adverse effect. However, traditional protective structures such as composite (Liu et al. 2018; Ma et al. 2019) and metal (Chen et al. 2019; Sun et al. 2020) sandwich structures can hardly avoid damage to some extent under strong impact loads. Some scholars have obtained ideal protection results by studying the bionic characteristics (A F J et al. 2015; Tran et al. 2014) and the characteristics of the fiber metal laminated structure (Langdon 2005; Ye et al. 2019) in terms of explosion energy absorption. Polyurea can effectively reduce the damage caused by the explosive impact due to its strong strain rate and phase change effect (Miao et al. 2019; Sarva et al. 2007; Yi et al. 2006; Wang et al. 2019). Moreover, polyurea elastomers are lightweight, tough, and have a high tensile ratio, for which it is widely used in the fields of architecture (Iqbal et al. 2018), navigation (Clark 2001), etc.

In the 1990s, the US Air Force Research Laboratory (Daniel 2008) first proposed the use of polyurea elastomer as an impact-resistant structure. Research shows that the structure

---

*Corresponding Author: jiaxin@mail.njust.edu.cn

coated with polyurea elastomer has significantly improved its explosion resistance (James et al. 2004). Wu Hecheng (Kathryn et al. 2013) studied the high-speed impact resistance of the composite structure of mother-of-pearl-like ceramics/polyurea. Through numerical simulation, it is found that compared with the pure ceramic beam, the composite structure sprayed with polyurea has better integrity, after impact, and better resistance to secondary damage. Samiee et al. (2013) compared the front and back sprayed polyurea with a bare steel plate under the same surface density and studied the dynamic response of the three under explosive load. Results show that when polyurea is used as the front explosion surface, the average plastic strain in the central area of the steel plate is smaller than when polyurea is used as the back explosion surface. The thickness of polyurea is an important factor for the explosion-proof effect, and the antiknock ability increases with the increase of polyurea thickness. Changhai (Samiee et al. 2013) studied the influence of good strength matching on the antiknock performance of polyurea-coated metal sheets. It is found that the increase in polyurea strength reduces the energy absorption of the kinetic energy of polyurea. In general, polyurea fragments have an important contribution to the total energy absorption of the structure.

Ceramic materials are being considered as good impact-resistant materials due to their good physical properties, such as high hardness, high strength, and low density. However, high brittleness and low toughness limit their application in the field of impact resistance. So far, there is little research on the explosion resistance of polyurea-coated ceramic composite structures at home and abroad, so this study selects ceramic/steel composite plates as the base material. To overcome the application defects of ceramics, glass fiber is used as a coating. ANSYS/LS-DYNA software is used to conduct numerical simulation on the explosion resistance of polyurea-coated ceramic/steel composite target plate and compared with single steel plate, steel plate/polyurea structure. The deformation characteristics of the target plate under the same explosion load are analyzed. The correctness of the numerical simulation is verified by experiments.

2 NUMERICAL SIMULATION

2.1 *Model establishment*

Based on the display dynamic analysis mode, ANSYS/LS-DYNA is a commonly used software in the field of explosion impact. ANSYS/LS-DYNA is used to simulate different structures. The models involved include TNT explosive, air, polyurea elastomer, ceramics, glass fiber, and steel plate. Glass fiber is used as the coating material of the ceramic layer, and the thickness of both sides is 1mm. The ceramic layer is made of hexagonal ceramic sheets with a side length of 50 mm. The composite target plate size is 300 × 300 mm square slab. Because of the symmetry of the antiknock problem of the composite plate, a quarter model is established. Symmetrical constraints are carried out on the symmetry plane, the target plate boundary is fully constrained, and the non-reflective boundary condition is used in the air domain. Single-side automatic contact is adopted between spliced ceramic sheets, and *CONTACT is adopted between glass fiber and ceramic layer_ TIED_ SURFACE_ TO_ SURFACE contact algorithm, common nodes are used between other layers to simulate the bonding between them. Eular grid is used for explosives and air, and Lagrange grid is used for polyurea, ceramics, glass fiber, adhesive layer and steel plate, and multi-material fluid-solid coupling algorithm *CONSTRAINED_ LAGRANG_ IN_ SOLID defines the coupling between the two meshes. The TNT charge of 50g is used for ignition. A space of 1mm is reserved between ceramic blocks. The geometric dimensions of four types of target plates of structures I, II, III, and IV are shown in Figure 1, and the simulation diagram is shown in Figure 2.

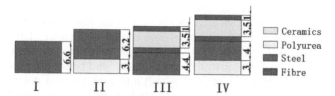

Figure 1. Geometric drawings of structures A, B, and C.

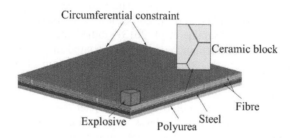

Figure 2. Simulation schematic diagram of structural C1/4 model.

2.2 *Material parameters*

Polyurea, as a kind of super viscoelastic material, has a remarkable strain rate effect. The Mooney Rivlin model is used as the material model, and the specific parameters are referred to (Chen et al. 2020). Steel plate is adopted for* MAT_ JOHNSON_ COOK (JC) material model, specific parameter reference (Liu et al. 2020). Ceramic materials are often used for *MAT_JOHNSON_ HOLMQUIST_ CERAMICS, or JH2 model; the material model parameters are shown in reference (Kathryn et al. 2013). Glass fiber adopts * MAT_COMPOSITE_SOLID_MODEL model, which is based on the Chang-Chang failure criterion. Some parameters of materials are shown in the literature (Xue et al. 2010).

2.3 *Analysis of numerical simulation results*

2.3.1 *Deformation analysis*

To intuitively compare the antiknock ability of four types of target plates under the same area density and explosive load, the 60μs time deformation diagram of four types of target plates is shown in Figure 3. It can be clearly seen that both type I and II steel plates have been damaged by punching, and the hole diameters are 29.6mm and 28mm, respectively. The steel plates at the center of types III and IV have large plastic deformation but no obvious cracks and holes; types II and IV polyurea elastomers on the back explosion surface break away from the steel plate under the action of a shock wave, forming fragments of different sizes. The ceramics, as the explosion-facing surface, failed in a large area, and the glass fiber layer coated with ceramics also broke. Comparatively speaking, types I and II

Figure 3. Schematic diagram of four types of structural deformation.

steel plates are penetrated, and the antiknock effect is poor. In contrast, types III and IV have no holes, which have certain integrity and superior antiknock ability. To clearly compare the anti-explosion ability of structures III and IV, several units at the center line of the steel plate back explosion surface are taken to analyze their radial deformation. It can be seen from the plastic deformation amount of the two obtained in Figure 4 that the displacement trend of types III and IV is consistent, and the radial displacement value of type III is always slightly greater than that of type IV. At 60μs, the maximum plastic deformation values of structures III and IV are 26.6mm and 24.6mm, respectively. To sum up, under the same surface density and explosive load, the composite structure with ceramics and polyurea has greater advantages in antiknock. The composite structure of ceramics as the front explosion surface has better anti-explosion ability than that of polyurea elastomer as the back explosion surface.

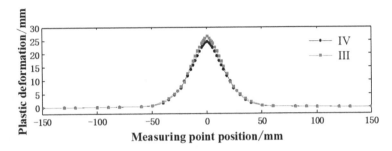

Figure 4. Plastic deformation diagram of structures III and IV.

2.3.2 Shock wave attenuation analysis

The explosion reaction can release a lot of energy in a short time, form continuous high-strength shock waves around, and damage the structure. The shock wave propagates in the form of a stress wave in the structure. Different structure configurations have different attenuation mechanisms for the shock wave. Based on the same explosion load, the smaller the back pressure value, the stronger the attenuation of the structure to the shock wave, and the better the protection effect. The numerical calculation for the stress value when the shock wave reaches the blasting face is 1781 Mpa. The back pressure values of structures I, II, III, and IV after stress wave attenuation are 45.5 Mpa, 39.3 Mpa, 2.0 Mpa, and 0.7 Mpa, respectively, which are I>III>II>IV from large to small. The back pressure values of structures II, III, and IV are respectively 86.37%, 43.96%, and 15.38% of that of structure I. It can be seen that when the area density is the same, the back pressure value of homogeneous steel plate structure is the largest, and the attenuation of shock wave when polyurea is sprayed on the back blast surface is better than that of homogeneous steel. The back pressure value of structure IV target plate is the smallest, and the shock wave attenuation effect is the best. The composite structure with ceramics as the front explosive surface has a stronger attenuation ability to the shock wave than the structure with the polyurea elastomer as the back explosive surface.

2.3.3 Energy absorption analysis

When the explosion acts on the target, part of the energy generated is converted into kinetic energy and internal energy of the target. To quantitatively analyze the energy absorption of different structures under the same load, the kinetic energy and internal energy time history curves of each material of four types of structures are made, as shown in Figures 7–10. It can be seen from the kinetic energy time history curves of the four types of steel plates in Figure 5(a) that the curves of types I and II rise sharply to the peak value and then drop

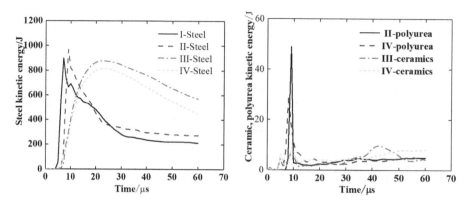

Figure 5. (a) Kinetic energy curve of steel; (b) Kinetic energy curve of polyurea and ceramics.

rapidly, with the same trend; types III and IV also decrease when reaching the peak, but the reduction span is large. The peak kinetic energy of steel plates of I, II, III, and IV structures are 896J, 965J, 876J, and 813J, respectively. The type II ceramics added have the most energy absorption, while the type IV ceramics as the structure of the front explosion face polyurea as the back explosion face have the least energy absorption. This is because of the failure of ceramics as the blasting face absorbs part of the energy and reduces the kinetic energy of the steel plate; The existence of types III and IV ceramics extends the action time, making the kinetic energy of steel plates have a certain time span. Type II has better energy absorption as far as kinetic energy absorption of the four types of structures is concerned.

It can be seen from Figure 5(b) that the kinetic energy curves of polyurea of types II and IV rapidly reach the peak value and then decrease sharply, where 9μs, 8μs reach the peak value of 48.9J and 28.9J, respectively. This means that the kinetic energy of polyurea in the ceramic-free composite structure increases more. The ceramic peak values in types III and IV are 9.8J and 12J, respectively, and the time history curve has certain fluctuations. This may be caused by the influence of the mesh size of ceramic materials, failure removal, crack formation, and propagation. According to the kinetic energy time history curve of steel plates, the existence of polyurea elastomers and ceramics can absorb energy and convert it into their own kinetic energy, which increases the energy absorption of the composite structure.

Figure 6(a) shows the internal energy time history curves of four types of steel plates. The trend of types I and II curves is consistent, which can increase rapidly in the initial stage of response, and at 13μs and 22μs, respectively, reached the peak value of 461J and 438J, and the internal energy value became stable after a period of time. The trend of types III and IV curves is consistent, showing a rising trend, with peaks of 1039J and 789J, respectively. The internal energy of types I and II steel plates continues to increase at the initial deformation stage of the steel plate due to the impact failure, and the plastic deformation of the steel plate decreases after the steel plate is sheared off, making the internal energy value stable. The increase of plastic deformation of types III and IV steel plates means more absorption of internal energy, so it shows a constant growth trend. It can be seen from the curve that the steel plate of type III has the largest internal energy, which means that it has the best energy absorption capacity.

Figure 6(b) shows the internal energy time history curves of four types of polyurea and ceramics. It can be seen that the internal energy of polyurea in the composite structure with or without ceramics as the blast face has a large difference. In type II, the polyurea elastomer quickly reaches the peak at the initial stage of response and then drops sharply. This is because the high pressure generated by contact explosion makes the polyurea elastomer respond quickly, and the internal energy increases. With the failure of a large number of

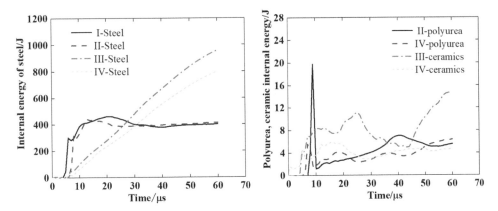

Figure 6. (a) Internal energy curve of steel; (b) Internal energy curve of polyurea and ceramics.

materials, the internal energy starts to decrease rapidly. In type IV, the internal energy of the polyurea elastomer changes little due to the small impact pressure. Both polyurea internal energy values fluctuate and increase to some extent after a sharp decrease. This is due to the deformation of the interaction between polyurea and steel plate, which makes the conversion between kinetic energy and internal energy occur. The internal energy of ceramic materials fluctuates greatly. It can be seen from the curve that the internal energy of ceramics in type III without polyurea spraying fluctuates more than that in type IV. This is because the existence of the polyurea elastomer reduces the deformation of the steel plate to a certain extent, weakens the role between the ceramic and the steel plate, and reduces the conversion between kinetic energy and internal energy. In short, the internal energy of polyurea elastomers and ceramics has increased to a certain extent, and they have a certain energy absorption effect.

The total energy absorption of different types of composite structures is shown in Figure 7. It can be seen that the homogeneous steel plate has the worst energy absorption effect, while the ceramic/steel composite structure has the best energy absorption effect. The total energy absorption of types II, III, and IV are 1.13 times, 2.08 times, and 2.64 times of type I, respectively. It means that the polyurea elastomer has a certain energy absorption capacity, and the addition of ceramics makes the overall energy absorption effect better. Considering the deformation, damage, energy absorption, and shock wave attenuation of the four types of structures, the ceramic steel polyurea composite structure has relatively good antiknock performance.

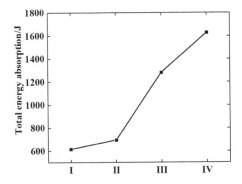

Figure 7. Total energy absorption of four structural types.

3 TEST STUDY

3.1 Test scheme

To verify the reliability of the numerical results, a proximity explosion test was carried out on the type IV structure. The layout of the test site is shown in Figure 8. The size of the structure used in the experiment is consistent with the numerical simulation model. The glass fiber-coated spliced ceramic sheet structure and the steel plate are bonded with AB adhesive (the adhesive layer thickness is as uniform as possible). Explosives are detonated by detonators, which are connected to the detonator through plastic detonators. Explosives are placed in the center of the upper surface of the target plate, constraining around the target plate.

Figure 8. Schematic diagram of test site setting and structure.

3.2 Comparison and analysis of test and simulation results

Table 1 shows the numerical simulation and test results of the composite structure. Figure 9 shows the comparison between the test and simulation results. It can be seen that the numerical simulation results are very close to the test results, which means that the results are reliable.

Table 1. Numerical simulation and test results of structure IV.

Type		Ceramics	Steel	Polyurea	Destruction
IV	Simulation	The diameter of the fragmentation area is about 100 mm	Large plastic deformation occurs in the center, and the bulge height is 24.6 mm	Damaged, area diameter about 87.5 mm	No break
	Test	The fragmentation area is a whole ceramic chip	Large plastic deformation occurs in the center, and the bulge height is 24 mm	Damaged, area diameter about 80 mm	
	Error%	0	2.5	9.4	

In the numerical simulation, the ceramic material, as the face of the explosion, forms a regular hexagon area with a length of 50 mm on one side after the end of the explosion, and the test results match with it, with an error of 0. In the test and simulation, the polyurea elastomer, as the material of the back burst surface, has failed. In the test, the diameter of the polyurea damaged area is about 80 mm, and the diameter of the simulated polyurea elastomer failure area is about 87.5 mm, with an error of 9.4%.

Under the strong impact, the steel plate in the test has a large plastic deformation at the center directly below the explosive, without cracks and holes. The numerical simulation shows that the steel plate has no cracks and holes, and the maximum plastic deformation is

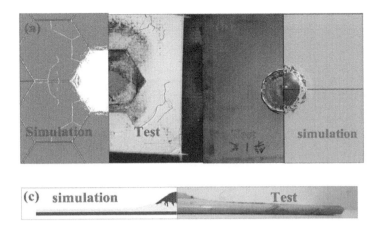

Figure 9. Type IV composite structure test and simulation results (Ceramics: (a), Polyurea: (b), Steel: (b)).

24.6 mm, with an error of 2.5% from the test of 24 mm. The comparison between the test and simulation results shows that the errors between the numerical simulation and the test are within the error range required by the project, which proves the authenticity of the numerical simulation results.

3.3 *Structural optimization research*

Through the comparison and analysis of the above simulation and experimental results, the addition of ceramics obviously weakens the damage to the structure and makes the structure play a good role in deformation, energy absorption, and shock wave attenuation. The target plate maintains high integrity and can be used for secondary antiknock. However, the steel plate directly under the explosive shows a large displacement and deformation, and a certain strength is required to resist deformation to prevent the steel plate from being damaged. Only the effects of ceramics, steel plates, and polyurea on the antiknock ability of composite structures are studied. Ceramics, steel plates, and polyurea are set as a single variable to study their deformation, energy absorption, back pressure, and specific energy absorption per unit mass (Wang & Lu 2013). The results are shown in Table 2.

It can be seen from Table 2 that when ceramics as a single variable increases from 2 mm to 8 mm, the displacement, energy absorption, and stress values of the target plate decrease to a certain extent (by 43.1%, 49.5%, and 24.7%, respectively). In comparison, the specific energy absorption per unit mass decreases by 66.6%. When the steel plate is a single variable, from 3 mm to 8 mm, the displacement, energy absorption, and stress of the target plate also decrease significantly, by 57.7%, 22.0%, and 19.5%, respectively, and the specific energy absorption per unit mass decreases by 58.9%. When the single variable of polyurea is increased from 1 mm to 8 mm, the displacement, energy absorption, and stress of the target plate also decrease to a certain extent (by 17.9%, 21.5%, 67.8%, respectively), and the specific energy absorption per unit mass decreases by 31.3%. By comparing 1-I, 2-i, and 3-i (i = 1−5), it can be seen that the increase in the thickness of ceramics, steel plates, and polyurea has a certain effect on deformation, energy absorption, and shock wave attenuation when the steel plate is of 3 mm thickness. The structural deformation is the largest, and the secondary antiknock effect is relatively poor. When the steel plate is of 8 mm thickness, the structural deformation is the minimum, indicating that the secondary anti-explosion effect will be better.

Table 2. Research results of antiknock performance of composite structures with different sizes[a].

Number	Ceramics (mm)	Steel (mm)	Polyurea (mm)	Quality (kg)	Displacement (mm)	Energy absorption (kJ)	Pressure (Mpa)	Mass specific energy absorption (kJ/kg)
1-1	2	4	3	4.12	28.8	2.00	0.797	0.485
1-2	3.5	4	3	4.65	24.6	1.62	0.702	0.348
1-3	5	4	3	5.18	20.2	1.38	0.691	0.266
1-4	6.5	4	3	5.71	17.6	1.22	0.647	0.214
1-5	8	4	3	6.24	16.4	1.01	0.600	0.162
2-1	3.5	3	3	3.94	29.8	1.82	0.750	0.462
2-2	3.5	4	3	4.65	24.6	1.62	0.702	0.348
2-3	3.5	5.5	3	5.71	18.5	1.58	0.650	0.277
2-4	3.5	7	3	6.76	14.1	1.37	0.634	0.203
2-5	3.5	8	3	7.47	12.6	1.42	0.604	0.190
3-1	3.5	4	1	4.46	25.7	1.91	0.934	0.428
3-2	3.5	4	3	4.65	24.6	1.67	0.702	0.359
3-3	3.5	4	5	4.83	23.0	1.62	0.458	0.335
3-4	3.5	4	6.5	4.97	22.0	1.56	0.369	0.314
3-5	3.5	4	8	5.11	21.1	1.50	0.301	0.294

[a]Note: 2 mm glass fiber is coated outside the ceramic layer; 0.1 mm adhesive layer between ceramic and steel plate; the specific energy absorption formula of unit mass $E_m = E/M$, where E is the total energy absorption of target plate and M is the total mass of target plate.

4 CONCLUSION

In this paper, the ceramic/steel/polyurea composite structure is used to compare the damage, energy consumption, and attenuation of the shock wave of a single steel plate and steel/polyurea structure under the same explosive load. The ANSYS/LS-DYNA software is used to conduct numerical simulations, and the test is used to verify the comparison.

The conclusions are summarized as follows:

1) When polyurea is coated on the back of the steel plate, the size of the hole is effectively reduced. Through simulation and testing, it is found that the polyurea has a ductile fracture in the process of resisting detonation impact.
2) The composite structure with ceramic materials can effectively resist the damage of detonation shock waves. Compared with the other two types of composite structures with ceramic front and polyurea back, the composite structure has the best attenuation effect on shock waves.
3) The energy consumption of the ceramic structure is higher than that of the other two types. The fracture failure of ceramics also consumes part of the energy. In the aspect of the antiknock effect, the composite structure with a ceramic front side and polyurea back side is the best.
4) The increase in the thickness of the ceramic, steel, and polyurea layer will strengthen the anti-explosion ability of the composite structure. In comparison, the increase in the thickness of steel plate has greater advantages in reducing structural deformation.

REFERENCES

A.F.J., A.L.S. and A.I.G., et al. A numerical Study of Bioinspired Nacre-like Composite Plates Under Blast Loading[J]. *Composite Structures*, 2015, 126:329–336.

B.J.Z., Ye Y. and Qin Q. On Dynamic Response of Rectangular Sandwich Plates with Fibre-metal Laminate Face-sheets Under Blast Loading[J]. *Thin-Walled Structures*, 2019, 144: 106288–106288.

Chen G., Pan Z. and Liu J., et al. Experimental and Numerical Analyses on the Dynamic Response of Aluminum Foam Core Sandwich Panels Subjected to Localized Air Blast Loading[J]. *Marine Structures*, 2019, 65:343–361.

Chen C., Wang X. and Hou H., et al. Effect of Strength Matching on Failure Characteristics of Polyurea Coated Thin Metal Plates Under Localized Air Blast Loading: Experiment and numerical analysis[J]. *Thin-Walled Structures*, 2020, 154:106819.

Clark G A., Burch I A. Emergent Technologies for the Royal Australian Navy's future afloat Support Force. 2001.

Collombet F., Lalbin X. and Lataillade, J.L. Impact Behavior of Laminated Composites: Physical Basis for Finite Element Analysis [J]. *Composites Science and Technology*, 1998, 58(3–4):463–478.

Daniel G., Linzell. State-of-the-Art Technological Developments in Concrete Computational Modeling. *31st annual Airport conference*, 2008: 325–331.

Iqbal N., Sharma P.K. and Kumar D., et al. Protective Polyurea Coatings for Enhanced Blast Survivability of concrete[J]. *Construction and Building Materials*, 2018, 175 (JUN.30): 682–690.

James S. Davidson, Jonathan R. Porter and Robert J. Dinan, Michael I. Hammons, James D. Connell. Explosive Testing of Polymer Retrofit Masonry Walls[J]. *Journal of Performance of Constructed Facilities*,2004,18(2).

Kathryn Ackland, Christopher Anderson and Tuan Duc Ngo. Deformation of Polyurea-coated Steel Plates Under Localised Blast Loading, *International Journal of Impact Engineering*, Volume 51, 2013, Pages 13–22, ISSN 0734-743X.

Langdon G.S., Cantwell W.J. and Nurick G.N. The Blast Response of Novel Thermoplastic-based Fibre-metal Laminates-some Preliminary Results and Observations[J]. *Composites Science & Technology*, 2005, 65(6):861–872.

Liu J., Liu J. and Mei J., et al. Investigation on Manufacturing and Mechanical Behavior of All-composite Sandwich Structure with Y-shaped cores[J]. *Composites Science & Technology*, 2018, 159:87–102.

Liu Q., Chen P. and Zhang Y., et al. Compressive Behavior and Constitutive Model of Polyurea at High Strain Rates and High Temperatures[J]. *Materials Today Communications*, 2019, 22:100834.

Liu R.H., Chen G.G. and Hou F., et al. Numerical Study of Projectile Penetrating Ceramic/steel Composite Target [J]. *Sichuan Journal of military Industry*, 2020, 041(003):183–188.

Ma X.A., Li X. and Li S.A., et al. Blast Response of Gradient Honeycomb Sandwich Panels with Basalt Fiber Metal Laminates as Skins[J]. *International Journal of Impact Engineering*, 2019, 123:126–139.

Miao Y., Zhang H., He H., et al. Mechanical Behaviors and Equivalent Configuration of a Polyurea under Wide Strain Rate Range[J]. *Composite Structures*, 2019, 222:110923.

Samiee A., Amirkhizi A.V. and Nemat-Nasser S. Numerical Study of the Effect of Polyurea on the Performance of Steel Plates Under Blast Loads[J]. *Mechanics of Materials*, 2013, 64(sep.):1–10.

Sarva S.S., Deschanel S. and Boyce M.C., et al. Stress-strain Behavior of a Polyurea and a Polyurethane From Low to High Strain Rates[J]. *Polymer*, 2007,48(8):2208–2213.

Sun G., Wang E. and J. Zhang, et al. Experimental Study on the Dynamic Responses of Foam Sandwich Panels with Different Facesheets and Core Gradients Subjected to Blast Impulse[J]. *International Journal of Impact Engineering*, 2020,135(Jan.):103327.1–103327.20.

Tran P., Ngo T.D. and Mendis P. Bio-inspired Composite Structures Subjected to Underwater Impulsive Loading[J]. *Computational Materials Science*, 2014, 82:134–139.

Wang H., Deng X. and Wu H., et al. Investigating the Dynamic Mechanical Behaviors of Polyurea Through Experimentation and Modeling [J]. *Defence Technology*, 2019, 15(6):10.

Wu H.C. Numerical Simulation Study on Impact Resistance of Ceramic/polyurea Composite Structure[D]. *East China Jiaotong University*, 2020.

Wang Y.X., Gu Y.X. and Sun M. Theoretical Calculation of Explosion Protection of Porous Composite Structure Under Impact Load [J]. *Acta Armamentarii*, 2006(2):375–379.

Wang Z.G., Lu Z.J. Experimental Evaluation of Energy Absorption Characteristics of Aluminum Honeycomb Heteroplane Compression [J]. *Journal of Central South University: Natural Science Edition*, 2013(3):6.

Xue. liang, Mock W. and Belytschkoa T. Penetration of DH-36 Steel Plates with and Without Polyurea Coating[J]. *Mechanics of Materials*, 2010,42(11):981–1003.

Yi J., Boyce M.C. and Lee G.F., et al. Large Deformation Rate-dependent Stress–strain Behavior of Polyurea and Polyurethanes [J]. *Polymer*, 2006, 47(1):319–329.

Tunnel stability analysis of karst cave location distribution in karst area

Xuzhe Zhang, Yucheng Zhao*, Quan Wang & Zhaoyu Dai
School of Civil Engineering, Shijiazhuang Railway University, Shijiazhuang, China

ABSTRACT: The stability of the tunnel is studied according to the different location distributions of karst caves in the karst area. Based on Midas/GTS finite element software, a three-dimensional coupling model of karst caverns, linings, and tunnels was established. The deformation of the monitoring position, the stress of the supporting structure, and the deformation of the surrounding rock are compared and analyzed. The influence of karst cave distribution on the distance between the karst cave and the tunnel is studied.

1 INTRODUCTION

With the advancement of urbanization and continuous economic development in China, promoting coordinated development between regions and reducing the spatial imbalance of population, economy, and resources, more and more tunnels are being built. Adverse geological conditions, such as karst and various hidden locations, which are not easily detected, bring great difficulties in tunnel construction. A large number of engineering examples show that the stability of surrounding rocks in karst areas is greatly influenced by cavity development and tunnel distance. Therefore, it is the main task of related workers to determine the location and size of cavities, understand their deformation laws, and design a reasonable construction plan. Randall (Randall et al. 2001) collated the morphology of karst in various periods. Grove and Meiman (2002) argued that karst landform development is the result of the combined effect of geological structure and intense dissolution. Fenart (Fenart et al. 1999) studied the mechanism of karst aquifer evolution by analyzing the accumulation in the karst cavities. Dogan's (Dogan & Yilmaz 2011) investigation of 19 karst collapse sites formed in a region of Turkey between 1977 and 2009 showed that thousands of deep wells were used for irrigation in local agricultural production, resulting in a 24 m drop in groundwater level, which became the main factor inducing karst as well as its collapse. Wang (Wang et al. 2019) takes Wuhan Metro Line 6 as the background and focuses on the magnitude and prediction of the influence of relevant geometric parameters and filling degree on the safety thickness when the hidden cavity is located above as well as below the tunnel through numerical simulation, while the safety critical distance of the cavity in the horizontal direction is initially calculated by combining the dichotomous method. Wang (2012) started his research on advanced forecasting in the field of tunneling. The research focused on the advanced forecasting techniques for different strata and the exploration methods to ensure the normal tunnel boring speed degree. Tang (Tang & Fu 1997) researched the influence law of hidden caverns on tunnels surrounding rock deformation and management.

---

*Corresponding Author: zyc36306@163.com

2 PROJECT OVERVIEW

Huang Dongpo Tunnel is located in the hilly area of Tongren, Guizhou Province, an area under the jurisdiction of the town of Chadian, with central mileage IDK23+312. The maximum depth of burial of the tunnel is about 31 m, and the total length is 664 m. The tunnel area belongs to the dissolution of tectonic hilly terrain. The topography of the area is undulating, and the ground elevation is 660 m $\sim$ 726 m. The tunnel inlet ground slope is steep, with a natural horizontal slope of about 30 degrees. The tunnel exit ground slope is also steep, and the natural horizontal slope is about 25 degrees. The CRD method was used to excavate the section from IDK23+196 to IDK23+256, which is prone to collapse and has karst cavities.

3 3D FINITE ELEMENT MODEL

In this study, the section IDK23+196~IDK23+256 has been selected, which is prone to collapse and has karst cavities. The tunnel span is 13.13 m, the tunnel structure height is 11.59 m, and the tunnel burial depth is 18.2 m. To fully analyze the construction influence range and the cavity influence range, as shown in Figure 1, the model is established by Midas/GTS in X (tunnel section aspect), Y (excavation direction), and Z (vertical direction) direction size of 120 m × 60 m × 70 m. When performing model calculations, the soil is simulated approximately as a continuum. The Moore–Coulomb principal structure model is used for mechanical calculations. The initial support uses plate units to simulate the initial spray concrete. The later measures, such as overrunning small conduit and large pipe shed reinforcement, are simulated by changing the soil parameters. To exclude the influence of the filler as well as the filling state, the original empty cavern, i.e., the empty field mechanical model, can be simulated. The tunnel is an unfilled cavern, and a spherical cavern with a circular cross-section is simulated. Since the cavity was formed before the excavation, the displacement due to the self-weight stress of the soil was cleared before the excavation was simulated.

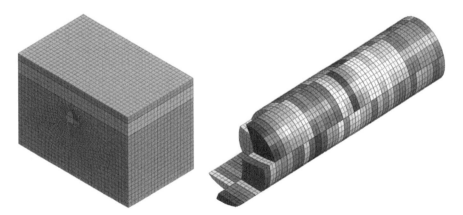

Figure 1. Numerical simulation of three-dimensional stereogram.

The model adopts Moore–Coulomb's intrinsic relationship for calculation, and the soil distribution is not uniform in the actual process. When the model adopts uniform soil for simulation, the three-dimensional soil is simulated by a solid unit. The initial support is simulated by 30 cm thick C30 concrete with reinforcement. The temporary support is steel support, which is simulated by plate units. Different reinforcement measures are simulated by changing the soil parameters. Each soil and support parameter is determined by referring to the relevant design specifications and data. The specific physical parameters are shown in Table 1.

Table 1. Table of physical parameters of the model.

Materials	Modulus of Elasticity E/MPa	Poisson's Ratio μ	Weight Capacity λ/(kN·m³)	Cohesion c/MPa	Friction Angle φ/(°)
Sandy Soil	30	0.35	18	20	28
Weathered Soil	40	0.31	20	25	30
Weathered Rock	300	0.25	23	100	35
Overrun Small Conduit	1680	0.21	25	80	35
Overrun Large Pipe Shed	5500	0.2	20	55	42
I-beam	20000	0.2	23.8	—	—
Initial Full Ring Steel Frame Support	33957	0.2	24	—	—
Anchor Rods	210000	0.3	78.5	—	—

4 MONITORING PROGRAM

(1) Geology and support observation of working face
(2) Measurement of vault subsidence and convergence
(3) Monitoring of surface settlement of overburden less than 40 m.

The arrangement of surface settlement observation piles is shown in Figure 2.

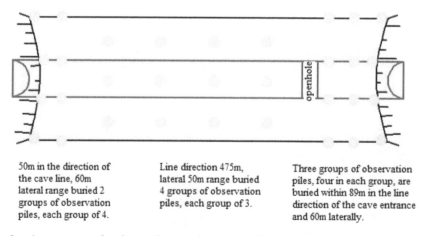

| 50m in the direction of the cave line, 60m lateral range buried 2 groups of observation piles, each group of 4. | Line direction 475m, lateral 50m range buried 4 groups of observation piles, each group of 3. | Three groups of observation piles, four in each group, are buried within 89m in the line direction of the cave entrance and 60m laterally. |

Figure 2. Arrangement of surface settlement observation pile.

5 NUMERICAL SIMULATION RESULTS AND MONITORING DATA COMPARISON ANALYSIS

5.1 *Comparison analysis between simulated and monitored values of surface settlement*

An analysis of Figure 3 shows that during the tunnel excavation, the monitored and simulated values have roughly the same trend. Also, the monitored values are larger than the

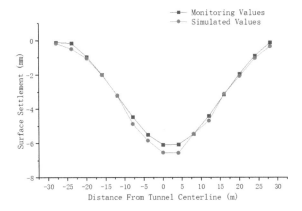

Figure 3. Comparison of surface settlement data between monitored and simulated values.

simulated values in individual places, which is related to the fact that the soil layer selected for the numerical simulation is more regular and does not contain any impurities.

5.2 Comparison analysis between tunnel arch settlement simulation and monitoring values

By analyzing Figures 4 and 5, it can be seen that the change trends of the monitored and simulated values are more or less the same. Their deformation values meet the construction deformation requirements and are much smaller than their specified values. On the whole, due to the slow construction process, the monitoring data is more stable, while the simulated value changes greatly. The maximum daily change rate of the monitoring value is 0.7 mm/d, while the maximum daily change rate of the simulated value reaches 2.4 mm/d, which is greater than the deformation convergence rate of 0.2 mm/d. It is necessary to strengthen the initial support strength and stiffness at the beginning to control the larger deformation. However, the final convergence rate is maintained at 0.1 mm/d. mm/d, which means the surrounding rock is basically stable.

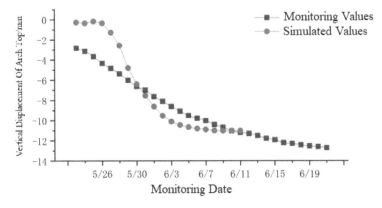

Figure 4. Comparison of data between monitoring and simulation values of the vault.

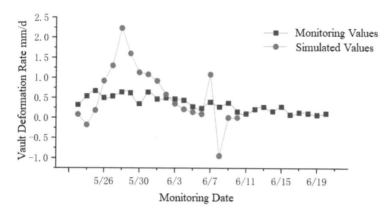

Figure 5. Comparison of monitoring value and simulated value of vault deformation rate data.

5.3 *Comparison analysis between simulated and detected values of tunnel headroom convergence*

By analyzing Figures 6 and 7, we can see that the trends of the monitored and simulated values are basically the same. The deformation of the arch is more consistent than that of the vault. The deformation value meets the requirements of construction deformation and is

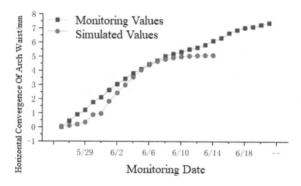

Figure 6. Comparison of monitoring and simulation values of arch waist level convergence data.

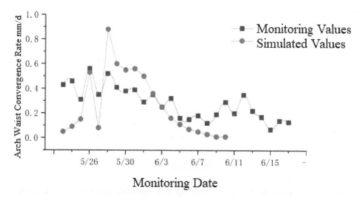

Figure 7. Comparison of monitoring and simulation values of arch waist level convergence rate.

much smaller than the specified value. Overall, due to the slow construction process on-site, its monitoring data is more stable, while the simulated values vary relatively more. The maximum daily change rate of the monitored value is 0.57 mm/d, while the maximum daily change rate of the simulated value reaches 0.91 mm/d, which is greater than the deformation convergence rate of 5 mm/d. The surrounding rock is in a state of rapid change, and measures need to be taken to strictly control the deformation. However, the final convergence rate tends to be about 0.1 mm/d. The surrounding rock is basically stable.

6 CONCLUSION

By comparing and analyzing the finite element simulation values with the actual monitoring values, it is found that some of the simulated values are larger than the actual monitoring values, which is greatly related to the construction progress, monitoring rate, and the simplification of numerical simulation. The numerical simulation and the actual detection data change in the same trend. The actual monitoring data is more moderate than the simulated value change, indicating that the safety of tunnel construction is largely ensured by managing the karst, improving the support strength, and reasonably controlling the excavation progress in the construction field.

REFERENCES

Dogan U., Yilmaz M. Natural and Induced Sinkholes of the Obruk Plateau and Karapınar-Hotamış Plain, Turkey[J]. *Journal of Asian Earth Sciences*, 2011, 40(2): 496–508.

Fenart P., Cat N.N., Drogue C., et al. Influence of Tectonics and Neotectonics on the Morphogenesis of Peak Karst of Halong Bay[J]. *Vietnam, Geodinamica Acta (Paris)*, 1999(12)3–4.

Grove C., Meiman J. Karst Aquifers at Atmospheric Carbon Sinks: An Evolving Global Network of Research Sites [C], US Geological Survey Karst Interest Group Proceedings. Shepherdstown, 2002, 20–22.

Randall C., Omdorff, Jack B. Proposed National Atlas Karst Map, Geological Survey Interest Group Proceeding, Water-resources Investigation[J], *Water-Resources Investigation Report*, 01–4011, 2001.

Tang C.A., Fu Y.F. A New Approach to Numerical Simulation of Source Development of Earthquake[J]. *Acta Seismologica Sinica*, 1997, 10(4): 425–434.

Wang, Gao., Liu, Wen., Li, Chen. Analysis on the Safe Distance Between Shield Tunnel Through Sand Stratum and Underlying Karst Cave[J]. *Geosystem Engineering*, 2019, 22(2).

Wang G. Application of Integrated Tunnel Geological Prediction Technologies to Highway Tunnels[J]. *Modern Tunnelling Technology*, 2012: 7.

Seismic analysis of structures with friction dampers

Xiangxiang Wu*
Shanghai Urban Construction Vocational College, Shanghai, China

Dawei Zhou & Yang Li
Shanghai Yingliang Construction Technology Co. Ltd., Shanghai, China

ABSTRACT: Friction dampers, a type of energy-dissipating device, are applied in a 14-story school building. To investigate the function of friction dampers, nonlinear seismic analysis is performed on the FE (finite element) model equipped with friction dampers. The analysis is done with seismic input of minor, moderate, and severe earthquakes. The analysis results show that the friction dampers start dissipating energy in minor earthquakes. In moderate and severe earthquakes, the energy dissipated is more. Therefore, the application of friction dampers significantly reduces the seismic response of the building and effectively protects the building structure from severe damage or collapse.

1 INTRODUCTION

In recent years, energy-dissipating and seismic isolation techniques have been widely applied in the high severity seismic zones (Feng et al. 2021; Wu et al. 2021; Zhou et al. 2021). Some engineering structures with such techniques have shown better seismic performance in the earthquake events, such as the Wenchuan earthquake or the Lushan earthquake. The energy-dissipating and seismic isolation techniques can effectively reduce the seismic response of the buildings and improve their earthquake resistance abilities (Zhou et al. 2021). Among these techniques, friction dampers, as a type of energy-dissipating device, are widely applied in engineering structures due to their stable technical features and easy applications (Wang et al. 2021; Ye et al. 2001). To investigate the function of friction dampers in high-rise buildings, the FE model of a 14-story school building with friction dampers is simulated, and seismic analysis is performed.

2 FRICTION DAMPERS

2.1 *Properties of friction dampers*

The friction damper converts the vibration energy of the building into thermal energy absorption by using the friction force opposite to the sliding direction generated on the contact surface when two contact objects are relatively displaced. The friction damper is installed onto the connection members extending from the upper and lower girders of the building. When the earthquake is coming, the friction damper can start dissipating energy, protecting the main structural members from seismic damage.

---

*Corresponding Author: wuxiangxiang@succ.edu.cn

The advantages of a friction damper are:

① It can improve the horizontal lateral stiffness of the structure to a certain extent so as to improve the overall seismic performance of the structure.
② In case of small horizontal displacements (minor earthquakes), it can quickly enter the yield section to provide effective additional damping for the structure so as to achieve the effect of energy dissipation.
③ The fatigue performance and ultimate deformation capacity are very strong, which can meet the deformation requirements under large earthquakes.

2.2 *Advantages of friction dampers*

A friction damper has the following characteristics:

(1) It is small in volume, so it can be hidden in the non-structural wall, which saves the occupied space and significantly improves the effective utilization of building space.
(2) The start-sliding displacement can be set small enough to make its energy dissipating capacity under minor earthquakes possible. Its stiffness after sliding is nearly zero. Therefore, its force remains almost unchanged under medium and large earthquakes, imposing little impact on its surrounding substructures.

3 APPLICATION OF FRICTION DAMPERS IN THIS PROJECT

3.1 *Project introduction*

The building has 14 stories on the ground and a two-story basement, with a total height of 58.5 m. The building structure is a concrete frame with a shear wall. The basic seismic acceleration of the design is 0.10 g. The isometric view of the FE model is shown in Figure 1.

3.2 *Arrangement of friction dampers*

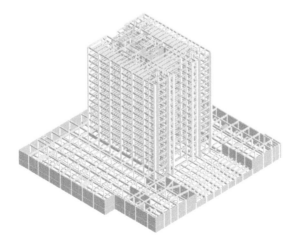

Figure 1. FE model of building.

The arrangement of friction dampers in the plan view is shown in Figure 2 (purple parts). The connection between the damper and the concrete frame is shown in Figure 3. The factors of dampers are set to ensure they will start dissipating energy during minor earthquake input.

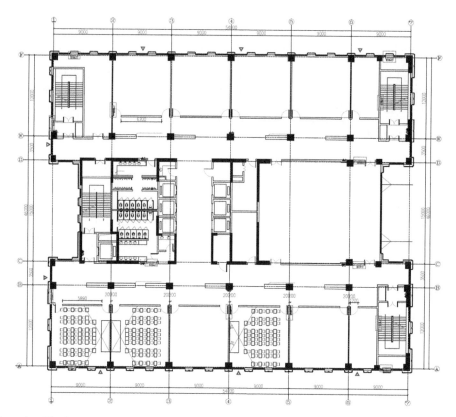

Figure 2. The layout of dampers in the plan view (purple part).

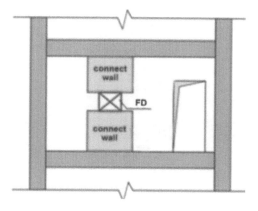

Figure 3. Connection of a friction damper.

4 SELECTION OF SEISMIC WAVES

The artificially simulated seismic waves that fit well with the design spectrum are selected as input for analysis. One of the artificial waves is shown in Figure 4 as an example. The comparison between its spectrum and the design spectrum is shown in Figure 5, which shows that the differences at the main periodic points are less than 10%. Seismic waves are selected

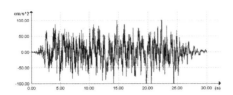

Figure 4. Seismic waves.

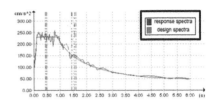

Figure 5. Comparison of input spectra and design spectra.

for two horizontal directions with intensity, as shown in Table 1. The seismic input in the vertical direction is not included in the analysis.

Table 1. Peak acceleration in three seismic levels (cm/s^2).

Direction	Minor earthquake	Moderate earthquake	Severe earthquake
Main Dir (X Dir)	35	100	200
Other Dir (Y Dir)	30	85	170

5 STRUCTURAL ANALYSIS

YJK software is used for analysis under small and moderate earthquakes, and SAUSG software (SAUSG Zeta) is used for analysis under large earthquakes. Both are with high calculation efficiency, stable solution, and meet the analysis and design requirements.

5.1 *Codes to be followed*

The code for seismic design of buildings (national code) and the technical standard for energy dissipation, vibration reduction, and isolation of buildings (Shanghai Standard) are followed. With the application of dampers, the aim of inter-story drift is strictly set (as listed in Table 2) to ensure that the main structural members won't yield during moderate seismic input.

Table 2. The target of the inter-story drift angle.

Seismic level	Target of inter-story drift
Minor	1/800
Moderate	1/300
Severe	1/120

5.2 *Analysis results under minor seismic input*

The average inter-story drift angle under minor seismic input is 1/903 in the X direction and 1/893 in the Y direction, which are both smaller than the aim drift angle in Table 2.

By checking the hysteresis curve of each damper, it can be seen that some dampers have started dissipating energy to a certain extent. The results show that the additional damping ratio due to dampers is 0.9% in the X direction and 0.7% in the Y direction.

5.3 Analysis results under moderate seismic input

The energy dissipation chart (as seen in Figure 6) shows the energy dissipated by each component under moderate seismic input. The pink part is the energy dissipated by dampers. It shows that a considerable part of the energy is dissipated by the dampers. The outcome data shows that the additional damping ratio due to dampers is 1.5% in the X direction and 1.4% in the Y direction. The dampers play an important role in protecting the main structure from damage, and the desired purpose is achieved.

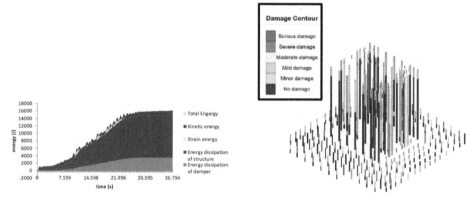

Figure 6. Energy dissipation under moderate input.

Figure 7. Column damage under moderate input.

It can be seen from the damaged contours that most columns are of no damage or minor damage, and more than 90% of the beams and shear walls are slightly damaged. The surrounding structural members don't yield. Column damage contour is shown as an example in Figure 7.

The inter-story drift angle with moderate seismic input (as shown in Figure 8) is smaller than 1/300 both in the X and Y direction, which meets the drift angle target in Table 2.

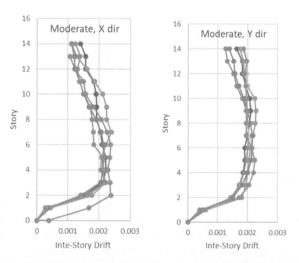

Figure 8. Inter-story drift angle in two horizontal directions with moderate seismic input.

5.4 Analysis results under severe seismic input

The energy dissipation chart in Figure 9) shows the energy dissipated by each component under severe seismic input. The dampers dissipate much seismic energy to protect the main structure from severe damage or collapse. On the other hand, the analysis outcome shows that the additional damping ratio due to the dampers is 3.3%.

The contour of damage shows that the overall damage to frame beams is slight; the damage to frame columns is minor or mild; some shear walls are moderately damaged. Figure 10 gives an example of a damaged contour of columns under severe seismic input.

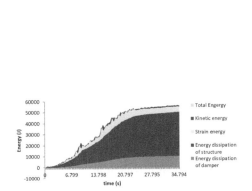

Figure 9. Energy dissipation under severe input.

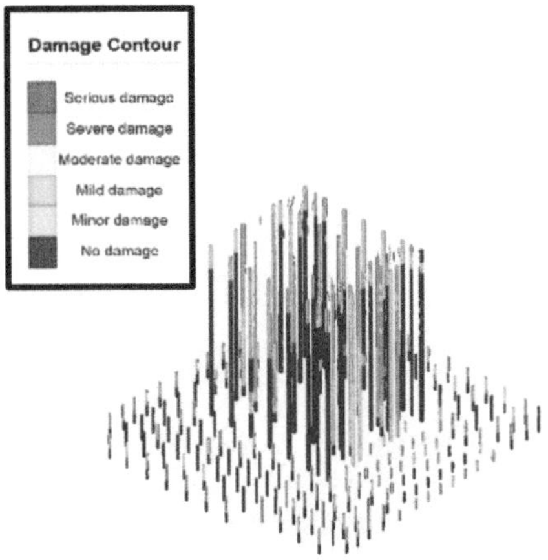

Figure 10. Column damage under severe input.

The inter-story drift angle under severe seismic input shown in Figure 11 is smaller than 1/120, both in the X and Y direction, which meets the drift target, as shown in Table 2. The seismic design targets under severe seismic input are therefore achieved.

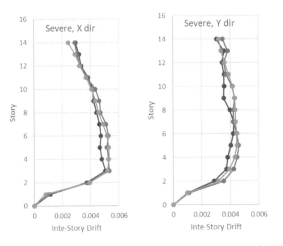

Figure 11. Inter-story drift angle in two horizontal directions under severe input.

6 CONCLUSION

Considering the structural seismic performance obtained in this paper, it is feasible to adopt the wall friction dampers in the seismic design of buildings of 10 to 20 stories. A total of 148 sets of friction dampers are set in this project. The selected seismic input waves conform to the design response specified in the code. Under minor earthquakes, the dampers start dissipating seismic energy, producing an additional damping ratio of 0.9% in the X direction and 0.7% in the Y direction. Under moderate and severe earthquakes, the dampers continue dissipating energy, producing greater additional damping ratios. The overall indicators of the structure, such as inter-story drift angle, component damage, etc., all meet the target requirements. Under the severe earthquake input, the overall damage to frame beams is slight; the damage to frame columns is minor or mild; some shear walls are moderately damaged; the substructure connected to the dampers does not yield. The dampers played a great role in protecting the main structural members from serious damage or collapse.

ACKNOWLEDGMENT

This project is financially supported by Shanghai Urban Construction College Project cjky202110.

REFERENCES

Code for Seismic Design of Buildings [S] (GB 50011-2010). Beijing: China, Architecture & Building Press, 2010. (In Chinese)

Feng, H., Zhou, F., Ge, H, Zhu, H., Zhou, L. (2021) Energy Dissipation Enhancement Through Multi-toggle Brace Damper Systems for Mitigating Dynamic Responses of Structures. [J] *Structures*, Volume 33: 2487–2499

Wang, C., Shen, T., Wei, J., Pan, Q. (2021), Energy-dissipated Design of an Office Building with Friction Dampers, *J. Shanxi Architecture*, Vol 47(12):43–45

Wu, X., Zhou, D., Huang, T. (2021) Structural Analysis of a School Building with Viscous Dampers Applied. *Journal of Physics: Conference Series* (JPCS) (ISSN:1742-6588), Vol. 2011(2021)012026

Ye, Z., Li, A., Chen, W. (2001), Study on Vibration Energy Dissipation Design of Structures with Fluid Friction Dampers. *Journal of Building Structures*, Vol.22(4):61–66

Zhou, Y., Liu, X., Wang, M. (2021) Experimental Study on Mechanical Properties of Two Types of Viscoelastic Dampers, *J. Engineering Mechanics*, Vol.38 Suppl:167–177

Seismic response analysis of shield tunnel considering the internal structure

Qiaozhen Fu & Xiwen Zhang*
School of Civil Engineering and Architecture, University of Jinan, Jinan, Shandong, China

ABSTRACT: In order to study the influence of the internal structure on the mechanical properties of the tunnel under earthquake action, this paper relies on a shield tunnel project across the river, and establishes a three-dimensional solid numerical calculation model by the finite difference software FlAC3D, considering the internal structure of the tunnel. This paper analyses the seismic response of the tunnel by the characteristics of the displacement of the tunnel, acceleration and stress. The results show that the displacement, acceleration and stress of the tunnel are improved when the internal structure is considered. The maximum transverse diameter deformation rate and vertical diameter deformation rate are greatly reduced when the internal structure is considered compared with when the internal structure is not considered. Considering the internal structure can effectively improve the seismic response characteristics of tunnels, which should be considered in seismic design.

1 INTRODUCTION

With the acceleration of urbanization and serious traffic congestion, underground space has become an important resource for urban construction and development (Chen & Shi 2022). The development of underground space is the only way to solve the problem of urbanization development, which is of great significance for realizing sustainable urban development. In 1863, the first underground railway in the world was built in London (Chen et al. 2016). So far, more and more cities in the world have built subways, and China is also in an important period of large-scale construction and rapid development of underground transportation (Li et al. 2020). In order to meet different functional requirements and traffic requirements, the tunnel engineering size will be more inclined to develop in large size, and the internal structure forms will gradually diversify, while the existence of the internal structure will certainly affect the stiffness and deformation characteristics of segments and other structures, and then affect the seismic performance of the structure (Yu et al. 2016). In addition, China is located between the Ring of the Pacific earthquake zone and the Eurasian earthquake zone, squeezed by the Pacific plate, the Indian Ocean plate and the Philippines plate, and has very active seismic activity, making it a country prone to earthquakes (Pan 2019). Many earthquake disasters occur every year, causing great casualties and property losses, and seriously hindering the modernization of the city (Wang et al. 2017). Seismic waves have three elements. Since natural seismic waves are greatly disturbed by the outside world, in order to ensure that the input conditions of seismic waves for dynamic calculation are met, many current studies use artificial seismic waves synthesized artificially as input seismic waves (Gu et al. 2022). The seismic response of underground buildings is different from that

---

*Corresponding Author: cea_zhangxw@ujn.edu.cn

of ground buildings, and its seismic response and structural damage are mainly caused by deformation. With the emergence of various forms of tunnel internal structures in order to meet different traffic requirements, compared with the single seismic problem of underground structures, taking the influence of internal structures into consideration has become a non-negligible content of the seismic design. At present, many scholars have studied the influence of tunnel internal structure on tunnel mechanical properties. For example, WU Huaqi (Wu et al. 2021) discusses several control points of fair-faced concrete in the inner structure of a double-layer shield tunnel from the design point of view based on the implementation case of the inner structure of a double-layer shield tunnel in the west section of Shanghai Beiheng Passage Project. Cui Jia (Cui et al. 2021) et al. investigated the internal structure forms of several typical tunnels and analyzed the prefabricated internal structure forms of shield tunnels and their influence on segment forces. Sun Wenhao (Sun et al. 2021) et al. took Wuhan Lianghu Tunnel (East Lake section) project as the background, established a three-dimensional solid calculation model with the large-scale finite element software ABAQUS, took the internal structure of the double-layer tunnel into account, and studied the longitudinal mechanical properties of the double-layer tunnel. A lot of research has also been done on the seismic resistance of tunnels. Taking the Luojiashan Tunnel of the Zhengzhou-Wanhigh-speed Railway as the engineering background, Yang Xuehai (Yang 2019) has done a lot of research on prefabricated bottom structure tunnels by means of numerical simulation. Li Siming (Li et al. 2021) analyzed the seismic response of shield tunnel through soil-rock changing strata, and the results showed that the peak deformation and stress of the shield tunnel occurred at the soil-rock junction, and the dynamic response at the soil-rock junction was obviously different under the same buried depth. Guo Zhengyang (Guo et al. 2021) et al., relying on the shield tunnel project of the Tianjin Z2 line and using the finite element software ABAQUS, established a three-dimensional beam-spring model to simulate the shield tunnel, analyzed the longitudinal aseismic performance of the shield tunnel, analyzed the corresponding results under four different circumstances, and drew a conclusion with the reference value in terms of aseismic design.

Nowadays, with the continuous development of tunnel engineering, the internal structure of the tunnel gradually occupies a non-negligible position in tunnel engineering. When an earthquake disaster occurs, considering the internal structure may lead to the deformation change of the contact site and the interaction between the internal structure of the tunnel and the segment structure, which will affect the original seismic response of the tunnel. Therefore, the seismic design of the tunnel should also take the internal structure into account. At present, the research on this aspect is relatively few. This study takes the internal structure of the tunnel into consideration, calculates the impact of the existence of the internal structure on the seismic response of the tunnel, analyzes and summarizes the response law of the tunnel containing the internal structure, and provides a theoretical basis for the relevant engineering seismic design.

2 NUMERICAL CALCULATION

The inner diameter of the tunnel segment is 13.9 m and the outer diameter of the tunnel segment is 15.2m. The thickness of the segment is 0.65 m, and the concrete strength grade is C60, while the other internal structure's concrete strength grade is C40.

Based on FLAC3D finite difference software, two tunnel models with and without internal structure were established to analyze the influence of internal structure on tunnel deformation under earthquake. The model size is (X × Y × Z) 120 m × 4 m × 120 m, and the free field boundary conditions and local damping are set. The model is shown in Figure 1.

The soil layer adopts Moore-Coulomb constitutive model, and the physical parameters are shown in Table 1. Tunnel segments and internal structure are regarded as elastic

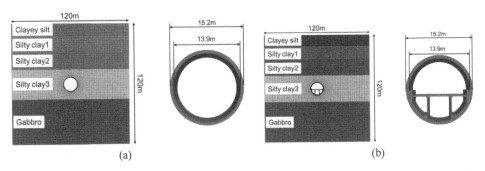

Figure 1. numerical analysis model. (a) No internal structure model (b) Internal structure model.

Table 1. Values of physical parameters of rock and soil layer.

Name of the soil	Density (g/cm^3)	Cohesion (kPa)	Internal friction angle (°)	Elastic modulus (MPa)	Poisson's ratio	Constitutive model
Clayey silt	1.846	20.5	15.6	4.6	0.3	M-C
Silty clay 1	1.897	18.7	10.4	5.6	0.3	M-C
Silty clay 2	2.009	35.4	16.0	7.3	0.3	M-C
Silty clay 3	2.020	40.5	24.6	8.2	0.3	M-C
Gabbro	2.244	40.0	38.0	s833.3	0.25	M-C

constitutive models, and the specific parameters of tunnel segments and internal structure are shown in Table 2.

In this paper, synthetic seismic wave is used as the seismic wave input condition for tunnel dynamic calculation. The peak acceleration of synthetic seismic wave is 0.1 g, which is input from the bottom of the model. The time-history curve of seismic acceleration is shown in Figure 2.

Table 2. Internal parameters of tunnel.

Structure	Concrete grade	Elastic modulus (MPa)	Poisson's ratio	Density (kg/m^3)	Constitutive model
Internal structure	C40	3.25e4	0.2	2400	Elastic
Segment	C60	3.60e4	0.2	2500	Elastic

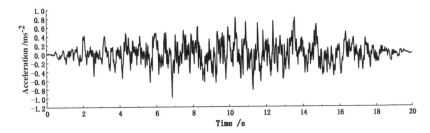

Figure 2. Time history curve of seismic wave acceleration.

3 RESULTS AND ANALYSIS

3.1 Transverse deformation analysis of tunnel structure

The time history curves of the lateral relative displacement of the left and right arch waist of the tunnel are obtained by subtracting the displacement of the left and right arch waist in the X direction. Figure 3 shows the time history curves of the relative displacement under the two working conditions considering the internal structure and without considering the internal structure.

When the seismic wave with a peak acceleration of 0.1 g is input from the bottom, the changing trend of the displacement time history curve of the left and right arch waist of the tunnel is roughly the same under the two working conditions, and the maximum displacement occurs around 13 s. It can be seen from Figures 3(a) and 3(b) that the maximum relative displacement of the tunnel occurs around 18 s without considering the internal structure, and the maximum relative displacement is about 0.0083 m, that is, the maximum diameter deformation rate is about 0.535‰. When considering the influence of internal structure, the maximum relative displacement occurs around 13 s, and the maximum relative displacement of the left and right arch waist is 0.0003 m, that is, the maximum diameter deformation rate is about 0.0197‰. The final relative displacement is 0.008 m without considering the internal structure. When considering the internal structure, the final relative displacement is 0.0002 m.

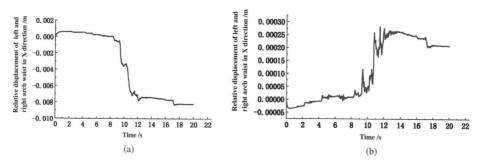

Figure 3. Relative displacement of left and right arch waist in X direction. (a) No internal structure model (b) Internal structure model.

The maximum displacement in the X direction of the tunnel occurs around 13 s. Through the analysis of the tunnel displacement cloud image at 13 s and the final tunnel displacement cloud image, it is found that the tunnel transverse diameter deformation at 13 s and the final tunnel transverse diameter deformation are greatly improved when the internal structure is taken into account, compared with that without the internal structure. It can be concluded that considering the internal structure greatly reduces the transverse diameter deformation of the tunnel.

3.2 Vertical deformation analysis of tunnel structure

The relative displacement is obtained by subtracting the vault displacement from the arch bottom displacement. The time-history curve of relative displacement in the Z direction of the arch bottom of the vault is shown in Figure 4.

Under the condition of no internal structure, the maximum displacement of the tunnel arch bottom is 0.0023 m, which is around 11 s. The maximum displacement of the vault occurs in 18 s, which is 0.0081 m. As can be seen in Figure 4(a), The maximum relative

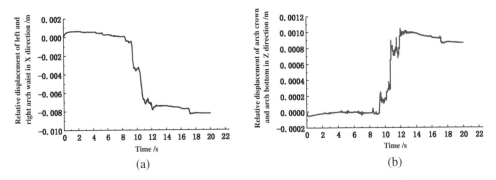

Figure 4. Relative displacement of left and right arch waist in Z direction. (a) No internal structure model (b) Internal structure model.

displacement of the arch bottom was 0.008 m, and the maximum vertical diameter deformation rate was 0.526‰.

When considering the internal structure, the maximum displacement of the tunnel vault and arch bottom is 0.0015 m and 0.0010 m, respectively. Figure 4(b) shows that the maximum vertical relative displacement of the arch bottom of the vault is 0.0010m, that is, when the internal structure is considered, the maximum vertical diameter deformation rate is 0.0568‰.

By comparing the vertical relative displacements of the vault and the arch bottom under two working conditions, it can be seen that considering the internal structure greatly improves the vertical deformation of the tunnel.

3.3 Analysis of acceleration

In the process of upward propagation of seismic waves, the amplification effect will be generated, and the peak acceleration at different monitoring positions of the tunnel will be counted as shown in Figure 5.

According to the analysis of Figure 5, with the decrease of the buried depth, the peak acceleration at each monitoring position in the stratum tends to gradually increase, and the

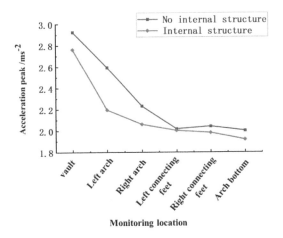

Figure 5. Comparison of peak acceleration at different monitoring positions under two working conditions.

amplification effect at the vault is the most significant, arch bottom vice versa. The peak acceleration of the vault is amplified by 2.93 times and 2.76 times, respectively. Compared with the tunnel without considering the internal structure, the amplification coefficient of the peak acceleration of the included structure is reduced to a certain extent, and the reduction effect is more obvious at the left and right arch waist and the left and right connecting foot. It can be shown that the internal structure of the tunnel can effectively reduce the acceleration of the tunnel segment under the action of an earthquake.

3.4 *Stress analysis*

Figure 6 shows the maximum principal stress cloud diagram. It can be seen that the internal structure changes the stress state of the segment, and the position of the left and right arch waist is the most obvious. The maximum principal stress at the connecting foot of the internal structure is also significantly reduced compared with the same position without the internal structure. The stress on the arch waist of the tunnel without internal structure is more disadvantageous, which should be paid attention to in the seismic design, and corresponding measures should be taken to prevent the damage when necessary. When the internal structure is considered, the maximum principal stress on the internal structure of the tunnel is larger and the minimum principal stress is smaller.

Figure 6. Cloud diagram of maximum principal stress. (a) No internal structure model (b) Internal structure model.

4 CONCLUSION

This paper mainly studies the seismic response of the tunnel under two working conditions, considering the internal structure and not considering the internal structure, and focuses on the deformation, acceleration and stress response of the internal structure of the tunnel under the action of low-intensity earthquake. The main conclusions of this paper are as follows:

(1) Under the action of seismic waves with a peak acceleration of 0.1 g, the seismic response of the tunnel considering the internal structure is significantly different from that without considering the internal structure.
(2) Considering the internal structure, compared with not considering the internal structure, the diameter deformation of the tunnel is greatly improved. The acceleration response, stress response, and displacement response are decreased as a whole.
(3) The existence of this type of internal structure can effectively reduce the tunnel diameter deformation. Among them, the relative displacement of the joints with the internal structure, such as the left and right arch waist, is more obvious.

In conclusion, the influence of the internal structure should be considered in the seismic design of the tunnel, and appropriate internal structure forms can be selected to improve the deformation of the parts of the tunnel with large deformation.

REFERENCES

Chen Y.H., Shi H.W. Research on the Natural Lighting Design of Underground Space Under the Guidance of the Concept of Urban Sustainable Development: A Case of the Underground Garage of the Central Park in the Eastern New Town of Ningbo [J]. *Urbanism and Architecture*, 2022, 19(06): 69–71.

Chen G.X., Chen S., Du X.L., Lu D.C., Qi C.Z. Review of Seismic Damage, Model Test, Available Design and Analysis Methods of Urban Underground Structures: Retrospect and Prospect [J]. *Journal of Disaster Prevention and Mitigation Engineering*, 2016, 36(01): 1–23.

Cui J., Zhang L., Guo W.Q. Assembled Internal Structure of Shield Tunnel and Its Influence on Force of Segmen [J]. *Building Technology Development*, 2021, 48(15): 96–98.

Gu K.S., Guo M.Z., Tang X.W., Wang T.C. Dynamic Response and Spectrum Characteristics of Anti-Dip Rock Slopes Under Earthquake [J]. *China Earthquake Engineering Journal*, 2022, 44(01): 62–71.

Guo Z.Y., Liu Y., Liang J.W., Li D.Q., Wang Z.K., Wu Z.Q. Longitudinal Seismic Analysis of Shield Tunnels in Complex Soft Soils [J]. *China Earthquake Engineering Journal*, 2021, 43(03): 687–692+703.

Li L.P., Cheng S., Zhang Y.H., Tu W.F. Opportunities and Challenges of Construction Safety in Underground Engineering Projects [J]. *Journal of Shandong University of Science and Technology (Natural Science)*, 2020, 39(04): 1–13.

Li S.M., Yu H.T., Xue G.Q., Xu L. Analysis on Seismic Response of Shield Tunnels Passing Through Soil-Rock Strata [J]. *Modern Tunnelling Technology*, 2021, 58(05): 65–72.

Pan J.C. *Seismic Response Analysis and Seismic Damping of Metro Shield Tunnel* [D]. Anhui University of Science and Technology, 2019.

Sun W.H., Feng K., Xiao M.Q., Wang J.Y., Guo W.Q., Lu X.Y. Influence of Internal Structure of Double Deck Tunnel on Longitudinal Mechanical Properties of Tunnel [J]. *Railway Engineering*, 2021, 61(05): 43–48.

Wang S.Y., Liang C., Ji X.X., Tang D.L., Zhang XW. Secondary Geological Disasters Induced by the Weining Earthquake on Oct. 9, 1948 in Guizhou [J]. *Guizhou Science*, 2017, 35(02): 33–35.

Wu H.Q., Sun W., Liu N. Several Control Points of Fair-Faced Concrete for Internal Structure of Double-Layer Shield Tunnel[J]. *China Municipal Engineering*, 2021(05): 108–111+122.

Yang X.H. *Seismic Response Analysis and Anti-seismic Measures of Tunnels with Assembled Bottom Structure* [D]. Beijing Jiaotong University, 2019.

Yu H.T., Li L.J., Cao C.Y., Duan K.P., Zhang F. Seismic Analysis of Shield Tunnel Considering Internal Prefabricated Structure [J]. *Chinese Journal of Underground Space and Engineering*, 2016, 12(S2): 834–840.

Analysis of vibration effect of pile foundation parameters on superstructure under blasting in a tunnel

Baofu Duan, Weishan Xu & Zhaowen Yu
Shandong University of Science and Technology, Shandong Provincial Civil Engineering Disaster Prevention and Mitigation Key Laboratory, Shandong Qingdao, China
Hubei Key Laboratory of Blasting Engineering of Jianghan University, Hubei Wuhan, China

Zongjun Sun*
Shandong University of Science and Technology, Shandong Provincial Civil Engineering Disaster Prevention and Mitigation Key Laboratory, Shandong Qingdao, China
Qingdao Ruihan Technology Group Co, Ltd, Shandong Qingdao, China

ABSTRACT: To study the influence of pile foundation parameters on superstructure in tunnel blasting, the three-dimensional calculation model of rock and soil mass, tunnel and superstructure were established by using MIDAS /GTS NX finite element simulation software, and the vibration law of superstructure was studied by selecting the parameters of pile foundation. The results show that when the pile lateral friction resistance value is within the range of [160,165], the vibration velocity of the structure increases by leaps and bounds with the increase of pile lateral friction resistance, in a certain range, the relationship between the stiffness of pile foundation material and the vibration velocity of superstructure shows an exponential growth trend.

1 INTRODUCTION

With the development of urban rail construction, the subway network is increasingly complicated. Due to the limitation of engineering conditions, many tunnels need to be excavated by blasting construction (China Journal of Highway and Transport 2015). Blasting construction not only brings great convenience to engineering construction but also brings great potential harm. The blasting seismic wave caused by the explosive explosions is likely to cause damage or even damage to buildings, which seriously threatens the safety of people's lives and property (Gong et al. 2015; Liu & Cao 2015). Therefore, it is of great significance to study the influence of pile foundation parameters on superstructure in tunnel blasting construction.

At present, domestic and foreign scholars aim at the vibration response of buildings under tunnel blasting, generally through on-site monitoring (Yu 2015; Wang et al. 2018; Wang et al. 2020), theoretical analysis (Wei & Chen 2011; Yang et al. 2009) and numerical simulation (Hu et al. 2016; WEI & Hao 2009) and other methods, and achieved certain results. Tunnel blasting is a complex problem of geotechnical dynamics, and field tests have great limitations. With the popularization of numerical simulation technology, numerical analysis methods are more and more applied in this field. For example, Shi Jiehong, etc. (Shi 2016) used the LS-DYNA dynamic finite element software, and it is found that the blasting

---

*Corresponding Author: 3519808967@qq.com

operation in the metro depot has a high impact on the risk of residential buildings, and the safety distance beyond 100 meters is suggested to carry out the blasting operation. Zhu Zebing et al (Zhu et al. 2010) used FLAC3D software to simulate the blasting vibration effect and compared it with the measured results to conduct research, effectively controlling the blasting ground vibration intensity, and blasting did not cause damage to the building. Jiang Nan; Zhang Yuqi; Zhou Chuanbo et al (Jiang et al. 2020) combined the small explosion test and numerical simulation, the prediction model of the vibration velocity of high-rise buildings related to the floors proposed, and the safety of high-rise building under the action of blasting vibration is analyzed and evaluated. Fei Honglu et al (Fei et al. 2018) used FLAC3D software to perform a numerical calculation on the vibration velocity peak of particles at each observation point in the field test. The monitoring result of underground blasting vibration velocity was smaller than that of the ground, and the peak value of vibration velocity was about 50%~64% of that of the ground, and it is proposed that the Sadowsky formula can be modified by the numerical simulation results to improve the prediction accuracy.

At present, the research on the disturbance of pile foundations and superstructures caused by tunnel blasting construction is more common among domestic and foreign researchers, while the research data on the vibration of superstructure caused by pile foundation parameters are less, and most of them stay in the research on the vibration response of single structure of pile foundation or superstructure caused by blasting vibration. So, in tunnel blasting construction, it is necessary to investigate the vibration response effect of pile foundation parameters on the superstructure.

2 PROJECT OVERVIEW

Shenzhen Urban Rail Transit Line 6 is located in Longhua New District and Futian District, with a total length of 5.45 km. In this paper, the Shen Mei section is selected as the research object, and the whole is in the north-south direction. The shield construction is carried out underground at the shield shaft at the Shilong intersection. The tunnel body of the shield-driven section in the Shenmei section is mainly made of slightly weathered granite and moderately weathered granite, the relationship between the tunnel and the upper building is shown below.

Figure 1. The relationship between the section tunnel and the upper building.

3 NUMERICAL SIMULATION

3.1 Calculation model

The model simplifies the upper building to a steel-concrete frame structure without considering the basement, each floor is 3 meters high and a total of 15 floors are selected. The foundation is piled raft foundation, the pile body is reinforced concrete material, and the pile foundation type is the end-bearing friction pile. Raft thickness is one meter, two meters outside pick, foundation depth is 3 meters, and piles are deep into the lightly weathered coarse-grained granite rock formation with 25 m length of the pile. The tunnel is 30 meters deep and has a diameter of 7.3 meters. The boundary of the numerical simulation of tunnel engineering is 3~5 times the diameter of the hole, and the length of the pile should be doubled in the vertical direction (Zhang et al. 2017). Combining the actual project and the operating capacity of the computer, the model size is determined to be 120 m×120 m×60 m, the rock stratum grid size is 2 m, and the tunnel grid size is 1 m, the model generates 173,560 units and 145,022 nodes in total. Models are shown in Figures 2 and 3.

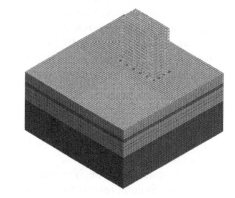

Figure 2. Mesh generation diagram of the finite element model.

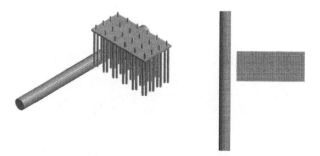

Figure 3. Internal structure diagram of finite element.

3.1.1 Foundation reaction coefficient

The foundation reaction coefficient is composed of the vertical foundation reaction coefficient and horizontal foundation reaction coefficient. The vertical foundation reaction force coefficient is:

$$k_v = k_{v0} \cdot \left(\frac{B_v}{30}\right)^{-3/4} \tag{1}$$

The horizontal foundation reaction coefficient is:

$$k_h = k_{h0} \cdot \left(\frac{B_h}{30}\right)^{-3/4} \tag{2}$$

where $k_{v0} = \frac{1}{30} \cdot \alpha \cdot E_0 = k_{h0}$, $B_v = \sqrt{A_h}$, $B_h = \sqrt{A_h}$, A_v and A_h are the vertical and horizontal sectional areas of the model, respectively. α is taken as 1.0, and E_0 is the elastic coefficient of the foundation.

3.1.2 *Viscous boundary of time history analysis*

The damping ratio of the model needs to be calculated for the viscoelastic boundary. The calculation formula is as follows:

$$P-\text{wave}: C_p = \rho \cdot A \cdot \sqrt{\frac{\lambda + 2G}{\rho}} = \gamma \cdot A \cdot \sqrt{\frac{\lambda + 2G}{\gamma \cdot 9.81}} = c_p \cdot A \tag{3}$$

$$S-\text{wave}: C_s = \rho \cdot A \cdot \sqrt{\frac{G}{\rho}} = \gamma \cdot A \cdot \sqrt{\frac{G}{\gamma \cdot 9.81}} = c_s \cdot A \tag{4}$$

In the formula: λ is the volume elasticity coefficient (kN/m^2); G is the shear elasticity coefficient (kN/m^2); E is the elastic coefficient (kN/m^2); A is the sectional area.

3.2 *Simplified loading of blasting loads*

Based on the Saint-Venant principle in mechanics, the equivalent load does not consider the shape of the blast hole when modeling. That is, it is not reflected in the model. And then the blasting load time history curve is equivalent, applied to the plane determined by the blast hole connecting the center line and the blast hole axis in the same row, the specific equivalent calculation process is shown in Figure 4 below, where P$_0$ is the blasting center load and Pe is the equivalent load.

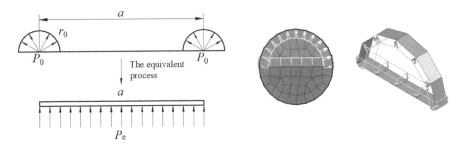

Figure 4. Schematic diagram of equivalent load and blasting load application.

3.3 *Monitoring data and numerical simulation analysis results*

Comparing the on-site detection data with the data calculated by simulation analysis, the vibration velocity was monitored in the Shuxiangmendi building near the upper part. The measuring points are shown in the red dots below.

Sort out the data collected by field monitoring and numerical model, as shown in the Table 1 below.

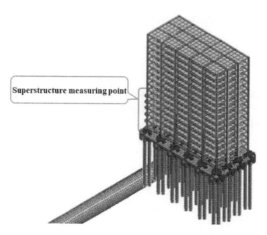

Figure 5. Monitoring points in blasting construction.

Table 1. Vibration velocities in the Z direction at each layer measurement point.

Floor number	Floor height (m)	Simulated vibration speed (cm/s)	Measured vibration velocity (cm/s)	Error
1	0	0.272	0.249	0.023
2	3	0.0364	0.021	0.0154
3	6	0.0142	0.007	0.0041
4	9	0.0151	0.011	0.0041
5	12	0.0146	0.008	0.0066
6	15	0.0142	0.006	0.0082

It can be seen from the above table that the numerical simulation results are similar to the measured results, and the maximum error between the on-site monitoring and numerical simulation results is 0.023 cm/s, to ensure the correctness of the numerical calculation model in the future.

4 INFLUENCE OF PILE FOUNDATION PARAMETERS ON VIBRATION RESPONSE OF SUPERSTRUCTURE

4.1 *Contact parameters of pile-soil interface*

Based on the end-bearing friction pile, the relationship between the pile-soil interface contact and the structural vibration response is explored by changing the pile-soil interface contact parameters, which are reflected in the numerical simulation as the difference of the final shear force in the pile-soil interface parameters. The final shear force values of 50 kN/m^2, 100 kN/m^2, 112 kN/m^2, 160 kN/m^2, 165 kN/m^2, 170 kN/m^2, 180 kN/m^2, 200 kN/m^2 and 300 kN/m^2 are respectively selected, and the vibration speed of the upper structure measurement point is taken as the characterization parameter of the vibration response. The relationship between different final shear forces and the vibration speed of the superstructure is shown Figure 6 below.

It can be found that the larger the final shear value, the larger the corresponding vibration speed of the superstructure, and the maximum vibration speed is within the safe range. The vibration velocity at the bottom of the structure is the largest. Taking the vibration velocity at the bottom of the structure as an example, it can be seen that the vibration

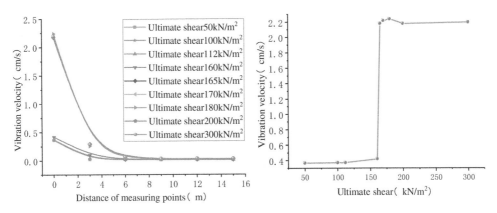

Figure 6. Relationship between final shear force and vibration response of superstructure.

velocity corresponding to different final shear forces varies greatly. There is no linear correspondence between the final shear force and the vibration speed of the superstructure, but with the increase of the final shear force, the vibration speed of the structure shows a jumping growth in the range between one of the small areas of the final shear force, and then with the increase of the final shear force, the vibration speed of the superstructure continues to stabilize. In this numerical simulation, the interval of the final shear force corresponding to the jump growth of the vibration speed is [160, 165]. Therefore, in the process of tunnel construction, the pile side friction affects the vibration response of the superstructure, and when the side friction is in a small area, the vibration response of the superstructure is particularly significant.

4.2 *Material rigidity of pile shaft*

Based on steel concrete piles, only the elastic modulus of the pile foundation material is changed, and the relationship between the elastic modulus and the vibration velocity of the superstructure is explored. The values of the elastic modulus of the pile foundation are 11 GPa, 38 GPa, 206 GPa, 250 GPa, 300 GPa, and 520 GPa. The vibration velocities of the superstructure corresponding to different pile foundation elastic moduli are shown in the Figure 7 below.

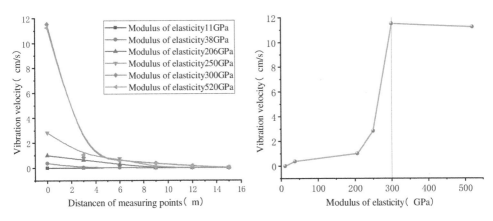

Figure 7. Vibration velocity of the superstructure corresponding to the elastic modulus of the pile foundation.

It can be found that there are large differences in the vibration speed curves of the superstructure. The larger the elastic modulus of the pile foundation material, the larger the corresponding vibration speed of the superstructure. When the elastic modulus increases to a certain value, the maximum vibration speed has exceeded the safe range. It can be seen from the above figure that with the increase of the elastic modulus of the pile shaft, the vibration speed of the superstructure measurement points increases exponentially, but when the vibration speed increases to a certain value, the vibration speed will not increase. When the soil layer medium is constant, the distance between the blasting centers and the blasting sources is constant, that is, the instantaneous input energy is constant, it is meaningless to increase the rigidity of the pile shaft again, and it will not have a corresponding impact on the superstructure, so the vibration speed will not be increased.

5 CONCLUSION

(1) The corresponding parameter of the pile side friction in the numerical simulation is the final shear force. According to the corresponding relationship between the final shear force and the superstructure vibration speed, when the final shear force is lower than 160 kN/m^2, as the final shear force increases, the structure vibration speed increases very slowly; when the final shear force is greater than 165 kN/m^2, with the increase of the final shear force, the vibration speed of the structure tends to be stable; when the final shear force value is in the interval [160, 165], with the increase of the final shear force, the structural vibration velocity exhibits a rapid increase.
(2) The corresponding parameter for the stiffness of the pile material in the numerical simulation is the elastic modulus. When the elastic modulus exceeds 300 MPa, as the elastic modulus increases, the vibration velocity of the superstructure does not increase; when the elastic modulus does not exceed 300 MPa, the relationship curve between elastic modulus and structural vibration velocity is similar to the exponential function curve.

ACKNOWLEDGMENTS

Funding: This research was funded by the Shandong Natural Science Foundation of China, grant number ZR2020ME096, and the work is supported by the Foundation of Hubei Key Laboratory of Blasting Engineering, grant number BL2021-24, and Qingdao West Coast New Area High-level Talent Team Project, grant number RCTD-JC-2019-06.

REFERENCES

Editorial Office of China Journal of Highway and Transport. Summary of Academic Research on Chinese Tunnel Engineering·2015 [J]. *China Journal of Highway and Transport*, 2015, 28(05): 1–65.
Fei Honglu, Wang Zhenda, Jiang Anjun, et al. Study on the Influence of Subway Tunnel Blasting Vibration on the Ground [J]. *Blasting*, 2018, 35(3): 68–73.
Gong Min, Wu Haojun, Meng Xiangdong, et al. Micro-Vibration Controlled Blasting Method and Vibration Analysis for Tunnel Excavation Under Dense Buildings [J]. *Explosion and Shock*, 2015, 35(3): 350–358.
Hu Huirong, Huang Huadong, Wang Xianyi, et al. Analysis of the Characteristics of the Amplification Effect of Tunnel Blasting Under Dynamic and Static Conditions [J]. *Tunnel Construction*, 2016, 36(7): 812–818.
Jiang Nan, Zhang Yuqi, Zhou, Chuanbo, Wu Tingyao, Zhu Bin. Influence of Blasting Vibration of MLEMC Shaft Foundation Pit on Adjacent High-Rise Frame Structure: A Case Study [J]. *Nutrients*, 2020, 13(19): 5140.

Liu Junwei, Cao Shengli. Non-Blasting Construction Technology of Underground Excavated Tunnel of Intercity Railway Through Ground Surface Building [J]. *Tunnel Construction*, 2015, (S2): 91–96.

Shi Jiehong. Research on Safety Control Distance and Analysis Method of Blasting Construction Around Structures [J]. *China Work Safety Science and Technology*, 2016, 12(11): 125–129.

Wang Lintai, Gao Wenxue, Zhang Facai, et al. Research on the Vibration Response of Buildings Under Blasting Earthquake [J]. *Acta Armamentarii*, 2018, 39(S1): 121–134.

Wang Songqing, Zhang Quanfeng, Wang Haibo, et al. Research on Blasting Vibration Control Technology of Wuhan Metro Section Tunnel Underpassing Building [J]. *Engineering Blasting*, 2020, 26(1): 85–90.

Wei Haixia, Chen Shihai. The Influence of the Three Elements of Blasting Seismic Waves on the Elastoplastic Seismic Response of Multi-Story Masonry Structures [J]. *Explosion and Shock*, 2011, 31(1): 55–61.

Wei Xueying, Hao Hong. Numerical derivation of homogenized dynamic masonry material properties with strain rate effects [J]. *International Journal of Impact Engineering*, 2009, 36(3): 522.

Yang Youfa, Liang Wenguang, Cao Jianliang. Time-History Prediction of Blasting Vibration Response of Frame Structure [J]. *Journal of Vibration and Shock*, 2009, 28(10): 147–149+178+231.

Yu Lei. The Impact of Blasting Vibration on the Safety of Multi-Storey Buildings[J]. *Journal of Railway Engineering Society*, 2015, (03): 86–89.

Zhu Zebing, Zhang Yongxing, Liu Xinrong, et al. The Effect of Blasting Excavation of A Super-Large Section Station Tunnel on Surface Buildings [J]. *Journal of Chongqing University*, 2010, (02): 110–116.

Zhang Zhiguo, Xu Chen, Gong Jianfei. Elastoplastic Solution to the Influence of Tunnel Excavation on the Deformation and Bearing Capacity of Adjacent Pile Foundations [J]. *Chinese Journal of Rock Mechanics and Engineering*, 2017, 36(01): 208–222.

Analysis of influencing factors of segment floating during the construction period of the shield tunnel

Yalong Jiang*, Fugen Lin, Yuchen Hu & Xin Tang
Institute of Geotechnical Engineering, School of Civil Engineering and Architecture, East China Jiaotong University, Nanchang, Jiangxi, P.R. China

ABSTRACT: During the construction period of the shield tunnel, the stratigraphic conditions, tunnel burial depth, and driving parameters will all have a specific impact on the segment floating. If the segment floating is too large, it may cause a safety problem for the shield tunnel. Relying on the tunnel project of Nanchang Metro Line 4 (Anfeng Station~Dongxin Station), we summarize and analyze the monitoring values of the segment floating when the shield passes through different strata using the mathematical statistics method. Then, the shield tunnel construction model is established by ABAQUS software, to study the influence of varying tunnel burial depths and main driving parameters on segment floating. Research results show that: (1) The amount of the segment floating is closely related to the stratigraphic conditions. When the shield machine is driving in different strata, the value of the segment floating will also change. (2) The greater the tunnel buried depth, the smaller the amount of the segment floating. When the buried depth changes from 15 m to 21 m, the amount of the segment floating basically shows a significant linear relationship with the buried depth. (3) With the increase of the three main driving parameters, the amount of the segment floating shows a linear increasing trend, and the influence of grouting pressure and mud silo pressure on segment floating is stronger than that caused by the total shield thrust.

1 INTRODUCTION

Subway construction has promoted the rapid development of the urban economy and greatly improved the living conditions of citizens. However, in the process of urban subway construction, affected by the stratum, tunnel burial depth, and driving parameters, shield tunnels often have different degrees of segment floating. In severe cases, it will lead to segment misalignment, cracking, axis offset, and even segment waterproofing structure failure and other diseases, which will affect the engineering quality of the tunnel lining structure.

At present, many scholars have carried out a lot of research on the problem of segment floating control. Shu et al. (2017a, 2017b) analyzed the measured data of segment floating during the construction of shield tunnels with variable composite strata. Wang et al. (2020) studied the influence of unfavorable geological conditions and loads on the segments floating during shield construction through numerical simulation. Kasper et al. (2006) used finite element software to establish models of tunnels, grouting layers, and soil components. They analyzed the characteristics of shield construction and the influence of material properties on segment floating. Jin et al. (2019) considered the extrusion effect of the shield machine shell, earth pressure, jack thrust, and grouting pressure when establishing the segment floating model. They found that if the segment floating is too large, it will be detrimental to the safety of the tunnel structure. Chen et al. (2014) based on the stress characteristics of the shield

---
*Corresponding Author: yalongjiang@whu.edu

tunnel segment during the construction stage, used the finite element software ABAQUS to establish a three-dimensional analysis model of the segment floating during the construction period, studied the law of floating deformation under the thrust of different jacks. Geng et al. (2021) analyzed the different stress states of the segment before and after the initial setting of the slurry. They proposed that adjusting the assembly height of the segment can alleviate the segment floating. To ensure the assembly quality and safety of segment lining structures in shield tunnels, it is necessary to study the factors that affect segment floating.

Relying on the tunnel project of Nanchang Metro Line 4 (Anfeng Station ∼ Dongxin Station), this paper summarizes and analyzes the monitoring values of the segment floating when the shield passes through different strata using mathematical statistics, and then uses numerical simulation methods to study the influence of different tunnel buried depth and main driving parameters on segment floating.

2 PROJECT GENERAL SITUATION

The thickness of the top of the tunnel of Nanchang Metro Line 4 (Anfeng Station ∼ Dongxin Station) is about 9.9∼25.8 m, and the buried depth of the groundwater level is 3.07 m∼11.2 m. The section tunnel successively passes through muddy clay, silty clay, gravelly soil, and strongly weathered and moderately weathered argillaceous siltstone on the west coast of Ganjiang River. The river crossing section passes through moderately weathered argillaceous siltstone on the whole area. The east coast successively passes through medium sand, coarse grit, sand gravel and moderately weathered argillaceous siltstone. The main part of the section tunnel is in moderately weathered argillaceous siltstone and passes through Ganjiang River, and the quality of surrounding rock in the stratum is mainly graded III.

3 MONITORING SCHEME

The method of manual measurement combined with automatic monitoring is used to measure each segment ring at a frequency of 2 times/d. The measurement point is located inside the bottom of the segment structure. The specific monitoring point is shown in Figure 1. The monitoring instruments mainly use an artificial total station and Leica small prism, among which the artificial total station is shown in Figure 2. The specific monitoring plan is: firstly, to use the total station to measure the coordinates of the bolt hole position H_1 at the bottom of the segment; then to measure the real-time coordinate H_2 of the shield machine site;

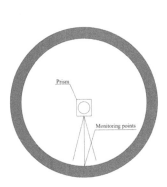

Figure 1. Diagram of segment floating monitoring points.

Figure 2. Leica TS30 total station with servo motor.

Figure 3. Field monitoring of segment floating.

finally, to calculate the difference between the segment coordinate H_1 and the real-time coordinate H_2 of the shield machine, namely, H_1-H_2 is the value of the segment floating. Figure 3 is the field monitoring of the segment floating.

4 MONITORING DATA ANALYSIS

To monitor the situation of the segment floating on the right line 1-897 ring of the tunnel between Anfeng Station and Dongxin Station, there are mainly three kinds of strata in this interval: 1) Full-section sand layer (No.1-77segment rings); 2) Upper-soft and lower-hard strata (No. 79~268 segment rings); 3) Moderately weathered argillaceous siltstone (Nos. 270~761 segment rings). Figure 2 shows the quantities of segments floating within the interval, as shown in Figure 4.

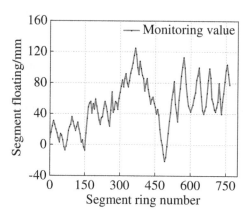

Figure 4. The quantities of segments floating within the interval.

When the shield machine is driving in the sand layer, the amount of the segments floating is not significant. The value is relatively stable, and the floating value is −7 mm~31 mm, in which the negative float means the segment sinks. And when the float is positive, it means the segment floats, and the average value of the segment floating is about 10.6 mm. When the shield machine is driving in the upper-soft and lower-hard strata, the segment floating value is −7 mm~68.6 mm, and the average value is about 33.8 mm. When the shield machine is driving in the moderately weathered argillaceous siltstone, the situation of the segment floating fluctuates obviously. The floating value is −21 mm~125 mm, and the average value reaches 66.7 mm, which shows that in the moderately weathered argillaceous siltstone, the amount of segment floating is not only unstable but also much more significant than that in the sand layer and the upper soft and lower hard strata. There will be a sudden increase of 15–25 mm in the floating amount at the separation between the upper-soft and lower-hard strata and the moderately weathered argillaceous siltstone. This is because when the shield tunnels in the hard rock formation, the slurry is lost, so there is no slurry filling the gap between the back of the segment wall and the surrounding rock, providing more floating space.

5 NUMERICAL SIMULATION RESEARCH

5.1 *Influence of tunnel buried depth on segment floating*

The model is built as a two-dimensional fluid-structure interaction model with a width of 40 m and a height of 50 m. The tunnel burial depths are 15 m, 18 m, and 21 m, respectively. The soil mass adopts the Mohr-Coulomb constitutive model, the element type is CPE4P, and the segment and grouting layer adopt the CPE4I solid element. The exact parameters of the soil layer are shown in Table 1. The detailed parameters of the model are shown in Table 2, and the soil model shown in Figure 5 is finally established. According to the reference (Geng et al. 2021), it can be seen that the buoyant force of about 300 kN will be applied to the segment when the segment comes out of the shield tail, converting it into a surface load of 212.31 kPa acting on the segment ring, the vertical displacement cloud map of different buried depths of the segment can be obtained. Taking 18 m as an example, the vertical displacement cloud map of the segment under the 18 m buried depth is shown in Figure 6.

Taking the vertical displacement of the node at the bottom of the arch as the amount of

Table 1. The exact parameters of the soil layer.

Soil layer Parameter value	Fill soils	Muddy clay	Sandy gravel	Strongly weathered argillaceous siltstone	Moderately weathered argillaceous siltstone
Soil thickness/m	2.0	2.5	8.0	1.0	36.5
Density/kg·m^{-3}	1860	1800	1880	2000	2440
Young's modulus/MPa	2.5	9.6	35	120	4000
Cohesion/kPa	10	13.13	1	38	150
Internal friction angle	10	6.77	34	33	39
Poisson's ratio	0.3	0.34	0.24	0.28	0.23
Permeability coefficient /m·s^{-1}	2×10^{-6}	2×10^{-6}	8×10^{-6}	2×10^{-6}	2×10^{-9}

Table 2. The detailed parameters of the model.

Part	Density /kg·m^{-3}	Young's modulus/MPa	Poisson's ratio
Segment	2500	30000	0.2
grout	1140	5	0.24
Shield	11986	210000	0.28

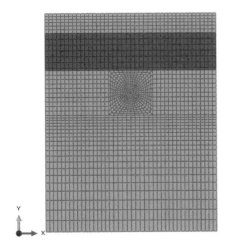

Figure 5. Soil model.

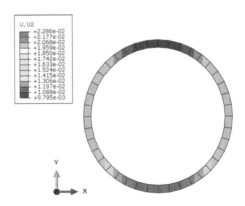

Figure 6. Nephogram of vertical displacement of the segment under 18m buried depth.

the segment floating, the finite element results are extracted and drawn to obtain the situation of the segment floating under different burial depths, as shown in Figure 7.

Figure 5 shows that when the tunnel burial depth is 15 m, the segment floating is

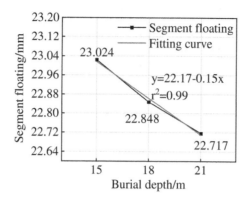

Figure 7. Variation of segment floating movement with buried depth.

23.024 mm; when the tunnel burial depth is 18 m, the segment floating is 22.848 mm; and when the tunnel burial depth is 21 m, the segment floating is 22.717 mm. The greater the burial depth of the tunnel, the smaller the amount of the segment floating. This is because the overlying soil provides a greater anti-buoyancy force to the segment. When the burial depth changes from 15 m to 21 m, the floating amount of the segment shows a significant linear relationship with the burial depth. The linear fitting of the amount of the segment floating versus the burial depth can obtain formula (1), and r^2 reaches 0.99, indicating that the fitting effect is good.

$$y = 27.17 - 0.15x \quad r^2 = 0.99 \tag{1}$$

5.2 Influence of main driving parameters on segment floating

To reduce the calculation cost, half of the shield tunnel model is taken, the tunnel burial depth is 21 m, and the overall size of the model is (24 × 42 × 50) m. Model soil parameters and material properties of all components are consistent with those in the 2D model. All components are divided into 28 blocks according to the driving direction of 1.5 m, and are numbered 1 to 28 in sequence. The load application method is determined according to the reference (Geng et al. 2021), and finally, the part model is shown in Figure 8~Figure 11.

Referring to the field driving parameters data, the finite element model under the corresponding working conditions is established by changing the main driving parameters (total shield thrust, grouting pressure, and slurry pressure), and the segment displacement under different working conditions is calculated. The vertical displacement data of the bottom of two ring segments are extracted (the 9th and 10th ring segments are selected in this paper), and the influence of different driving parameters on the segment floating is drawn, as shown in Figures 12 to 14.

It can be seen from the simulation results that with the increase of the three main driving parameters, the amount of the segment floating will show a linear growth trend. When the total shield thrust increases from 1.49 MPa to 2.23 MPa, the float of the ninth ring segment increases from 11.61 mm to 11.70 mm, an increase of 0.8%; the float of the tenth ring segment increases from 13.61 mm to 13.70 mm, an increase of 0.7%. When the grouting

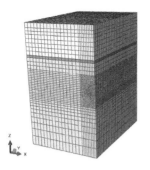

Figure 8. Soil.

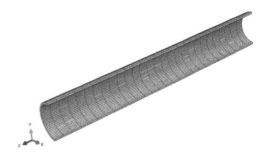

Figure 9. Segment.

Figure 10. Grouting layer.

Figure 11. Shield machine.

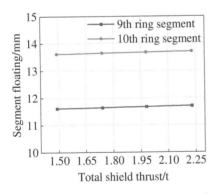

Figure 12. Floating movement of segments in the 9th and 10th rings under different total shield thrust.

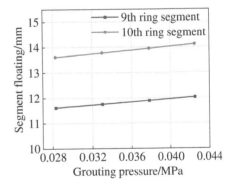

Figure 13. Floating movement of segments in the 9th and 10th rings under different grouting pressure.

pressure increases from 0.0283 MPa to 0.0425 MPa, the float of the ninth ring segment increases from 11.61 mm to 12.03 mm, an increase of 3.6%; the float of the tenth ring segment increases from 13.61 mm to 14.11 mm, an increase of 3.7%. When the slurry pressure increases from 0.15 MPa to 0.2 MPa, the float of the ninth ring segment increases from 11.61 mm to 12.03 mm, an increase of 3.6%; the float of the tenth ring segment increases from 13.61 mm to 14.10 mm, an increase of 3.7%. It can be seen that the degree of the segment floating is more sensitive to the grouting pressure and the slurry pressure, and is the

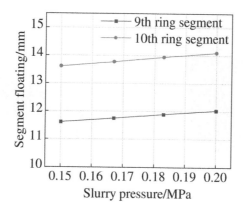

Figure 14. Floating movement of segments in the 9th and 10th rings under different slurry pressure.

least susceptible to the total shield thrust. That is, the influence of the total shield thrust on the segment floating is lower than that caused by the grouting pressure and slurry pressure.

6 CONCLUSIONS

Taking Nanchang Metro Line 4 (Anfeng Station to Dongxin Station) as the engineering background, the method of field monitoring and numerical simulation is used to study the influence of different strata, different tunnel burial depths, and three main driving parameters on segment floating. The main conclusions are as follows:

- The amount of the segment floating is closely related to the stratum where it is located. When the shield machine is driving in the sand layer, the average value of the segment floating is 10.6 mm. When the shield machine is driving in the upper-soft and lower-hard strata, the average value of the segment floating is 33.8 mm; When excavating in moderately weathered argillaceous siltstone, the average floating of segments is 66.7 mm, and the overall floating amount is not only much more significant than that in the sand layer and the upper-soft and lower-hard strata but also unstable and fluctuating.
- When the buried depth changes from 15 m to 21 m, the amount of the segment floating shows a significant linear relationship with the buried depth.
- With the increase of the three main driving parameters, the amount of the segment floating will also increase. The influence of grouting pressure and slurry pressure on segment floating is more substantial than that caused by the total shield thrust.

ACKNOWLEDGMENTS

The research work is supported by the National Natural Science Foundation of China (42267022), the Natural Science Foundation of Jiangxi Province (2021BAB204012), the State Key Laboratory of Performance Monitoring and Protecting of Rail Transit Infrastructure Foundation (No.HJGZ2021102).

REFERENCES

Chen D., Liu Z., Liu J.Y. (2021) State-of-Art and Prospects for Intelligent Construction Technology for Railway Shield Tunneling. *Tunnel Construction*, **41**: 923–932.

Chen J., Feng XY., Wei H. (2021) Statistics on Underwater Tunnels in China. *Tunnel Construction*, **41**:483–516.

Chen R.P., Liu Y., Liu S.X. (2014) Characteristics of Upward Moving for Lining During Shield Tunnelling Construction. *Journal of Zhejiang University (Engineering Science)*, **48**(06): 1068–1074.

Geng D.X., Hu Y.C., Jiang Y.L. (2021) Modified Calculation Model for Segment Floating in Slurry Shield Tunnel. *Journal of Performance of Constructed Facilities*, **35**(5): 1–11.

He C., Feng B., Fang Y. (2015) Review and Prospects on Constructing Technologies of Metro Tunnels Using Shield Tunnelling Method. *Journal of Southwest Jiao tong University*, **50**: 97–109.

Jin H., Yu K.W., Zhou S.H. (2019) Performance Assessment of Shield Tunnel Damaged by Shield Shell Extrusion During Construction. *International Journal of Civil Engineering*, **17**: 1015–1027.

Kasper T., Meschke G. (2006) On the Influence of Face Pressure, Grouting Pressure and TBM Design in Soft Ground Tunnelling. *Tunnelling and Underground Space Technology incorporating Trenchless Technology Research*, **21**: 160–171.

Shu Y., Zhou S.H., Ji C. (2017a) Analysis of Shield Tunnel Segment Uplift Data and Uplift Value Forecast During Tunnel Construction in Variable Composite Formation. *Chinese Journal of Rock Mechanics and Engineering*, **36**: 3464–3474.

Shu Y., Ji C., Zhou S.H. (2017b) Prediction for Shield Tunnel Segment Uplift Considering the Effect of Stratum Permeability. *Chinese Journal of Rock Mechanics and Engineering*, **36**: 3516–3524.

Wang S., Liu C., Shao Z. (2020) Experimental Study on Damage Evolution Characteristics of Segment Structure of Shield Tunnel with Cracks Based on Acoustic Emission Information. *Engineering Failure Analysis*, **118**: 104899.

Technical analysis of lattice super-high pier in bridge construction

Mei Guo* & Renan Jiang

JiLin Communications Polytechnic, Changchun City, Jilin Province, China

ABSTRACT: In bridge construction projects in complex areas, to improve the mechanical properties of bridge structures, lattice super-high buttress is a common supporting technology system in construction. To further explore the application of this technology, in this study, the key steps, technical measures, and management measures in the application of lattice super-high pier technology mode are expounded, summarized, and summarized in combination with practical engineering cases. On the whole, in this study, the demand for concrete in the support mode with steel pipe piles as the main structure is greatly reduced, and the applied "cotton pole analysis model" has unique innovations, which can provide some references for similar projects in the future.

1 INTRODUCTION

In the selection of a temporary support system, lattice super-high buttress technology has a relatively high application ratio because of its strong mechanical properties, adaptability and economy, etc. In recent years, researchers have paid more attention to this and carried out a lot of research. Tian Jun et al. combined with the actual engineering case, adopted the support measures of steel pipe combined support and cable wind rope, and adopted the auxiliary support method, which reduced the cost by about 15% while meeting the mechanical requirements. Zhang Ziwei and others adopted the *pushing erection method* to construct the lattice super-high pier technology and applied finite element analysis software to analyze the construction effect. He believes that in this construction mode, the stress and deformation of steel beams should be monitored and the application of monitoring technology should be strengthened to ensure the construction quality. On the whole, the above research has made outstanding progress. However, due to many differences in the actual conditions of various projects, it is still necessary to carry out the design and construction work based on the concept and method of adapting to local conditions.

2 PROJECT OVERVIEW

A newly-built railway bridge is designed with a continuous steel truss bridge structure. The total length of the bridge is 800 m, and the overall weight of the steel structure is about 20,000 tons. The whole bridge is supported by two main piers and two side piers. Because the bridge construction area is located in a mountainous area, the terrain is relatively rugged, and there is a river with a large drop in the construction area. All these factors lead to a further increase in the difficulty of this construction. On the whole, in this construction, there are the following difficulties: (1) The height of the super-high temporary buttress is high, and

---

*Corresponding Author: gmei@jljy.edu.cn

the requirements for bearing load are also high, which makes the buttress design the most difficult point of this project; (2) The construction site belongs to a typical canyon area in the mountains. Due to the limitation of construction conditions, the super-high pier must be set on a steep slope, which makes the construction deployment difficult; (3) In this project, the hoisting height of super-high piers is high, so how to select the corresponding hoisting equipment becomes a difficult problem; (4) This super-high pier needs to be welded at high altitude, and its quality control is difficult; (5) In the construction of super-high piers, there is a lot of aerial work, which requires high safety protection.

Considering the above situation, to improve the construction efficiency and quality at the same time, the construction unit decided to set up two lattice super-high piers after comprehensive research combined with the design drawings. The height of the left buttress is set to 111.5 m, and the height of the right buttress is set to 133.2 m.

3 MECHANICAL ANALYSIS OF LATTICE SUPER-HIGH PIER CONSTRUCTION

3.1 Construction of mechanical analysis model

After deciding to adopt the construction mode of a lattice super-high pier, the technicians first determined the analysis method. Considering that the contact element method in classical elasticity theory has the limitation of too much calculation in its application, after research, it is decided to adopt a brand-new mechanical analysis model-"Cotton Bar Element Analysis Model" to analyze lattice super-high piers (Figure 1).

As can be seen from Figure 1, the analysis model established this time is mainly divided into upper and lower units. The main function of the upper unit is to adjust the elastic modulus and unit length. Considering that this super-high pier is mainly made of steel, and the adjustment range of its elastic modulus is low, it is called a "cotton pole unit". In this unit, L1 represents the actual length of the pad, and L2 represents the assumed length of the "cotton stalk" unit.

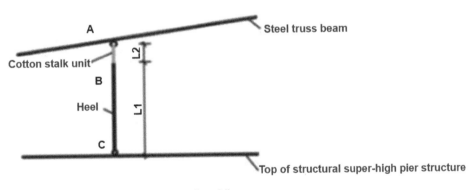

Figure 1. Analysis model of "cotton stalk unit".

Further analysis based on this figure shows that in this model, the lower unit releases the constraint of the lower beam end and is hinged with the lower structure. On the contrary, the "cotton stalk unit" mainly faces the upper structure and is hinged with the upper structure. In the whole structure, the upper and lower units are in a rigid connection state, and the force transmitted from the steel truss to the buttress is expressed in the form of axial force.

On the other hand, considering that the lattice super-high pier will have a certain displacement when the ambient temperature changes, therefore, in the calculation process, the researchers integrated the displacement influencing factors and adjusted the length of the cotton pole unit. If the buttress rises, the unit length will be shortened correspondingly, and vice versa. Then, the researchers further applied APDL language in finite element analysis software to make secondary feedback adjustments to the cotton stalk unit, to improve the accuracy of mechanical analysis.

3.2 Mechanical analysis results

According to the theoretical analysis, the main stress nodes of this lattice super-high pier are located in the interval of cantilever erection behind the pier towards the main pier. Before the construction of the main pier of the steel truss bridge, the load value of the supporting pier is the highest, corresponding to the most unfavorable working condition. To avoid the problem of "pier-beam void", researchers used finite element software to iteratively analyze and feedback on the mechanical model. The analysis results show that the supporting reaction force of the pier is basically proportional to the cantilever erection length. The longer the cantilever length is, the greater the supporting reaction force is, and the highest value is about 23,500 kN.

On the other hand, the researchers also analyzed the stress field changes of super-high piers. Considering that the super-high pier is exposed to the wind load, the researchers fully consider the influencing factors in this aspect and use the finite element software to analyze the maximum Von Mises stress of the pier. The results show that the Von Mises stress of the pier also increases with the increase of cantilever length, and the maximum value is 133 MPa, and the maximum value appears at the bottom of the steel column of the pier. On the whole, the above two values are all within the mechanical properties of this super-high pier, which proves that the structural super-high pier technology adopted this time has high feasibility.

4 KEY POINTS OF THE CONSTRUCTION PROCESS OF LATTICE SUPER-HIGH PIER

4.1 Foundation construction part

In this construction operation, the lattice super-high pier is constructed with a single pile and single bearing platform structure, the pile foundation is constructed with manual hole-digging cast-in-place pile, and the design of pile well lock and retaining wall is adopted. In the design process, all steel bars of HPB300 are used for design, and other parameters are shown in Table 1.

After the reinforcement material is determined, the concrete with strength grade C20 is used to pour the lock and the retaining wall, in which the concrete pouring of the retaining wall adopts the sectional pouring mode, and the pouring height of each section is controlled to be 1 m.

Table 1. Main parameters of lock mouth and retaining wall reinforcement (unit: mm).

Project	Parameter
Diameter of the locking ring to the main reinforcement	12.0
Diameter of the lock stirrup	8.0
Diameter of the retaining ring to the main reinforcement	8.0
Diameter of stirrup for retaining wall	8.0

4.2 Layout of lifting equipment

In the layout of lifting equipment, the engineering staff comprehensively considered the actual situation on the spot, and adopted the bridge deck girder crane and the 220 t truck crane to jointly install the lattice super-high pier adopted this time. Among them, the 220 t truck crane is used for installation within the 0–60 m height range of the buttress, and a bridge deck girder crane is used for installation over 60 m.

4.3 Installation of steel column

In the process of steel column installation, the engineering personnel first determine the hoisting scheme as a whole. Considering the actual situation, this hoisting shall be carried out in the order from far to near, with priority given to hoisting the far-away axis members. After determining the above scheme process, organize the construction personnel to start the installation work in the following aspects.

First of all, the truck crane is used to hoist the steel column. To improve the stability of the steel column, before the hoisting starts, the staff uses two traction ropes to fix the steel column. Secondly, after the steel column is lifted to a predetermined height, the staff uses the positioning plate to connect and fix the upper and lower parts of the steel column. Then, a steel liner is arranged in the steel column with no groove at the lower part to ensure that the steel liner is completely attached to the inner wall of the butt joint steel pipe, and the final butt joint of the steel column is carried out. Thirdly, after the steel column is hoisted in place, the engineers correct the verticality and elevation of the steel column. Two total stations are used in the correction process, and the two total stations are arranged orthogonally on the longitudinal and transverse axes to realize simultaneous observation and control. According to the observation results, the staff adjusts the steel column with a jack and wedge iron to ensure that the crosshair position of the instrument basically coincides with the edge of the steel column, thus ensuring that its verticality deviation meets the requirements. After the above steps are completed, the construction personnel shall weld the steel cushion block and weld it firmly. After welding, adjust the steel column. In this link, the jack is used to adjust to ensure that the amount of weld misalignment falls below 1.5 mm, and the groove clearance is controlled within 2 mm. Finally, the steel column is welded, and the welding is carried out by two operators synchronously and symmetrically. The welding parameters are shown in Table 2.

Table 2. Main welding parameters.

Welding process	Welding current/A	Electric arc/V	Welding speed (cm/min)	The linear energy (kJ/cm)
SMAW	110~130	20~22	6~10	15~24

After the welding is completed, the jack is removed to prevent the upper steel column from tilting due to the influence of high temperature during the welding process.

4.4 Installation of web members

After the steel columns are installed and accepted, the web members are hoisted. The overall hoisting sequence of web members is "horizontal web members-inclined web members", and the installation operation is carried out from bottom to top. Specifically, this link is mainly divided into the following steps.

First of all, steel wire rope is used to tie the web members. In this link, the engineers control the left and right symmetry of the sling, control the angle between the sling and the horizontal plane to be 45, and then tie the link. At the same time, additional welding stops are added to prevent the sling from slipping. Then the engineers use lifting equipment to lift the web member, control the crane to climb the bar forward and the lifting of the hook to ensure that the corresponding positions of the web member and the steel column are aligned, and control the elevation position to meet the requirements.

Secondly, after the web member is hoisted in place, the staff will fine-tune the elevation and axis of the web member, and handle the joint of the web member by manual gas cutting to ensure that it meets the actual needs. Then, the inspectors further check the elevation and axis of the web member, and after determining that the deviation between the elevation and axis is within the required range, the inspectors weld the web member and the steel column. When the weld height reaches more than 35% of the designed weld height, the operator will loosen the hook and continue to weld according to the design requirements. At the same time, in the welding process between the web joint and the steel column, the operator added an inner lining plate in the web member to avoid the problem of the excessively staggered joint.

4.5 *Installation of connection system*

After the web bar is installed and accepted, the construction department will further install the connection system. In this construction, the connection system is installed in the mode of connecting double-angle steel and steel columns. Among them, the angle steel is made of prefabricated components, which are perforated in advance. After all the materials enter the market and pass the inspection, the operator will start the installation of the connection system. This link is divided into the following steps.

First of all, it is the wire rope binding link, which is similar to the wire rope binding in the installation of web members. Engineers still control the left and right symmetry of slings, and control the binding link after the angle between the slings and the horizontal plane is 45. At the same time, additional welding stops are added to prevent the slings from slipping.

After the sling is ready, the operator controls the crane hook to coincide with the center of the upper chord of the connecting system and starts the lifting operation of the connecting system. In this link, the crane climbs the rod forward, the lifting hook is matched with each other, and the connecting tie bar is controlled to correspond to the elevation position of the steel column, to realize the alignment operation between them. In this link, the operator still uses two traction ropes for auxiliary control to avoid the unstable swing of the connecting system.

Finally, after the hoisting of the connecting tie bar is completed, the operator will further fine-tune the position of the connecting tie bar to ensure that the bolt holes of the connecting system correspond to those of the steel column and that each hole is bolted tightly. Therefore, the operator first bolted the two bolts at the diagonal of the splice plate, and then bolted the rest of the bolts to complete the final installation of the connection system.

4.6 *Installation of the distribution beam*

In the installation process of the distribution beam, the construction department adopts a beam erecting crane for installation, and the installation sequence is from bottom to top. Before the start of hoisting, the engineering personnel shall conduct a comprehensive inspection of the state of the girder erection cranes, slings and slings. After the inspection is correct, the symmetrical hoisting method shall be adopted to slowly hoist the distribution beam to the top of the super-high temporary buttress. To realize the accurate installation of the distribution beam, in this link, the staff takes the following two control measures.

One is the control of the measurement link of the buttress. In the measurement process, aiming at the same steel column, two total stations are used to measure at the same time, and the two total stations are arranged orthogonally on the longitudinal and transverse axes, to simultaneously observe and control the verticality and installation elevation of the steel column. On this basis, the engineers arranged settlement observation points and sensor equipment at several key nodes of the pier to monitor the settlement of the pier in real-time. To ensure the deformation can be effectively controlled, engineers use finite element analysis software to predict the change of settlement and take corresponding corrective measures.

The second is "linear control". In the process of manufacturing components, technical personnel will determine the geometric axis of the pier processing in combination with the design drawings of the pier, and check the geometric axis of the pier processing after each section of processing is completed. After the hoisting of each section is completed, total station equipment shall be used to accurately measure the verticality and top surface table of the temporary buttress, and according to the measurement results, the correction value for the next stage installation shall be determined. In this link, surveyors calculate the inclination angle of the column top plane according to the verticality deviation of each segment and then calculate the correction value of steel column length according to the elevation difference of every two adjacent segments. At the same time, the correction value of each point at the end of the steel column is calculated according to the plane inclination angle of the column top. Considering that the temperature has a significant influence on the parameters of temporary piers, in this link, all the measurement and observation time is set at around 05:00 in the morning, and the influence of temperature factors is controlled at a relatively low level.

4.7 *Installation acceptance link*

After all the work in the above links is completed, the engineering supervisors will check and accept the installation quality of each section. After the installation, the overall situation of the whole lattice super-high pier will be accepted.

5 RELEVANT MANAGEMENT MEASURES IN THE APPLICATION OF LATTICE SUPER-HIGH PIER TECHNOLOGY

5.1 *Quality control measures*

To ensure the orderly construction of lattice super-high piers, managers pay more attention to quality control measures. On the one hand, to ensure effective quality control of raw materials, relevant units have strengthened the inspection of raw materials before admission. The first is to check the quality certification documents and inspection reports of raw materials. For some steel materials with large thicknesses, flaw detection tests and other methods can be used to carry out inspection work on these materials and judge whether the performance parameters meet the requirements. Only when the performance parameters of raw materials meet the requirements, can they be admitted.

On the other hand, due to more bolting work in this construction. Therefore, technicians take the following two measures to control the quality of bolted connections: (1) to control installation nodes, mainly using manual wrenches or temporary bolts for fastening, and calculate the specific number of bolts inserted into each node. (2) ensure that all bolts are installed in the same direction when installing bolts, and tighten the bolts several times.

5.2 *Safety measures for aerial work*

Considering that a large number of safety operations are involved in the construction of this super-high pier, and there are many risk factors in aerial work, the construction unit takes the following measures to ensure safety: (1) Professional technical training is provided for aerial workers, and they can only take up their posts after passing the examination and training; (2) Regular physical examination is conducted for aerial workers, and stop aerial work immediately after finding diseases unsuitable for aerial work; (3) Adequate safety facilities are equipped with, in addition to infrastructure such as safety belts, temporary fixing tools, etc., to ensure the safe installation of high-rise components.

6 CONCLUSION

On the whole, in bridge construction, the construction of lattice super-high pier technology is complicated work, involving many key technologies and management measures. Therefore, in the actual construction process, attention should be paid to the work of each link, and the related work should be summarized in time, to provide a reference for subsequent similar projects.

FUND PROJECT

Jilin Research Project of Vocational Education: Research on the path of promoting the reform of "Three Education" in the course of bridge construction technique in higher vocational colleges. Project No.: 2021XHY119.

REFERENCES

Cheng Xianxun. *Mechanical Analysis and Construction Monitoring Technology of Steel Truss Bridge and Super-High Temporary Pier Structure* [D]. Hefei University of Technology, 2021.

Huang Yuhuan. *Mechanical Analysis of the Construction Process of Long-Span Steel Truss Bridge And Super-High Pier* [D]. Hefei University of Technology, 2020.

Li Minzi, Han Tao. Construction Technology of Long-Span Super-High Formwork Curved Beam [J]. *Building Technique Development*, 2021,48(18): 69–71.

Peng Qin-feng, Tao Zhong, Xie Huang-dong, et al. Key Technologies of Temporary Buttress Construction of Super-High Steel Structure of Super-Large Bridge [J]. *Construction Technology (English and Chinese)*, 2022,51(03): 125–128.

Tian Jun. Key Construction Technologies of Steel Box Girder Lattice Temporary Buttress and Steel Tower Support Frame [J]. *Sichuan Cement*, 2022(09): 201–202+213.

Zhang Ziwei, Rao Qi, Xu Canchao, et al. Simulation Analysis and Technology of Jacking Erection of Lattice Steel-Concrete Composite Beams [C]//. *Proceedings of 2021 National Civil Engineering Construction Technology Exchange Conference* (Vol.2)., 2021: 365–368.

Zhang Xiaoping, Construction Technology of Super-High Formwork Support System of New Sliding Buckle Frame. Zhejiang, *Zhongtian Construction Group Co., Ltd.*, 2020-09-06.

Research on adit rock mass mechanics test of Baihetan Hydropower Station

Di Liu & An Liu*
PowerChina Huadong Engineering Corporation Limited, China

ABSTRACT: The dam has high requirements on the quality of the rock mass of the dam foundation and both banks. As the foundation rock mass supports the arch dam structure, it should have sufficient bearing capacity and deformation stability. In this paper, through the field rock mass deformation test, the mechanical properties of the rock mass of the adit PD9012 in the dam site area are understood to provide the required rock mass deformation parameters for the design. The best value of the horizontal and vertical deformation modulus test is respectively 22.93 GPa/ 10.83 GPa.

1 INTRODUCTION

Baihetan Hydropower Station is located in Ningnan County, Sichuan Province and Qiaojia County, Yunnan Province, at the lower reaches of the Jinsha River, 45 km away from Qiaojia County, adjacent to the Wudongde Cascade on the upper side, Xiluodu Cascade on the lower side, 195 km away from Xiluodu Hydropower Station, with a controlled drainage area of 430300 km^2, accounting for 91.0% of the Jinsha River drainage area. The dam site is about 260 km from Kunming, 400 km from Chongqing, Chengdu and Guiyang, and 1850 km from Shanghai in East China (Jiang et al. 2017).

Baihetan Hydropower Station is a concrete double-curvature arch dam with an underground powerhouse arranged on both banks. The dam height is 289 m, the flood discharge is large, and the seismic intensity is high. When the arch dam bears various basic loads, the maximum compressive stress at the downstream arch end is nearly 10.0 MPa, the maximum tensile stress at the upstream arch end is more than 1.0 MPa, and the total thrust is up to 14 million tons. The dam has high requirements for the quality of the rock mass of the dam foundation and both banks (Meng et al. 2016). As the foundation rock mass supports the arch dam structure, it should have sufficient bearing capacity and deformation stability. The rock mass mechanical parameters of the dam foundation are important input parameters for dam engineering analysis. Understanding the various rules of its deformation parameters is of great significance for engineering geological evaluation and analysis and design of dam engineering (Jin et al. 2010).

The reasonable determination of rock mass mechanical parameters has been a major problem in rock mechanics for many years (Xu et al. 2017). Studying rock mass mechanical parameters from the perspective of tests is one of the most effective methods. However, due to the size effect, the rock mass mechanical parameters obtained from laboratory tests in geotechnical engineering cannot be directly applied to engineering practice (Zhu et al. 2011). Therefore, it is necessary to directly carry out in-situ rock mass deformation test research.

Based on this, to obtain the rock mechanical parameters of the right bank area of the dam site of Baihetan Hydropower Station, a group of rock mass deformation tests was conducted in

---

*Corresponding Author: liu_a2@hdec.com

this area. The deformation test points are arranged at the depth of 30~35 m of PD9012 adit, one horizontal point and one vertical point, of which the horizontal point is arranged on the downstream adit wall as required. The specific site layout of each pilot is shown in Figure 1.

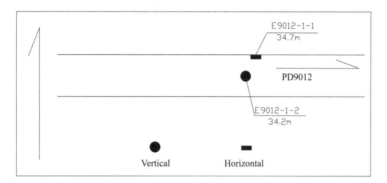

Figure 1. Location map of site rock mass deformation test pilot for PD9012 adit of Baihetan Hydropower Station.

2 ENGINEERING GEOLOGICAL CONDITIONS

Adit location: 34.40 m downstream the right bank of Survey IX line, with an elevation of 772.46 m and a depth of 151.10 m; overall tunnel direction: E; Section size: 2 × 2 m, the shape of the tunnel wall is still regular. The tunnel is 0~39.00 m deep, and the overlying rock is 5.00~115.00 m thick, which is a steep slope with an inclination of 70°, 39.00–151.10 m, 115.00–285.00 m thick, and is a medium slope terrain with an inclination of 45°.

2.1 Formation lithology

The cave depth of 0~151.10 m is the basalt of the Emeishan Formation of Upper Permian (P2β51), in which the tunnel depth is 2.00 m. It is cinereous aphanitic basalt, columnar joints are slightly developed, and the joint spacing is 5~15 cm. 2.00~11.70 m is purplish red variegated breccia lava, with breccia diameter of 1~5 cm, and some of them are larger than 5 cm, composed of almond basalt, calcareous, fused and cemented; 11.70~61.00 m is grayish green almond bearing basalt, composed of calcite and a small amount of chlorite, with the content of about 5%; 61.00~85.50 m is grayish green almond basalt. The diameter of an almond is 1~3 mm, with a small amount of up to 6 mm. The composition is chlorite and a small amount of calcite, with a content of 20~25%; 85.50~85.80 m is purplish red tuff; 85.50~151.10 m is grayish black and grayish green columnar jointed basalt with a diameter of 15~40 cm, a dip angle of 75° and extremely developed micro cracks.

2.2 Rock mass structure

The tunnel depth is 0~4.00 m, with mosaic structure; 4.00~13.00 m sub-block structure; 13.00~70.00 m columnar mosaic structure, 70.00~79.00 m, sub blocky mosaic structure, 79.00~85.50 m blocky structure; 85.50~85.80 m, cataclastic structure; 85.80~151.10 m, columnar mosaic structure.

2.3 Weathering of rock mass

The upper section of weak weathering is 0~23.00 m deep; 23.00~128.00 m is the lower section of weakly weathered; 128.00~151.00 m is slightly new rock mass.

2.4 Unloading of rock mass

The tunnel depth of 0~6.00 m is a strong unloading section; 6.00~112.00 m is the weak unloading section. Concerning the engineering geological classification standard of dam foundation rock mass, the adit depth of 0~6.00 m is classified as Class IV; 6.00~23.00 m is Class III 2; 23.00~85.50 m is Class III 1; 85.50 ~ 85.80 m is Class IV; 85.80~112.00 m is Class III 2; 112.00~128.00m is Class III 1; 128.00~151.10 m is Class II 2.

To supplement and find out the geological conditions such as the rock mass quality of the right bank arch dam abutment dam foundation and the stability of the dam abutment, it is revealed that four gently inclined interlaminar and internal staggered zones and two faults are developed, but the scale is small, which is basically consistent with the original exploration results, and the inclination is opposite to the slope direction, so there is no stability problem on the slope.

3 ROCK MASS DEFORMATION TEST

According to the actual situation of the site, a group of rock mass deformation tests with 2 points in total were conducted in the PD9012 adit of Baihetan Hydropower Station. One horizontal pilot and one vertical pilot are arranged for the rock mass deformation test of this group, and the horizontal point is arranged at the downstream tunnel wall.

3.1 Pilot layout

According to the requirements of geologists and the actual situation of the site, the rock mass deformation test of this group is arranged at 30~35 m of adit PD9012 by site survey. The lithology of the pilot site is Class III1 cinereous aphanitic basalt in the lower weakly weathered section.

3.2 Sample preparation

(1) The loose rock block within Φ 100 cm in the area around the pilot is cleared first, taking the center of the pilot as the center, the test surface is chiseled and polished within Φ 60 cm, with a fluctuation difference of 0.5 cm. The test surface of the tunnel wall pilot is vertical to the ground. The fault pilot requires that the test surface is vertical to the fault plane. The distance from the edge of the bearing plate at each point to the tunnel wall is greater than 1.5 times the diameter of the bearing plate.
(2) The loose rock blocks at the back seat of the pilot are removed, and then the rock surface within the scope of Φ 50 cm shall be chiseled (or ground) and parallel to the test surface of the pilot, and the location shall be on the same lead line or horizontal line with the pilot.
(3) A layer of high-strength mortar with a thickness of ≤ 0.5 cm on the test surface is plastered, then the pressure-bearing plate is pressed and the level or verticality of the pressure-bearing plate is calibrated with a level ruler. After the high-strength mortar reaches a certain strength (generally curing for 5–7 days), the pressure test starts.

3.3 Geological description of the sample

After the preparation of the test surface, the geological description of the test surface is generally carried out by engineering geologists with the assistance of the test personnel. The description content is as follows:

(1) Number, position, orientation, depth, floor elevation, section shape and size of the test chamber;
(2) Pilot number, location and size;

(3) Rock name, structure, main mineral composition and color;
(4) The exposed position, occurrence, width, extension, continuity, density, and relationship with stress direction of bedding, schistosity, cleavage, joint, fissure and other structural planes and fault zones, the genetic type, mechanical properties, roughness, nature and composition of fillings, softening and marginalization of various structural planes, and the contact relationship between rock veins and surrounding rock.
(5) Weathering degree and characteristics;
(6) Exposed position and seepage volume of seepage water;
(7) Rock burst and rock deformation.

3.4 Test equipment and test methods

(1) The rigid bearing plate method is adopted for the test, and the diameter of the rigid bearing plate is Φ 504.6 mm, each piece is 60 mm thick, with an area of 2000 cm^2;
(2) See Figure 2 for the installation of the jack, dowel column and other equipment;
(3) 3000 kN jack and manual high-pressure oil pump are used for loading;
(4) The maximum test pressure is 10.0 MPa, and the test is applied in five levels of 2, 4, 6, 8 and 10 MPa;
(5) The pressurization method is step by step one cycle method;
(6) Four dial indicators are symmetrically arranged in four directions of the pressure-bearing plate to measure the rock mass deformation. The dial indicators are fixed on the reference beam. Before loading, they are measured every 10 minutes. The load is applied after the readings of the four dial indicators remain unchanged for three consecutive times (this reading is the initial measured value of each gauge). The deformation value shall be measured immediately after loading, and then every 10 minutes until the deformation is stable, and the next level of load shall be added;
(7) Stability standard: when the ratio of the difference between two adjacent readings of four gauges on the pressure plate and the difference between the first deformation reading under the same pressure level and the last deformation reading under the previous pressure level is less than 5, the deformation is considered stable;
(8) Pressure relief: step by step pressure relief is adopted. The process pressure is required to be read once. After zero reduction, it is required to read once immediately. After

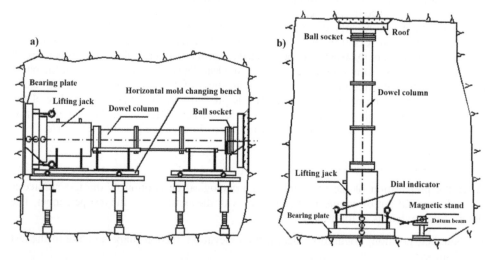

Figure 2. Installation diagram of deformation modulus test equipment: a) horizontal; b) Vertical.

10 minutes, it is required to read again. The stability standard is the same as that of pressurization.

3.5 Arrangement of test results

(1) The total deformation and elastic deformation of rock mass are calculated according to the average value of 4 dial indicators;
(2) The pressure deformation relationship curve with pressure p(MPa) is drawn as the ordinate and deformation W (mm) as the abscissa;
(3) The deformation (elastic) modulus of rock mass is calculated according to the following formula:

$$E = \frac{\pi}{4} \cdot \frac{(1-\mu^2)pD}{W} \qquad (1)$$

Where:

E – deformation modulus or elastic modulus, MPa; The deformation modulus E_0 is calculated by substituting the full deformation W_0 into the formula. When the elastic deformation W_e is substituted into the formula, it is the elastic modulus E_e;
W – surface deformation of rock mass, cm;
p – pressure calculated according to unit area of pressure bearing surface, MPa;
D – Diameter of bearing plate, cm;
μ – Poisson's ratio of rock mass (0.25 for rock mass).

3.6 Analysis of test results

Two rock mass deformation modulus tests have been carried out for the adit PD9012 of Baihetan Hydropower Station, including one horizontal point and one vertical point. The deformation modulus test results of each test point are shown in Table 1.

The rock mass at the test site is grayish aphanitic basalt in the lower weakly weathered section, belonging to Class III1 rock mass. The test pressure deformation relationship curve of each test point is shown in Figure 3, and the recommended values of deformation modulus are described as follows:

(1) The recommended deformation modulus E0/of the horizontal pilot E9012-1-1 is 22.93 GPa.
(2) The recommended deformation modulus E0/of the vertical pilot E9012-1-2 is 10.83 GPa.

Table 1. Summary of Field rock deformation modulus test results of PD9012 Adit on the right bank of baihetan hydropower station.

Test No.	Position	Force Direction		Under the action of pressure p (MPa) E_e(GPa) and E_0(GPa)					E0' (GPa)
			p	2	4	6	8	10	
E9012-1-1	34.7 m	Horizontal	E_e	24.77	31.62	31.40	32.66	34.72	22.93
			E_0	19.05	21.54	22.75	23.22	22.93	
E9012-1-2	34.2 m	Vertical	E_e	16.15	17.48	15.81	16.07	15.88	10.83
			E_0	12.38	12.08	11.09	11.09	10.83	

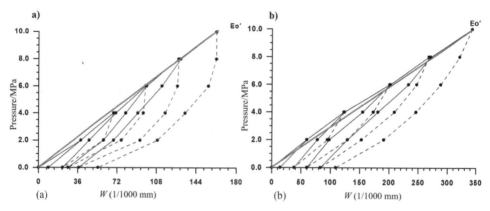

Figure 3. Pressure deformation curve: (a) E9012-1-1; (b) E9012-1-2.

4 CONCLUSION

According to the requirements of the test assignment and the actual situation of the site, the rock mass deformation test was carried out in the adit PD9012 of Baihetan Hydropower Station. The main conclusions are as follows: Type III1 cinereous aphanitic basalt in the lower weakly weathered section: the best value of the horizontal deformation modulus test is 22.93 GPa; The best value of vertical deformation modulus test is 10.83 GPa.

REFERENCES

Jiang Q., Fan Y., Feng X., Li Y., Pei S., Liu G. Unloading Fracture of Hard Rock Under High Stress: Case Study of Basalt Crack Observation in the Underground Powerhouse of Baihetan Hydropower Station. *Journal of Rock Mechanics and Engineering*, 2017,36 (05): 1076–1087.
Jin C., Feng X., Zhang C. Research and Analysis on Initial Ground Stress Field of Baihetan Hydropower Station. *Geotechnical Mechanics*, 2010, 31(03): 845–850+855.
Meng G., Fan Y., Jiang Y., He W, Pan Y., Li Y. Research on Key Rock Mechanics Problems and Engineering Countermeasures of the Huge Underground Caverns of Baihetan Hydropower Station. *Journal of Rock Mechanics and Engineering*, 2016,35(12): 2549–2560.
Xu N., Li T., Dai F., Li B., Fan Y., Xu J. Stability Analysis of Rock Slope on the Left Bank of Baihetan Hydropower Station Based on Discrete Element Simulation and Microseismic Monitoring. *Geotechnical Mechanics*, 2017, 38(08): 2358–2367.
Zhu Y., Zhu H., Shi A., Meng G. Stability Analysis of Complex Blocks of Baihetan Hydropower Station Based on Discrete Element Method. *Journal of Rock Mechanics and Engineering*, 2011, 30(10): 2068–2075.

Key construction technology of complex "Xumishan" spatial bending-torsion aluminum alloy structure

Xiaoyang Sun*, Feng Yang* & Hai Zhao*
China Construction Eighth Engineering Division Corp., LTD, Nanjing, China

ABSTRACT: With the rapid development of China's social economy and the continuous progress of science and technology, the application of Aluminum Alloy Structure to achieve super-high, special-shaped and large-scale complex space art modeling has been increasingly widely used. Combined with the example of Xumishan space bending and torsion aluminum alloy structure in the Guanyinshengtan project, this paper introduces the construction technology of complex space bending and torsion aluminum alloy structure, including the production of aluminum alloy rod, overall unloading of aluminum alloy structure, structure installation, stress and strain monitoring and other technology or process innovation application. It provides reference cases for similar engineering construction in the future.

1 INTRODUCTION

Mount Putuo Guanyinshengtan project is located in the south of the Baishan scenic area in the Zhujiajian scenic area of Zhoushan City, Zhejiang Province. The architectural form comes from the Pilu Guanyin statue worshipped by the Puji temple in Putuo Mountain. The overall layout is a building complex of "one master and two slaves". The project includes the altar, shancai building, Longnv building and square. The total building area is 61900 m². After finishing the project, this place will be the holy land of Buddha worship, the place of practice, the Dharma promotion center and the believer service base (Qiao et al. 2015). It is necessary not only to create the handed-down works of Buddhist architecture but also to become the center of Guanyin culture in Putuo Mountain and the "cultural landmark" of modern Buddhism.

The structural decoration integration project of Xumi mountain inside the altar is an axially symmetrical multi-curved space art building with a net height of about 55.8 m. The "Spatial bending-torsion multi-curved-surface aluminum-alloy shell structure" at the upper part of Xumi mountain is in a bottle-neck shape, as shown in Figure 1. The structural plane is circular,

Figure 1. Spatial bending-torsion multi-curved-surface aluminum-alloy shell structure.

---

*Corresponding Authors: 25728791@qq.com, 206A0638@cscec.com and 206A0638@cscec.com

with a total height of 32.7 m, a bottom diameter of 18.17 m, a middle diameter of 7.6 m, a top diameter of 21.65 m, and more than 7000 aluminum alloy structural members and 587 glasses.

2 TECHNICAL DIFFICULTIES AND APPLICATION OF NEW TECHNOLOGIES

2.1 Technical difficulties

1) The aluminum alloy structures of the project are all composed of special-shaped hyperboloid members, which are difficult to process and manufacture, and require high precision for installation.
2) It is difficult to control the precision of sectional pre-assembly of ultra-high aluminum alloy structure with annular section. The stress distribution during pre-assembly is complex, so controlling the overall settlement and avoiding the deformation of components will be difficult.
3) The indoor installation space is small and the use of large lifting machinery and transportation equipment is limited. The transportation and hoisting of aluminum alloy rods and glass are difficult; The structure of decorative joints is complex and the installation is cumbersome. Therefore, the point is how to innovate and apply the construction technology of aluminum alloy structures to ensure installation quality.

2.2 Application of new technologies

Based on the conventional aluminum alloy construction technology combined with the digital and scientific concept, the following new technologies and processes are adopted considering the actual situation of the project: 1) Manufacturing of aluminum alloy structural members; 2) Sectional pre-assembly technology of aluminum alloy structure; 3) Space dot matrix measurement and positioning technology; 4) Installation of foundation steel platform; 5) Erection technology of high-altitude turnbuckle; 6) Installation technology of aluminum alloy structural members; 7) Unloading of aluminum alloy structure; 8) Installation of multi-curved fireproof glass and decorative aluminum plate; 9) Stress and strain monitoring technology (Sun 2016).

3 KEY TECHNOLOGIES

3.1 Construction process flow

The construction process flow is shown in Figure 2.

3.2 Manufacturing of spatial bending-torsion aluminum alloy structural members

3.2.1 Detailed design of aluminum alloy structure
BIM technology is used to establish a three-dimensional building information model of aluminum alloy structures. The study aided by the BIM technology mainly focuses on deepening the research on the joint parts between aluminum alloy structure and another form of structure, deepening the design from the perspective of completion effect and subsequent construction, and clarifying the shape and functional requirements of complex members used in the project. The complex members include a hyperbolic aluminum alloy rod, hyperbolic aluminum alloy gusset plate, hyperbolic cross-shaped connecting plate and ring groove rivet fastener, as shown in Figure 3.

3.2.2 Manufacturing of aluminum alloy structural members
To realize the bending and torsion effect of aluminum alloy structure, the existing heavy-duty CNC three-roller machine was customized. By controlling the manufacturing angle of

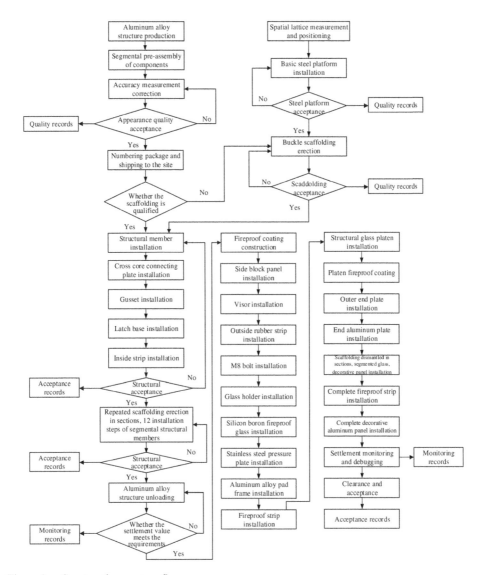

Figure 2. Construction process flow.

Figure 3. Connection of aluminum alloy structure.

Figure 4. Bending and torsion devices. (a) Bending and torsion equipment and (b) Bending and torsion tooling.

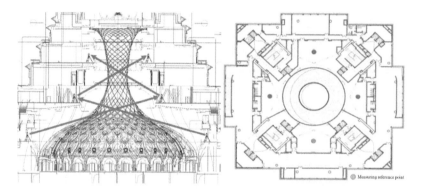

Figure 5. Spatial dot matrix measurement and positioning technology.

the bending and torsion, the rod needed large bending and torsion angle will be bent and twisted at least 10 times, and then the bending and torsion correction shall be carried out. The above process can simplify the processing of the curved surface, and ensure the curvature requirements of the aluminum alloy rod.

3.3 Installation technology of aluminum alloy structure

3.3.1 Spatial dot matrix measurement and positioning technology

According to the structural rules of the Xumishan aluminum alloy structure, a three-dimensional spatial dot matrix system is established. The construction method of a "multi-point layered survey" is adopted. Four survey control points on the third floor are used as survey reference points, and a horizontal control network is arranged. The elevation control network is arranged by using the elevation control lines of each floor, which solves the problem that the GPS positioning instrument cannot be used to arrange survey control reference points indoors.

3.3.2 Installation technology of aluminum alloy structural members

1) Division of aluminum alloy member

As shown in Figure 6, aluminum alloy rods are divided into 12 sections for construction according to curvature and height. Scaffolds are erected and installed layer by layer, and

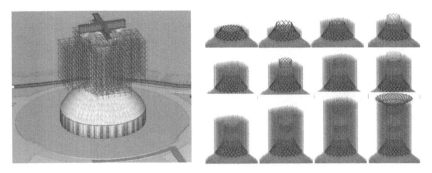

Figure 6. Installation of aluminum alloy structural members.

the scaffolds are also installed 12 times (Ouyang et al. 2016). The whole process of dynamic construction is ensured and the construction efficiency is improved.

2) Installation process of aluminum alloy rod
The "full hall scaffold high-altitude bulk method" is adopted to install the aluminum alloy structure from the bottom to the top. To avoid the accumulation of installation errors, the aluminum alloy members shall be corrected once after each span is finished. The scaffold shall be erected while installing the aluminum alloy structure, and the scaffold also shall be erected throughout the installation of the latticed shell. In addition, when installing bending-torsion aluminum alloy members and gusset plates in each section, the erection of the scaffold shall always be faster than the installation (Liu et al. 2018).

(1) Installation of aluminum alloy members and cross-core connecting plate: As shown in Figure 7, a cross-core connecting plate is developed to connect the aluminum alloy structural members. Using this plate can avoid secondary opening on the members and meet the strength requirements of the connection of the Xumishan aluminum alloy structure with large space and complex shape.

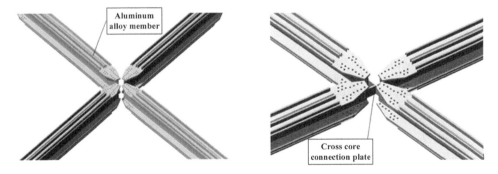

Figure 7. Installation of aluminum alloy members and cross-core connecting plate.

(2) Installation of aluminum alloy gusset plate and maintenance pin base: As shown in Figure 8, after the installation of a horizontal circle of aluminum alloy members and cross core connecting plates is completed, the aluminum alloy gusset plates are installed outside the cross core connecting plates. The aluminum alloy gusset plate is locked and connected with special M10 bolts of aluminum alloy. Furthermore, the maintenance pin

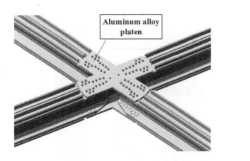
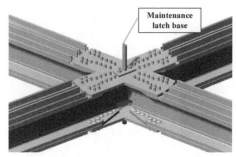

Figure 8. Installation of aluminum alloy gusset plate and maintenance pin base.

base is fixed with bolts outside the aluminum alloy gusset plate for decoration aluminum plate installation. The aluminum alloy gusset plate, countersunk Huck stainless steel screw, decorative plate and aluminum alloy members are all extruded.

(3) Installation of inner EPDM rubber strip: As shown in Figure 9, the EPDM rubber strips with heat resistance and weather resistance are installed and fixed on the inner side of the Xumishan aluminum alloy structure to ensure that the connectivity, sealing, decoration, safety and energy saving performance meet the design requirements (China Construction Industry Press, 2010).

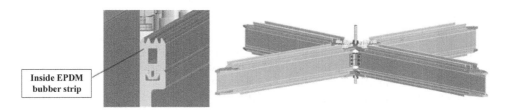

Figure 9. Installation of inner EPDM rubber strip.

3.3.3 *Unloading technolgoy of the aluminum alloy structure*

During unloading, the digital visual displacement monitoring system is used to monitor the real-time displacement of four symmetrical points on the outer ring of the top of the Xumishan aluminum alloy structure. The unloading time is 1 month. The displacement monitoring of all points of the Xumishan aluminum alloy structure shows that the unloaded Xumishan aluminum alloy structure meets the design requirements because the overall settlement is about 40 mm, which is less than the design predetermined value by 100 mm (Wang & Yan 2013).

3.3.4 *Stress and strain monitoring technology*

Figure 10 shows the curves of the structural stress change with time from the sectional assembly of the Xumishan aluminum alloy structure to the settlement stability of the aluminum alloy structure, which intuitively reflects the stress change of the monitoring points in each stage of the structural unloading process. The curves show that the stress of the aluminum alloy structure increases slowly in a step shape at the beginning of each round installation of aluminum alloy members (or glasses), indicating that the stress of the aluminum alloy structure increases slowly without local abnormal mutation and the stress, strain of the whole aluminum alloy structure are normal (Yang et al. 2016).

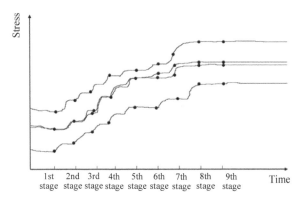

Figure 10. Stress-time curve of monitoring points.

4 CONCLUSION

As the holy place of Buddha worship in Putuo Mountain, the Guanyin altar in Putuo Mountain is a masterpiece of Buddhist architecture and a "cultural landmark" of modern Buddhist concepts. To ensure the smooth implementation of the "Xumishan" spatial bending-torsion aluminum-alloy structure project, which is an important indoor building of the Guanyin altar in Putuo Mountain, the difficulties in the manufacturing and installation of bending-torsion aluminum alloy members have been solved through the application of new technologies such as deepening design with BIM, three-dimensional space lattice measurement technology, segmented pre-assembly of aluminum alloy structures, and unloading technology of aluminum alloy structures. The use of these new technologies reduces material loss and installation costs and shortens the installation period. This project provides a reference for the construction of such projects.

ACKNOWLEDGMENTS

The Science and Technology Development Project of China Construction Eighth Bureau (No. 2016-27) is greatly appreciated for the financial support.

REFERENCES

Liu Q., Li B.Y., Qin Z.H., et al. Study on Plastic Forming Process of Aluminum Alloy Spatially Shaped Thin-Walled Components [J]. *Forging & Stamping Technology*, 2018,43(11): 42–47.

Ouyang Y.W., Sun X.Y., Chen B., et al. Installation Technology of Large-span Multi Curvature Irregular Curved Surface Aluminum Alloy Roof [J]. *Construction Technology*, 2016,45(14): 58–63.

Qiao G.Y., Sun X.Y., Zhang S., et al. Key Construction Technology for the Large-span Ellipsoidal Aluminum Alloy Dome Structure Under Complex Environment[J]. *Construction Technology*, 2015(8): 5.

Sun X.Y. Construction Technology of Large-span Hyperbolic Surface Cable-type Special Shaped Hollow Aluminum Plate [J]. *Construction Technology*, 2016,45(13):1–5.

Technical Regulations for Spatial Grid Structure: JGJ7-2010 [S]. Beijing: China Construction Industry Press, 2010.

Wang W., Yan P. Experimental Study on Mechanical Behavior of Concrete Filled Steel Tubular Intersecting Connections. *Journal of Building Structures*, 2013,34(S1): 123–127.

Yang G., Cao WL, Dong H.Y., et al. Compressive Behavior of Specially-Shaped Multi-Cell Mega-Bifurcated Concrete Filled Steel Tubular Columns. *Journal of Building Structures*, 2016, 37(05): 57–68.

The feasibility analysis of a new prefabricated steel structure system

Rong Xing
Civil Engineering College of Taiyuan University of Technology, Taiyuan, China
Shanxi Vocational University of Engineering Science and Technology, Taiyuan, China

Junjie Zhang
Architectural Design and Research Insitute Co., Ltd. of Taiyuan University of Technology, Taiyuan, China

Honggang Lei*
Civil Engineering College of Taiyuan University of Technology, Taiyuan, China

ABSTRACT: Prefabricated steel structure building system is becoming gradually mature, and accepted by the public day by day. To make this system more and more widely applicated, this paper uses YJK software to establish a finite element model of a new type of prefabricated steel structure system based on the existing solid abdominal I-beam research content. It is concluded that the overall performance of the new system is better, and what's more, it is more economical and practical than a solid abdominal I-beam system.

1 INTRODUCTION

In the process of promoting prefabricated buildings, the type of prefabricated steel structure system is gradually becoming more and more rich and perfect. Its common system has a frame structure system, frame-support structure system (Shi 2020; Wang 2020), steel frame-shear wall (core tube) system, mixed frame-shear wall (core tube) system (Liu 2019), suspension-floor frame structure system (Yuan & Liu 1996), cylinder structure (Ding 2019; Hao et al. 2018; Yang 2016; Yuan et al. 2005), the application of these systems is mature. With the progress of the times, the social need is constantly updating, prefabricated steel structure system is constantly updating. Based on the common and special steel structure systems which have penetrated research, a new type of square pipe column-truss beam support frame system has occurred, the beam-column of the node the system makes use of welding or full bolted connection, the composition of the new system is shown in Figure 1.

2 BUILDING OF STRUCTURE SYSTEM

2.1 *System component and constituent materials*

The top chord's material of the truss beam of this system is of narrow flange H type steel; the materials of the lower chord and the web member are all made of square steel pipe; corner column supports' materials are also square steel pipe, as not only may save the quantity of steel and reduce dead weight, but also may be easy to pipeline penetration and installation in

---

*Corresponding Author: lhgang168@126.com

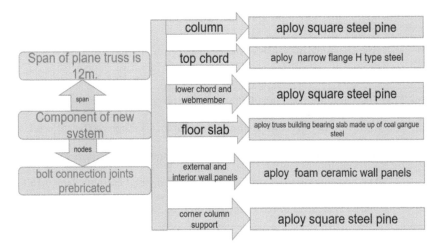

Figure 1. Detail of composition of the new system.

the relatively wide internal space. To a certain extent, the square steel pipe column has improved the overall torsion and anti-overturning performance. And what's more, compared with the concrete component, the component section of square steel pipe column is smaller; based on the reasonable design it is reasonable to hide columns in the wall when installing; to a large extent, the utilization rate of building space is improved (Chen et al. 2004). The main load-bearing member is made up of Q355B low-alloy high-strength steel, and the floor slab adopts a truss building bearing slab made up of coal gangue steel. External wall panels and interior wall panels adopt foam ceramic wall panels. Through solid waste utilization researching new building materials is conducive to energy conservation and environmental protection. The selection of floor forced reinforcement should conform to the Code for Design of Steel Structure (China Engineering Construction Standardization Association organizing translation, 2006).

2.2 *Load value*

Load value is crucial for the force analysis of the new system. The common load of the new system is shown in Table 1 below.

Table 1. Load types and standard values.

Functional partition	The standard values of floor dead load (kN/m^2)	The standard values of floor live load (kN/m^2)
office	5.0	2.0
hotel, apartment	5.0	2.0
fire evacuation staircase	7.0	3.5
equipment room, elevator machine room	4.5	7.0
non-accessible roof	6.0	0.5
accessible roof	6.0	2.0
cistern room	4.5	10.0
refuge floor room	4.5	5.0

Note: The partition wall is made up of light material, and the floor dead load includes the partition wall load.

The value of special loads as wind load and earthquake action is shown in Table 2 below.

Table 2. Load types and standard values of special load.

Name	Value
wind load	1. The basic air pressure of the tower is taken at 0.40 kN/m^2, and the ground roughness is of class C; 2. To the overall structure analysis, the wind load's shape coefficient is taken at 1.3.
earthquake action	1. According to the seismic intensity 8, the design's basic seismic acceleration value is 0.20 g; 2. The maximum impact coefficient of the horizontal earthquake is 0.16 (multiple earthquakes). The structural damping ratio is taken at 0.05; 3. The seismic grade of the frame is level 1.

Note: The basic wind pressure is taken at the wind pressure value of 50 years of recurrence period; the reference of seismic action is the "Classification Standard for Seismic Fortification of Construction Engineering" GB50223-2008.

2.3 *Member section*

Detailed analysis and comparison of the studied solid abdominal I-type cross-section steel beam frame system and the new truss beam system are shown in Tables 3 and 4 below.

Table 3. Component section of the beam member.

Region	Member name		Section dimension (mm)
office	column	1st~5th floor	~1100 × 1100 × 20
		6th~10th floor	~1000 × 1000 × 20
		11th~2th floor	~950 × 950 × 20
	Solid abdominal main beam		~12 × 800 × 400 × 20 × 40
	Secondary beam		HN594 × 302
			HN400 × 200
	Corner pillar support		~100 × 100 × 10

Table 4. Component section of truss beam.

Region	Member name		Section dimension (mm)
office	column	1st~5th floor	~1100 × 1100 × 20
		6th~1th floor	1000 × 1000 × 20
		11th~20th floor	950 × 950 × 20
	truss main beam	upper chord	HN200 × 100
		web member	~50 × 50 × 5
		lower chord	~80 × 80 × 5
	Secondary beam		HN594 × 302/HN400 × 200
	Corner column support		~100 × 100 × 10

3 PERFORMANCE ANALYSIS OF THE STRUCTURE SYSTEM

3.1 *Model building*

The truss beam structure system adopts the layout plan of Figure 2, with a span of 12 m; the main beam adopts a truss; the truss girder and column adopt a rigid connection; the truss web and chord adopt a hinge connection. According to the determined building and structure scheme, the structural scheme is modeled by the way of the YJK software. The finite element model is shown in Figure 3.

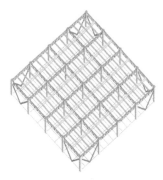

Figure 2 The structural plan of the office building.

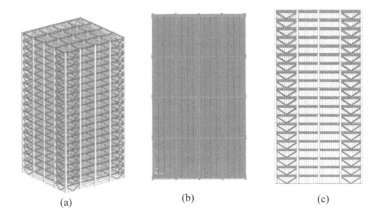

Figure 3. Pure steel frame model of the system. (a) Integral model, (b) Structure plan of standard level and (c) Elevation.

3.2 *Elasticity analysis under frequent earthquakes*

3.2.1 *Cycle and vibration type*

The top three vibration types of the truss beam structure system are analyzed and calculated by the YJK 4.1.1 software, as shown in Figure 4.

And the top 15 orders for the vibration period, translation coefficient, and torsion coefficient in the X/Y direction are obtained. According to the vibration pattern, the first vibration type is the translation in the Y direction, the second vibration type is the translation in the X direction, and the third vibration type is the torsion in the Z direction. In the two directions T1 = T2 = 2.92 s, the structure has the same dynamic properties. The ratio of the first torsion period to the

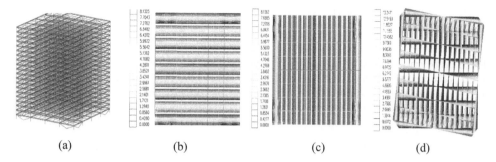

Figure 4. The first three-order vibration shape diagram. (a) Integral model, (b) The first vibration type, (c) the second vibration type, (d) the third vibration type.

first translation period of the structure is 1.7908/2.9229 = 0.61 < 0.85, indicating that the structure will not twist too early. In the first 15 order modes, the quality coefficient of translation vibration in structure's X and Y direction is 96.75% and 95.86% respectively, meeting the requirement of more than 90% that the specification stipulates.

3.2.2 *Shear-to-weight ratio and maximum interlayer displacement*

Under the action of the horizontal earthquake, the shear weight ratio of each layer of the structure in both directions is almost the same, and it is more than 3.2%, indicating that the horizontal seismic shear force obtained by vibration type decomposition reaction spectroscopy can meet the standard requirements.

Under earthquake action, the value of the maximum interlayer displacement angle in the X and Y directions appears on the 11th floor, where the maximum interlayer displacement angle in the X direction is 1/499, and the maximum interlayer displacement angle in the Y direction is 1/499, which meets the requirement that is less than 1/250.

3.2.3 *Interlayer stiffness ratio*

The minimum stiffness ratio of the floor in the X direction is also close to that in the Y direction, so it can be seen that there is no irregular lateral stiffness. The minimum stiffness ratio of the first floor in the X direction is 2.7001, and the minimum stiffness ratio in the Y direction is 2.6995, which is the maximum of all floors. So, the ground as the embedded end of the structure provides the larger lateral stiffness of the structure on the first floor, improves the reliability of the structure, and increases the service life of the structure, thus meeting the safety and durability requirements of residential and office.

4 CONCLUSION

4.1 *Cycle and vibration comparison*

Table 5. The table of structure period analysis.

System	Cycle	Cycle ratio	Quality participation coefficient
solid abdominal I-type steel beam	TI=3.02s	0.62<0.85	95.91%
	T2=3.02s		96.80%
truss beam	TI=2.92s	0.61<0.85	96.75%
	T2=2.92s		95.86%

4.2 The comparison of maximum interlayer displacement (angle)

Table 6. Two systems maximum displacement of floor and displacement angle of the structure.

System / Load type		solid abdominal I-type steel beam		truss beam	
		maximum interlayer displacement / mm	maximum interlayer displacement angle	maximum interlayer displacement / mm	maximum interlayer displacement angle
Seismic action	X	8.70	1/483	8.40	1/500
	Y	8.70	1/483	8.41	1/500

4.3 The comparison of steel quantity

Table 7. The steel weight of the two systems.

The system of the structure	Column (t)	Beam (t)	Support (t)	Total steel quantity (t)
truss beam	1987.698	2430.674	318.828	4737.200
solid abdominal I-type steel beam	1987.698	3146.869	318.828	5453.395
ratio	1.00	1.00	0.77	0.87

From the above conclusions, the self-vibration period for different vibration types of the truss beam structure system is significantly reduced; it is indicated that the truss beam structure system has superior lateral resistance performance. The overall performance of the truss beam is significantly better than the solid abdominal I-type steel beam structure system. The steel quantity of the beam in the truss beam structure system is only 77.24% of the solid abdominal I-type steel beam structure system, and the economy is more obvious. So, it can be reasonably applied to the large-span and large-space high-rise structure, which has a superior promotion prospect.

ACKNOWLEDGMENT

We are grateful for the support from the National Science Foundation (51578387).

REFERENCES

Chen C. C., Lin C. C., Tsai C. L. 2004. Evaluation of Reinforced Connections Between Steel Beams and Box Columns. *Engineering Structures*. 26(13): 1889–1904.

China Engineering Construction Standardization Association organizing translation 2006.09. *Code for Design of Steel Structure GB50017-2003 English Version* = CODE FOR DESIGN OF STEEL STRUCTURES. (Beijing: China Construction Industry Press).

Ding X. 2019. *Seismic Performance Analysis of Fully Assembled Steel Modular Frame Barrel Structure System*. (China: Harbin Institute of Technology).

Hao Y., Lou Y., and Du Xiuli, et al. 2018. Analysis of Shear Lag of Steel Frame Bundled Tube Structure and Floor Damage Assessment. *Journal of Architectural Architecture*. 02.61–71.

Liu C. 2019. *Research on Combined Shear Wall System of Steel Pipe Concrete*. (China: Tsinghua University).

Shi F. 2020. *The Research on Seismic Performance of New SMA Supporting Steel Frame Structure.* (China: Guangzhou University).
Wang Y. 2020. *The Research on Seismic Properties of Double Corner Supporting Steel Frame Structure.* (China: Agricultural University Of Hebei).
Yuan F. S., Liu Y. X. 1996. New Seismic Structure System of Interlayer Suspension Floor Cover. *Journal of Nanjing Institute of Architecture and Engineering.* 03.29–34.73.
Yang Z. Y. 2016. *Comparative Analysis on Steel Structure System of High-rise Residential Building.* (China: Guizhou University).
Yuan Z. J., Zhang W. Y., and Zhang Y. C. 2005. Finite Element Analysis on Performance of Giant Truss Cylinder Structure. *Journal of Harbin Jianzhu University.* Issue **5**: 30–35.

Experimental method for verifying truck weight limits of highway bridges

Songhui Li* & Di Liu*
Shandong University of Science and Technology, Qingdao, China

ABSTRACT: Based on the assumption that the external load effect of the test beam is similar to the vehicle load effect, a test verification method for the load limit value of highway bridges is proposed. First, based on the field static load test data of three 13m full-size reinforced concrete hollow slabs, the imposed load limit of the test beam is determined according to the mid-span deflection, as well as the crack and strain limit of tensile reinforcement of the reinforced concrete beam. Then, the resistance standard value is calculated according to the design size of the test beam and the standard value of material strength. The dead load effect and vehicle load effect corresponding to different live dead load ratios are determined according to the design expression of flexural member-bearing capacity in Code 85 and Code 04. Finally, according to the bridge load limiting coefficient under different safety levels and the midspan bending moment generated by the maximum allowable external load, the theoretical load limiting value and the test load limiting value corresponding to different live and dead load ratios are calculated. The results show that the ratio between the safety test limit value and the theoretical limit value calculated according to different load limit levels is greater than 1.0, which verifies the rationality of the proposed simplified analysis model and the limit value of the bridge load limit.

1 GENERAL INSTRUCTION

In recent years, China's highway transportation industry has developed rapidly, and both the freight flow and the total weight of single vehicles have increased significantly, which has brought great challenges to the safety management of highway bridges (Zhang 2014). However, due to the influence of the fleet load of the original 85 specification, there are certain misunderstandings in the industry regarding the value of bridge load limit (JTG D62-2004), which leads to great differences in bridge load limit standards between different provinces and cities and brings great difficulties to highway transportation and overweight vehicle management. Therefore, it is necessary to systematically study the bridge load limit theory and formulate the bridge load limit standard at the national level.

Aiming at the load limit standard of reinforced concrete medium and small-span bridges, the author has carried out research work for more than ten years. Based on an in-depth analysis of the influence law of load limit coefficient on bridge safety (Li 2016), three different bridge load limit analysis models have been proposed by Li and Wang (Li 2013) (Li 2014) (Wang 2010), and the physical significance of bridge load limit value has been theoretically clarified. Using the load limit analysis model proposed by Li (Li 2013), the author compiles the load limit analysis program and calculates the load limit values of Code 85 bridges and Code 04 bridges. On this basis, a load limit analysis model considering the actual resistance level is proposed to solve the problem of load limit value of low-grade bridges or

---

*Corresponding Authors: songhui.l@163.com and 303553699@qq.com

resistance attenuation bridges (Li 2016). However, the proposed load limit needs to be verified by field or indoor test data, whether it has sufficient safety reserves or not.

Compared with the theoretical research of load limiting, the existing bridge load limiting test results at home and abroad mostly focus on the analysis of the mechanical performance of overloaded bridges (Sun 2005; Wang 2010; Zhang 2011), mainly studying the deflection and crack change law of in-service bridges under different loading ranges and comparing the relationship between the measured value of bearing capacity and the design value. Since the above research did not clarify the structural overload criteria and failed to establish the quantitative relationship between overload and main performance indicators, it is difficult to solve the problem of the load limit value of bridges.

Given the shortcomings of the existing research, the following is a simplified analytical model of the highway bridge load limit proposed by Li (Li 2014). First, an experimental verification method for the value of the bridge load limit is established. Furthermore, the self-compiled load limiting analysis program is applied to calculate the load limiting value of the reinforced concrete medium and small span, which simply supported beam bridges designed according to the code for the design of highway reinforced concrete, prestressed concrete bridges and culverts (JTJ 023-85) (hereinafter referred to as Code 85), the code for design of highway reinforced concrete, and prestressed concrete bridges and culverts (JTG D62-2004) (hereinafter referred to as Code 04). On this basis, the safe loading value and maximum allowable loading value of the test beam are determined based on the field loading test data of the reinforced concrete slab (beam) and the index limits of deflection, maximum crack width, and reinforcement strain of reinforced concrete beam specified in the 04 specifications. Finally, based on the standard value of bending resistance of the test beam, the test load limit value corresponding to different design live dead load ratios are calculated. By comparing with the theoretical load limit value, the safety and engineering applicability of the proposed load limit analysis model and load limit value is verified.

2 GETTING STARTED

According to the simplified analysis model of load limiting, the value of the reinforced concrete medium, and small span simply supported beam bridges (Li 2013), if the variability of vehicle load effect is not considered, the structural function can be expressed as:

$$Z = g(R, S_G) = R - S_G - S_{Qth} \tag{1}$$

Where R and S_G are the random variables of resistance and dead load effect respectively, and S_{Qth} is the standard value of vehicle load effect.

For a bridge structure or member, if the probability density functions of resistance and dead load effect are $f_R(r)$ and $f_{SG}(S_G)$, respectively, the failure probability PF of the structure can be expressed as:

$$p_f = P(Z < 0) = \iint_{r < s_G + s_{Qth}} f_R(r) f_{SG}(s_G) dr ds_G \tag{2}$$

On the contrary, if the allowable failure probability of the structure is given, the only corresponding limit value of vehicle load effect S_{Qth} can be inversely calculated according to Formula (2), which is also called the load effect limit value of the bridge. Therefore, the bridge load limit analysis process is the reverse calculation of the S_{Qth} process according to the allowable failure probability of the structure.

In actual analysis, let $S_{Qth} = \zeta_q S_{Qk}$, where S_{Qk} is the standard value effect of vehicle load adopted in the design, and ζ_q is the bridge load limiting coefficient. In this way, the bridge load limit analysis is transformed into the process of calculating the load limit coefficient according to the allowable failure probability.

According to the code for the design of highway reinforced concrete and prestressed concrete, bridges and culverts (JTJ 023-85) (hereinafter referred to as Code 85), and bridge design code (JTG D62-2004) (hereinafter referred to as Code 04) bearing capacity limit state design expression (JTG D62-2004, JTJ 023-85), if only the action of dead load and vehicle load is considered, the relationship between the standard value of flexural member resistance and the combined design value of action effect can be expressed as:

$$R_k = \gamma_R(\gamma_1 S_{Gk} + \gamma_2 S_{Qk}) \quad (3)$$

or

$$R_k = \gamma_0 \gamma_R(\gamma_G S_{Gk} + \gamma_Q S_{Qk}) \quad (4)$$

Where, R_k is the standard value of resistance calculated according to the standard value of material performance and geometric parameters specified in the specification; γ_R is the partial coefficient of resistance, and the bridge designed according to Code 85 can be regarded as the partial coefficient of comprehensive resistance; γ_0 is the importance coefficient of the bridge structure, which is related to the design safety level of the bridge. Level 1, level 2, and level 3 take 1.1, 1.0, and 0.9 respectively; γ_1, γ_2 are the partial coefficients of the dead load effect and vehicle load effect respectively. For a simply supported beam bridge, $\gamma 1$ takes 1.2, and γ_2 takes 1.4; S_{Gk} is the standard value effect of dead load; S_{Qk} is the standard value effect of vehicle load calculated according to Specification 85 or 04.

In Equations (3) and (4), γ_R is only the ratio to the live dead load adopted in the design ρ (=S_{Qk}/S_{Gk}) related (Li 2013, Li 2014). Therefore, if the resistance standard value R_k of the member is known, for any given design live dead load ratio ρ, the corresponding dead load effect S_{Gk} and vehicle load effect S_{Qk} can be determined. The calculation formula is:

$$S_{Gk} = \frac{R_k}{\gamma_R(\gamma_G + \gamma_Q \rho)}, \quad S_{Qk} = \rho S_{Gk} \quad (5)$$

or

$$S_{Qk} = \frac{\rho R_k}{\gamma_R(\gamma_G + \gamma_Q \rho)}, \quad S_{Gk} = S_{Qk}/\rho \quad (6)$$

According to Formula (3) or Formula (4), the ratio with different resistance standard values and design live dead load can be determined as ρ, with corresponding dead load standard value effect S_{Gk} and vehicle load standard value effect S_{Qk}. Similarly, S_{Gk} and S_{Qk} values expressed according to the design formula of Specification 04 can be obtained.

When the load limit value of the actual bridge is verified by the reinforced concrete slab beam loading test, the standard value of resistance can be calculated according to the actual size of the test slab (beam) and the measured value of material strength. For different live dead load ratios of different structures, the corresponding S_{Gk} and S_{Qk} can be calculated according to Formula (5) or Formula (6).

The load limiting coefficient is only related to the live dead load ratio adopted in the design and has nothing to do with the specific values of S_{Gk} and S_{Qk}. The specific geometric dimensions and material parameters of the bridge structure may not be involved in the calculation process.

3 GETTING STARTED

3.1 *Calculation of theoretical limit value*

When calculating S_{Gk} or S_{Qk} according to Formula (3) and Formula (4), it is necessary to determine the partial coefficient of resistance corresponding to different live dead load ratios γ_R.

According to Code 85, the partial factor of the resistance of reinforced concrete medium and small span simply supported beam bridges is the safety factor of concrete or reinforcement γ_3 and load effect improvement coefficient γ_5. The calculation results of the product can be seen in Table 1.

Table 1. Resistance factors correspond to the 85 edition of bridge design specifications.

ρ	$S_{Qk}/(S_{Gk} + S_{Qk})$ /%	γ_3	γ_5	$\gamma_R = \gamma_3\gamma_5$
0.1	9	1.25	1.05	1.3125
0.25	20	1.25	1.05	1.3125
0.5	33	1.25	1.03	1.2875
1.0	50	1.25	1.00	1.2500
1.5	60	1.25	1.00	1.2500
2.5	71	1.25	1.00	1.2500

For the reinforced concrete simply supported beam bridge with medium and small spans designed according to Code 04, if the structural importance factor γ_0 is taken into account, S_{Gk} and S_{Qk} corresponding to different resistance standard values and design live dead load ratios can also be determined according to Formula (5) or Formula (6). According to the unified standard, the partial coefficients of resistance correspond to the design safety Class I, Class II, and Class III bridges ($=\gamma_0\gamma_R$) are 1.2379, 1.1254, and 1.0129, respectively (GB/T 50283-1999).

According to the simplified analysis model of bridge load limitation (Wang 2010), the load limitation coefficients corresponding to the design of bridges in Codes 85 and 04 can be calculated. The calculation results of the load-limiting coefficient ξ_q under the first, second, and third grades of the load-limiting safety grade are given in Table 2.

Table 2. Resistance factors correspond to the 85 edition of bridge design specifications.

Load limiting Safety level	ρ	85 specifications (JTG D62-2004)	04 specifications (first level)	04 specifications (second level)	04 specifications (third level)
First level	0.1	0.601	–	–	–
	0.25	0.946	0.650	0.203	–
	0.5	0.997	0.877	0.605	0.332
	1.0	1.007	0.987	0.802	0.617
	1.5	1.039	1.022	0.866	0.710
	2.5	1.064	1.049	0.916	0.783
Second level	0.1	1.450	0.765	–	–
	0.25	1.333	1.018	0.541	–
	0.5	1.227	1.099	0.808	0.517
	1.0	1.158	1.137	0.939	0.741
	1.5	1.166	1.148	0.981	0.814
	2.5	1.171	1.156	1.014	0.871
Third level	0.1	2.353	1.621	0.514	–
	0.25	1.746	1.409	0.899	0.388
	0.5	1.473	1.336	1.024	0.713
	1.0	1.319	1.297	1.085	0.872
	1.5	1.302	1.283	1.104	0.924
	2.5	1.287	1.271	1.118	0.965

By comparing the theoretical limit value of the mid-span bending moment of the test beam with the test limit value, the rationality of the load limiting coefficient and the load limiting analysis model can be indirectly verified.

3.2 Verification steps

The test verification of the load limit value is the comparison process between the theoretical limit value and the test limit value. By comparing the theoretical limit value and the test limit value of the test beam, the rationality of the load limit analysis model can be effectively tested. The following is an example of reinforced concrete flexural members to illustrate the test verification method of load-limiting value. The basic steps are as follows:

(1) The allowable imposed load effect of the test beam is determined according to the yield strain of the reinforcement, the mid-span deflection limit, and the crack width limit of the test beam.
(2) Given different live dead load ratios ρ, the corresponding resistance partial coefficient γ_R can be checked from Table 1 or the unified standard.
(3) According to the standard value R_k of bending resistance of the test beam, the corresponding standard value effect S_{Gk} of dead load and standard value effect S_{Qk} of vehicle load can be calculated according to Formula (5) or (6).
(4) The calculated S_{Gk} and S_{qk} are used to simulate the dead load effect and vehicle load effect adopted in bridge design, and the corresponding load limiting coefficient ζ_q is selected according to Table 2. The theoretical limit value $\zeta_q S_{Qk}$ can be calculated.
(5) The required dead load counterweight value can be determined according to the size of S_{Gk}. If the self-weight effect of the test beam is less than S_{Gk}, the applied load can be taken as the dead load counterweight, and the difference between the applied load effect and the dead load counterweight effect is the load limiting test value of the test beam; Similarly, if the self-weight effect of the test beam is greater than S_{Gk}, the sum of the self-weight effect of some test beams and the effect of applied load is taken as the load limiting test value S_Q, except for the test beam.
(6) The size of $\zeta_q S_{Qk}$ and S_Q can be compared. If the ratio of S_Q, exp to $\zeta_q S_{Qk}$ is greater than 1.0, it is considered that the theoretical limit value is partial to safety.

It should be noted that the load limiting factor ζ_q is only related to the live dead load ratio adopted in the design, and has nothing to do with the specific value of the standard value effect of vehicle load and the standard value effect of dead load. Therefore, the rationality of the load limit value of the bridge designed according to Code 85 and Code 04 can be indirectly verified by constructing different live dead load ratios. At the same time, considering the small variability of the dead load effect, part of the applied load is directly taken as the dead load counterweight in the analysis process to achieve the purpose of simulating different live dead load ratios.

4 TEST VERIFICATION

4.1 Test overview

A channel bridge of the Shugang expressway was built in 1995. The superstructure is a 13m fabricated hinged hollow slab, and the original design load is qichao-20 and gua-120. Due to the antifreeze mixed with chlorinated salt during construction, the unqualified construction mixing water, and the corrosion of chloride ions in the external environment, only 11 years after the opening to traffic, most of the main board bottoms had serious diseases such as reinforcement corrosion and concrete protective layer falling off, so it was decided to demolish and rebuild. To verify the engineering applicability of the proposed bridge load limiting simplified analysis model, three intact hollow slabs were selected for field loading tests and the specimen numbers were G1, G2, and G3 respectively.

In the test, the weight is loaded symmetrically at two points and the spacing between loading points is 2m. To simulate the stress state of the in-service hollow slab bridge as much as possible, the fulcrum of all specimens adopts circular plate rubber bearings. Due to the limitation of on-site support conditions, the calculated span used for test beams G1 and G2 is 11.4m and the calculated span used for the G3 test beam is 12.3m. The dead weight bending moments at midspan is 168.9kN·m and 196.6kN·m respectively.

4.2 *Experimental data*

This test mainly determines the maximum loading value corresponding to different failure criteria. The observation items include the strain of tensile reinforcement, concrete strain, deflection of the beam, crack width and crack development of the mid-span section.

The measured mid-span section crack width and mid-span deflection of each test beam under different graded loads can be seen in Tables 3 and 4.

Table 3. Crack widths at midspan sections of box beams under various loads.

Classification load/kN	Midspan bending moment/kN·m		G1 Width/mm	G2 Width/mm	G3 Width/mm	Classification load/kN
	$L = 11.4$m	$L = 12.3$m				
0.0	0.0	0.0	0.02	0.02	0.03	0.0
80.4	188.9	207.0	0.04	0.04	0.06	80.4
134.0	314.9	345.1	0.08	0.06	0.12	134.0
187.6	440.9	483.1	0.10	0.10	0.16	187.6
240.0	564.0	618.0	0.14	0.14	0.22	240.0
288.0	676.8	741.6	0.2	0.18	0.28	288.0
314.0	737.9	808.6	0.24	0.24	0.30	314.0
343.0	806.1	883.2	0.30	0.26	0.34	343.0

Table 4. Deflections at midspan sections of box beams under various loads.

Classification load/kN	Midspan bending moment/kN·m		Midspan deflection/mm			Classification load/kN	Midspan bending moment/kN·m		Midspan deflection/mm		
	$L =$ 11.4m	$L =$ 12.3m	G1	G2	G3		$L =$ 11.4m	$L =$ 12.3m	G1	G2	G3
0.0	0.0	0.0	0	0	0	0.0	0.0	0.0	0	0	0
80.4	188.9	207.0	2.07	1.8	3.18	80.4	188.9	207.0	2.07	1.8	3.18
134.0	314.9	345.1	4.39	4.19	5.6	134.0	314.9	345.1	4.39	4.19	5.6
187.6	440.9	483.1	7.12	7.08	8.94	187.6	440.9	483.1	7.12	7.08	8.94
240.0	564.0	618.0	9.69	10.13	11.61	240.0	564.0	618.0	9.69	10.13	11.61
288.0	676.8	741.6	12.31	14.13	16.14	288.0	676.8	741.6	12.31	14.13	16.14
314.0	737.9	808.6	13.87	16.3	19.42	314.0	737.9	808.6	13.87	16.3	19.42
343.0	806.1	883.2	16.2	19.2	24.03	343.0	806.1	883.2	16.2	19.2	24.03
371.0	871.9	955.3	18.85	22.22	43.97	371.0	871.9	955.3	18.85	22.22	43.97

4.3 *Bending moment limit*

According to the current bridge design code, the maximum allowable loading value of the test beam under different safety levels can be determined according to the deflection, crack width, and yield strain limit of tensile reinforcement. Among them, the load value corresponding to the deflection and crack limits is the safe load value that the test beam can bear, while the load value corresponding to the yield strain limit of tensile reinforcement is the maximum load value that the test beam can bear.

After analysis, the crack limit is taken as 0.2mm according to the Class II environment, and the mid-span deflection limit is determined according to L/600, where L is the calculated span used in the test. The deflection limit corresponding to test beams G1 and G2 is 19mm, and the deflection limit corresponding to test beam G3 is 20.5mm. For the yield moment, the yield strain of tensile reinforcement is calculated according to the standard value of strength, which is 1675 $\mu\varepsilon$.

According to Tables 3 and 4, the loading values corresponding to the limit values of each index can be calculated by the interpolation method. Since the calculated span of the G3 test beam is different from that of the G1 and G2 test beams, the loading limits under different control indexes are uniformly converted into the mid-span section load effect to facilitate comparison, as shown in Table 5.

Table 5. Flexural capacities at midspan sections of box beams /kN·m.

Control index	Test beam Number		
	G1	G2	G3
Crack width limit	676.8	697.2	573.0
Mid-span deflection limit	875.1	801.6	826.0

If analyzed according to the normal service limit state, the allowable loading value of the test beam is controlled by the crack limit value and the bending moment limit of the midspan section corresponding to the crack limit value of the G3 test beam is the smallest. Therefore, 573.0kN·m is taken as the allowable bending moment value of the test beam.

Next, based on the mid-span bending moment limit of the test beam, and taking different design live dead load ratios as the basic parameters, the difference between the theoretical limit value of the test beam and the test value is compared to test the rationality of the proposed load limit analysis model and load limit coefficient.

5 TEST VERIFICATION

5.1 *Comparison of bridge load limits in Code 85*

The load limit values of 85 specification bridges under safety Levels I, II, and III are verified according to the serviceability limit state. The load limit in this state is controlled by the G3 beam. The dead load self-weight bending moment of the test beam is 196.6kN·m, and the maximum loading bending moment corresponding to the G3 beam is 573.0kN·m. For the convenience of distinction, the load limit calculated according to this value may be defined as the load limit value of the safety test. The calculation results can be seen in Table 6.

The calculation results show that for each load-limiting safety level, with the increase of live dead load ratio, the ratio of load-limiting test value to load-limiting theoretical value gradually decreases and tends to be stable. The ratio of the theoretical limit value of the mid-span bending moment determined according to the first and second levels of structural safety

Table 6. Comparison between theoretical and safe experimental weight limits. (85 Edition).

Load limiting Grade	ρ	S_{QK}/kN·m	S_{GK}/kN·m	Dead load counter-weight/kN·m	ζ_q	Theory Limit value $\zeta_q S_{QK}$/kN·m	Test Limit value $S_{Q,exp}$/kN·m	$\dfrac{S_{Q,exp}}{\zeta_q S_{QK}}$
First level	0.10	59.4	593.5	396.9	0.601	35.7	176.1	4.94
	0.25	128.3	513.1	316.5	0.945	121.2	256.5	2.12
	0.50	213.4	426.7	230.1	0.997	212.7	342.9	1.61
	1.00	321.2	321.2	124.6	1.007	323.4	448.4	1.39
	1.50	379.6	253.1	56.5	1.039	394.4	516.5	1.31
	2.5	444.2	177.7	−18.9	1.063	472.2	591.9	1.25
Second level	0.10	59.4	593.5	396.9	1.450	86.1	176.1	2.05
	0.25	128.3	513.1	316.5	1.333	171.0	256.5	1.50
	0.5	213.4	426.7	230.1	1.227	261.8	342.9	1.31
	1.00	321.2	321.2	124.6	1.158	371.9	448.4	1.21
	1.50	379.6	253.1	56.5	1.166	442.6	516.5	1.17
	2.5	444.2	177.7	−18.9	1.171	520.2	591.9	1.14
Third level	0.10	59.4	593.5	396.9	2.353	139.7	176.1	1.26
	0.25	128.3	513.1	316.5	1.746	224.0	256.5	1.15
	0.5	213.4	426.7	230.1	1.473	314.3	342.9	1.09
	1.00	321.2	321.2	124.6	1.319	423.7	448.4	1.06
	1.50	379.6	253.1	56.5	1.302	494.2	516.5	1.05
	2.5	444.2	177.7	−18.9	1.287	571.7	591.9	1.04

to the test limit value is greater than 1.14. It can be considered that the Code 85 bridge is partial to safety according to the first and second levels of load limit, which also confirms the rationality of the simplified analysis model according to the bridge load limit.

For the load limiting coefficient determined according to the third level of load limiting grade, the load limiting ratio is mostly close to 1.0. To further test the engineering practicability of the lower limit coefficient of the load limiting grade, the theoretical and experimental load limiting values are calculated by using the yield moment in the middle of the test beam span to test whether the corresponding load limiting coefficient has sufficient safety reserves.

It is worth noting that when analyzing according to the yield moment, the yield moment of G2 beam920.7kN·m is taken as the basis for calculation, and the dead load moment of the main beam corresponding to the beam is taken as 168.9kN·m. Accordingly, the limit value calculated according to this value is the maximum test limit value. The load limit analysis results calculated according to the third level of load limit level can be seen in Table 7.

Table 7. Comparison between theoretical and safe experimental weight limits. (85 Edition).

ρ	S_{QK}/kN·m	S_{GK}/kN·m	Dead load counter-weight/kN·m	ζ_q	Theory Limit value $\zeta_q S_{QK}$/kN·m	Test Limit value $S_{Q,exp}$/kN·m	$\dfrac{S_{Q,exp}}{\zeta_q S_{QK}}$
0.10	59.4	593.5	424.6	2.353	139.7	496.1	3.55
0.25	128.3	513.1	344.2	1.746	224.0	576.5	2.57
0.5	213.4	426.7	257.8	1.473	314.3	662.9	2.11
1.00	321.2	321.2	152.3	1.319	423.7	768.4	1.81
1.50	379.6	253.1	84.2	1.302	494.2	836.5	1.69
2.5	444.2	177.7	8.8	1.287	571.7	911.9	1.60

The calculation results show that although the load limit ratio under the three-level load limit level is close to 1.0, there is still enough safety reserve to calculate the load limit value according to the load limit coefficient of this level, and the ratio of the test value to the theoretical value is greater than 1.60. Therefore, the Code 85 bridge can calculate the load limit value according to the three-level load limit level.

5.2 *Comparison of bridge load limits in Code 04*

The following is the verification and analysis of load limit values for bridges with design safety levels of Class I, Class II, and Class III. The dead load self-weight bending moment and the maximum loading bending moment are the same. Specific calculation results can be seen in Tables 8–10.

Table 8. Comparison between theoretical and safe experimental weight limits. (Safety Class I).

Load limiting Grade	ρ	$S_{QK}/$ kN·m	$S_{GK}/$ kN·m	Dead load counter weight/kN·m	ζ_q	Theory Limit value $\zeta_q S_{QK}$/kN·m	Test Limit value $S_{Q,exp}$/kN·m	$\dfrac{S_{Q,exp}}{\zeta_q S_{QK}}$
First level	0.10	62.9	629.3	432.7	–	–	140.3	–
	0.25	136.0	544.0	347.4	0.650	88.4	225.6	2.55
	0.50	221.9	443.8	247.2	0.877	194.6	325.8	1.67
	1.00	324.3	324.3	127.7	0.987	320.1	445.3	1.39
	1.50	383.3	255.5	58.9	1.022	391.7	514.1	1.31
	2.5	448.6	179.4	−17.2	1.049	470.5	590.2	1.25
Second level	0.10	62.9	629.3	432.7	0.765	48.1	140.3	2.91
	0.25	136.0	544.0	347.4	1.018	138.5	225.6	1.63
	0.5	221.9	443.8	247.2	1.099	243.9	325.8	1.34
	1.00	324.3	324.3	127.7	1.137	368.8	445.3	1.21
	1.50	383.3	255.5	58.9	1.148	440.0	514.1	1.17
	2.5	448.6	179.4	−17.2	1.156	518.5	590.2	1.14
Third level	0.10	62.9	629.3	432.7	1.621	102.0	140.3	1.38
	0.25	136.0	544.0	347.4	1.409	191.6	225.6	1.18
	0.5	221.9	443.8	247.2	1.336	296.5	325.8	1.10
	1.00	324.3	324.3	127.7	1.297	420.7	445.3	1.06
	1.50	383.3	255.5	58.9	1.283	491.8	514.1	1.05
	2.5	448.6	179.4	−17.2	1.271	570.1	590.2	1.04

The above tables respectively verify the load limiting coefficients of bridges with design safety Classes I, II, and III in Code 04. The results show that the ratio of theoretical and experimental bending moment limits corresponding to different live dead load ratios is greater than 1.0. Therefore, no matter the design safety level of reinforced concrete medium and small span simply supported beam bridge designed according to Code 04 is Level I, Level II, or Level III, the load limit value determined according to Table 2 load limit coefficient table can meet the requirements of normal service limit state. However, considering that the load limit ratio under the third load limit level is close to 1.0, and considering the need for durability, it is not recommended to formulate the bridge load limit standard according to the third load limit level, and the load limit value under this level is only for the approval of overweight vehicles.

To test the safety of a Grade III bridge with a load limit, the yield moment is also used to calculate the corresponding load limit ratio, and the corresponding load limit ratio is shown in Table 11.

It can be seen from Table 11 that the ratio of the maximum allowable load limit value calculated according to the three-level load limit coefficient of the load limit level to the theoretical load limit value is greater than 1.60, and the structure has sufficient safety reserves under the action of corresponding vehicle loads.

Table 9. Comparison between theoretical and safe experimental weight limits. (Safety Class II).

Load limiting Grade	ρ	S_{QK}/ kN·m	S_{GK}/ kN·m	Dead load counter weight/kN·m	ζ_q	Theory Limit value $\zeta_q S_{QK}$/kN·m	Test Limit value $S_{Q,exp}$/kN·m	$\dfrac{S_{Q,exp}}{\zeta_q S_{QK}}$
First level	0.10	69.2	692.2	495.6	–	–	77.4	–
	0.25	149.6	598.4	401.8	0.203	30.4	171.2	5.64
	0.50	244.1	488.2	291.6	0.605	147.7	281.4	1.91
	1.00	356.8	356.8	160.2	0.802	286.1	412.8	1.44
	1.50	421.6	281.1	84.5	0.866	365.1	488.5	1.34
	2.5	493.4	197.4	0.8	0.916	451.9	572.2	1.27
Second level	0.10	69.2	692.2	495.6	–	–	77.4	–
	0.25	149.6	598.4	401.8	0.541	80.9	171.2	2.11
	0.5	244.1	488.2	291.6	0.808	197.2	281.4	1.43
	1.00	356.8	356.8	160.2	0.939	335.0	412.8	1.23
	1.50	421.6	281.1	84.5	0.981	413.6	488.5	1.18
	2.5	493.4	197.4	0.8	1.014	500.3	572.2	1.14
Third level	0.10	69.2	692.2	495.6	0.514	35.6	77.4	2.17
	0.25	149.6	598.4	401.8	0.899	134.5	171.2	1.27
	0.5	244.1	488.2	291.6	1.024	250.0	281.4	1.13
	1.00	356.8	356.8	160.2	1.085	387.1	412.8	1.07
	1.50	421.6	281.1	84.5	1.104	465.5	488.5	1.05
	2.5	493.4	197.4	0.8	1.118	551.6	572.2	1.04

Table 10. Comparison between theoretical and safe experimental weight limits. (Safety Class III).

Load limiting Grade	ρ	S_{QK}/ kN·m	S_{GK}/ kN·m	Dead load counter weight/kN·m	ζ_q	Theory Limit value $\zeta_q S_{QK}$/kN·m	Test Limit value $S_{Q,exp}$/kN·m	$\dfrac{S_{Q,exp}}{\zeta_q S_{QK}}$
First level	0.10	76.9	769.1	572.5	–	–	0.5	–
	0.25	166.2	664.9	468.3	–	–	104.7	–
	0.50	271.2	542.4	345.8	0.332	90.0	227.2	2.52
	1.00	396.4	396.4	199.8	0.617	244.6	373.2	1.53
	1.50	468.5	312.3	115.7	0.71	332.6	457.3	1.37
	2.5	548.2	219.3	22.7	0.783	429.2	550.3	1.28
Second level	0.10	76.9	769.1	572.5	–	–	0.5	–
	0.25	166.2	664.9	468.3	–	–	104.7	–
	0.5	271.2	542.4	345.8	0.517	140.2	227.2	1.62
	1.00	396.4	396.4	199.8	0.741	293.7	373.2	1.27
	1.50	468.5	312.3	115.7	0.814	381.3	457.3	1.20
	2.5	548.2	219.3	22.7	0.871	477.5	550.3	1.15
Third level	0.10	76.9	769.1	572.5	–	–	0.5	–
	0.25	166.2	664.9	468.3	0.388	64.5	104.7	1.62
	0.5	271.2	542.4	345.8	0.713	193.4	227.2	1.17
	1.00	396.4	396.4	199.8	0.872	345.6	373.2	1.08
	1.50	468.5	312.3	115.7	0.924	432.9	457.3	1.06
	2.5	548.2	219.3	22.7	0.965	529.0	550.3	1.04

Table 11. Comparison between theoretical and maximum experimental weight limits. (04 Edition).

Design security Grade	ρ					
	0.10	0.25	0.50	1.00	1.50	2.50
First level	4.51	2.85	2.18	1.82	1.70	1.60
Second level	–	3.65	2.41	1.89	1.74	1.62
Third level	–	6.58	2.83	2.01	1.80	1.65

6 CONCLUSION

After analysis, the following conclusions can be drawn:

(1) The ratio of the test limit value to the theoretical limit value of the reinforced concrete medium and small span simply supported beam bridges designed by Code 85 and Code 04 is greater than 1.0. The proposed simplified analysis model of bridge load limit can be used to determine the load limit value of highway bridges.
(2) Under the safety level of load limit, if the live dead load ratio is greater than or equal to 1.0, the ratio of the calculated test limit value and the theoretical limit value of bridges with different design safety levels according to Code 85 and Code 04 is very close, and the verification results are in good consistency.
(3) When the design live dead load ratio is greater than or equal to 1.0, the ratio of the safe load limit test value to the theoretical value corresponding to the third level of load limit level is close to 1.0. From the perspective of durability, it is not recommended to set the load limit standard according to the third level of structural safety level.
(4) If the maximum load limit test value is calculated based on the yield moment, the ratio between it and the theoretical load limit value is greater than 1.60, which shows that although the three-level load limit according to the load limit level will affect the durability of the structure, the structure still has sufficient safety reserves, and this level of load limit value can be used as a reference for the approval of over limit vehicles.

REFERENCES

BR-MCEB-2-I2 *Manual for Condition Evaluation of Bridges (2nd Edition with 2001 and 2003 Interim Revisions)* [S]. Washington DC: American Association of State Highway and Transportation Officials, 2000.
GB/T 50283-1999 *Unified Standard for Reliability Design of Highway Engineering Structures* [S]. Beijing: China Plan Press, 1999.
JTG B01-2003 *Technical Standard of Highway Engineering* [S]. Beijing: China Communications Press, 2004.
JTG D62-2004 *Code for the Design of Highway Reinforced Concrete and Prestressed Concrete Bridges and Culverts* [S], Beijing: China Communications Press, 2004.
JTJ 023-85 *Code for Design of Highway Reinforced Concrete and Prestressed Concrete Bridges and Culvers* [S]. Beijing: China Communications Press, 1985.
Li Songhui, Jiang Hanwan. Analytical Model for Determining Weight Limits of Highway Bridges with Various Resistance Levels [J]. *China Civil Engineering Journal*, 2016, 49(6):76–83. (In Chinese)
Li Songhui, Jiang Hanwan. Influence Law of Different Weight Limit Levels on Highway Bridges [J]. *China Journal of Highway and Transport*, 2016, 29(3):82–88. (In Chinese)
Li Songhui. Analytical Approach for Determining Bridge Weight Limits Based on Truncated Distributions of Live Load Effect [J]. *Engineering Mechanics*, 2014, 31(2):117–124.
Li Songhui. Reliability-Based Analytical Model for Determining the Truck Weight Limits on Highway Bridges [J]. *China Civil Engineering Journal*, 2013, 46(9): 83–90.
LRFR-1 *Guide Manual for Condition Evaluation and Load and Resistance Factor Rating (Lrfr) Of Highway Bridges* [S]. Washington DC: American Association of State Highway and Transportation Officials, 2003.
MBE-2 *The Manual for Bridge Evaluation* [S]. Washington, DC: American Association of State Highway and Transportation Officials, 2011.
Roschke PN, Pruski KR. Overload and Ultimate Load Behavior of Post-Tensioned Slab Bridge [J]. *Journal of Bridge Engineering*, 2000, 5(2): 148–155.
Sun Xiaoyan, Huang Chengkui, Zhao Guofan. Experimental Study of The Influence of Truck Overloads on The Flexural Performance of Bridge Members [J]. *China Civil Engineering Journal*, 2005, 38(6): 35–40.
Wang Songgen, Li Songhui. Reliability-Based Vehicular Weight Limits of Highway Bridges [J]. *Engineering Mechanics*, 2010, 27(10):162–166.
Zhang Jianren, Peng Hui, Cai CS. Field study of overload behavior of an existing reinforced concrete bridge under simulated vehicle loads [J]. *Journal of Bridge Engineering*, 2011, 16(2), 226–237.
Zhang Xigang. *Research on Vehicular loading standard of Highway Bridges* [M]. Beijing: Beijing: China Communications Press, 2014.

Wind vibration response analysis of derrick steel structure based on time domain analysis method

Dongying Han*
School of Vehicle and Energy, Yanshan University, Hebei, China

Nian Liu*
China North Industries Group Jiangshan Heavy Industry Research Institute Co., Ltd., Jiangshan, China

Guoqing Zhu*, Yan Huang*, Xinjun Yang* & Liming Zheng*
School of Vehicle and Energy, Yanshan University, Hebei, China

ABSTRACT: Aiming at the safety problem of derrick steel structure under wind load, the wind vibration response analysis of derrick steel structure is carried out. The linear filtering method is used to simulate the wind load on the derrick steel structure. The algorithm is simple and the simulation results are accurate. The finite element model of the ZJ70 derrick steel structure is established, and the dynamic response analysis of the derrick steel structure under wind load under different working conditions is carried out. The results show that: First, compared with the equivalent static wind load analysis, the response peak of the derrick steel structure under the wind vibration time domain analysis is much larger than that under the equivalent static load analysis. Second, under the condition of preservation equipment, the stress response of the middle-inclined bar member of the ZJ70 derrick steel structure exceeds the allowable value of the specification, and the strength of the ZJ70 derrick steel structure under this condition does not meet the requirements. Third, the wind level of wind load has a great influence on the safety of the derrick steel structure, and the middle and low parts of the ZJ70 derrick steel structure are sensitive to the change in wind load strength. Fourth, the overall damage and local damage of the derrick steel structure have a great influence on its wind resistance safety, and the local damage is not easy to be directly found by observation.

1 GENERAL INSTRUCTION

As the main bearing structure of oilfield production, its slender structural characteristics make the wind load greatly affect the safety of the derrick steel structure. In recent years, the time domain analysis method has been widely used in the calculation of wind vibration response. Lou Wenjuan et al. (Lou et al. 2013) compared the wind tunnel test of the aero-elastic model with the numerical analysis results and proved that the wind vibration response of the transmission tower obtained by the numerical analysis in the time domain has a certain reference value. Wang Hui et al. (Wang et al. 2020) proposed a structural damage detection method based on time-domain response correlation analysis and data fusion and verified the effectiveness of the method through experimental research. Liqiang et al. (An et al. 2017) established a transmission tower-line system model in ANSYS and calculated the wind load effect of the model under the action of typhoons. It was concluded that the

---

*Corresponding Authors: dongying.han@163.com, whiteliu4466@163.com, 1026600545@163.com, 295788857@qq.com, yangxj@ysu.edu.cn and zheng_liming_happy@126.com

response of the model under the typhoon extreme state was much larger than that under the static wind equivalent effect. Huang Guosheng (Huang et al. 2014) used the Shiyuan typhoon wind speed spectrum and harmonic superposition method to simulate the wind field of the transmission tower. Therefore, based on the time domain analysis method, a wind field simulation method based on the AR model is proposed. Based on the basic principle of time series analysis, the wind load is numerically simulated, and then the nonlinear analysis of the derrick steel structure under wind load is carried out to obtain the response time history curve of the structure. The response results of time domain analysis and static analysis are compared to verify the necessity of wind vibration response analysis of the derrick steel structure. The structural response data under different wind loads and damage are analyzed, and the influence of different factors on the safety of derrick steel structures under wind load is summarized. The results show that the algorithm is simple and efficient, and the simulation results are accurate.

2 BASIC THEORY OF WIND LOAD SIMULATION

The p-order auto-regressive (AR) model of the $V(X, Y, Z, t)$ column vector of the fluctuating wind speed time history is shown at M space points:

$$V(X, Y, Z, t) = \sum_{k=1}^{p} \psi_k V(X, Y, Z, t - kt) + N(t) \qquad (1)$$

In the formula: X, Y, Z are the coordinate vector matrix; p is the AR model order; Δt is the time step; ψ_k is the AR model autoregressive coefficient matrix $k = 1, 2, 3, \cdots, p$; $N(t)$ is the independent random process vector $N(t) = L \cdot n(t)$; L is a lower triangular matrix of order m; $n(t)$ is a normal random distribution of $(0,1)$ independent of each other. By multiplying both sides of Equation (1) to the right $N(t)^T$ at the same time and taking the mathematical expectation, we get:

$$R(0) = \sum_{k=1}^{p} \psi_k R(k\Delta t) + R_N \qquad (2)$$

$R(j\Delta t)$ is the cross-correlation function matrix of $m \times m$ order spatial pulsating wind speed:

$$R(j\Delta t) = \begin{bmatrix} R_{11}(j\Delta t) & R_{12}(j\Delta t) & \cdots & R_{1m}(j\Delta t) \\ R_{21}(j\Delta t) & R_{22}(j\Delta t) & \cdots & R_{2m}(j\Delta t) \\ \vdots & \vdots & \ddots & \vdots \\ R_{m1}(j\Delta t) & R_{m2}(j\Delta t) & \cdots & R_{mm}(j\Delta t) \end{bmatrix} \qquad (3)$$

From Formula (2), it can be known that $R_N = R(0) - \sum_{K=1}^{P} \psi_k R_u(k\Delta t)$, and the lower triangular matrix L is obtained by the Cholesky decomposition of R_N. The pulsating wind speed of M space points at time $j\Delta t$ can be obtained:

$$\left\{ \begin{bmatrix} V^1(j\Delta t) \\ V^2(j\Delta t) \\ \vdots \\ V^m(j\Delta t) \end{bmatrix} = \sum_{k=1}^{p} [\psi_k] \begin{bmatrix} V^1[(j-k)\Delta t] \\ V^2[(j-k)\Delta t] \\ \vdots \\ V^m[(j-k)\Delta t] \end{bmatrix} + \begin{bmatrix} N^1(j\Delta t) \\ N^2(j\Delta t) \\ \vdots \\ N^m(j\Delta t) \end{bmatrix} \right. \qquad (4)$$
$$(j\Delta t = 0 - T, k \leq j)$$

3 WIND VIBRATION RESPONSE OF DERRICK STEEL STRUCTURE UNDER MAINTENANCE EQUIPMENT CONDITION

3.1 Establishment of finite element model of derrick steel structure

The ZJ70 type derrick steel structure finite element model is 52.8m high and made of Q345 steel. According to the different bearing directions, the rods of the derrick steel structure are classified into horizontal rods, inclined rods, and vertical rods, as shown in Figure 1.

3.2 Security equipment condition

According to the requirements in API SPEC 4F, the working conditions of the security equipment are as follows in the 15-level bushel wind, that is, the average wind speed at a height of 10m within 10 minutes is 47.8m/s, and all the steel roots of the ZJ70 derrick are put down to ensure the derrick. The steel structure is not damaged. When calculating the wind load of the derrick steel structure, the average wind speed $\bar{v}_{10}$ in 10 minutes should be converted into the average wind speed V_s in 2 minutes.

3.3 Response analysis of derrick steel structure under equivalent static wind load

The displacement and stress response results are shown in Figures 2 and 3. In Figure 2, the displacement at the top position of the derrick steel structure is the largest, and the maximum displacement value is 0.355m in combination with the diagram. In Figure 3, a positive value indicates that the unit is pulling, and a negative value indicates that the unit is in a compression state. The highest stress values are in the middle-inclined rods. The tensile stress value of the unit at position MX in the figure is the highest, which is 111MPa, and the compressive stress value of the unit at position MN is the highest value, which is 101MPa. According to the AISC specification, the horizontal angle limit of the derrick steel structure under wind load is 1/50, the tensile allowable stress of the rod is 207MPa, and the compressive allowable stress is 200MPa, so the derrick steel structure under static analysis is in the security equipment in a safe state.

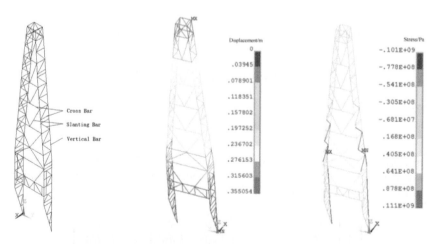

Figure 1. Schematic diagram of ZJ70 type derrick steel structure.

Figure 2. Structural displacement under equivalent static wind load.

Figure 3. Structural stress under equivalent static wind load.

3.4 Time domain analysis of wind vibration response

Taking the simulated wind pressure as the load, the dynamic analysis under the 15-level wind load is carried out, and the wind load is applied in the X direction in Figure 1. The displacement response of the top is extracted, and the time history curve is drawn. As shown in Figure 4 (a), the maximum displacement of the top of the ZJ70 derrick steel structure is $0.74m$, which meets the displacement angle limit requirements in the AISC specification. The axial stress value of the inclined rod with the highest stress value can be extracted and the time history curve can be drawn in Figure 3. As shown in Figure 4 (b), the average axial stress of the inclined rod is $67.133 MPa$, and the maximum axial stress is $283.79 MPa$, which exceeds the allowable tensile force. Therefore, the ZJ70 derrick steel structure under the wind load of grade 15 is outside the safe range.

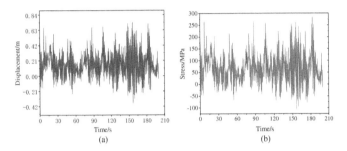

Figure 4. (a) Displacement time-history curve. (b) Stress response time history curve.

In Figures 5 and 6, it can be seen that under the wind pressure of the security equipment, the stress response value of the horizontal bar axis is small, while the stress response value of the vertical bar and the oblique bar axis is large. As the main load-bearing member, the overall axial stress response decreases with the increase in height. The axial stress value of the inclined rod unit in the middle of the derrick steel structure is significantly higher than that of other parts.

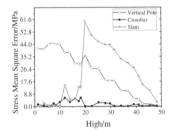 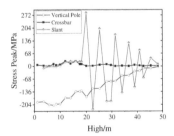

Figure 5. Mean square error of element stress at different heights.

Figure 6. Peak stress of element at different heights.

4 INFLUENCE OF DIFFERENT FACTORS ON WIND VIBRATION RESPONSE OF DERRICK STEEL STRUCTURE

4.1 Wind level

The average wind speed under different winds at a height of $10m$ in 2 minutes is shown in Table 1:

Table 1. Wind speed values under different wind levels.

Wind level	7	9	11	13	15
Beaufort wind speed (m/s)	16	22	30	40	47.8
Calculate wind speed (m/s)	17.3	24.1	33.2	44.5	53.4

The calculated wind speed in Table 1 is taken as the average wind speed at the reference height, the displacement at the top of the derrick steel structure, and the stress of the vertical and inclined rods are extracted for analysis. Figure 7 shows the mean and peak values of the top displacement of the derrick steel structure under different wind loads. Under the wind load of Level 7, the average displacement of the top of the derrick steel structure is $0.0131m$, and the peak value is $0.0539m$. Based on the wind-induced response, the average displacement increases by 78% ~ 378%, and the peak value increases by 79% ~ 468% for every 2 levels of wind increase. Figures 8 and 9 show the peak stress of different members. The results show that for the bottom bar, the increase in load wind level has a great influence on the stress of the vertical bar, while for the middle bar, the increase in load wind level will lead to a large increase in the stress of the vertical bar and inclined bar.

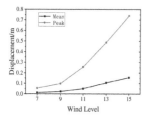
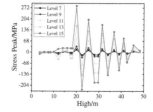

Figure 7. Displacement response for different wind levels.

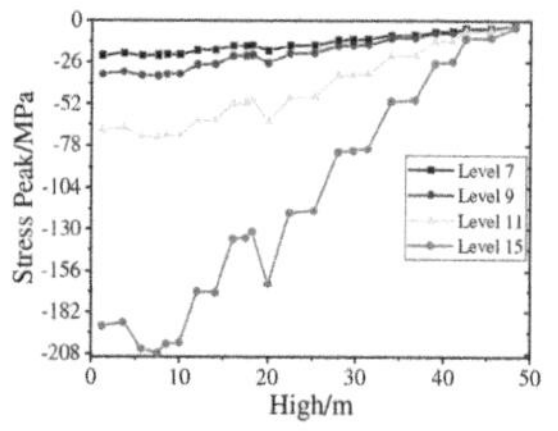

Figure 8. Peak stress response of vertical poles at different wind levels.

Figure 9. Peak stress of inclined bar under different wind forces.

4.2 Damage

4.2.1 Overall damage

The overall damage simulation of the derrick steel structure model is carried out, and the damage degree is 0~20%. The wind vibration analysis is carried out by applying a 13-level leeward load. It can be seen from Figure 10 that the maximum displacement response at the top is $0.503m$ when there is no damage. When the overall damage degree reaches 20%, the maximum displacement at the top increases to $0.893m$, increasing by 77.5%. Therefore, the overall damage has a greater impact on the safety of the derrick steel structure under wind load. Figures 11 and 12 show the peak stress of each vertical and inclined bar under damage. As the main component bearing the gravity of the derrick, the bottom pole of the derrick is greatly affected by the damage. When the overall damage is 20%, the axial stress of the bottom multiple vertical rods exceeds the allowable compressive stress of the derrick steel structure rods, which is unsafe. When the overall damage degree reaches 15%, the axial stress of the inclined rod at the height of $26m$ reaches $237MPa$, which exceeds the allowable tensile stress.

4.2.2 Local damage

It is known from the previous article that under the wind load, the stress of the inclined rod is the largest at the height of $26m$. Therefore, the inclined rod is selected as the damaged

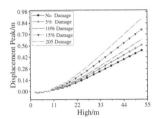 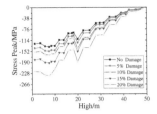 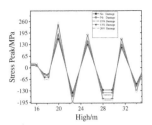

Figure 10. The peak displacement of each node of the derrick under different damage degrees.

Figure 11. The peak stress of vertical rod unit under different damage degrees.

Figure 12. The peak stress of inclined rod unit under different damage degrees.

component when analyzing the wind vibration response of the derrick steel structure under local damage. The damage degree is 0–20%, a 13-level back-wind load is applied, and the wind vibration response analysis is carried out. As shown in Figure 13, the damage degree of the column increases by 2.5%, the average displacement response of the derrick top increases by only 0.02%, and the peak displacement response increases by 0.06%. It is shown in Figure 14 that as the damage degree of the member element deepens, its own stress response peak value increases greatly. Under the 13-level backwind load, the stress peak value of the member without damage is $159.6 MPa$. When the damage degree reaches 20%, the peak stress is $224 MPa$, which exceeds the allowable tensile stress. The damage of the local element of the derrick steel structure has a weak influence on the overall displacement response but has a greater influence on its stress value.

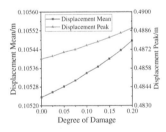

Figure 13. Displacement response under different damage degrees.

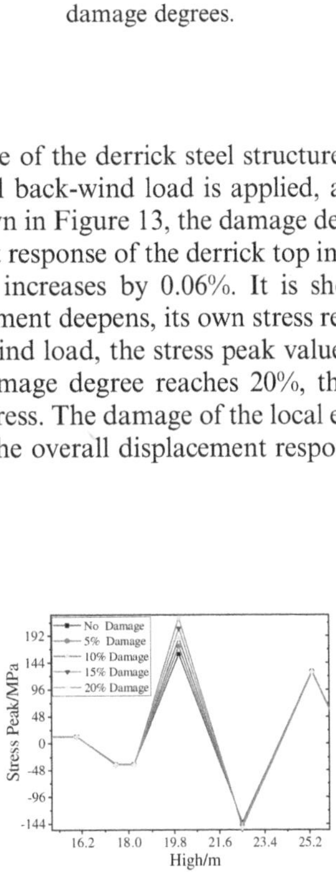

Figure 14. Peak stress response under different damage levels.

5 CONCLUSIONS

(1) The peak value of the response of the derrick steel structure in the time domain analysis of wind vibration is much larger than the response value of the equivalent static load analysis, so it is necessary to analyze the wind vibration response of the derrick steel structure in the time domain.

(2) Under the condition of the security equipment, the middle and lower part of the steel structure of the ZJ70 derrick is relatively stressed. Among them, the stress value of the inclined rod in the middle is the highest, which has exceeded the allowable stress value of the specification. It is in an unsafe state, which shows that the strength of the middle

structure should be strengthened in the design of the derrick steel structure to ensure the safety of the derrick during service.
(3) The wind grade of the wind load has a great influence on the wind vibration response of the derrick steel structure. Therefore, during the service process of the derrick steel structure, it is necessary to pay close attention to the changes in the local environmental wind and adjust the working state of the derrick steel structure in advance according to the wind level prediction.
(4) The damage to the rod will greatly reduce the safety of the derrick steel structure under the action of wind load, especially the local damage that is not easy to detect. Therefore, it is of great significance to strengthen the investigation of local damage to the derrick steel structure and deal with the damaged rods in time, which is of great significance to improve the wind resistance safety of the derrick steel structure.

ACKNOWLEDGMENTS

The studies were funded by the National Natural Science Foundation of China (Grant number 51875500), the Natural Science Foundation of Hebei Province (Grant number E2020203147), Other projects of Hebei Province Department of Science and Technology (Grant number 216Z4301G).

REFERENCES

An Liqiang, Zhang Zhiqiang, Huang Renmou, et al. Dynamic Response Analysis of A Transmission Tower-Line System Under Typhoon [J]. *Journal of Vibration and Shock*, 2017, 36(23): 255–262.
Hu Xue-lian, Li Zheng-liang, Yan Zhi-tao. Simulation of Wind Loading for Large-Span Bridge Structures [J]. *Journal of Chongqing Jianzhu University*, 2005, 27(03): 63–67.
Huang Guo-sheng, Liu Shutang, Han Lin-tian. Numerical Simulation of Fluctuating Wind Fields for Transmission Tower Based on Shiyuan Typhoon Wind Speed Spectrum [J]. *South China Journal of Seismology*, 2014, 34(S1): 71–75.
Lou Wenjuan, Xia Liang, Jiang Ying, et al. Wind-Induced Response and Wind Load Factor of Transmission Tower Under Terrain Wind Field and Typhoon Wind Field [J]. *Journal of Vibration and Shock*, 2013, 32 (06): 13–17.
Wang Hui, Wang Le, Tian Run-ze. Structural Damage Detection Using Correlation Functions of Time Domain Vibration Responses and Data Fusion [J]. *Engineering Mechanics*, 2020, 37(09): 30–37+111.
Zhang Liang-Chen, Lyu Ling-Yi. Study on Wind Vibration Coefficient of Annular Open-Web Cable-Truss Structures Based On Ar Model [J]. *Engineering Construction*, 2018, 50(02): 1–5.

Study on overall and local mechanical properties of tower-beam longitudinal restraint of three tower cable-stayed bridge

Junhao Tong*
Guangdong Provincial Highway Construction Co., Ltd., Guangzhou, China

Pan Wu* & Yaoyu Zhu*
China Communications Construction Company Highway Bridges National Engineering Research Center Co., Ltd., Beijing, China

ABSTRACT: The Huangmaohai Bridge will become the world's largest single-column tower three-pylon cable-stayed bridge with double cable planes. The longitudinal restraint system and connection structure of the tower beam have a great influence on the static and dynamic performance of the structure. Based on the Huangmaohai Bridge, this paper has studied the superiority of the longitudinal elastic restraint system in the middle tower beam of the Huangmaohai Bridge through the overall finite element model and analyzed the influence of elastic cable stiffness. The Huangmaohai Bridge finally connects the elastic cables to the tower and the transverse connection box respectively. The stability of the transverse connection box under the maximum cable force and the most unfavorable cable replacement condition is verified by establishing a local finite element model. The results show that: ① The longitudinal elastic restraint system has better vertical and longitudinal stiffness, and the force is more reasonable under E2 earthquake; ② When the elastic cable stiffness of the middle tower is greater than 5e5kN/m, the beam end displacement and the bending moment of the middle tower bottom change slightly; ③ Under the combined response of initial cable force and E2 earthquake and the most unfavorable condition of cable replacement, the stress level of the transverse connection box is low and the structure is safe.

1 INTRODUCTION

The longitudinal restraint system of the cable-stayed bridge tower beam has a great influence on the static and dynamic performance of the structure. The connection structure between the tower and beam is the core structure that reflects the characteristics of the cable-stayed bridge structure system and realizes the normal operation of the system. It is also the key content of cable-stayed bridge design. At present, in the cable-stayed bridges that have been built at home and abroad, there are four main types of tower-beam longitudinal connection systems commonly used: tower-beam consolidation rigid restraint system, the tower-beam full floating system without longitudinal restraint, tower-beam longitudinal damping connection system, and tower-beam longitudinal elastic restraint system.

The first three tower-beam longitudinal connection systems have their advantages and disadvantages: ① The rigid constraint system of tower-beam consolidation can effectively reduce the longitudinal displacement of the beam end and the tower top, but the structure of the intersection of the tower and the beam is more complicated, and it will increase the

*Corresponding Authors: 871829267@qq.com, 826094622@qq.com and zhuyaoy@sina.com

longitudinal response of the tower bottom under earthquake and temperature effects; ② The longitudinal stiffness of the structural system is low and the displacement of beam end is large under earthquake, although the Tower-beam without longitudinal constraint fully floating system avoids the consolidation node of tower-beam and improves the bending moment of main tower bottom under earthquake; ③ The longitudinal damping connection system between tower and beam can improve the dynamic performance of the main bridge, but cannot effectively enhance the longitudinal stiffness of the structure.

In summary, some scholars have carried out some research based on the longitudinal elastic constraint system of tower-beam, but the overall research is less. So, this paper will be conducted based on the Huangmaohai Bridge (single-column tower double-plane three-tower cable-stayed bridge), and the study on overall and local mechanical properties of tower-beam longitudinal restraint of three towers cable-stayed bridge.

2 ENGINEERING SITUATION

The Huangmaohai cross-sea channel project is located in Guangdong Province, with a total length of 31.26 km, connecting Zhuhai city and Jiangmen city. It is a key project in the west extension of the Hong Kong-Zhuhai-Macao Bridge. The span arrangement of the Huangmaohai Bridge is (100 + 280 + 2 × 720 + 280 + 100) m. After completion, it will become the world's largest single-column tower three-pylon cable-stayed bridge with double cable planes. The main beam of the bridge adopts a separated steel box beam, which is composed of two steel box beams and a transverse connection box. The steel box beam is 4m high, as shown in Figure 1. The longitudinal restraint system and connection structure of the three-tower cable-stayed bridge are one of the key design contents of the bridge, which will be studied in this paper.

Figure 1. The Huangmaohai bridge.

3 ANALYZES OF OVERALL MECHANICAL PROPERTIES

3.1 *Overall finite element model*

The static and dynamic calculation model of the Huangmaohai Bridge is established by MIDAS finite element software. The tower, main beam, and pier are simulated by the space beam element, the cable is simulated by the cable element in tension space truss element only, and 6 degrees of freedom springs are set at the bottom of the tower and the bottom of the pier. The design load is highway-I, the lateral is considered as 8 lanes, and the E2 seismic response is calculated concerning the seismic safety assessment report. The peak acceleration of the ground motion is 0.177g. The overall finite element model is shown in Figure 2.

Figure 2. Overall finite element model of the Huangmaohai bridge.

3.2 *Comparison of different longitudinal restraint systems*

As shown in Table 1, the finite element model is used to compare the vertical stiffness and longitudinal stiffness of the structure of the longitudinal elastic restraint system (elastic cable stiffness K = 1e5kN/m) of the middle tower beam with the vertical stiffness and longitudinal stiffness of the tower beam consolidation system and the full floating system of the tower beam, and the bending moment of the tower bottom of the middle side tower. The calculation results show that: ① The full floating system of the tower beam will cause the vertical and longitudinal stiffness of the bridge to be too low to meet the requirements of the specification; ② The tower-beam consolidation system has better vertical and longitudinal stiffness, but the bending moment of tower bottom is larger under E2 earthquake. The bending moment of the middle tower bottom is 5902MN·m, the bending moment of the side tower bottom is 1864 MN·m, and the bending moment of the middle tower bottom is 3.17 times of the side tower bottom, which is unreasonable; ③ The longitudinal elastic restraint system (elastic cable stiffness K = 1e5kN/m) has good vertical and longitudinal stiffness, and the bending moment of the middle tower bottom is 4618MN·m. The bending moment of the edge tower bottom is 1626 MN·m, and the bending moment of the middle tower bottom is 2.84 times the bending moment of the edge tower bottom, which is more reasonable. In summary, the longitudinal elastic constraint system is the most reasonable, so the longitudinal elastic constraint scheme is finally adopted at the middle tower of the Huangmaohai Bridge.

Table 1. Comparison of structural overall stiffness under different longitudinal restraint systems.

Mechanical property index		Longitudinal restraint systems		
		The full floating restraint system	Elastic restraint system (K = 1e5kN/m)	Rigid restraint system
Vertical stiffness under live load	Maximum deflection (m)	0.785	0.726	0.718
	Minimum deflection (m)	−1.152	−1.110	−1.103
	The ratio of deflection under the span	372	392	395
Longitudinal stiffness under E2 seismic response	The maximum displacement of beam end (m)	1.481	0.514	0.206
	Minimum displacement of beam end (m)	−1.521	−0.536	−0.242
A bending moment under E2 seismic response	Bottom of the middle tower (kN·m)	3548780	4618280	5901850
	Bottom of the side tower (kN·m)	1534660	1626290	1864000

3.3 Influence of elastic cable stiffness

Based on the whole model of the elastic longitudinal restraint system, the influence of elastic cable stiffness on the static and dynamic performance of the structure is further studied when the elastic cable stiffness increases from 1e4kN/m to 1e8kN/m. As shown in Figures 3 and 4 below, with the increase of the stiffness of the elastic cable, beam end displacement gradually decreases, the bending moment at the bottom of the side tower changes slightly, and the bending moment at the bottom of the middle tower gradually increases. When the elastic cable stiffness of the middle tower is greater than 5e5kN/m, beam end displacement and the bending moment at the bottom of the middle tower change slightly. Considering the static response and seismic response, the stiffness of the elastic cable of the middle tower of the Huangmaohai Bridge is determined to be 7.8e5kN/m. A total of 16 elastic cables are used in the tower and beam, and the initial maximum cable force of a single elastic cable is 4800kN.

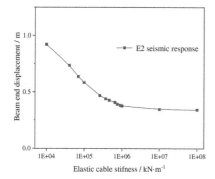
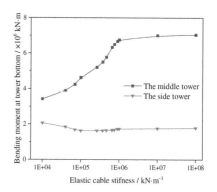

Figure 3. Variation of beam end displacement under different elastic cable stiffness.

Figure 4. Variation of bending moment at tower bottom under different elastic cable stiffness.

4 ANALYZES OF LOCAL MECHANICAL PROPERTIES

The longitudinal elastic constraint system is finally adopted at the middle tower beam of the Huangmaohai Bridge. As shown in Figure 5, one end of the elastic cable is anchored on the cable tower through the conventional steel anchor beam, and the other end is anchored on the transverse connection box of the main beam in two batches through the transverse connection box. To ensure that the cable is still in the tension state under the most unfavorable working conditions, the initial maximum cable force of each elastic cable is 4800kN, and the maximum cable force of the elastic cable is 8690kN under the combined response of the initial tension and E2 earthquake. The cable force of elastic cable has little influence on

Figure 5. 3D schematic diagram of longitudinal elastic cable restraint system of tower beam.

the tower, but it may have some influence on the transverse connection box of the main beam. Therefore, it is necessary to establish a local finite element model to analyze the stability of the transverse connection box under elastic cable force.

4.1 Local finite element model

The ABAQUS finite element software is used to establish the local model of the transverse connection box of the Huangmaohai Bridge, as is shown in Figure 6. There is a stiffened partition with a human-shaped hole in the box, and the steel plate adopts the S4R four-node curved surface reduced integral shell element to specify the corresponding thickness. Fixing the two ends of the transverse connection box, the initial cable force of the elastic cable is 4800kN at the anchorage position of each elastic cable, and the maximum cable force of a single elastic cable under the combined response of the initial tension and E2 earthquake is 8690kN.

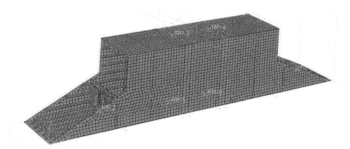

Figure 6. Local finite element model of the transverse connection box.

4.2 Analyzes of local model results

Figure 7 shows that under the combined response of initial tension and E2 earthquake (the maximum tension of a single elastic cable is 8690kN), the stress program of the transverse connection box shows that the maximum stress of the steel plate is only 166MPa, which is mainly distributed near the anchorage point of the elastic cable and the top edge of the web. In general, the stress level of the steel plate of the transverse connection box is low and the structure is safe.

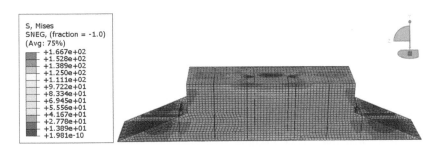

Figure 7. Stress distribution of transverse connection box under maximum design tension.

Figure 8 is the stress distribution at the initial yield. It can be found that most of the roof of the box reaches the yield stress of 355MPa, and the corresponding elastic cable tension reaches 19630kN. The safety factor of the transverse connection box reaches 2.25, which is safer.

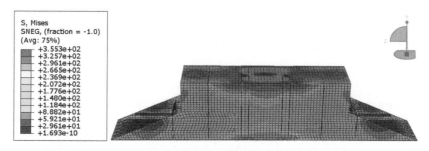

Figure 8. Stress distribution of transverse connection box under steel plate yield.

4.3 *Checking calculation of elastic cable replacement*

After the operation state of the elastic cable of the tower beam is degraded for some time, the elastic cable needs to be replaced. When the elastic cable is replaced, the elastic cable force of the top plate and the bottom plate of the transverse connection box may be asymmetric. Therefore, the influence of the most unfavorable working condition of replacing the elastic cable of the bottom plate at the same time on the force of the transverse connection box is considered.

Figure 9 is the stress distribution of the top plate of the transverse connection box under the most unfavorable working condition of the simultaneous replacement of the elastic cable of the bottom plate. It can be found that the stress of the transverse connection box is generally small, and the maximum stress of the top plate and the bottom plate is 91MPa, indicating that the elastic cable replacement has little effect on the transverse connection box. The elastic cable can be replaced arbitrarily in the subsequent operation period.

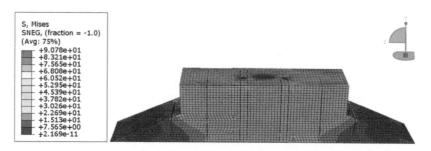

Figure 9. Stress distribution of transverse connection box top plate roof under elastic cable replacement condition.

5 CONCLUSIONS

Based on the Huangmaohai bridge, this paper studied the overall and local mechanical properties of three tower cable-stayed bridges with longitudinal elastic restraints of tower-beam, and drew the following conclusions:

(1) Compared with the full-floating system of tower-beam and the full-floating system of tower-beam, the longitudinal elastic restraint system of the middle tower beam has better vertical and longitudinal stiffness. The force is more reasonable under the response to the E2 earthquake, which is the final longitudinal restraint scheme of the middle tower and beam of Huangmaohai Bridge;

(2) With the increase of the stiffness of the elastic cable, beam end displacement decreases gradually, the bending moment at the bottom of the side tower changes slightly, and the bending moment at the bottom of the middle tower increases gradually. However, when the stiffness of the elastic cable in the middle tower is greater than 5e5kN/m, beam end displacement and the bending moment at the bottom of the middle tower change slightly. Finally, the stiffness of the elastic cable in the middle tower of Huangmaohai Bridge is 7.8e5kN/m, and the initial tension of a single elastic cable is 4800kN;

(3) The local finite element analysis of the transverse connection box based on the anchored elastic cable is carried out. Under the combined response of the initial cable force and the E2 earthquake (the maximum tension of the elastic cable is 8690kN), the maximum stress of the transverse connection box is only 166MPa, and the maximum stress appears on the top plate of the transverse connection box;

(4) The most unfavorable replacement conditions of the elastic cable during the operation period are checked and analyzed. The results show that the replacement of the elastic cable has little effect on the transverse connection box, and the elastic cable can be replaced at will during the subsequent operation period.

REFERENCES

Chen, Y. H. (2013). Analysis of The Static and Dynamic Characteristics of Steel Arch Tower Cable-Stayed Bridge Affected By Tower Beam Connection Mode. *Journal of Highway and Transportation Research and Development* 9(01): 102–106.

Huang, Y. F. (2022). Study on Restraint System of Double Tower Cable-Stayed Bridge with High Pier in the Mountainous Area. *Value Engineering* 41(13): 94–96.

Jiang, X. P. (2020). Longitudinal Seismic Response of Multi-Pylon Partially Cable-Stayed Bridge. *Highway* 65 (02):90–95.

Jiao, C. Y. (2009). Influence of Tower Beam Connection Mode on Seismic Response of Long-Span Cable-Stayed Bridges. *Journal of Vibration and Shock* 28(10): 179–184+233–234.

Wang, Z. W. (2021). Research on The Design of Longitudinal Restraint System of A Long-Span Three-Tower Cable-Stayed Bridge. *World Bridges* 49(4): 42–48.

Xu, L. P. (2003). The Structural System Analyzes for Super-Long Span Cable-Stayed Bridges 31(4): 400–403.

Xu, X. D. (2020). Study on Seismic Reduction Effect of A Three-Pylon Cable-Stayed Bridge with Super-High Pier Using Elastic Cables. *World Earthquake Engineering* 36(2): 138–145.

Zhang, Y. L. (2011). Influence of Longitudinal Elastic Constraints on Dynamic Characteristics and Seismic Response of Railway Cable-Stayed Bridge. *Journal of Railway Science and Engineering* 8(2): 21–26.

Zheng, F. L. (2020). Influence of Pylon-Beam Connection on Seismic Response of The Multi-Pylon Cable-Stayed Bridge. *Journal of Highway and Transportation Research and Development* 37(08): 58–65.

Zhou, K. (2018). Influence of Tower-Beam Connection Method on Longitudinal Seismic Response of Multi-Tower Cable-Stayed Bridges. *Structural Engineers* 34(S1): 31–36.

Damage and permeability of surrounding rocks during blasting excavation of tunnels in karst areas

Wen Hao Li*
School of Safety Science and Emergency Management, Wuhan University of Technology, Wuhan, China

Deng Xing Qu
School of Civil Engineering and Architecture, Wuhan University of Technology, Wuhan, China

Qi Yue Zhang
School of Safety Science and Emergency Management, Wuhan University of Technology, Wuhan, China

Ben Yang
China Gezhouba Group No.1 Engineering Co., Ltd., Yichang, China

ABSTRACT: This research aimed to tackle practical engineering problems including high geo stress, high seepage pressure, and dynamic disturbance in the blasting excavation of the Yuelongmen tunnel. Numerical simulation software COMSOL Multiphysics was used to establish a numerical model based on the Mohr-Coulomb failure criterion and the theory of continuum damage mechanics. On this basis, the Spatio-temporal evolution of the stress field, seepage field, and damage field of the tunnel under different seepage water pressures and fault conditions during blasting excavation was investigated. The results show that the seepage water pressure has significant influences on water inflow in the tunnel while exerting small influences on the severity of damage to the surrounding rocks. As the fault-zone length increase, damage to the surrounding rock and water inflow in the tunnel both changes significantly and the fault-zone thickness exerts larger influences on water inflow to the tunnel. If the fault is located above the tunnel, the seepage water pressure has the greatest seepage effect on the tunnel and water inflow in the tunnel is also maximized.

1 INTRODUCTION

The development and use of underground spaces have received much attention from geotechnical engineers. Tunnels are a common form of utilization of underground space in geotechnical engineering, and long and deep tunnels crossing regions in a complex geological environment are crucial to controlling engineering works and their safety (Hadi et al. 2018) (Przemyslaw Bukowski 2011). Deep tunnels constructed in karst areas are characterized by high geo stress and seepage pressure that are different from shallow engineering works. Under the effects of blasting excavation and mechanical vibration, disasters including rock damage, even collapse, and water inflow may be triggered in the construction process of such tunnels. Blasting excavation changes the original mechanical equilibrium of surrounding rocks and triggers the instantaneous release of the energy stored in the rocks. Disasters including collapse and water inflow in tunnels are the results because of the coupling of

---

*Corresponding Author: 375213316@qq.com

multiple factors, which will cause serious engineering disasters once treated inappropriately and even cause secondary environmental disasters (Golian et al. 2018; Odintsev et al. 2016; Xie et al. 2019).

In the construction process, blasting excavation and mechanical vibration may cause irreversible damage to nearby rocks. Apart from blasting shock waves, surrounding rocks are also influenced by stress redistribution due to the instantaneous release of initial geo stress (Sunny Murmu et al. 2018). Different blasting modes exert different influences on the evolution of stress and damage fields in surrounding rocks. Yang et al. (Yang et al. 2018, 2016) and Umit Ozer, et al. established different numerical models to evaluate the influences of different blasting patterns on stress distribution and damage evolution. Aiming at different damaged forms of rocks, different scholars have proposed various damage models. For example, Taylor et al. constructed the TCK model and Grady et al. (1980) established the GK model. Blasting excavation causes pre-existing fractures in rocks to further develop, propagate, coalesce, and produce new fractures, which weakens the strength of rocks and causes non-negligible risks to engineering construction (Abierdi et al. 2020; Ömer Aydan 2013; Simha 1986; Siren et al. 2015, Xu et al. 2019, Yang et al. 2019). Abierdi et al. (2020) and Yang et al. (2019) separately investigated failure modes of rocks and fracture development during blasting excavation at the tunnel scale and triaxial compression at the sample scale. Many scholars have performed in-depth research on the results of single-blasting excavation. However, multi-deck continuous blasting is generally used in engineering practice (Hamis Salum Amiri et al. 2019; Hu et al. 2015; Ji et al. 2021). While causing pre-existing fractures in surrounding rock to develop, blasting excavation also facilitates the generation of new fractures, thus changing the distribution of the seepage field therein. Blasting excavation changes the stress and seepage fields in nearby rocks and the coupling mechanism of the two is one of the foci in current geotechnical engineering, for which numerical simulation software FLAC, ABAQUS, and COMSOL Multiphysics can be the main choices. Due to the complex and diverse geological conditions in China, construction under different geological conditions faces different difficulties. To solve water inflow during the excavation of tunnels, some researchers established fluid-structure interaction models of tunnels considering many different conditions including the geo stress, seepage pressure, complex fractures, and size and location of karst caves.

At present, high geo-stress environments can be simulated using existing equipment, while existing equipment is unable to simulate the high seepage pressure, which is one of the research foci in geotechnical mechanics. The research took the practical engineering problems including high geo stress, high seepage pressure, and dynamic disturbance faced in the blasting excavation process of the Yuelongmen tunnel as the background. The relevant numerical model was established using the numerical simulation software COMSOL Multiphysics, the Mohr-Coulomb failure criterion, and the theory of continuum damage mechanics.

2 ENGINEERING OVERVIEW OF THE YUELONGMEN TUNNEL

2.1 *Engineering overview*

The Yuelongmen tunnel is located between Gaochuan and Maoxian railway stations, comprising left and right tunnels that are separately 19,981 and 20,042 m long. The two tunnels were constructed separately, with a maximum burial depth of 1445 m. The entrance of the tunnel adjoins the two-lane bridge in Gaochuan County, where the tracks at Gaochuan railway station separately extend into the left and right tunnels. The tunnels are designed to have a single-free-face upslope. Due to being located in an area of so complex geological conditions, the construction of the Yuelongmen tunnel faced a series of engineering problems including seismically active fault zones, fault fractured zones, rock bursts, high geo temperature, large deformation of soft rocks with high geo stress, noxious gas, and water-rich karst structures. Therefore, it is a tunnel project at extremely high risk of several types of geological accidents.

2.2 Geo stress test

Among the mainstream methods for testing stress, the common ones are hydraulic fracturing and stress relieving methods. According to comparisons of previous tests, the hydraulic fracturing method (giving the best effect) was applied to test the geo stress at the Yuelongmen tunnel. Boreholes for measuring the geo stress were arranged at D2K100 +059.4 and their diameter and depth were 56 mm and 35 m, respectively. The test results are listed in Table 1.

Table 1. Measured geo stresses.

Depth of test section (m)	Rupture pressure (MPa)	Retention pressure (Mpa)	Close pressure (Mpa)	Head pressure (Mpa)	Maximum horizontal principal stress (Mpa)	Minimum horizontal principal stress (Mpa)	Vertical stress (Mpa)	The Maximum horizontal principal stress direction
29.1	16.89	16.16	13.40	0.29	24.62	13.69	27.76	N35.0°E
17.9	16.92	13.24	11.39	0.28	21.49	11.67	27.73	
26.8	12.56	11.84	10.56	0.27	20.38	10.83	27.70	
25.7	14.04	11.39	10.35	0.26	20.17	10.61	27.67	N14.7°E
23.2	12.72	11.11	10.09	0.23	19.62	10.32	27.60	N43.5°E

2.3 Field water inflow

Groundwater at the Yuelongmen tunnel mainly comes from pore water and karstic water, which mainly comes from atmospheric precipitation and fracture water in rocks. Field prediction shows that the normal water inflow to the Yuelongmen tunnel is 1.19×10^5 m^3/d and the maximum water inflow is 1.79×10^5 m^3/d. Water inflow to some tunnel sections is shown in Table 2.

Table 2. Water inflow and lithology in some tunnel sections.

Depth of test section	Stratigraphic lithology	Total stratigraphic length (m)	Surge capacity Q (m^3/d)
D2K96+770~D2K97+000	Sandstone, apatite	230	2344
D2K97+000~D2K98+100	Slate, Shale	1100	3716
D2K98+100~D2K98+780	Diabase	680	2776
D2K100+540~D2K101+600	Slate, Shale	1060	3802
D2K105+065~D2K105+170	Faults	105	3769
D2K105+170~D2K107+220	Slate, Shale	2050	8757
D2K107+220~D2K107+270	Sandstone, apatite	50	575
D2K110+150~D2K110+994	Chimei Rock	844	389

3 NUMERICAL SIMULATIONS OF THE DAMAGE AND WATER INFLOW MECHANISM IN THE ROCK SURROUNDING THE TUNNEL

3.1 Damage and seepage mechanism of surrounding rocks under blasting excavation

3.1.1 Evolution equation of damage

In the blasting excavation process, surrounding rocks of the tunnel were subject to plastic strain under coupled action of the geo stress, blast load, and pore pressure. The equivalent plastic strain is expressed as follows (Jia et al. 2009):

$$\bar{\varepsilon}_p = \frac{\sqrt{2}}{3}\sqrt{(\varepsilon_{p1}-\varepsilon_{p2})^2 + (\varepsilon_{p2}-\varepsilon_{p3})^2 + (\varepsilon_{p3}-\varepsilon_{p1})^2} \qquad (1)$$

where ε_{p1}, ε_{p2}, and ε_{p3} separately denote the principal plastic strains in three directions.

The equivalent plastic strain is normalized, that is:

$$D = A_1 e^{-\bar{\varepsilon}_{pm}/a} + B_1 \qquad (2)$$

where $\bar{\varepsilon}_{pm}$ represents the normalized equivalent plastic strain; a, A_1, and B_1 refer to material parameters.

3.1.2 Evolution equation of seepage

Changes in the permeability coefficient of surrounding rocks of the tunnel in the blasting excavation process are supposed to be mainly caused by the closure of pores and local damage. The Kozeny-Carman model is a seepage model in which the permeability varies due to changes in the porosity, and it is expressed as follows:

$$K_p = c \frac{n^3}{(1-n)^2} \qquad (3)$$

where K_p represents changes in the permeability coefficient caused by the closure of pores; c is a constant applied to such mesostructures; n denotes the porosity.

Supposing that rock particles are incompressible, then the volumetric strain therein is:

$$\varepsilon_v = \frac{\Delta V}{V_0} = \frac{\Delta V_p}{V_0} = \frac{V_s \Delta e}{V_s(1+e_0)} = \frac{\Delta e}{1+e_0} = \frac{e-e_0}{1+e_0} \qquad (4)$$

where V_p and V_s separately represent the volumes of pores and rock particles; e_0 and Δe denote the initial void ratio and changes thereof, respectively.

The following relationship between the porosity and volumetric strain is obtained according to the relationship between the void ratio and porosity:

$$k = k_0 \left[\frac{1}{n_0}(1+\varepsilon_v)^3 - \frac{1-n_0}{n_0}(1+\varepsilon_v)^{-1/3} \right]^3 \qquad (5)$$

where n_0 denotes the initial porosity.

The permeability coefficient of surrounding rocks changes constantly during the excavation of tunnels. In terms of changes in permeability caused by local damage, damaged surrounding rocks have a lower strength. By referring to the cubic law for seepage, the evolution of the permeability of the surrounding rocks is:

$$k = (1-D)k^M + Dk^N \left(1+\varepsilon_V^{pF}\right)^3 \qquad (6)$$

where D represents the damage variable; k^M and k^N separately denote the coefficients of permeability of undamaged and damaged rocks; ε_V^{pF} is the plastic volumetric strain in the damaged surrounding rocks.

3.2 Dimensions and parameters of the numerical model

Combining with the engineering and geological data of the Yuelongmen tunnel, a three-dimensional model measuring 100 m × 100 m × 100 m (length × width × height) was established. The tunnel was designed to have a cross-section in the shape of a straight-sided arch: the upper part was a semicircle with a diameter of 10 m and the lower part was a rectangle with a length of 10 m and width of 7 m. The initial footage for excavation of the tunnel was 30 m and the tunnel was excavated over 2 m after a single blasting operation. Three irregular parameterized surfaces were imported at the defined interfaces to simulate faults and ten different elliptical models were inserted randomly in the model to simulate fractures. The established model is displayed in Figure 1.

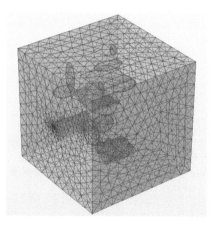

Figure 1. Numerical model.

The numerical model was selected according to geological reports, field samples, and laboratory test results, to calculate the selected parameters. The Hoek-Brown strength criterion was selected. Table 3 lists the initial physicomechanical parameters of the rocks.

Table 3. Mechanical and seepage parameters of rocks.

Rock character	Uniaxial compressive strength (MPa)	Young's modulus (Gpa)	Poisson's ratio (v)	Density (kg/m³)	Initial porosity	Initial permeability	Fluid density (kg/m³)	Fluid compressibility
Slate	75	30	0.25	2 700	0.1	1.2e-9	1 000	4e-10

3.3 Curves of blast load and determination of peak blast load

3.3.1 Curves of blast load

The commonly used blast load curves include parabolic and triangular ones. The two types of blast load curves demonstrate different characteristics and are displayed in Figure 2. For the convenience of calculation, the triangular blast load curve was selected. The calculation takes 7 ms, and t_0 and t_1 separately represent 1 and 6 ms, as shown in Figure 2.

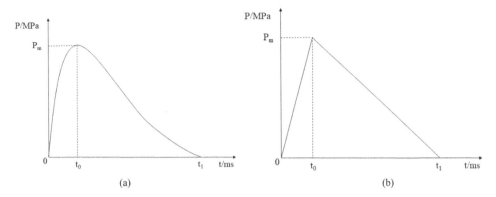

Figure 2. Blast load curves. (a) Parabolic blast load. (b) Triangular blast load.

3.3.2 *Peak blast load*

It is blasting shock waves that mainly act on blasthole walls during blasting. Supposing that blast load acts on blasthole walls in a form vertical to the free face, then the initial peak pressure produced by shock waves on blasthole walls is determined by Yang et al. (Yang et al. 2010):

$$P = \frac{\rho_e D^2}{2(\gamma + 1)} \left(\frac{d_i}{d_j}\right)^{2\gamma} \left(\frac{l_i}{l_j}\right) n \qquad (7)$$

The equivalent peak pressure is:

$$P_0 = \frac{d_i}{a} P \qquad (8)$$

where D represents the detonation velocity (set to 3600 m/s here); d_i and d_j separately denote the cartridge diameter and blast hole diameter, which are 0.03 m and 0.04 m, respectively; l_i and l_j separately represent the charge length and blast hole length and are separately 2.0 m and 2.5 m; γ is the isentropic exponent of explosive gas, which is 3 here; n is a constant in the range of 8 to 11; a refers to the hole spacing. The equivalent pressure is calculated to be about 60 MPa.

4 DAMAGE AND PERMEABILITY OF SURROUNDING ROCKS DURING BLASTING EXCAVATION OF THE TUNNEL

4.1 *Blasting excavation of the tunnel*

Rocks in nature have been solidified under their deadweight, while a few rocks, such as artificial backfill, have not been solidified under their deadweight when they will still settle naturally over time. In the model, the settlement caused by the deadweight stress of soil is not considered the better to align calculated data with actual conditions.

4.2 *Influences of seepage water pressure on damage and permeability of surrounding rocks*

To more intuitively study the results, five monitoring points were selected: the right arch foot, right spandrel, vault, left spandrel and left arch foot. Keeping other conditions

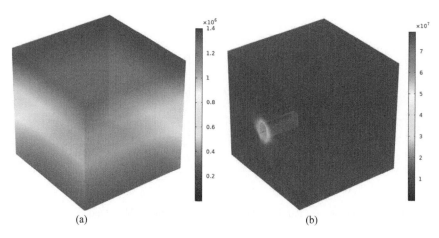

Figure 3. Stress distributions before and after excavation of the tunnel. (a) Stress distribution before excavation. (b) Stress distribution after excavation.

unchanged, the seepage water pressure was set to be 1, 2, 4, 6, and 10 MPa to study changes in damage to the surrounding rocks, Darcy's velocity, and water inflow under different seepage water pressures. The calculated results are displayed in Figures 4 and 5.

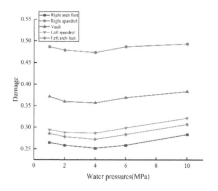
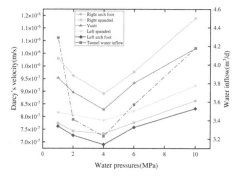

Figure 4. Changes in damage at monitoring points under different water pressures.

Figure 5. Darcy's velocity and water inflow in the tunnel at monitoring points under different seepage water pressures.

It can be seen from Figure 4 that damage at various monitoring points first decreases, then increases albeit with a narrow range with the gradual increase in the water pressure. This is because the presence of fissure water pressure in the early stage of increasing the water pressure can offset some of the impacts from the blast load; since the seepage water pressure is low, damage to the tunnel face is a little less severe. As the seepage water pressure increases constantly and reaches the critical value, the damage to the tunnel face begins to increase slightly under the coupled action of the seepage water pressure and blast load. Figure 5 shows that with the constant increase in the water pressure, Darcy's velocity at various monitoring points first decreases, then increases; it falls to a minimum under a seepage water pressure of 4 MPa and then begins to grow rapidly. This is because the increase in the seepage water pressure in fractures accelerates the flow velocity of fracture water from surrounding rocks and therefore leads to constantly increasing water inflow to the tunnel. The

Darcy velocity and water inflow in the tunnel at monitoring points are significantly influenced by changes in the seepage water pressure. *In-situ* monitoring data show that the average water inflow, horizontal geo stress, and vertical geo stress at chainages D2K100 +540 to D2K101+600 are 3.59 m^3/d, 20 to 25 MPa, and 27 MPa. When calculating the water inflow, the lateral pressure coefficient is set to 1. The numerical calculation yields a result of about 3.87 m^3/d using the model. Considering the complex lithology in the engineering site and that simulated fracture development does not completely conform to the practical situation, it is regarded that the measured water inflow in the tunnel is consistent with the calculated result. This finding further verifies the correctness of the model.

4.3 *Influences of the fault range on damage and permeability of surrounding rocks*

While keeping other conditions unchanged, the fault-zone length was set to 10, 20, 30, 40, and 50 m to study changes in the damage to the surrounding rock, Darcy's velocity, and water inflow to the tunnel under different fault-zone lengths. The results are shown in Figures 6 and 7.

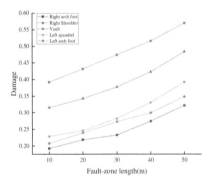

Figure 6. Changes in damage at monitoring points under different fault-zone lengths.

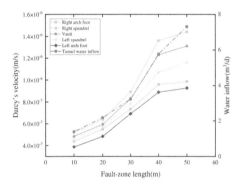

Figure 7. Darcy's velocity and water inflow at monitoring points under different fault-zone lengths.

As shown in Figure 6, damage at various monitoring points increases at a gradually rising rate with the increase in the fault-zone length, and the maximum damage always occurs at the right spandrel. This is because the seepage water pressure in faults exerts a more significant seepage effect on the surrounding rock with the increase in the fault-zone length. Under the coupled action of geo stress and blast load, surrounding rock is more likely to be damaged. In addition, many fractures are distributed near the right spandrel in the model, an area more affected by seepage, so damage thereat is the most severe. Figure 7 shows that, as the fault-zone length increases, Darcy's velocity and water inflow to the tunnel at various monitoring points also increase constantly. The rate of change of the Darcy velocity at various monitoring points gradually decreases while that of water inflow increases constantly. This is because, as the fault-zone length increases, the water content in the nearby fractures also increases, and fracture water in surrounding rocks flows out at a higher speed under the impact of the blast load, which results in the increased Darcy's velocity at various points. Meanwhile, the permeability coefficient of surrounding rocks also rises due to increasing damage. As a result, the water inflow to the whole tunnel constantly grows under the coupled action of the stress field, seepage field, and damage field.

4.4 Influences of relative locations of faults on damage and permeability of surrounding rocks

Figure 8 illustrates that the damage at various monitoring points is always maximized when the faults are located above the tunnel; while the damage at various monitoring points is always at a minimum when the faults are beneath the tunnel. This is because changes in the location of the faults affect the stress distribution in the surrounding rocks and cause changes in both the extent and severity of the damage at various monitoring points under the effects of blasting shock and geo stress. As shown in Figure 9, Darcy's velocity and water inflow at various monitoring points are maximized when the faults are above the tunnel, while they are minimized at various monitoring points if the faults are beneath the tunnel. The reason for this is that surrounding rock is damaged under the blast load and the permeability at various points also changes, leading to a more significant seepage effect. This causes Darcy's velocity and water inflow at various monitoring points to increase constantly. If the faults are located above the tunnel, the tunnel is subject to a greater influence from the seepage water pressure and water stored in the faults and fractures flows out more rapidly under the blast load. If the faults are located beneath the tunnel, the seepage water pressure exerts a weaker seepage effect on the tunnel due to gravity, so Darcy's velocity and the water inflow at various monitoring points also decrease accordingly.

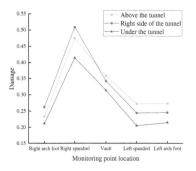
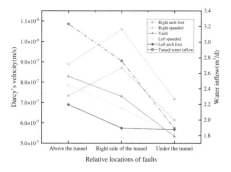

Figure 8. Changes in damage at monitoring points when the faults are at different locations.

Figure 9. Darcy's velocity and water inflow to the tunnel at monitoring points when the faults are at different locations.

5 CONCLUSION

This research aimed at solving practical engineering problems including the high geo stress, high seepage pressure, and dynamic disturbance during the blasting excavation of the Yuelongmen tunnel. The main conclusions are as follows:

1. As the blast load increases, damage first occurs around the tunnel face and the rate of change of the amount and severity of the damage to the tunnel face decreases upon reaching the peak blast load. Finally, the damage is concentrated on the spandrel.
2. With the constant increase in the seepage water pressure, fracture water flows out from surrounding rocks at a higher speed under the coupled action of blast load and geo stress. As a result, water inflow to the tunnel increases constantly, while the peak damage at monitoring points changes slightly.
3. The enlargement in the fault-zone length results in more abundant fracture water in the surrounding rocks, more significant damage to the surrounding rocks, and greater water inflow to the tunnel. Moreover, damage at various monitoring points also increases with increasing fault-zone length.

4. Changes in the relative location of faults also alter the stress distribution in the surrounding rock. Under the influences of the blast load, the permeability of the surrounding rocks also changes. The seepage effect of the seepage water pressure is the most significant on the tunnel and the water inflow to the tunnel is also maximized when the faults are located above the tunnel.

REFERENCES

Abierdi, et al. Laboratory Model Tests and DEM Simulations of Unloading-Induced Tunnel Failure Mechanism. *Cmc-Comput Mater Con* (2020); 63(2): 825–844.

Graday D. E., et al. Continuum Modeling of Explosive Fracture in Oil Shale. *International Journal of Rock Mechanics and Mining Sciences and Geomechanics Abstracts* (1980); 17(3): 147–157.

Hadi F., et al. Water Flow Into Tunnels in Discontinuous Rock: A Short Critical Review of the Analytical Solution of the Art. *Bulletin of Engineering Geology and the Environment* (2018).

Hamis Salum Amiri, et al. Optimising Blast Pulls and Controlling Blast-induced Excavation Damage Zone in Tunneling Through Varied Rock Classes. *Tunn Undergr Sp Tech* (2019); 85:307–318.

Hu Ying Guo, et al. Numerical Simulation of the Complete Rock Blasting Response by SPH–DAM–FEM Approach. *Simulation Modelling Practice and Theory* (2015); 56:55–68.

Ji Ling, et al. Numerical Studies on the Cumulative Damage Effects and Safety Criterion of a Large Cross-section Tunnel Induced by Single and Multiple Full-Scale Blasting. *Rock Mech Rock Eng* (2021); 54(12): 6393–6411.

Ji Ling, et al. Modeling Study of Cumulative Damage Effects and Safety Criterion of Surrounding Rock Under Multiple Full-face Blasting of a Large Cross-section Tunnel. *Int J Rock Mech Min* (2021); 147.

Jia Shan Po, et al. Research on Seepage-stress Coupling Damage Model of Boom Clay During Tunneling. *Rock and Soil Mechanics* (2009); 30(01):19–26.

Golian M., et al. Prediction of Water Inflow to Mechanized Tunnels During Tunnel-boring-machine Advance Using Numerical Simulation. *Hydrogeol J* (2018); 26(8):2827–2851.

Odintsev V. N., et al. Water Inrush in Mines as a Consequence of Spontaneous Hydrofracture. *Journal of Mining Science* (2016); 51(3):423–434.

Ömer Aydan. In Situ Stress Inference from Damage Around Blasted Holes. *Geosystem Engineering* (2013); 16 (1): 83–91.

Przemyslaw Bukowski. Water Hazard Assessment in Active Shafts in Upper Silesian Coal Basin Mines. *Mine Water Environ* (2011); 30(4):302–311.

Sunny Murmu, et al. Empirical and Probabilistic Analysis of Blast-induced Ground Vibrations. *Int J Rock Mech Min* (2018); 103:267–274.

Simha K. Studies on Explosively Driven Cracks Under Confining In-situ Stresses. *International Journal of Rock Mechanics and Mining Sciences and Geomechanics Abstracts* (1986); 23:84–84.

Siren T., et al. Considerations and Observations of Stress-induced and Construction-induced Excavation Damage Zone in Crystalline Rock. *Int J Rock Mech Min* (2015); 73:165–174.

Taylor L. M. Microcrack-induced Damage Accumulation in the Brittle Rock Under Dynamic Loading. *Computer Methods in Applied Mechanics and Engineering* (1986); 55:301–320.

Umit Ozer, et al. An Analytical and Applied to Blast Approach to Facilitate Rock-socketed Bored Pile Excavation. *Arab J Geosci* (2020); 13(12).

Xie He Ping, et al. Research Review of the State Key Research Development Program of China: Deep Rock Mechanics and Mining Theory. *Journal of China Coal Society* (2019); 44:1283–1305.

Xu Jin Hui, et al. Dynamic Characteristics and Safety Criterion of Deep Rock Mine Opening Under Blast Loading. *Int J Rock Mech Min* (2019); 119:156–167.

Yang Jian Hua, et al. Dynamic Stress Adjustment and Rock Damage During Blasting Excavation in a Deep-buried Circular Tunnel. *Tunn Undergr Sp Tech* (2018); 71:591–604.

Yang Jian Hua, et al. A Study on the Vibration Frequency of Blasting Excavation in Highly Stressed Rock Masses. *Rock Mech Rock Eng* (2016); 49(7):2825–2843.

Yang Sheng Qi, et al. Failure Behavior and Crack Evolution Mechanism of a Non-persistent jointed Rock Mass Containing a Circular Hole. *Int J Rock Mech Min* (2019); 114:101–121.

Yang Jian Hua, et al. Determination of the Blasting Load Variation in Borehole [C]// *The Second National Conference on Engineering Security and Protection.* (2010). 73–777.

Risk factors affecting tunnel collapse and treatment

Lipeng Yang*
Monitoring Company, Shenyang Metallurgical Inspection Engineering Co., Ltd., Liaoning, China

Tong Wang
Hebei GEO University, Shijiazhuang, China

ABSTRACT: Tunnel engineering has been advancing in the direction of growth in the past ten years. With the improvement in the country's comprehensive strength, more and more investments have been made in disaster prevention and governance. For tunnel engineering, the entire process includes an initial survey, design, construction, prevention, and treatment. However, disasters cannot be avoided. What can be done is minimize unnecessary losses and reduce costs. This article analyzes the identification and treatment of tunnel tower defense risk factors.

1 INTRODUCTION

The identification of the influencing factors of tunnel collapse, which refers to the factors that may have an impact on the safety of the tunnel, is based on the collection of relevant data before the construction of the site. It is important to determine the influencing factors from their source, their characteristics, and their degree of impact on the project, and make appropriate judgment, classification, and identification (Medina et al. 2020).

The principles of selection are as follows:

One has to integrate the whole and the part in the identification of the influencing factors. From a broader point of view, it is divided into the relationship between the whole and the parts, and the indicators are divided into first and second levels (Lei et al. 2019).

First, as a whole, the first-level impact indicators have an inclusive and causal relationship with the second-level indicators. When analyzing, the second-level and the first-level analysis should be combined. Second, it is important to distinguish the primary and secondary risks in the process of analyzing and studying the risks that affect the collapse (Zhang et al. 2020). Solving prominently representative factors and eliminating indicators with small impacts should be focused on (Luo & Wang 2022). The selection of indicators should be sufficient, and at the same time, it should be adapted to the tunnel project.

The workforce and material resources should be arranged as a whole, and resources should be allocated to improve the efficiency of project construction. Third, choose independent factors as much as possible. Related factors should be considered together to avoid overlapping between factors to avoid duplication of work (Yue et al. 2021). Therefore, according to the identification principle of influencing factors, combined with tunnel engineering in Hebei, geological conditions, and historical disasters, the factors affecting the collapse of tunnel engineering have been identified. The reasons that affect tunnel collapse should be summarized and identified based on the literature on tunnel collapse in the area.

---

*Corresponding Author: ylp8246@163.com

The following various reasons combined lead to the occurrence of tunnel collapse accidents:

1. Engineering geological conditions; it fundamentally affects the safety of the tunnel project.
2. Support design.
3. Excavation plan.
4. Other factors include (a) poor understanding of the construction process by the construction people, (b) the method used in the design not adopted in strict accordance with the requirements, (c) the unpredictability of the tunnel construction, (d) the unfavorable situations that may occur in future.

2 IDENTIFICATION OF FACTORS AFFECTING TUNNEL COLLAPSE

2.1 *Engineering geological conditions*

1) Rock structure lithology: The rock mass structure is divided into loose, fractured, layered, and block structure based on specifications. The influence of lithology is obvious, and the quality of lithology affects the stability of the tunnel. Especially when minerals, such as montmorillonite or kaolin, exist, it will swell in contact with water and may cause disasters. In addition, soft rock has lower strength than hard rock, and the possibility of disasters is more.

   The rock mass is mixed with structural surfaces of different types, shapes, and properties, such as bedding surfaces and weak interlayers. When the structural surface is small, or the degree of influence is not high, the structure determines its strength. Most landslides will move and fall along the weak surface.

2) Strength of surrounding rock: The strength of the surrounding rock is classified according to the standard uniaxial saturated compressive strength. The level III surrounding rock has no support, and it may easily collapse when the blasting vibration is too much. If the arch of the grade IV surrounding rock is unsupported, the sidewall instability will occur, resulting in relatively large landslides. In grade V surrounding rock, the arch foot stress releases quickly, and may easily cause instability and failure of the side walls. The grade VI surrounding rock has poor self-stability, and it is extremely easy to excavate the deformation that caused the collapse. When cracks are well-developed in weaker rock masses and contain groundwater, the flow of mud and sand reduces the friction between the rock blocks, and collapses and roof failures are likely to occur in shallow buried structures.

3) Special bad geology: When the tunnel passes through the fault fracture zone, shallow buried section with weak coverage, sections with an abundant water source, swelling soil layer, quicksand layer, severely biased section, and fillings in the cracks, such as silty debris, it is prone to collapse.

4) Groundwater: When the tunnel traverses the fault fracture zone, as long as the surrounding rock is dry, it is easy to overcome the difficulties even if the rock mass is broken. But when it is in an aquifer or when there is water seepage in the surrounding rock, the impact is obvious, especially when there are well-developed joints and weak interlayers in the surrounding rock. When the tunnel is covered with a loess layer, it will collapse once it encounters rain during the construction period.

5) Burial depth: As the excavation depth of the tunnel increases, the collapse will be reduced to a certain extent, and finally, a stable collapsed arch will be formed. As the buried depth increases, the stress on the arch foot and the arch bottom also increases. If it is considered that the cave span is about 20 m, the depth exceeds double the cave span, and the overlying rock and soil tend to be stable when the range increases. Being buried in a

stable state and under the same conditions as surrounding rock and fissures, the range of landslides is affected by the burial depth. Most of the geologically poor sections encountered during tunnel excavation are due to the existence of faults, which will have weak interlayers. When water or forces above the bearing range are encountered, other disasters will be triggered. Therefore, for this type of section, the scope and breadth of possible disasters should be forecast in advance.

2.2 Support design

1) Design plan: During the preliminary geological survey of the tunnel, the changes in the geological structure in the history of the area and the distribution range of unfavorable geological zones, such as faults, karst caves, etc., were not known. The uncertainty and inaccuracy of advanced geological forecasts show that there design and actual engineering errors.
2) Support method: For engineering, timely and effective support is important. The selection of support methods must be carried out by standards, and there must be certain feasible emergency measures in the face of emergencies. For example, the support is not in place, the support of the weak surrounding rock is not implemented in time, the construction continues without being completely enclosed, and the corresponding load cannot be withstood. If the time between the primary lining and the secondary lining is too long, and the surrounding rock-bearing capacity is not enough to continue to support the upper structure, it will cause deformation or settlement, greatly increasing the risk of tunnel collapse.
3) Excavation span: When the tunnel ground diameter is greater than or equal to the distance between the excavated section and the initial lining structure, it is difficult for the surrounding rock to lose stability. But when this distance is greater than twice the tunnel diameter, if the secondary lining structure has not been completed, the surrounding rock will be deformed sharply, and the initial support resistance will not be able to withstand the effect of the moving force, leading to the collapse of the highway tunnel. If the construction measures are not changed on time and if there is no timely and effective plan when the actual situation is encountered, the measures do not conform to the design. If the interval between the first and second linings in the construction is long, the stratum is exposed for too long, and the strength of the rock mass is insufficient, it would meet the force condition.

2.3 Excavation plan

1) Blasting disturbance: Blasting prolongs the cracks in the rocks, opens the cracks, and even produces new cracks. The waves generated by blasting reflect on some interfaces and create tensile stress. When the tunnel face is excavated, the forward blasting disturbs the subsequent support section and easily leads to a collapse.
2) Monitoring and quality: Failure to perform monitoring and measurement by regulations, or failure to update the results promptly, will result in delayed feedback and decision-making errors. If the support time does not meet the standard requirements, the construction is started in a hurry and continues to catch up with the construction period, it makes the support development not in place and the bearing capacity insufficient, resulting in frequent accidents. Sometimes low-quality support materials are used to save costs or design changes are brought in to reap greater benefits.
3) Construction method: Drilling and blasting (traditional mining method and new Austrian method), road header, immersed pipe, jacking, open-cut, etc., are methods involving construction. The main method used in engineering is the new Austrian tunneling method (NATM), introduced in the country in the 1960s. NATM has become the most important

method of construction of underground projects in China. The core idea of the method is that the surrounding rock and supporting structure in the tunnel excavation section deform together and bear the load.

3 PREVENTION AND TREATMENT OF FACTORS AFFECTING TUNNEL COLLAPSE

Based on the previous tunnel risk ranking results, the overall recommendations for the prevention and control of engineering landslide disasters in the entire Hebei area are as follows:

1. First, we must consider the entire Hebei Province. According to the comprehensive analysis of the regions where different types of disasters are prone to occur, prevention and control planning should be carried out in consideration of sudden natural factors, such as precipitation and earthquakes. At the same time, the period should be referred to for rainfall.

   There are east-west cracks in the regional structure of Hebei Province, and the northwestern region is located in a plating zone. It is recommended that such areas be screened out for prevention and control according to the number of historical disasters. The disasters that have been classified in the previous section should be controlled according to the status quo of development, the trend of change, and the degree of crisis. As highway tunnel areas are mostly rock structures, which is likely to cause human and land injuries during blasting and excavation, and tunnels are mostly located in places with little population flow, it is recommended to protect human and land safety in construction.

3.1 *Tunnel collapse prevention*

1) Prevention in the process of tunnel design: The method of excavation and support can be used for doing a good job of geological survey when designing the tunnel, selecting the unfavorable section, combining the actual geological conditions of the section to be excavated, and designing an implementable plan.
2) Prevention during tunnel construction and advanced geological forecasting: In tunnel construction, in addition to monitoring the surrounding rock, it is also necessary to analyze the surrounding rock by geological advance forecasting to increase the excavation and support measures to minimize risks and to do a good job of water treatment.

In landslide tunnels, disasters caused by water as an inducing factor should not be underestimated. To avoid the dangers caused by groundwater, drainage channels should be set up to drain the water in time. The surrounding rocks should be measured by the requirements of the code. Based on the dynamic monitoring data, pre-judging the deformation in advance to prevent trouble, carefully choosing the excavation and blasting methods, and controlling the number of explosives are good ways to avoid large impacts and disturbances on the surrounding rock. This helps in adopting advanced support and construction support that match the type of surrounding rock in the fault fracture zone. In handling the collapsed section, the construction quality must be ensured, and the design must be strictly followed.

The supervision report requests the next step instructions – to pay attention to the precursors of the collapse approach. The precursors of the collapse are mainly the sprayed concrete that has cracks and expands. When the excavation passes through the bad section, it is important to not only do a good job in supporting work but also not rushing to work. We must prevent and drain the water. The original bad section is more prone to disasters.

After encountering water, the risk of disasters will increase. At the same time, to ensure the quality and safety of construction, step-by-step construction must be steady.

3.2 Treatment of tunnel collapse

Based on the related landslide literature, landslide management may be divided into two stages:

1) Initial treatment measures for landslides: When the collapse has just occurred, the collapsed body should be closed and reinforced at the same time to avoid secondary collapse. According to the Platts correlation method, the height of the collapsed body is in direct proportion to the width of the excavation. Through the relationship between the two, it is inferred whether the collapse will continue to occur based on the height of the collapse. If the ground surface sinks in the landslide, the first thing to do is to drain the water to avoid secondary disasters caused by the water. After the water treatment is completed and landed on the backfill surface and compact, a secondary lining is made at the back end of the collapsed body for consolidation to prevent disturbance during the treatment of the collapsed body and adverse effects on the remaining parts.
2) Post-treatment measures for collapsed body self-stabilization: After taking measures to stabilize the collapse at the initial stage of the collapse, the following methods should be used to continue construction and strengthen real-time monitoring.

The pipe shed method is commonly used as an auxiliary method. This method is suitable for Quaternary soil layers and grade IV surrounding rocks with severe weathering, well-developed joints, and poor geological conditions. There are two types – small pipe sheds and large pipe sheds. After the collapsed body and surrounding rock masses are reinforced, the pipes are driven into the pipe shed for treatment based on the reserved deformation.

Flower tubes are mostly used in the project. This kind of tube shed has a better compaction effect and long length, which can reduce unnecessary waste.

The catheter grouting method is suitable for small landslides and loose soil layers. Grouting is used to reinforce the landslides. Arch and concrete are used for construction.

The angle of extrapolation during construction is different, which should be determined according to the actual situation. Step excavation method – joint excavation by workforce and machine, as far as possible without disturbing the surrounding rock, is a relatively safe and effective method, which can not only ensure the safety of construction but also increase the speed of construction.

It can be seen from the literature statistical data in Table 1 that for small landslides, the number of times of backfilling method is the most, the landslide volume is small, and the cavity can be backfilled and compacted. For medium-sized landslides, the pipe grouting method is used the most times. Dense and loose rock masses are excavated after the rock mass and the loose parts are consolidated and stabilized. When the landslides are relatively large, safety is the main choice when choosing treatment methods. After the step method is chosen, the excavation is gradually advanced.

Table 1. Comparison of governance methods.

Approach	Backfill	Void method	Step method	Catheter grouting method	Pipe shed method
Small landslide	49	6	10	29	6
Medium landslide	14	14	3	48	21
Large landslide	8	4	41	4	33

4 PREVENTION OF FACTORS AFFECTING TUNNEL COLLAPSE

4.1 *Prevention of engineering geological conditions*

Engineering geological conditions are the foundation of tunnel construction. The geological environment and evolution process of the tunnel site area, including the number of earthquakes, magmatic activity, and historical structure, are of great significance to the safety of tunnel engineering. The focus of prevention and control in engineering geological conditions is surrounding rock, groundwater, and unfavorable geology.

1) Rock structure: The rock mass structure plays a major role in the surrounding rock, and most of the landslides occur along the weak structural plane. For rock masses with good integrity, shotcrete and bolt support are used. The role of the anchor rod is to strengthen the part because most of this type of rock mass is caused by the falling of key blocks. For broken and loose rock masses, especially when encountering groundwater, there is no self-stabilization ability. Therefore, the pre-grouting reinforcement, drainage, and pressure reduction method is adopted. The disturbance of surrounding rock should be reduced during excavation. After excavation, support with bolt and shotcrete or hang grid, or adopt an advanced bolt and pipe shed.
2) Surrounding rock: The prevention and control technologies for surrounding rock mostly used are methods to improve the formation, pre-grouting, small pipe grouting, horizontal mixing pile, first protection and digging, small pipe shed and large pipe shed, inserting plate, closed loose body, sprayed concrete closed, sprayed steel fiber concrete closed, and inserted board drainage method.
3) Groundwater: According to its occurrence status, groundwater in the country is categorized into the plain basin, loess, karst, and bedrock mountain groundwater. For tunnels, two conditions trigger disasters. One is the existence of water-bearing structures, including faults and intrusive water-bearing structures, etc., and the other is karst areas, which have their special geological structures and high-pressure and water-rich structures. For Hebei Province, due to geological conditions, there is no karst area, so the disaster is caused by the existence of bedrock fissure water in the rock. Rain causes an abundance of groundwater, causing undesirable disasters. The principle of groundwater treatment is based on diversion while reducing the supply of surface water to the water source in the cave. When there are many water outlets, and the amount of water inflow is large, according to the distribution of fissure water, the principle of grouting can be used for partition injection.
4) Special bad geology fault zones, lithological abrupt sections, weak joint interbedded zones, and discontinuous penetrating zones are all bad geological sections that can lead to landslides. For the fault fracture zone, methods such as advanced bolts, pipe sheds, and surrounding rock grouting are used. Small pipe and large pipe sheds are used for support, and grouting is used to form a compact structure to ensure stability and collapse after excavation. For squeezing faults, the arc-shaped heading method of retaining the core soil is generally used. When blasting, the plan with a small disturbance should be selected. The rock and soil are loose, so the lining support should be done on time. If the deformation is found, the plan should be changed in time. CD or CRD method has a good degree of control.

4.2 *Other control*

Other influencing factors, which include human management, cognition, and precipitation, rank second among the three major indicators, with the understanding of geological data and support design. Factors such as earthquakes have a greater degree of influence because such sudden disasters are unavoidable, leading to fatal losses for the project. Therefore, while it is necessary to do a good job in advance, that is forecasting and geological survey work, to understand the historical disaster situation and the law of disaster occurrence, it is also important to take preventive measures for sudden disasters from being caught off guard.

5 CONCLUSION

In this paper, the influencing factors of tunnel collapse are identified and analyzed. After the identification of the influencing factors, prevention and treatment can be carried out in a targeted manner, which includes prevention in the early stage and treatment in the later stage. Prevention and control measures suitable for tunnel tower defense should be formulated. Targeted prevention involving reducing unnecessary measures and implementing economical and reasonable methods should be adopted.

REFERENCES

Medina, J. J. Hernández-Gómez, C. R. Torres-San Miguel et al. (2020) Prototype of a Computer Vision-Based CubeSat Detection System for Laser Communications[J]. *International Journal of Aeronautical and Space Sciences* (Pre-publish).

Jiliang Zhang, Honglin Ran, Xiaojun Pan, et al. (2020) Outage analysis of wireless-powered relaying FSO-RF systems with nonlinear energy harvesting[J]. *Optics Communications*, 477.

Wenchao Luo, Junwei Wang. (2022) Optimization study of surrounding rock classification indexes based on drilling parameters [J]. *Engineering survey*, 50 (10): 19–23 + 30.

Yao Lei, Xiaoming Li, Lizhong Zhang. (2019) Experimental Study on PAT System for Long-Distance Laser Communications Between Fixed-Wing Aircrafts[J]. *Photonic Sensors*, 9(2).

Zhai Yue, Fandong Meng, Qu Lu, Yunsheng Zhang, Gao Huan. (2021) Comprehensive analysis of the risk factors of "Japanese" shaped underground ring tunnel construction [J]. *Science, Technology and Engineering*, 21 (30): 13196–13202.

Collapse risk assessment of highway tunnel construction based on game theory and mutation progression method

Yingwei Ren, Boxun Wang* & Rong Ma
School of Civil Engineering and Architecture, Shandong University of Science and Technology, Qingdao, Shandong, China

Haiyan Wang
CIECC Overseas Consulting Co., Ltd, Beijing, China

ABSTRACT: In order to make the risk assessment of highway tunnel collapse more scientific and reduce the risk of collapse accidents during tunnel construction, this paper proposed a collapse risk assessment model for highway tunnel construction based on game theory and the mutation progression method. Firstly, an index system of highway tunnel construction collapse risk assessment was established, and then a comprehensive weighting of the indexes was determined based on the entropy and CRITIC methods, combined with game theory ideas, and finally, the mutation level method was applied to the risk assessment. The model was compared with the results of the survey and the fuzzy hierarchical analysis. The results show that the model is more consistent with the actual situation and can provide a reference for the risk assessment of highway tunnel construction collapse, so as to effectively reduce the occurrence of tunnel collapse accidents.

1 INTRODUCTION

In the context of the new era, with the high-quality development of China's economy and the enhancement of comprehensive national power, significant progress has been made in the construction of highways. Tunnels, as an important part of highways, are widely used in countries around the world for their convenience. Tunnel engineering has the characteristics of a long construction period, large investment, influenced by the environment, complex construction technology, etc., and is prone to safety accidents (Huang 2006). Collapse is the most common safety accident in tunnel construction, which can lead to serious consequences such as casualties, economic losses, and schedule delays.

In recent years, many scholars have deeply explored the risk of tunnel collapse through experimental demonstration, theoretical analysis, investigation, and research. Zhang et al. (2012) studied the risk of Mingshan tunnel construction collapse, and obtain the important factors of inducing tunnel collapse; Yang et al. (2015) and Guan et al. (2018) used the cloud model as a basis to study and determine the collapse risk level, showing that it was feasible to apply the cloud model to the evaluation of tunnel collapse risk level; Liu et al. (2020) provided scientific theoretical support for determining the collapse risk level of road tunnels by constructing an entropy power-improved grey correlation model; Chen et al. (2019) analyzed the characteristics of collapse accidents to establish a tunnel collapse risk evaluation model based on rough set conditional information entropy, which provided a

---

*Corresponding Author: 896573683@qq.com

feasible research idea for the evaluation of collapse risk in mountainous tunnels; Nan et al. (2017) investigated and counted the tunnel construction accidents, conducted risk source identified, and established a tunnel collapse accident tree model using the accident tree analysis method to derive an analysis method for risk events leading to tunnel collapse, which effectively reduced the probability of collapse accidents after combining the analysis with engineering examples and taking measures.

In the research of the above scholars, most of them use the hierarchical analysis method, fuzzy evaluation method, and other methods. The hierarchical analysis method and accident tree model rely too much on human subjective judgment, which is easy to cause errors. Rough set theory and cloud model need to determine the objective weight based on a large number of basic data of accidents. Therefore, in order to avoid excessive human subjectivity and ensure the rationality of the evaluation results, this paper proposes a risk assessment method based on the game theory combination of empowerment and mutation series method, in order to provide a new method for the tunnel collapse risk assessment work, which can effectively reduce the occurrence of collapse accidents.

2 BASIC THEORY AND METHODS

2.1 *Entropy weight method*

The entropy weight method is based on the information entropy principle and determines the variability of each index value (Luo 2021). The basic idea is that the lower the information entropy of an indicator, the greater the variability of the indicator, the more information it provides, the greater its role in the evaluation process, and therefore the higher its weight; conversely, the lower its weight. The calculation steps are as follows (Ma 2021):

(1) Construct a judgment matrix of m schemes, n indicators $X(x_{ij})_{m \times n}$, $i = 1, 2, \ldots, m; j = 1, 2, \ldots, n$:

$$X = \begin{pmatrix} x_{11} & x_{12} & \cdots & x_{1n} \\ x_{21} & x_{22} & \cdots & x_{2n} \\ \vdots & \vdots & & \vdots \\ x_{m1} & x_{m2} & \cdots & x_{mn} \end{pmatrix} \quad (1)$$

(2) Standardize X to get the evaluation matrix $P(P_{ij})_{m \times n}$:

$$P_{ij} = \frac{x_{ij}}{\sum_{i=1}^{m} x_{ij}}, \quad i = 1, 2, \ldots m; j = 1, 2, \ldots, n \quad (2)$$

(3) Determine the entropy of each index. The entropy of the *j*th index is:

$$H_j = -\frac{1}{\ln m}\left(\sum_{i=1}^{m} p_{ij} \ln p_{ij}\right), j = 1, 2, \ldots, n \quad (3)$$

Where H_j = entropy value of the *j*th index; m = number of evaluation schemes.

(4) Calculate the entropy right α: The entropy right of the *j*th index is:

$$\omega_j = \frac{1 - H_j}{\sum_{j=1}^{n}(1 - H_j)}, j = 1, 2, \ldots, n \quad (4)$$

The weight $\alpha = (\alpha_1, \alpha_2, \ldots, \alpha_n)$ of all indicators can be calculated according to the above formula.

2.2 CRITIC method

The CRITIC method is an objective weighting method that uses the amount of information about each indicator and the degree of correlation between the indicators to determine their weight (Xie 2020). The coefficient of variation and correlation coefficient are calculated to measure the discriminative power and the size of the conflict, respectively, and thus reflect the amount of information and the degree of correlation. The calculation process is as follows (Deng 2019):

(1) Standardize the original evaluation index matrix X in Eq. (1) to obtain the standardization matrix $X^*(x^*_{ij})$:
  ① Calculate the mean value of the *jth* indicator $\bar{x}_j$:

$$\bar{x}_j = \frac{1}{m}\sum_{i=1}^{m} x_{ij} \tag{5}$$

  ② Calculate the standard deviation S_j of the *jth* indicator:

$$S_j = \sqrt{\frac{1}{m}\sum_{i=1}^{m}(x_{ij} - \bar{x}_j)^2} \tag{6}$$

  ③ Calculate the elements x^*_{ij} of the normalized matrix X^*:

$$x^*_{ij} = \frac{x_{ij} - \bar{x}_j}{S_j}, i = 1, 2, \ldots, m; j = 1, 2, \ldots, n \tag{7}$$

(2) Calculate the coefficient of variation of the indicator v_j:

$$v_j = \frac{S_j}{\bar{x}_j} \tag{8}$$

(3) Calculate the independence coefficient of the indicator η:
  Calculate the correlation coefficients between the assessment indicators and obtain the correlation coefficient matrix $R = (rkl)_{n \times n}$,

$$r_{kl} = \frac{\sum_{i=1}^{m}(x^*_{ik} - \bar{x}^*_k)(x^*_{il} - \bar{x}^*_l)}{\sqrt{\sum_{i=1}^{m}(x^*_{ik} - \bar{x}^*_k)^2}\sqrt{\sum_{i=1}^{m}(x^*_{il} - \bar{x}^*_l)^2}}, r_{kl} = r_{lk}(k = 1, 2, \ldots, n; l = k+1, \ldots, n) \tag{9}$$

Where x^*_{ik} = standardized values of the *kth* and *lth* assessment indicator of the *ith* assessment scenario in the standardization matrix X^*, x^*_{il} = standardized values of the *lth* assessment indicator of the *ith* assessment scenario in the standardization matrix X^*, $\bar{x}^*_k$ = average values of the *kth* assessment indicators of the standardization matrix, $\bar{x}^*_l$ = average values of the *lth* assessment indicators of the standardization matrix.
  Calculate the independence coefficient for each assessment indicator η:

$$\eta_j = \sum_{k=1}^{n}(1 - r_{kj}), j = 1, 2, \ldots, n \tag{10}$$

(4) Calculate the weights for each indicator $\beta = (\beta_1, \beta_2, \ldots, \beta_n)$:

$$C_j = v_j \eta_j, \beta_j = \frac{C_j}{\sum_{j=1}^{n} C_j}, j = 1, 2, \ldots, n \tag{11}$$

2.3 Game-theoretic portfolio empowerment method

The game-theoretic-based combination weighting method is based on Nash equilibrium and seeks compromise and agreement between the two weightings, which can effectively avoid the preference for a particular weighting method and more objectively reflect the contribution of the two weighting methods to the combined weighting (Wang 2020). The combined weights and the two sets of weights are formed into a linear combination according to Eq. (12).

$$W = \lambda_1 \alpha + \lambda_2 \beta \tag{12}$$

where W = vector of all possible combined weights; λ_1, λ_2 = weight assignment coefficients.

According to game theoretical ideas, to find the optimal weights, the combined weight vector W needs to minimize the deviation from the two sets of weights, which is modeled as:

$$\min \| W - \alpha - \beta \|_2 \tag{13}$$

Based on the differential properties of the matrix, the optimal first-order derivative of Eq. (13) can be obtained as.

$$\begin{bmatrix} \alpha \alpha^T & \alpha \beta^T \\ \beta \alpha^T & \beta \beta^T \end{bmatrix} \begin{bmatrix} \lambda_1^* \\ \lambda_2^* \end{bmatrix} = \begin{bmatrix} \alpha \alpha^T \\ \beta \beta^T \end{bmatrix} \tag{14}$$

Solving for Eq. (14) gives the optimal weight assignment coefficients λ_1^*, λ_2^*, which are then normalized as follows:

$$\lambda_i' = \frac{\lambda_i^*}{\lambda_1^* + \lambda_2^*}, i = 1, 2 \tag{15}$$

Finally, the combined weights are obtained from game theory and the formula is:

$$W' = \lambda_1' \alpha + \lambda_2' \beta \tag{16}$$

2.4 Mutation progression method

The mutation progression method is a comprehensive evaluation method based on the mutation theory, which can scientifically understand the mechanism of the occurrence of discontinuities and make predictions. And tunnel collapse is a sudden situation with obvious discontinuity, so the two are a good match (Hu 2014). The steps of the method are as follows:

(1) Multi-level decomposition of evaluation objects to form a hierarchical structure.
(2) Determine the mutation model of the evaluation index system, the common primary mutation model is shown in Table 1. $v(x)$ in the assortment indicates the potential function of the state variable x, and the coefficients $a \sim d$ indicate the control variables of the state variable. When an evaluation index is decomposed into $1 \sim 5$ indicators, the

corresponding ones are folded into mutation, pointed mutation, dovetail mutation, butterfly mutation, and shed mutation respectively.

(3) Deriving a normalization formula. The normalization formula can transform the control variables of different qualitative states in the mutation system into the same qualitative state that can be easily compared, in order to find the total mutation affiliation value that can characterize the state of the system for evaluation. The normalization formulae for several commonly used mutation models are as follows:

$$\text{Folded mutations: } x_a = \sqrt{a} \qquad (17)$$

$$\text{Pointed mutations: } x_a = \sqrt{a}, x_b = \sqrt[3]{b} \qquad (18)$$

$$\text{Dovetail mutations: } x_a = \sqrt{a}, x_b = \sqrt[3]{b}, x_c = \sqrt[4]{c} \qquad (19)$$

$$\text{Butterfly mutations: } x_a = \sqrt{a}, x_b = \sqrt[3]{b}, x_c = \sqrt[4]{c}, x_d = \sqrt[5]{d} \qquad (20)$$

$$\text{Shed mutations: } x_a = \sqrt{a}, x_b = \sqrt[3]{b}, x_c = \sqrt[4]{c}, x_d = \sqrt[5]{d}, x_e = \sqrt[6]{e} \qquad (21)$$

(4) Rank the importance of the control variables. Taking the spike-point mutation model as an example, the expression of the potential function shows that the control variable a is the primary factor, the control variable b is the secondary factor, i.e., the importance of the control variable $a > b$, and so on: the order of importance is $a > b > c > d > e$.

Table 1. Elementary catastrophe model.

Mutations Models	Status variables	Control Variables	Potential functions
Folded type	1	1	$V(x) = x^3 + ax$
Pointed type	1	2	$V(x) = x^4 + ax^2 + bx$
Dovetail type	1	3	$V(x) = x^5 + ax^3 + bx^2 + cx$
Butterfly type	1	4	$V(x) = x^6 + ax^4 + bx^3 + cx^2 + dx$
Shed type	1	5	$V(x) = x^7 + ax^5 + bx^4 + cx^3 + dx^2 + ex$

(5) Comprehensive analysis. If the control variables in the mutation system are not correlated with each other, i.e., the principle of non-complementarity is satisfied, then the minimum value of the mutation affiliation value of each control variable is taken; if the control variables in the system are correlated, i.e., the principle of complementarity is satisfied, then the average of the mutation affiliation values of the control variables is taken (Zhang 2020).

3 COLLAPSE RISK ASSESSMENT MODEL OF TUNNEL CONSTRUCTION BASED ON GAME THEORY AND MUTATION PROGRESSION METHOD

3.1 Analysis of tunnel collapse risk elements

By collecting investigation reports of accidents and relevant literature (Liu 2019; Wen 2021), nearly 100 tunnel construction collapse accidents that occurred in China in the past 20 years were analyzed, and the influencing factors of tunnel collapse were grouped into the following four categories, taking into account the mechanism of collapse accidents.

(1) Natural environmental factors. The natural environment includes the following aspects: excessive precipitation will seep into the pit, making the soil softer, heavier, and less resistant to shear, which will easily cause collapse accidents; the higher the seismic intensity, the more difficult the tunnel construction will be, and the greater the probability of collapse accidents; when the depth of the tunnel is shallow, the surrounding rock cannot form a natural arch with sufficient curvature after excavation, which is less stable and more prone to collapse accidents (Hou 2018).
(2) Engineering geological factors. Geological conditions are mainly manifested in five aspects: surrounding rock condition, deviated topography, fault rupture zone, rock weathering degree, and groundwater level. In general, the higher the level of the surrounding rock, the worse its stability; fault rupture zone rock strength is lower, and weathering resistance is weak, with strong water permeability; bias pressure often has a greater impact on the stability of the shallow buried area of the tunnel entrance, and is very likely to cause collapse accidents.
(3) Survey and design factors. Survey design and construction process have a direct relationship with the occurrence of safety accidents, survey unit qualification level, survey investment, etc. They will affect the degree of detail and accuracy of the ground survey data, and whether the parameters of the design fully consider the geological and hydrological situation of the tunnel site area and construction conditions. And whether the support program is perfect and reasonable will also directly affect the safety of tunnel construction, usually, the larger the tunnel excavation span. In general, the larger the span of the tunnel excavation, the more complex the tunnel structure and forces, and the more likely it is that the surrounding rock will become unstable and collapse.
(4) Construction level factors. In the tunnel construction process, the choice of excavation method is practical and feasible, whether the construction personnel have undergone safety training and education, whether the construction site has adopted scientific and reasonable management measures, or whether the monitoring and measurement are in place and the possibility and consequences of collapse accidents have a great relationship.

3.2 Indicator system construction and risk classification

Combined with the above analysis, the natural environmental factors, engineering geological factors, survey and design factors, and construction level factors as the primary indicators, with the average annual precipitation, seismic intensity, burial depth, surrounding rock conditions, and other 15 factors as secondary indicators, to build road tunnel collapse risk assessment index system. See Table 2.

The collapse risk level was divided into 5 evaluation levels from high to low, namely high risk, high risk, medium risk, and low risk, corresponding to the affiliation intervals of (0, 0.2), (0.2, 0.4), (0.4, 0.6), (0.6, 0.8) and (0.8, 1).

As the normalization operation in the mutation level method is in the form often open squarely, the upward step-by-step calculation will make the mutation affiliation value of the total target high. Besides, as the indicator values are relatively close, it is difficult to make an accurate comparison, so this paper uses the score transformation method to correct the total indicator mutation affiliation value (Liu 2010). The transformation relationship is shown in Table 3.

3.3 Calculation of indicator weights

According to the entropy and CRITIC method $\alpha\beta$, i.e., through Eqs. (1) to (11), the weights of the secondary indicators $\alpha = (\alpha_1, \alpha_2, \ldots, \alpha_{15})$ and $\beta = (\beta_1, \beta_2, \ldots, \beta_{15})$ are calculated. The weights W' of the primary indicators are obtained by taking the average of the weights of the corresponding secondary indicators and then normalizing them. Then, according to the size of the combined weight, each risk indicator is ranked in importance.

Table 2. Collapse risk assessment index system of highway tunnel construction.

Total indicators	Tier 1 indicators	Secondary indicators
Highway tunnel construction collapse risk A	Natural Environment B_1	Average annual precipitation C_1 Depth of burial C_2 Seismic intensity C_3
	Engineering Geology B_2	Enclosure Class C_4 Degree of rock weathering C_5 Fault rupture zone C_6 Groundwater C_7 Bias C_8
	Survey and Design B_3	Excavation span C_9 Geological Survey Information C_{10} Parametric Design C_{11} Support Scheme C_{12}
	Construction level B_4	Excavation method C_{13} Security Management C_{14} Monitoring and measurement C_{15}

Table 3. X-Y correspondence table.

x	0.10	0.20	0.30	0.40	0.50	0.55	0.60	0.65
y	0.9008	0.9286	0.9456	0.9581	0.9679	0.9722	0.9762	0.9798
x	0.70	0.75	0.80	0.85	0.90	0.95	0.97	1.00
y	0.9832	0.9864	0.9895	0.9923	0.9950	0.9976	0.9985	1.0000

Note: 1) y is the calculated value, and x is the transformed value. After the transformation, it grades the risk corresponding to the risk membership value at all levels;
2) When the calculated value y is less than 0.9008 and the transformed value x is less than 0.1, that is, the risk level is level I.

3.4 Risk level determination

The risk analysis is carried out using the mutation level method, where the data of the risk indicators are dimensionless to obtain the initial mutation affiliation value, and the total mutation affiliation value is calculated step by step upwards according to the normalization formula and evaluation principles, and finally, the risk level of highway tunnel construction collapse is determined.

4 ENGINEERING APPLICATION ANALYSIS

4.1 Project overview

In order to verify the reasonableness and validity of the assessment method, a tunnel construction collapse risk assessment was carried out based on a particular section of each of the five road tunnels (S_1, S_2, S_3, S_4, and S_5) with a higher risk of survey results. The secondary risk indicators are divided into those that can be directly obtained from geological survey data, design documents, and construction organization documents, including average annual precipitation C_1, depth of burial C_2, seismic intensity C_3, rock level C_4, fault rupture zone C_6, deflection C_8, excavation span C_9, and those that need to be converted into specific values based on the data, including the degree of rock weathering C_5, groundwater C_7, geological survey C_{10}, parameter design C_{11}, support scheme C_{12}, excavation method C_{13}, safety management C_{14}, monitoring, and measurement C_{15}. These indicators are scored by experts according to a percentage system, and the higher the score, the higher the risk value.

4.2 Risk assessment

According to the risk assessment model based on game theory combination of weighting and mutation level method for tunnel construction collapse risk assessment of five tunnel samples, the weights of each risk indicator were calculated by applying Eqs. (1) to (21) and ranked in order of importance according to the weights, and the results are shown in Table 5.

Dimensionless processing of risk indicators. In order to solve the problem of comparability of the raw data for the secondary risk indicators, the data in Table 4 need to be normalized before the evaluation begins to find the initial mutation affiliation value of the indicators so that it lies between 0 and 1, using the following formula.

$$X_i = \frac{R_i}{R_{max}} \quad (22)$$

$$r_i = R_{max} + R_{min} - R_i \quad (23)$$

Where X_i = initial mutation affiliation function value obtained by conducting normalization, R_i = original data of the indicator, R_{max} = maximum values, and R_{min} = minimum values.

For positive indicators directly normalized using Eq. (22), for negative indicators were first transformed using Eq. (23) and later processed using Eq. (22), the initial mutation affiliation values obtained after processing are shown in Table 6.

Taking the S_1 tunnel as an example, the natural environmental factor B_1 consists of three secondary indicators C_1, C_2, and C_3, which constitute a dovetail mutation, according to the dovetail mutation normalization formula and the principle of non-complementarity.

$$B_1 = \min\left(\sqrt[3]{0.522}, \sqrt[4]{0.494}, \sqrt{0.857}\right) = \min(0.806, 0.838, 0.926) = 0.806 \quad (24)$$

Table 4. Statistics of five tunnels.

	S_1	S_2	S_3	S_4	S_5
Average annual precipitation C_1/mm	897.20	523.20	396.00	1103.90	650.00
Depth of burial C_2/m	60.02	45.70	85.40	23.50	121.40
Seismic intensity C_3/magnitude	6.00	7.00	7.00	5.00	7.00
Rock envelop class C_4/grade	5.00	5.00	4.00	5.00	4.00
Degree of rock weathering C_5	65.00	63.00	56.00	78.00	55.00
Fault rupture zone C_6/m	45.00	18.00	25.00	13.00	20.00
Groundwater C_7	23.00	15.00	17.00	12.00	43.00
Bias C_8/°	40.00	20.00	33.00	20.00	35.00
Excavation span C_9/m	12.75	10.65	11.00	13.00	10.25
Geological Survey Information C_{10}	20.00	17.00	33.00	25.00	22.00
Parametric Design C_{11}	18.00	25.00	29.00	33.00	19.00
Support Scheme C_{12}	15.00	17.00	23.00	21.00	27.00
Excavation method C_{13}	42.00	40.00	55.00	34.00	47.00
Safety Management C_{14}	15.00	18.00	22.00	43.00	32.00
Monitoring and measurement C_{15}	25.00	15.00	18.00	30.00	17.00

Table 5. Weight value and ranking of risk indicators.

Risk indicators	Entropy method weights α	CRITIC method weights β	Combined weights W'	Importance Sort by
C_1	0.0597	0.0552	0.0578	2
C_2	0.0430	0.0599	0.0500	3
C_3	0.0399	0.0928	0.0619	1
C_4	0.0784	0.0892	0.0829	1
C_5	0.0809	0.0572	0.0710	4
C_6	0.0734	0.0639	0.0694	5
C_7	0.0880	0.0720	0.0813	2
C_8	0.0798	0.0754	0.0780	3
C_9	0.0701	0.0611	0.0664	2
C_{10}	0.0615	0.0578	0.0600	4
C_{11}	0.0696	0.0658	0.0680	1
C_{12}	0.0566	0.0673	0.0611	3
C_{13}	0.0511	0.0686	0.0585	3
C_{14}	0.0735	0.0597	0.0678	1
C_{15}	0.0743	0.0541	0.0659	2
B_1	0.184	0.262	0.192	4
B_2	0.310	0.27	0.306	1
B_3	0.250	0.238	0.249	3
B_4	0.256	0.23	0.253	2

Table 6. Mutation membership value of risk indicator.

	S_1	S_2	S_3	S_4	S_5
C_1	0.522	0.884	1.000	0.359	0.769
C_2	0.494	0.376	0.703	0.194	1.000
C_3	0.857	0.714	0.714	1.000	0.714
C_4	0.800	0.800	1.000	0.800	1.000
C_5	0.872	0.897	0.987	0.705	1.000
C_6	0.289	0.889	0.733	1.000	0.844
C_7	0.744	0.930	0.883	1.000	0.279
C_8	0.500	1.000	0.675	1.000	0.625
C_9	0.807	0.969	0.942	0.788	1.000
C_{10}	0.909	1.000	0.515	0.757	0.848
C_{11}	1.000	0.787	0.667	0.545	0.969
C_{12}	1.000	0.925	0.703	0.778	0.556
C_{13}	0.854	0.890	0.618	1.000	0.763
C_{14}	1.000	0.930	0.837	0.449	0.604
C_{15}	0.667	1.000	0.900	0.500	0.933
B_1	0.806	0.783	0.845	0.664	0.845
B_2	0.813	0.894	0.906	0.932	0.653
B_3	0.931	0.887	0.816	0.738	0.863
B_4	0.873	0.964	0.886	0.670	0.777
A	0.902	0.945	0.950	0.875	0.810

The engineering geological factor B_2 consists of five secondary indicators C_4, C_5, C_6, C_7, and C_8, which constitute the shed type mutation, according to the shed mutation normalization formula and the principle of non-complementarity.

$$B_2 = \min(\sqrt{0.800}, \sqrt[5]{0.872}, \sqrt[6]{0.289}, \sqrt[4]{0.744}, \sqrt[4]{0.500})$$
$$= \min(0.894, 0.972, 0.813, 0.906, 0.840) = 0.813 \qquad (25)$$

The survey and design factor B_3 consists of four secondary indicators, C_9, C_{10}, C_{11}, and C_{12}, which constitute the butterfly mutation, according to the butterfly mutation normalization formula and the principle of non-complementarity.

$$B_3 = \min\left(\sqrt[3]{0.807}, \sqrt[5]{0.909}, \sqrt[4]{1}, \sqrt[4]{1}\right) = \min(0.931, 0.981, 1, 1) = 0.931 \qquad (26)$$

The construction level factor B_4 consists of three secondary indicators C_{13}, C_{14}, and C_{15}, which constitute dovetail mutations, according to the dovetail mutation normalization formula and the principle of non-complementarity.

$$B_4 = \min\left(\sqrt[4]{0.854}, \sqrt{1}, \sqrt[3]{0.667}\right) = \min(0.961, 1, 0.873) = 0.873 \qquad (27)$$

The total indicator road tunnel collapse risk A consists of four primary indicators B_1, B_2, B_3, and B_4, which constitute a butterfly mutation, according to the butterfly mutation normalization formula and the principle of non-complementarity.

$$A = \min\left(\sqrt[5]{0.806}, \sqrt{0.813}, \sqrt[4]{0.931}, \sqrt[3]{0.873}\right) = \min(0.957, 0.902, 0.982, 0.955) = 0.902 \quad (28)$$

Similarly, the mutation affiliation values of each risk indicator for the four tunnels S_2, S_3, S_4, and S_5 can be obtained from Table 6. According to Table 4, the total mutation affiliation values of S_1, S_2, S_3, S_4, and S_5 tunnel construction collapse risk were calculated by interpolation method as 0.104, 0.297, 0.335, 0.097 and 0.089, which belong to risk level I, risk level II, risk level II, risk level I and risk level I respectively.

4.3 Analysis of tunnel collapse risk elements

To verify the reasonableness of this assessment model, the risk assessment of S_1-S_5 was carried out using fuzzy hierarchical analysis (Hu 2021), and the results were compared with the results of the method in this paper as shown in Figure 1.

As can be seen from Figure 1: the conclusions obtained by the two methods are generally consistent, but the fuzzy hierarchical analysis method requires experts to score the

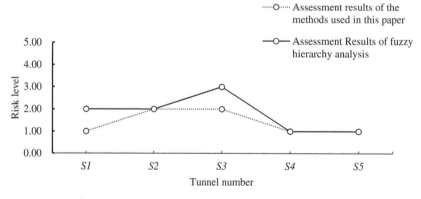

Figure 1. Results comparison.

importance of indicators. Besides, it also requires a subjective judgment of experts in determining the subordination value. So, it can be said that the calculation results are based on experts' experience and knowledge accumulation, and therefore the method is subjective and not objective enough easily leading to a larger error in the assessment results. The method used in this paper to establish the model does not rely much on expert judgment, and the weights are only used to judge the importance of the indicators and are not directly involved in the calculation, so the error is smaller.

As can be seen from Table 6, the secondary indicators for the S_1 tunnel are, in descending order of risk, natural environmental factors, engineering geological factors, construction level factors, and survey, and design factors, and the first two factors have lower values of a sudden change in affiliation, which, when combined with their relevant information, indicate that the S_1 tunnel area has higher average annual precipitation, a higher degree of rock weathering, and the existence of large fracture zones, etc., which are affected by these undesirable factors and its collapse. The risk level for tunnel S_2 is higher due to the following factors in descending order of risk: natural environmental factors, survey, and design factors, engineering geological factors, construction level factors, higher surrounding rock level, higher seismic intensity, lack of consideration of certain special conditions in the parameter design, etc. The risk level for tunnel S_3 is higher due to the following factors in descending order of risk: survey and design factors, natural environmental factors, construction level factors, engineering geological factors, etc. For the S_4 tunnel, the risk level is from natural environment factor, construction level factor, survey design factor, and engineering geology factor, as there is a lot of precipitation in the tunnel area, while the depth of burial is shallow, the surrounding rock level is high, and the parameter design has some defects. For the S_5 tunnel, the risk factors are, in descending order of risk, engineering and geological factors, construction level factors, natural environmental factors, and survey and design factors, with a slightly water-rich area, high seismic intensity, high envelope rock level, severe deviations, and general safety of construction methods.

In view of the above problems, measures should be taken to prevent them, such as the use of overrunning pipe shed support or overrunning small conduit for the V-grade section of the tunnel, and other overrunning support schemes according to the surrounding rock conditions, and reinforced lining for the cavern entrance section; the implementation of the whole process of overrunning forecasting program to accurately identify the geological conditions under construction to ensure construction safety. During construction, dripping water may occur in the tunnel chamber, and then cause dampness in the tunnel. The speed of excavation should be strictly controlled and forbidden to avoid blindly pursuing progress and violating the rules of construction.

5 CONCLUSION

The road tunnel construction collapse risk assessment index system proposed in this paper includes four primary indicators, including natural environmental factors, engineering geological factors, survey and design factors, and construction level factors, as well as 15 secondary indicators, including average annual precipitation, depth of burial, seismic intensity and rock level. The index system is more appropriate to the actual situation, and the content is more complete, more scientific, and reasonable.

The mutation level method is a dynamic and comprehensive evaluation method, which not only can make up for the shortcomings of static evaluation methods, but also can scientifically understand the mechanism of occurrence of discontinuous phenomena and make predictions. Tunnel collapse is a sudden situation with obvious discontinuity, and the two match well. The entropy power method and CRITIC method are used to judge the importance ranking of indicators, which makes the improved mutation level method more scientific and reasonable, and it is applied to road tunnel collapse risk assessment, and the assessment results are compared with the survey results and fuzzy hierarchy analysis method, which verifies the applicability of the method in and can provide a reference for road tunnel construction collapse risk assessment.

There is a certain degree of subjectivity in the evaluation of risk indicators in this paper, the risk assessment model constructed still has limitations, such as the application of the "complementary" and "non-complementary" principles in the mutation level method, and the value of some indicators depends on the subjective judgment of experts. Thus, further research is needed.

REFERENCES

Chen Wu, Zhang Guohua, Wang Hao, et al. (2019). Risk Assessment of Mountain Tunnel Collapse Based on Rough Set Conditional Information Entropy[J]. *Geotechnical Mechanics*, 40(09): 3549–3558.

Deng Wenzhou, Liu Tangzhi, Pu Wenpei (2019). Risk Management of Mountain Highway Based on Critical Fuzzy Comprehensive Analysis Method[J]. *Science, Technology and Engineering*, v.19; No.498(29):339–343.

Guan Xiaoji (2018). Evaluation Method of Tunnel Collapse Risk Grade Based on Extension Connection Cloud Model[J]. *Science and technology for Work Safety in China*, 14(11): 186–192.

Hou Yanjuan, Zhang Dingli, Li Ao (2018). Analysis and Control of Tunnel Construction Collapse Accident [J]. *Modern Tunnel Technology*, v.55; No.378(01):45–52.

Hu Changming, Gong Shaorui, Zhang Chaohui, et al. (2014). Risk Prediction of Mountain Tunnel Collapse Based on Catastrophe Theory[J]. *Journal of Xi'an University of architecture and technology (Natural Science Edition)*, 46(01): 10–15.

Hu Junming, Huang Hongzhong, Huang Peng (2021). Reliability Allocation of Industrial Robots Based on Fuzzy Analytic Hierarchy Process[J]. *Science, Technology and Engineering*, V.21; No.562(21):8965–8969.

Huang Hongwei (2006). Research Progress on Risk Management in the Tunnel and Underground Engineering Construction[J]. *Journal of Underground Space and Engineering*, (01): 13–20.

Liu Can, Zheng Bangyou, Li Zheng, et al. (2020). Risk Assessment of Highway Tunnel Collapse Based on Entropy Weight Improved Grey Correlation Model[J]. *Science, Technology and Engineering*, v.20; No.520 (15):6292–6297.

Liu Ke (2019). *Research on Safety Risk Assessment and Application of Highway Tunnel Construction* [D]. Tian Jin: Hebei University of technology.

Liu Wenyuan, Tao Juan (2010). Offshore Ship Navigation Safety Evaluation Based on Catastrophe Theory[J]. *Chinese Journal of Safety Science*, 20 (10): 113–118.

Luo Wenjie, Li Mingjie, Xiao Ziliang (2021). Evaluation Method of GA-BP Neural Network Programming Ability Based on Entropy Weight Deviation[J]. *Science, Technology and Engineering*, V.21; No.551 (10):4117–4123.

Ma Sha, Liu Congcong, Zhang Runhua, et al. (2021). Suitability Evaluation of Underground Space Based on Entropy Weight Analytic Hierarchy Process[J]. *Science, Technology and Engineering*, V.21; No.564 (23):10013–10020.

Nan Yuhong, Zhao Song, Yang Zhengmao (2017). Risk Assessment of Tunnel Collapse Based on Fault Tree [J]. *Road Construction Machinery and Construction Mechanization*, 34(07): 106–110.

Wang Jinxiang, Zhao Shuen, Yang Qizhi, et al. (2020). Vehicle Collision Risk Situation Assessment Based on Game Theory Combined Weighted TOPSIS method[J]. *Science, Technology And Engineering*, v.20; No.513(08):3315–3322.

Wen Yanfang, Chen Jingpei (2021). Study on Coupling Mechanism of Collapse Risk in Subway Tunnel Construction[J]. *Journal of Underground Space and Engineering*, v.17; No.130(03): 943–952.

Xie Xuebin, Li Dexuan, Kong Lingyan, et al. (2020). Rockburst Tendency Grade Prediction Model Based on CRITIC-XGB Algorithm[J]. *Journal of Rock Mechanics And Engineering*, 39(10): 1975–1982.

Yang Guang, Liu Dunwen, Chu Fujiao, et al. (2015). Tunnel Collapse Risk Grade Evaluation Based on Cloud Model[J]. *Science and Technology for Work Safety in China*, 11(06): 95–101.

Zhang Sheng, Chen Xiuhe, Wang Bo, et al. (2012). Suggestions on Risk Assessment and Control Measures for the Collapse of Mingtangshan Tunnel[J]. *Journal of Underground Space and Engineering*, v.8; No.62 (S1):1567–1570 + 1620.

Zhang Yibing, Liu Handong, Yang Jihong, et al. (2020). Risk Analysis of Reservoir Slope Stability Based on Improved Catastrophe Theory – Take the Right Dam Abutment Slope of Qianping Reservoir as an example[J]. *Science, Technology and Engineering*, v.20; No.513(08): 3246–3251.

Study on construction technology optimization of post-cast strip in underground engineering

Yaohui Shen*
Department of Civil Engineering, Xiamen University Tan Kah Kee College, Zhangzhou, Fujian Province, China

ABSTRACT: Based on the engineering case, this paper discusses the advantages and disadvantages of the post-cast strip construction technology of underground engineering and puts forward the optimization measures for the post-cast strip construction of underground engineering in combination with the research results in other fields, to solve the common engineering problems at present.

1 INSTRUCTION

In high-rise or large-scale buildings, it is often necessary to set up underground garages and shear walls. High-rise buildings often have podium structures, and the foundation needs to be designed as a whole. Due to the settlement difference between the high-rise buildings and podium structures, the two parts need to be temporarily disconnected by the post-pouring belt during construction. After the main structure of the high-rise building is completed and most of the settlement has been completed, the concrete of the connecting part is poured to connect the high and low floors as a whole. In addition, if the area of the building is too large, the structure will expand and contract due to temperature change, and the concrete will crack due to plastic shrinkage during hardening, so the expansion post-pouring belt needs to be set. In the construction of existing projects, when it is necessary to reserve settlement post pouring belt or expansion post pouring belt, especially in the construction of foundation works and external walls with underground water, in order to prevent underground water from gushing out during construction, a concrete base plate or guide wall shall be added at the post pouring belt and a water stop shall be set. The base plate or guide wall shall be poured at the same time as the foundation base plate during construction, and the post-pouring belt shall be reserved. When there is a water stop, the precipitation can be stopped as long as the superstructure load can resist the buoyancy of groundwater. After the main body is completed, the concrete at the post-pouring belt shall be poured. However, after the completion of the project, the main structure may still be subject to settlement and contraction, leading to the occurrence of cracks, especially at the post-pouring belt, thus causing groundwater leakage (Gao 2019; Li 2022). Therefore, it is of great significance to study and optimize the construction technology of the post-cast strip in underground engineering.

2 ENGINEERING CASE

Yinhong home living Plaza project is located in Yuelu District, Changsha City, with a building area of about 210000 square meters. It consists of a shopping mall with 2 floors

---

*Corresponding Author: yhshen@xujc.com

underground and 7 floors above ground, a commercial residential tower with 3 floors underground and 28 floors above ground, and an apartment office building with 3 floors underground and 26 floors above ground. The shopping mall is about 100000 square meters, which is a reinforced concrete frame structure. The commercial residential tower is about 54000 square meters, which is a reinforced concrete frame shear wall structure. The fire resistance rating is grade I, the design life is 50 years, and the seismic fortification intensity is 6 degrees. As the underground garages of the three buildings are connected, the height difference and load difference are large. The foundation slab of the underground garages of the tower and the podium is provided with a settlement post-pouring belt, expansion post-pouring belt, and reinforcement belt, which divides the basement into 13 blocks. The settlement post-pouring belt shall be constructed after the main structure of the building is capped. The expansion post-pouring belt can be constructed after the concrete age on both sides reaches 45 days.

During construction, horizontal construction joints shall be set at the height of 450 mm from the shear wall to the basement ground, and 3 × 300 folded water stop steel plate to prevent water seepage; The foundation slab and external wall shall be provided with water stop post pouring belt according to the design requirements. On both sides of the foundation slab post-cast belt or on both sides of the external wall post-cast belt, the 3 × 300 folded edge water stop steel plate, and reinforced wire mesh isolation mesh belt shall be set along both sides of the post-cast belt. The relevant structures are shown in the following Figures (1) to (3).

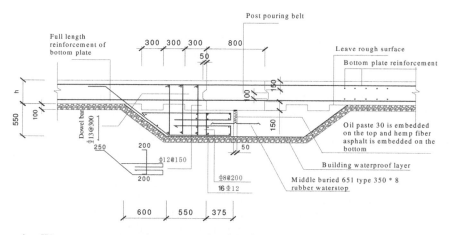

Figure 1. Water stop structure of post-cast strip of the basement floor.

The structure of the post-cast strip of the baseplate of the basement is shown in Figure 1. The water stop layer is composed of a 100 mm thick plain concrete cushion and a waterproof layer, which is constructed to the same elevation as the waterproof protection layer of the baseplate. This can play a waterproof role before the concrete pouring of the post-pouring belt. When the load of the superstructure is greater than the buoyancy of the groundwater, the precipitation can be stopped, which can shorten the dewatering time of the foundation pit, reduce the construction cost and protect the environment. The post-pouring belt is provided with a guide wall, which is poured simultaneously with the foundation slab. The reinforcement structure of the guide wall is shown in Figure 1 50 mm expansion joint shall be reserved in the middle of the guide wall, and a rubber water stop shall be set, embedded with oil paste and hemp asphalt. The guide wall and the bottom plate are concreted at the same time. When settlement and expansion deformation occur in the later stage, the displacement is mainly concentrated at the expansion joint of the guide wall. The main disadvantage of

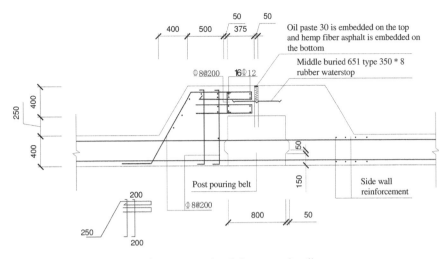

Figure 2. Water stop structure of post-cast strip of the external wall.

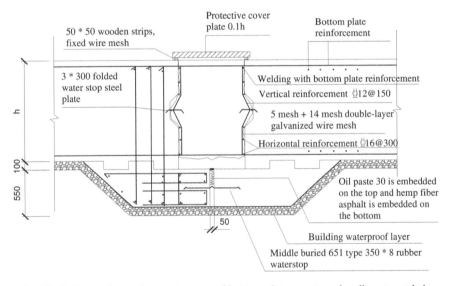

Figure 3. Steel wire mesh membrane structure of bottom plate or external wall post-cast belt.

this structure is that when the waterproof layer of the bottom plate, the rubber water stops, oil paste and asphalt are damaged due to aging, stress deformation, and other reasons, the groundwater will permeate to the post-pouring belt under the pressure, and the water stop will not play a waterproof role.

As shown in Figure 2, the water stop structure of the post-pouring belt of the outer wall has only three waterproof lines of the rubber water stop, oil paste, and asphalt, which is one less line than the bottom plate, and there is no plain concrete cushion. Other structures are similar to the bottom plate. Considering that the groundwater pressure of the external wall is not as great as that of the bottom plate, it is also possible to reduce the waterproof layer. However, the exterior wall also has the hidden danger of water seepage similar to the floor.

As shown in Figure 3, before the concrete pouring of the bottom plate, the reinforcement at the post-pouring belt shall be installed, and the folded water stop steel plate shall be welded firmly. Bind 5-mesh and 14-mesh double-layer galvanized wire mesh on both sides of the post-pouring belt and pour the concrete off the bottom plate after it is fixed firmly. For the settlement post-pouring belt between the high-rise building and the podium structure, concrete shall be poured after the main structure of the building is capped. For other expansion post-pouring belts, concrete shall be poured after the concrete age on both sides reaches 45 days. The double-layer wire mesh can increase the roughness of the contact surface, the adhesion between the concrete of the post-cast belt, and the concrete of the bottom plate, as well as reduce the occurrence of cracks. However, the wire mesh is vertically arranged, which is not very effective in reducing the transverse shrinkage deformation. The shrinkage crack is mainly caused by the accumulation of shrinkage deformation of the floor in the horizontal direction. After the building is capped, the building may still be subject to settlement and shrinkage deformation. Therefore, cracks and water seepage in the post-pouring belt are common.

After the concrete age of the bottom plate meets the requirements or after the main body is capped, before the concrete pouring of the post-pouring belt, the sundries of the formwork shall be removed, the surface scum and loose concrete shall be removed, the rust and cement slurry of the reinforcement and the water stop steel plate shall be removed, the solid and rough surface shall be exposed, washed, and the accumulated water shall be removed. The water shall be sprayed with a spray to make it fully wet. Before the second pouring of the post-pouring belt and construction joint concrete, the mortar shall be connected at the junction of the new and old concrete. The mortar shall be consistent with the concrete strength grade. For the post-pouring belt of the basement floor, a layer of 150 mm-200 mm thick mortar shall be laid before the concrete pouring. When vibrating the concrete, it shall be fully vibrated and compacted to make the new and old concrete closely combine. The post-pouring belt is poured with expansive concrete, and the strength grade is one grade higher than the original concrete. The concrete shall be vibrated and compacted to avoid missing vibration and excessive vibration. The inclined surface shall be used for layered construction. The vibration shall start from the lower end of the pouring layer and then gradually move up to ensure the quality of the concrete. After the initial setting of concrete, cover it with plastic film and a layer of straw bags. After the hydration heat peak, after the concrete temperature drops, the film can be spread, covered with straw bags, and watered for curing. The total curing time of the two kinds of curing is not less than 14 days.

After the concrete construction of the post-cast belt is completed, the protection of the post-cast belt shall be strengthened in time. Then, the scattered concrete, laitance on reinforcement and building materials shall be cleaned up. Along the length direction of the post pouring belt, using 2.5-inch iron nails to nail the 18 mm thick plywood on the 80 mm * 60 mm square wood, it forms a whole along the length direction to cover the post pouring belt.

After the formwork of the post-pouring belt is removed, if there is water seepage in the construction joint during the construction process, fs-888 single liquid polyurethane foam leakage plugging agent with excellent waterproof effect shall be used for grouting to prevent water seepage.

3 PROCESS IMPROVEMENT

As shown in Figure 4, the steel wire mesh shall be arranged on the upper and lower surfaces of the bottom plate or the inner and outer sides of the outer wall. Other construction measures shall not be changed. Some research results show that steel wire mesh can reduce the shrinkage crack of concrete by 20% - 50% (Shao 2007) and can also enhance the strength of concrete (Mohamed 2020; Raiyani 2021; Zhao 2020). The original construction waterproof measures include multiple waterproof lines such as base plate waterproof layer, rubber water stop, oil paste, and asphalt. The double-layer galvanized wire mesh is added on both sides of the post-pouring belt to increase the adhesive force of the concrete. The expansive concrete

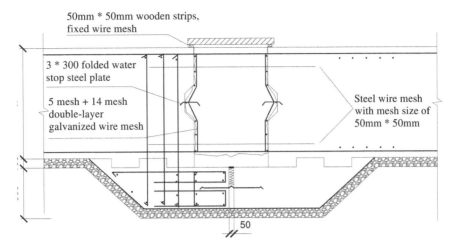

Figure 4. Optimization and improvement measures for post-cast steel wire mesh membrane of bottom plate or outer wall.

with a higher strength than the original concrete is also used for pouring. The existing technical measures have been used as much as possible. If the latest research results can be applied to the optimization of construction technology, it will be perfect. By increasing the transverse wire mesh, the shrinkage crack of concrete can be reduced, and the strength of concrete can be enhanced so that the post-cast belt can better resist the additional stress caused by uneven settlement and shrinkage, thus avoiding water seepage.

4 CONCLUSION

At present, there are many corresponding measures for the construction method of the post-cast strip in the project, but there are still construction defects due to the restrictions of the construction conditions on the site. This paper combines the latest research success to optimize the construction process, which can better solve the problems in the current construction.

REFERENCES

Abu Maraq Mohamed A., Tayeh Bassam A., Ziara Mohamed M., Alyousef Rayed. Flexural Behavior of Rc Beams Strengthened With Steel Wire Mesh and Self-Compacting Concrete Jacketing – Experimental Investigation and Test Results[J]. *Journal of Materials Research and Technology* 2020.

Gao Gao, Xu Ziguo, Yan Feng. Crack Prevention Analysis and Design of a Super long Basement Concrete Structure [J]. *Architectural Science*, 2019,35 (9): 134–141.

Hua Zhao. Axial Compressive Behaviour of Concrete Strengthened with Steel Rings, Wire Mesh and Modified High Strength Mortar (MHSM)[J]. *Construction and Building Materials Volume* 250, Issue C. 2020.

Li Zihua, Xiao Xinyu, Liu Dong, Zhang Guojun, Chen Jiakun Crack Control Technology of Super Large Structure Concrete [J] *Cement and Concrete Production*, 2022,6:1–5.

Raiyani Sunil D., Patel Paresh V. Torsional Strengthening of Reinforced Concrete Hollow Beam Using Stainless Steel Wire Mesh [J]. *Proceedings of the Institution of Civil Engineers – Structures and Buildings* 2021. pp. 1–38.

Yutian Shao, Amir Mirmiran. Control of Plastic Shrinkage Cracking of Concrete with Carbon Fiber-Reinforced Polymer Grids[J]. *Journal of Materials in Civil Engineering* Volume 19, Issue 5. 2007. pp. 441–444.

Study on the conception, static and dynamic performance of the ribbed flat grid

Lanchao Jiang*
Beijing Jiaotong University, Beijing, China

Ri Chol Nam* & Jon Ming Nam*
Pyongyang Communications and Transportation University, Pyongyang, Democratic People's Republic of Korea

ABSTRACT: In recent years, the space structure has made great progress in China, especially the flat grid, which is the fastest-growing and most widely used. Inspired by the reinforced concrete ribbed floor system, this paper conceives a ribbed flat grid structure, which changes the traditional double-layer grid structure into a local three-layer grid structure, and arranges the ribs according to certain rules. It is a new type of grid structure system, which can not only apply to the local prestressed structure but also reduce the overall structure height of the grid structure. In order to study the mechanical characteristics of the common grid structure and the ribbed grid structure system, taking the surrounding support pyramid system as an example, the MST2020 spatial structure analysis software was used to compare and analyze the internal force, stress ratio, structural deformation, structural weight and other parameters of a common grid structure, the ribbed grid structure, the number of ribs, height, type and other design parameters of the ribbed grid structure was set, and finally, finite element calculation was carried out. The structural advantages and engineering feasibility of large-span stiffened steel space truss are verified. Compared with the ordinary space truss, the stiffened space truss has higher structural stiffness, reasonable internal force distribution, and better economic performance.

1 STRUCTURE CONCEPTION

In the development of long-span space structures, the proportion of steel grid structures is increasing, and it can play a great role in the sustainable construction of many projects. To ensure that future work can be carried out in accordance with the expected goals, we should adhere to the design process of the steel grid structure, select the correct ways and methods, and make effective arrangements in different functions, to continuously create higher value in the process of making up for a series of deficiencies. On the other hand, the reliability and feasibility of steel grid structure design also need to be greatly improved. For some specially shaped steel roofs, it is an efficient solution to arrange ribbed grid structures in key stress parts and heavy-duty long-span roofs.

*Corresponding Authors: lcjiang@bjtu.edu.cn, 17239007@bjtu.edu.cn and 17239008@bjtu.edu.cn

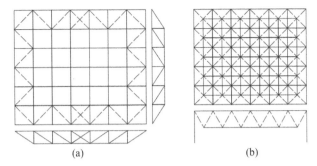

Figure 1. Grid structure form. (a) Cross truss system grid (b) quadrangular pyramid system grid.

1.1 *The present situation of steel grid structure design*

The grid structure is a type of spatial structure which is composed of multiple members connected by nodes according to a certain grid form.

Space truss structures can be divided into four categories according to different compositions: cross truss system, triangular cone system, quadrangular cone system, hexagonal cone system space truss, etc. The common form is the quadrangular pyramid system grid.

In the process of engineering construction and innovation, China is constantly integrating with long-span spatial structures. The grid structure is characterized by high stability, and it still has very good results for all kinds of highly complex projects (Jia & Yan 2021). The development and utilization of the steel grid structure have high value for the operation of long-span space structures, and the overall development trend is very clear. The analysis shows that the current situation of the steel grid structure is mainly reflected in the following aspects: first, it can be industrialized and mass-produced, which has a very free architectural shape, flexibility in construction, and can use small components to form a large space, so it has the characteristics of lightweight and has been highly welcomed in the industry (Li & Li 2007; Wan et al. 2015; Yin 1989). However, it has the problems of excess strength, high manufacturing accuracy requirements, and high cost, especially in the consideration of details, we must strengthen all kinds of tests and research. So, in the system of the steel grid structure, we also need to continue to improve, reduce redundant operation modes, and ensure that the follow-up work can be completed according to higher standards. Second, the design of the steel grid structure must be continuously optimized. Even for small and medium-sized projects, it is necessary to put forward schemes with high reliability and feasibility. The advantage of this operation is that it can continuously achieve higher benefits in making up for a series of deficiencies (Shen 2008; Wang 1991).

1.2 *The idea of the ribbed grid structure*

Inspired by the reinforced concrete ribbed floor system in order to optimize the stress transmission path and stiffness distribution form of the flat plate grid structure, a certain number of truss ribs are arranged in the flat plate grid structure; so, the ribbed grid structure is a structural system with multiple rows of ribs connected under the grid structure, and each row of ribs extends along the span direction of the grid structure (Wang 2021).

In the past ten years, China has been very active in discussing new types of space truss structures. The pipe truss and grid combined structure similar to the ribbed grid structure technology has been applied in the Huadian Jiangling coal yard project (Li & Liu 1989). This structure can significantly reduce on-site welding and shorten the construction period. In addition, the steel structure has been applied more and more because of its light weight and good stability. The structure aims to solve the problem of large spacing of the pipe trusses, but it is still in its infancy. Therefore, the design parameters of the ribbed space truss structure need to be further studied.

LI Jingyun and others (Li 1992; Yin 2089; Wan et al. 2015) carried out a relatively systematic study on the form, stress characteristics, and structural stiffness of the local three-layer grid. The local three-layer grid has certain similarities with the ribbed grid proposed in this paper. The research shows that the steel consumption of the two types of grids is relatively close and the stiffness of the local three-layer grid is better than that of the ordinary flat grid. However, the static and dynamic characteristics of ribbed space grids and ordinary space grids have not been studied in depth.

This paper analyzes the influence of static and dynamic characteristics of the ribbed flat surface grid structure.

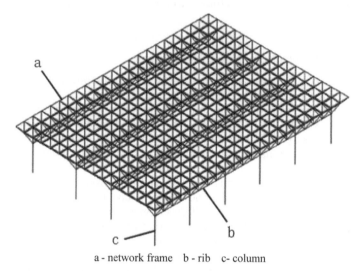

a - network frame   b - rib   c - column

Figure 2. Ribbed grid structure model.

2 COMPUTATIONAL MODEL

2.1 Model overview

An ordinary grid model without ribs and a ribbed grid model are established respectively. The ordinary grid structure adopts the form of an upright quadrangle pyramid, the span of the steel grid structure is 84 m, the plane size of the grid structure is 84 m × 60 m, the height of the non-ribbed grid structure is 3.3 m, and the maximum height of the ribbed grid structure is 6.6 m. The support method adopts peripheral support, and the support is located at the upper chord node. The bar material used for the grid is Q355 steel, the elastic modulus is 2.06×105 N/mm^2, the unit weight is 78.5 kg/mm^3, the Poisson's ratio is 0.3, the section is a hollow circular pipe, the node is bolt ball node, the grid adopts peripheral support, which is fixed and hinged. Each hinged member is the basic unit of grid structure calculation, and each member only bears axial force, not bending moment and shear force.

2.2 Model geometric parameters

2.2.1 Non-ribbed grid
1) Grid height: h = 3.3 m
2) Network size: the network size of upper and lower chords is 4 m × 4 m
3) The layout plan of the grid is as follows:

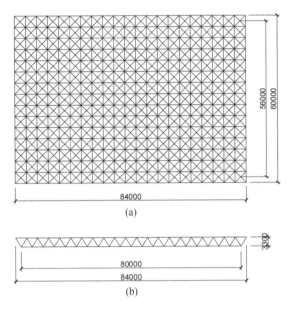

Figure 3. Structural model of normal square pyramid space truss. (a) Plan (b) elevation.

2.2.2 Ribbed grid

1) Grid height: $h_1 = 3.3$ m
2) Network size: the network size of upper and lower chords is 4 m × 4 m
3) Height of rib: $h_2 = 3.3$ m
4) The layout plan of the grid is as follows:

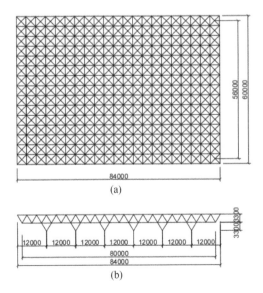

Figure 4. Structural model of square quadrangular pyramid ribbed grid structure. (a) Plan (b) elevation.

2.3 Load parameters

The support condition of the grid is fixed hinge, that is, $D_{x,y,z} = 1$, $R_{x,y,z} = 0$, the surface load of the grid is 0.3 kN/mm² dead load, 0.5 kN/mm² live load, 0.2 kN/m² wind load.

The working conditions considered in the calculation model are dead load, live load, wind load, temperature load, and earthquake action. The maximum temperature is 42°C, the minimum temperature is −22°C, and the closure temperature is 20°C. During the calculation, the temperature action is ±40°C, the seismic fortification intensity is 8 degrees, the seismic acceleration value is 0.20 g, and the site type is class III. because the snow load is relatively small compared with the live load, the snow load is not considered in the analysis, At the same time, it is also necessary to consider the changes caused by temperature working conditions. The combination of 16 working conditions composed of 6 load working conditions including dead load, live load, wind load, earthquake load, and temperature action is considered, as shown in Table 1.

Table 1. Load case combination.

№	Load case combination
1	1.3DEAD
2	1.3DEAD+1.5LIVE
3	1.3DEAD+1.5WIND
4	1.3DEAD+1.5LIVE+0.9WIND
5	1.3DEAD+1.5TEMPERATURE (+42)
6	1.3DEAD+1.5LIVE+0.9TEMPERATUR (+42)
7	1.3DEAD+1.05LIVE+1.5TEMPERATURE (+42)
8	1.3DEAD+1.5WIND+0.9TEMPERATURE (+42)
9	1.3DEAD+0.9WIND+1.5TEMPERATURE (+42)
10	1.3DEAD+1.5TEMPERATURE (−22)
11	1.3DEAD+1.5LIVE+0.9TEMPERATURE (−22)
12	1.3DEAD+1.05LIVE+1.5TEMPERATURE (−22)
13	1.3DEAD+1.5WIND+0.9TEMPERATURE (−22)
14	1.3DEAD+0.9WIND+1.5TEMPERATURE (−22)
15	1.3DEAD+1.3QUAKE
16	1.3DEAD+0.65LIVE+1.3QUAKE

3 STATIC CHARACTERISTIC ANALYSIS

3.1 The internal force of member

The first lines of paragraphs are indented 4 mm (0.16") except for paragraphs after a heading or a blank line (First paragraph tag). Equations are indented 12 mm (0.47") (Formula tag).

To facilitate data comparison, the two grid frames use the same member section. Select the unit with the largest internal force in the grid, and the internal force value is shown in Table 2.

Table 2. Internal force value of the member.

Grid form	Internal force	
	Max (pull rod)	Min (Pressure bar)
Non-ribbed grid	1462.8	−1469.6
Ribbed grid	1628.7	−1090.9

It can be seen from the data in the table that the maximum tension of the non-ribbed grid is 165.9 kn smaller than that of the ribbed grid, and the maximum pressure of the ribbed grid is 379.7 kn smaller than that of the non-ribbed grid.

3.2 *Deformation (disturbance) comparison*

After MSTCAD simulation analysis, the deformation of the grid is shown in Figures 5 and 6. The deformation of the ribbed grid is more uniform than that of the non-ribbed grid, and the deflection of the ribbed grid is smaller. Due to the upward displacement of the ribbed grid, the downward displacement of the load on the grid is partially offset, which increases the stiffness of the grid.

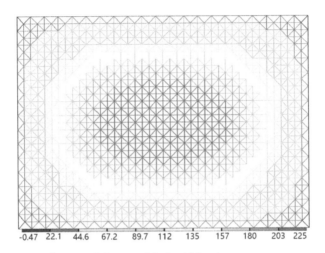

Figure 5. Displacement diagram of the non-ribbed grid (mm).

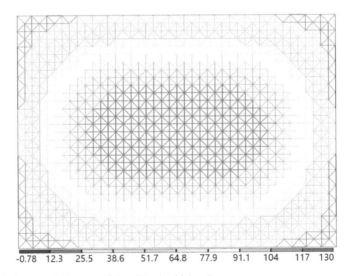

Figure 6. Displacement diagram of the ribbed grid (mm).

It can be seen from the above figure that the stiffening significantly reduces the deflection of the grid. From the comparison of the maximum deflection points, the non-ribbed grid is 225 mm, while the ribbed grid is only 130 mm. The largest reduction is 43%.

3.3 Comparison of member stress

3.3.1 Stress comparison of the top chord

The stiffened grid and the non-stiffened grid are matched with members with the same section. It can be seen from Figures 7 and 8 that the compressive stress distribution of the stiffened grid is more uniform than that of the non-stiffened grid, and the stress of most members is smaller than that of the non-stiffened grid. The uniformity of stress value distribution makes the mechanical performance of more bars play. Because the stiffened part increases the bending stiffness of the structure, the stress value decreases. With the decrease in stress value, the steel consumption of the rod is reduced, and the cost is reduced.

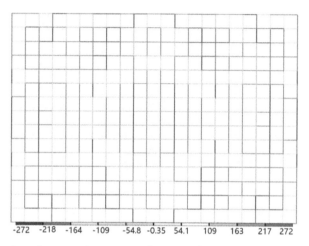

Figure 7. Distribution of compressive stress on the upper chord of non-ribbed space truss (Mpa).

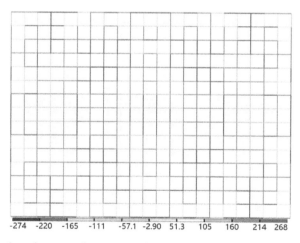

Figure 8. Distribution of compressive stress on the upper chord of ribbed space truss (Mpa).

3.3.2 Stress comparison of the bottom chord

It can be seen from Figures 9 and 10 that the maximum stress of the non-ribbed grid is 271.1 n/mm². The maximum stress of stiffened grid is 267.9 n/mm².

There is little difference in the uniformity of stress distribution between the two grid structures, but as for the bottom chord, the stress of the members of the ribbed grid structure

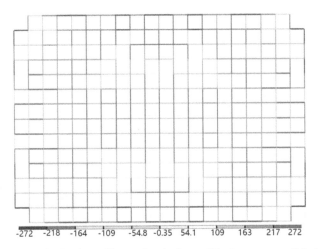

Figure 9. Tensile stress distribution of lower chord of non-ribbed space truss (Mpa).

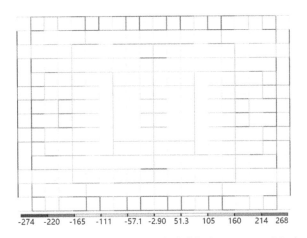

Figure 10. Tensile stress distribution of lower chord of ribbed space truss (Mpa).

is basically smaller than that of the non-ribbed members, which reduces the cross-sectional area of the members and reduces the amount of steel used.

3.4 *The reaction force of support and self-weight of the member*

3.4.1 *Bearing reaction comparison*
Take the reaction force of the support on the upper chord plane of the ribbed grid and compare it with the corresponding non-ribbed grid. The results are shown in Figures 11 and 12. It can be seen from the figure that the maximum bearing reaction of a non-ribbed space frame is 899 kn, and the maximum bearing reaction of a ribbed space frame is 754 kn. Using

Figure 11. The reaction force of non-ribbed grid support (kN).

Figure 12. The reaction force of ribbed grid support (kn).

the support reaction on this side, it can be seen that the distribution of the support reaction of the ribbed grid structure is relatively uniform, which reduces the steel consumption of most of the support nodes and simplifies the design and construction of the grid structure.

3.4.2 *Bearing reaction comparison*
After the optimization design of the two structures, the total steel consumption of the non-ribbed grid is 174.08 T, and the total steel consumption of the ribbed grid is 156.48 T, which can save 17.6 T of steel.

4 DYNAMIC CHARACTERISTIC ANALYSIS

To understand the dynamic characteristics of the non-stiffened grid and the stiffened grid, the general finite element software SAP2000 is used for modal analysis, and the first 20 vibration modes of the non-stiffened grid and the stiffened grid are taken respectively. The first 20 natural vibration periods are shown in Table 3.

It can be seen that the natural vibration period of non-ribbed space grids is significantly longer than that of ribbed space grids. According to the design seismic acceleration response spectrum, the seismic force borne by ribbed space grids is significantly lower than that of non-ribbed space grids.

Table 3. First 20 natural vibration periods of two types of grid structures.

Mode order	Non-ribbed grid/s	Ribbed grid/s
1	0.646	0.493
2	0.479	0.438
3	0.457	0.410
4	0.321	0.301
5	0.259	0.260
6	0.222	0.178
7	0.176	0.169
8	0.167	0.150
9	0.154	0.147
10	0.127	0.122
11	0.113	0.107
12	0.111	0.105
13	0.097	0.094
14	0.07	0.091
15	0.074	0.073
16	0.073	0.065
17	0.056	0.060
18	0.046	0.043
19	0.037	0.032
20	0.035	0.032

5 INFLUENCE ANALYSIS OF STRUCTURAL DESIGN PARAMETERS

5.1 Influence analysis of space truss span

The span of the space truss is an important parameter in the design of a space truss. A span of 60m is designed × 84 m, 72 m × 96 m, 84 m × 108 m, 96 m × the distance between the ribs of the four 120m grid structures remains the same, but the number of ribs increases from 6 to 9.

Under the same load, the maximum displacement and structural self-weight of the four grid structures are calculated by using the full gravity design, as shown in Table 4.

Table 4. Design parameters of different grid spans.

Grid span/m	Number of ribs	Maximum displacement/mm	Structure weight/t
60×84	6	130	156.48
72×96	7	195	228.67
84×108	8	263	330.69
96×120	9	355	469.14

It can be seen from the data in the table that the longer the span of the grid, the greater the deformation and self-weight of the structure.

5.2 Analysis of the influence of the number of stiffeners

The number of stiffeners of stiffened space grids increased from zero to eight, and the full stress method was used to calculate the node displacement and weight of five space grids.

First, it can be seen from the maximum displacement curve in Figure 13 that the more the number of ribs of the grid, the smaller the maximum displacement of the grid. There is a significant gap between the displacement of the grid without ribs and that with two ribs. After the number of ribs exceeds four, the maximum displacement of the grid does not change significantly.

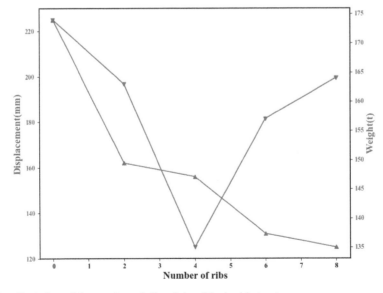

Figure 13. Variation of the number of ribs of the ribbed grid structure.

The self-weight of the grid is the largest when there is no rib, and the smallest when there are four ribs.

5.3 Influence analysis of rib height

By changing the height of the ribs, the displacement and weight of seven kinds of ribbed space grids with different rib heights are compared.

It can be seen from Figure 14 that the greater the height of the rib, the smaller the maximum displacement of the grid. The higher the height of the rib, the self-weight of the grid increases to a certain extent, but it is not obvious.

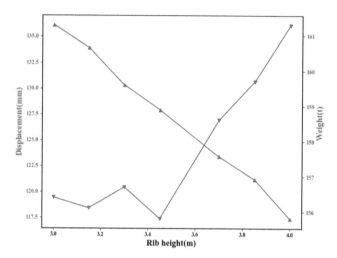

Figure 14. Height change diagram of ribbed grid rib.

5.4 Influence analysis of rib height

To find a good-stiffened form, two forms of stiffened grid structures are built with MSTCAD software, and the structural static analysis of the pyramid ribbed grid and plane truss ribbed grid is carried out with the full stress method.

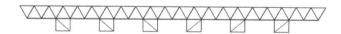

Figure 15. Ribbed grid structure in Truss Form.

From the calculation results in Table 5, the displacement and the self-weight difference between the two kinds of grid structures are small, so two kinds of grid structures can be used in the construction process.

Table 5. Design parameters of different rib forms.

Grid form	Maximum displacement/mm	Structure weight/t
Pyramid rib	130	156.48
Plane truss rib	127	157.96

6 CONCLUSION

In this paper, MSTCAD is used to calculate and analyze the design parameters of a long-span perimeter-supported quadrangular pyramid ribbed steel grid. It can be seen from the analysis that after the arrangement of ribs, the deflection of the grid structure is significantly reduced, and the internal force of the members is redistributed. The internal force of most members decreases. The reduction of the reaction force of the ribbed grid support also simplifies the design and construction of the substructure. In short, through the above static analysis, it can be seen that the long-span perimeter supported quadrangular pyramid ribbed steel grid structure is a kind of large-span grid structure with convenient construction, good mechanical performance, no increase in indoor redundant space of buildings, and low cost.

REFERENCES

Jia Jianbo, Yan Falin (2021). Structural characteristics of multi-layer grid and its application in steel roof of high-speed railway station. *J. Building Structure*. 51 (S1), 531–534.

Li Haiwang, Li Weiwen (2007). Dynamic Response and Failure Mode Analysis of Three-Story Grid Bridge Under Earthquake. *J. Sci-tech Information Development & Economy* 17(7),151–153.

Li Jingyun, (1992). Comparison of Local Three-Layer Network Frame and Flat Plate Network Frame Design Scheme. C. *Proceedings of the Sixth Academic Conference on Spatial Structure*. Beijing. 724–726.

Li Qingjian, Liu Xuewu (1989). Technological Innovation and Engineering Application of Industrial Building Spatial Grid Structure EB/OL.

Shen Zuyan, (2008). *Design of Building Steel Structure*. M. Beijing: China Architecture and Architecture Press. 2–17.

Wan Xiang, Yin Zhiwei, Shao Qingliang (2015). Design and Characteristic Analysis of Local Three-Layer Grid in a College Basketball Gymnasium. *J. Sichuan Building Science*. 41(1), 531–534.

Wan Xiang, Yin Zhiwei, Wang Jian (2015). Analysis of the Whole Work of a Local Three-Layer Grid Frame and its Undersupport Structure. *J. Spatial Structures*. 21(3), 49–53.

Wang Bocheng. (1991). *Research and Application of Segmental Hoisting Technology for Large Span Steel Grid Structure*. D. Chongqing university.

Wang Lijun (2021). *The Utility Model Relates to a Ribbed Grid Structure System*. P. 201711170198.

Yin Dewang, (1989). Application and Research of Three-Layer and Multi-Layer Grid. *Journal of Taiyuan University of Technology*. 20(1), 103–110.

Yin Deyu, (1989). Application and Research of Three-layer and Multi-layer grid. *Journal of Taiyuan University of Technology*. 20(1), 103–110.

Study on the shape of flushing system of diversion pool in urban drainage network

Bocheng Liu
University of Jinan, Jinan, China
China Institute of Water Resources and Hydropower Research, China

Guangning Li*
China Institute of Water Resources and Hydropower Research, China

Yancheng Han
University of Jinan, Jinan, China

Zhensheng Chen
Vacuum Flushing (Tian jin) Smart Water Systems Co. LTD., China

Guoqin Dou
China Institute of Water Resources and Hydropower Research, China

ABSTRACT: To solve the storage pool water in a large number of silts, suspended solids, and heavy metals in bottom sediment serious problems, combined with the bottom sediments of anti-scouring feature, it is expected to change the outlet of the negative pressure water flushing system size, water diversion chamber, flushing corridor, and the length of the slope, and make it more reasonable, thus to achieve more efficient desilting effects in flushing water. By using CFD software to carry out a three-dimensional numerical simulation of the scour flow in the flushing system and combining it with the relevant physical scour test, the flow velocity and flow pattern of the bottom water flow were obtained, the appropriate hydraulic indexes were established to evaluate the scour capacity of the flow, and the optimization suggestions of the flushing system were given for the design of this kind of engineering.

1 INTRODUCTION

Combined sewer overflow pollution (CSOs) (Li 2004) can pose a serious threat to the environment and drinking water safety in water bodies, and urban drainage network storage tanks can effectively intercept rainwater and combined sewerage generated during heavy rainfall. In Japan, rainwater storage ponds were set up in the 1960s to treat and utilize rainwater. In the last century, Germany, the United States, and the United Kingdom also built a large number of storage ponds in urban stormwater pollution control measures (Yang et al. 2008, Tang et al. 2003). In the 1990s, Shanghai was the first city in China to propose pollutant control techniques for combined flow drainage systems (Huang et al. 2008), and in recent years more and more cities in China have been building storage ponds to alleviate environmental pollution problems in urban water bodies. Sediments can accumulate at the bottom of a storage pond, and when the sediment volume is too thick, it will occupy the storage pond space and reduce the effective storage capacity. The accumulation of sediment over a long period of time also causes corrosion and damage to the bottom of the pond, and

---

*Corresponding Author: lgnchina@163.com

the odor emitted has a negative impact on the surrounding environment. The most convenient and energy-efficient method is to use a hydraulic device to create a jet at the bottom of the pond, using the kinetic energy of the water to destroy and carry away the silt from the bottom. One of the practical applications of this method is the diversion flushing system (Liu et al. 2015), the process of which consists of two steps: water storage and drainage flushing, where the proper design of the diversion chamber, outlet body, and flushing corridor is very important to ensure flushing efficiency. This paper investigates the physical characteristics of sediments in the urban drainage network, such as their flushing resistance, and studies and utilized the body shape of the flushing system for the regulating pond of the urban drainage network to achieve efficient flushing of sediments, which can provide a new solution to alleviate the siltation problem of urban drainage pipes.

2 FLUSHING RESISTANCE OF URBAN PIPELINE SEDIMENTS

Based on the physical and chemical properties of pipeline sediments, Ahyerre (Ahyerre M et al. 2001) classified them into three categories: underlying coarse-grained sediment (GBS), organic layer (OL), and biofilm (Biofilm). The literature shows (Ma 2014) that the biofilm cover can be destroyed by experimental measurements at a shear force of 0.5 N/m^2, i.e., a flow velocity of approximately 0.5 m/s, with a certain erosion rate for sediments with a biofilm cover. It can be seen that when the flow velocity is greater than 0.5m/s the main consideration can be the pushing effect of the water on the sediment. In the Water Supply and Drainage Design Manual (Beijing Municipal Engineering Design and Research Institute 2004), the concept of stopping flow velocity of pipeline sediment is proposed based on the concept of sediment dynamics, and the range is given as 0.35~0.40 m/s, and the lifting flow velocity is 2.4 times of the stopping flow velocity. The Beijing Municipal Engineering Design Institute made a large number of observations on the completed sewage pipes in Beijing in 1965 and obtained the non-siltation flow velocity generally in the range of 0.4~0.5 m/s, which is similar to the above stop flow velocity values. Shi Shan (Shi 2015) found that the flow velocities of pipe sections with sedimentation were basically below 0.65 m/s when investigating the sewerage network in Xi'an, while sedimentation did not occur in pipes with flow velocities above 0.7 m/s. Combined with the technical characteristics of the large amount and thickness of sediment in the storage pond and the short flushing time, this paper takes a flow rate greater than 2.0 m/s as the effective flushing flow rate. At the same time, this paper refers to the flushing water flow of the sedimentation pond in the Hydraulic Design Manual (Dong et al. 2014). To introduce $q_s = (1.1 \sim 1.25)v_s h_{cp}$ (q_s is the single width flow rate; v_s is the flushing flow rate; h_{cp} is the average water depth when flushing sand), in order to reflect the flushing capacity of the water flow more intuitively.

3 3D NUMERICAL SIMULATION

3.1 Diversion flushing system body type scheme

The diversion flushing system consists of a diversion chamber, a flow outlet, a flushing gallery, and a tailing basin, as shown in Figure 1 (a). A comparative study was carried out for different outlet body types (Figure 1 (b), (c), and (d)), gallery length, gallery bottom slope, and chamber volume. Body type I has the outlet sloping upwards, body type II is based on body type I and adjusts the outlet direction to slope 30° downstream (Gao 2017); body type III adjusts the outlet direction to slope 30° downstream and further changes the body shape of the diversion chamber in the hope of improving the efficiency of water utilization, and the flow outlets all use rectangular nozzles with the same width as the bottom of the pond. The flushing gallery lengths for the comparison scenarios are 20 m, 40 m, 60 m, 80 m, and 100 m respectively, and the gallery bottom slopes are 0.5%, 1%, 2%, 3%, 4%, and

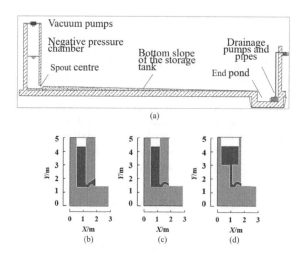

Figure 1. Schematic representation of the structure of the negative pressure diversion flushing system and the body shape of the water outlet. (a) Illustration of the structure of the diversion flushing system (b) Body Type I (c) Body Type II (d) Body Type III.

5% respectively. The volume of the diversion chamber is 18 m³ (standard volume), 36 m³ (2x volume), and 54 m³ (3x volume) respectively, increasing the volume by vertical expansion with no change in height.

The volume of the basic scheme is 18 m³, the height difference between the liquid level of the supply tank and the center of the outlet is fixed at 4.5 m, the height of the outlet is 0.3 m, the flushing gallery is 30 m long, the bottom slope is 1%, the outlet is Body I, the tailing tank is set at the downstream end and the width of each of the above parts is 4 m.

3.2 Control equations, calculation region, and boundary conditions

The control equations and solution methods used are the same as those in the literature (Li et al. 2016), including the continuity equation, momentum equation, and k-ε equation. The model is solved using the finite volume method in second-order windward format, the pressure-velocity coupling is solved using the pressure correction method, the discrete equations are solved using the GMRES method, the time difference is solved using the fully implicit format, and the VOF (The Volume of Fluid) method is used for the free water surface. The calculation area includes the negative pressure diversion chamber, the flow outlet, the flushing gallery, the tailing basin, etc. The 3D digital model is shown in Figure 2. In the calculation, the initial water body is set within the model diversion chamber, the flow outlet boundary is free outflow, the solid wall boundary meets the no-slip condition, and the water surface is a free surface. The model uses a nested and encrypted rectangular grid with grid sizes ranging from 0.01 m to 0.1 m, with a total grid size of approximately 1.2 million.

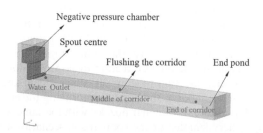

Figure 2. Three-dimensional model of the negative pressure diversion flushing system.

3.3 Model validation

Numerical calculations and field flushing tests (geometric scale 1:1) were carried out for the basic scheme and the results were compared. The field flushing test was designed at a 1:1 scale, as shown in Figure 3, with the same body shape and dimensions as the calculated scheme. The flow velocity process lines at the middle and end of the flushing corridor were extracted and compared. The flow velocity values and the flow velocity change process were basically the same, with a maximum deviation of 6.5%, as shown in Figure 4, indicating that the numerical calculation method used in this paper is reliable.

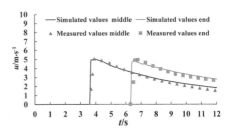

Figure 4. Comparison of test measurements with numerically calculated flow rate processes.

Figure 3. Flushing test site.

4 CALCULATION RESULTS AND ANALYSIS

4.1 Water outlet body type

Numerical calculations were carried out for water outlet body type I, body type II, and body type III. The flow velocity and effective flushing time at the three locations of the water outlet, the middle of the flushing gallery, and the end of the flushing gallery were compared and analyzed, as shown in Figure 5. The results show that the maximum flow velocity at the center of the outlet of Body Type I is 2.4 m/s and the effective flushing time is 0.6 s. The maximum flow velocities at the middle and end of the flushing gallery are 5.0 m/s and 4.8 m/s respectively, and the water can be flushed effectively for about 7.5 s. The maximum flow velocity at the center of the outlet of Body Type II is 5.3 m/s, which is 121% higher than that of Body Type I. The effective flushing time is 3.8 s. The maximum flow velocities at the middle and end of the flushing gallery are 6.0 m/s and 5.6 m/s respectively, and the water can be flushed effectively for about 8 s. The maximum flow velocity at the center of the outlet of body type III is 4.8 m/s, with an effective flushing time of 4.5 s. The maximum flow velocities in the middle and at the end of the flushing gallery are 5.7 m/s and 5.2 m/s respectively, with an effective flushing time of about 7.5 s.

It can be seen that the flow velocity at the outlet of Body I is relatively small and the flushing capacity is poor. The reason for this is that the outlet direction of Body Type I is inclined upwards, so the flushing water is shot up into the air and will consume part of the potential energy conversion, so the maximum flow velocity is less than the other two body types. Body type II has the highest maximum flow velocity of the three options at the outlet, in the middle, and at the end. Body type II, by changing the outlet to a 30° downward jet, the potential energy of the water flow can be converted into kinetic energy more directly and effectively, which acts on the gallery floor to achieve higher flow velocities and a significant increase in flushing capacity, indicating that changing the direction of the flow outlet can

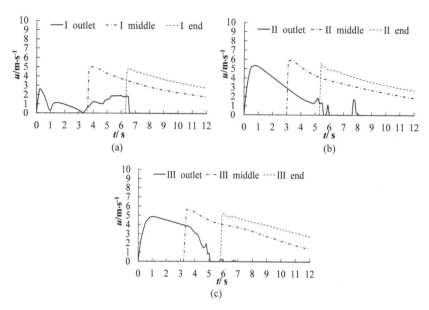

Figure 5. Comparison of flow rates for the three body types. (a) Body Type I Flow Rate (b) Body Type II Flow Rate (c) Body Type III Flow Rate.

effectively increase the flow velocity at the outlet and enhance the instantaneous flushing capacity. Body Type III raises the position of the entire water body, increasing the potential energy and lengthening the path from the diversion chamber to the outlet, enabling the high flow rate to be sustained for a long and relatively smooth period of time. Body Type III is more difficult to construct, but the results show that raising the water body and increasing the potential energy is an effective way to increase flushing capacity, so if space is available in the upper part of the water body, an increase in the volume of the upper part can be considered. On balance, Body Type II was chosen as the basis for a comparative study of different flushing gallery lengths, bottom slopes, and diversion chamber volumes.

4.2 Length of flushing gallery

Based on Body Type II, a comparative study was carried out for the bottom slope of 1% and the flushing gallery lengths of 20 m, 40 m, 60 m, 80 m, and 100 m respectively. Of the five corridor lengths, only the end section in the 100 m corridor did not reach the effective flushing velocity of 2 m/s. The rest of the sections met and exceeded the effective flushing velocity. As both the body and outlet chamber size are the same for each body type, the outlet flow velocity process is essentially the same, with the difference being the variation in flow velocity and single-width flow rate q in the middle and end of the corridor, as shown in Figure 6.

The results show that the maximum flow velocities in the middle and end of the flushing gallery are 6.50 m/s and 5.68 m/s respectively for the 20 m corridor length scheme, and the water flow can be maintained for about 7 s with a maximum single width flow rate q of 1.07 $m^3/(s·m)$ and 1.05 $m^3/(s·m)$ respectively. The maximum flow velocities in the middle and end of the flushing corridor are 5.12 m/s and 4.16 m/s respectively for the 40 m corridor length option, and the flow can be effectively flushed for about 9 s with a maximum single width flow rate q of 0.90 $m^3/(s·m)$ and 0.57 $m^3/(s·m)$ respectively. The maximum flow velocities in the middle and end of the flushing corridor for the 60m length option are 4.81 m/s and 3.22 m/s, respectively, and the flow can be maintained for about 10 s with a maximum single width flow rate q of 0.66 $m^3/(s·m)$ and 0.39 $m^3/(s·m)$, respectively. The single-width

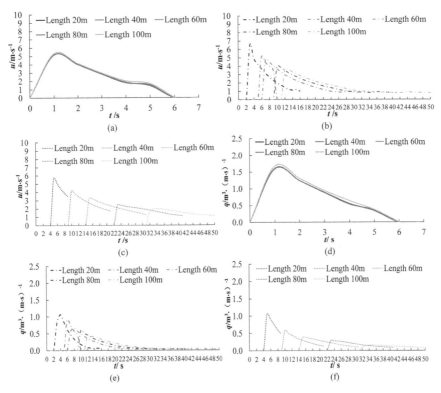

Figure 6. Hydraulic parameters for different length options of the flushing corridor. (a) Water outlet flow rate (b) Flow rate in the middle (c) Flow rate at the end (d) q at water outlet (e) q in the Middle (f) q at the end.

flow rate q for the 60 m end section reaches 0.36 m³/(s·m) for only 1.5 s. The maximum single-width flow rates q for the 80 m and 100 m corridor lengths were only 0.28 m³/(s·m) and 0.15 m³/(s·m), which is not a good flushing effect.

This shows that the corridor length has a strong influence on the flushing effect, as the corridor length increases, the flow rate decreases in the middle and end of the corridor, and the single-width flow rate q also decreases in turn. On the whole, good flow rates are maintained in the flushing corridor for corridor lengths ≤ 60 m. For corridor lengths > 60, all flushing indicators in the flushing corridor are unsatisfactory.

4.3 The bottom slope of the flushing gallery

Based on Body Type II, calculations were carried out for a flushing gallery length of 30 m and bottom slopes of 0.5%, 1%, 2%, 3%, 4%, and 5% respectively. As the chamber body and outlet body are the same for all body types, the outlet flow velocity process is essentially the same, the difference being the variation in flow velocity and single width flow rate q in the middle and end of the gallery, as shown in Figure 7.

The results show that the maximum flow velocity in the middle and end of the flushing gallery for the 0.5% bottom slope of the corridor are 5.48 m/s and 4.47 m/s respectively, and the water flow can be maintained for about 9 s. The maximum single-width flow rate q is 0.83 m³/(s·m) and 0.61 m³/(s·m) respectively. The maximum flow velocities at the middle and end of the flushing corridor are 5.56 m/s and 4.70 m/s respectively for the 1% bottom slope of the corridor, and the flow can be maintained for about 10 s. The maximum single-width flow rates q is 0.80 m³/(s·m) and 0.75 m³/(s·m) respectively. The maximum flow velocities at the middle and end of the flushing corridor are 6.08 m/s and 4.90 m/s respectively for the 2% bottom slope of

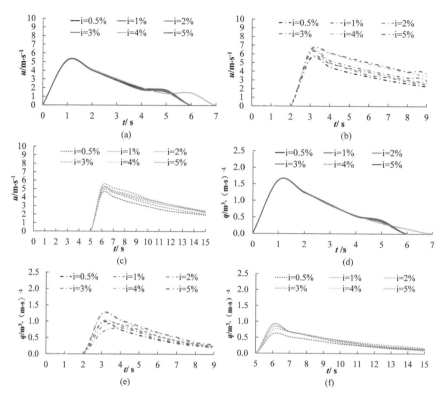

Figure 7. Hydraulic parameters for different bottom slope options for the flushing corridor. (a) Water outlet flow rate (b) Flow rate in the middle (c) Flow rate at the end (d) q at the water outlet (e) q in the Middle (f) q at the end.

the corridor, and the flow can be maintained for about 11 s. The maximum single-width flow rates q is 0.91m³/(s·m) and 0.73 m³/(s·m) respectively. The maximum flow velocities at the middle and end of the flushing corridor are 6.20m/s and 5.00m/s respectively for the 3% bottom slope of the corridor, and the flow can be maintained for about 10 s. The maximum single-width flow rates q is 0.98m³/(s·m) and 0.82 m³/(s·m) respectively. The maximum flow velocities at the middle and end of the flushing corridor are 6.32 m/s and 5.14 m/s respectively for the 4% bottom slope of the corridor, and the flow can be maintained for about 10 s. The maximum single-width flow rates q is 1.21 m³/(s·m) and 0.88 m³/(s·m) respectively. The maximum flow velocities at the middle and end of the flushing corridor for the 5% bottom slope are 6.48m/s and 5.35m/s, respectively, and the flow is maintained for about 9 s. The maximum single-width flow rates q is 1.23 m³/(s·m) and 0.91 m³/(s·m), respectively.

It can be seen that the maximum flow rate for each slope option can achieve an effective flushing flow rate. The higher the bottom slope J of the flushing gallery, the higher the flow rate in each section of the flushing gallery and the longer the effective flushing time; the single width flow rate q increases with the increase of J, i.e., the flushing capacity is stronger and more stable.

4.4 *Diversion chamber volume*

Based on Body Type II, with a flushing gallery length of 30m and a bottom slope of 1% respectively, calculations were carried out for scenarios with a diversion chamber volume of 18 m³ (standard water volume), 36 m³ (2x water volume), and 54 m³ (3x water volume) respectively. The flow rates and single-width flow processes for flushing are shown in Figure 8.

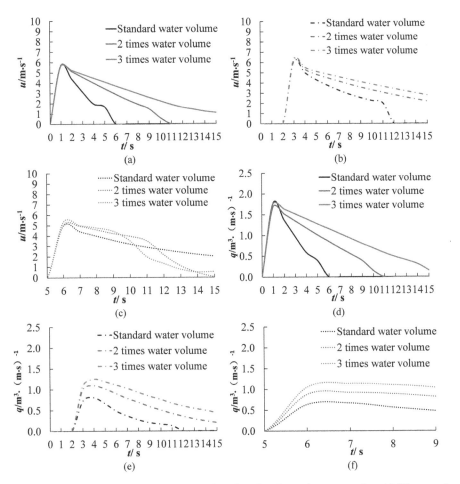

Figure 8. Hydraulic parameters for different diversion chamber volume scenarios. (a) Water outlet flow rate (b) Flow rate in the middle (c) Flow rate at the end (d) q at the water outlet (e) q in the Middle (f) q at the end.

The results show that the maximum flow velocities in the middle and end of the flushing gallery for the 18m³ chamber volume scenario are 6.11 m/s and 4.95 m/s respectively, and the water can be effectively flushed for about 10s with maximum single-width flow rates q of 0.80 m³/(s·m) and 0.67 m³/(s·m) respectively. The maximum flow velocities in the middle and end of the flushing corridor for the 36 m³ chamber option are 6.24 m/s and 5.02 m/s, respectively, and the flow can be maintained for about 11.5 s with a maximum single width flow rate q of 1.09 m³/(s·m) and 0.92m³/(s·m), respectively. The maximum flow velocities in the middle and end of the flushing corridor for the 54 m³ chamber volume scenario are 6.35 m/s and 5.32 m/s, respectively, and the flow is maintained for about 12s with a maximum single width flow rate q of 1.25 m³/(s·m) and 1.40 m³/(s·m), respectively.

This shows that as the volume of the chamber becomes larger, the decay of the flow rate becomes significantly slower and the effective flushing flow rate is maintained for a longer period of time. The larger the volume of the chamber, the larger the q-maximum, the stronger the pushing capacity of the flow, and the longer and more stable the strong flushing state is maintained. The flushing volume is mainly determined by the energy difference between the upstream and downstream (Guo et al. 2004), and a higher liquid level in the reservoir can increase the flushing volume of the sediment.

5 CONCLUSIONS AND RECOMMENDATIONS

Through the three-dimensional numerical calculation of the flushing process of the transfer pond and the analysis of its hydraulic elements, the following conclusions were drawn.

(1) The target flushing object sources, classification, and flushing resistance characteristics were investigated, and the flow rate and single-width flow rate were selected as evaluation indicators.
(2) The silt surface is smooth, the bottom is consolidated, and the flushing water flow needs to form a certain flushing angle with the sediment surface, which can be oriented at 30° downwards diagonally from the bottom slope.
(3) The length of the corridor should not be too long (≤ 60 m is appropriate for the body type in the text), as the length of the corridor increases, the flow velocity and single-width flow rate q in the corridor decrease in turn.
(4) The larger the slope J of the corridor floor is, the larger the velocity in the corridor will be. Single flow q will also increase with the increase. That is enhanced flushing ability.
(5) The larger the volume of the diversion chamber, the longer the time to maintain the effective flushing flow rate, the larger the q value, the stronger the ability of the water to push, raising the water body to increase the potential energy is an effective way to enhance the flushing capacity, in the engineering design should try to increase the volume of the diversion chamber, and try to increase the volume of the upper water body.

REFERENCES

Ahyerre M., Chebbo G., Saad M. (2001). Nature and Dynamics of Water Sediment Interface in Combined Sewers [J]. *Environ Eng*, 127(3):233–239.

Beijing Municipal Engineering Design and Research Institute (2004). *Water Supply and Drainage Design Manual* [M]. China Construction Industry Press.

Dong Anjian, Li Xianshe et al. (2014). General Institute of Water Resources and Hydropower Planning and Design, (2014). Ministry of Water Resources. *Handbook of Hydraulic Design* [M]. China Water Conservancy and Hydropower Press.

Gao Yaping. (2017). Experimental Study of Planar Oblique Jet Erosion of Sand Beds with Different Slopes [J]. *South-North Water Diversion and Water Conservancy Science and Technology*, (03):147–154.

Guo Q., Fan C.Y., Raghaven R. (2004). Raghaven R. Gate and Vacuum Flushing of Sewer Sediment: Laboratory Testing [J]. *Journal of Hydraulic Engineering*, 130(5):463–466.

Huang M. & Chen H. et al. (2008). Analysis of the Design and Operational Efficiency of Rainwater Storage Tanks in Chengdu Road, Shanghai [J]. *China Water Supply and Drainage*, 24(18):33–36.

Li Chunguang (2004). Setting and Suggestions for Combined Sewerage System Storage Tanks [J]. *Shanghai Construction Science and Technology*, 19–20.

Li Shuntao, Geng Bo et al. (2016). CFD Simulation Analysis of Sand Flushing Process in Sedimentation ponds [J]. *Hydropower Energy Science*, 34(11):104–108.

Liu Z.H. & Li H.Z. (2015). Hydraulic Scouring Technology for Drainage Pipe Sediments [J]. *Science and Technology Innovation and Application* (02);1–3.

Ma Yan. (2014). *Study on the Erosion Transport Characteristics and Laws of Sediment in Drainage System Pipes* [D]. Zhejiang University.

Shi Shan. (2015). *Study on Sediment Formation Pattern and Influencing Factors of Sewage Network in Xi'an City* [D]. Xi'an University of Architecture and Technology.

Tang J. & Cao F. et al. (2003). Introduction of Drainage Pipes in Germany [J]. *Water Supply and Drainage*, 29 (5):4–9.

Yang Xue & Che Wu et al. (2008). Control and Management of Overflow Pollution in Combined Flow System Pipelines at Home and Abroad [J]. *China Water Supply and Drainage*, (16):7–11.

Research on characteristics and quality defects of aluminum molds in high-rise buildings

Jichao Xie*

Kunming University of Science and Technology Oxbridge College, Kunming, China

ABSTRACT: With the change in the international energy situation, a fresh round of technological revolution and industrial transformation, "Aluminum mold" as a kind of green template, is used by more and more construction industry. Through comparative analysis of the characteristics and limitations of aluminum mold, the common quality problems of aluminum mold in high-rise buildings are analyzed, such as the potting, porosity, peeling, slurry leakage, structural deformation, and empty drum in the process of mold removal, and the corresponding prevention measures are put forward to provide a reference for the construction management of aluminum mold in high-rise buildings.

1 INTRODUCTION

The application of aluminum formwork has gradually replaced wood formwork and steel formwork in construction since 2000. Aluminum dies have become widely used with the promotion of green energy saving, assembly types, and industrial building concepts. Aluminum alloy formwork is well suited for steep-rise residential construction projects due to its lightweight, lengthy turnaround times, excellent construction quality, less construction waste, no reliance on lifting equipment, and the advantage of shortening the overall construction period (Liu 2020; Shi 2016; Zhang 2020).

At present, the construction technology, construction points, and advantages and disadvantages of the aluminum mold are mainly focused on in the research of aluminum mold in our country: Xie Jianglu 2017 studied the application of aluminum alloy formwork in the green construction of super high-rise buildings. Zhao Xueqi 2020 studied the advantages and limitations of aluminum mold. Fang Sanling 2018 conducted an example analysis and comparison of the aluminum alloy template and wood template; Lu Liang 2017 studied the comprehensive value of aluminum alloy formwork in construction engineering; Wang Xuanyi 2020 studied the construction technology of aluminum formwork in high-rise buildings.

In this paper, based on the study of the characteristics of aluminum mold, a series of common quality defects of aluminum mold in high-rise building construction, such as concrete forming quality, pouring slurry leakage, formwork structure deformation, bay window forming quality, are analyzed, and the related prevention measures are proposed.

2 ALUMINUM MOLD OVERVIEW

2.1 *Introduction of aluminum mold and its support system*

Aluminum template and aluminum alloy template, according to ≪Aluminum alloy template≫ (JG/T 522-2017) in the standard provisions, for which aluminum alloy profile is the main material, is suitable for concrete engineering template. The aluminum alloy

---
*Corresponding Author: 957088335@qq.com

formwork system is used in the concrete pouring process, and the classification of aluminum alloy formwork is shown in Table 1 below:

Table 1. Classification of aluminum alloy templates.

Number	Classification form	Classification
1	Structure form	Plane template
		The angle of the template
		Component
2	To form	Pull rod formwork
		Pull type template
3	General form	Standard template
		Nonstandard form

Aluminum formwork can be combined and assembled into different sizes and relatively complex overall formwork, which is the preferred system formwork for assembly and industrial construction. The aluminum alloy template system consists of a plate surface system machined from an extruded aluminum alloy profile with high-strength steel support and fastening system, coupled with a high-quality hardware bolt and attachment system to support the use of general accessories. The slab system is a bearing plate directly in contact with the cast-in-place concrete. Supporting and fastening systems are temporary structures used to support and stabilize formwork, concrete, and construction loads. An accessory system is a supporting tool for mold installation and removal.

Roughly 80 percent of the modules in the aluminum formwork system can be recycled in additional projects, and 20 percent of the modules can only be recycled in one class of standard floors. As a result, aluminum formwork systems are suitable for ultra-high-rise buildings with a significant degree of standardization.

2.2 *Characteristics of aluminum mold – compared with wood mold and steel mold*

In recent years, the domestic construction field has been rapidly growing and developing at an average annual rate of roughly 20%. Aluminum formwork has the advantages of construction, environmental protection, and efficiency.

Table 2. Comparison and analysis of aluminum alloy characteristics with wood die and steel die.

Template	Wooden template	Steel template	Aluminum template
Material	Wooden	Steel	Aluminum alloy
Weight (kg/m^2)	10.5 (18 mm)	60–80 (2.3–6 mm)	20–25 (3–4 mm)
Bearing capacity (kN/m^2)	30	30–60	60
Lifting equipment	Manual assembly, partial use	Coordinated use (4 persons to coordinate installation)	Professional manual installation (coordinated by 2 persons)
	Aluminum mold light weight, bearing capacity and steel mold, need 2 special installations		
Turns	5–8	30–80	100–300
Construction time limit for a project	6, 7 days a floor	8 days a floor	4 days a floor
Time to dismantle	Floor board 168 h	–	Column and shear wall 24 h, floor plate 36 h

(*continued*)

Table 2. Continued

Template	Wooden template	Steel template	Aluminum template
Efficiency of construction	10–15 m²/day	15–25 m²/day	20–50 m²/day
	Aluminum mold high turnover rate, a fast construction period, and high efficiency		
Concrete forming and surface effect	Rough surface and vulnerability to quality problems	Smooth surface, precision is not easy to control, an improper connection is prone to quality problems	The surface is smooth and smooth, forming quality is good, need to use a release agent
Subject to external influences	Exposure to water leads to deformation of die expansion	Long-term use is easy to rust	No deformation, no corrosion
Environmental protection	Waste wood, construction site-generated waste mold and nails and other garbage	Basically, no construction waste	Basically, no construction waste
Fire safety	Flammable, fire	Non–flammable, poor fire resistance	Non–flammable, poor fire resistance
	Aluminum mold high precision, environmental protection		
Specifications	Large size can be on–site processing size	Serialization and industrialization	Serialization and industrialization (customized by manufacturers)
Application area	Wall, beam, pillars, and plate	Wall, beam, pillars, and plate	All components
	Aluminum mold can be industrialized construction of whole parts		

In addition, aluminum dies have the advantage of low carbon emissions and elevated recycling value.

3 LIMITATIONS OF ALUMINUM MOLD

Currently, aluminum molding in high-rise buildings has the following limitations:

(1) Aluminum mold construction needs to be deepened: The first step is to deepen the drawing – commonly, starting from the third layer, component size, optimization of wall stacking, deepening lintel, structural column, staircase, expansion joint, mechanical and electrical, weight concrete exterior wall, etc., to ensure the accuracy of construction. Secondly, the additional design is carried out from the aspects of deformation, bearing capacity, material properties, back bending, vertical rod, and arch lifting of beam and plate of aluminum mold (Wang 2020), so as not to affect the construction quality.
(2) The aluminum mold system has a high degree of serialization, but its applicability has certain limitations: it has certain advantages in the construction process of standardized buildings, especially when the standard floor is more than 30 floors (Lin 2014), so it does not have the advantage of low cost in the middle and high-rise buildings or the projects with additional flat apartments, and the underground and conversion floors of another building are also not applicable.

4 COMMON QUALITY FAULTS, CAUSES, AND PREVENTION MEASURES OF ALUMINUM MOLD IN HIGH-RISE CONSTRUCTION

According to the "2020 Aluminum Formwork Industry Development Survey Report" by the China Formwork and Scaffolding Industry Association: in recent years, the construction speed of Chinese construction has increased by 20% per year. Based on fully borrowing the development of a foreign aluminum mold system, Chinese aluminum mold is deepening and extending new technical characteristics. China's aluminum mold industry is on the rise, the high-rise housing has reached 30% of the market share, but there is still room for a continuous rise. More and more enterprises and industries pay attention to the application and development of aluminum mold. According to the relevant construction materials and literature, the common quality problems, causes, and prevention measures of aluminum mold in high-rise construction are summarized.

4.1 *There are pockmarks, pores, and peeling on the concrete surface*

The surface of the concrete is liable to be pockmarked by the oxidation of the aluminum and concrete, and the principal materials of the aluminum mold. The sealing of the aluminum mold is excellent. When the mold is removed, the pores will be generated on the concrete surface, which will affect the use function and beauty of the structure, and even lead to the putty bubble phenomenon in the later decoration project.

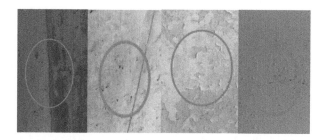

Figure 1. Concrete surface pockmark, air hole, peeling, putty bubble diagram.

Table 3. Causes and prevention measures of concrete pockmarks, pores, peeling, and putty bubbles.

Common causes of concrete forming problems	Prevention and control measures
The template is not clean	Residue must be removed after mold removal
Improper use of release agent and uneven coating	Reasonable use of release agent (mixing quality standards), brushing evenly
Concrete leakage vibration, less vibration, vibration is not in place	Layered casting with concrete vibrating in place at each position
Concrete water ash is relatively large, or too much fly ash	Control water–cement ratio during trial mixing

If there are pores in the concrete surface, the putty scraping process can also be improved during the putty-making process-coarse putty can seal the base or bubble back pressure.

4.2 Slurry

Aluminum mold in the installation and use of the process will have a template between the joint is not strict-gap, resulting in concrete pouring leakage, and even concrete surface honeycomb, leakage reinforcement, holes, etc., leakage will affect the beauty and even durability of the structure.

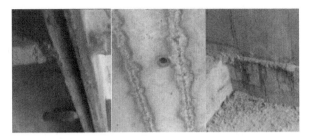

Figure 2. Schematic diagram of joint, leakage, and root rot of aluminum mold.

Table 4. Aluminum mold joint, slurry leakage, root rot causes, prevention, and control measures.

Slurry leakage reason	Prevention and control measures
Template deformation or size deviation	Correct and replace the unqualified template
The template is not installed vertically	Control the ground leveling and calibrate the template
Template cement paste is not cleaned	After removing the mold, remove the cement slurry in time, and pay attention to the use of a release agent
Pin piece is not fastened in place	Tighten the pins and pins at the joints after the die is finished
Premature removal or application of construction load	During construction, the vertical formwork is removed after 12 h, and the beam and plate bottom molds are removed after 48 h, but the supporting system is retained

The support of the joint part of the beam, column, wall, and board should be ensured to be firm, the joint should be tight, the dislocation should be corrected in time, and the mortar should be used to block the gap before pouring.

4.3 Formwork and structural deformation

Aluminum formwork typically has a turnover rate of between 100 and 300, but after repeated use, the formwork becomes misshapen, causing deformities in the concrete after the molding is removed. Or aluminum molding plates, which are highly common and lack overall stiffness, can also cause structural deformations.

Figure 3. Diagram of aluminum mold deformation and small plate.

Table 5. Reason and prevention measures of deformation of aluminum mold and structure.

Deformation reasons	Prevention and control measures
Pouring wall column concrete too fast or at one time too high	Uniform feeding and strict control of the pouring height, especially the door and window openings to ensure that the concrete vibration is compacted, and prevent excessive vibration caused by formwork deformation
The stiffening stiffness of the formwork is insufficient	According to the requirements of adding pins, pins, back, pull screw, etc., to ensure the stiffness of the support phantom system
Small plate too much	Focus on the template deepening for non-standard parts and the lower part of the negative corner of the wall panel
Too much deformation at the top of the wall	When deepening, add a pull-screw on the top of the wall or add a pull-screw on the top of the wall

In addition, attention should also be paid to the acceptance and material inspection before the entrance of the aluminum mold and construction.

4.4 Bay window bubble, mold expansion, mold explosion, board top plastering empty drum

In the construction of the bay window, there are commonly bubbles, mold expansion, and mold explosion, resulting in deformation of the bay window, and even an empty drum when plastering the bay window roof.

Figure 4. Schematic diagram of the opening and reinforcement of the bay window aluminum die.

Table 6. Reason for bay window deformation, bubble, mold expansion, and prevention measures.

Reason	Prevention and control measures
No pressure groove is designed on the top of the bay window	When deepening the drawing, pay attention to the design of the pressure groove on the top of the bay window
The bay window is not properly strengthened, and the floating causes deformation	Reasonable reinforcement of bay window template
The pressure is high during pouring concrete	The bay window panel is arranged on the pull screw, the back flange is set for reinforcement, and the top support is set up and down for the back top; Bay window panels are perforated to reduce air bubbles

In view of the deformation and plastering of the bay window after the removal of the mold, the bay window without toning and deformation of the ceiling should be plastered, and the pressure frame is 5 mm; the ceiling base plaster before the use of beat slurry process and do a fine job before and after maintenance work.

5 CONCLUSION

Compared with wood mold and steel mold, aluminum mold has great advantages in terms of the material itself, finished product quality, and environmental protection, but it has certain limitations in the number of building layers, optimal design, and even professional personnel.

In view of the field application of aluminum mold in high-rise buildings, common quality defects are found, such as poking, air holes, peeling, leakage, structural deformation, and hollow drum in the process of mold removal, and the corresponding prevention measures are put forward to provide a reference for the efficient application of aluminum mold in the later stage.

ACKNOWLEDGMENT

This paper is one of the results of the research on "aluminum mold + climbing frame + full penetration" construction in the construction management of high-rise buildings, which is a scientific research fund project of the Yunnan Education Department.

REFERENCES

2020 Aluminum Formwork Industry Development Survey Report.

Aluminum Alloy Formwork. JC/T 522-2017

Hong L. Z. (2014). *Research on Construction Quality Control of Real Estate Development Projects Based on Aluminum Alloy Building Formwork System*. (03):137

Jiang L. (2020). *Study on Application of Aluminum Formwork in High-rise Residential Building Engineering*. South China University of Technology.

Liang L. (2017). *Study on the Comprehensive Value of Aluminum Alloy Formwork in Construction*. Taiyuan University of Technology.

Lu S. C. (2016). *Research and Application of Aluminum Formwork Construction Technology in Super Tall Buildings*. Anhui University of Science & Technology.

Ling F. S. (2018). *Quality Control of Aluminum Alloy Formwork in Building Engineering Application*.12:34.

Lu X. J. (2017). *Application of Aluminum Alloy Template Technology in Green Construction of Super Tall Buildings*. (7):67–69

Lin Z. Z. (2020). *Research on Efficient Construction Based on the Construction Mode of "Aluminum Mold + Climbing Frame + Interspersed"*. (3):74–76

Qi Z. X, Tao T. J., Xin C. H. (2020). *Analysis of Advantages and Limitations of Aluminum Alloy Template System and Research on its Application Technology*. (12):869–872

Yi W. X. (2020). *Research and Application of Aluminum Formwork Construction Technology in High-Rise Buildings*. Jilin university.

Zhong W. Z. (2020). *Study on the Comparative Application of Steel and Wood Formwork System and Aluminum Alloy Formwork System*. Southwest University of Science and Technology.

Application of green low-carbon prefabrication technology in specific buildings

Lin Xu[†]
Guangzhou RBS Architectural Engineering Design Consultant Co., Ltd, China

Jing Yang[†]
Zhuhai Hengqin Rong Sheng Project Management Consulting Co., Ltd, China

Zhe Li[†], Jinxiu Wang[†], Dan Li*, Chunhong Zhang* & Linfang Mo*
PLA Naval Medical Center, China

Hongyu Xue & Wei Yang
Shanghai RBS Architectural Engineering Design Consultant Co., Ltd, China

Linhai Peng
Guangzhou RBS Architectural Engineering Design Consultant Co., Ltd, China

Yiwen Zeng
PLA Naval Medical University, China

ABSTRACT: This paper reviewed the development, status, direction, and research results of prefabricated building technology for specific applications in certain buildings. It was necessary and demanding to apply low-carbon green prefabricated building technology in such specific buildings. The value to study the low-carbon prefabricated building technology application was evaluated and clarified. The new technology of prefabricated building was generalized and introduced with respect to characteristics and application potential. From the example of a specific canteen building design on basis of modular design combined with prefabrication, green and low-carbon environmental protection measures, use of new material, and new energy technology, it was concluded that low-carbon prefabricated building technology can be successfully applied in the specified building construction. As testified in the canteen design, a low-carbon intelligent canteen can be practically implemented with appropriate study and suitable technology. This study also provided a good example and reference for further application of prefabricated buildings in the specified building construction.

1 STATUS OF PREFABRICATED BUILDING AND BARRACK BUILDING

During the "13th Five-Year Plan" period, China launched green building promotion. The green building standards (Zhang 2021) shall be fully implemented in government public housing, government-invested institutional buildings, and large-scale public buildings in province-level cities across the country. Prefabricated buildings not only comply with the green and low-carbon building requirements but also significantly expedite the construction

---

*Corresponding Authors: 13030734@qq.com, 956308151@qq.com and molinfang@126.com
[†] 1st: Joint Author

progress meanwhile reduce maintenance costs. The General Office of the State Council issued the *"Guiding Opinions on Vigorously Developing Prefabricated Building" (State Office Issued [2016] No.71)*. This issue stipulated eight tasks and required to make a proportion of prefabricated buildings of 30% among newly built buildings nationwide in ten years. *"14th Five-Year Building Energy Efficiency and Green Building Development Plan"* (Mohurd 2022) and *"14th Five-Year Construction Industry Development Plan"* (Mohurd 2022) further proposed to vigorously develop prefabricated buildings and improve the quality in developing green buildings.

In 2020, the proportion of prefabricated buildings in China reached more than 20%. Since then, the level of construction industrialization has been further developed along with the rapid expansion of industrial scale and favorable policy environment. The technical specification and standard system for prefabricated buildings have been completed. The application of prefabricated buildings has been increased constantly. Awareness of green building, low-carbon environmental protection, and sustainable development are getting more received. As a result, more matured development of prefabricated building technology has been achieved.

The concept of green environmental protection has become a "new normal" so far. Recently, with continued cultural development in science and technology, the demand for a quality life is escalating. In response to the "Plastic Ban" and "Plastic Restriction" order stipulated by the country, the requirements of white pollution reduction, carbon reduction, and low carbon have been established in different fields such as from green new material packaging technology for food to environmental protection and energy savings in all trades, more importantly for barrack building construction.

In March 2018, the Ministry of Housing and Urban-Rural Development of the PRC signed a cooperation agreement to jointly develop dismountable barracks. Wang Zhenggang (Wang Li & Wang 2020) summarized the construction principles and characteristics of such specific buildings and made a preliminary exploration of the development of prefabricated assembly in such buildings. They concluded that prefabrication technology can meet the demand of barrack construction and will have a wide application prospect in this field.

From 2018 to 2021, with more promotion and application of prefabricated buildings in institutional construction projects, prefabricated building technology began to be more applied in barrack building construction. However, it is still underdeveloped mainly as a perspective and proposal. It has not yet formed systematic model technologies and standards for various functional barrack buildings.

Zhao Qingyang (Zhao Liu et al. 2022), Dong Baoping (Dong 2021), and Xu Shiming (Xu Tang and Tang et al. 2022) summarized the characteristics, development status, and driving factors of prefabricated buildings. However, specific buildings, such as barracks, were not included.

With reference to *"the 14th Five-Year Building Energy Saving and Green Building Development Plan"*, the characteristics of barrack building shall include integrated functions, comprehensive facilities, convenient use, safety and hygiene, energy saving, environmental protection, intelligence with comfort, high flexibility (multi-function), and in varied forms (protective camouflage).

2 INTRODUCTION OF NEW PREFABRICATED BUILDING TECHNOLOGY

2.1 *Overview of architectural technology*

2.1.1 *The integration of architectural design, structural engineering, mechanical & electrical engineering, and building fitting-out*

As a result of the revolution in the construction industry, prefabricated building initiated new industrialized construction methods of "the integration of architectural design, structural engineering, mechanical & electrical engineering, and building fitting-out" via "the integration of design, production, and assembly" (Ye Zhou et al. 2017). Meanwhile, by intensively integrating informatization and industrialization of the industry, the prefabricated building also played a key role in promoting green, industrialization, and informatization construction. On the premise of fully understanding the needs of building functions, integrated construction was introduced to ensure the final product of the building as such to meet the architectural requirements (Zhang Zhou et al. 2021). The integration of "design, production, and assembly" can be achieved, and the problems of uncoordinated design, production, and assembly can be eliminated with modular building design, selection of the most suitable materials and equipment, application of low-carbon technology, and modernized information technology. This integrated product maximized the integration of advanced civilian technology and can be quickly applied to all kinds of specified areas of the building.

2.1.2 *Integration and application of energy-saving materials and renewable energy*

(1) Building components prefabricated in a factory can be quickly assembled and put into use on-site.
The main structure of the building adopts a steel structure or other forms. The external wall, thermal insulation layer, and interior finishing are formed in one go. Plumbing, HVAC, electrical, mechanical, and other intelligent equipment pipelines, as well as interior finishes and ceilings, are formed holistically. Prefabricated furniture should be fixed as much as possible to complete at the same time as interior fitting-out.
(2) Supply of renewable energy and application of rainwater recycling and purification system.
To solve the problem of power shortage in areas far from the city center, a local area network of renewable energy shall be built to provide safe, reliable, clean, and efficient power services (Zhang 2016). Sustainable ecological barrack construction shall be developed by application of new energy-saving and environment-friendly materials and technologies, renewable energy, recycling of cold and heat energy, water resources, etc. Flexibility in barrack building site selection and low energy consumption during operation can then be implemented.

2.1.3 *Application of intelligent and digital technology*

The intelligent building project shall be composed of the entire barrack environment, intelligent systems, and management, providing a comfortable, safe, and convenient living and working environment, and also saving manpower, material, and financial resources for the operation and maintenance of the barrack development (Zhong 2019). These intelligent and digital technologies have been widely used in civilian projects. Adopting part of such technologies and integrating them into prefabricated specific building construction can achieve intelligent building control in one step.

With the application of sensor technology, graphics and image technology, computer and modern communication technology, the building automation system implements fully automatic monitoring and control of the equipment and facilities of the building, including electric power, air conditioning, elevator, chiller, heating station, water supply and drainage, fire protection system, security system, access control system and so on. In addition, the system also has functions in controlling and monitoring the on and off operations of power equipment, displaying the operating status of equipment, alarms for the equipment in an abnormal condition, and controlling the energy-saving of a power plant.

2.1.4 *Application of building information modeling (BIM) technology*

BIM technology integrates the design of all disciplines into one 3D model to simulate construction and preassembly processes before the production of modules in a factory and on-site installation, which can significantly minimize discrepancies and conflicts in the design and optimize construction plans. BIM technology has distinct features of visualization, simulation, and coordination. It can integrate advanced digital technologies to provide a platform of full life span service for the construction industry and build a data foundation for refined operational management (Xia & Li 2022). Those digital technologies include but are not limited to the Internet of Things (IoT), Artificial Intelligence (AI), Virtual Reality (VR), Big Data, Cloud Computing, and Blockchain.

2.2 *Overview of structural engineering technology*

Compared with other countries, although prefabricated building structure in China started slightly behind, it has developed rapidly and has been widely used in civilian buildings in the country because of national policy encouragement and a demanding market. A full range of prefabricated structure technology and three key prefabricated construction technologies for mainstream materials of concrete and steel to suit a variety of building functions have been developed.

2.2.1 *Prefabricated concrete structure technology*

The prefabricated concrete technology includes the traditional PC system and the new PI system (T/CECS 2021). The traditional PC system was mostly applied in residential buildings. The disadvantage was that the structural joints were less reliable than those of cast-in-situ construction. Their transportation and lifting were difficult. However, the new PI system integrated the formwork and the steel reinforcement cage into a holistic prefabricated part, easy for transport and lifting, meanwhile achieving the same seismic performance as cast-in-situ construction, as shown in Figure 1. The PI was suitable for all types of buildings. Infill materials can be changed according to specific building functions to meet the diverse need of building functions.

Figure 1. PI structural system.

2.2.2 *Prefabricated steel-concrete composite structure technology*

The prefabricated steel-concrete composite structure technology was mainly used in the integration of industrialized design and construction of multi-story, high-rise, and super high-rise steel structures. At present, it was commonly applied for super high-rise residential buildings with the steel bundled tube structural system (Jiang Xie et al. 2018). The SPI system was applied for super high-rise prefabricated buildings (Xu Peng & Feng 2020), as

Figure 2. SPI structural system.

shown in Figure 2. These systems were of high stiffness and good seismic performance. They can be applied to construct super high-rise buildings and various types of buildings that required high strength and large deformation capacity. They can also be combined with other prefabricated structures to form a hybrid lateral force-resisting system.

2.2.3 *Prefabricated steel structure technology*
The steel structure system has a wide range of applications in various industrial and civilian buildings including lightweight steel structures, factory steel structures, multi-story, and high-rise steel structures due to the steel structure being factory manufactured, assembled on site with high accuracy of finished products, and high production efficiency, fast installation on-site, short construction period, etc. However, the steel structure required careful corrosion protection and fire resistance protection. Special attention shall be given to issues such as durability under complex climatic conditions.

2.2.4 *Three key technologies of prefabricated structure construction*
The prefabricated structure needs to consider the project specifics and application environment with the following three key technologies:

(1) *The technology of component disassembly, transportation, and packaging.*
   The concept in the design stage was to divide the entire building into units for transportation and lifting which required discretization, standardization, and modularization of structural components, as shown in Figure 3. This technology facilitated

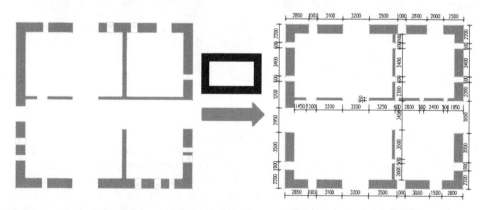

Figure 3. Discretization of the structural wall.

Figure 4. Transportation and lifting of the unit component.

transportation and lifting, more industrialization of components production in the factory, and less work on-site so that fast installation is possible, as shown in Figure 4.

(2) *The technology of quick installation of joints.*

The traditional prefabricated structure required a lot of mechanical equipment and complex joint connection technology. However, for some sites with complex environments and constraint conditions for construction, such as islands, mountain areas, deserts, etc., to develop a new type of assembly scheme was necessary. The new scheme shall satisfy the requirements of lightweight structural components, simplification of joint connections, miniaturization of installation equipment, and fast construction.

(3) *The technology of modern construction and installation.*

Based on the application of advanced technologies such as numerical control, visual microcomputer, laser synchronous tracking and navigation, BIM and other technologies, on-site real-time monitoring, and construction systems shall be researched and developed. The system shall be seamlessly coordinated with other disciplines such as architecture, and mechanical and electrical engineering through BIM technology, to integrate all disciplines and to further facilitate the implementation of design and construction integration.

3 OVERVIEW OF LOW-CARBON ENVIRONMENTAL PROTECTION MATERIALS TECHNOLOGY

According to statistics, most of the carbon emissions in cities came from the activities in the building operation stage. Therefore, it is of great social significance and economic value to reduce and evaluate the carbon consumption in the building operation stage, and also to design and construct low-carbon and zero-carbon buildings. The use of new ecological materials can significantly reduce carbon emissions in the building operation stage, which has become an important direction for research.

For example, SiO_2 aerogel is a green and environment-friendly material with the features of heat resistance, heat insulation, no decomposition at high temperatures, and no harmful gas emission. It has properties of acoustic delay, sound insulation at high temperatures, acoustic resistance coupling, gas filtration, transparent and heat insulation, etc. The energy stone with negative ions is a new type of green and environment-friendly material that can naturally and continuously release negative ions without consuming additional energy, and not generating ozone and active oxygen. The use of this material can improve air quality, increase human comfort and enhance human immunity (Zhang & Liu 2021).

In the future, with low-carbon buildings as a trend, new materials and technologies for energy saving and emission reduction, new green and environment-friendly materials with negative ions will have great market potential.

4 RESEARCH AND APPLICATION ON LOW-CARBON PREFABRICATED SMART CANTEEN

This section takes the prefabricated smart canteen as an example to discuss a number of mature civilian technologies applied to the barracks following modularization and integration.

4.1 Architectural layout and space research of prefabricated smart canteen

The characteristic of the canteen is that a large number of people dining with a relatively constant number of people, a concentrated dining time, and a large instantaneous flow of people. The prefabricated smart canteen has many advantages such as: greatly accelerated construction speed, less staff with high efficiency in daily operation, increasing the satisfaction of dining people with food through big data analysis by increasing food variety and reducing food waste.

4.2 Research on prefabricated structure scheme of smart canteen

According to the architectural function of the prefabricated smart canteen, combined with the characteristics of the prefabricated structural system and the three key technologies of construction, the structural system can be implemented with the design approach of modularization, integration, and full assembly. For example, the box-type modular assembly structure can be used to produce the structure in a factory and integrate the functions of buildings, MEP equipment, etc., Then transported to the construction site through the container truck. To ensure rapid assembly on site, intelligent and information technology installation will be adopted. Furthermore, by applying quick connection of joint, ordinary staff can complete the site installation without special training.

4.3 Application of low-carbon design and technology in canteen

Low-carbon design and technology can reduce energy consumption by adopting the following: new thermal insulation and fire-proof materials, local sunroof design to allow for natural lighting ventilation and cooling, roof equipped with solar panels to achieve self-power supply, rainwater collecting system, water recycling, etc.

5 CONCLUSIONS AND FUTURE PROSPECT

(1) This paper illustrated the development process, direction, staged achievement, and standard of prefabricated building technology in China. The current application of prefabricated buildings in the barrack construction field was elaborated. Compared with the civilian construction field, prefabricated building technology is still in its early development in barrack building construction. Relevant technical standards and systematic production have not been fully established.
(2) The feasibility and necessity of promoting prefabricated building technology to barrack construction have been illustrated by comparing the specified building construction with the civilian building. The study to apply low-carbon green prefabricated building technology by considering the characteristics of barrack buildings demonstrated both strategic significance and economic value.
(3) This paper introduced the development status of prefabricated building technology and assembly structure technology. Also, the advantages, disadvantages, and application prospects of various assembly technologies were studied. It was concluded that applying

low-carbon environmentally friendly materials could significantly reduce carbon consumption during building operations.
(4) From the example of smart canteen building design on basis of modular design combined with prefabrication, green and low-carbon environmental protection measures, use of new material, and new energy technology, it was concluded that integration of architectural, structural, MEP and interior fitting-out can be achieved. Low-carbon prefabricated building technology can be successfully applied in specific barrack building construction. This study also provided a good example and reference for further application of prefabricated buildings in specific barrack construction.

REFERENCES

Baoping Dong (2021). Domestic Development Status of Prefabricated Building Management[J]. *Building Development*, 5(6), 106–107.

Cunjing Xia and Chuanhai Li (2022). Research on the Deep Application of BIM Technology and Intelligent Building Concept in Construction Engineering [J] *Industrial Construction*,52(5), 1.

Haowen Ye, Chong Zhou, Zesen Fan, Chengwei Liu (2017). Thinking and Application of Integrated Digital Construction to Prefabricated Building[J]. *Journal of Engineering Management*,31(5), 86–89.

Jianguo Zhang (2021). "13th Five-Year Plan" Achievements in Building Energy Efficiency and Low Carbon Development, and the Study on the Development Route of "14th Five-Year Plan": The Integration of Prefabricated Building Technology and Near Zero Energy Consumption Building has Become a Hot spot in the Industry[J]. *Energy of China*, 43(6), 8.

Li Zhang, Jieling Liu (2021). Application of New Composite Building Material SiO2 Aerogel in Civil Engineering[J]. *Chemical Engineer*, 6, 52–55.

Lin Xu, Linhai Peng, Jian Feng (2020). *Special Technical Research Report on the Industrial Integration Technology (SPI) of the Super High-Rise Steel Structure in the Hengqin Headquarters Building Phase II* [R].

Lin Zhang, Wenjie Zhou, Cheng Ji, et al. (2021). Research Status and Prospect of the Prefabricated Construction and Integrated Decoration[J]. *Shanxi Architecture*, 47(3),184–187.

Ministry of Housing and Urban-Rural Development of the People's Republic of China (2022). "14th Five-Year" Building Energy Efficiency and Green Building Development Plan [R].

Ministry of Housing and Urban-Rural Development of the People's Republic of China (2022). "14th Five-Year" Construction Industry Development Plan[R].

Qingyang Zhao, Chang Liu and Lizhu Tian (2022). The Characteristics and Development of Prefabricated Building[J]. *Prefabricated Components and Application*, 154, 27–30.

Shiming Xu, Qiubo Tang and Lei Tang et al. (2022). *Research on the Current Situation and Development Driving Factors of Prefabricated Buildings*[J]. 11(2), 176–180.

T/CECS 949 (2021). *Technical Specification for Assembly Integrated Concrete Structures with Reinforcement Cage and Framwork*[C].

Tao Zhong (2019). Application of Intelligent Technology in Building Engineering[J]. *Equipment Technology*, 136.

Zhenggang Wang, Zhen Li and Wei Wang (2020). Thinking on the Application of Assembled Building Technology in Barracks[J]. *Sichuan Building Science*, 46(5), 105–110.

Zhishun Zhang (2016). The Key Technology and Realization Way of Ecological and Sustainable Architectural Design of the Plateau and Alpine barracks[J]. *Architecture Technology*, 47(12), 95–97.

Zhiwu Jiang, Yousheng Xie, Rong Fu, et al (2018). Design of High-rise Steel Pipe Bundle Concrete Combined Structure Residential Building[J]. *Building Structure*, 48, 84:87.

Construction engineering quality management based on BIM and Big Data

Shu Zong*

Jinqiao College of Kunming University of Technology, Kunming, Yunnan, China

ABSTRACT: In the process of the development of the construction industry, the safety of construction projects has always been the focus of attention, and it is also an important criterion for construction enterprises. Due to various factors, safety accidents frequently occur in construction projects, and the QM methods of construction projects need to be improved. Therefore, this paper studies the quality management (QM) of construction engineering (CE) based on BIM and big data (BD). This paper first briefly describes the BIM BD collaboration platform and the BD quality platform and then gives a detailed explanation and analysis of the research status of engineering quality grade assessment, hierarchical method, subject quality responsibility weight distribution, and finally, QM. The simulation research and analysis of the path are carried out, and relevant strategies are put forward for the QM of CE.

1 INTRODUCTION

Due to the increasingly fierce market competition in the construction industry, small and medium-sized enterprises urgently need a standardized quality control system to improve their core competitiveness (Al-Ali A R et al. 2018). The traditional management mode is mainly the quality evaluation mode of "whoever builds, who tests". Some construction units blindly pursue the construction progress, ignoring the importance of quality evaluation (Andre J C et al. 2018). The establishment of the BIM BD quality control system can improve the knowledge level of employees within the enterprise, avoid the same mistakes appearing many times in different projects and different people, and is conducive to the inheritance of experience (Al-Salim A M et al. 2018). Through the combination of BIM technology, various quality data required for QM of construction projects can be obtained quickly, timely, and accurately, providing technical support for front-line construction personnel (Ashiru A R et al. 2021).

With the development of the construction industry, a large number of researchers have conducted in-depth research on construction engineering quality management (CEQM), BIM, and BD, and have achieved good results. For example, experts and scholars such as Goel A use BIM information integration technology to integrate different business data of construction projects from the design stage to the construction stage and use the integrated BIM data as the basis for data cleaning and text segmentation of text BD. Cluster analysis mines the associations between data structures and converts the integrated unstructured data into structured data (Goel A et al. 2020). Hofmann E and other researchers use the support vector machine and BP neural network fitting algorithm to build an expert decision model for the problems existing in the quality evaluation of construction projects and apply the expert decision model to the specific construction project quality evaluation (Hofmann E & Rutschmann E 2018). Although researchers have done a lot of research on the quality of

---

*Corresponding Author: 928288652@qq.com

construction projects, there are still many problems in QM, so it is necessary to continue to study the quality of construction projects.

With the development of BIM and BD, this paper conducts in-depth research on CEQM based on BIM and BD technology. The structure of this paper can be roughly divided into three parts: The first part is a brief description of the BIM BD platform, which includes two aspects of the BIM BD collaboration platform and the BD quality platform. The second part is about the analysis of the current research status of CEQM, it analyzes the example of the layered method, and finally analyzes the status quo of the weight distribution of the quality responsibility of each subject of the construction project in detail. The third part is the analysis of the research results of the QM of the construction project, which mainly includes two aspects, the first aspect is the simulation research and analysis of the QM path, and the second aspect is the CEQM suggestion and strategy.

2 BIM BD PLATFORM

2.1 *BIM BD collaboration platform*

The quality information of construction projects can be generated through the QM system, making full use of modern computers and information technology to generate an information database, and processing the construction quality information in the whole process of QM implementation, which can greatly improve the level of information management (Hasan A et al. 2021). However, the QM systems may be independent and can only process information within their respective scope of work, and there may be incompatibility problems between them, resulting in inefficient processing of information and data (Lamba K & Singh S P 2018). Therefore, on this basis, a BD collaboration platform is established, through which information can be used, transmitted, and shared faster and more efficiently. The structure of the BIM BD collaboration platform is shown in Figure 1.

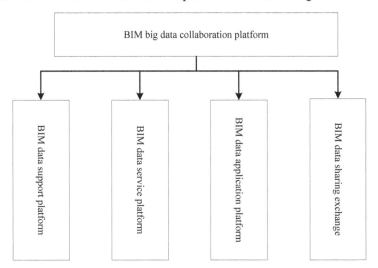

Figure 1. Structure diagram of the BIM BD collaboration platform.

2.2 *BD quality platform*

The BD quality platform is to feed back some quality problems and solutions with a high probability of occurrence in previous projects to project managers in a timely manner, reminding them to pay attention to potential quality risks. At the same time, new quality issues and control measures are added to the BD quality platform to provide a reference for future project management. By establishing a BIM quality model and comparing it with the actual on-site engineering, the quality problems that have occurred are summarized and fed back to the large quality database. By establishing a large database of BIM quality, construction enterprises can improve the knowledge level of internal personnel, avoid the same mistakes appearing repeatedly in different construction projects and different people, further improve the efficiency of experience inheritance, and reduce the frequent flow of personnel in construction enterprises and the losses to the enterprise (Ogunrinde O et al. 2020). On the other hand, through the combination of BIM, various quality data required for QM can be obtained quickly, timely, and accurately, providing technical support for front-line construction personnel (Quezon E T et al. 2021). Figure 2 is a schematic diagram of the functions of the BD quality platform.

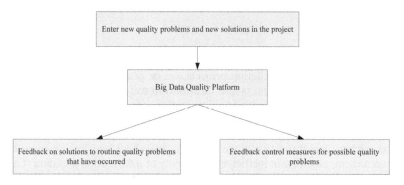

Figure 2. Functional diagram of BD quality platform.

3 RESEARCH STATUS ANALYSIS

3.1 *Project quality rating*

Inspection batches, sub-projects, sub-projects, and unit projects are set up in the BIM big data collaboration platform, and they are interrelated and progressive. The specific construction project quality grade assessment is carried out in accordance with the rules, and the assessed project quality grade can be directly displayed on the platform. In addition to the evaluation of each stage, construction technology and material quality control, equipment operation and personnel safety management, quality defect and accident handling, project information management and release, and online assessment of project grades are interrelated. The information interaction of each function is as follows: Figure 3, the whole

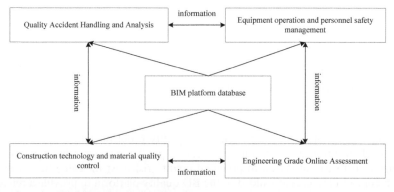

Figure 3. Functional information interaction diagram.

process and each link of the construction project are recorded on the big data collaboration platform. The big data collaboration platform scores according to each link, and then generates a comprehensive score according to the weight of each link, and finally evaluates the project quality level according to the setting of the score segment.

3.2 *Hierarchical method*

The basic principle of the layered method is the process of extracting the quality status of construction projects from the many factors that affect the quality of the project and then determining the source of the problem. The hierarchical control method for the quality of binding steel bars is analyzed as follows: in a steel bar group, there are workers in operation A, workers in operation B, workers in operation C, and workers in operation D to bind steel bars. We selected 100 lashing points, of which 33 were identified as unqualified, and the unqualified lashing points reached 30% of the total. Through the analysis of the stratified method survey results in Table 1, it can be seen that it is the lashing quality of the workers in operation D that leads to the decline of the overall level. Therefore, in the construction process of the construction project, it is necessary to combine the relevant specifications and the summary of previous construction experience and focus on the training of worker D for reinforcing steel bar binding.

Table 1. Statistical table of hierarchical data.

Worker	Sampling Points	Failed Points	Individual Failure Rate (%)	Percentage of Total Failure Points (%)
A	25	2	8	6
B	25	5	20	15
C	25	8	32	24
D	25	18	72	55
Total	100	33		100

3.3 *Weight distribution of subject quality responsibility*

The data relating to the QM of construction projects are processed by the method of counting analysis. The counting analysis method is a method to count and measure the number of times each subject is selected as "the most important" and "the least important". Its calculation formula is:

$$F_a = \frac{\sum_{b=1}^{x} F_{ab}}{x} \quad (1)$$

$$F = \sum_{a=1}^{5} F_a \quad (2)$$

Among them, F represents the final importance score of a single subject, F_a represents the importance score of a single subject in each group, F_{ab} represents the choice of the b-th respondent in the a-th group, and x represents the number of respondents. The value of Fab is most responsible = 1; least responsible = -1; no option = 0.

Arrange the subjects in descending order, and the result is shown in Figure 4. As can be seen from Figure 4, among all the subjects, the construction unit is the most important subject in the construction quality governance path, with a weighted score of 9.64; the

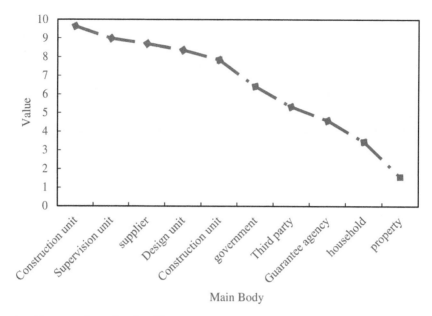

Figure 4. The score chart of each subject.

property has the lowest influence, with a weighted score of only 1.58. According to the score, each subject can be divided into 6 levels. Among them, property influence is the lowest, followed by users, insurance institutions are ranked in the middle and lower reaches, third-party institutions and the government are ranked in the middle and upper reaches, construction units, design units, material suppliers and supervision units have significantly increased their influence compared to the other entities mentioned above. The impact of the construction unit on the quality of the project is the most direct and the most important.

4 ANALYSIS OF RESEARCH RESULTS

4.1 Simulation research of QM path

To further study the influence of different subjects on building quality, this paper changes the input value of each subject to the highest and lowest values while keeping the weight coefficient unchanged and observes the impact range of the value change on the output results. To express the impact of each subject's change more intuitively on the system, this paper calculates the change difference caused by each subject and obtains the results shown in Figure 5.

As can be seen from Figure 5, each subsystem is still in a relatively independent state. In the overall system, the supervisory body can supervise and constrain the quality behavior of the construction-responsible body, thereby improving the building quality, but the construction-responsibility body Quality changes are hardly responsive to the regulatory system, mainly because the feedback system is not yet functioning effectively. On the one hand, there is a lack of effective feedback channels for the quality problems found by the feedback subjects. For example, the service quality of the property has a certain degree of independence. Although a good property can greatly improve the user's living experience and improve user satisfaction, there is a lack of processes and channels for feedback on problems found in the use process, which can neither provide a basis for early design

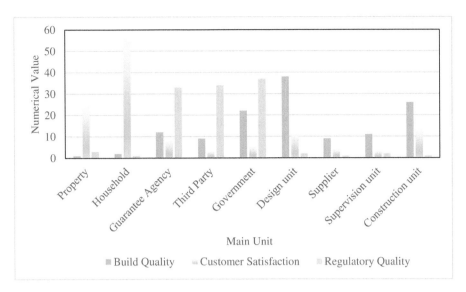

Figure 5. Governance path simulation result diagram.

optimization nor improve the pertinence of supervision during the construction process and provide assistance for the improvement of building quality at the macro level.

According to the simulation results shown in Figure 5, in terms of building quality, the proportion of the government is 22, ranking third among all entities. The government's influence on building quality is second only to the design and construction units. In terms of quality supervision, the government's ratio is 38, which is the largest ratio among all subjects. Therefore, the government's supervision of construction quality is a key part of the entire supervision path. The government's attitude towards quality supervision and management directly affects the development trend of the entire industry and affects the progress of the industry. The government's influence on construction quality at the macro level is even higher than that of construction enterprises to a certain extent.

4.2 *CEQM strategy*

(1) We should understand and master the comprehensive development trend of the quality and safety of construction projects, regularly carry out inspections and assessments at different stages of construction projects, track the quality and safety of different stages, and then obtain the development trend of the quality and safety of the project.
(2) It is attempted to establish a comprehensive network of regional overall construction quality and safety conditions. Through the comprehensive evaluation of project quality and safety at each stage, the client can keep abreast of the progress of the project and have a clear understanding of the quality and safety of the project.
(3) Commitment is made to give full play to the role of the BIM BD collaboration platform in the QM of construction projects. The BIM BD collaboration platform can realize data accumulation and statistics, timely feedback on building quality problems, and provide continuous and targeted optimization ideas for builders and regulators.
(4) Efforts are intensified to strengthen the promotion and popularization of socialized supervision, introduce the stakeholders, public, media, and other social forces into the supervision of market entities, realize diversified supervision, and strengthen the public's understanding of their construction quality rights and interests.

5 CONCLUSION

The use of BIM and BD technology is conducive to the healthy development of the construction industry. Therefore, this paper conducts in-depth research on CEQM based on BIM and BD. Through the research and analysis of construction QM based on BIM and BD, this paper finds that the impact of construction units on the quality of construction projects is the most direct and most important. The influence of the government on construction quality is second only to the design and construction units. The supervision of the construction project is a key link in the entire construction project management path. There are many deficiencies in this paper, which need to be improved, but the research on construction QM based on BIM and BD technology is conducive to the development of QM, which can avoid the occurrence of some safety accidents and promote the healthy development of the construction industry.

REFERENCES

Al-Ali A.R. & Zualkernan I.A. & Rashid M. (2018). A Smart Home Energy Management System Using IoT and Big Data Analytics Approach[J]. *IEEE Transactions on Consumer Electronics*, 63(4):426–434.

Al-Salim A.M. & El-Gorashi T. & Lawey A.Q. (2018). Greening Big Data Networks: The Impact of Veracity [J]. *Iet Optoelectronics*, 12(3):126–135.

Andre J C & Antoniu G & Asch M. (2018). Big Data and Extreme-Scale Computing: Pathways to Convergence – Toward a Shaping Strategy for a Future Software and Data Ecosystem for Scientific Inquiry[J]. *International Journal of High Performance Computing Applications*, 32(4):435–479.

Ashiru A.R. & Aule T.T. & Anifowose K. (2021). The Use of Total Quality Management (TQM) Principles for Construction Projects in Nigerian Tertiary Institutions[J]. *IOSR Journal of Mechanical and Civil Engineering*, 17(3):7–12.

Goel A. & Ganesh L.S. & Kaur A. (2020). Social Sustainability Considerations in Construction Project Feasibility Study: A Stakeholder Salience Perspective [J]. *Engineering Construction & Architectural Management*, 27(7):1429–1459.

Hasan A. & Ahn S. & Rameezdeen R. (2021). Investigation into Post-adoption Usage of Mobile ICTs in Australian Construction Projects[J]. *Engineering, Construction and Architectural Management*, 28(1): 351–371.

Hofmann E. & Rutschmann E. (2018). Big Data Analytics and Demand Forecasting in Supply Chains: A Conceptual Analysis [J]. *The International Journal of Logistics Management*, 29(2):739–766.

Lamba K & Singh S P. (2018). Modeling Big Data Enablers for Operations and Supply Chain Management [J]. *The International Journal of Logistics Management*, 29(2):629–658.

Ogunrinde O. & Amirkhanian A. & Corley M. (2020). Effect of Nighttime Construction on Quality of Asphalt Paving [J]. *Journal of Construction Engineering and Management*, 146(9):1–7.

Quezon E T & Getu A & Genie T. (2021). Factors Affecting Supervision Practice of Public Building Construction Projects in Dire Dawa Administration [J]. *American Journal of Civil Engineering and Architecture*, 9(4):134–141.

ns Research on key technology of prefabricated building construction

Nana Liu, Yingjia Wang* & Ting Hu
Chongqing College of Architecture and Technology, Chongqing, China

ABSTRACT: With the development of modern industrial technology, the construction of houses can be made in batches like machine production. Prefabricated building components are shipped to the site and assembled. The emergence of prefabricated buildings is an important direction for the transformation and development of the construction industry to intelligence, industrialization and information technology. With the increasing novelty of prefabricated buildings, people also put forward higher requirements for construction technology. This paper mainly discusses the key technology of prefabricated building construction from the supporting technology of prefabricated building, node waterproof technology, hoisting construction technology, speed and other aspects, so as to provide a theoretical and practical basis for prefabricated building construction.

1 GENERAL INSTRUCTIONS

As a labor-intensive industry, there are still many problems in the development of traditional construction, such as large consumption of resources, serious environmental pollution, low labor efficiency, and less use of emerging technologies. In the past, the construction industry was a labor-intensive industry with "crowd tactics" as the core. With the improvement of people's living standards, more and more young people no longer choose the construction industry, and the aging of construction employees is gradually prominent.

With the transformation and development of the construction industry, prefabricated buildings have been widely recognized. Under the constant updating of the building style, the overall design of the building structure has correspondingly put forward higher requirements. In order to meet such high requirements, the construction technology of prefabricated buildings needs to be improved at any time. This paper mainly focuses on the key technologies used in the construction process of prefabricated buildings, technical analysis, functional introduction and the rationality of the application of building technology in different construction occasions.

This paper mainly discusses the key technology of prefabricated building construction from the supporting technology of prefabricated building, node waterproof technology, hoisting construction technology, speed and other aspects, so as to provide a theoretical and practical basis for prefabricated building construction.

2 COMPARISON BETWEEN PREFABRICATED BUILDING AND TRADITIONAL BUILDING CONSTRUCTION PROCESS

The traditional way of building built mainly adopts the scene mode, using a certain technology, in accordance with the design chart of the whole building construction, and

---

*Corresponding Author: liunana@sina.com

prefabricated buildings, which will complete the construction of the original site job transferring to a specific manufacturing factory for each part of the built building and the building structure and accessories to the construction site, By effectively connecting different building structures and fittings, the final assembly plant building (Li 2013).

Prefabricated buildings began to attract interest in the early 20th century and were finally realized in the 1960s. Britain, France, the Soviet Union and other countries made the first attempt. Because of the speed of construction and the low cost of production, prefabricated buildings quickly spread around the world. The early prefabricated buildings were rigid and monotonous in appearance. Later improvements in design have increased flexibility and variety, allowing prefabricated buildings to be built not only in batches, but in a variety of styles. There is a kind of mobile home in the United States, which is a relatively advanced prefabricated building. Each residential unit is like a large trailer, which can be used as long as it is pulled to the site by a special car and hoisted by a crane onto the floor mat to connect with the embedded water channel, power supply and telephone system. The mobile home has heating, bathroom, kitchen, dining room, bedroom and other facilities. Mobile homes can either stand alone as a unit or be connected to each other.

The construction organization design and management system of prefabricated concrete construction still have some shortcomings, which are still carried out according to the traditional construction techniques. In order to promote the progress of prefabricated concrete buildings in China, a set of sound construction organization design and management systems should be urgently developed.

As shown in Table 1, the modern construction industry represented by prefabricated concrete structures is fundamentally different from the traditional construction industry (Zhao 2020).

Table 1. Differences between the modern construction industry and traditional construction industry.

Compare the content of	Prefabricated building	Traditional architecture
Industry structure	Whole process industrial chain	Mainly in the construction stage
Industrial organization	Intensive, integration, coordination management	Low-end competition and lack of cooperation
Production organization	Whole process, integrated management	Disjointed at all stages
The production technology	Standardization and integration	Independent of each other
Means of production	Factory, assembly, information	"Crowd tactics"
Factors of production	Harmonious and unified	On its own input
Production target	Pursue the benefits of the whole industrial chain	Pursue the interests of all departments

3 KEY TECHNOLOGY OF PREFABRICATED BUILDING CONSTRUCTION

3.1 Supporting technologies

In building construction, the support system is one of the important links in the construction of prefabricated concrete buildings. Support technology is a very important technology. Whether the complete implementation of the supporting technology is related to the stability of the external wall structure.

The wall panel support system is composed of horizontally connected harmoniously adjustable adjusting rods (see Figure 1), with a pre-embedded nut tube at the upper end and a steel plate support at the lower end. Steel plate is a common component of the support system. Compared with other types of bottom support, the steel plate support is stronger. The use of support technology also needs to be judged according to the situation of the

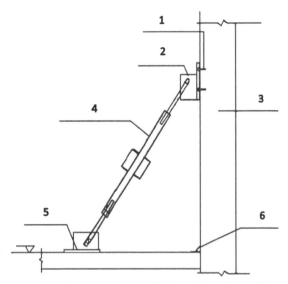

1-Embedded nut tube, 2-Connection support, 3-Prefabricated exterior wall panels, 4-Flower basket adjusting rod, 5-Bottom steel bearing, 6-Bottom limit connector

Figure 1. Schematic of prefabricated exterior wall supports.

construction site, if the weight is too heavy or some overhanging members have used two groups of horizontal connection between the two ends of the setting which also needs to impose three groups of an adjustable screw to ensure that the construction process can be carried out safely.

Because there is no specially designed support frame for assembly type in the industry, most construction units are experienced construction, and the spacing between support rods is usually taken as the empirical value of 1m. Too many support bars will cause the construction to face congestion which affects the construction speed and increases the economic cost; The installation of the support rod layout is too little, and cannot ensure the safety of construction (Long 2016). In the erection of horizontal component support, not only the construction speed but also the construction safety and economy should be taken into account, so it is particularly important to design a reasonable support spacing of laminated plate.

For the construction process of PC component installation, it is necessary to measure the wall according to the given horizontal height and control axis, and mark the measured horizontal elevation line and axis according to the standard, and paste the seal strip according to the measured data and the mark made on the wall. When placing the PC on the ground, at least four people below are connected to ensure that the PC can be placed in the correct position. Before stabilizing the PC on the ground, you can measure the horizontal line of the PC board to reduce the handling of the PC board. When the PC board is perpendicular to the ground, first of all, the longitude and weft measuring instrument is used to measure the vertical line, and then the oblique support bracket is installed through the oblique support and vertical adjustment of the board's verticality, to ensure that the PC version can be well perpendicular to the ground (Lu 2021).

3.2 *Node waterproofing technology*

In the design of prefabricated buildings, waterproof treatment is the most critical content. The effect of waterproof treatment on the final construction quality and even later service life of the project has a decisive role. Prefabricated building is in fact in accordance with a certain sequence and process of the components of reasonable production and assembly, in the assembly joint, a certain gap inevitably exists, which is prone to leakage phenomenon, seriously affecting the quality of construction in the late use of the process of safety. Nowadays in China, it is the common waterproof sealing technology, its application principle is effective seam sealing treatment, achieving the impermeable effect. To continuously improve the waterproof sealing effect, it is necessary to optimize the waterproof treatment process, reasonably design the drainage path, and prevent the outside water from seeping into the room (Gong 2019; Liu 2016).

The node waterproofing technology of prefabricated buildings is far more effective than traditional building waterproofing measures. The waterproofing of traditional building gaps is often difficult to be solved, and also needs to consider the outer waterproof layer and set a reasonable drainage path. In prefabricated buildings, different node drainage systems are set according to the different structures of each building, which can effectively solve the problem of water leakage and seepage of the building.

3.3 Hoisting construction technology

In prefabricated concrete construction, prefabricated component hoisting is the leading process. Before lifting prefabricated components, the construction and installation personnel need to communicate with the hoisting personnel in detail to ensure that the lifting operation can be carried out in an orderly manner. Large components can be hoisted by means of open parts such as balconies or Windows. In the process of lifting components, staff need to be highly focused, and pay attention to handling with care and to avoid the damage of components and other problems. Hoisting operations can be carried out through man-machine cooperation (Zhang 2017). Once the staff finds any deviation in the lifting position, it is necessary to adjust the lifting position the first time. The adjustment operation shall be carried out according to the following process: first, after the component lifting is completed, the floor commanding staff shall issue instructions to the tower crane operator, and the tower crane operator shall carry out work according to the instructions; After initial positioning, the horizontal direction of prefabricated components is aligned with the control line; Secondly, when the prefabricated components arrive near the floor, manual assistance can be adopted to accurately position the components. Among them, there may be some errors in the vertical and verticality of the components. Finally, workers can use special tools such as jacks to further adjust the position of prefabricated components.

In the process of building lifting construction, it is mainly to transport the building materials from point A to point B. The position of lifting does not have much influence on the lifting process, but the placement point needs a high degree of accuracy (Fu 2013). In general, the construction of the hoisting system needs to be controlled at the millimeter level, and the error of some materials after hoisting and positioning must be controlled within 2 to 5 mm. Therefore, the requirements for the erection of the hoisting platform will be higher than other technical requirements (as shown in Figure 2).

3.3.1 Speed matching technology for hoisting

The stretching of the lifting platform also needs to take into account the different situations in the construction process. In order to control the emergency situation better in the construction process and improve the construction efficiency of the lifting platform, it is also necessary to design the speed matching technology of the lifting platform. The lifting platform is actually a process of lifting, transporting and lowering the material, so the speed of its rise and fall determines the efficiency of the whole lifting process. The whole operation process of the hoisting platform is faced with different working states such as no-load operation, load operation, ascending operation, parallel operation and descending operation. In this process, speed often determines the overall efficiency of the whole working process (Ma 2016).

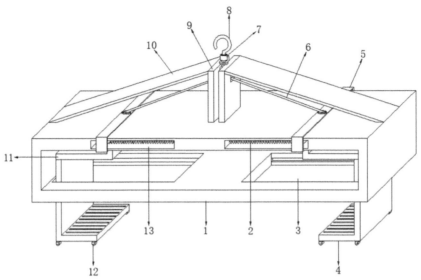

1-Frame, 2-scale plate, 3- limit slot, 4-clamping plate, 5-movable enclosure frame, 6-hanging rope, 7-fastener, 8-hook, 9-support plate, 11-mounting plate, 12-universal wheel, 13-through slot.

Figure 2. Prefabricated lifting platform.

At present, the lifting platform used in the construction of prefabricated buildings has a maximum bearing capacity of 16 tons and a maximum lifting height of 60 meters. Therefore, for such a large weight and height, the lifting speed can be set to five-speed gears: 5m/min, 16m/min, 45m/min, 75m/min and 100m/min. The lifting mechanism of the lifting platform can be tested according to the five-speed gears, and the optimal lifting speed can be selected at last.

3.3.2 *Research on micro control technology*
Prefabricated building is the symbol of construction industrialization and also the process of technological innovation and development, and lifting technology is also developing towards micro-control. In the process of prefabricated building construction, the traditional hoisting technology encounters some problems, such as low automation, low accuracy, low construction efficiency and high labor intensity. In order to better build prefabricated buildings and realize the intelligent and refined development of building construction, micro-motion control is currently used in the hoisting technology of matching buildings. The working principle of micro-motion control is actually to click the micro-motion button on the lifting system to enter the "micro-motion control state", which realizes the minimum starting point of the lifting platform at about 1cm, which is one-tenth of the traditional lifting technology and ensures the advancement and accuracy of the technology. The hoisting construction process is also faced with an important problem, that is, how to lift components safely and how to place them quickly, stably and accurately in the exact location, which directly affects the efficiency of installation.

Component hoisting in place can ensure the displacement of three directions (x, Y, z) in the three-dimensional space. In the three-dimensional space, the horizontal direction (x, Y) can be lifted to the specified position by the horizontal movement of the lifting platform, or the component can be placed in the position by manual righting during the falling process. The commonly used method for displacement in the Z direction is frequency conversion control because the movement in the Z direction is moving up and down in the lifting process and the minimum step distance cannot be less than 10cm. The use of micro-motion control technology can effectively realize the accurate alignment of components and docking of building components, so as to reduce the number of debugging and debugging processes, and finally improve the efficiency of prefabricated building construction.

4 TECHNICAL IMPROVEMENT MANAGEMENT OF CONSTRUCTION PERSONNEL

Prefabricated building is the product of modernization construction, its realization requires the construction personnel to master certain construction skills and theoretical knowledge, including node waterproof technology, lifting technology and support technology and other professional skills, to ensure the safety and stability of the entire construction site and to achieve the building function. To improve the skill quality of employees, a relatively strict management mode should be formulated, skills training should be carried out regularly, online and offline courses should be taught, and excellent technical workers should be invited to summarize experience and share work skills. Both theoretical knowledge and operational skills should be mastered by employees. The learning of skills not only requires regular training courses, but also tests the student's learning situation after learning to understand the mastery of work skills of construction personnel, so as to better complete the construction task of prefabricated buildings.

5 CONCLUSION

The prefabricated building has greatly changed the construction method of traditional buildings, optimized the construction process and improved construction efficiency. At the same time, the construction quality is greatly improved, and the overall stability and use function of the building is guaranteed. In particular, technical leaps have been made in the support technology, node waterproof and lifting technology.

REFERENCES

Fu Shuqing (2013). Research on Construction Technology of High-rise Housing [J]. *Management and Technology of Small and Medium-sized Enterprises*, 2013(25):2.

Gong Xiaoyu, Yuan Jianang (2019). Design and Construction of Prefabricated Waterproofing Joints [J], *Housing*, 2019.09(116–117).

Li Chao (2022). Discussion on the Application of Prefabricated Building Construction Technology [J]. *China Building Metal Structure*, 222, 03(54–55).

Li Lihong, Geng Bohui, Qi Baobo, et al. (2013). Comparison and Empirical Study on the Cost of Prefabricated Construction Engineering and Cast-in-place Construction Engineering [J]. *Construction Economy*, 2013(9):4.

Liu Xiaohui (2017). Discussion on Construction Technology of Prefabricated Building [J]. *Housing*, 2017(17):2.

Long Libo, Ma Yueqiang, Zhao Bo, et al. (2016). Research on the Innovation of Prefabricated Building Construction Technology and its Supporting Equipment [J]. *Building Construction*, 2016,38(3):3.

Lu Kaifeng (2021). *Research on Key Construction Technology of Prefabricated Concrete Residence*. 2021.03.

Ma Zhiming (2016). Research on Construction Technology of Civil Engineering Building [J]. *Engineering Technology* (Cited version), *2016: 00208–00208*.

Zhang Yang (2017). Research on Construction Technology of Prefabricated Building [J]. *Decoration and Decoration World*, 2017,000(024):278. (in Chinese)

Zhao Bensheng (2020). Research and Application of Key Technology of Prefabricated Building *Construction Based on Intelligent Construction*, 2020.10.

Effect of clay on microstructure and mechanical properties of Guilin red clay

Liangyu Wang
China MCC5 Group Corp. Ltd, Chengdu, China

Qingye Shi, Bai Yang* & Junhuan Chen
School of Architecture and Transportation Engineering, Guilin University of Electronic Technology, Guilin, Guangxi, China

Shujiang Li
China MCC5 Group Corp. Ltd, Chengdu, China

ABSTRACT: In order to explore the effect of clay on the shear strength and microstructure of red clay, the consolidated undrained test and scanning electron microscope test were carried out by preparing red clay samples with different clay contents. The results show that the clay has an obvious influence on the cohesion and internal friction angle of the red clay. The cohesion increases with the increase of the clay, and the internal friction angle decreases with the increase of the clay. Through the analysis of the microstructure, the particle distribution of the soil is uneven, the size is different and complex, and the clay will fill the pores of the red clay. The higher the clay content is, the finer the soil sample looks, the clay particles will closely connect the particles. The conclusions of this paper have certain practical significance for the study of engineering properties of red clay and provide some help for the treatment of red clay.

1 INTRODUCTION

Red clay is a kind of special soil, which is more and more used in engineering construction as the building foundation and building medium (Meng 2014). With the development of the social economy, a large number of projects are faced with the special physical characteristics of red clay (Dong 2019). Many scholars at home and abroad have studied red clay more and more extensively. In the study of the microstructure properties of red clay, Zhang et al. (2007) studied the microstructure and chemical composition of Kunming red clay by using an X-ray fluorescence spectrometer and scanning electron microscope and found that the high liquid-plastic limit of red clay is due to the strong hydration ability of the cement-free iron oxide in the red clay. The unit particles in the red clay cause the soil to have high water content. In order to study the relationship between the permeability coefficient and the number of dry and wet cycles, Chen Ran et al. (2020) conducted dry and wet cycles on the red clay after vacuum saturation and constant temperature drying and observed the distribution characteristics of pores under different cycle times under microscopic conditions. In the study of the mechanical and engineering properties of red clay, Tan et al. (2001) proposed a soil cementation structure model to explain the irreversibility

---

*Corresponding Author: ayangbai@163.com

of red clay in engineering mechanical properties in detail, and explained the special engineering properties of red clay through cementation disaggregation. Tan et al. (2014) took compacted red clay as the research object, proposed the limited expansion method, and explored the influence mechanism of different states on the expansion force of red clay from the microscopic perspective. Fang Wei et al. (2008) studied the stress-strain, consolidation deformation and expansion-contraction deformation of Wuguang red clay through a triaxial test, consolidation test and expansion test, and found that the red clay has the deformation characteristics of over consolidation, solid but not dense, consolidation reverse section, medium compressibility, small expansion and strong contraction. Huang Xiang et al. (2016) conducted undrained triaxial tests on red clay to discuss the conditions for the formation of shear bands. The test results show that the shear band can be divided into four types in the triaxial test: single type, sub single type, double slit T type and multi-slit type. Meng et al. (2018) carried out dynamic triaxial tests on red clay in Xinxiang, Jiangxi Province under different physical states and stress states to study the dynamic stress failure vibration order relationship, dynamic cohesion and dynamic internal friction angle, dynamic strain dynamic elastic modulus relationship and damping ratio under different water content and consolidation confining pressure. Dong (2018), Zhao (2018, 2017), Pu (2019), Luo Wenjun (2020), et al. all conducted shear strength tests on red clay under different water content, It is found that the shear strength and cohesion of red clay decrease with the increase of water content. The recommended values of cohesion and internal friction angle are given.

Clay is an important component of red clay, with a large content in red clay. Kaolinite is the main mineral component of clay in Guilin red clay, and its content is up to 90% (Xiao 2016, Cui 2021). Lu et al. (2012) took soil samples from the Yanshan campus of Guilin University of technology and measured the clay content of 61.27% through the clean water settlement test. Wei (2005) found that the clay content was generally more than 50% through the basic geotechnical test on the red clay. In terms of Engineering geological properties, Guilin red clay has the characteristics of weak swelling and weak disintegration. Wang et al. (2021) conducted fracture tests on red clay with different clay content to explore the influence and development law of clay on fissures. In this paper, by adding the different proportions of clay into the red clay with low clay content, the scanning electron microscope test and triaxial shear test were carried out on the red clay with different clay content to explore the changes in microstructure and shear strength parameters of clay in the red clay.

2 EXPERIMENT

2.1 Basic physical indicators

The soil for this test is taken from a construction site in Guilin. A series of geotechnical foundation tests are carried out on undisturbed soil. The basic physical indexes are shown in Table 1.

Table 1. Basic physical indexes.

Water content (%)	Density (g/cm^3)	Proportion	Liquid limit (%)	Plastic limit (%)	Dry density (g/cm^3)
35	1.93	2.74	54	32	1.4

2.2 Preparation of clay particles

The particle gradation of the soil was measured by the particle analysis test. The particle composition of the soil used in this test is shown in Table 2.

Table 2. Particle composition (%).

≤ 0.075 mm	≤ 0.059 mm	≤ 0.045 mm	≤ 0.033 mm	≤ 0.021 mm	≤ 0.009 mm	≤ 0.005 mm	≤ 0.001 mm
96.30	85.54	75.48	69.88	61.30	53.14	42.30	40.82

Particles with D ≤ 0.005mm are called clay particles (Tang 1980). Therefore, the clay content of the soil used in this test is 42.3%. Then, clay particles are prepared by using a self-made clay extraction device based on the principle of clean water sedimentation (Wang 2021). Then, clay particles are added to the soil with low clay content in a gradient of 5%, and then scanning electron microscope test and consolidated undrained triaxial test are carried out. The prepared clay particles are shown in Figure 1.

Figure 1. Clay particles of red clay.

2.3 Test scheme

Triaxial shear test: the dry density of the soil sample is 1.4g/cm^3, the water content is 35%, the soil samples are mixed with clay particles in the proportion of 0%, 5%, 10% and 15%, and then the triaxial sample with a volume of 96.01mm^3 is prepared. The consolidated undrained shear under different confining pressures is carried out to analyze the influence of clay on shear strength.

Scanning electron microscope test (SEM): The microstructure of clay was observed by SEM, and the red clay after adding clay was observed by SEM, so as to observe and analyze the morphology of clay in the structure of red clay. The scanning electron microscope test was carried out on the red clay samples after the triaxial test to observe the changes in soil structure in the soil samples.

3 ANALYSIS OF TEST RESULTS

3.1 Triaxial shear test

The consolidated undrained shear test is conducted on red clay with different added clay content. The test results are shown in Figure 2.

It can be seen from Figure 2 that the shear peak of red clay with 0% added clay content is 83.6 to 548.5kPa, the shear peak of red clay with 5% added clay content is 101.8 to 569.2kPa, the shear peak of red clay with 10% added clay content is 165.3 to 507.1kPa, and the shear

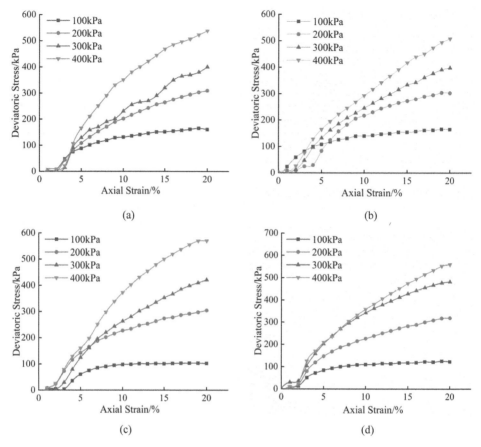

Figure 2. Stress-Strain curves of different added clay content. (a) Added clay content of 0% (b) Added clay content of 5% (c) Added clay content of 10% (d) Added clay content of 15%.

peak of red clay with 15% added clay content is 124.8 to 558.8kPa; No matter what the clay content is, at the low confining pressure of 100kPa, the stress-strain curve rises first, and it basically keeps horizontal development in the later stage, and the soil is plastically deformed and reaches the hardening state. At the stage of high confining pressure of 200kPa, 300kPa and 400kPa, the stress-strain curve of red clay is elastically deformed before the inflection point, and then the stress-strain curve keeps rising, and the soil mass reaches a strong plastic state.

The clay content has an influence on the stress-strain curve of the soil. When the clay content increases, the deviatoric stress of the red clay increases first and then decreases with the increase of the confining pressure. However, the inflection point of some curves occurs later, which is caused by the uneven distribution of clay in the soil during the soil preparation process, The abrupt change of some curves in the later stage is also due to the uneven distribution of clay particles affecting the test. The different content of clay particles will have a great impact on the stress and strain of soil.

3.2 *Effect of clay particles on cohesion and internal friction angle of red clay*

The effective stress index of the soil sample is obtained by drawing the effective stress molar circle, and the corresponding curve is drawn by the obtained cohesion and internal friction angle, as shown in Figure 3.

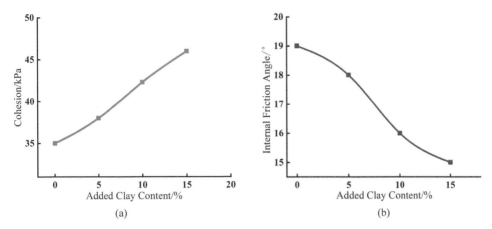

Figure 3. Effect of clay particles on shear strength parameters of red clay. (a) Effect of clay particles on cohesion of red clay (b) Effect of clay particles on internal friction angle of red clay.

As shown in Figure 3 (a), different proportions of clay have an obvious influence on the cohesion of red clay, and the clay content has a great influence on the value of cohesion c. the cohesion increases gradually with the increase of clay content of red clay. When a small amount of clay is added, the change of cohesion is small; With the continuous addition of clay particles, the cohesion will continue to rise. This is because the clay particles are composed of secondary minerals and double accompanying oxides. There are always bound water films with different thicknesses in the red clay. When the clay particles are added, the double accompanying oxides will react with the bound water film to make the soil more closely connected, so the cohesion of the soil will increase.

As shown in Figure 3 (b), the influence of clay particles on the internal friction angle of red clay is also obvious. Compared with natural red clay, the content of clay particles has a great influence on the internal friction angle A of red clay, and the internal friction angle decreases gradually with the increase of the content of clay particles. The internal friction angle decreases from 18°–22° to 15°–16°. The change in internal friction angle is related to the roughness of soil particles. With the increase of clay particles, the clay particles will fill the pores of the soil, so that the surface roughness of soil particles decreases, and the internal friction angle decreases.

3.3 *Microstructure of clay particles*

In this test, clay is a very important part and plays an important role in red clay. In order to explore the distribution of clay in the red clay, the prepared clay was tested by scanning electron microscopy. Through observing the microstructure of clay, the influence of clay on the physical and mechanical properties of red clay was analyzed from a microscopic perspective. The microstructure of clay particles is shown in Figure 4.

According to Figure 4, the clay particles are magnified by 200 times. At this time, the clay particles are like many fine powder particles laid together. The concave part in the figure is conductive adhesive, and there is no distribution of clay particles. From the 200 times SEM image, it can be roughly seen that the pores and clay particles are densely stacked together. The magnification is 1000 times. At this time, small solid particles in the clay particles can be seen. The distribution of the clay particles is closely and evenly stacked by particles of different sizes, dense and soft. When the magnification is 5000 times, there are many pores in the clay particles, and the presence of crystals can be seen. These are the minerals in the clay particles, which are occluded and stacked together. When the magnification is 10000 times,

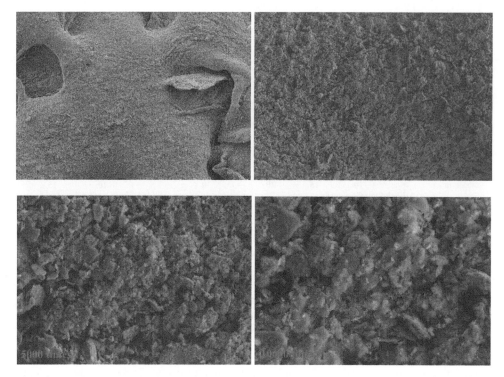

Figure 4. Microstructure of clay particles.

this layer of "plush" material is attached to the surface of clay particles. Through these "plush" materials, the soil particles can be closely connected, thus reducing the porosity of soil and increasing its strength of the soil.

3.4 *Microstructure analysis of red clay with different clay content*

Scanning electron microscope (SEM) tests were carried out on red clay with different added clay content to observe its microstructure and analyze the role of clay in red clay.

Figure 5 is the SEM image of red clay with an added clay content of 0%. It can be seen from the images with a magnification of 200 times and 1000 times that the shape and size of particles are different and complex. Some particles are very large and some particles are as small as powder. It can be seen from the images with magnifications of 5000 times and 10000 times that the intergranular pores are more and larger, the pores lack clay particles, and there are fine crystals on the surface of the particles, which are the mineral substances contained in the red clay. Moreover, it can be seen from the images that the connection force between particles is weak, the structural strength is not high, but the sensitivity is high, and the compressibility of the soil is good.

Figures 6, 7 and 8 are SEM images of remolded red clay with the added clay content of 5%, 10% and 15% under different magnifications. Through the image with a magnification of 200 times, it can be seen that the particles in the remolded red clay mixed with clay particles are unevenly distributed, different in size, and complex. It can be seen from the image with a magnification of 1000 times that the pores of red clay are filled with clay particles locally. The higher the content of clay particles is, the finer and denser the soil sample looks, and it looks like the soil sample is adsorbed together. This is because the clay particles connect the large and small red clay particles. It can be seen from the images with

Figure 5. Added clay content of 0%.

Figure 6. Added clay content of 5%.

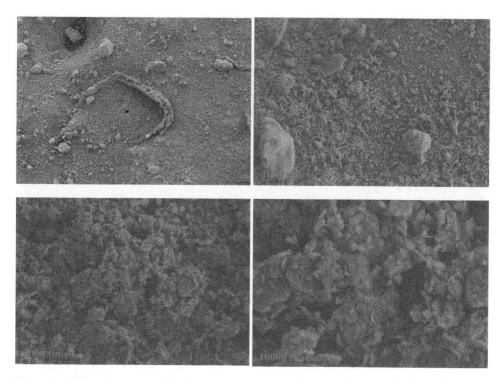

Figure 7. Added clay content of 10%.

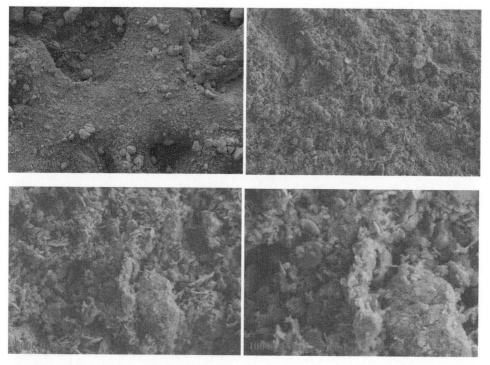

Figure 8. Added clay content of 15%.

452

magnifications of 5000 and 10000 that with the increase of clay content, the pores of soil samples become smaller and smaller. This is because clay particles fill the pores of soil and form aggregates. In the images with a magnification of 10000 times, it can be seen that the pores connected by these aggregates in the form of surface-surface contact, edge-surface contact and edge-edge contact are also decreasing.

3.5 *Microstructure analysis of red clay with different clay content in a shear test*

In this section, the structural changes of soil samples after compression are analyzed from the microscopic perspective. The soil samples selected in this section are SEM images of different clay content after the 300 confining pressure test. In the scanning electron microscope test, the soil samples must be dry soil. Therefore, the soil samples are dried after the test, and then the scanning test is conducted to explore the structural changes of the soil samples after shearing. In this section, the images with magnifications of 200 times and 10000 times are selected for comparative analysis, and the microstructure of soil samples between local large magnification and high magnification is analyzed, as shown in Figures 9, 10, 11 and 12.

Figure 9. Added clay content of 0%.

Figure 10. Added clay content of 5%.

Figures 9, 10, 11 and 12 are SEM images of samples with different added clay content after shearing. From the images with low magnification, it can be seen that the samples after shearing are more compact, the particles are all tightly bound together, and the surface of the soil sample has full concavity and graininess; It can be seen from the image with high

Figure 11. Added clay content of 10%.

Figure 12. Added clay content of 15%.

magnification that the cemented connection of the agglomerates before compression is destroyed to form larger agglomerates. The particles and agglomerates exist in the form of surface-to-surface connections. The size of the agglomerates is also different with different clay content, and the agglomerates with large clay content are large, this is because the increase of clay particles enhances the bonding force between soil particles. There are large and small pores between these agglomerates, which are left by the loss of water, and few individual flake particles can be observed. The particles basically exist in the form of aggregates, and there are pores of different sizes between each aggregate.

4 CONCLUSION

Through the triaxial test and scanning electron microscope test of red clay with different added clay content, it is found that clay has an obvious influence on the cohesion and internal friction angle of red clay. The cohesion increases with the increase of clay, and the internal friction angle is the opposite. Through the analysis of microstructure, after the addition of clay particles, the particle distribution of the soil is uneven, different in size and complicated. Clay particles will fill the pores of red clay. The higher the content of clay particles is, the finer the soil sample looks. The clay particles will closely connect the particles. After shearing, the soil particles are more compact, forming aggregates of different sizes, making the pores of the soil continuously decrease.

Red clay is widely distributed in Guilin area. The influence of clay particles on Guilin red clay is obtained through the above tests, which has a little practical significance for the study of engineering properties, and provides some help for the treatment of special soil such as red clay.

ACKNOWLEDGMENTS

This research was supported by Research Basic Ability Improvement Project of Young and Middle-aged Teachers in Guangxi of China (No.2020ky05036); Foundation for Science and Technology Base and Talents of Guangxi Provence of China (GUIKEAD21220051).

REFERENCES

Chen R., Wang X.Y., Wu S.M., Zhang M. (2020). Experiment on Saturated Permeability Coefficient and Pore Characteristics of Guilin Red Clay Under Dry Wet Cycle [J] *Subgrade Engineering*, (05): 94–98.

Cui Y.S. (2021). *Experimental Study on Fracture Development of Guilin Red Clay with Different Clay Content* [D] Guilin University of Technology.

Dong J. (2019). *Experimental Study on the Effect of Lignin Fiber on the Physical and Mechanical Properties of Guiyang red clay* [D] Guizhou University.

Dong J.Y., Zhao Y.W. (2018). Study on Shear Strength Characteristics of High and Low Liquid Plastic Limit Red Clay Under Different Water Content [J] *Journal of North China University of Water Resources and Hydropower* (Natural Science Edition), 39 (03): 84–87.

Fang W., Yang G.L., Yu D.M. (2008). Study on Deformation Characteristics of Red Clay in Wuhan Guangzhou Passenger Dedicated Line [J] *Journal of Railway Engineering*, (09): 13–20.

Huang X., Chen X.J., Chen L., Xiao G.Y. (2016). Experimental Study on Shear Zone of Guilin Remolded red clay [J] *Geotechnical Foundation*, 30 (02): 255–258. + 280.

Luo W.J., Wang H.Y., Liu H.Q., Liu M.X. (2020). Experimental Study on Shear Strength of Red Clay with Different Water Content [J] *Journal of East China Jiaotong University*, 37 (01): 119–126.

Lv H.B., Zeng Z.T., Yin G.Q., Zhao Y.L. (2012). Mineral Composition Analysis of Guangxi Red Clay [J] *Journal of Engineering Geology*, 20 (05): 651–656.

Meng G.L., Liu Z.K., Lei Y. (2014). Research Status and Prospect of Red Clay [J] *Subgrade Engineering*, (04): 7–11.

Meng X.Y., Xiong F., Wu K., Liu R.Y., Wei W. (2018). Experimental Study on Dynamic Characteristics of Jiangxi Remolded Red Clay Under Cyclic Load [J] *Science, Technology And Engineering*, 18 (25): 84–90. MENG.

Pu L.X. (2019). Effect of Water Content on Shear Strength of Red Clay in Different States [J] *Journal of Guizhou University* (Natural Science Edition), 36 (04): 93–100.

Tan L.R., Kong L.W. (2001). Basic Characteristics and Microstructure Model of a Certain Type of Red Clay [J] *Journal of Geotechnical Engineering*, (04): 458–462.

Tan Y.Z., Hu X.J., Yu B., Liu X.L., Wan P. and Zhang X.H. (2014). Study on Constant Volume Expansion Force and Meso Mechanism of Compacted Red Clay [J] *Geotechnical Mechanics*, 35 (03): 653–658.

Tang D.X., Sun S.W. (1980). Engineering geotechnical [M] Changchun: Geological Publishing House Wei F C (2005). Material Composition and Engineering Geological Properties of Guilin Red Clay [J] *Journal of Jiangxi Normal University* (Natural Science Edition), (05): 88–92.

Wang L.Y., Liu B.C., Yang B., Cui Y.S. (2021). Experimental Study on Water Loss Cracking of Red Clay with Different Clay Content [J] *Journal of Wuhan University of Technology*, 43 (07): 61–68.

Xiao S.D. (2016). *Study on Disintegration Law of Red Clay in Guilin area* [D] Guangxi: Guilin University of Technology.

Zhang H.Y., Zhang Y.S., Peng Y.L., Fu L.N. (2007). Discussion on the Basic Characteristics of Kunming Red Clay and the Influence Mechanism of Engineering Effect [J] *Journal of Yunnan Agricultural University*, (04): 615–617 + 622.

Zhao Y.W. (2017). Discussion on Research Methods of Shear Strength Characteristics of Red Clay Under Different Water Content [J] *Journal of Kaifeng University*, 31 (03): 95–96.

Zhao Y.W. (2018). Discussion on the Method of Shear Strength Characteristics of Red Clay Under Different Water Content [J] *Nanfang Agricultural Machinery*, 49 (05): 174.

Comparative analysis of domestic and foreign tunnel maintenance codes

JingTao Yu* & Ming Pang*
China Road & Bridge Corporation, Beijing, China

ZiXuan Chen* & MuKe Shi*
University of Science and Technology Beijing, Beijing, China

ABSTRACT: With the development of the economy and society, there has been a massive increase in infrastructure construction, and road construction has shown a gradual growth trend in terms of quantity and scale. As an important part of transport infrastructure, the maintenance management and regulation of tunnels are particularly important. In this paper, the tunnel maintenance systems of developed countries such as the United States, the United Kingdom and France are selected for an in-depth study, briefly summarising the inspection categories, frequencies and assessment criteria in each country's specifications. On this basis, the similarities and differences between domestic and foreign inspection systems and assessment methods are analysed and compared, providing a reference for the development of maintenance work in domestic operational highway tunnels, which is of great importance to Chinese enterprises participating in major infrastructure construction projects abroad. It is of great importance for Chinese enterprises to participate in major infrastructure projects abroad.

1 INSTRUCTION

In recent years, with the development of society and the improvement of people's living standards, vehicles have become more and more popular, which in turn has promoted the rapid development of highways, of which the construction of highway tunnels has become an integral part (Wang et al. 2013). Tunnel is a common structure of the expressway, but also the expressway bottleneck section and the key section of daily management. By the end of 2020, China has operated 16,798 railway tunnels with a total length of 19,630km, 2,746 railway tunnels under construction with a total length of about 6,083km, and 6,395 planned railway tunnels with a total length of 16,325km (Tian et al. 2021). The construction of expressway tunnels has greatly shortened people's travel distance, but the current maintenance of highway tunnels is not enough attention. In recent years, the number of accidents in the expressway tunnel has increased year by year, causing huge losses to the safety of people's lives and property.

Therefore, in the process of highway use, only timely prediction and detection of minor tunnel diseases can ensure the usual traffic safety of tunnels by killing dangerous accidents in the cradle (Sun 2021). Although many studies have been done on the characteristics of road surface and tunnel traffic accidents in China, the characteristics of road surface and long-up

---

*Corresponding Authors: yujingtao1999@163.com, 651319970@qq.com, sdczx1998@163.com and shimuke324@163.com

tunnel accidents on the same road section are rarely studied. How to ensure the safe operation of tunnels on a specific traffic road has become a hot concern (Gao 2021). In this paper, we compare tunnel maintenance techniques and evaluation methods at home and abroad, learn from the successful experiences of various countries, recognize the shortcomings of tunnel maintenance specifications, improve and perfect tunnel evaluation methods and maintenance systems, enhance tunnel management and maintenance, and guarantee the safety of people's lives and property.

2 OVERVIEW OF THE TESTING CODES DURING THE FOREIGN OPERATION PERIOD

2.1 The United States Code

The US Tunnel Testing Code is promulgated and developed by the Federal Highway Administration (FHWA) and the Federal Transportation Administration (FTA) and maintains the National Tunnel Inspection Standard (NTIS) established by the FHWA. In addition, the Tunnel Operation and Maintenance Inspection and Evaluation Manual Highway and Rail Transit Tunnel Inspection Manual (TOMIE) provide consistent guidance for the operation, maintenance, inspection and evaluation of tunnels, and instructions for the National Tunnel List (SNTI) of submission of inventory and inspection data to the FHWA. The Highway and Rail Transit Tunnel Inspection Manual (HRTTIM) issued in 2005 contains inspection procedures for civil, structural, and functional systems in highway and traffic tunnels.

The United States divides tunnel operation period testing into initial inspection, periodic inspection, damage inspection, in-depth inspection, and special inspection, in addition to daily, weekly, or monthly hiking inspections by tunnel operation and maintenance personnel. Among them, the detection frequency requirements of initial inspections, routine inspections and in-depth inspections are shown in Table 1, and the frequency of special and damage inspections is at the discretion of the tunnel owner.

Table 1. American tunnel detection frequency.

Type of Activity	Applications	Margin
Initial inspection	The new tunnel	Before opening it to the public
	The existing tunnel	Within 24 months from the effective date of the NTIS
Routine inspection	Implied condition	Every 24 months during the tunnel life cycle
	Approval of written reasons	A maximum extension of 48 months may be allowed
Depth testing	Complex tunnels and specific structural and functional systems	The level and frequency were determined by the project manager

The results of the above types of inspections will be graded by using the evaluation system in Table 2 for testing tunnel components, with each component being assigned a numerical rating from 0 to 9, with 0 being the worst condition and 9 being the best condition, on the basis of which the condition status of the tunnel will be determined according to the grading results and percentage of each unit, divided into four states, CS1, CS2, CS3, and CS4, representing good, fair, poor, and severe, respectively.

Table 2. US tunnel testing status rating.

Ratings	Description
9	Newly completed buildings
8	Good condition - no defects found
7	Good condition - individual defects found, but no repairs required
6	Status between 5 and 7
5	Good condition - minor repairs required, minor, moderate and isolated serious defects, but the structure still functions as originally designed
4	Status between 3 and 5
3	Poor condition - requires extensive repairs, has serious defects, and the structure does not function as originally designed
2	Critical condition - requires immediate major repairs to keep the structure open to road or rail traffic
1	Crisis situation - immediate closure is required and a study should be conducted to determine the feasibility of repairing the structure
0	Critical situation - structure is closed and cannot be repaired

2.2 French norms

The CETU series of documents published and distributed by the French Centre for Tunnelling Research (CETU), "Guide to civil engineering inspection of road tunnels", which is more applied in practical engineering inspection, provides detailed provisions on the categories of tunnel inspection, the names of various types of defects, their causes and means of maintenance. In addition, the Technical Instructions for Monitoring and Maintenance of Engineering Structures (ITSEOA) Volumes 0 and 40, as a supplement to the abovementioned specifications, provide specific testing procedures and technical means for various types of testing items.

Some of the tunnel inspection programs specified in the French specifications are ongoing throughout the project cycle, while others are cyclical or one-off, related to specific events in the service life of the project. Among them, the cyclical ones include annual inspections, evaluation visits, periodic detailed inspections (IDP) and detailed inspections of parts of the works, and the one-off ones include initial detailed inspections (IDI), specific inspections at the end of the contract guarantee, and actions related to unforeseen events. In addition, the first IDP is required 3 years after the IDI, the 2nd IDP no later than 9 years after the completion of the work, and in general, consecutive IDPs every 6 years.

The French tunnels are evaluated by using the IQOA method, which is based on a detailed inspection and is divided into two parts: "civil work" and "water". The works are evaluated by dividing them into areas of the same size, and the evaluation levels and descriptions are shown in Tables 3 and 4.

2.3 UK norms

The UK Inspection and Recording of Highway Tunnel Systems (CS 452) is mainly applied to tunnel inspection and evaluation. In practical engineering applications, the highway inspection specifications Record of Completion, Operation and Maintenance of Highway Structures (CG 302) and Inspection of Highway Structures (CS 450) are often used in conjunction with CS 452 to provide specific requirements and recommendations for the inspection and recording of all structures.

The inspections carried out periodically are divided into surface inspections, general inspections, major inspections, special inspections and safety inspections. Surface inspections are regular, informal visual inspections; general inspections are visual inspections of all

Table 3. Classification table of civil engineering conditions of IQOA tunnels.

Level 1	Area surfaces are in good condition. Level 1 areas require only routine maintenance and regular preventive professional maintenance.
Level 2	The area has minor deficiencies (on structures within the zone of influence or on civil engineering equipment) that do not affect the stability of the structure or do not reflect the instability of the area. Class 2 areas may require specialized non-emergency remedial maintenance in addition to the maintenance specified for Class 1 areas.
Level 2E	Areas presenting Level 2 defects (on the structure or within the zone of influence) that may evolve into major defects and may jeopardize the stability of the structure or areas where civil engineering equipment is severely damaged or its stability may be compromised. Level 2E areas require special monitoring and emergency remedial specialized repairs, in addition to the maintenance required for Level 1 areas, in order to prevent rapid and greater failure of the structure, or to repair damaged civil engineering equipment. The index "E" reflects the evolving condition of the area.
Level 3	Observed defects indicate structural changes or potential compromise of the stability of the area in question. Level 3 areas require non-emergency protection, repair or reinforcement, and judgment must be made quickly.
Level 3U	The observed defects indicate a significant level of degradation and that the overall stability of the area is threatened in the short to medium term. Class 3U areas require urgent repairs to ensure the durability of the structure or to stop the rapid evolution of defects that threaten the stability of the structure. Generally, surveys and monitoring must be conducted prior to construction to ensure that the work is well adapted to local geotechnical conditions, which are often not well known. The index "U" indicates the urgency of the action to be taken.

Table 4. Classification table of IQOA tunnel water flow status.

Level 1	Areas with no visible water flow or areas where water droplets or wet spots are found only on the roadway or sidewalk. Class 1 areas require only routine maintenance and specialized preventive maintenance for drainage and sewerage systems.
Level 2	Low-intensity water flow zones. • drip injection (without regard to flow rate). • localized puddles $\leq$ 5mm in thickness. • Wet stains on the pavement. • Water film formed by continuous flow, left from the pavement, with a thickness of less than 1 mm. In addition to the interventions specified for Level 1 areas, Level 2 areas must be subject to regular monitoring by the management
Level 3	High-intensity water flow zone. • ormation of a continuous flowing film of water falling from the pavement with a thickness of more than 1 mm. • continuous flow of water falling on the pavement (without regard to the flow rate). • Puddles with an area of more than 10 m^2 or a thickness of more than 5 mm. The rating is 3 when the intensity of the flow at the inlet or the volume of water coming from the diffuse surface is large. This requires specific work to communicate to the management the safety measures to be carried out.

accessible parts of road tunnels and electromechanical equipment; major inspections are further detailed inspections of the above-mentioned parts; special inspections include careful examination of specific areas with defects. Among them, the general inspection shall be carried out no later than one year after the due date of the last general or major inspection of the electromechanical equipment and related systems, and the major inspection shall be carried out at intervals agreed by the supervisory organization and not later than three years from the date of the last major inspection.

2.4 *Chinese norms*

The current tunnel maintenance specification in China is mainly the Technical Specification for Highway Tunnel Maintenance (JTG H12-2015), which stipulates that the maintenance of civil structures includes daily inspection, cleaning, structural inspection and technical condition assessment, maintenance and repair, and disease treatment. Among them, the inspection of civil structures includes frequent inspection, regular inspection, emergency inspection and special inspection, which are general qualitative inspection, comprehensive condition inspection, detailed inspection after an accident and targeted inspection of cavern entrance, cavern door, lining, road surface, drainage facilities, etc. respectively. When abnormal conditions are found in the tunnel during regular inspections and the reasons for their generation and details are unknown, regular inspections or special inspections should be done. The frequency of regular inspection is 1 time/month, 1 time/2 months, 1 time/quarter, corresponding to the first, second and third maintenance level of tunnel works, respectively, and the periodicity of regular inspection should be determined according to the technical condition of the tunnel, preferably once a year, with a maximum of 1 time in 3 years.

The technical condition rating of the civil structure of China's tunnels is specified in detail in the Technical Specification for Maintenance of Road Tunnels. Based on the inspection results of each sub-item, the technical condition rating of the civil structure (JGCI) is calculated according to the formula given in the specification and the weighting table of each sub-item, and the technical condition is divided into five categories according to the rating classification threshold values, as shown in Table 5. The evaluation classification boundary values are shown in Table 5. When the cavity entrance, cavity door, lining, pavement and ceiling and pre-buried items of the assessment of the status value of 3 or 4, the technical condition of the civil structure should be directly rated as 4 or 5; when the tunnel appears in the specification of the 7 types of a special critical situation, its technical condition assessment should be rated as 5 types of tunnels.

Table 5. Civil structure technical condition assessment classification boundary value.

Technical condition score	Classification of civil structure technical condition assessment				
	Class 1	Class 2	Class 3	Class 4	Class 5
JGCI	≥ 85	$\geq 70, <85$	$\geq 55, <70$	$\geq 40, <55$	<40

3 COMPARATIVE ANALYSIS OF DOMESTIC AND FOREIGN TUNNEL INSPECTION SPECIFICATIONS

3.1 *Comparison of testing systems*

To summarise the domestic and international specifications for tunnel maintenance and inspection, as each country's specifications are developed on the basis of internationally accepted norms, taking into account specific project experience, there are differences in the inspection categories, but the classification methods and ideas are basically the same. In China, the regular inspection, periodic inspection, emergency inspection and special inspection in tunnel maintenance inspection can roughly correspond to the initial inspection, periodic inspection, damage inspection and depth inspection in the American code, while the French code divides tunnel maintenance inspection into initial detailed inspection (IDI) and periodic detailed inspection (IDP), but its content is roughly the same as in China.

Translated with www.DeepL.com/Translator (free version) in addition to the detection categories, the content of testing in each country is also roughly the same, occasionally there are specific items or testing techniques, frequency, determination methods and other inconsistencies, such as China's specific inspection results based on cracks, lining, water seepage and other projects, it is divided into two categories of general abnormalities, serious abnormalities, the corresponding description to qualitative judgment, while the results of the U.S. inspections to determine the quantitative description of the main clearly specifies the specific limits to distinguish between mild, moderate, serious three categories of cases.

3.2 *Comparison of assessment methods*

Chinese specification classifies the technical condition rating of tunnel civil structures into 1, 2, 3, 4 and 5 categories. First, the condition value of each sub-technical condition of the tunnel civil structure is evaluated hole by hole and section by section, and on this basis, the technical condition of each sub-technical condition is determined, and then the technical condition rating of the civil structure is evaluated. Similar to China, the tunnel rating in the U.S. code uses unit-level inspection technology, grading each tunnel unit according to SNTI, and determining the condition status of the tunnel according to the grading results and percentage of each unit, divided into four states, CS1, CS2, CS3, and CS4, representing good, average, poor, and severe, respectively. While France uses a slightly different method, where the IQOA assessment method is used by cutting the structure into tubes and segments, or by partitioning the project into areas of the same size and then evaluating the project into five categories: 1, 2, 2E, 3, and 3U.

When comparing the assessment methods of tunnels in different countries, although the specific assessment methods and principles are different, the assessment ideas are basically the same, all of them are divided into several sections or graded for each structural element of the tunnel project, and on this basis, the final rating of the tunnel project is based on different weights.

4 CONCLUSION

This paper briefly introduces the methods of testing and evaluation of tunnels in the tunnel maintenance code of the United States, France, Great Britain, China and other countries, and obtains the following conclusions:

(1) The tunnel maintenance and detection system at home and abroad are roughly the same, and their classification method is basically corresponding to the idea and detection content. There are occasionally specific items or different places in detection technology, frequency and determination methods. China's description is mainly based on qualitative judgment, while the judgment of various inspection results in the United States is mainly based on quantitative description.
(2) In terms of evaluation methods, the norms of various countries are generally the same, and the tunnel engineering is first divided into several sections or the structural parts are evaluated and graded first, but the segmentation method is slightly different. The

Chinese Code evaluates tunnel civil structures hole by hole and section; the US Code grades each tunnel unit according to SNTI; The IQOA evaluation method used in France cuts the tunnel into tubes and sections, or divides the works into areas of the same size.
(3) The difference between domestic and foreign tunnel maintenance specifications is very small, but the United Kingdom, France, the United States and other countries have been in use for a long time, and a large number of tunnel projects in urgent need of maintenance phase, and therefore has accumulated a wealth of tunnel maintenance engineering cases and experience technology, while China's tunnel maintenance experience is not enough and lack a large number of cases to test the applicability of the relevant specifications, which need to gradually accumulate and improve in engineering practice.

REFERENCES

Centre d'études des tunnel (2015). *Guide de l'inspection du genie civil des tunnels routiers. Livre 1: du desordre a l'analyse*[S]. de l'analyse à la cotation, Janvier.

Centre d'études des tunnel (2012). *Fascicule 40: Tunnels-Genie Civil Et Equipements-Centre D'etudes Des Tunnels*[S]. de l'analyse à la cotation, Janvier.

Federal Highway Administration (2019). *National Tunnel Inspection Standard* [S]. U.S. Department of Transportation, Washington, DC.

Federal Highway Administration and Federal Transit Administration (2005). *Tunnel Operations, Maintenance, Inspection, and Evaluation (TOMIE) Manual* [S]. Washington: U.S. Department of transport action.

Federal Highway Administration and Federal Transit Administration (2005). *Highway and Rail Transit Tunnel Inspection Manual* [S]. Washington: U.S. Department of transport action.

Gao L., Rao F.Q., Xi Y.J., Liao Z.P. Risk analysis and Risk control Countermeasures [J]. *Highway Traffic Technology*, 2021, 37 (06): 148–152+158.

Highways England (1995). *Inspection and Records for Road Tunnels:* CS 452 [S]. London: The highways agency.

Highways England (1995). As-built, Operational and Maintenance Records for Highway Structures: CG 302 [S]. London: the highways agency.

Highways England (1995). *Inspection of Highway Structures:* CS 450 [S]. London: The highways agency.

Ministry of Transport of the People's Republic of China(2015). *Technical Specification for Highway Tunnel Maintenance: JTG H12-2015* [S]. Beijing: People's Traffic Publishing House.

Sun, J (2022). Study on the Management Measures of Highway Bridge and Tunnel Maintenance[J]. *Bulk Cement*, 01, 42–44.

Tian S.M., Wang W., Gong J.F. Development and Prospect of Railway Tunnels in China (including statistics of railway tunnels in China by the end of 2020)[J]. *Tunnel Construction*, 2021, 41(2): 308.

Wang S., Lin Z., Chen J. (2013). Discussion on Traffic Safety in Mountain Highway Tunnels[J]. *Highway Traffic Technology*, 16(06), 137–143.

Intelligent monitoring and engineering technique optimization

intelligent monitoring and engineering
methods for environment

Research on the optimization of joint dispatching between the quay crane and AGV in automated container terminal

Ji Zhang*

Dalian Vocational & Technical College, Ganjingzi District, Dalian City, Liaoning Province, China

ABSTRACT: With the continuous development of the global economy and the rapid leap of trade globalization, port terminals, as a transit platform connecting waterway transportation and other modes of transportation, play a vital role in the development of global trade. In recent years, with the rapid growth of container terminal throughput, the automation construction of container terminals has received extensive attention from countries around the world. In this paper, by analyzing the scheduling mode of quay cranes and AGVs that affect the loading and unloading efficiency of automated terminals, in order to achieve the purpose of synchronously optimizing the scheduling of quay cranes and AGVs, the coordinated scheduling of quay cranes and AGVs is selected as the optimization object. Using the scheduling mode of the operation surface, considering the constraints of continuous tasks, equipment resources, operation time, and the floating state of ships, and taking the minimization of the maximum completion time of all tasks as the scheduling optimization goal, a coordinated scheduling optimization model of quay crane and AGV is established. The optimization model is closer to the actual working situation and has better practical significance.

1 INTRODUCTION

1.1 Overview of automated container terminals

Automated container terminal (Automated Container Terminal, ACT) refers to a container terminal in which several basic working links of the terminal operation function can achieve automatic operation. The main working links of a container terminal are the two-way horizontal transportation between the shore and the yard. The stacking operation of the yard bridge in the yard, etc. The automated container terminal is composed of regional elements such as the terminal front operation area, the AGV horizontal operation area, and the yard operation area. Figure 1 is a schematic diagram of the layout of an automated container terminal.

In the front-end operation area of the terminal, the double-trolley quay crane is responsible for loading and unloading containers for the container ships in the port, realizing the conversion of container shipping and land transportation. The AGV horizontal operation area is mainly used by AGV to realize the two-way transportation operation between the quay crane and the yard area. The yard operation area consists of a certain number of superimposed areas, and each yard area is equipped with a grounding box and a certain number of ARMGs (Automatic Rail-Mounted Gantry Crane).

---

*Corresponding Author: 524756405@qq.com

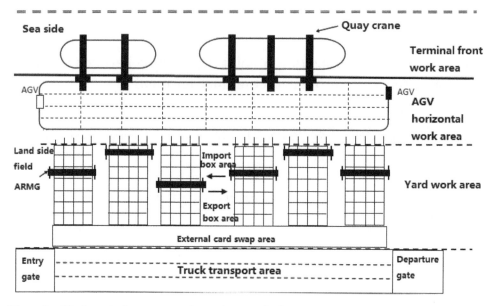

Figure 1. The layout of an automated container terminal.

1.2 Literature review

At present, most of the research on the optimal scheduling of automated terminal operations is to study the scheduling optimization of single equipment in the automated terminal. Although such research can achieve the optimal scheduling of an intermediate operation link of the terminal to a certain extent, the overall optimization of scheduling the sex is not strong and the integration is not enough. Moreover, in the process of researching the coordinated scheduling of quay cranes and AGVs, the loading and unloading types and operation sequences of quay cranes are often set as known, so that the coordinated scheduling between the two pieces of equipment is still fixed to the scheduling of a single equipment. Therefore, in the process of co-scheduling the quay crane and AGV, the restrictions on the scheduling mode and operation sequence of the quay crane should be minimized. The integration of the problem increases the applicability and operability of the scheduling.

Although the existing research optimizes the scheduling of a single piece of equipment to a certain extent or optimizes the coordinated scheduling of two or more pieces of equipment. In the process of researching the problem, the influence of the floating state of the ship on the efficiency of ship loading and unloading is not considered. A practical factor, or it does not take into account the problem of the floating state of the ship during the loading and unloading operations (Luo & Wu 2015). In summary, based on the ship-related theory, in the process of researching the coordinated optimal scheduling of quay cranes and AGVs, this paper considers the cooperative scheduling problem of quay cranes and AGVs, and considers the balance state of ships in loading and unloading operations. By adding ship floating constraints, the research problem is closer to the actual work.

2 ANALYSIS OF AUTOMATIC CONTAINER TERMINAL QUAY CRANE AND AGV DISPATCH MODE

2.1 Crane operation mode

The quay crane operation mode is divided into single-cycle (independent loading and unloading) and double-cycle operation modes. The single-cycle operation mode refers to the cycle process of loading and unloading a container by the quay crane, and one container is loaded or unloaded individually. Containers provide services. The double-cycle operation mode of the quay crane (Yang 2019) is that in a round-trip cycle operation process, the quay crane first lifts an inlet box from the ship and unloads it to the shore, and before the quay crane trolley returns to the ship side for the next cycle operation, an export box is hoisted from shore and loaded onto the ship. The quay cranes carry out mixed loading and unloading of unloading and packing respectively in the same round-trip cycle process. In the same way, in a working cycle, after picking up an export box from the land side of the terminal for loading, the quay crane can pick up an import box from the ship and unload it at the front of the terminal. The double-cycle operation mode of the quay crane enables the quay crane to realize the secondary heavy load in one round-trip cycle. Compared with the single-cycle mode, the number of containers loaded and unloaded by the quay-side crane in the double-cycle mode is twice the number of quay cranes in circulation mode.

2.2 AGV scheduling mode

The operation form of cooperation between quay cranes and AGVs is an important part of the loading and unloading transportation process (Journal of Guangxi University (Natural Science Edition), 2016). An appropriate number of quay cranes and a certain number of AGVs work together to shorten the waiting time between equipment, so that quay cranes can efficiently complete the handling of the container. The scheduling mode of AGV includes two modes: traditional work line mode and work surface scheduling mode.

2.2.1 Traditional job line scheduling mode

If the number of AGVs equipped is 6, the quay crane and the AGVs that only serve them form a group of operation lines. Simply put in a specific loading and unloading process, a fixed number of AGVs only serve one quay crane for container transportation. This scheduling mode is called the traditional operation line scheduling mode.

2.2.2 Job plane scheduling mode

The scheduling mode of the operation surface means that the number of AGVs serving a certain quay crane is not fixed, and the AGVs are centralized as a shared resource library for the entire terminal operation area for shared use by all quay cranes. The increase in the number of quay cranes served by AGVs in the terminal makes the transportation operations of AGVs more frequent because the operation mode of the original point-to-point single task line has changed to a working mode that serves multiple quay cranes and forms a working surface. In this way, the AGV can be scheduled in one working face, but the working face is generally not allowed to cross, that is, the operation process between each quay crane does not interfere with each other. By rationally arranging AGV resources, the terminal can achieve optimal scheduling in the operation surface, which can form a collaborative operation mode among the quay crane, AGV, and ARM. And optimize the intermediate hub link between the three, which can reasonably and effectively improve the work efficiency of the terminal.

2.3 Optimal scheduling mode of quay crane and AGV

Through the above analysis of the operation mode of the quay crane and the scheduling mode of the AGV, it is clarified that in the coordinated scheduling process of the automatic terminal quay crane and the AGV, the operation mode of the quay crane should give priority to the double-cycle operation mode to improve the equipment utilization of the quay crane. The AGV adopts the method of working surface scheduling, that is, one AGV serves more than one quay crane, and can serve multiple quay cranes. For example, an AGV directly supplies another quay crane after completing the transportation requirements of one quay crane. Therefore, it can avoid waiting for each other between devices (Liang & Lin 2019), and can effectively improve the single-machine heavy load rate of AGV. In the actual work process, by effectively scheduling AGVs, the waiting time between various devices can be shortened, thereby improving work efficiency.

3 OPTIMAL MODELLING OF COORDINATED SCHEDULING BETWEEN AUTOMATED TERMINAL QUAY CRANES AND AGVS

3.1 Problem description

The complete operation of an automated container terminal includes a series of closely linked technological processes (Angeloudis & Bell 2010). When the ship is docked at the corresponding berth of the terminal, the quay crane starts to load and unload the container, the AGV is responsible for the transportation of the container between the shore and the yard, and the automatic rail crane completes the sorting of the container in the yard and the loading and unloading of the trailer. The card transports the container from the yard to the destination of the cargo. In view of the fact that the berth resource of the terminal is an important factor in determining the scale of the terminal operation, when scheduling and optimizing the terminal loading and unloading system, the available loading and unloading equipment resources are limited. Scheduling to improve dock loading and unloading efficiency. Based on this, this paper minimizes the maximum completion time of all loading and unloading tasks as the scheduling optimization goal, establishes an operation process based on container tasks, and comprehensively considers the constraints of continuous tasks, equipment resources, operation time, and ship floating state. Scheduling optimization models.

3.2 Model assumptions

The loading and unloading efficiency of the quay crane directly determines the time of the ship in port. Therefore, when arranging other equipment such as AGV and ARMG, the number of them should not be lower than the loading and unloading efficiency of the quay crane, so as to avoid the situation that the quay crane waits for AGV. In this paper, an optimization mathematical model considering the constraints of the ship's floating state is established. In order to simplify the research, the following basic assumptions should be made:

3.2.1 Container ships are divided into multiple bays along the length of the ship from the bow to the stern and numbered in ascending order;
3.2.2 The loading and unloading types of containers are known, and this paper considers the mixed loading and unloading of containers;
3.2.3 One container task can only be completed by one quay crane, and one container can only be transported by one AGV;
3.2.4 Since the operation time of the quay crane is much longer than the time that the quay crane moves between different bay levels, the moving time of the quay crane between the bay levels is not counted in this study;
3.2.5 The time for AGVs to wait for the automated rail crane (ARMG) operation in the yard is zero (the yard is equipped with an AGV companion to achieve this). This article only considers two container areas in the yard, namely the inlet container area and the exit container area;
3.2.6 The traffic congestion of the AGV is not considered;
3.2.7 Failures of operating equipment such as quay cranes, AGVs, and automated rail cranes are not considered;
3.2.8 It is assumed that each quay crane serves a fixed number of bays.

3.3 Parameter calibration

3.3.1 Set parameters. B: *The set of bay positions, arranged from the bow to the stern along the length of the ship, numbered {1,...,b..., B};*
Y: The set of horizontal coordinates of container tasks, which is determined by the serial number of the column where the container is located, and the range is {-Y,...,yi..., Y};
N: the set of all loading and unloading tasks, numbered {1,...,i,...,N}, $i \in N$;
K: The set of quay cranes, bay positions, and AGVs, numbered {1,...,k...,K};
B_y: The set of completed Bay task numbers y, .$y \in B$.

3.3.2 Symbol parameters. o_{ij} : *The jth operation of task i (operation means that each container task occupies equipment resources. If task i is an unloading task, operation j is the bay position-quay crane-AGV in sequence. If task i is a loading task, operation j is in sequence for AGV-quay crane-bay);*
p_{kij}: The duration of the operation on k;
M: A sufficiently large number;
C_{max}: The time when the task is all completed;
CI_i: The time when all operations of task i are completed;
S_{kij}: The start time of the operation on k;
E_{kij}: The end time of the operation on k;
D: The total weight of the ship after removing the container, that is, the displacement of the empty ship;
g: The average weight of a single container;
N_b: The total number of containers in bay b;
θ: The maximum allowable heel angle of the ship;
ΔG: The allowable offset of the ship's center of gravity along the length of the ship.

3.4 Decision variables

$$x_{kij} = \begin{cases} 1, \text{ if } o_{ij} \text{ is operated by } k \\ 0, \text{ else} \end{cases}$$

$$y_{kijgh} = \begin{cases} 1, \text{If the operation } o_{ij} \text{ is performed on } k \text{ first, then the operation } o_{gh} \text{ is performed} \\ 0, \text{ else} \end{cases}$$

3.5 Model construction

The scheduling optimization objective of the terminal loading and unloading system is to minimize the maximum completion time of all tasks, and the objective function is shown in formula (1):

$$f = \min C_{max} \tag{1}$$

$$C_{max} \geq CI_i, \forall i \in N \tag{2}$$

$$CI_i = E_{ki3}, \forall i \in N \tag{3}$$

$$\sum_{k \in K} x_{kij} = 1, \forall i \in N, j \in \{1,2,3\} \tag{4}$$

$$E_{kij} \geq S_{kij} + p_{kij} - (1 - x_{kij}) * M, \forall i \in N, j \in \{1,2,3\}, \forall k \in K \tag{5}$$

$$S_{kij} \geq E_{kgh} - y_{kijgh} * M, \forall i, g \in N, j \in \{1,2,3\}, h \in \{1,2,3\}, \forall k \in K \tag{6}$$

$$S_{kij} \geq E_{kgh} - (1 - y_{kijgh}) * M, \forall i, g \in N, j \in \{1,2,3\}, h \in \{1,2,3\}, \forall k \in K \tag{7}$$

$$\sum_{k \in K} S_{kij} \geq \sum_{k \in K} E_{kij-1}, \forall i \in N, j \in \{1,2,3\} \tag{8}$$

$$S_{kij} + E_{kij} \leq x_{kij} * M, \forall i \in N, j \in \{1,2,3\}, \forall k \in K \tag{9}$$

$$-\tan\theta \leq \frac{\sum_{i \in N} g y_i}{GM \bullet (D+g)} \leq \tan\theta, y_i \in Y \tag{10}$$

$$-\Delta G \leq \frac{\sum_{b \in By}(b - \frac{1+|B|}{2})N_b \bullet g}{D + \sum_{b \in By} N_b \bullet g} \leq \Delta G \tag{11}$$

$$C_{\max} \geq 0 \tag{12}$$

$$S_{kij} \geq 0, \forall i \in N, j \in \{1,2,3\}, \forall k \in K \tag{13}$$

$$E_{kij} \geq 0, \forall i \in N, j \in \{1,2,3\}, \forall k \in K \tag{14}$$

$$x_{kij} \in \{0,1\}, \forall i \in N, j \in \{1,2,3\}, \forall k \in K \tag{15}$$

$$y_{kijgh} \in \{0,1\}, \forall i, g \in N, j \in \{1,2,3\}, h \in \{1,2,3\}, \forall k \in K \tag{16}$$

Constraint (2) means that the entire process cannot be completed earlier than all tasks, so that the entire process is completed in time equal to the last task. Constraint (3) represents a value constraint. Constraints (4) make it only possible to occupy a certain equipment resource (bay, quay crane, AGV) corresponding to the jth operation. Constraint (5) states that if an operation is performed on k, its end time on k is equal to the start time plus the operation time. Constraint (6) and constraint (7) together indicate that at the same time, any bay position, quay crane, or AGV can only perform one operation, that is, only one container can be transported. Constraint (8) means that the start time of the subsequent operation of the task must be greater than or equal to the end time of the previous operation, which ensures that the task order is satisfied. Constraint (9) states that if the operation does not choose k, its start time and end time on k are both 0. Constraint (10) indicates the maximum allowable range of the ship's heel angle, ensuring that after the ship completes each container task, the ship's heel is within the allowable range. Constraint (11) is expressed as a constraint condition for the ship to trim, ensuring that after the quay crane completes each bay position task, the displacement of the ship's center of gravity along the ship's length direction caused by the loading or unloading task is within the allowable range. Constraints (12) to (16) are variable value constraints.

4 CONCLUSION

This paper analyzes the composition of the automatic container terminal loading and unloading system and the operation process of each loading and unloading equipment. The double-cycle operation mode is selected through the quay crane, and the AGV adopts the operation surface scheduling mode. The scheduling optimization goal is to minimize the maximum completion time of all tasks. A mathematical model for the coordinated scheduling of quay cranes and AGVs, which comprehensively considers the constraints of continuous tasks, equipment resources, operating time, and ship floating state, is presented. The research on the dispatching of quay cranes and AGVs supplements relevant theoretical knowledge and provides a certain reference for further research in this field in the future.

REFERENCES

Angeloudis P., Bell M.G.H. An uncertainty-aware AGV assignment algorithm for automated container terminals[J]. *Transportation Research Part E: Logs and Transportation Review*, 2010,46(3):354–366.

AGV scheduling and configuration problems in automated container terminals under uncertain environment [J]. *Journal of Guangxi University* (Natural Science Edition), 2016, 41(2): 589–597.

Luo J, Wu Y. Modelling of dual-cycle strategy for container storage and vehicle scheduling problems at automated container terminals[J]. *Transportation Research Part E Logs & Transportation Review*, 2015,79:49–64.

Liang Chengji, Lin Yang. Research on Coordinated Scheduling Problem of Double-trolley Quay Crane and AGV in Automated Wharf [J]. *Computer Engineering and Application*, 2019, 55(10): 256–263.

Yang Xue. *Optimization of L-AGV Task Scheduling in Automated Terminals Under The "Double Cycle" Mode* [D]. Jilin University, 2019.

Comprehensive evaluation method and application of foundation pit construction safety based on measured data

Jianfeng Liu*
CCCC Sihang Engineering Research Institute Co., Ltd., Guangzhou Guangdong, China
CCCC Transportation Fundamental Engineering Environmental Protection and Safety Key Laboratory, Guangzhou, Guangdong, China
Southern Marine Science and Engineering Guangdong Laboratory, Guangdong, China

Kunpeng Wu*, Junxing Luo*, Hongxing Zhou* & Qingchang Qiu*
CCC Sihang Engineering Research Institute Co., Ltd., Guangzhou, Guangdong, China

ABSTRACT: The fuzzy analytic hierarchy process and set pair analysis theory are introduced to establish a comprehensive safety risk assessment method to scientifically carry out the safety risk assessment of foundation pit engineering construction process. Based on the monitoring data of the actual project, a comprehensive safety risk assessment of a foundation pit engineering construction process is carried out. This method can effectively reflect the evolution process of safety risk in the construction process of a foundation pit and determine the risk source in combination with construction records and monitoring data, providing significance guidance for engineering construction. It also verifies the rationality and practicability of the engineering application of the comprehensive risk assessment method and provides a useful reference for risk assessment of similar deep foundation pit construction.

1 INTRODUCTION

In recent years, foundation pit projects have shown a deep and large development trend, and the risks faced by construction are getting higher and higher (Wang 2020). To ensure the safety of the foundation pit engineering system, several monitoring points and monitoring items are usually set up in their key parts. Construction safety risk assessment is carried out by comparing the monitoring value of a monitoring item at a single monitoring point and its early warning limit. However, due to the complexity and particularity of the foundation pit engineering site and the existing geotechnical environment, different monitoring projects contain different safety risks in different foundation pit projects. At the same time, the monitoring data is also affected by measurement noise, external interference, and sensors. Traditional safety risk assessment methods, due to the influence of performance and other factors, frequently cannot directly reflect the safety risk status of foundation pits and are prone to false alarms and excessive warnings, which impede construction progress and increase the project's economic loss. Therefore, tracking the safety monitoring data on-site and carrying out safety evaluation scientifically along with the construction process has important theoretical value and practical significance in engineering (Ye 2018; Xia 2016). This paper intends to introduce fuzzy uncertainty theories, such as the fuzzy analytic

*Corresponding Authors: ljianfeng@cccc4.com, wkunpeng@cccc4.com, ljunxing@cccc4.com, zhongxing@cccc4.com and qqingchang@cccc4.com

hierarchy process and set-pair analysis method, into the safety risk assessment of foundation pit engineering (Xu 1988; Zhao 2000). The evaluation results will be more scientific and reasonable in providing a new way of thinking for the safety evaluation of foundation pit engineering.

2 PROJECT OVERVIEW

The support structure of the basement pit project in the C area of a construction project in Guangzhou is a 1.0-m thick underground diaphragm wall + beam plate support. The basement covers an area of about 29,000 m^2 (including the connecting section of areas B and C). The support structure of the foundation pit is about 130 m in the north-south direction and 300 m in the east-west direction. The excavation depth of the C area is about 14.70 m, and two supports are provided. The excavation depth of the connecting section between B and C is about 11.60 m. The surrounding environment of the foundation pit is shown in Figure 1(a). The foundation pit is close to the Pearl River, and the water level of the site is affected by the tide. There are many buildings, municipal roads, and municipal pipelines around. The geological survey report shows that the embedded solid soil layer at the pit

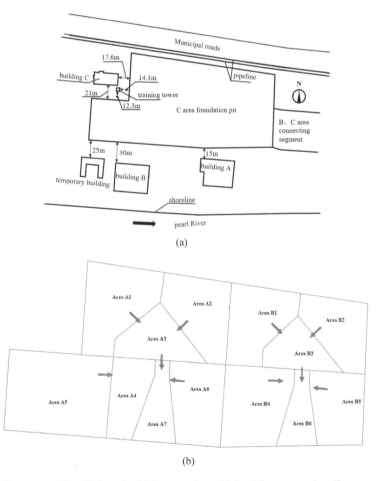

Figure 1. Overview of foundation pit. (a) Surroundings (b) Partition excavation diagram.

bottom of the retaining wall is mainly fully weathered and strongly weathered silty mudstone, which is a weakly permeable soil layer. Groundwater is mainly found in pores in sandy soil. In the construction process, the water collection and open drainage method are adopted to carry out precipitation in the pit, and the environmental and design safety level is Class 1. Figure 1(b) shows the earthwork excavation map of the foundation pit zone. The sequence of excavation from all sides to the middle and from north to south is adopted, and the original excavation ramp position is finally excavated.

3 COMPREHENSIVE SECURITY RISK ASSESSMENT

3.1 Construction of evaluation index system

The foundation pit project in Zone C has the characteristic of being deep, large, and close to the water. It adopts a beam-plate support structure, and the force transmission path is complicated. There are existing buildings, municipal roads, and municipal pipelines (rainwater pipes, sewage pipes, power cable pipes) in the surrounding main affected areas. The groundwater level is shallow and fluctuates with the tide of the Pearl River and regional weather. This paper refers to the experience of similar projects through expert consultation, combined with the requirements of relevant technical specifications (MOHURD 2019), to establish the safety risk evaluation index system of this project, as shown in Table 1.

Table 1. Safety risk evaluation index system of foundation pit construction.

Primary indicator	Secondary indicators (Monitoring items)	Three-level indicator (Monitoring indicators)
Foundation pit support structure (I_1)	Deep horizontal displacement of the building envelope (I_{11})	Cumulative value (I_{111}) rate of change (I_{112})
	Horizontal displacement of crown beam (I_{12})	Cumulative value (I_{121}) Rate of change (I_{122})
	Vertical displacement of crown beam (I_{13})	Cumulative value (I_{131}) rate of change (I_{132})
	Support shaft force (I_{14})	Cumulative value (I_{141})
	Column settlement (I_{15})	Cumulative value (I_{151}) rate of change (I_{152})
Surroundings (I_2)	Subsidence (I_{21})	Cumulative value (I_{211}) Rate of change (I_{212})
	Building settlement (I_{22})	Cumulative value (I_{221}) Rate of change (I_{222})
	Leaning building (I_{23})	Cumulative value (I_{231})
	Pipeline settlement (I_{24})	Cumulative value (I_{241}) rate of change (I_{242})
	Water level outside the pit (I_{25})	Cumulative value (I_{251}) Rate of change (I_{252})

3.2 Determination of indicator weights

The foundation pit project is a high-risk project. To fully consider the opinions of multiple parties and take into account the efficiency of group decision-making, this paper invites a total of five experts in this field related to engineering construction, using the triangular fuzzy analytic hierarchy process and 1-9 scale. The criteria compare and score the importance of evaluation indicators at different levels (Xu 1988). At the same time, the set-pair analysis theory is introduced to synthesize the opinions of many experts. Due to space limitations, this paper takes the secondary index corresponding to the foundation pit support structure as an example. After discussion by the expert group, it is believed that the importance order of the secondary index in this example is: $I_{11} > I_{12} > I_{14} > I_{15} > I_{13}$.

The importance ranking of the triangular fuzzy judgment matrix $X^{(1)} = [x_{ij}]_{n \times n}$ is as follows:

$$\begin{bmatrix} (1.00\ 1.00\ 1.00) & (2.00\ 2.00\ 3.00) & (3.50\ 4.00\ 4.50) & (4.50\ 5.00\ 6.00) & (5.50\ 6.00\ 7.00) \\ (0.33\ 0.50\ 0.50) & (1.00\ 1.00\ 1.00) & (2.50\ 3.00\ 3.00) & (3.50\ 4.00\ 5.00) & (4.00\ 5.00\ 5.00) \\ (0.22\ 0.25\ 0.29) & (0.33\ 0.33\ 0.40) & (1.00\ 1.00\ 1.00) & (2.50\ 3.00\ 3.00) & (3.50\ 4.00\ 4.50) \\ (0.17\ 0.20\ 0.22) & (0.20\ 0.25\ 0.29) & (0.33\ 0.33\ 0.40) & (1.00\ 1.00\ 1.00) & (2.50\ 3.00\ 4.00) \\ (0.14\ 0.17\ 0.18) & (0.20\ 0.20\ 0.25) & (0.22\ 0.25\ 0.29) & (0.33\ 0.33\ 0.40) & (1.00\ 1.00\ 1.00) \end{bmatrix}$$

$X^{(1)}$ is calculated to get the conversion judgment matrix $Y^{(1)} = [y_{ij}]_{n \times n}$ (Fan 2005). The consistency test is carried out to obtain the consistency ratio $CR^{(1)} = 0.057 < 0.1$, which meets the consistency requirements. Similarly, the consistency ratios of the other four experts are 0.040, 0.043, 0.049, and 0.004, respectively.

$$Y^{(1)} = \begin{bmatrix} 1.00 & 2.17 & 4.00 & 5.08 & 6.08 \\ 0.47 & 1.00 & 2.92 & 4.08 & 4.83 \\ 0.25 & 0.34 & 1.00 & 2.92 & 4.00 \\ 0.20 & 0.25 & 0.34 & 1.00 & 3.08 \\ 0.17 & 0.21 & 0.25 & 0.33 & 1.00 \end{bmatrix}$$

Using the geometric mean method, the transformation judgment matrix of the five experts is used to synthesize the judgment matrix $Z = [z_{ij}]_{n \times n}$ of the expert group (Li 2009). The consistency ratio is calculated to be 0.037, which is less than 0.1, which meets the consistency requirements and ensures the experts' internal consistency of opinion.

$$Z = \begin{bmatrix} 1.00 & 1.82 & 5.83 & 2.95 & 4.78 \\ 0.55 & 1.00 & 4.75 & 2.25 & 4.15 \\ 0.17 & 0.21 & 1.00 & 0.26 & 0.41 \\ 0.34 & 0.44 & 3.82 & 1.00 & 3.16 \\ 0.21 & 0.24 & 2.45 & 0.32 & 1.00 \end{bmatrix}$$

The set-pair analysis method is introduced to synthesize the opinions of the five experts. The comprehensive judgment matrix of the five experts is transformed into a set-pair judgment matrix U using Formula 1 as follows:

$$U = A + \alpha B = \begin{bmatrix} a_{11} & a_{12} & \cdots & a_{1n} \\ a_{21} & a_{22} & \cdots & a_{2n} \\ \vdots & \vdots & \vdots & \vdots \\ a_{n1} & a_{n2} & \cdots & a_{nn} \end{bmatrix} + \alpha \begin{bmatrix} b_{11} & b_{12} & \cdots & b_{1n} \\ b_{21} & b_{22} & \cdots & b_{2n} \\ \vdots & \vdots & \vdots & \vdots \\ b_{n1} & b_{n2} & \cdots & b_{nn} \end{bmatrix} \quad (1)$$

In the formulas: $a_{ij} = \begin{cases} \min_r\{y_{ij}\}, y_{ij} \geq 1 \\ \max_r\{y_{ij}\}, y_{ij} \leq 1 \end{cases}; b_{ij} = \begin{cases} \left|\max_r\{y_{ij}\} - \min_r\{y_{ij}\}\right|, y_{ij} \geq 1 \\ -\left|\max_r\{y_{ij}\} - \min_r\{y_{ij}\}\right|, y_{ij} \leq 1 \end{cases}$

$$U = \begin{bmatrix} 1.00 & 1.83 & 3.04 & 4.54 & 5.61 \\ 0.57 & 1.00 & 2.33 & 4.04 & 4.58 \\ 0.37 & 0.49 & 1.00 & 3.29 & 3.63 \\ 0.22 & 0.26 & 0.32 & 1.00 & 2.29 \\ 0.18 & 0.22 & 0.28 & 0.50 & 1.00 \end{bmatrix}$$

Since the set pair judgment matrix U is generally not consistent, the compatibility judgment matrix $D = [d_{ij}]_{n \times n}$ is obtained after the compatibility processing [Ye 2006] as follows:

$$D = \begin{bmatrix} 1.033 & 1.491 & 2.460 & 5.313 & 7.870 \\ 0.728 & 1.051 & 1.733 & 3.743 & 5.545 \\ 0.447 & 0.645 & 1.064 & 2.298 & 3.404 \\ 0.204 & 0.295 & 0.486 & 1.051 & 1.556 \\ 0.136 & 0.196 & 0.324 & 0.699 & 1.068 \end{bmatrix}$$

Finally, the weight of each secondary indicator corresponding to I1 can be obtained as: $W_1 = (w_{11}, w_{12}, w_{13}, w_{14}, w_{15})^T = (0.4055, 0.2857, 0.0533, 0.1754, 0.0533)^T$.

In the same way, the weights of the evaluation indicators at all levels can be obtained as shown in Table 2:

Table 2. Weight distribution table of foundation pit safety risk evaluation index.

Primary indicator	Secondary indicators (Monitoring items)	Three-level indicator (Monitoring indicators)
w_1: 0.5599	w_{11}: 0.4055	w_{111}: 0.6127
		w_{112}: 0.3873
	w_{12}: 0.2857	w_{121}: 0.5906
		w_{122}: 0.4094
	w_{13}: 0.0533	w_{131}: 0.6194
		w_{132}: 0.3806
	w_{14}: 0.1754	w_{141}: 1.0000
	w_{15}: 0.0533	w_{151}: 0.6378
		w_{152}: 0.3622
w_2: 4401	w_{21}: 0.1276	w_{211}: 0.5837
		w_{212}: 0.4163
	w_{22}: 0.1986	w_{221}: 0.5798
		w_{222}: 0.4202
	w_{23}: 0.3351	w_{231}: 1.0000
	w_{24}: 0.2623	w_{241}: 0.5919
		w_{242}: 0.4081
	w_{25}: 0.0761	w_{251}: 0.6191
		w_{252}: 0.3809

3.3 Classification of safety risk levels and evaluation criteria

This paper uses the commonly used five-level classification method to establish the safety risk level classification standard and foundation pit construction safety risk based on the accumulated value and change rate alarm value of the monitoring data stipulated in the relevant specifications or design documents through literature research, engineering analogy, and other means. (Xia 2016; MOHURD 2019). Evaluation, acceptance criteria, and countermeasures are shown in Table 3.

3.4 Contact number calculation and comprehensive safety evaluation analysis

3.4.1 Contact number calculation

In this paper, the evaluation level and the evaluation index are composed of set pairs. The five-element connection number expression of the bottom evaluation index is established by using the same-difference-inverse hierarchy method (Pan 2010; Qin 2010). Finally, the safety

Table 3. Classification criteria and evaluation of safety risk levels for monitoring projects.

Risk level	Monitoring value/Alarm value	Risk status	Risk assessment and acceptance criteria	Countermeasures
I	<0.6	Safety	Very low risk; negligible	Risk treatment measures, routine management, and review are not required.
II	0.6~0.7	Track	Low risk; tolerable	No need to take risk treatment measures, attract attention, and routinely monitor and manage.
III	0.7~0.8	Warning	Medium risk; acceptable	Attention should be paid to formulating detailed risk prevention, treatment, and monitoring measures.
IV	0.8~1.0	Alarm	High risk; partially acceptable	Take risk treatment measures to reduce the risk to three levels as much as possible, and strengthen monitoring and warning.
V	≥1.0	Danger	Very high risk; refuse to accept	Pay high attention and take measures to avoid and transfer; otherwise reduce the risk to at least Level IV at any cost.

risk assessment level is determined by referring to Qin's research. The specific steps are as follows:

(1) The five-element connection coefficient expression of the bottom evaluation index:

$$\mu_{mk} = \begin{cases} 1 + 0i_1 + 0i_2 + 0i_3 + 0j & c_{mk} \in [0, s_{(1,2)k}) \\ \frac{1}{2}\frac{c_{mk} - s_{(2,3)k}}{s_{(1,2)k} - s_{(2,3)k}} + \frac{1}{2}i_1 + \frac{1}{2}\frac{s_{(1,2)k} - c_{mk}}{s_{(1,2)k} - s_{(2,3)k}}i_2 + 0i_3 + 0j & c_{mk} \in [s_{(1,2)k}, s_{(2,3)k}) \\ 0 + \frac{1}{2}\frac{c_{mk} - s_{(3,4)k}}{s_{(2,3)k} - s_{(3,4)k}}i_1 + \frac{1}{2}i_2 + \frac{1}{2}\frac{s_{(2,3)k} - c_{mk}}{s_{(2,3)k} - s_{(3,4)k}}i_3 + 0j & c_{mk} \in [s_{(2,3)k}, s_{(3,4)k}) \\ 0 + 0i_1 + \frac{1}{2}\frac{c_{mk} - s_{(4,5)k}}{s_{(3,4)k} - s_{(4,5)k}}i_2 + \frac{1}{2}i_3 + \frac{1}{2}\frac{s_{(3,4)k} - c_{mk}}{s_{(3,4)k} - s_{(4,5)k}}j & c_{mk} \in [s_{(3,4)k}, s_{(4,5)k}) \\ 0 + 0i_1 + 0i_2 + 0i_3 + 1j & c_{mk} \in [s_{(4,5)k}, +\infty) \end{cases} \quad (2)$$

In the formula: m is the mth evaluation unit; k is the kth evaluation index; c_{mk} is the sample value of the evaluation index; $s_{(x,y)k}$ is the first evaluation index; The limit value of the x, y is the grade standard; i_1, i_2, i_3 are the difference coefficients, let i_1, i_2, i_3 be 0.5, 0, −0.5, respectively. This paper refers to the cumulative value and change rate of the monitoring data.

Equation 3 can be expressed as a general formula:

$$\mu_{mk} = r_{mk1} + r_{mk2}i_1 + r_{mk3}i_2 + r_{mk4}i_3 + r_{mk5}j \quad (3)$$

In the formula: $r_{mk1}, r_{mk2}, r_{mk3}, r_{mk4}, r_{mk5}$ are the contact number components.

(2) Calculation of comprehensive evaluation connection number of subsystems:

$$\mu_m = r_{m1} + r_{m2}i_1 + r_{m3}i_2 + r_{m4}i_3 + r_{m5}j \quad (4)$$

$$r_{ml} = \sum_{p=1}^{k} w_{mp} r_{mpl} \quad (5)$$

In the formula: r_{ml} is the connection degree component of the subsystem index relative to the security risk level; w_{mpl} is the evaluation index weight of the secondary subsystem obtained in Section 3.2.

(3) Calculation of the comprehensive evaluation connection number of the total system of the foundation pit:

$$\mu = r_1 + r_2 i_1 + r_3 i_2 + r_4 i_3 + r_5 j \tag{6}$$

$$r_l = \sum_{m=1}^{k} w_l r_{ml} \tag{7}$$

In the formula: r_l is the connection degree component of the system relative to the security risk level; w_{mp} is the evaluation index weight of the Level I subsystem obtained in Section 3.2.

(4) Evaluation of safety risk level:

According to the principle of equal division, the $[-1,1]$ interval is divided into five sub-intervals, namely (0.6, 1), (0.2, 0.6), (−0.2, 0.2), (−0.6, −0.2), and (−1, −0.6), corresponding to the safety risk Levels I, II, III, IV, and V, respectively.

According to the above calculation method, the maximum value of the monitoring indicators of each monitoring project in the weekly monitoring report is used as the basic data for the foundation pit safety risk assessment. The connection number of the safety risk assessment index is calculated, and the first-level safety risk assessment index of the foundation pit are drawn. The change in the main value of the system's contact number and the number of monitoring periods are shown in Figure 2. The comprehensive safety risk assessment of the foundation pit involves a total of 19 consecutive weekly reports, and the monitoring time period is 2020.11.10–2021.3.22.

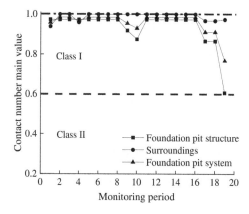

Figure 2. Changes in the safety risk indicators of foundation pits.

3.4.2 Comprehensive safety evaluation analysis

It can be seen from Figure 2 that the main value of the connection number of the foundation pit changes with the construction process. Among them, the safety of the foundation pit can be seen to decrease in Phases 8–10 and 16–19, and the possible reasons are analyzed according to the site construction records and monitoring data. During monitoring period from 8 to 10, the second layer of earthwork was excavated in the A6 and A7 areas, and the excavation depth is about 9 m. At this time, part of the second support was not poured, and the intersection of the two diagonal supports was the enclosure structure corresponding to the A7 area. As the horizontal displacement increases, the safety of the foundation pit decreases. During the 11th monitoring period, the second support was all poured and reached the design strength. The soil in the A1 and A2 areas continues to be excavated downwards, which reduces the horizontal displacement of the enclosure structure at the corresponding position in the A7 area and improves the safety of the foundation pit. During the monitoring period from 16 to 19, after heavy rain and high tide, the water and soil pressure behind the wall increased. At the same time, the third layer of soil was excavated in the foundation pit. The comprehensive security risk level of the pit system is on the rise.

During this monitoring period, the overall safety of the surrounding environment of the foundation pit is higher than that of the foundation pit itself. The main value of the connection number of the surrounding environment changes little with the construction. The change law of the main value of the connection number of the body is basically the same, indicating that the foundation pit body is the main part that affects the safety of the foundation pit system during the monitoring period. More attention should be paid to the safety of the foundation pit body during the construction process in the future. In addition, during this monitoring period, the safety risk level of the foundation pit has always been at Level I, indicating that the current site construction has good control over the safety of the foundation pit system.

4 CONCLUSION

The fuzzy AHP (analytical hierarchy process) theory and set-pair analysis theory are introduced into the construction safety risk assessment of foundation pit engineering. Based on the on-site monitoring data, it can effectively reflect the safety risk changes during the project construction process. It can indicate the risk source and provide a good source of risk for the site when combined with on-site records and monitoring data. A resource for construction safety management personnel is available.

REFERENCES

Anonymous. Regulations on the Safety Management of Sub-item Projects With High Risk (Order No. 37 of the Ministry of Housing and Urban-Rural Development) [J]. *Ministry of Housing and Urban-Rural Development*, 2018.

Fan Weigang, Hou Lihong. Improvement of Analytic Hierarchy Process [J]. *Library and Information Guide*, 2005, 15(4):153–154.

Li Lingjuan, Dou Kun. Research on Consistency of Judgment Matrix in Analytic Hierarchy Process [J]. *Computer Technology and Development*, 2009(10):137–139.

Ministry of Housing and Urban-Rural Development of the People's Republic of China, State Administration for Market Regulation. *Technical Specification for Monitoring of Construction Foundation Pit Engineering* (GB 50497-2019) [S]. Beijing: China Planning Press, 2019.

Pan Zhengwei, Jin Juliang, Wu Kaiya, et al. System Comprehensive Evaluation Model Based on Link Function and Its Application [J]. *Practice and Understanding of Mathematics*, 2010, 40(023):40–47.

Qin Zhihai, Qin Peng. Fuzzy Hierarchy and Set Pair Analysis Coupling Model for High Slope Stability Evaluation [J]. *Chinese Journal of Geotechnical Engineering*, 2010, 32(005):706–711.

Wu Jianjun, Cai Yao, Liu Zhengjiang. Set Pair Analysis of Index Weights in Comprehensive Safety Evaluation [J]. *China Navigation*, 2010, 33(003):60–63.

Wang Weidong, Ding Wenqi, Yang Xiuren, et al. Foundation Pit Engineering and Underground Engineering: High Efficiency and Energy Saving, Low Environmental Impact and Sustainable Development New Technology [J]. *Chinese Journal of Civil Engineering*, 2020(7):78–98.

Xia Yuanyou, Chen Chunshu, Chen Jinpei, et al. Dynamic Risk Assessment of Deep Foundation Pit Construction Based on On-site Monitoring [J]. *Chinese Journal of Underground Space and Engineering*, 2016(12): 1378–1384.

Xu Shubai. *Practical Decision-Making Methods: The Principles of Analytic Hierarchy Process* [M]. Tianjin: Tianjin University Press, 1988.

Ye Yicheng, Ke Lihua, Huang Deyu. *System Comprehensive Evaluation Technology and Its Application* [M]. Beijing: Metallurgical Industry Press, 2006.

Zhao Keqin. *Set Pair Analysis and Its Preliminary Application* [M]. Zhejiang: Zhejiang Science and Technology Press, 2000.

Fast building method of bridge OpenSees dynamic model based on Python language

Guangping Zhu* & Hujun Lei*
FuJian University of Technology, China

ABSTRACT: The OpenSees is widely used in the field of bridge earthquake resistance. The low efficiency of modelling is one of the bottlenecks restricting its popularization. In order to quickly establish the dynamic model of the OpenSees full bridge, the conversion formula between the β angle of the Midas Civil beam element and the axial vector of the OpenSees beam element is derived based on the basic concept of β angle. Based on Python language, a fast method for establishing the dynamic model of the whole bridge of OpenSees is established, and the corresponding ANSYS model is generated and verified. Taking Dashengguan Yangtze River Bridge in Nanjing as an example, the dynamic responses of structures under three typical seismic waves are compared based on the two generated dynamic models. The results show that the automatically generated OpenSees model is completely consistent with the first 10 natural frequencies of the Midas Civil model, and the maximum error of the first 20 natural frequencies is 4.5%. Further comparing the displacement, velocity and acceleration of the vault nodes, the maximum amplitude error is 9.8%, which comes from the calculation error of the software itself. Using the MTO conversion program, a one-click generation of complex bridge OpenSees dynamic model can be realized. The research results of this paper can be used to quickly establish the dynamic model of the whole bridge of OpenSees.

1 GENERAL INSTRUCTIONS

OpenSees is the abbreviation of Open System for Earthquake Engineering Simulation, which is a software platform developed by the Pacific Earthquake Engineering Research Center (PEER) as a simulation research and application of structural and geotechnical engineering systems. OpenSees is widely used in the dynamic analysis of bridge structures due to its strong dynamic analysis capability, small program storage and open-source property. OpenSees has no graphical user interface and has a large workload of modelling and repetitive work, which is one of the difficulties that limit its popularization and application.

In previous research, most scholars mainly rely on the OpenSees software platform to study specific problems or secondary development, and seldom pay attention to the modeling efficiency of OpenSees. Chen Xuewei and others proposed to use the ETABS as a modeling tool for building structures and compiled a conversion program ETO (ETABS to OpenSees), which can realize the rapid conversion of the ETABS model to OpenSees model. For the full bridge finite element model of bridge, it is very important to be able to quickly model and improve the research efficiency.

Firstly, based on the basic concepts of the β angle of beam element in Midas Civil and the axial vector of beam element in OpenSees, this paper deduces the conversion formula from β angle to axial vector. On this basis, the conversion program from Midas Civil to OpenSees is compiled based on Python language. MTO and verified the validity. We use Dashengguan Yangtze River Bridge as an example, typical seismic waves are selected as the input, and the

*Corresponding Authors: 2742417564@qq.com and leihujun@yeah.net

differences in the dynamic responses of the Midas Civil model and the automatically generated OpenSees model are compared in depth to judge the applicability of the conversion program MTO to complex bridges. The research results in this paper can be used for the rapid establishment of the OpenSees full-bridge dynamic finite element model.

2 DERIVATION OF BEAM ELEMENT AXIAL VECTOR CONVERSION FORMULA

For full bridge finite element models of bridge structures, the most used element types are beam and rod elements. In Midas Civil, beam elements require beta angles to define the element cross section. According to the Midas Civil technical documentation, the global coordinate system is defined as XYZ, and the local coordinate system of the unit is xyz, X, Y, Z, which represents the coordinate system parallel to the global coordinate system through the beam element. There are two cases:

(1) If the x-axis of the element coordinate system is parallel to the Z-axis of the global coordinate system. It is a vertical member. The angle β is the angle between the X-axis of the global coordinate system and the z-axis of the element coordinate system.
(2) If the x-axis of the unit coordinate system is not parallel to the Z-axis of the global coordinate system, and the component is horizontal or inclined. The β angle is the angle between the plane formed by the Z-axis of the global coordinate system (GCS) and the xz-axis of the unit coordinate system.

In OpenSees, the local coordinate axis of the beam element is defined by the axial vector. The axis vector is the normalized direction vector of the axis of the local coordinate axis 2 of the element relative to the orthogonal coordinate system.

Therefore, before data conversion, the conversion relationship between the β angle of the beam element in Midas Civil and the axial vector of the beam element in OpenSees should be deduced first. The N1 and N2 node coordinates and β angle of a beam element in Midas Civil are known. The x, y, and z coordinates of the N1 node are x_1, y_1, z_1 respectively. The N2 node is x_2, y_2, z_2. The length components of the beam element dx, dy, dz, dL, dxy are respectively:

$$\begin{cases} dx = x_2 - x_1, dy = y_2 - y_1, dz = z_2 - z_1 \\ dL = \sqrt{dx^2 + dy^2 + dz^2} \\ dxy = \sqrt{dx^2 + dy^2} \end{cases} \quad (1)$$

Let the axial vector of the beam element in OpenSees be $(vecxzX, vecxzY, vecxzZ)$, the angle between the beam element and the X-Y plane is $yota$, and the angle between the beam element's projection on the X-Y plane and the X axis is $seta$. Then by definition:

$$yota = \arcsin\left(\frac{dz}{dL}\right) \quad (2)$$

Case 1: When $dxy = 0.0$, the beam element is parallel to the Z axis. At this time, the calculation formula of the axial vector in OpenSees can be obtained according to the definition of the β angle in Midas Civil, which is divided into:

① When $dz > 0.0$, the three components of the axial vector are:

$$\begin{cases} vecxzX = \cos(\beta) \\ vecxzY = \sin(\beta) \\ vecxzZ = 0.0 \end{cases} \quad (3)$$

② When $dz < 0.0$, the three components of the axial vector are:

$$\begin{cases} vecxzX = \cos(\beta) \\ vecxzY = -\sin(\beta) \\ vecxzZ = 0.0 \end{cases} \quad (4)$$

Case 2: When $dz < 0.0$, the beam element is not parallel to the Z axis. According to the definition of beta angle in Midas Civil:

$$\begin{cases} seta = \arcsin\left(\dfrac{dy}{dxy}\right) & dx \geq 0.0 \\ seta = \arcsin\left(\dfrac{dy}{dxy}\right) + \pi dx < 0.0, dy \geq 0.0 \\ seta = \arcsin\left(\dfrac{dy}{dxy}\right) - \pi dx < 0.0, dy < 0.0 \end{cases} \quad (5)$$

Then, according to the geometric relationship, the conversion formula of β angle and OpenSees axis vector in Midas Civil can be obtained as:

$$\begin{cases} vecxzX = \sin(seta)\sin(\beta) \\ \qquad - \sin(yota)\cos(seta)\cos(\beta) \\ vecxzY = -\cos(seta)\sin(\beta) \\ \qquad - \sin(yota)\sin(seta)\cos(\beta) \\ vecxzZ = \cos(yota)\cos(\beta) \end{cases} \quad (6)$$

According to the above formulas (1)~(6), the axial vector of the corresponding beam element in OpenSees can be calculated by the N1 and N2 node coordinates and β angle of the beam element in Midas Civil.

3 RAPID MODELLING METHOD AND VERIFICATION

3.1 Rapid modelling method

Python programming language is concise, easy to read and highly extensible. It is one of the most popular programming languages in the world. Based on the derived axial vector calculation formula, Python language is used to read the MCT file of Midas Civil software, and then according to the keywords in the MCT file to accurately locate and format the output, the rapid establishment of the OpenSees dynamic model can be realized. Based on this, this paper compiled the Midas to OpenSees program (referred to as MTO).

3.2 Method validation

Taking a simple frame structure as an example, a Midas Civil finite element model is established. Among them, the beams, columns, and inclined rods are all H-shaped sections, the beams and columns are H300 × 400 sections, 400 mm high, 300 mm wide, and 25mm thick; the inclined rods are H200 × 200 sections, 200 mm high, 200 mm wide and 12 mm thick, see Figure 1(a). Using the MTO program, the OpenSees model is automatically generated, and the corresponding ANSYS model is generated at the same time. The ANSYS model is shown in Figure 1(b).

The calculation results of the Midas Civil model and the ANSYS model are compared, and the first 10-order natural frequencies are shown in Table 1. By comparison, it can be found that for this simple frame structure, the first 10-order natural frequencies of the generated OpenSees model and ANSYS model are completely consistent with the calculation results of Midas Civil. This shows that the rapid establishment method of the OpenSees dynamic model based on Python language and the compiled MTO program is effective.

Table 1. Comparison of natural frequencies of frame structures.

Order	Midas	OpenSees	ANSYS
1	5.706	5.706	5.706
2	15.415	15.415	15.415
3	16.404	16.404	16.404
4	23.088	23.088	23.088
5	25.871	25.871	25.871
6	32.981	32.981	32.981
7	34.393	34.393	34.393
8	36.026	36.026	36.026
9	36.401	36.401	36.401
10	36.877	36.877	36.877

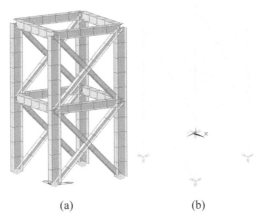

Figure 1. Finite element model of frame structure. (a) Midas model (b) ANSYS model.

4 EXAMPLES

Taking the Beijing-Shanghai (Beijing-Shanghai) high-speed railway Nanjing Dashengguan Yangtze River Bridge as an example, the main bridge is a continuous steel truss arch of (108 + 192 + 2 × 336 + 192 + 108) m, located between the 4#~10# piers. The established Midas Civil full bridge finite element model is shown in Figure 2(a). Among them, there are 3062 nodes, 7389 beam elements, and 6 springs at the bottom of the pier to simplify the simulation of the foundation stiffness.

The corresponding OpenSees model and ANSYS model are generated with one click of the MTO program. The ANSYS model is shown in Figure 2(b). At the same time, the first 20-order natural frequencies of different models are calculated, and the calculation results are compared in Figure 3. The natural frequencies of the OpenSees model and the ANSYS model are exactly the same; compared with the Midas Civil model, the first 10 natural frequencies of the OpenSees model and the ANSYS model are exactly the same, and the maximum relative error of the first 20 natural frequencies is 4.5%, which appears in 18th order. This shows that the MTO program can realize the rapid conversion of the complex bridge Midas Civil model to the OpenSees model and the ANSYS model.

Figure 2. Finite element model of the whole bridge of Dashengguan Yangtze River Bridge. (a) Midas model (b) ANSYS model

In order to further verify the conversion effect of the MTO program, select the typical seismic wave EL Centro seismic wave (1940, El Centro Site, 180 Deg) as input, the original seismic wave data are all from the seismic wave database of Midas Civil 2020 software, and the seismic wave acceleration time history is shown in Figure 4.

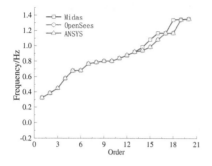

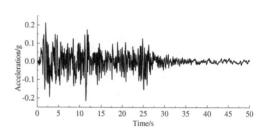

Figure 3. Comparison of natural frequencies of Bridge.

Figure 4. EL Centro.

The seismic waves were input along the longitudinal and lateral directions of the bridge respectively, and the differences between the calculation results of the OpenSees model and the Midas Civil model were compared. In the calculation, both models use the direct integration method, the calculation time is 50 s, and the step size is 0.01 s; the damping adopts the Rayleigh damping model and is based on the first-order natural frequency of Dashengguan Yangtze River Bridge of 0.322 Hz and the 10th-order natural vibration frequency. The vibration frequency is 0.801Hz and the damping ratio is 3%. The time-history comparison of displacement, velocity, and acceleration of the left vault node when different seismic waves are input along the longitudinal and lateral directions of the bridge is shown in Figures 5 and 6, respectively.

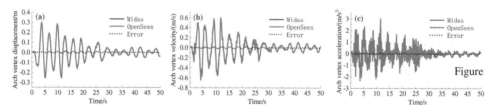

Figure 5. Comparison of longitudinal displacement, velocity, and acceleration responses of vault joints under different seismic waves: (a)~(c) ELCentro.

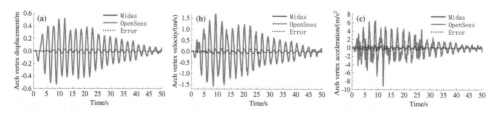

Figure 6. Comparison of lateral displacement, velocity, and acceleration responses of vault joints under different seismic waves: (a)~(c) EL Centro.

Figures 5 and 6 compared the Midas Civil model. The automatically generated OpenSees model is completely consistent with the time-history waveforms of displacement, velocity and acceleration of the vault nodes, and the errors are all around 0.0. Among them, the errors of nodal displacement, nodal velocity and nodal acceleration tend to increase in turn. In addition, in general, the error between the OpenSees model and the Midas Civil model is smaller when seismic waves are input longitudinally along the bridge than when the seismic waves are input laterally along the bridge. The displacement, velocity, and acceleration amplitudes of the vault nodes when different seismic waves are input along the longitudinal and lateral directions of the bridge are further compared, as shown in Table 2.

Table 2. Comparison of dynamic response amplitudes of vault nodes.

Response metrics	Longitudinal			Lateral		
	Midas	OpenSees	Relative error	Midas	OpenSees	Relative error
Displacement	0.297	0.300	1.0%	0.496	0.529	6.6%
Velocity	0.636	0.646	1.5%	1.612	1.699	5.4%
Acceleration	2.947	3.045	3.3%	8.434	9.112	8.0%

From Table 2, it can be obtained: (1) When the seismic wave is input along the longitudinal direction of the bridge, the maximum relative error of the node displacement, velocity, and acceleration amplitude between the OpenSees model and the Midas Civil model is 3.3%, while the seismic wave is input along the transverse bridge direction. When, the maximum relative error of the two models is 8.0%; (2) Under the action of different seismic waves, the relative errors of nodal displacements, velocities and acceleration amplitudes calculated by the OpenSees model, and the Midas Civil model are different.

Analysis of the reasons shows that when the Midas Civil model is converted to the OpenSees model, although the MTO program can generate completely consistent data such as nodes, elements and boundaries, the processing methods of boundary conditions in the two models, the set of sparse matrices, and the freedom of nodes. There are differences in the number of degrees, convergence conditions, etc. The above error comes from the calculation error of the software itself, not the conversion error of the MTO program. The MTO program can automatically generate the OpenSees dynamic model of complex bridges, and the accuracy meets the calculation requirements.

5 CONCLUSION

This paper firstly deduces the conversion formula between the β angle of the Midas Civil beam element and the axial vector of the OpenSees beam element. On this basis, based on the Python language, the rapid establishment of the OpenSees full-bridge dynamic model is studied. The main conclusions are as follows:

(1) Based on the derived conversion formula, the correct conversion between the β angle of the Midas Civil beam element and the axial vector of the OpenSees beam element can be realized when the element node coordinates and β angle are known.
(2) The MTO conversion program based on the Python programming language can realize the rapid establishment of the OpenSees full-bridge dynamic model, and the calculation accuracy can meet the requirements.

(3) For complex bridges, the difference between the calculation results of the Midas Civil model and the generated OpenSees model is caused by the calculation error of the software, and the MTO conversion program is still valid.

ACKNOWLEDGEMENTS

This research was supported by the Natural Science Foundation of Fujian Province (Grant No. 2020J01883) and the Natural Science Foundation of China (Grant No.51878173).

REFERENCES

Cheng Xuewei, Lin Zhe. *Structural Nonlinear Analysis Program OpenSEES (2nd Edition) Theory and Tutorial* [M]. China Construction Industry Press, 2020.
Jia Junfeng, Tan Yuqing, Bai Yulei, et al. Seismic Performance Analysis of Self-centering Precast RC bridge columns based on OpenSees [J]. *Journal of Basic Science and Engineering*, 2022, 30(2): 328–340.
Jin Zhenglu, Zheng Guang, Liu Shaoqian, et al. Numerical Analysis for Seismic Behavior of Segmental Assembly Piers Based on OpenSees [J]. *Science Technology and Engineering*, 2022, 22(6): 2383–2393.
Lei Hujun. *Non-uniform Seismic Excitation of the Following Vehicle-Track-Bridge Coupled Vibration and Driving Safety Research* [D]. Chengdu: Southwest Jiaotong University, 2014.
Liu Huanyun, Yu Zhiwu, Guo Wei, et al. A Rapid Simulation Technique of the Train-Track-Bridge Interaction Based on OpenSEES [J]. *Journal of Railway Science and Engineering*, 2021, 18(4): 957–965.
Wang Jingxuan, Wang Wenda, Wei Guoqiang. Numerical Simulation on Mechanical Performance of Concrete Filled Steel Tubular Structures Based on OpenSees plateform [J]. *Journal of Disaster Prevention and Mitigation Engineering*, 2014, 34(5): 613–618 + 631.
Yan Dong, Jiang Guanlu, Liu Xianfeng, et al. Experimental Research on Liquefaction Behavior of Saturated Silt for Beijing-Shanghai High-speed Railway [J]. *Rock and Soil Mechanics*, 2008, 29(12): 3337–3341.
Zhao Jingang, Du Bin, Zhan Yulin, et al. Comparison of Constitutive Concrete Models in OpenSees for Hysteretic Behavior of Structures [J]. *Journal of Guilin University of Technology*, 2017, 37(1): 59–67.
Zhao Xuqing. Implementation and application of nonlinear dynamic Davidenkov model of soil in OpenSees [J]. *Journal of Civil Engineering and Management*, 2015, 32(2): 19–22 + 73.
Zhou Linlu, Su Lei, Qiu Zhijian, et al. Comparison of Four Constitutive Models for Sand Based on OpenSees [J]. *China Earthquake Engineering Journal*, 2022, 44(1): 128–135 + 151.

Dynamic displacement calculation of constrained steel plate based on rigid plastic model under blast load

Guangqing Hu* & Taochun Yang*
School of Civil Engineering and Architecture, University of Jinan, Jinan, China

ABSTRACT: Based on the equivalent single degree of freedom method and energy method, a simplified calculation method for the maximum elastic-plastic displacement of steel plates with different constraints under uniform blast load is proposed in this paper. The calculation results are compared with the numerical simulation results obtained by the finite element analysis software Abaqus. The results show that the error between the simplified calculation method proposed in this paper and the experimental value is within 8%, proving that the rigid-plastic model has high accuracy in calculating the maximum displacement of different restrained steel plates, and can be used to evaluate the anti-explosion ability of steel plates.

1 INTRODUCTION

It is of great engineering significance to study the dynamic response and failure mode of structural members under blast load. In recent years, many scholars have done a lot of research on this. Among them, Low analyzed the possible failure modes of reinforced concrete slabs constrained at both ends under different blast loads, different stiffness, and other factors (Low 2002); Nurick analyzed the dynamic response of clamped square plate under uniform blast load and local blast load using LS-DYNA software (Nurick 2005); Wang Wei obtained three failure modes of the four-side fixed reinforced concrete floor under different charge quantities and different scale distances (Wang 2010). However, these analyses are mainly based on the constraints at both ends, and there are few studies on the dynamic response and failure mode of steel plates under blast loads under other different constraints.

Tables of transfer coefficients of beam and plate members under partial restraints are given by the American blast-resistant design manual TM5-1300 (TM5-1300 1990). However, the transfer coefficients of plate members under different constraints are not complete, which makes the dynamic analysis of plate members under blast load difficult Based on the equivalent single-degree-of-freedom method and energy method, a simplified calculation method for the maximum vertical displacement of different constrained steel plates under blast load is derived. The transfer coefficient of steel plate under multilateral constraints not involved in the literature is given. The accuracy of the simplified calculation model is verified by the finite element software Abaqus.

2 THEORETICAL ANALYSIS OF SIMPLIFIED CALCULATION MODEL OF STEEL PLATE UNDER BLAST LOAD

Norris and Biggs proposed an equivalent single-degree-of-freedom theory in the 1950s (Biggs 1964; Norris 1959). It is an analysis method that focuses on the response of structural

---

*Corresponding Authors: 1612865277@qq.com and yangtaochun@126.com

members. To simplify the theoretical analysis, this paper only considers the bending deformation of steel plates without considering the shear deformation.

2.1 Simplified calculation model of hinged steel plate at both ends under blast load

2.1.1 Elastic stage

Figure 1 shows the schematic diagram of the steel plate and the deformation diagram in the elastic stage. The dotted line indicates hinged restraint at both ends of the steel plate. The distributed mass of the steel plate is m, length is a, width is b, and rotational stiffness is EI; uniform blast load q is applied on the surface of the steel plate.

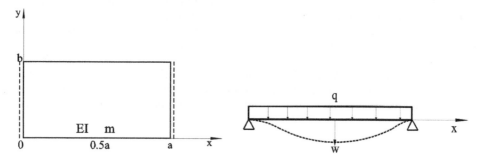

Figure 1. Bending member-elastic stage.

According to the approximate differential equation of the deflection surface and its integral, the deflection surface equation of steel plate expressed by the maximum vertical displacement w_{max} can be obtained:

$$w = \frac{w_{max}}{5a^4}\left(16x^4 - 32x^3 a + 16xa^3\right) \quad (1)$$

When the steel plate is deformed in Equation 1, the work done by the external load (uniform blast load q) is:

$$W = q\int_0^b dy \int_0^a w\,dx = \frac{q^2 a^5 b}{120 EI} = \frac{16}{25} qabw_{max} \quad (2)$$

The kinetic energy of the steel plate is:

$$K_E = \frac{1}{2} m \int_0^b dy \int_0^a (\dot{w})^2 dx = 0.25 mab\dot{w}_{max}^2 \quad (3)$$

The strain energy of steel plate is:

$$u = \frac{1}{2} EI \int_0^b dy \int_0^a (w'')^2 dx = \frac{q^2 a^5 b}{240 EI} = \frac{3072 EIb}{125 a^3} w_{max}^2 \quad (4)$$

Letting the external work, kinetic energy, and strain energy of the steel plate in Figure 1 and the single degree of freedom system be equal, the relevant parameters of the equivalent single degree of freedom system in the elastic state can be obtained respectively:

$$\begin{cases} \text{Equivalent load:} \ F^e = 0.64qab \\ \text{Equivalent mass:} \ M^e = 0.50mab \end{cases} \quad \begin{cases} \text{Load coefficient:} \ K_L = \dfrac{F_e}{F} = 0.64 \\ \text{Mass coefficient:} \ K_m = \dfrac{M_e}{M_t} = 0.50 \end{cases} \quad (5)$$

$$\text{Equivalent stiffness:} \ K^e = \dfrac{49.152EIb}{a^3} \quad \text{Stiffness coefficient:} \ K_E = \dfrac{k_e}{k_0} = 0.64$$

2.1.2 Plastic stage

It is assumed that the plates between the yield lines of the steel plate are plane, and the yield line divides the steel plate into Plate ① and Plate ②. Its deformation is shown in Figure 2.

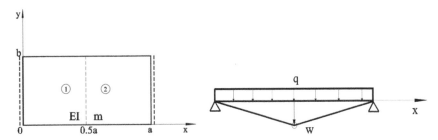

Figure 2. Bending member-plastic state.

① and ② have plate deformation of the same shape. According to the points (0,0,0), (0, b,0), (a/2,0,w_{max}) on Plate ①, the deflection surface equation of the steel plate can be obtained:

$$w = Z = \dfrac{2w_{max}}{a} x \left(0 \leq x \leq \dfrac{a}{2}\right) \quad (6)$$

When the steel plate is deformed in Equation 6, the work done by the external load (uniform blast load q) is:

$$W = q \int_0^b dy \int_0^{\frac{a}{2}} w dx \cdot 2 = \dfrac{abqw_{max}}{2} \quad (7)$$

The kinetic energy of the steel plate is:

$$K_E = \dfrac{1}{2}m \int_0^b dy \int_0^{\frac{a}{2}} \dot{w}^2 dx \cdot 2 = \dfrac{mab\dot{w}_{max}^2}{6} \quad (8)$$

The strain energy of steel plate is:

$$u = M_P \cdot 2\theta = \dfrac{4M_P w_{max}}{a} = \sigma_s \cdot \dfrac{bh^2}{a} w_{max} \quad (9)$$

Letting the external work, kinetic energy, and strain energy of the steel plate in Figure 2 and the single degree of freedom system be equal, the relevant parameters of the equivalent single degree of freedom system in the plastic stage can be obtained respectively:

$$\begin{cases} \text{Equivalent load: } F^P = \dfrac{abq}{2} \\ \text{Equivalent mass: } M^P = \dfrac{abm}{3} \\ \text{Equivalent stiffness: } R^P = \sigma_s \cdot \dfrac{bh^2}{a} \end{cases} \begin{cases} \text{Load coefficient: } K_L = \dfrac{F_P}{F} = 0.5 \\ \text{Mass coefficient: } K_M = \dfrac{M_P}{M_t} = \dfrac{1}{3} \\ \text{Stiffness coefficient: } K_R = \dfrac{R_P}{R_0} = 0.5 \end{cases} \quad (10)$$

2.2 Simplified calculation model of different constrained steel plates under explosion

The steel plate usually undergoes large plastic deformation, and the elastic deformation can be almost ignored under blast load. Therefore, this paper assumes that the steel plate is a rigid plastic material. Table 1 shows the equivalent single-degree-of-freedom model transfer coefficient of the steel plate under different constraints. The dotted line in the figure represents the plastic hinge line appearing in the plastic deformation of the steel plate.

Table 1. Simplified calculation model parameters of different constrained steel plates under explosion.

Model	Constraint condition	Deformation stage	Load coefficient K_L	Mass coefficient K_M	Equivalent resistance R
	Two-end hinge	plastic	0.500	0.333	$\dfrac{b}{a}h^2\sigma_s$
	Three-end hinge	plastic	0.333	0.167	$\dfrac{a^2 + 4b^2}{4ab}h^2\sigma_s$
	Four-end hinge	plastic	$\dfrac{3a-b}{6a}$	$\dfrac{2a-b}{6a}$	$\dfrac{a+b}{b}h^2\sigma_s$
	Two-end fixed	plastic	0.500	0.333	$\dfrac{2bh^2}{a}\sigma_s$
	Three-end fixed	plastic	0.333	0.167	$\dfrac{a^2+4b^2}{2ab}h^2\sigma_s$
	Four-end fixed	plastic	$\dfrac{3a-b}{6a}$	$\dfrac{2a-b}{6a}$	$\dfrac{2a+2b}{b}h^2\sigma_s$

3 DYNAMIC DISPLACEMENT RESPONSE OF DIFFERENT CONSTRAINED STEEL PLATES UNDER BLAST LOAD

3.1 Establishment of finite element model

In finite element analysis, the square steel plate is taken as the research object. The material model of steel is the isotropic model, and the material constitutive model is the rigid plastic model. The static yield strength f_y of the steel plate is 244.8 MPa, the elastic modulus E_s is 2.07×10^{11} pa, and the Poisson's ratio v is 0.3 (Yuen 2008). The steel plate is established by an S4R shell element in Abaqus with a mesh size of 20 mm. The blast load is simplified as an inverted triangular load and evenly distributed on the surface of the steel plate. The pressure-time curve of the blast load is shown in Figure 3.

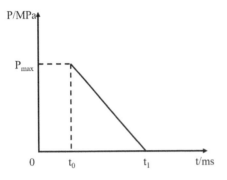

Figure 3. Pressure-time curve of blast load.   Figure 4. Steel plate test device.

3.2 *Validation of finite element model*

Yuen and Nurick carried out a series of experimental studies on the dynamic displacement response of square low-carbon steel plates under blast load (Yuen 2008). The test device is shown in Figure 4. In this paper, some experimental data are selected for numerical simulation. The experimental results are compared with the simulation results to verify the accuracy of the material model. The test object is a $700 \times 700 \times 3$ mm steel plate, the scale distance is 0.83 m/kg$^{1/3}$, the peak value of the explosion load is 1.496 Mpa, the duration of overpressure is 3.53 ms, and the 100 mm around the plate is fixed.

Figure 5 shows the maximum residual deformation of the steel plate simulated by the finite element method under the above conditions. Figure 6 shows the displacement time history curve of the maximum displacement point of the steel plate under the above conditions. The dotted line in the figure represents the maximum displacement value of the steel plate measured by the test in the literature. Among them, the finite element simulation value is 38.84 mm, the experimentally measured value is 40.70 mm, and the error is 4.6%,

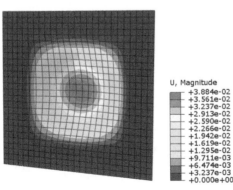

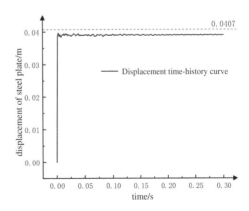

Figure 5. Residual deformation of steel plate.   Figure 6. Displacement time-history curve.

indicating that the finite element analysis method given in this paper meets the accuracy requirements.

3.3 *Dynamic displacement response of different constrained steel plates under blast load*

Based on Abaqus software, the finite element model of steel plate under different constraints is established to verify the accuracy of the simplified calculation model. Six steel plates with different boundary conditions were established in Abaqus. The geometric size of the steel plate was 700 × 700 × 3 mm. The blast overpressure peak was 1.5 MPa, the overpressure duration was 3.0 ms, and the 100 mm around the plate is fixed.

Table 4 shows the maximum displacement of the steel plate calculated by the finite element simulation and equivalent SDOF model. It can be seen from Table 4 that the error between the calculated value of the equivalent SDOF simplified model proposed in this paper and the finite element simulation value is within 10%. It indicates that the equivalent simplified calculation model proposed in this paper has high calculation accuracy and can be used to evaluate the anti-explosion ability of the steel plate.

Table 4. Values of finite element simulation and SDOF model.

Constraint condition	Simulation value/mm	Calculated values/mm	Error
Two-end hinge	72.84	76.74	5.6%
Three-end hinge	81.16	77.34	−4.7%
Four-end hinge	49.35	52.33	4.8%
Two-end fixed	71.92	68.46	−4.2%
Three-end fixed	79.52	81.33	3.5%
Four-end fixed	46.49	51.52	5.8%

4 CONCLUSION

Based on theoretical analysis and numerical simulation, the maximum dynamic displacement response of different restrained steel plates under blast load was analyzed in this paper, and a simplified model for calculating the maximum displacement of different restrained steel plates was proposed.

The specific conclusions are as follows:

(1) Under the same blast load, the maximum vertical displacement of the steel plate is the three-sided constraint > two-sided constraint > four-sided constraint, and hinged constraint > fixed constraint.
(2) The simplified calculation model given in this paper has high calculation accuracy under blast load and can be used to evaluate the anti-explosion ability of steel plates under different constraints.

REFERENCES

Biggs J. M. (1964). *Introduction to Structural Dynamics.* New York: McGraw-Hill Book Company, 23–32.
Chung Kim Yuen S.; Nurick G. N.; Verster W.; Jacob N.; Vara A. R.; Balden V. H.; Bwalya D.; Govender R. A.; Pittermann M. (2008). *Deformation of Mild Steel Plates Subjected to Large-scale Explosions.* 35(8), 684–703.

Hsin Yu Low; Hong Hao (2002). *Reliability Analysis of Direct Shear and Flexural Failure Modes of RC Slabs Under Explosive Loading*, 24(2), 189–198.

Langdon G. S.; Chung Kim Yuen S.; Nurick G. N. (2005). *Experimental and Numerical Studies on the Response of Quadrangular Stiffened Plates*. Part II: localized blast loading, 31(1), 85–111.

Norris G. H., Hansen R. J., et al. (1959). *Structural Design for Dynamic Loads*. McGraw-Hill Book Co, New York.

U.S. Army Corps of Engineer Headquarters (1990). TM5-1300 Structures to Resist the Effects of Accidental Explosions: UFC3-340-02. *Department of the Army, the Navy, and the Air Force*.

Wang Wei, Zhang Xiang, Lu Fangyun, Lin Hualing, Jiang Zengrong (2010). Analysis of Failure Mode and Anti-explosion Performance of Reinforced Concrete Floor Under Blast Loading. *Army Journal*, 31(S1): 102–106.

Analysis of measured data of upper span tunnel foundation pit engineering

Ting Bao*
Zhejiang Mingsui Technology Co., Ltd., China

Xiaobo Sun*
China Railway 14th Bureau Group Mege Shield Construction Engineering Co., Ltd., China

Jin Pang*, Lingchao Shou* & Lifeng Wang*
Zhejiang Mingsui Technology Co., Ltd., China

ABSTRACT: Based on the measured data of the foundation pit excavation project of the cross-operating tunnel in Hangzhou, the influence of the foundation pit excavation on the envelope structure, the surrounding environment and the underlying tunnel structure is analyzed. The results show that the surface subsidence curve after the wall is of groove type during the foundation pit excavation, the maximum surface subsidence is about 0.117%H, the maximum support shaft force is about 1151 kN, the overall displacement of the ground connecting wall is within 5 mm, the settlement change of the lower tunnel bed is within 10 mm, the horizontal displacement change is within 3 mm, and the selection of envelope structure is reasonable.

1 INTRODUCTION

In recent years, with the rapid development of urban rail transit in my country, the number and scale of deep foundation pit projects in subway stations have doubled, and many tunnel foundation pit projects with upper-span operations have inevitably appeared. Such projects where the foundation pit is close to the tunnel are often difficult to construct, and the deformation control of the subway itself is very strict. If there is a lack of reasonable research and analysis, and corresponding measures are taken, the deformation of the subway will be too large, which will threaten the safety of the subway operation. Effective monitoring of foundation pits and tunnels has become a necessary means to ensure construction safety (Bi 2022; Xu 2021; Ying 2014).

The deformation problems during the excavation of the foundation pit include the deformation of the foundation pit itself and the deformation around the foundation pit, which affect the safety and stability of the enclosure structure itself. Wang Weidong et al. (Wang 2020) took an ultra-deep large foundation pit in Shanghai as the engineering background, and analyzed the deformation characteristics of the foundation pit in the soft soil layer in the implementation of the subdivisional crossover; Tong Jianjun et al. (Tong 2015) collected nine subway station foundation pits in Chengdu area. Based on the monitoring data, the envelope of land subsidence and its mathematical expression are given, and the similarities and differences of the envelope of surface subsidence around the foundation pit in Chengdu and Shanghai are summarized; Zhu Wenjun et al. (Zhu 2022) studied

---

*Corresponding Authors: 1542528435@qq.com, 114938824@qq.com, pangj00@163.com, slczust@163.com and wanglfzust@163.com

Changzhou from the perspective of statistical analysis. The law of surface settlement and deformation outside the foundation pits of 38 subway stations; Liu Nianwu et al. (Liu 2019) carried out systematic monitoring and analysis in combination with a deep foundation pit project in a deep soft clay area. The displacement value accounts for about 20% of the lateral displacement value of the soil during the excavation of the foundation pit.

In this paper, relying on the foundation pit project of an upper-span operating tunnel in Hangzhou, and taking the actual monitoring data as the criterion, this paper analyzes the engineering properties of surface settlement, supporting axial force, deformation of the ground connecting wall and the deformation of the tunnel below during the excavation process of the foundation pit, in order to provide a basis for the construction of the upper-span tunnel in Hangzhou and guidance on cross-tunnel excavation design.

2 ENGINEERING BACKGROUND

2.1 Project overview

The Genshan East Road crossing the river tunnel project is designed as an urban expressway, including the river crossing tunnel, comprehensive pipe gallery, ground connecting line project and supporting ancillary projects. The total length of the project is 4612.26 m (including the ground connecting line road), and the total length of the tunnel is 4462.26 m (including the shield section of about 3210 m). Among them, the open-cut section near YK0 + 700 on the Xiasha side is orthogonal to the existing tunnel, and the foundation pit of the open-cut section is shown in Figure 1. It is located about 6 m below the proposed tunnel project.

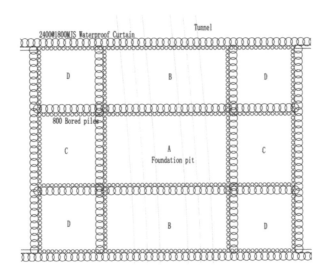

Figure 1. Position relationship between foundation pit and Metro Tunnel.

2.2 Engineering geological conditions

The minimum clear distance between the open-cut tunnel floor and the operating subway tunnel is about 5.2 m, and the soil layers in the interval are, from top to bottom, miscellaneous fill, sandy silt, silt, sandy silt and silty clay. The mechanical properties of each layer of soil are shown in Table 1. The subway shield tunnel is located in the silt, and is not affected by groundwater within the construction area of the site.

Table 1. Distribution table of soil layer.

Solum	The name of soil layer	Top-of-layer elevation (m)	Bottom burial depth (m)	Lift height (m)
1	Miscellaneous fill	0.20~4.10	5.39~10.29	0.20~4.10
2	Sandy silt	0.40~12.60	−12.45~5.62	2.90~18.90
3	Silt	1.20~14.20	−13.93~−1.86	13.30~25.30
4	Sandy silt	0.90~5.50	−0.17~4.81	3.00~9.70
5	Silty clay	0.90~6.00	−22.79~−18.70	27.20~32.50
6	Silly clay	0.60~12.20	−26.59~−16.58	22.90~41.80
7	Silty clay	1.50~9.10	−28.60~−20.92	29.80~41.50

2.3 *Foundation pit design scheme*

The foundation pit enclosure piles use bored cast-in-place piles with a diameter of 800 mm and a spacing of 1000 mm, of which double-row piles are used directly above the subway, and the waterproof curtain is 2400@1800MJS rotary jet piles. A bored cast-in-place pile with a diameter of 1200 mm is installed in the foundation pit as an anti-uplift pile. The vertical distance between the bottom of the enclosure pile directly above the shield tunnel and the top of the tunnel is greater than 2.0 m, and the elevation of the bottom of the remaining enclosure piles is −21.2 m. The two sides of the shield tunnel are reinforced with portal bodies. The reinforcement method is MJS rotary jet piles. The depth of the foundation pit is reinforced to −21.2 m. The sub-pits are separated by bored piles, and the overall construction sequence is to excavate areas A, B, C and D of the foundation pit above in sequence.

3 ENGINEERING CHARACTER ANALYSIS

3.1 *Surface settlement and deformation law after the wall*

Figures 2 and 3 are the surface settlement maps of the TDB8 and TDB9 measuring points of the foundation pit at different distances from the foundation pit. It can be seen from the

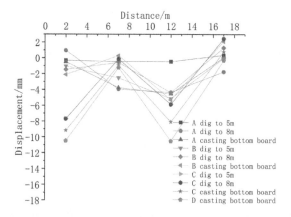

Figure 2. TDB8.

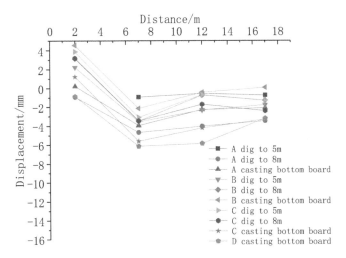

Figure 3. TDB9.

figures that the surface settlement value of the TDB8 measuring point on the north side of the foundation pit is larger than that of the TDB9 measuring point, and the maximum value appears at the measuring point TDB8-3, which is about 12 m away from the foundation pit, the appearance position is about 1.2H (H is the excavation depth), and the maximum settlement value is 10.54 mm, which decreases with the increase of the distance. There is almost no subsidence on the surface, and the excavation of the foundation pit has little impact on this place. The surface settlement shape curve behind the foundation pit wall is "grooved".

Figure 4 shows the relationship between the maximum value of the surface settlement behind the wall at each measuring point and the excavation depth. With the increase of the excavation depth of the foundation pit, the surface settlement also gradually increases, which is roughly proportional. The maximum surface settlement is about 0.117%H, and the average value is about 0.064%H. From the statistical results, it can be seen that the foundation pit has good control over the surrounding surface settlement.

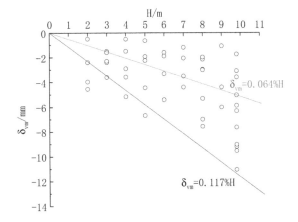

Figure 4. Relationship between maximum surface subsidence and excavation depth.

3.2 Analysis of support axial force monitoring data

Figures 5 and 6 show the change of the foundation pit support axial force in area A with time. The axial force of the first concrete support increases rapidly in the early stage of excavation. When area A is excavated to 5m, the axial force value is small and basically kept within 300 kN. That is to say, the effect of concrete support in the early stage of excavation is limited. When the foundation pit is excavated to the bottom elevation of the second support on October 25, 2021, after the second steel support is erected and begins to bear the force, the first support shaft The force enters a plateau and remains between 300–400 kN. As the foundation pit continued to be excavated, the first support axial force gradually increased, and when the excavation reached the bottom, the support axial force reached a maximum of about 567 kN. The second support axial force occurred on November 26, 2021 (measurement point TZL7-2), at about 1151 kN. At this time, the foundation pit in Area A has been excavated to the bottom. After the construction of the basement floor is completed, the axial force is stable and has a downward trend, indicating that the support effectively controls the deformation of the foundation pit.

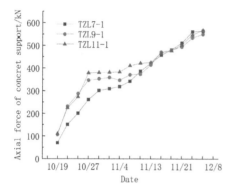 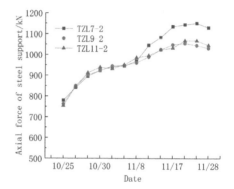

Figure 5. First concrete support time curve.   Figure 6. second steel support time curve.

3.3 Analysis of the horizontal displacement data of the ground connection wall

The monitoring section of TCX4 is located on the south side of the foundation pit, and the horizontal displacement data of the connecting wall of the monitoring section under different excavation conditions were selected, as shown in Figure 7. It can be seen from the figure that when area A is excavated for 5 m, the top of the ground connecting wall has shifted to the outside of the pit. With the continuous excavation of the foundation pit, the displacement of the ground connecting the wall to the outside of the pit increases. When the excavation depth of the foundation pit in Zone B reaches 8 m, the ground connecting walls within the upper 8 m range of the foundation pit are all displaced outward, and the maximum displacement is 4.99 mm.

The monitoring section of TCX1 is located on the west side of the foundation pit, and the horizontal displacement data of the connecting wall of the monitoring section under different excavation conditions were selected, as shown in Figure 8. It can be seen from the figure that the excavation of the C area causes the ground connecting wall within the upper 20 m of the foundation pit to move out of the pit. When the bottom plate of the C area is poured, the maximum displacement of the ground connecting wall occurs at a depth of 5.5m, and the maximum displacement value is 3.71 mm. The excavation of the D area leads to a decrease in the outward displacement of the ground connecting wall.

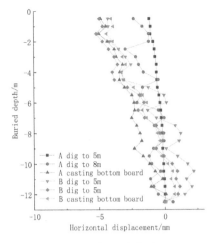
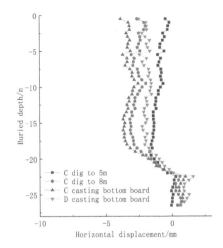

Figure 7. TCX4 horizontal displacement diagram.

Figure 8. TCX1 horizontal displacement diagram.

According to the displacement data of the ground connecting wall, it can be seen that the overall displacement of the ground connecting wall is within 5 mm, indicating that the design of the enclosure structure is reasonable and the deformation of the foundation pit is within a reasonable range.

3.4 Analysis of tunnel deformation data

Figure 9 shows the time-history curve of the tunnel bed settlement and horizontal displacement at some monitoring points. The curve in the figure is stepped, reflecting the excavation process of each subsection of the foundation pit. The variation of the tunnel bed settlement is within 10 mm, and the variation of the horizontal displacement is within 3 mm. within. Among them, the development section is the superposition of the cumulative impact of earthwork excavation and excavation in the previous stage; the flat section is the working condition of erecting supports and pouring the bottom plate; the steeply increasing section corresponds to the long-term exposure of the soil at the bottom of the pit, which causes the bottom of the pit to bulge,

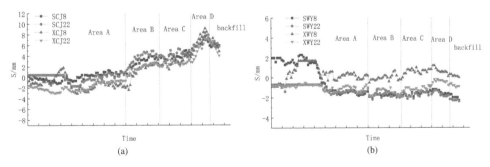

Figure 9. Tunnel monitoring time course curve. (a) Roadbed settlement time chart (b) Horizontal displacement time chart.

and the tunnel is affected. Under working conditions, after the excavation and backfilling are completed, the change in the settlement of the track bed decreases.

4 CONCLUSION

During the excavation of the foundation pit, the surface settlement curve behind the wall is a "groove type". The increase of the excavation depth of the foundation pit gradually increases the surface settlement, which is roughly proportional. The maximum surface settlement of the foundation pit is about 0.117%H, and the average value is about 0.064%H, which can control the surrounding surface settlement well.

The maximum axial force of the first concrete support is about 567 kN, and the maximum axial force of the second steel support is about 1151 kN; the overall displacement of the ground connecting wall is within 5 mm, the design of the foundation pit enclosure is reasonable, and the foundation pit itself Small deformation.

The track bed settlement and horizontal displacement change of the tunnel below the foundation pit are stepped. Corresponding to the excavation process of each zone of the foundation pit, the track bed settlement change is within 10 mm, and the horizontal displacement change is within 3 mm, which meets the requirements of subway protection.

During the long-term exposure of the soil at the bottom of the pit, a large deformation of the tunnel occurs. It is recommended to speed up the construction process in the actual project to avoid the uplift of the bottom of the pit caused by the long-term exposure of the soil at the bottom of the pit, which is conducive to the protection of tunnel deformation.

REFERENCES

Bi Shuqi, Gan Binlin, Liang Yahua, Yang Yonghua, Li Huan, Cui Hao, Bian Chao. Measurement Analysis of Foundation Pit Excavation on Existing Close-Lying Tunnel [J]. *Science, Technology and Engineering*, 2022, 22 (03): 1198–1204.

Liu Nianwu, Chen Yitian, Gong Xiaonan, Yu Jitao. Study on Deformation Characteristics of Foundation Pit and Adjacent Buildings of Subway Station Caused by Deep Excavation of Soft Soil [J]. *Rock and Soil Mechanics*, 2019, 40(04): 1515–1525 + 1576.

Tong Jianjun, Wang Mingnian, Yu Li, Liu Dagang, Xu Rui. Study on the Surface Settlement Law Around the Deep Foundation Pit of Chengdu Metro Station [J]. *Hydrogeology and Engineering Geology*, 2015, 42 (03): 97–101.

Wang Weidong, Xu Zhonghua, Zong Ludan, Chen Yongcai. Analysis of Ultra-deep Large Foundation Pit Project in Shanghai International Financial Center [J]. *Building Structure*, 2020, 50 (18): 126–135.

Xu Sifa, Zhou Qihui, Zheng Wenhao, Zhu Yongqiang, Wang Zhe. Analysis of the Influence of Foundation Pit Construction on the Deformation of Adjacent Operating Tunnels [J]. *Journal of Geotechnical Engineering*, 2021, 43 (05): 804–812.

Ying Hongwei, Sun Wei, Lu Mengjun, Chen Dong. Study on the Measured Properties of a Deep Soft Soil Foundation Pit in a Complex Environment [J]. *Geotechnical Journal*, 2014, 36 (S2): 424–430.

Zhu Wenjun, Zhang Siyuan, Tong Liyuan, Yudi, Liu Song. Research on surface subsidence deformation law of Changzhou Metro Station based on measured data [J]. *Construction Technology* (Chinese and English), 2022, 51 (07): 15–18.

Analysis of the spatiotemporal pattern of the gathering and dispersal of evacuated people in large and medium-sized cities after strong earthquakes

Weiyong Shen, Dongping Li*, Jingfei Yin, Yibo Jian, Xinhui Miao & Di Yao
Zhejiang Earthquake Agency, Hangzhou, Zhejiang, China

ABSTRACT: Studying the spatiotemporal patterns of the gathering and dispersal of evacuated people in the community after an earthquake can provide practical support for the government's emergency response and management and provide a more accurate basis for earthquake relief and assessment of the timely restoration of social order. On December 22, 2021, a series of 4.2-magnitude (M4.2) earthquakes struck the Tianning District of Changzhou City, Jiangsu Province, China, causing enormous social impact in the local area. Using 81.69 million mobile phone location records collected after the earthquake, this study performed a quantitative spatiotemporal analysis of people's locations after the earthquake under different scales and obtained the optimal bandwidth of kernel density estimation (KDE) for each study area. This study obtained the flow pattern of evacuated people at the community level after a nondestructive earthquake, which provides a quantitative basis for the planning and reduction of disaster shelters and has good research and application value for improving the capacity of earthquake prevention and disaster mitigation.

1 INTRODUCTION

At 21:46 on December 22, 2021, in the Tianning District, Changzhou City, Jiangsu Province, China, a 4.2-magnitude (M4.2) earthquake occurred (31.75°N, 120.00°E), with a focal depth of 10 km. The earthquake did not cause direct losses, but it was felt in a large area of Eastern China. In particular, the earthquake epicenter is located in the center of a large city, and the earthquake was felt strongly in the urban area of Changzhou City. This study used the wireless Internet location data obtained during the earthquake to study the changes in people flow at different scales, used high-resolution real-time mobile terminal location data to analyze the spatiotemporal patterns of the gathering and dispersal of evacuated people at the community level after the earthquake, and obtained satisfactory results.

After an earthquake, the clear understanding of the spatiotemporal distribution characteristics of people after an earthquake can not only guide the formulation of corresponding measures and plans but also provide theoretical and technical support to the government for the successful handling of post-earthquake emergency response. In the past, it was difficult to obtain the real-time distribution of people and changes after an earthquake by traditional technical means, resulting in the inability to quickly obtain an accurate picture of the earthquake disaster. It was thus difficult for the government to respond quickly, and relief

---
*Corresponding Author: lidongping@zjdz.gov.cn

efforts were ultimately inefficient and disorganized. In the current technical environment, when the geographical location of a mobile terminal user is within a set range, the spatial information of the user can be counted and described, and the current number of people in a geographical space can be calculated through a differentiated model. Mobile phone location data can be applied in many application scenarios, i.e., the mobile phone location data can be used to estimate the real-time distribution of people in the earthquake area, and changes in mobile phone location data can be used to obtain the damage to mobile communications infrastructure caused by strong earthquakes, which can lead to an estimation of the extremely affected area and the damage caused by the earthquake. Many scholars have performed in-depth studies on the relationship between mobile phone big data and the distribution of natural disasters. Gao et al. applied the people data obtained from mobile phone location data in the field of emergency disaster relief and used the obtained people data to assist in studying the disaster situation. Based on several destructive earthquakes, such as the 7.0-magnitude Jiuzhaigou earthquake, Pang et al. proposed a method for selecting earthquake disaster indicators based on mobile phone location data. Using the acquired 1-hour post-earthquake mobile phone location data, Li et al. performed a quantitative spatiotemporal analysis on the distribution and flow of people in the earthquake-stricken area of Jiuzhaigou, Sichuan. Xia used telephone signal data to construct a people density model and estimated the real-time location of people in an earthquake-stricken area, which can be used to support earthquake relief efforts. Calabrese F proposed an algorithm to analyze the collected mobile phone location data to estimate people's travel demand based on the starting point and destination of the individual's movement; this algorithm can provide a key service for traffic management and emergency response. Takahiro Y used large-scale mobile data such as mobile phone data to observe people's movement patterns within and between cities with high spatiotemporal granularity and studied how people flow can promote post-disaster recovery. In recent years, the maturity of technologies based on location services, cloud computing, and low-latency network transmission has enabled data managers to dynamically obtain accurate wireless Internet terminal location information, and the use of wireless Internet terminals to estimate the real-time location of people has become a reliable technical tool.

2 ACQUISITION AND TECHNICAL PROCESSING OF SPATIOTEMPORAL DATA

2.1 Data source

With the development of the Internet and communication technology, location big data, such as smartphone positioning, social media check-in, and picture geotagging, provide new data resources that can quickly perceive dynamic changes in people's locations. The data used in this study constitute the massive amount of app data of each mobile phone. Currently, the software development kit (SDK) has been installed on more than billions of terminals, with hundreds of millions online simultaneously. Based on many apps and mobile terminals, the location information reported by a mobile phone is used to analyze the geographical location of the user. The location information comes from the application programming interface (API) of the mobile phone and mainly includes global positioning system (GPS) data from the mobile phone, WIFI data, and base station signal data. Based on the ratio of covered local mobile terminals to actual people, the real-time number of people is obtained by performing model correction. In this study, since Geohash grids with precisions between 5 and 8 were used, the data used have different precision. In this study, the physical quantity based on the number of Internet terminals ultimately refers to each individual holding a mobile smart terminal.

2.2 Study area and data range

In this study, the densely populated area around the epicenter of the M4.2 Tianning earthquake in Jiangsu Province on December 22 was used as the macroscopic study area. In this study, data collection was performed in 20 periods with a total of 83.204 million sampling points, involving a total of 14 prefectures and cities in the central and southern parts of Jiangsu Province, the northern part of Shanghai City, and the eastern part of Anhui Province, which include a total of 81.69 million people. The data were collected using the Geohash6 grid. In addition, to conduct the community-level microscopic analysis of people's movement, study areas in Changzhou City within a theoretical intensity of six due to the M4.2 earthquake were selected for fine analysis. As shown in Figure 1, the first study area is located in the middle of the Tianning District and is closest to the epicenter, with a total area of 9.2 km^2; the data were collected using the Geohash8 grid, with a total of 1593 sampling points and 113,670 people. The Geohash8 grid has a resolution of approximately 20*30 m, which can more intuitively reflect the people flow at the community level and can also facilitate the analysis of the phenomenon of dense people flow moving from indoors to outdoors.

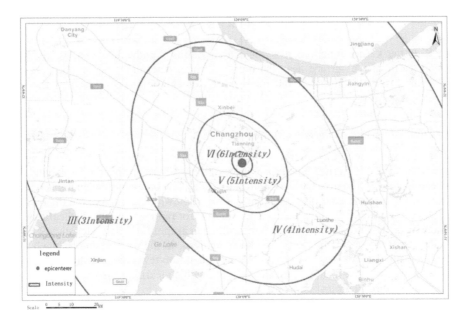

Figure 1. Schematic diagram of the study area for the M4.2 Tianning earthquake.

Data collection included residential, commercial, and mixed commercial and residential areas in the urban area. High-precision sampling data (Geohash8) were used for the fine analysis areas, and slightly lower-precision data (Geohash6) were used for the macroscopic study area. The people's information in different periods was extracted. Between 21:40 and 23:20 on December 22, the data were collected every 10 minutes. For each earthquake, the data collection was performed for ten periods so that the post-earthquake dynamic distribution of people could be more comprehensively reflected. To facilitate comparison, ten sets of data from the same period of the previous day were obtained.

In addition, to compare and analyze the data, 3797 three-dimensional (3D) models of the main buildings in the two fine analysis areas and 272 pieces of Weibo data related to the earthquake were collected.

3 DATA PROCESSING TECHNIQUES AND METHODS

3.1 Distribution of people using Kernel Density Estimation (KDE)

Since the collected data are summary point data, it is necessary to use a model to simulate the distribution of people. In this study, KDE was used. The KDE is derived from the first law of geography, that is, everything is related, close things are more closely related, and the closer the location to the core element, the greater the density expansion value, and reflects the characteristics of the spatial heterogeneity and intensity attenuation with the distance. The center point of the Geohash grid is used as the sample point. Due to the spatial or central clustering characteristics of people, the distance from the center is inversely proportional to the people's density under normal circumstances, and the distribution of people within the grid can be simulated by calculating the attenuation of the point value density. KDE is a relatively suitable method for simulating the distribution of people. In the calculation of kernel density, as the radius increases, the density value does not change significantly, and the neighborhood contains more point values. The formula to calculate KDE is the following:

$$f(x) = \sum_{i=1}^{n} \frac{1}{h^2} K\left(\frac{x - x_i}{h}\right) \qquad (1)$$

where K is the kernel function; h is the search radius (bandwidth), that is, the spatial extension width of the surface near the point x, which determines the smoothness of the kernel density surface; and n is the number of element points contained in the search bandwidth at point x.

3.2 KDE analysis and bandwidth selection

In KDE analysis, the choice of the variance in the kernel function, i.e., the bandwidth (neighborhood) of the kernel function, is a key technical issue, and different bandwidths could lead to large differences in the final fitting results. The bandwidth selection is related to the scale of the geographical phenomenon. A small bandwidth can make the fluctuation of sampling points more obvious, which is more suitable for representing the fine features of the local density distribution, while a large bandwidth can amplify the characteristics of hotspot areas under macroscopic conditions. For KDE analysis, in theory, the smaller the h (bandwidth), the better the results; however, if h is too small, too few points in the neighborhood could be included in the fitting. With the help of machine learning theory, cross-validation can be used to select the most appropriate bandwidth.

Figure 2 shows a random sample drawn from the standard normal distribution. The green curve is the probability density of 25 m. In contrast, the red curve is the probability density curve obtained by using an excessively small bandwidth of 100 m, and the steep twists and turns can be easily noticed. The blue curve is too smooth because it uses a bandwidth of 200 m, which masks most of the basic structure of the data.

When the sampling points are determined, for calculating the density of the sampling points under ideal conditions, the probability density is basically consistent with the kernel calculation. An error function is defined; then, by minimizing the error function, an approximate direction for the selection of h can be provided. The mean integrated squared error (MISE) function is selected; the definition of MISE is the following:

$$\text{MISE}(h) = E \int (\widehat{f}(x) - f(x))^2 dx \qquad (2)$$

MISE (h) is the accumulation of the local mean squared error at each point x, which can also be understood as the overall average error concerning the sampling density. A feature of

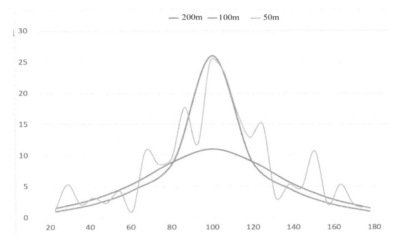

Figure 2. Comparison of KDE analysis at bandwidths of 25 m, 100 m, and 200 m using the same Geohash8 data.

KDE is to fully mine the information of the data itself, which can avoid subjectively setting the empirical route; then, the estimation of the minimum error of the sample data is performed. In this study, the bandwidth was calculated based on the distance between the adjacent point elements.

$$R = \sum_{i=1}^{n} 1 \sum_{j=1}^{n} 1 \frac{d_{ij}}{kn} \quad (3)$$

where d_{ij} refers to the average distance between the point element i and the k number of other adjacent point elements J. The larger K is, the larger the bandwidth and the smoother the obtained kernel density surface. According to Equation (3), the K value was continuously adjusted, and the optimal bandwidths for cities, towns, and rural areas were finally selected (Table 1).

Table 1. Comparison experiments with different bandwidths to determine parameter settings.

Type	Optimal bandwidth (m)	Original default bandwidth (m)
Rural area	70	100
Town	62	100
Residential area	31	100
Commercial district	25	100
Contiguous open space	100	200

4 ANALYSIS OF THE STUDY AREA AND PEOPLE FLOW

4.1 *Analysis of macroscale people flow variation at the epicenter*

After an earthquake disaster occurs, communication is necessary for asking for help from the outside world or reporting safety among relatives and friends. For nondestructive earthquakes, the overall communication volume and activity in the affected area should be increased significantly under the premise that the base stations are working normally, resulting in many mobile phones being used. The closer to the epicenter, the greater the deviation of the actual moved people data from the historical moved people data in the affected area, especially for late-night earthquakes. For example, On the morning of July 12, 2020, there was an M5.1 earthquake in Guye; the number of mobile terminals in residential communities near the macroscopic epicenter increased by 7% within 30 minutes. In the data analysis, the KDE analysis and rasterization were performed on point data, i.e., the density analysis results of a 10-minute time slice before the earthquake and a 10-minute time slice after the earthquake were used to calculate the grid difference. In the past, a large amount of digital elevation model (DEM) data was used in the grid calculation, and the grid calculator can be used to directly obtain the data difference between the two grids. In this study, the grid calculation was used to analyze the people density difference between the two periods. The change in the people before and after the earthquake was calculated. In Equation (4), m is the people before the earthquake, n is the people after the earthquake, s is the degree of people gathering, s_i is the average daily variation within a week, and c is the random value.

$$s = \frac{m-n}{n} - s_i + c \tag{4}$$

The largest s value (degree of people gathering) indicates the most severely affected area. In the macroscale calculation, the Geohash6 data were used, and the calculation results can be used to objectively infer the range and extent of the people gathering caused by the earthquake; under certain circumstances, the distribution range of the macro-epicenter can be inferred.

Figure 3 shows that the overall people flow changes at the macroscopic scale are discrete, but the number of active terminals near the epicenters of the two earthquakes increases significantly, which is particularly evident in the urban area of Changzhou with an intensity of VI after the M4.2 magnitude earthquake. The increase in the terminal volumes in the urban areas of Jiangyin and Wuxi, which are approximately 30 km away from the epicenter,

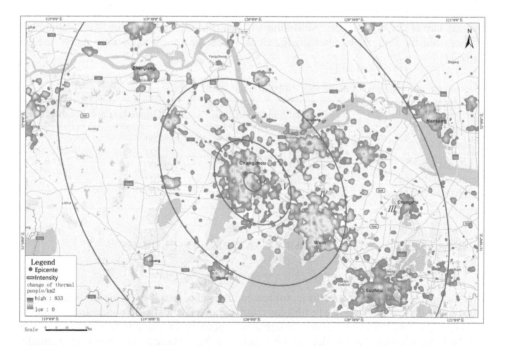

Figure 3. Results of 20-minute slice grid data difference the macro scale of Jiangsu Tianning 4.2 earthquake.

is also relatively significant. Terminal volumes in areas over 50 km away from the epicenter show a slight overall increase, with an alternating distribution of increase and decrease.

4.2 *Analysis of people's spatial changes at the community level*

The typical densely populated areas were selected. The study area is a well-planned new area, buildings in the area are mainly new high-rise buildings, and there are small parks and open spaces. The villages in the city mostly have old and low-rise buildings, but there are also some new high-rise communities, factories and warehouses, and two large parks. The two study areas are located within the area of a theoretical intensity of VI. Because the Geohash6 (1000*600 m) data are clearly not accurate enough to reflect the people flow at the community level, the higher-precision Geohash8 data (20 m*30 m) were used, and Equation 4 was used to perform the subtraction of the grid density in the two time periods. In the study areas, crowd gathering occurs when many people move out of the core of the residential area; the crowd gathering places are the main entrances and exits of the community, the internal roads of the community, and the community square. However, in the main traffic lines, especially the city squares and parks that are relatively far away from the residential areas, no crowd gathering occurs, indicating that the residents respond to the earthquake but soon calm down.

The people flow at the community level after an earthquake simulated using the high-precision location data is close to the actual situation and can be used to objectively infer the range and extent of people gathering caused by the earthquake. The distribution of the earthquake impact can be determined by overlaying high-resolution remote sensing images with the increase and decrease in mobile phone data from the same area. In the two study areas, there are dramatic changes in people's locations in the high-rise communities, such as the Changfa Dihao Community (mainly the 35-story high-rise buildings), and Tianning Wuyue Plaza (mainly the 36-story high-rise buildings). For the Tianning Wuyue Plaza, the high-rise buildings and commercial areas are mixed, and the community floor area ratio is high. Due to the amplification effect of high-rise buildings on earthquake impacts, the changes in people's locations are even more dramatic. In other areas with multistory or single-story buildings, drastic changes in people's locations rarely occur (Figure 4).

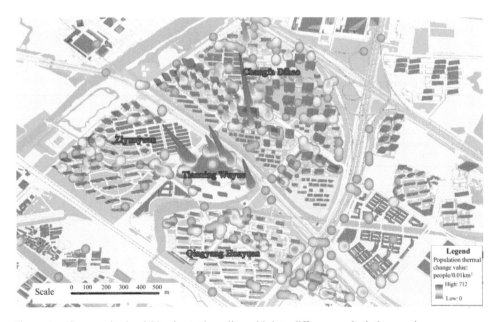

Figure 4. Community level 20-minute time slice grid data difference calculation results.

Table 2. Analysis of microscale people flow changes in typical areas.

Community	21:40 (people)	22:00 (people)	Sample point	Ratio of dots where density changes by more than 5%	The proportion of the terminal increase	Sampling unit properties	Main building floors
Around Tianning Wuyue	11286	12038	152	59.02%	6.67%	Commercial and residential area	4, 12
Around Qingyang Huayuan	12906	13699	170	57.06%	6.17%	Residential area	12, 30
Around Ziyunyuan	14499	15069	156	57.69%	5.76%	Residential area	28, 4
Changfa Dihao and Xiangyi Zijun	10441	11038	148	53.38%	5.72%	Residential area	33

The selected study areas all have large parks and squares, such as Rose Garden Park. Although the surrounding squares of Tianning Wuyue Plaza are densely populated with commercial and residential areas, there is basically no gathering of people and no significant increase in the number of people (Figure 4, Table 2).

Table 2 shows that the average density change in 56.37% of the areas exceeds 5%, which indicates that people are evacuated in more than half of the areas. Areas with a reduced number of people caused by people escaping and areas with an increased number of people due to gathering of people are both included, which indicates that people in the study areas generally respond to an earthquake; thus, the people locations or mobile phone signals change. The increase in the total number of terminals can be interpreted as an increase in the number of mobile phones switched on due to earthquake wake-up calls. The increase occurs in sampling points in all study areas and is more pronounced in areas with highly concentrated high-rise buildings such as Wanda Plaza in the Lunan District. Because some people rush out of the building and do not take their mobile phones when an earthquake

Table 3. The destination of population flow.

	Maximum move distance (m)	Average move distance (m)	Move destination					
			Road	Community Road	Exit Road	Community open space/ square	Regional Plaza	Large park
Around Tianning Wuyue	109	49	16.76%	19.51%	14.25%	43.30%	5.01%	1.17%
Around Qingyang Huayuan	121	63	7.96%	37.54%	18.21%	31.54%	2.36%	2.39%
Around Ziyunyuan	120	51	11.11%	29.18%	18.03%	37.65%	0.53%	3.50%
Changfa Dihao and Xiangyi Zijun	131	57	20.31%	22.79%	17.45%	39.15%	0.26%	0.00%

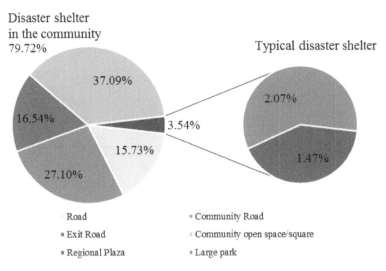

Figure 5. Schematic diagram of the people flow after an earthquake.

occurs, there is no location change for the abandoned mobile phones. Therefore, the actual data could be larger than the monitored data.

A significant people gathering occurs in the study areas, which is the phenomenon that after an earthquake, people respond to the earthquake and move from residential areas or areas with buildings to relatively open spaces. in some relatively open spaces, such as community squares, community roads, and entrances and exits of the community, many people gather. Similarly, within a community, there are significant people outflow in the areas with dense buildings, and a clustering analysis of the people who move inside the community was performed.

4.3 Time-series analysis of changes in people in the affected area

After an earthquake, the distribution of people near the epicenter experiences a dynamic change. A certain number of people start to gather in relatively open spaces, such as residential squares. Eight typical open spaces in the two study areas were selected for analysis. After an earthquake, people gathering occurs in these typical areas, and the number and location of the people are always changing. In this study, beginning at 21:40 on December 22, 2021, the people data in 10 periods were analyzed (each period = 10 minutes), the blue curve is the change on December 22, 2021, and the red curve is the people flow on December 21, 2021, the day before the earthquake occurred.

At 21:50, people start to gather in several typical open spaces. At 22:00, the number of people in some areas slowly increased. At approximately 22:10 (20 minutes after the earthquake), people gathering reaches its peak and starts to return to normal. At approximately 22:40, the level of people gathering basically returns to normal (i.e., it is consistent with the level of people gathering on the previous day). Figure 6 shows people gathering occurs in several typical open spaces at 21:50. In most areas, the level of people gathering basically returns to normal at approximately 22:50 (i.e., it is essentially consistent with that on the previous day), and people in areas with high-rise buildings feel the earthquake more strongly than those in areas with low-rise buildings. Figure 6 shows that since the well-planned building complex in the study area has reasonable open spaces, the gathering time of evacuated people is significantly lower than that of people in areas with a greater number of old buildings (Figure 6).

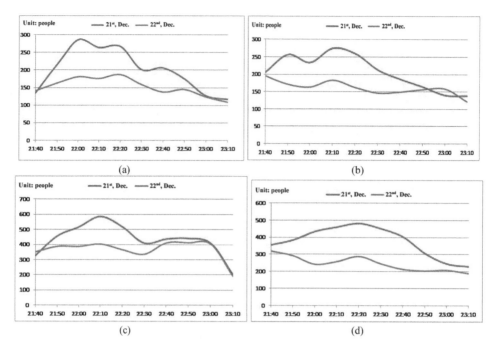

Figure 6. Time series analysis of regional population change. (a) The east side of Tianning Wuyue (b) Entrance and exit of Qingyang Huayuan (c) Ziyunyuan open space (d) Around Changfa Dihao.

5 CONCLUSIONS AND PROSPECTS

(1) The data analysis found that there are no large-scale and long-term people gatherings after an earthquake, and most people complete the gathering-dispersal process within the community. The level of the spatiotemporal gathering of people is much smaller than expected, the entire gathering-dispersal process requires between 20 and 50 minutes to complete, and the average moving distance is only approximately 50 m. The rapid dissemination of post-earthquake warning information through media channels, such as the Internet, TV, and radio, enables residents in the affected areas to be informed of the situation promptly, and the high level of public awareness of the earthquake and more rational earthquake avoidance behavior also contribute to the rapid restoration of social order after an earthquake.

(2) The earthquake avoidance behavior of people in this study indicates that compared with constructing large-scale disaster shelters, more attention should be given to the planning of small and medium-sized shelters at the community level, which can satisfy the emergency disaster avoidance needs that are the most urgent, effective, and practical needs of people after an earthquake. Because an M4.2 earthquake is a nondestructive earthquake, the traditional shelters, such as large squares and parks, do not play a role in people's actual earthquake avoidance, while open spaces at the community level become the main places for people to gather. These places effectively relieve the pressure due to earthquake avoidance and avoid a great disruption to traffic and social order. As a microcosm of cities in eastern China, Changzhou has a dense population, concentrated social wealth, dense urban buildings, and many high-rise and ultra-high-rise buildings. For historical reasons, a considerable number of cities still have many old buildings and residential areas that have not been well planned and have no emergency evacuation locations. Once earthquakes and induced fires occur, serious casualties and social

disorder could be easily caused. Community-level disaster shelters can effectively support the needs for emergency disaster sheltering.

(3) Big data provide an unprecedented social perception means to understand the evolution of natural disasters from a human perspective. This study performed quantitative analysis on the spatiotemporal patterns of post-earthquake people gathering and dispersal and studied the post-earthquake spatiotemporal changes in people locations under different scales (urban level to community level to building units). New technologies and methods are emerging, and with the support of high-precision data, community-level quantitative spatiotemporal analysis of people after an earthquake can be realized, which can provide practical and reliable micro-level data for government disaster relief within a few hours of an earthquake. For the M4.2 earthquake that occurred in Changzhou, with the support of wireless Internet terminal data, this study obtained the spatiotemporal dynamic changes of people after the earthquake and performed multi-source data verification; therefore, the obtained data are close to the actual situation.

REFERENCES

Becker, J.S., Paton, D, Johnston, D.M., Ronan, et al. (2017) The Role of Prior Experience in Informing and Motivating Earthquake Preparedness. *Int J Disaster Risk Reduct*. 22: 179–193.

Calka B., Bielecka E. (2019) Reliability Analysis of Landscan Gridded Population Data. The Case Study of Poland. *ISPRS Int J Geo Inf* 8(5):222–240.

Calabrese F., Di Lorenzo G., Liu L. (2011) Estimating Origin-Destination Flows Using Opportunistically Collected Mobile Phone Location Data From One Million Users In Boston Metropolitan Area. *IEEE Perv Comput*. 10(4) 36–44.

Freire S (2010) Modeling of Spatiotemporal Distribution of Urban Population at High Resolution–Value for Risk Assessment and Emergency Management. In: *Geographic Information and Cartography for Risk and Crisis Management*. Springer, pp. 53–67.

Samardjieva E (2002) Estimation of the Expected Number of Casualties Caused by Strong Earthquakes. *Bull Seismol Soc Am* 92(6):2310–2322.

Pang X., Nie G., Zhang X., et al. (2019) Selection of Earthquake Disaster Index Based on Mobile Phone Position Data. *Earthquake Research in China*.

Zhang, X. B., Kelly, S., Ahmad, K., Ieee, (2016) *The Slandail Monitor: Real-Time Processing And Visualisation of Social Media Data for Emergency Management*. In: Proceedings of 2016 11th International Conference on Availability, Reliability and Security. Ieee, New York, pp. 786–791.

Study on the regulation of pentamode lattice ring structure on impact stress wave

Bangyi Han & Zhenhua Zhang
College of Navel Architecture & Ocean, Naval University of Engineering, Wuhan, China

ABSTRACT: The pentamode lattice ring structure as an impact protection structure can bend the propagation path of the impact stress wave, which is completely different from the mechanism of the traditional impact protection structure. In order to study the regulation effect of the pentamode lattice ring structure on the impact stress wave, a pentamode lattice ring structure test model is designed and made. Through the impact test and simulation, it is found that the peak strain of the inner-ring front shock surface of the pentamode lattice ring structure is 92.5% of that of the inner-ring rear shock surface. Compared with the equal mass solid ring, the peak strain on the inner-ring front shock surface of the pentamode lattice ring structure is reduced by 44%, while the peak strain on the inner-ring rear shock surface is increased by 25%. It is found that the normal vector direction distribution of the dispersion surface isoline can accurately reflect the energy flow characteristics in the material. The relevant conclusions can provide some reference for the application of pentamode materials in the field of impact protection.

1 INTRODUCTION

Collision shock widely exists in daily life. At present, the protection against collision shock is mainly based on the energy dissipation theory, that is, the energy of collision shock is absorbed through the deformation of the structure, and conventional materials have gradually failed to meet the protection needs. As a new type of metamaterial, pentamode materials have the characteristics of both fluid and solid. Through the design of pentamode materials, the control of stress waves can be realized, which may provide a new solution for the collision protection of the structure.

In 1995, Milton and Cherkaev (1995) first proposed the concept of pentamode materials. Only one eigenvalue of its stiffness matrix is not zero, that is, it can only bear one state of stress. The shear modulus of the ideal pentamode materials is zero, but it will retain a certain shear strength for practical use. Norris (2008) proposed the idea of pentamode materials acoustic cloak, which can bend the propagation path of sound waves and make the target object "disappear" in the sound field. Scandrett et al. (2010, 2011) proved the feasibility of using layered pentamode materials to build an acoustic cloak, which made the coordinate transformation theory break through the limitation of continuously changing anisotropic materials, and made it possible to realize the acoustic cloak. Zafiris (2014) studied a kind of pentamode materials based on layered rod structure through simulation and found that it is close to the ideal pentamode materials in a certain frequency range. The kind of pentamode materials can effectively regulate sound waves and has broad prospects in the field of elastodynamics. Chen et al. (2015) designed a pentamode acoustic cloak made of single-phase solid materials and verified the acoustic stealth effect by simulation. However, due to the resonance caused by weak shear modulus, the cloak has a stealth effect only in part of the frequency range. Li et al. (2019) designed a pentamode spherical acoustic cloak. The study found that the stronger the anisotropy of pentamode materials, the better the stealth effect.

At present, the application of pentamode materials in the field of stress wave regulation mainly focuses on acoustic cloaking. The application potential in other fields, such as impact cloaking, mechanical cloaking, vibration reduction, and isolation, has not been effectively explored. Only a few scientific researchers are carrying out relevant research, and it is still in its infancy (Amendola et al. 2016a, 2016b, 2017; Lymperopoulos & Theotokoglou 2022). In order to explore the regulation effect of the pentamode lattice ring structure on the impact stress wave, this paper designs and makes a pentamode lattice ring structure, and carries out the impact test and simulation. By calculating the dispersion characteristics of cells, we can obtain the normal vector direction distribution of the dispersion surface isoline, and on this basis, we can research the relationship between the normal vector direction distribution and the energy flow vector. The relevant conclusions can provide some theoretical support for the application of pentamode materials in the field of collision protection.

2 DESIGN AND MANUFACTURE OF PENTAMODE LATTICE RING STRUCTURE

2.1 Design theory of pentamode lattice ring structure

2.1.1 Coordinate transformation theory

The dynamic process of wave propagation is described by the following equation

$$\begin{cases} \sigma = -p\mathbf{S} \\ \rho \cdot \dot{v} = \nabla \cdot \sigma \end{cases}. \tag{1}$$

From this, we can get

$$\begin{cases} k\mathbf{S} = k_0 \det \mathbf{AP} \\ \rho^{-1} S = \rho_0^{-1} \dfrac{P^{-1} A A^T}{\det A} \end{cases}, \tag{2}$$

where S and P are general symmetric matrices, A is the transformation matrix, k_0 and are the bulk modulus and density before a coordinate transformation, and k and ρ are the bulk modulus and density after a coordinate transformation. The coordinate mapping relationship of the two-dimensional cylindrical structure can be expressed as

$$r' = \sqrt{\dfrac{R_2^2 - \delta^2}{R_2^2 - R_1^2} r^2 - \dfrac{R_1^2 - \delta^2}{R_2^2 - R_1^2} R_2^2}, \tag{3}$$

$$\theta' = \theta, \tag{4}$$

where r and θ are the polar coordinates before coordinate transformation, r' and θ' are the polar coordinates after a coordinate transformation, R_1 and R_2 are the inner and outer radius of the stealth structure respectively, δ is a small value. The existence of δ can effectively avoid the singularity of material parameters, so that the region of $0 < r < \delta$ is mapped to $0 < r' < R_1$. Substitute Equation (3) into Equation (2) to obtain

$$\rho = \rho_0 \frac{R_2^2 - \delta^2}{R_2^2 - R_1^2}, \tag{5}$$

$$k_r = k_0 \frac{r^2(R_2^2 - \delta^2) - (R_1^2 - \delta^2)R_2^2}{r^2(R_2^2 - \delta^2)}, \tag{6}$$

$$k_\theta = k_0 \frac{r^2(R_2^2 - \delta^2)}{r^2(R_2^2 - \delta^2) - (R_1^2 - \delta^2)R_2^2}, \tag{7}$$

where k_r is the radial bulk modulus of the ring, k_θ is the tangential bulk modulus of the ring.

2.1.2 Numerical homogenization of complex microstructure materials

For the microstructure with a complex cell shape, it is almost impossible to obtain the equivalent mechanical parameters by analytical solution, so it is necessary to use the numerical homogenization method for material equivalent transformation. For equivalent materials, we usually need to obtain their equivalent density and equivalent stiffness matrix. Under the condition of the long wave, there is generally no resonance phenomenon in microstructure cells, so the equivalent density of cells can be expressed as

$$\rho^{eff} = m_{cell}/V_{cell}, \tag{8}$$

where m_{cell} is the total mass of cells, and V_{cell} is the space occupied by the cell. For two-dimensional microstructures, the equivalent stiffness matrix can be expressed as

$$C = \begin{bmatrix} K_x^{eff} & K_{xy}^{eff} & 0 \\ K_{xy}^{eff} & K_y^{eff} & 0 \\ 0 & 0 & G_{xy}^{eff} \end{bmatrix}. \tag{9}$$

Substituting the equivalent stiffness matrix into the dynamic equation, we can obtain

$$C_{Lx} = \sqrt{K_x^{eff}/\rho^{eff}}, \tag{10}$$

$$C_{Ly} = \sqrt{K_y^{eff}/\rho^{eff}}, \tag{11}$$

$$C_T = \sqrt{G_{xy}^{eff}/\rho^{eff}}, \tag{12}$$

$$C_{qL}^2 = \frac{1}{4\rho^{eff}} \left(K_x^{eff} + K_y^{eff} + 2G_{xy}^{eff} + \sqrt{\left(K_x^{eff} - K_y^{eff}\right)^2 + 4\left(K_{xy}^{eff} + G_{xy}^{eff}\right)^2} \right), \tag{13}$$

$$C_{qT}^2 = \frac{1}{4\rho^{eff}} \left(K_x^{eff} + K_y^{eff} + 2G_{xy}^{eff} - \sqrt{\left(K_x^{eff} - K_y^{eff}\right)^2 + 4\left(K_{xy}^{eff} + G_{xy}^{eff}\right)^2} \right), \tag{14}$$

where C_{Lx} is the phase velocity of a longitudinal wave in the x direction, C_{Ly} is the phase velocity of a longitudinal wave in the y direction, C_T is shear wave phase velocity, C_{qL} is the longitudinal wave phase velocity in the 45° direction, C_{qT} is the shear wave phase velocity in the 45° direction. Solving Equations (10~14), we can obtain

$$K_x^{eff} = \rho^{eff} C_{Lx}^2, \tag{15}$$

$$K_y^{eff} = \rho^{eff} C_{Ly}^2, \tag{16}$$

$$G_{xy}^{eff} = \rho^{eff} C_T^2, \tag{17}$$

$$K_{xy}^{eff} = \rho^{eff} \left(\sqrt{\left(C_{qL}^2 - C_{qT}^2\right)^2 - \left(C_{Lx}^2 - C_{Ly}^2\right)^2/4} - C_T^2 \right). \tag{18}$$

The phase velocities in all directions of the cell can be obtained through the dispersion surface. On this basis, we can obtain the cell equivalent stiffness matrix and realize the numerical homogenization of complex microstructures.

2.2 Parameters of pentamode lattice ring structure

Based on the coordinate transformation theory and numerical homogenization method, we design the cell structure size of each layer and make equivalent parameters basically meet the requirements of coordinate transformation. On this basis, we make the physical model by using additive manufacturing technology. The inner radius of the ring R_1 = 45 mm, the outer radius R_2 = 92 mm, and the cell wall thickness t = 0.75 mm. The base material is PC material, and its material parameters are elastic modulus E = 2000 MPa, density ρ = 1200 kg/m³, and Poisson's ratio ν = 0.4. The cell size and equivalent parameters of each layer in the pentamode lattice ring structure are shown in Figure 1 and Table 1. The similarity between the designed material and the ideal pentamode materials is mainly determined by the following indicators

$$\pi = \frac{|K_{XY}|}{\sqrt{K_X K_Y}}, \tag{19}$$

$$\mu = \frac{G}{\sqrt{K_X K_Y}}. \tag{20}$$

The pentamode characteristic parameters of ideal pentamode materials are π = 1 and μ = 0. At this time, the material is a complex fluid with solid characteristics. But in reality, π = 1 and μ = 0 can't be achieved, π and μ can only tend to 1 and 0 respectively. From Table 1, it is

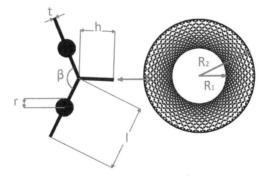

Figure 1. Pentamode lattice ring structure diagram.

Table 1. Pentamode lattice ring structure parameters.

Number of layers	β (°)	l (mm)	h (mm)	r (mm)	ρ^{eff} (kg/m³)	K_x^{eff} (Pa)	K_y^{eff} (Pa)	K_{xy}^{eff} (Pa)	G_{xy}^{eff} (Pa)	π	μ
1	152	3.14	0.85	0.75	723.50	1.22E9	3.87E7	1.56E8	1.10E8	0.717	0.507
2	151	3.25	1.04	0.75	634.72	1.04E9	4.19E7	1.54E8	8.53E7	0.736	0.409
3	145	3.47	1.16	0.75	537.60	8.23E8	5.13E7	1.65E8	7.25E7	0.804	0.353
4	143	3.67	1.29	0.875	483.86	7.47E8	5.49E7	1.69E8	6.10E7	0.833	0.301
5	141	3.88	1.42	0.875	493.69	6.48E8	5.61E7	1.63E8	5.02E7	0.856	0.263
6	139	4.11	1.57	0.875	444.07	5.62E8	5.72E7	1.57E8	4.12E7	0.875	0.230
7	137	4.37	1.73	1	450.18	5.05E8	6.03E7	1.56E8	3.40E7	0.895	0.195
8	135	4.65	1.90	1	403.67	4.36E8	6.09E7	1.48E8	2.76E7	0.910	0.169
9	134	4.96	2.09	1	364.39	3.85E8	5.88E7	1.39E8	2.20E7	0.921	0.146
10	133	5.30	2.28	1.125	383.98	3.53E8	5.83E7	1.34E8	1.79E7	0.932	0.125
11	131	5.67	2.50	1.125	342.36	3.02E8	5.87E7	1.26E8	1.43E7	0.942	0.107
12	130	6.08	2.74	1.125	307.23	2.67E8	5.62E7	1.16E8	1.13E7	0.950	0.092
13	130	6.39	2.98	1.125	287.65	2.35E8	5.10E7	1.04E8	8.80E6	0.954	0.080

not difficult to find that the pentamode characteristic parameters of the outer layer cell are significantly better than those of the inner layer cell in the pentamode lattice ring structure, which is mainly limited by the manufacturing process. The rod slenderness ratio of the inner layer cell is too large, which reduces the pentamode characteristic of the inner layer cell.

In the table, the pentamode lattice ring structure has 1–13 layers from the inner layer to the outer layer.

3 IMPACT TEST OF PENTAMODE LATTICE RING STRUCTURE

3.1 Test conditions

We select three measuring points in the inner ring, which are located at the top (front shock surface), side and bottom (rear shock surface), and paste a strain gauge along the tangential at each of the measuring points. We fix the support together with the test model on the iron frame with bolts, place a 50 g steel column at the height of 20 cm, 30 cm and 40 cm directly above the test model, corresponding to working conditions 1, 2 and 3, and impact the test model with a static free fall. The test device is shown in Figure 2.

3.2 Test results and analysis

After repeated tests many times to eliminate accidental errors, we select the test data of the strain peak section, and draw the strain time history curve of the three measuring points with MATLAB software. The strain time history curve of the test model under different working conditions is shown in Figure 3, and the strain peak of each measuring point is shown in Table 2.

Under different working conditions, $\varepsilon_1/\varepsilon_3$ of the pentamode lattice ring structure is stable at about 92.5%. The peak strain on the inner-ring front shock surface is significantly smaller than that on the inner-ring rear shock surface, which is significantly different from the conventional ring structure. After the impact of the conventional ring, the peak strain on the inner-ring front shock surface will be much larger than that on the inner-ring rear shock surface. So the pentamode lattice ring structure designed in this paper has an obvious regulation effect on the impact stress wave.

In the table, ε_i is the peak strain of measuring point i, and unit is $\mu\varepsilon$.

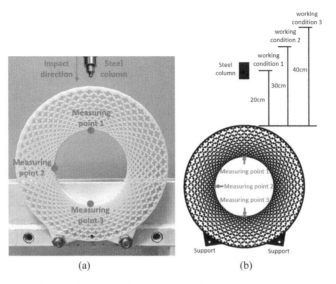

Figure 2. Test setup diagram. (a) Physical map (b) Sketch map.

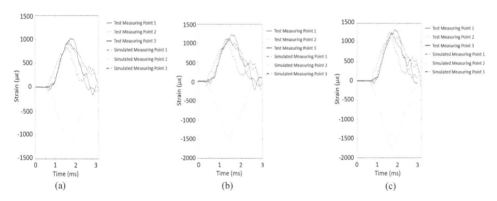

Figure 3. Strain time history curve of the pentamode lattice ring structure. (a) Working condition 1 (b) Working condition 2 (c) Working condition 3.

Table 2. Peak strain of test and simulation measuring points.

Working condition	Pentamode lattice ring structure (test)				Pentamode lattice ring structure (simulated)				Equal mass solid ring (simulated)			
	ε_1	ε_2	ε_3	$\varepsilon_1/\varepsilon_3$	ε_1	ε_2	ε_3	$\varepsilon_1/\varepsilon_3$	ε_1	ε_2	ε_3	$\varepsilon_1/\varepsilon_3$
1	813.6	−1225.9	880.2	92.4%	855.5	−1021.3	915.5	93.4%	1525	−3103	701.3	217.45%
2	1046.4	−1425.7	1136.9	92.0%	1033.4	−1247.4	1105.8	93.5%	1854	−3661	881.4	210.35%
3	1164.2	−1611.8	1254.7	92.8%	1182.5	−1423.2	1267	93.3%	2129	−4378	1002	212.48%

4 SIMULATION OF THE PENTAMODE LATTICE RING STRUCTURE UNDER IMPACT

4.1 Simulation modeling

We use MSC.Patran software to conduct pre-processing modeling of the pentamode lattice ring structure. In order to highlight the protective performance of the pentamode lattice ring structure, we establish an equal mass solid ring made of the same base material as a comparison. The simulation conditions of the two models are completely consistent with the test conditions. The inner radius of the equal mass solid ring is equal to the inner radius of the pentamode lattice ring structure, as shown in Figure 4.

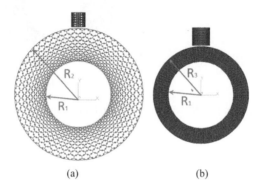

Figure 4. Simulation modeling diagram. (a) The pentamode lattice ring structure (b) The equal mass solid ring.

4.2 Strain result analysis

Using MSC.Dytran software, we conduct post-processing calculation on the model and output the strain values at the three measuring points in Figure 2. We compare the simulation results of the pentamode lattice ring structure with the test results, which are shown in Figure 3. From the overall change trend, the test results are very consistent with the simulation results, and the error of peak strain between test results and simulation results is within 15%. $\varepsilon_1/\varepsilon_3$ of simulation is stable at 93.5%, and the error with the test is within 10%. The simulation peak strain of the pentamode lattice ring structure is shown in Table 2.

We compare the simulation strain results of the equal mass solid ring and the pentamode lattice ring structure, as shown in Figure 5. The simulation peak strain of equal-mass solid ring is shown in Table 2. The peak strain on the inner-ring front shock surface of the equal mass solid ring is significantly greater than that on the inner-ring rear shock surface, and $\varepsilon_1/\varepsilon_3$ of the equal mass solid ring is up to 210%, which is much larger than $\varepsilon_1/\varepsilon_3$ of the pentamode lattice ring structure. Compared with the equal mass solid ring, the peak strain of measuring point 1 in the pentamode lattice ring structure decreased by 44%, the peak strain of measuring point 2 decreased by 65%, but the peak strain of measuring point 3 increased by 25%. This shows that the pentamode lattice ring structure can effectively reduce the impact on the inner-ring front shock surface and protect the internal structure.

4.3 Stress result analysis

MSC.Patran can help visually see the stress distribution in the model at each time by reading the calculation result file. The stress distribution under the three working conditions is

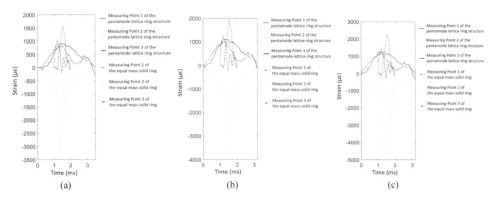

Figure 5. Simulated strain time history curve. (a) Working condition 1 (b) Working condition 2 (c) Working condition 3.

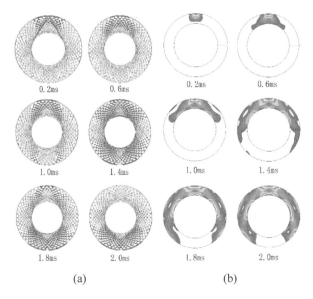

Figure 6. Simulated stress distribution. (a) The pentamode lattice ring structure (b) The equal mass solid ring.

roughly similar. Then we select the stress distribution results of working condition 1 for analysis, as shown in Figure 6.

It can be clearly seen from Figure 6 that after the pentamode lattice ring structure is impacted, the stress tends to diverge from both sides, and the impact stress wave is effectively "guided". The stress distribution of equal mass solid rings is obviously different. After being impacted, the stress concentration on the inner-ring front shock surface is obvious. Therefore, the pentamode lattice ring structure can adjust and control the stress wave generated by the impact, and effectively reduce the stress concentration on the inner-ring front shock surface. The frequency spectrum analysis of the stress at the impact points of the pentamode lattice ring structure and the equal mass solid ring is carried out. The main frequency range of the stress wave generated by the collision is 0–100 Hz, as shown in Figure 7.

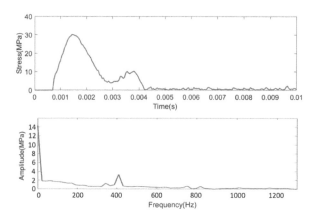

Figure 7. Spectrum analysis of impact stress.

5 CELL DISPERSION CHARACTERISTICS ANALYSIS

5.1 *Dispersion surface analysis*

The dispersion surface determines the direction and size of the group velocity, and the group velocity is the specific embodiment of the energy flow vector. Therefore, the propagation characteristics of energy in the structure can be obtained by analyzing the dispersion surface of pentamode lattice cells. The normal vector direction distribution of a frequency isoline reflects the main direction of energy transmission at that frequency, that is, if the angle of the normal vector of the isoline falls in a propagation direction more, the energy transmitted in that direction will be more. By using COMSOL software to sweep the irreducible Brillouin zone of the pentamode lattice cell, we can obtain the dispersion characteristics of the cell, as shown in Figure 8.

Figure 8 (a) shows the dispersion surface of the 13th layer cell in the pentamode lattice ring structure. The first six-order dispersion surfaces have been calculated. The first-order dispersion surface is the shear wave dispersion surface, the second-order dispersion surface is the longitudinal wave dispersion surface, and other higher-order dispersion surfaces are the dispersion surfaces of the longitudinal wave and shear wave coupling at extremely high frequencies. For frequencies below 4000 Hz, the shear wave velocity of cells in the pentamode lattice ring structure is less than 1/4 of the longitudinal wave velocity. According to the stress wave energy calculation formula, for the stress wave generated by the same vibration source, the energy contained in the shear wave and the longitudinal wave is proportional to the square of wave velocity. Therefore, the energy contained in the shear wave is less than 1/16 of the longitudinal wave, and the second-order dispersion surface is the main research object in this section. Figure 8(b) shows the isoline of the second-order dispersion surface. It is not difficult to find that within 4000 Hz, the change of frequency has little effect on the shape of the isoline. On this basis, calculate the normal vector of the isoline below 4000 Hz, as shown in Figure 8(c), and count the direction distribution of the normal vector, as shown in Figure 8(d). The normal vector direction distribution has obvious anisotropy and has a bifurcation angle α. We take the angle corresponding to N/2 as the boundary of the bifurcation angle, where N is the maximum magnitude of the normal vector direction distribution, then the normal vector bifurcation angle of the 13th layer cell can be read out $\alpha = 106°$. We calculate the normal vector bifurcation angle α of each layer cell in the pentamode lattice ring structure below 4000 Hz, as shown in Table 3.

5.2 *Energy flow characteristics analysis*

Based on the numerical homogenization method of complex microstructure materials, the microstructure materials can be equivalent to homogeneous materials in COMSOL

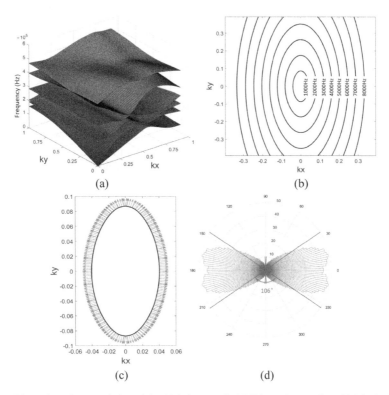

Figure 8. Dispersion characteristics of the 13th layer cell. (a) Dispersion surface (b) The isoline of the dispersion surface (c) The normal vector of the isoline (d) Normal vector direction distribution.

software. We apply a half sine impact load to the rectangular equivalent material, and conduct frequency spectrum analysis on the load. The main frequency distribution range of the load is below 5000 Hz. The mechanical energy distribution field in the equivalent homogeneous material is output, as shown in Figure 9, and there is an obvious bifurcation angle α'. The mechanical energy flow bifurcation angle α' in the equivalent homogeneous material of each layer cell is shown in Table 3. An isotropic PC material model is set up for comparison, through simulation we calculate its normal vector direction distribution and mechanical energy distribution field, as shown in Figure 10.

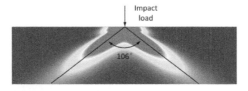

Figure 9. The mechanical energy distribution field in the equivalent homogeneous material of 13th layer cell.

In a comparison of normal vector bifurcation angles α and mechanical energy flow bifurcation angle α', it is found that the two angles are very consistent, and the error is within 2 degrees. The isoline normal vectors of isotropic material are evenly distributed in all directions, without normal vector bifurcation angle. Besides that, the mechanical energy also spreads in a spherical form, without mechanical energy flow bifurcation angle. Therefore, the normal vector direction distribution of the isoline can well reflect the energy flow direction in the material. The greater the normal vector bifurcation angle α of the material, the better the effect of regulating the impact stress wave.

Based on the energy flow characteristics of each layer cell, we predict the energy flow trajectory in the pentamode lattice ring structure, as shown in Figure 11(a). We establish an equivalent homogenized pentamode lattice ring structure model in COMSOL software, apply a half-sine impact load and output the mechanical energy distribution field, as shown in Figure 11(b). From Figure 11, it can find that the energy flow trajectory predicted is very consistent with the energy flow trajectory obtained by simulation. On this basis, we can locally strengthen the pentamode lattice ring structure by predicting the energy flow trajectory to avoid local failure.

Table 3. Normal vector bifurcation angle and mechanical energy flow bifurcation angle.

bifurcation angle	Number of layers												
	1	2	3	4	5	6	7	8	9	10	11	12	13
α	119°	122°	123°	126°	118°	116°	114°	112°	112°	111°	110°	107°	106°
α'	117°	122°	122°	125°	119°	117°	115°	111°	111°	110°	108°	106°	106°

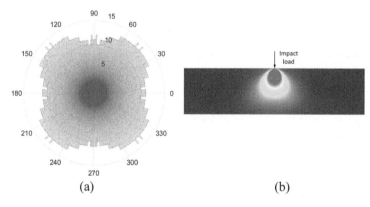

Figure 10. Dispersion characteristics and energy flow characteristics of isotropic materials. (a) Normal vector direction distribution (b) Mechanical energy distribution field.

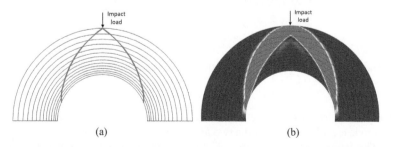

Figure 11. Energy flow trajectory. (a) Predicted energy flow trajectory (b) Simulated energy flow trajectory.

6 CONCLUSION

Based on the coordinate transformation theory and additive manufacturing technology, this paper designs and makes a pentamode lattice ring structure test model, and reveals the dynamic response of the pentamode lattice ring structure under impact through impact test, impact simulation and dispersion characteristics analysis. The specific conclusions are as follows:

(1) Through impact test, it is found that $\varepsilon_1/\varepsilon_3$, the ratio of the peak strain on the inner-ring front shock surface and the inner-ring rear shock surface of the pentamode lattice ring structure is stable at about 92.5%, which is obviously different from the conventional ring. This shows that the pentamode lattice ring structure has an obvious control effect on the impact stress wave, which can prevent the structure of the inner-ring front shock surface from being damaged first.

(2) The dynamic response process of the pentamode lattice ring structure under impact is revealed through simulation, and the error between the simulation results and the test results is within a reasonable range. Compared with the equal mass solid ring, the peak strain of the pentamode lattice ring structure on the inner-ring front shock surface decreases by 44%, and the peak strain of the inner-ring rear shock surface increases by 25%. $\varepsilon_1/\varepsilon_3$ of the pentamode lattice ring structure is only 93.5%, while $\varepsilon_1/\varepsilon_3$ of the equal mass solid ring is as high as 210%. From the point of stress distribution, the stress in the pentamode lattice ring structure has an obvious divergent trend on both sides, which can effectively avoid the stress concentration on the impact face.

(3) Through the analysis of the cell dispersion characteristics, it is found that the normal vector direction distribution of the dispersion surface isoline can accurately reflect the energy flow characteristics in the material. The error between the normal vector bifurcation angle and mechanical energy flow bifurcation angle doesn't exceed 2°. The normal vector bifurcation angle of each layer cell can accurately predict the energy flow trajectory in the pentamode lattice ring structure, so the pentamode lattice ring structure can be locally strengthened to avoid local damage.

REFERENCES

Aravantinos-Zafiris N., Sigalas M. M., Economou E. N. Elastodynamic Behavior of The Three Dimensional Layer-By-Layer Metamaterial Structure [J]. *Journal of Applied Physics*, 2014, 116(13).

Amendola A., Carpentieri G., Feo L., et al. Bending Dominated Response of Layered Mechanical Metamaterials Alternating Pentamode Lattices and Confinement Plates [J]. *Composite Structures*, 2016a, 157: 71–77.

Amendola A., Smith C. J., Goodall R., et al. Experimental Response of Additively Manufactured Metallic Pentamode Materials Confined Between Stiffening Plates [J]. *Composite Structures*, 2016b, 142: 254–262.

Amendola A., Benzoni G., Fraternali F. Non-Linear Elastic Response of Layered Structures, Alternating Pentamode Lattices and Confinement Plates [J]. *Composites Part B: Engineering*, 2017, 115: 117–123.

Chen Y., Liu X., Hu G. Latticed Pentamode Acoustic Cloak [J]. *Sci Rep*, 2015, 5: 15745.

Li Q, Vipperman J. S. Three-Dimensional Pentamode Acoustic Metamaterials With Hexagonal Unit Cells [J]. *The Journal of the Acoustical Society of America*, 2019, 145(3): 1372–1377.

Lymperopoulos P. N., Theotokoglou E. E. Computational Analysis of Pentamode Metamaterials for Antiseismic Design [J]. *Procedia Structural Integrity*, 2020, 26: 263–268.

Milton G. W., Cherkaev A. V. Which Elasticity Tensors are Realizable? [J]. *Journal of Engineering, Materials and Technology*, 1995, 117: 483–493.

Norris A. N. Acoustic Cloaking Theory [J]. *Proceedings of the Royal Society A: Mathematical, Physical and Engineering Sciences*, 2008, 464(2097): 2411–2434.

Scandrett C. L., Boisvert J. E., Howarth T. R. Acoustic Cloaking Using Layered Pentamode Materials [J]. *Journal of the Acoustical Society of America*, 2010, 127(5): 2856.

Scandrett C. L., Boisvert J. E., Howarth T. R. Broadband Optimization of A Pentamode-Layered Spherical Acoustic Waveguide [J]. *Wave Motion*, 2011, 48(6): 505–514.

Integrated launching technology of the beam-arch combination system

Jian Li
CCCCSHEC Fourth Engineering Company Ltd, Wuhu Anhui, China

Hanjin Yu*
Cccc WuHan Harbour Engineering Design and Research Co Ltd, Wuhan Hubei, China

ABSTRACT: The first-beam-then-arch method, commonly used in the beam-arch combination system, has the problems of a long construction period, high cost and high requirements in the construction environment. Relying on the rapid reconstruction project of Beijing Road in Suqian, in view of the problems existing in the current process, an integrated launching technology that can be used for the beam-arch combination system is proposed, which realizes the efficient construction of beam-arch bridges across the river and under high traffic pressure. The ANSYS software is used to analyze the force of the beam-arch combined launching platform in the whole launching process. And the key control parameters and calculation methods in the construction process are studied. The results show that the structural strength and stiffness of the beam-arch combination integrated launching platform can meet the construction requirements under different combinations of loads and working conditions, which provides a technical reference for the integrated construction of the beam-arch combination system.

1 INTRODUCTION

The total length of the main line of the rapid reconstruction project of Beijing Road in Suqian is about 3.726 km. Among the six main bridges, the main bridge and the north-south approach bridge of the newly-built Beijing-Hangzhou Canal Bridge are the key points. The main bridge is generally a bottom-bearing tied arch bridge, which mainly includes steel box girders, mid-span arch ribs, wind bracing, suspension rods, tie rods and deck ties. The total length of the main bridge is 252.5 m, and the span arrangement is (40 + 172.5 + 40) m.

The residential and enterprises around the main bridge of the Beijing Hangzhou canal are densely distributed, and there are many pedestrians. Furthermore, it is necessary to meet the requirements of ensuring the passage of vehicles and pedestrians across the entire line, and the road clearance must not be compressed. In addition, the existing roads in the bridge area have large traffic flow, fast driving speed, and high pressure on traffic diversion.

2 RESEARCH ON THE INTEGRATED LAUNCHING CONSTRUCTION SCHEME

The main bridge of the Beijing-Hangzhou Canal is a beam-arch combination system. Compared with the traditional beam-and-arch construction scheme (see Figure 1(a)), the

---

*Corresponding Author: 15002776933@163.com

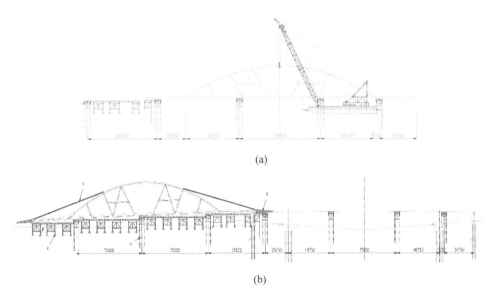

Figure 1. Construction scheme of the beam-arch combination system. (a) Beam-first-then-arch method (b) Integrated launching method of beam-arch combination.

beam-arch combination integrated launching construction scheme (see Figure 1(b)) can not only complete the construction in less period, but also save the project cost to a certain extent, and has lower requirements for the construction environment (Xu et al. 2013). Therefore, it is a more suitable construction scheme for the main bridge of the Beijing-Hangzhou Canal. According to the characteristics of the Beijing-Hangzhou Canal main bridge project and the beam-arch integrated launching construction, the unique design of this construction scheme is as follows:

2.1 Temporary piers

The temporary piers are used as launching piers during launching construction, which are mainly used to bear the load of launching construction. The temporary pier is mainly composed of the steel pipe pile foundation, steel pipe pile, connection system between piles, distribution beam at the top of the pile, etc. The foundation vertical load of the temporary pier is designed according to the maximum support reaction force when the steel box girder is pushed, the self-weight of the temporary pier and the unevenness of the force (Yang et al. 2014).

2.2 Guide beams

The design length of the guide beam is 45 m, and it consists of 6 sections (Mróz et al. 1985). Usually, the main beam of the guide beam will be first lifted and placed on the temporary piers in sections. Eventually, the horizontal links are installed to connect it as a whole.

2.3 Temporary support for arch rib

In order to meet the arch rib assembly requirements of the upper structure, the temporary support is arranged on the steel box girder as a support platform for arch rib assembly (Dimande et al. 2012).

2.4 Arrangement of A-type braces

When the arch rib structure is assembled, the temporary support needs to be removed. In order to ensure the overall stability of the arch rib structure during the jacking process, A-type bracing is set as a temporary support structure (Shan & Shan 2014). At the same time, the temporary cable is installed, and the welding process refers to the design requirements.

3 CONSTRUCTION PROCESS AND SIMULATION ANALYSIS

3.1 Construction process

The overall construction steps (Zhou et al. 2013) of the large-span beam-arch combination launching are as follows:

1) Install the guide beam and the first 89.4 m steel box girder segment to the design position, and complete the welding between the guide beam and the steel box girder (see Figure 2(a)).
2) Use the L3, L4 and L5 temporary piers to launch the steel box girder forward 79 m as a whole to reach the position shown in the figure (see Figure 2(b)).
3) Assemble the 141 m steel box girder on the assembly platform. In the meanwhile, complete the assembling of arch rib, remove the arch rib bracket, and install the arch rib temporary support rod and the temporary head cable (see Figure 2(c)).
4) Launch 65 m, the head of the guide beam reaches the maximum cantilever state, the maximum cantilever of the head is 70 m, and the maximum cantilever of the tail is 40 m (see Figure 2(d)).
5) Continue to launch 35 m, and the remaining segments of the steel box girder are assembled at the tail (see Figure 2(e)).
6) Start the launching system to launch the steel box girder into the designed place as a whole. During the launching process, when the guide beam exceeds the L9 temporary pier, it can be removed section by section. After completing the precise adjustment of the

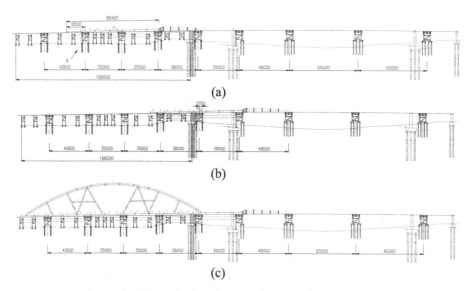

Figure 2. Construction steps of large-span beam-arch integrated launching. (a) Step 1 (b) Step 2 (c) Step 3 (d) Step 4 (e) Step 5 (f) Step 6.

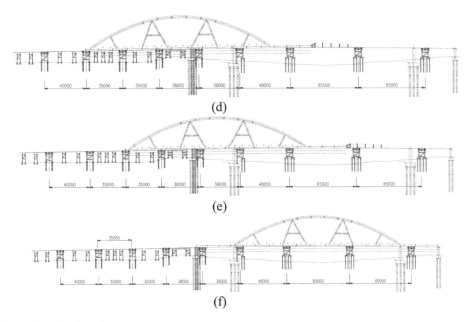

Figure 2. Continued.

steel box girder, drop the steel beam onto the support to complete the consolidation of the support and the steel box girder. After the steel box girder launching construction is completed, the temporary support rod of the arch rib can be removed and other engineering construction can be carried out (see Figure 2(f)).

3.2 *Analysis of working conditions*

In the process of launching construction, the position of the fulcrum of the bottom steel box girder is constantly changing, and the stress state of the beam-arch combination system is also changing constantly. These lead to large differences between the load results. Therefore, it is necessary to calculate and analyze the strength of the temporary structure, the force of the temporary structure and the local stability of the steel box girder. Among them, the most dangerous working condition of the beam-arch combination in each stage of construction is the maximum cantilever state at the front and rear at the same time. In step 4 of Figure 2(d), the head of the guide beam reaches the maximum cantilever state, the maximum cantilever of the head is 65 m, and the maximum cantilever of the tail is 40 m.

3.3 *Analysis model and calculation results*

ANSYS finite element analysis software is used to calculate and analyze the strength and fulcrum reaction force of the beam-arch composite structure during the launching process. The material properties are ideal for elastoplastic constitutive models. The wind loads are imposed on the full-bridge model in the form of acceleration (International Journal of Safety and Security Engineering 2020). For self-weight loads such as steel box girder, arch bridge and guide beam, a partial factor of 1.3 is considered, and a partial factor of 1.5 is considered for the wind load. In the calculation process, each launching fulcrum is set with vertical constraints (Y direction), and the tail launching fulcrum is set with full constraints, but the lateral rotation constraints (ROTX) are released to ensure the convergence of the calculation.

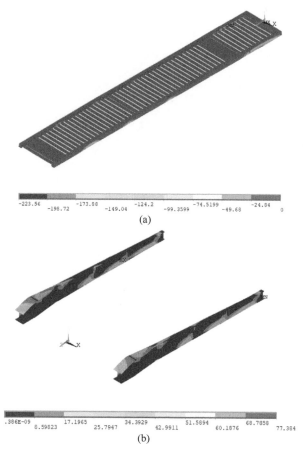

Figure 3. Comprehensive stress cloud diagram of the structure under the most unfavorable working condition. (a) Cloud map of compressive stress of the main beam (b) Cloud map of tensile stress of guide beam.

It can be seen from the finite element analysis results that the overall maximum compressive stress of the steel box girder is $f_{max} = -224\ MPa\ < f = 295\ MPa$. The maximum tensile stress of the guide beam is $f_{max} = 77\ MPa\ < f = 295\ MPa$. The maximum tensile stress of the arch rib is $f_{max} = 71\ MPa\ < f = 295\ MPa$. The maximum compressive stress of the arch rib temporary support rod is $f_{max} = -91\ MPa\ < f = 295\ MPa$. And the overall downward maximum vertical displacement is 426 mm. In general, the structural strength and stiffness meet the requirements, and the structure is safe and reliable. Therefore, the arch-beam integrated structure meets the launching requirements.

4 CONCLUSIONS

The main bridge of the Beijing-Hangzhou Canal in the Suqian Beijing Road Rapid Reconstruction Project adopts the beam-arch combination system. The communities and enterprises within the bridge range are densely distributed, and the traffic guidance and reform pressure are relatively large. Through the research of the scheme, the synchronous launching scheme of the long-span beam-arch combination system can greatly save the construction period while reducing the investment of the temporary structure. And it is better in terms of comprehensive cost and quality control. Through further detailed design and finite element analysis, the force of the beam-arch integrated structure meets the requirements, and the scheme is safe and feasible, which can provide a reference for similar projects.

REFERENCES

A Dimande, Mario Pimentel, Carlos Felix, et al. Monitoring System for Execution Control Applied to a Steel Arch Footbridge. *Structures and Infrastructure Engineering*. 8(3), 277–294 (2012).

A Practical and Safe Optimization Method for Temporary Cable Layout on the Upper Beam of Beam-Arch Composite Rigid Frame Bridge. *International Journal of Safety and Security Engineering*. 10(1), 89–95 (2020).

Jianrong Yang, Yu Bai, Xiaodong Yang, et al. Analysis and Experiment Study of Continuous Beam Arch Composite Bridge. *Advanced Materials Research*. 915–916: 3–6 (2014).

Lige Xu Yili Wang, Changjie Xu. Jacking Technology for a Simply Supported Girder Bridge. *Applied Mechanics and Materials*. 477–478: 675–680 (2013).

Rongxiang Shan, Yuhui Shan. The Confirmation of Closure Jacking Force in Continuous Rigid Frame Bridge. *Applied Mechanics and Materials*. 638–640: 987–993 (2014).

Zenon Mróz, M. P. Kamat, R. H. Plaut. Sensitivity Analysis and Optimal Design of Nonlinear Beams and Plates. *Journal of Structural Mechanics*. 13(3), 245–266 (1985).

Zhixiang Zhou, Yanmei Gao, Chengjun Li, et al. Exploration of Super-Large Span Steel Truss-Fabricated Concrete Composite Continuous Rigid Frame Bridges. *IABSE Symposium Report*. 101(14), (2013).

Failure probability assessment of Fujian earth buildings under multi-disaster coupling

Shangbin Pan, Ziyi Liang & Qiang Liu*
College of Harbour and Coastal Engineering, Jimei University, Xiamen, China

ABSTRACT: A multi-hazard risk assessment method for Fujian Hakka earth buildings under the coupling action of strong rainfall (flood) scouring and earthquake is proposed. Based on various failure states of Fujian earth buildings' structural performance, the failure risk assessment of Fujian earth buildings under the coupling effect of heavy rainfall (flood) scouring and earthquake is conducted. The research results show that the failure probability of Fujian earth-building structures under each failure state is greatly improved when considering the strong rainfall flood scouring and earthquake coupling effects compared with only considering the earthquake effects, and the impact of strong rainfall flood scouring on the seismic performance of structures cannot be ignored. The evaluation method is simple and reliable, which is of great significance for Fujian Hakka earth buildings in multi-disaster emergency response and disaster loss reduction.

1 INTRODUCTION

After Fujian Hakka earth buildings were listed in the world cultural heritage, the problems of development, utilization, protection and reinforcement of earth buildings have become increasingly prominent. As Fujian Hakka earth buildings are located in the coastal area of Fujian Province, where earthquakes and typhoons frequently occur, flood disasters caused by earthquakes, typhoons and heavy rainfall over the past few hundred years have caused damage and destruction of the earth buildings to varying degrees, thus reducing the whole life disaster resistance and reliability of the structure (Qiu 2019; Yi 2020; Zhou 2015).

However, in the traditional research on disaster resistance of Fujian earth building structures, the research on evaluation methods of disaster resistance of Fujian earth building structures is mainly focused on a single disaster (earthquake or wind disaster or heavy rainfall, etc.) (Yi 2019), while the impact of multi disaster coupling on the disaster resistance of Fujian earth building structures throughout their life cycle is ignored. At present, in the research of structural performance evaluation and design methods under multiple disasters, domestic and foreign scholars mainly focus on disaster load combination method and performance-based multi-disaster coupling disaster-resistant design and evaluation method. Petrini et al. proposed a new generation of performance-based multi-hazard coupling disaster-resistant design and evaluation method, which is relatively complex to solve and rarely used in actual research (Petrini 2012); Li Hongnan et al. studied the seismic performance of structures based on performance degradation under multi-hazard coupling action (Li 2019). Lv Guanghui et al. proposed a probability method to consider the failure risk assessment of masonry structures under flood and earthquake (Lv 2020). These results provide a theoretical basis for the design of multi-hazard coupling disaster resistance design and evaluation method of Fujian earth buildings.

---

*Corresponding Author: liutanq007@aliyun.com

Based on the above research results, this paper proposes a failure risk assessment method for Fujian Hakka earth buildings under the combined action of strong rainfall, flood scouring and earthquake. According to the risk assessment theory, the risk assessment of the earth-building structure is carried out according to the reliability index. Therefore, a more reasonable and reliable method is provided for the risk assessment of Fujian earth buildings under the coupling of multiple disasters.

2 TWO DISASTER RISK MODELS

2.1 Hazard model of heavy rainfall scouring

The risk model of heavy rainfall scouring refers to the curve of the probability that the structure will exceed a certain scouring depth in a certain period of time. The probability model of scour depth conforms to the lognormal distribution, and its probability density function can be expressed as (Lv 2020):

$$f_H(h) = \frac{1}{\sqrt{2\pi}\sigma h} \exp\left[-\frac{(\ln h - \mu)^2}{2\sigma^2}\right] \tag{1}$$

Where h is the scouring depth, μ is the logarithmic mean value of scouring depth, σ is the logarithmic standard deviation of scouring depth.

2.2 Seismic hazard model

The seismic hazard risk model is used to describe the curve of the probability that the structure location exceeds a certain earthquake intensity within a certain period of time. In this paper, the seismic hazard risk model established by Gao Xiaowang et al. is adopted (Gao 1986). The probability distribution of seismic intensity conforms to the extreme value type III distribution, and the expression is:

$$F_s(im) = \exp\left[-\left(\frac{w - im}{w - \varepsilon}\right)^k\right] \tag{2}$$

The derivation of the above equation leads to the following probability density function:

$$f_s(im) = \frac{k(\omega - im)^{k-1}}{(\omega - \varepsilon)^k} \exp\left[-\left(\frac{\omega - im}{\omega - \varepsilon}\right)^k\right] \tag{3}$$

Where *im* is the seismic intensity; ω is the upper limit value of seismic intensity, taking $\omega = 12$; k is the shape parameter; ε is a multi-value intensity.

According to China's seismic fortification thought, the fortification intensity is 10% of the earthquake intensity with a probability of exceeding 50 years. It is assumed that the fortification intensity is 7 degrees. Equation (3) can further obtain:

$$f_s(im) = \frac{8.33(12 - im)^{8.33-1}}{(12 - 5.45)^{8.33}} \exp\left[-\left(\frac{12 - im}{12 - 5.45}\right)^{8.33}\right] \tag{4}$$

The relationship between seismic intensity and PGA is:

$$im = 3.73\lg(PGA) - 1.23 \tag{5}$$

Where GPA is the seismic peak acceleration, the unit is Gal.

3 VULNERABILITY ANALYSIS OF FUJIAN EARTH BUILDING

3.1 Structural vulnerability analysis method under earthquake action

The seismic vulnerability of structures refers to the probability that the structural damage exceeds a specified value when a certain strength value of ground motion occurs. The seismic vulnerability of structures can be expressed as:

$$f_R(a) = P[EDP \geq LS/IM = a] = \int_V^\infty f_{R/IM}[r/a]dr \quad (6)$$

Where $f_R(a)$ is seismic vulnerability; $P[EDP \geq LS/IM = a]$ is the probability that the damage of the structure exceeds a specified value; EDP (engineering demand parameter) is the engineering demand parameter, LS (limit state) represents the failure state of a specified structure, and IM (intensity measure) represents the seismic intensity coefficient; $f_{R/IM}[r/a]$ is the conditional probability density that a certain engineering demand parameter EDP of the structure reaches or exceeds a specified failure state when the ground motion intensity IM = a.

Assuming that the relationship between EDP and IM follows the lognormal distribution, the average demand parameter of the structure is:

$$\widehat{EDP} = b(IM)^c \quad (7)$$

Where coefficients c and b are correlation coefficients.

Further calculating the seismic vulnerability $f_R(a)$ of the structure reaching or exceeding the failure state LS, and conducting curve fitting through statistical methods to obtain a smooth "seismic vulnerability curve", the equation expression is:

$$P = [EDP \geq LS/IM] = 1 - \Phi\left(\frac{\ln LS_m - \ln(bIM^c)}{\sqrt{\beta_{LS}^2 + \beta_{EDP}^2}}\right) \quad (8)$$

Where Φ is the cumulative density function of standard normal distribution, and LS_m is the mean value of logarithmic normal distribution under a failure state. β_{LS} and β_{EDP} are the logarithmic standard deviation of structural capacity and demand respectively.

If only earthquake action is considered, the probability of failure of the structure (conditional probability) is:

$$P_b = \sum_{i=0}^{+\infty} P(F/PGA_i)P(PGA_i) \quad (9)$$

Where $P(PGA_i)$ is the probability of earthquake intensity; $P(F/PGA_i)$ is the probability of exceeding the structure under the given ground motion intensity, which is determined by Equation (8).

Considering that the ground motion intensity is continuous, Equation (9) can be expressed as:

$$P_b = \int_0^{+\infty} F_k(PGA)f_s(PGA)dPGA \quad (10)$$

Where $F_k(PGA)$ is the seismic vulnerability function of the structure; $f_s(PGA)$ is the probability function of an earthquake disaster, which can be obtained from Equations (4) and (5).

3.2 Vulnerability analysis under coupling action of earthquake and strong rainfall scouring

In addition to considering the vulnerability of the structure, the probability of disasters (heavy rainfall scouring) should also be considered in the failure assessment of the structure. Considering that the structure reaches a certain failure state under the jth scouring depth, the probability of exceeding is:

$$P_b = \int_0^{+\infty} F_k(PGA, h_j) f(PGA, h_j) dPGA \tag{11}$$

Considering that the probability of structure failure beyond the continuous scouring depth is:

$$P_b = \int_0^{+\infty} \int_0^{+\infty} F_k(PGA, h) f(PGA, h) dPGA dh \tag{12}$$

Where $F_k(PGA, h)$ is the vulnerability surface function under the combined action of flood scouring and earthquake; $f(PGA, h)$ is the probability density function under the combined action of flood scouring and earthquake, where $f(PGA, h) = f_s(PGA) f_H(h)$.

The transcendental probability of structural failure is further expressed as:

$$P_b = \int_0^{+\infty} \int_0^{+\infty} F_k(PGA, h) f_s(PGA) f_H(h) dPGA dh \tag{13}$$

The two-dimensional vulnerability surface under the combined action of heavy rainfall, flood scouring and earthquake is obtained from multiple seismic vulnerability curves under different scouring depths using the linear interpolation method. Since the vulnerability surface is not easy to solve, the discrete probability solution (13) can be used:

$$P_b = \sum_{i=0}^{+\infty} \sum_{i=0}^{+\infty} P(PGA_i, h_j) P(PGA_i) P(h_j) \tag{14}$$

Where $P(PGA_i, h_j)$ refers to the probability of exceeding conditions corresponding to and below the vulnerable surface; $P(PGA_i) P(h_j)$ refers to the disaster probability density function under corresponding intensity.

3.3 Failure risk assessment

The failure probability of each failure state is the probability of the occurrence of each failure state risk, and the loss ratio C of the structure is introduced as the loss caused after the risk occurs, then the cumulative failure risk probability of the structure (Yin 1996):

$$P_f = \sum_{i=1}^{5} C_i P_{f_i} \tag{15}$$

Where C_i is the loss ratio of the structure in different damage states. According to the research results of Yin Zhiqian et al., the loss ratio of Fujian earth building structure in this paper is 0 in the intact state, 5% in slight damage, 10% in medium damage, 40% in severe damage and 60% in collapse (Yin 2020).

P_{f_i} (i = 1, 2, 3, 4, 5) correspond to the failure probability under the five states of structural integrity, slight damage, medium damage, severe damage and collapse respectively.

Finally, the failure probability of the structure under all performance states can be obtained from Equation (15). The structural reliability index can be obtained from the relationship between the failure probability and the reliability index in the reliability theory to check whether it meets the specified target reliability index, so as to evaluate the structure (Mohurd 2018).

4 APPLICATION STUDIES

4.1 Fujian earth buildings structure model

This paper takes the three-story square earth building structure as the research object. The structure has a floor height of 3.3 m, a length of 15.4 m, a single-side section width of 5.6 m, and a corridor width of 1.2 m. The single side section is shown in Figure 1. Consider the door and window openings, green tile wood roofs and floors, among which, the earth building walls are compacted with clay, and the strength simulation is plain concrete with the strength of C40. The structural foundation is a stone foundation, and the foundation wall is built with stones and mortar. The soil spring layer is used to simulate the scouring depth of a strong rainfall flood. A layer of spring is arranged every 0.1 m depth, and the scouring is simulated by removing the soil spring layer above the scouring surface. The seismic fortification intensity is 7 degrees. For Class II site, the seismic grade is Grade III, the design earthquake group is Group III, the site characteristic period is 0.45 s, and the structural safety grade is Grade II. The multi-hazard risk assessment of a square earth building structure under the combined action of strong rainfall, flood scouring and earthquake is carried out.

4.2 Calculation results and analysis

4.2.1 Performance index of Fujian earth building structure

Before the vulnerability analysis of the Fujian earth building structure, it is necessary to define the damage state of the Fujian earth building structure and quantify the indicators of each damage state. In this paper, the maximum inter-story displacement is used as the performance indicator of the structure. According to the recommendations of the reference (Gao 2019; Su 2013), the damage indicators of the Fujian earth building structure in this paper are shown in Table 1.

4.2.2 Vulnerability analysis under coupling action of earthquake and strong rainfall scouring

The measured ground motions of 15 Class II sites are selected to analyze the vulnerability of the structure, and the peak seismic acceleration range is 0.01 g~0.5 g. Six kinds of scouring depths (0.1 m, 0.2 m, 0.3 m, 0.4 m, 0.5 m, 0.6 m) need to be considered in the calculation process, and a total of 90 working conditions need to be calculated. The nonlinear dynamic time history analysis of the structure is carried out.

Due to space limitation, according to the seismic vulnerability curves of the above six kinds of strong rainfall flood scour depths, the vulnerability surface of Fujian earth building structure under severe damage conditions under different scour depths and PGA conditions is obtained through MATLAB linear interpolation, as shown in Figure 2.

It can be seen from Figure 2 that with the increase of the scouring depth of heavy rainfall, the seismic vulnerability of this Fujian earth building structure gradually increases, which has a significant impact on serious damage; In addition, when the scour depth reaches or exceeds the critical value (0.2 m~0.3 m), the impact of strong rainfall scour depth on the seismic vulnerability of severely damaged states increases rapidly. Therefore, the impact of heavy rainfall and flood scouring on the performance of Fujian earth buildings cannot be ignored.

In addition, considering the earthquake risk and heavy rainfall flood scouring risk model, the failure transcendental probability of each failure state of the Fujian earth building in the

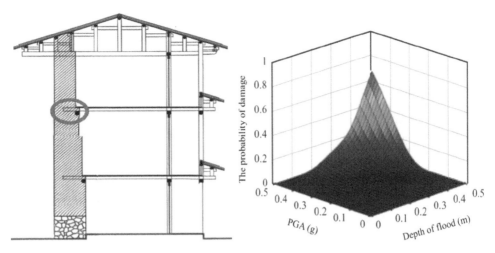

Figure 1. Cross section of earth buildings.   Figure 2. Vulnerability surface.

50-year design reference period is calculated according to Equations (14) and (15), and the failure probability of each performance is obtained by subtracting. The results are shown in Table 2.

It can be seen from Table 2 that in the 50-year design reference period, the failure probability of the structure under the four failure states decreases in turn; the probability of minor damage to the structure is 1.462%, the probability of moderate damage is 1.101%, the probability of severe damage is 0.082%, and the probability of collapse is 0.0056%. The cumulative failure probability of the structure is 0.219%, and the reliability index is 3.312, which is greater than the target reliability index of 3.2. Therefore, Fujian earth building structure meets the reliability requirements.

Table 1. Damage index of Fujian earth building structure.

Destructive state	Intact	Minor damage	Moderate damage	Serious damage	Collapse
Inter storey displacement angle limit (θ)	θ ≤ 1/921	1/921 < θ ≤ 1/424	1/424 < θ ≤ 1/124	1/124 < θ ≤ 1/58	θ > 1/58

Table 2. Comparison of failure probability values.

Destructive state	Minor damage	Moderate damage	Serious damage	Collapse	Cumulative
Coupling action	1.462%	1.101%	0.082%	0.0056%	0.219%
earthquake action	1.265%	0.916%	0.062%	0.0039%	0.182%
growth rate	15.6%	20.2%	32.1%	43.5%	16.9%

5 CONCLUSION

A failure risk assessment method for Fujian earth building structure under the combined action of strong rainfall, flood scouring and earthquake is proposed. Research showed that considering the flood scouring and earthquake coupling effects of strong rainfall, compared with considering only the earthquake effect, the failure probability of Fujian earth-building structures in each failure state has been greatly improved, and with the aggravation of the damage, the impact of scouring on the structural failure probability increases. Therefore, the impact of strong rainfall and flood scouring on the seismic performance of structures cannot be ignored.

ACKNOWLEDGMENTS

This work was financially supported by the Natural Scientific Research Foundation of Fujian Province (Grant No. 2021J01856) and the Fujian University Student Science and Technology Innovation Project (Grant No. 2021100390063).

REFERENCES

Gao, D.F., Zhang H., Wu, D.Y. (2019). Study on the Seismic Vulnerability of Chinese Multi Story Wood Structure Ancient Buildings Based on Multiple Seismic Intensity Parameters. *Earthquake Engineering and Engineering Vibration*. 39 (3), 41–52

Gao, X., Bao, Aibin. (1986). Determination of Seismic Fortification Standard by Probability Method. *Journal of Building Structure*. 7 (2), 55–63.

Li, H.N., Zheng, X.H., Li, C. (2019). Research Progress in Life Cycle Performance Analysis and Design Theory of High-performance Structures Against Multiple Disasters. *Journal of Building Structure*. 40 (2), 60–73.

Lv, G.H., Li, G., Li, H.N. (2020). Failure Risk Assessment of Masonry Structures Under Combined Action of Flood Scouring and Earthquake. *Journal of Disaster Prevention and Reduction Engineering*, 11, 1–9.

Mohurd. (2018). *Unified Standard for Reliability Design of Engineering Structures: GB 50068–2018*. Beijing: China Architecture Press.

Petrini F, Palmeri A. (2012). Performance-based Design of Bridge Structures Subjected to Multiple Hazards: A Review [C]/Biondini F, Frangopold M. Proceedings of the Sixth International Conference on Bridge Maintenance, *Safety and Management*, IABMAS 2012. London: Taylor & Francis Group. 12, 2040–2047.

Qiu, X., Zhao, H., You, R.Z, Zhuo, M.J., Lin, P.F. (2019). Investigation on Seismic Performance of Masonry Structures in Rural Areas of Southern Fujian [J]. *Sichuan Building Materials*. 45,81–84.

Su, Q.W., Xu, H., Wu, H., et al. (2013). Discussion on Storey Displacement Angle of Brick Masonry Structure. *Journal of Civil Engineering*. S1, 111–116.

Yi, L., Botao Yin, et al. (2019). Wind-rain Erosion of Fujian Tulou Hakka Earth Buildings. *Sustainable Cities and Society*. 50, 101666.

Yi, L., Meng, Q.Y., et al. (2020). Degradation of Rammed Earth Under Wind-driven Rain: The case of Fujian Tulou, China. *Construction and Building Materials*. 261, 119989.

Yin, Z.Q. (1996). *Prediction Method of Earthquake Disasters and Losses*. Beijing: Seismological Publishing House. 225–226.

Zhou, W.Q, Wang, S.B. (2015). Investigation and Vulnerability Analysis of Group Buildings in Southern Fujian. *Journal of Fuzhou University (Natural Science Edition)*. 43, 123–127.

Structural response prediction based on blind Kriging model

Zhixue Li & Jieshan Liu
CCCC Fourth Harbor Engineering Institute Co., Ltd, Guangzhou, Guangdong, China

Hejie Gao
CCCC Guanglian Expressway Investment Development Co., Ltd, Qingyuan, Guangdong, China

Yangwen Chen, Chunyi Yu, Wenrui Zhang & Zheng Yang*
South China University of Technology, Guangzhou, Guangdong, China

ABSTRACT: When conducting finite element numerical analysis on complex structures, huge calculations are always required, and a complex implicit relationship between structural output and structural input is inevitable, which brings challenges to structural performance evaluation and safety management. At present, the Kriging model, as a kind of data-driven proxy model, has shown an excellent ability for structural response prediction. However, when the Kriging model is used to predict the structural response, the standard Kriging model is generally adopted by the existing methods, that is, the basic function of the regression term is set as the standard form in advance without distinction. This will lead to some unnecessary computations, and in many cases, it will have an adverse effect. Given this, this study proposes a novel structural response prediction method based on the Blind Kriging model. It is used to replace the traditional Kriging model, by which the important factor variables are automatically selected and the non-important factor variables are discarded, and thereby the efficiency of predicting structural response can be greatly improved. Finally, a case study has been conducted to verify the effectiveness of the proposed method is higher and the prediction error is about 1/3 of the standard method. The effectiveness and superiority of the proposed method are verified by comparison with the other two traditional Kriging models. The proposed method can further promote the development and application of the Kriging model in proxy prediction of structural response.

1 INTRODUCTION

Proxy model methods, which essentially belong to the category of machine learning, i.e. machine learning of proxy models by training samples first, establish a one-to-one mapping relationship between excitation and response to achieve the purpose of replacing the numerical model with a proxy model, thus significantly improving computational efficiency. Depending on the agent model, the following agent model methods have been used for structural response prediction: 1) polynomial response surface method (Ren & Chen 2008, 2010; Zhang et al. 2015); 2) neural network methods (Han et al. 2011; He et al. 2008; Luo et al. 2021; Xu & Chen 2021; Zheng et al. 2021); 3) support vector machine methods (Jin & Yuan 2007; Li & Lu 2009; Yi et al. 2021); and 4) Kriging methods (Han 2016; Jia et al. 2013; Kaymaz 2005; Lin et al. 2021; Sacks et al. 1989; Sakata et al. 2003; Romero et al. 2004; Xiao et al. 2021). Among them, the polynomial response surface method is not suitable for

---

*Corresponding Author: ctyz@scut.edu.cn

problems with a high degree of non-linearity and its form is also more stereotypical; the neural network method is now widely used (especially with the emergence of deep learning networks), but its training sample requirements are high, while the computational efficiency and robustness aspects need to be continuously explored and researched; the support vector machine method can overcome the structure selection difficulties of neural networks, local extrema and other problems. The support vector machine method can overcome the problems of difficult structure selection and local extremes of neural networks but is easily limited by the dimensionality of the variable parameters, and its effect cannot be guaranteed if the variable parameters are large. The Kriging method is an optimal linear spatial estimation and interpolation method, proposed by Krige in 1951 (1951) and gradually evolved into a mathematical model in later years. Kriging is a linear regression technique that aims to minimize the estimate of the fitted covariance model or variance function. It uses the relationship between spatial and Euclidean distances (and assigns weights) to estimate the unsampled data values. The Kriging algorithm is described as the Best Linear Unbiased Estimator (BLUE) or Best Linear Unbiased Predictor (BLUP). Compared to the other methods mentioned above, Kriging can achieve better approximation accuracy and can handle simple or complex, linear or non-linear, and low-dimensional or high-dimensional problems. Secondly, Kriging can predict the uncertainty of unknown points and its basis functions usually have adjustable parameters, and Kriging can guarantee the smoothness of the fitted model with high computational efficiency and accuracy.

The existing Kriging-based structural response prediction methods are basically based on the traditional standard Kriging model, and no corresponding improvements have been made to it according to the actual engineering situation, mainly in the following two aspects: 1) Firstly, in the Kriging model, the determination of the regression basis function depends on the complexity of the actual engineering problem, and the existing methods basically adopt the standard Kriging 2) Secondly, in real structures (e.g. large span bridges), there are many structural parameters, some of which have a large effect on the response (i.e. high sensitivity) and some of which have a small effect on the response (some are even negligible). In the case of Kriging models, it is not the case that the more complex the regression model is (e.g. considering all parameters), the better the predictive performance of the Kriging model will be. This depends mainly on the importance of the selected factor variables (i.e. the action terms related to the structural parameters, including cross terms), and those non-important factor variables may in turn affect the accuracy and pull down the computational efficiency.

In summary, this study concludes that it is inappropriate to set the basis functions in the regression terms of the traditional standard Kriging model to be known, but that they should be optimized and confirmed by primary and secondary screening techniques, and therefore a blind Kriging model is used. It can automatically select important variables and discard non-important variables, which can significantly improve the efficiency of structural response prediction.

2 GENERAL INTRODUCTION TO THE STANDARD KRIGING MODEL

The general expression for the Kerriging model is (Kaymaz 2005; Sacks et al. 1989):

$$Y(\boldsymbol{x}) = f(\boldsymbol{x})'\boldsymbol{\mu} + Z(\boldsymbol{x}) = [f_1(x) \cdots f_p(x)]\boldsymbol{\mu} + Z(\boldsymbol{x}) \qquad (1)$$

where $f(\boldsymbol{x})$, $\boldsymbol{\mu}$ and p are the regression polynomial function, the regression coefficients, and the number of terms of the regression function, respectively. $Z(\boldsymbol{x})$ is a stochastic process that follows a normal distribution $N(0, \sigma_z^2)$, and the covariance is not zero. The covariance matrix of $Z(\boldsymbol{x})$ is as follows:

$$E[Z(\boldsymbol{x}_1)Z(\boldsymbol{x}_2)] = \sigma_z^2 R(\theta, \boldsymbol{x}_1, \boldsymbol{x}_2) \qquad (2)$$

With a training sample set $x = [x_1, \cdots x_m]$, $x_i \in \Re^n$ (m is the sample points, n is the number of parameters) and its response set $y = [y_1, \cdots, y_m]$, $y_i = g(x_i) \in \Re$, the predicted value, and root mean squared error (MSE) of any one prediction point x_{new} is:

$$\hat{y} = f'\hat{\mu} + r'R^{-1}(y - F\hat{\mu}) \tag{3}$$

$$\hat{\sigma}^2 = \hat{\sigma}_z^2\left(1 + v'(F'R^{-1}F)^{-1}v - r'R^{-1}r\right) \tag{4}$$

where $F = (F_{ij})_{m \times p}$ is the polynomial function matrix, $F_{ij} = f_j(x_i)$, $i = 1, \cdots, m; j = 1, \cdots, p$; $R = (R_{ij})_{m \times m}$ is the correlation function matrix between the training samples, $R_{ij} = R(\theta, x_i, x_j) i, j = 1, \cdots, m$; $r = r(x_{new}) = [R(\theta, x_1, x_{new}), \cdots, R(\theta, x_m, x_{new})]$ is the correlation matrix between the predicted sample points and the training sample points. $v = F'R^{-1}r - f$. $\hat{\mu}$ and $\hat{\sigma}_z^2$ are the estimates of μ and σ_z^2, obtained by generalized least squares.

$$\hat{\mu} = (F'R^{-1}F)^{-1}F'R^{-1}Y \tag{5}$$

$$\hat{\sigma}_z^2 = \frac{1}{m}(Y - F\hat{\mu})'R^{-1}(Y - F\hat{\mu}) \tag{6}$$

The determination of the parameters $\hat{\mu}$ and $\hat{\sigma}_z^2$ depends on the value of θ. Therefore, the optimal value of θ needs to be determined first. The maximum θ-value of the following logarithmic likelihood function can generally be obtained using the maximum likelihood estimation method.

$$\theta = \arg\min_{\theta}\left\{|R|^{\frac{1}{m}}\hat{\sigma}_z^2\right\} \tag{7}$$

3 AN IMPROVED KRIGING MODEL: THE BLIND KRIGING MODEL

The previous equation (1) corresponds to standard Kriging, where the basis function $f(x)'$ is known, i.e. predetermined, in the regression model. Here the standard Kriging is adapted so that the basis functions $f(x)'$ are no longer assumed to be known. Instead, they will be automatically identified by some data analysis. Since here the basis functions $f(x)'$ are not known in advance, they are referred to as blind Kriging. The blind Kriging model is written as:

$$Y(x) = f_k(x)'\mu_k + Z(x) = \sum_{i=0}^{k} f_k(x)\mu_i + Z(x) \quad Z(x) \sim N(0, \sigma_k^2 r) \tag{8}$$

Where $f_k(x)' = [f_1(x), \cdots, f_k(x)]$, $\mu_k = [\mu_0, \cdots, \mu_k]$, and k s unknown. At this point, the Kriging forecast can be expressed as:

$$\hat{y}(x) = f_k(x)'\hat{\mu}_k + r'R^{-1}(y - F_k\hat{\mu}_k) \tag{9}$$

Where, $F_k = (f_k(x_1), \cdots, f_k(x_m))'$, and $\hat{\mu}_k = (F_k'R^{-1}F_k)^{-1}F_k'R^{-1}y$.

3.1 Determination of $f_k(x)$

The most important step in a blind Kriging model is to determine $f_k(x)$, i.e., to perform variable selection. In complex structures (e.g. bridges), this problem will be particularly complex (because of the huge number of candidate parameters), and searching for all these variables to obtain the best model is almost impossible. A Bayesian variable selection algorithm to select the important factor variables (hereafter referred to as variables) is a reasonable approach. Candidate variables include primary terms, secondary terms, and two-factor crossover terms. The two-factor crossover terms include primary-primary, primary-quadratic, and secondary-quadratic crossover terms, and the total number of terms is $2n^2$ (n is the number of variables). The Bayesian variable selection technique is illustrated using the low-order term as an example. First, the variables are regularized to [1.0, 3.0] (other intervals are of course possible) to obtain the following orthogonal polynomial basis:

$$\begin{cases} x_j^l = \sqrt{\frac{3}{2}}(x_j - 2) \\ x_j^q = \frac{\sqrt{2}}{2}\left[3(x_j - 2)^2 - 2\right] \end{cases} \quad j = 1, 2, \ldots n \tag{10}$$

When x_j takes values 1, 2, and 3, the primary term x_j^l and the second term x_j^q will have the same length, which is $\sqrt{3}$. Therefore, these variables can be used to define the two-factor crossover term instead. For example, the primary-quadratic crossover term between x_1 and x_2 can be defined as $x_1^l x_2^q$.

Assume that $u_1, \ldots, u_t$ is a candidate variable and consider a simplified version of the model in equation (9): $y(x) = \sum_{i=0}^{k} \mu_i v_i + \sum_{i=0}^{t} \beta_i u_i$, where $u_0 = 1$, $v_1, \ldots, v_k \in \{u_1, \ldots, u_t\}$ is the important variable responsible for explaining most of the variation in the response. Now take the example of the two-factor x_1, x_2. At this point $y(x)$ can be expressed as: $y(x) = \sum_{i=0}^{k} \mu_i v_i + \sum_{i=0}^{8} \beta_i u_i$, where $u_0 = 1$, $u_1 = x_1^l$, $u_2 = x_1^q$, $u_3 = x_2^l$, $u_4 = x_2^q$, $u_5 = x_1^l x_2^l$, $u_6 = x_1^l x_2^q$, $u_7 = x_1^q x_2^l$, $u_8 = x_1^q x_2^q$. A dense valuation of all β_i is not possible when $t > m - 1$. However, this can be solved using a Bayesian updating method, and to do so, a prior distribution of $\beta = (\beta_0, \beta_1, \ldots \beta_t)$ needs to be assumed:

$$\boldsymbol{\beta} \sim \mathcal{N}(0, \tau_k^2 \psi) \rightarrow \begin{cases} \beta_0 \sim \mathcal{N}(0, \tau_k^2) \\ \beta_1 \sim \mathcal{N}(0, \tau_k^2 \psi_1^l) \\ \beta_2 \sim \mathcal{N}(0, \tau_k^2 \psi_1^q) \\ \vdots \\ \beta_t \sim \mathcal{N}(0, \tau_k^2 \psi_1^q \psi_2^q \cdots \psi_p^q) \end{cases} \tag{11}$$

where $\tau_k^2 \psi$ is the a priori covariance matrix, and ψ is the diagonal matrix of $(t+1) \times (t+1)$. ψ can be determined as follows: It is assumed that the correlation function in the Kriging model is a product-correlation structure, given by $r(h) = \prod_{j=1}^{p} r_j(h_j)$. $l_{ij} = 1$ if β_i contains the primary effects of the factor j and $l_{ij} = 0$ otherwise. Similarly, β_i contains a quadratic effect of the factor j, $q_{ij} = 1$, otherwise $q_{ij} = 0$. Thus the ith diagonal element of ψ is $\prod_{j=1}^{p} \left(\psi_j^l\right)^{l_{ij}} \left(\psi_j^q\right)^{q_{ij}}$, where $\psi_j^l = \frac{3 - 3r_j(2)}{3 + 4r_j(1) + 2r_j(2)}$, $\psi_j^q = \frac{3 - 4r_j(1) + r_j(2)}{3 + 4r_j(1) + 2r_j(2)}$. Assume that $Z(x)$ in equation (8) follows a Gaussian process. Then, based on Bayesian theory, the posterior mean of β can be approximated as:

$$\hat{\boldsymbol{\beta}} = \frac{\tau_k^2}{\sigma_k^2} \psi \boldsymbol{U}' \boldsymbol{R}^{-1}(\boldsymbol{y} - \boldsymbol{F}_k \hat{\boldsymbol{\mu}}_k) \tag{12}$$

Where $\boldsymbol{U}$ is the model matrix, in the form of an orthogonal polynomial basis at level 3, for example, is:

$$\boldsymbol{U} = \begin{pmatrix} 1 & -\sqrt{3/2} & \sqrt{1/2} \\ 1 & 0 & -\sqrt{2} \\ 1 & \sqrt{3/2} & \sqrt{1/2} \end{pmatrix} \tag{13}$$

If a variable has a large absolute coefficient, it is considered important. Thus, at each step $k = 0, 1, 2 \ldots$, those variables with the largest $\widehat{\beta}_i$ can be selected as important. Without loss of generality, we can set $\tau_k^2/\sigma_k^2 = 1$ in equation (12) so that the calculation can be further simplified.

3.2 *The optimal value of k*

In this Bayesian forward variable selection strategy, there is still an important question to be addressed: when should one stop adding terms to the mean component? In other words, what is the optimal value of k? The difficulty in choosing k is that the Kriging prediction model interpolates the data to give a perfect fit (with a prediction error of zero in all cases) regardless of its value, which makes it impossible to use traditional model selection criteria (such as the C_p-statistic criterion) in regression analysis. Fortunately, however, it is possible to use the cross-validation method as a selection criterion for the best k-value. The basic idea of cross-validation is to divide the data into two groups, one for model training and the other for testing the predictive performance of the model. Here the leave-one-out method of the cross-validation method is used, which selects only one set of samples for validation and uses the rest for training the model. The details are as follows:

$\widehat{y}_{(i)}(x_i)$ is taken as the prediction after removing the i-th data point. Then, the leave-one-out method cross-validation error is defined as:

$$loo_i = y_i - \widehat{y}_{(i)}(x_i) \tag{14}$$

This gives a validation prediction error of:

$$CVPE(k) = \sqrt{\frac{1}{n}\sum_{i=1}^{n} loo_i^2} \tag{15}$$

Next to solve:

$$k = \arg\min_{k} CVPE(k) \tag{16}$$

The optimal value of k is obtained.

3.3 *Determination of undetermined parameters*

After the important variables are selected, the undetermined parameters (parameters $\boldsymbol{\theta}$, σ_k^2, $\widehat{\boldsymbol{\mu}}_k$) in the blind Kriging model are determined. First, it is assumed that the correlation function $r(x)$ is the gaussian kernel function:

$$r(x) = \exp\left(-\sum_{j=1}^{p} \theta_j x_j^2\right) \tag{17}$$

$\boldsymbol{\theta} = [\theta_1, \theta_2, \ldots \theta_p]$ is defined. Parameters $\boldsymbol{\theta}$, σ_k^2, $\widehat{\boldsymbol{\mu}}_k$ can be estimated by maximum likelihood. Because the model was chosen based on cross-validation criteria, it seems more appropriate to use the same criteria for estimation. However, many empirical studies show that maximum likelihood estimation is superior to estimation based on cross validation. On the assumption that $Z(x)$ follows the Gaussian process, the negative value of the logarithmic likelihood function can be expressed as:

$$NLH = \frac{m}{2}\log(2\pi) + \frac{m}{2}\log(\sigma_k^2) + \frac{1}{2}\log(|\boldsymbol{R}|) + \frac{1}{2\sigma_k^2}(y - \boldsymbol{F}_k\widehat{\boldsymbol{\mu}}_k)'\boldsymbol{R}^{-1}(y - \boldsymbol{F}_k\widehat{\boldsymbol{\mu}}_k) \tag{18}$$

For the moment, let's say θ is given. *NLH* is minimized to σ_k^2 and $\widehat{\mu}_k$, and we will get:

$$\widehat{\mu}_k = \left(F_k' R^{-1} F_k\right)^{-1} F_k' R^{-1} y \tag{19}$$

$$\sigma_k^2 = \frac{1}{m}(y - F_k \widehat{\mu}_k)' R^{-1} (y - F_k \widehat{\mu}_k) \tag{20}$$

At this time, the corresponding minimum value of *NLH* is:

$$NLH = \frac{m}{2}[1 + \log(2\pi)] + \frac{1}{2}\left[m\log(\sigma_k^2) + \log(|R|)\right] \tag{21}$$

Now, the case where θ is unknown is considered. It can also be estimated by minimizing *NLH* in Equation (21). However, minimization is not an easy task (finding a global minimum is not easy). θ can be estimated only when $k = 0$:

$$\widehat{\theta} = \arg\min_{\theta} \left[m\log(\sigma_0^2) + \log(|R|)\right] \tag{22}$$

The process of establishing the blind Kriging model can be summarized as follows: at the beginning, $k = 0$ is set, $\widehat{\mu}_0$, σ_0^2 and $\widehat{\theta}$ are firstly calculated through Eq. (19), (20) and (22), and $\widehat{\beta}$ is obtained through Eq. (13). The variable corresponding to the maximum value of $|\widehat{\beta}_i|$ is selected as the first important variable v_1. The above process when $k = 1, 2, \cdots$ is repeated until the validation prediction error *CVPE* in the leave one method is minimized.

3.4 The flow path of the proposed method

Step 1: The initial parameters involved in the structural response are determined as the output of the blind Kriging model.

Step 2: Latin hypercube experimental design method is used to obtain the training sample x of structural parameter variables. The number of training sample points is generally $N \times n$, N is 5–10, and n is the number of variables (here $n = 4$).

Step 3: The finite element numerical model was established to obtain the structural response corresponding to each group of parameter samples (static response data was used in this study), and the response was used as the input y of the blind Kriging model, thus obtaining the training sample set with one-to-one mapping relationship [x,y].

Step 4: According to the training sample set [x, y], the blind Kriging model is established.

1) The blind Kriging model framework is determined, in which the basis function $f(x)$ in its regression model is set as unknown (Equation (8)); The training sample set [x,y] is introduced into the model.
2) The iterative strategy and Bayesian variable selection algorithm are used to determine the important variables in the Kriging model. Firstly, $k = 0$ is set, and the initial model is established by calculating the undetermined parameters in the blind Kriging model through Equations (19), (20) and (22) (the main purpose is to conveniently determine the optimal value of $\widehat{\theta}$).
3) Then the subsequent iteration process is carried out, and k increases one by one. In each iteration step, the blind Kriging model is transformed into a model based on the variable representation of orthogonal polynomial basis, and the bayesian updating method was used to determine the undetermined coefficient $\widehat{\beta}_i$ of the candidate variable (Equation (12)), and the variable corresponding to the maximum value of $|\widehat{\beta}_i|$ was selected as the important variable. Meanwhile, undetermined coefficients $\widehat{\mu}_k$ and σ_k^2 of the model corresponding to each step are obtained according to Equations (19) and (20).

4) The validation prediction error $CVPE$ corresponding to each iteration step (i.e., each k) is calculated by using the method of leaving one in the cross-validation method (Equation (15)).
5) Steps 3) –4) when $k = 1, 2, \ldots$ are repeated until the validation prediction error $CVPE$ in the leave one method reaches the minimum (Equation (16)), and the optimal value k is obtained: k_{opt}.
6) According to $\hat{\mu}_{k_{opt}}$ and $\sigma^2_{k_{opt}}$ corresponding to final k_{opt} and $\hat{\theta}$ in (2), the final establishment of the blind Kriging model is completed.

Step 5: The blind Kriging prediction model is obtained for structural response prediction. The overall flow chart is shown in Figure 1.

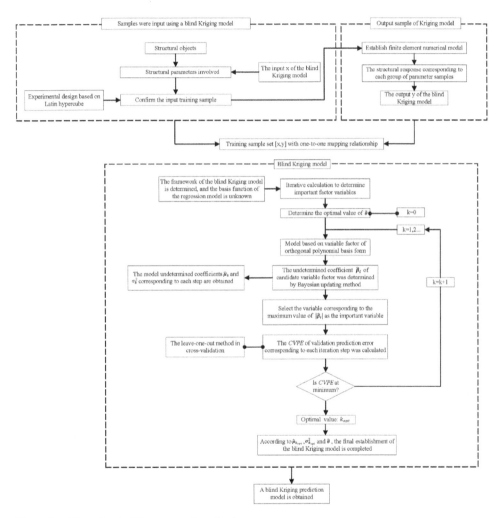

Figure 1. Flow chart of the proposed method.

4 AN EXAMPLE

The example object is the frame structure of three spans 20 floors in Figure 2. The initial parameters involved in the response are set as beam-column sectional area $A_1, A_2, A_3, A_4, A_5,$

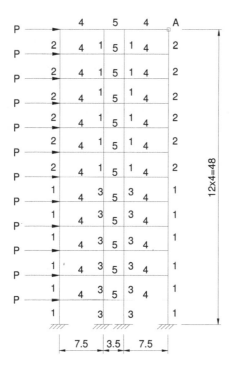

Figure 2. Three-span twelve-storey plane frame (unit: m).

and horizontal external load P. The relation between the bending moment of inertia and the section area of the beam-column section is $I_i = a_i A_i^2, i = 1, 2, 3, 4, 5$ (see Table 1 for the magnitude of a_i). The numerical statistics of the initial parameters involved in the response are shown in Table 1. Structural elastic modulus is an unrelated parameter, and its value is set as $E = 2.0 \times 10^7 \text{kPa}$. The structural response to be predicted is the horizontal displacement u_A of structural node A.

Table 1. Characteristics of parameter variables.

Parameters involved	Variable parameter	value	a_i
$A_1(\text{m}^2)$	x_1	0.25	0.08333
$A_2(\text{m}^2)$	x_2	0.16	0.08333
$A_3(\text{m}^2)$	x_3	0.36	0.08333
$A_4(\text{m}^2)$	x_4	0.2	0.2667
$A_5(\text{m}^2)$	x_5	0.15	0.2
P(kN)	x_6	30	/

$CVPE(0) = 1.4536$ is obtained by standard Kriging calculation (equation 16). By using the proposed method, four variable parameters x_4^I, x_5^I, x_6^I and $x_4^I x_6^I$ are successively selected through calculation, and the corresponding $CVPE(4) = 1.1208$, which is 29.6% higher than that of standard kriging. Finally, the kriging prediction model is as follows:

$$\widehat{u}_A = \widehat{y}(x) = -0.4151 - 0.02 x_4^I - 0.1105 x_5^I - 0.0176 x_6^I - 0.006 x_4^I x_6^I + r' R^{-1}(Y - F_4 \widehat{\mu}_4)$$

(23)

where $x_j^l = \sqrt{\frac{3}{2}}(x_{jl} - 2)$. To verify the effectiveness of the proposed method, 100 sets of data were added for testing, and RMSPE (Root Mean Squared Prediction Error) was used to evaluate the Prediction effect:

$$RMSPE = \sqrt{\frac{1}{n}\sum_{i=1}^{n}(y_i - \widehat{y}_i)^2} \quad (24)$$

The RMSPE results are shown in the following table:

Table 2. Comparison of RMSPE results.

Method	RMSPE
Standard Kriging method	0.9822
Van Krigen (second order multiple regression model) method	0.6633
The method proposed in this paper	0.3002

Based on the minimal RMSPE results in Table 2, it can be seen that the proposed method correctly selects important factor variables and the prediction model obtained is accurate. As can be seen from Table 2, compared with standard Kriging and Van Kriging, the proposed method achieves the best prediction performance (RMSPE = 0.3002). Standard Kriging only uses constant to describe the overall trend of the model, so it is impossible to achieve accurate prediction (RMSPE = 0.9822). However, due to the interference of a large number of non-important (or redundant) variables, the prediction performance of the Van Kriging model becomes poor (RMSPE = 0.6633), indicating that the more complex the regression model is, the better the prediction performance of the Kriging model will be, and the non-important factor variables will harm the prediction accuracy. Therefore, importance should be attached to the selection of important factor variables in Kriging modeling.

5 CONCLUSION

(1) In this study, a structural response prediction method based on the blind Kriging model is proposed, in which the blind Kriging model based on the Bayesian variable selection algorithm and cross-validation method is adopted. In this way, the important variables can be selected automatically and the non-important variables can be discarded, and the actual situation is effectively combined (the structural response often depends on some important factors, but not all of them), to achieve the slimming of the Kriging model and further ensure the accuracy and efficiency of the structural response. The proposed method is used to predict the structural response of multi-span high-rise structures successfully.
(2) The effectiveness and superiority of the proposed method are verified by comparison with the other two traditional Kriging models. The comparison results also show that the regression model setting in the Kriging model should not adopt a rigid standard form or complex form, but should make an automatic selection and trade-off according to the importance of factor variables, to give the optimal predictive value of the structural response. The proposed method can further promote the development and application of the Kriging model in proxy prediction of structural response.
(3) Kriging model can be used not only for structural response prediction (forward calculation problem) but also for structural parameter identification (inverse calculation problem). Therefore, future research will focus on the problem of structural parameter identification based on the blind Kriging model.

REFERENCES

Han F., Zhong D. and Wang J. 2011 Model Updating Based on Radial Basis Function Neural Network *J. Wuhan Univ. Sci. Tech.* **34(2)** 115–18.

Han Z. 2016 Kriging Surrogate Model and Its Application to Design Optimization A Review of Recent Progress *Acta Aeronautica et Astronautica Sinica.* **37(11)** 3197–225.

He H., Yan W. and Wang Z. 2008 Stepwise Model Updating Method Based on Substructures and GA–ANN *Engr. Mech.* **25(4)** 99–105.

Jia B., Yu X., Yan Q. and Chen Z. 2013 Bridge Seismic Reliability Analysis Based on Improved Kriging Response Surface Method *J. Vibration and Shock.* **32(16)** 82–7.

Jin W. and Yuan X. 2007 Response Surface Method Based on LS–SVM for Structural Reliability Analysis *J. Zhejiang Univ. (Engr. Sci.).* **41(1)** 44–7.

Kaymaz I. 2005 Application of Kriging Method to Structural Reliability Problems *Structural Safety.* **27(2)** 133–51.

Krige 1951 G. A. Statistical Approach to Some Mine Valuations and Allied Problems at the Witwatersrand *J. Chem. Metal.l Min. Soc.* **52(6)** 119–39.

Li H. and Lu Z. 2009 A Support Vector Machine Response Surface Method for Structural Reliability Analysis *Chinese Journal of Computational Mechanics.* **26(2)** 199–203.

Luo Y., Wang L., Liao F. and Liu J. 2021 Vibration–based Structural Damage Identification by 1–Dimensional Convolutional Neural Network *Earthq. Engr. & Engr. Dynamics.* **41(4)** 145–56.

Lin C., Ma Y., Xiao T. and Liu L. 2021 Multi–objective Efficient Global Optimization Method and Its Application Based on PLS Method and Kriging Model *Systems Engineering – Theory & Practice.* **41(7)** 1855–67.

Ren W. and Chen H. 2008 Response–surface Based on Finite Element Model Updating of Bridge Structure *China Civil Engr. J.* **41(12)** 73–8.

Ren W., Chen H. 2010 Finite Element Model Updating in Structural Dynamics by Using the Response Surface Method *Engineering Structures.* **32(8)** 2455–65.

Romero V. J., Swiler L. P., Giunta A 2004 A Construction of Response Surfaces Based on Progressive–lattice–sampling Experimental Designs with Application to Uncertainty Propagation *Structural Safety.* **26(2)** 201–19.

Sacks J., Schiller S. B., Welch W. J. 1989 Design for Computer Experiments *Technometrics.* **31(1)** 41–7.

Sakata S., Ashida F., Zako M. 2003 Structural Optimization Using Kriging Approximation *Computer Methods in Applied Mechanics & Engineering.* **192(7/8)** 923–39.

Xu Z and Chen J 2021 Neural Network Algorithm for Nonlinear Structural Seismic Response *Engr. Mech.* **38(9)** 133–45.

Xiao T., Ma Y. and Lin C. 2021 Ensemble Kriging Modeling Technique for Quality Design *Computer Integrated Manufacturing Systems.* **27(7)** 2023–34.

Yi W., Wang M., Tong J., Zhao S., Li J., Gui D. and Zhang X. 2021 Inhomogeneity Identification Method for Surrounding Rock of Large–section Rock Tunnel Face Based on Support Vector Machine *China Railway Science.* **42(5)** 112–22.

Zhang S., Gao F. and Zhang W. 2015 Bridge Finite Element Model Updating Based on Combination Function Response Surfaces *J. Chongqing Jiaotong Univ. (Nat. Sci.).* **34(5)** 18–24.

Zheng Q., Zhou G. and Liu D. 2021 Method of Modeling Temperature–displacement Correlation for Long–span Arch Bridges Based on Long Short–term Memory Neural Networks *Engineering Mechanics.* **38(4)** 68–79.

Research on the digital twin intelligent building management platform in communication industry existing office buildings

Weiwei Kou*
China Mobile Group Gansu Co. LTD, Lanzhou Gansu, China

Shuaihua Ye*
*Key Laboratory of Disaster Mitigation in Civil Engineering of Gansu Province,
Lanzhou University of Technology, Lanzhou, China*

Shigang Guo*
Gansu Engineering Design Research Institute Co. Ltd, Lanzhou Gansu, China

ABSTRACT: Aiming at the problems of difficult control and high energy consumption of existing office buildings in the communication industry, based on the integration and application of BIM, big data, artificial intelligence and other technologies and digital twins, this paper explores the construction theory and mode of intelligent building management platform based on the original various management platforms. Relying on the transformation project of an intelligent building management platform of an office building in the communication industry in Lanzhou, the research theory is applied in the actual existing buildings, and good results have been achieved. The application results show that: (1) according to the completed drawings and the actual situation of the site, the professional models and functional modules can be presented better by professional evaluation, data collation, clear standards and platform construction, etc, implementation building a smart building management platform; (2) It can realize the sharing and communication between the original management platform and the data of all kinds of sensors in the field and the model, achieve the effect of digital twinning intelligence management, and initially achieve the goals of energy saving and emission reduction and cost reduction and efficiency enhancement in the existing buildings. It is of certain reference significance to the construction of the following practical digital twinning intelligent building management platform project.

1 INTRODUCTION

Digital Twins are the digitization of the real world (Grieves M 2005, 2006). Since the concept was put forward, it has been progressively integrated with all walks of life. The concept of digital twin cities has gained wide attention from all walks of life in China, setting off a wave of research and construction. The typical characteristics of digital twin cities, such as global view, Precise mapping, simulation, virtual and real interaction, and intelligent intervention, are accelerating the development of cities. With the rapid development of BIM, Big Data, Artificial Intelligence, information, and communication technologies, combined with the practice of nearly 500 pilot cities in China, nowadays, society already could build digital twin cities (Qi 2018; M Sc Juhás P 2017; Li 2019). Fan Zheng (Fan 2021) et al. explored the deeper application scenarios of digital twin technology from the construction of digital platforms in urban operation centers. Li Yun (2021) et al. analyzed the development status

---

*Corresponding Authors: 13893424981@139.com, yeshuaihua@163.com and 164283167@qq.com

of Digital Twins related engineering standards at home and abroad and highlighted the achievements of standardization work in the field of Digital Twins in smart cities in China. Some scholars have studied Digital Twins solutions for older buildings. Li Gang (Li 2021) designed the application of digital twin technology in the operation, control and interaction of energy management in smart parks. Liu Yun (Liu 2020) et al. tried to reconstruct the existing office buildings with green wisdom.

Combined with the implementation of domestic digital twin cities, the new cities represented by Xiong'an New Area have the inherent advantages of building digital twin cities, while more cities in China are not new cities, and a large number of existing buildings already exist in cities. How to solve the construction problem of such digital twin cities is even more pressing. Some cities all over the country have sought the opportunity to try to carry out the planning and construction of digital twin cities based on communities. For example, Guiyang took the lead in proposing to build a digital twin city from the small urban ecosystems such as the Huaguoyuan Super – large Community Governance and Shubo Avenue. From the above examples, it can also be seen that from local to global is the proper way for the construction of digital twin cities in non-new cities. Buildings are the foundation of the city. Creating a digital twin intelligent building management platform for them is the basis for creating a city information model platform and realizing digital twin cities.

In summary, it is not difficult to see that existing studies on digital twin cities or buildings mainly focus on overall planning, technology research, standard formulation, new cities or existing community renovation, etc. While there are few studies on existing buildings, especially on existing buildings in the communication industry, namely data centers and their supporting office buildings. Based on the example of the digital twin intelligent building management platform transformation of an existing communication industry office building in Lanzhou, the research theory of this paper is applied to the actual existing building, which has achieved good results. And it also has certain reference significance for the subsequent construction of the actual digital twin intelligent building management platform project.

2 EXISTING BUILDING TECHNOLOGY IMPLEMENTATION PLAN IN THE COMMUNICATION INDUSTRY

The existing office buildings in the communication industry have the characteristics of a lot of equipment, pipelines, systems and monitoring points, complex system operation, high professionalism and safety requirements for external services(DB62/T3150-2018). In order to make the buildings in the communication industry operate, maintain and manage safely and reliably, it is necessary to determine the specific technical implementation plan and build a digital twin intelligent building management platform for existing buildings to control them, to improve management efficiency, share various information resources and reduce operating costs.

2.1 Determination of the implementation requirements

According to the pre-construction requirements and planning, various professional systems of the previous buildings have been running for some time. It is necessary to jointly evaluate the overall performance of the building with the original professional construction departments, post-maintenance departments, and renovation demand departments before the transformation. The specific contents are shown in Table 1.

According to the professional evaluation report or maintenance report, combined with the needs, planning content and investment situation of the renovation demand department, the implementation leading department should find out the existing problems during the operation and maintenance of the existing buildings and the direction to be upgraded, and determine the implementation demand plan.

Table 1. Table of contents of professional evaluation.

Evaluation content	Concrete contents	Remark
Structural evaluation	Main structure assessment according to requirements	Professional report
Non-structural evaluation	Assessment of components such as exterior walls, interior walls, parapet walls, curtain walls, ceilings, doors and windows, canopy, drains and equipment supports, as required	Professional or maintenance reports
Installation professional evaluation	This part is an important part. According to the needs and characteristics of the project, the lighting, air conditioning, fresh air, conference, power supply, environment (temperature, humidity, smoke, gas, water leakage, battery, UPS, distribution cabinet, oil machine, transformer, current, oil tank oil level), security (monitoring, alarm, access control) and other systems are evaluated	Professional or maintenance reports

2.2 *Data collection and collation*

According to the preliminary implementation of the needs of the program, the implementation lead department will first collect all kinds of engineering materials, mainly including as-built drawings and transformation drawings of various disciplines, other professional engineering system drawings and completion data. Secondly, after the above data collection and collation is completed, the implementation lead department should also cooperate with the original professional construction departments, post-maintenance departments and renovation demand departments to conduct full-scale on-site inspections on key parts and spot check on non-key parts in combination with the implementation demand plan, and revise the contents with differences. Finally, the leading implementation department will sort out, number, and save the final project data according to the standards.

2.3 *Determination of the implementation criteria*

According to the actual situation of each profession and each system in the final engineering data, the BIM model implementation standard and platform data standard in the implementation process are determined, and the most important one is the BIM model implementation standard. The implementation standard of the BIM model prescribes the fineness and color standard of the BIM model, which can ensure the creation of a unified and coordinated professional BIM model, and make the data of each stage of the BIM model accurate and complete, implement data sharing.

BIM model fineness: the model serves for each special task during the whole life of the construction project. The content and information requirements of the model elements are different for all kinds of special tasks and should be based on the requirements of different stages, develop different model fineness requirements and recommend local standards at the project site. In this paper, the application case is in the stage of operation and maintenance and adopts the model fineness of the operation and maintenance stage in the local standard of Gansu Province (DB62/T3150-2018). Other projects can choose the appropriate model fineness according to their stage and specific requirements.

BIM model color standard: To ensure the needs of model maintenance and data sharing, the color standard of the model is unified, in which the architecture and structure specialty are determined according to the completion map and the actual situation of the site, and the color standard of the installation of professional related components is implemented according to the relevant provisions of the local standard of Gansu Province (DB62/T3150-2018).

2.4 Determination of the data source

According to the final implementation requirements scheme, the data sharing scheme is determined by combining the data interface protocol type and sensor type of each system on site. The interface protocol of the system is preferentially used to realize data transmission and feedback. For the system that needs to be modified or added sensors, the sensor that can wirelessly transmit data is preferentially selected, so that the data can be transmitted to the data-sharing platform through the LAN laid in the building in the future. The purpose of multi-source heterogeneous data integration, exchange, and sharing between physical entities and virtual models is achieved, and the real-time interaction and iterative optimization of physical space and information space are realized.

2.5 Platform construction

Relying on the One City (Smart City Operation and Management Platform) base of our company's affiliated company, a digital twin smart building management platform for existing buildings is built. Firstly, the general IFC standard is used to connect the three-dimensional visualization model with the management platform. Secondly, the actual point information location of each system is located in the three-dimensional visualization interface of the operation and maintenance platform. Then, the data is converted through the monitoring or sensing equipment at the end of the point, and the data is uploaded to each subsystem of the data storage center of the data-sharing platform for data analysis. Finally, the data analyzed by each subsystem is uploaded to the integrated operation and maintenance platform through the API interface. Based on the analysis results, the instructions can be fed back to the end of each point to achieve digital twin intelligent management of existing buildings.

3 PROJECT EXAMPLE

3.1 General situation of office dispatching room project

A 42,000-square-meter office Gross leasable area in Anning District, Lanzhou, covers an area of about 13,000 square meters, main office scheduling room mainly, including office scheduling rooms, computer rooms, meeting rooms, equipment rooms, a total of about 375, about 750 sets of various types of the air-conditioning terminal, Elevator 4, network Point 1797, Fiber Point 527, telephone point 582.

4 ENGINEERING MODELING

4.1 Whole scene modeling

To quickly obtain the real model of the engineering example for the 3D display of the platform, this paper adopts the UAV to do the tilt photography. In this paper, the real scene modeling of the office scheduling room is carried out by using the Context Capture 3D scene modeling software of Bentley series software. The effect of real-scene modeling is detailed in Figure 1.

Figure 1. Scene modeling diagram.

It can be seen from the real scene modeling diagram that although the local area forming effect is not perfect, it can meet the requirements of the platform 3D display after the following lightweight processing.

4.2 *Office scheduling room modeling*

Because the construction time of the office dispatching room is earlier, BIM modeling has not been carried out in the early stage, the model is re-modeled according to the requirements of Lod600 in the operation stage according to the collected engineering data, as detailed in the Figures 2–6.

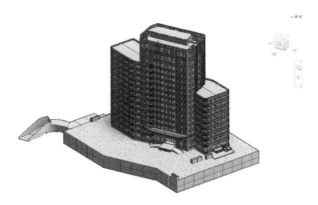

Figure 2. Models of buildings and structures.

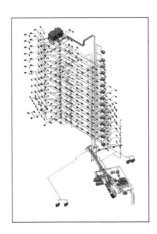

Figure 3. Air conditioning model.

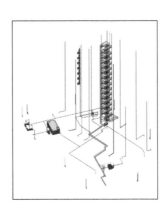

Figure 4. Water supply.

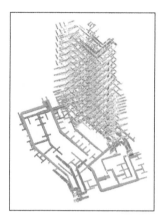

Figure 5. Ventilation model.

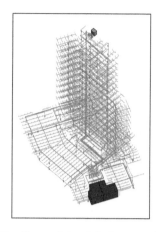

Figure 6. fire model and drainage model.

5 DATA COLLECTION

5.1 *The original system*

According to the actual situation of the existing buildings in the communication industry, to facilitate the management of various professional systems, most of the professional systems have built their management systems, which have data collection and remote control modules at the end of the equipment. However, the data of each system is not shared. The management platform can integrate, exchange, share and interact the data of each system to realize the intelligent operation of the existing buildings.

5.2 *Adding sensors*

According to the specific content of the implementation requirement plan, it is necessary to add sensors to the equipment without a management system and data collection module, to analyze the health and safety of building structures. So the related sensors for structural health monitoring are added. The common sensors are shown in Figure 7.

Figure 7. Common sensors.

6 SET UP THE PLATFORM

6.1 *The overall architecture*

Considering the need for long-term development, this paper establishes the whole solution framework with the park as the unit, and the construction content includes the management content of the digital twin intelligent building, for office scheduling room operation and maintenance characteristics and needs, focusing on data exchange systems, efficient operation, as detailed in Figure 8.

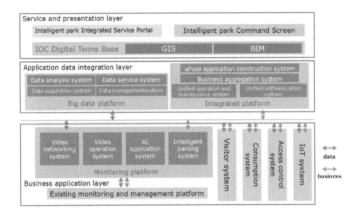

Figure 8. Office scheduling housing overall solution architecture.

6.2 *Platform function*

Through engineering modeling, system reconstruction, and platform construction, the digital twinning intelligent building management platform of the existing buildings in the communication industry has been completed in the management platform of the intelligent park of the communication industry, which mainly includes 5 functional modules, namely, comprehensive situation, security situation, personnel analysis, vehicle analysis and energy analysis, as detailed in Figures 9–13.

The comprehensive situation module can show information on building infrastructure (including pipelines), equipment operation, video data, OA office and other elements, combined with the sharing and interaction of real-time data such as personnel visitors, vehicle operation, video monitoring, energy consumption, etc., to assist the park management to master the real-time operation situation of the park.

The security posture module can support the access of video monitoring in the park, and it can predict, warn and prevent security events through big data and video AI analysis,

Figure 9. Integrated situation interface.

Figure 10. Security posture interface.

Figure 11. Personnel analysis interface.

Figure 12. Vehicle analysis interface.

Figure 13. Energy analysis interface.

including vehicle violations, personnel crossing the boundary, illegal invasion, fire alarm, perimeter alarm, etc. Special functions such as environmental alarm, APP usage statistics and alarm can also be added according to the specific needs of the park to ensure the safety of park personnel, assets, network and data.

The personnel analysis module can show the overall personnel situation of the park through the data provided by personnel, visitors, access control, time and attendance, elevator, monitoring and other systems, according to the relevant data model and data analysis established by the back-end, and realize the functions of personnel statistics, real-time display of security personnel on duty, stranger AI identification in specific areas, etc. It also realizes the functions of personnel gathering and mask-wearing identification in combination with the epidemic situation.

The vehicle analysis module in the collection of the park square, above-ground, and underground parking space data is based on the realization of parking data analysis, traffic flow analysis and other functions, to grasp the illegal parking, illegal, parking demand situation, the construction of an orderly parking environment, improve the management level, and reduce operating costs.

The energy analysis module is an important content of the platform, through real-time monitoring and collection of data of energy-using units such as air conditioning, fresh air, strong power, power, water supply and drainage, natural gas, various professional systems (such as conference, intelligent lighting, elevator, waste heat recovery and other systems) and server room area, energy consumption statistics and comparative analysis can be conducted by layer, by item, by year (quarter, month, week, day), etc., to diagnose energy-using problems. For example, the platform has realized the linkage between air quality and fresh air system, the linkage between personnel and vehicle dynamics and intelligent lighting system, the linkage between flooding system and alarm system, the linkage between the gas alarm system and gas valve, and the linkage between temperature and humidity of server room and air conditioning system, to achieve the goal of reducing cost and increasing efficiency and safe operation of energy consumption control in the park.

6.3 *Platform operation effect*

(1) The operation of this platform minimizes the number of operation and maintenance managers. The office dispatching rooms managed by this platform can reduce the number of owners and property management personnel by about 5 per year, saving labor costs by about 250,000 yuan per year.
(2) Through the platform's intelligent control of operation equipment, combined with the application of geothermal utilization, intelligent lighting and other measures, the office dispatching room can save energy consumption of about 1,904,761 kWh per year, and the total cost savings will be about 1.2 million yuan according to the average annual electricity price of electricity consumption classification.

7 CONCLUSION

In this paper, through the study of the digital twin intelligent building management platform for existing buildings in the communications industry, the following conclusions are drawn.

1) According to the completed drawings and the actual situation on the spot, we can present the professional models and functional modules and realize the construction of the intelligent building management platform through professional evaluation, data collation, clear standards (modeling, data collection) and the construction of the platform.
2) It can realize the sharing and interworking of data and models of all kinds of original management platforms and all kinds of sensors on the spot (replacement, addition and transformation), and achieve the effect of digital twinning intelligence management. The goal of energy-saving and emission-reducing, cost-saving and efficiency-increasing of the existing buildings has been preliminarily achieved, which is of certain reference significance for the subsequent construction of the actual digital twin intelligent building management platform project.

ACKNOWLEDGMENTS

This work was supported by the Science and technology planning project from the Department of Housing and Urban-Rural Development of Gansu Province "Digital Twin City Basic Research-Digital Twin Intelligent Building Management Platform"(JK2020-15).

REFERENCES

Fan, Z. (2021). Digital Platform Application of Urban Operation Center Based on Digital Twin. *Acta Electronica Sinica*, 50(10), 286–287.
Gansu Provincial Department of Housing and Urban-Rural Development, Gansu Provincial Bureau of Quality and Technical Supervision (2018) DB62/T 3150–2018, *Building Information Model (BIM) Application Standard*. Beijing: China Construction Industry Press.
Grieves, M. Product Lifecycle Management: The New Paradigm for Enterprises [J]. *International Journal of Product Development (S1477-9056)*, 2005, 2(1/2): 71.
Grieves, M. *Product Lifecycle Management - Driving The Next Generation of Lean Thinking* [M]. New York: McGraw-Hill Companies, 2006.

Kou, W.W., Chen, C.l., & Zhang, K. et al. (2017). *Comprehensive Application of BIM Technology in China Mobile (Gansu) Data Center Project*. Chinese Institute of Cartography. Proceedings of the Third National BIM Academic Conference, 155–160.

Kou, W.W., & Chen, C.L. (2018). Research on the Application of BIM Technology in the Whole Life Cycle of Construction Projects. *Construction Quality*, 36(9), 42–46.

Liu, E., & Pan, X. (2020). *Green Wisdom Renovation of Existing Office Buildings - Practice of Green Renovation of Lanzhou Jian Research Building*. 2020 International Green Building and Building Energy Efficiency Conference Proceedings, 292–295.

Li, X., Liu, X., & Wan, X.X. (2019) A review of digital twin applications and security development. *Journal of System Simulation*, 31(3), 385–392.

Li, Y. (2021). Exploration and Standardization of Digital Twin Technology Application in Smart Cities. *Information Technology & Standardization*, 10, 13–19.

Li, G., Wang, M., & Zuo, Z.B., et al. (2021). Exploring the Application of Digital Twin-based Energy Management System for Smart Parks. *Science and Technology & Innovation*, 18, 51–52.

M Sc Juhás P., M Sc Molnár K. Key Components of the Architecture of Cyber-Physical Manufacturing Systems [J]. *International Scientific Journal "Industry 4.0"* (S2643–8582), 2017, 2(5): 205–207.

Qi, Q, Tao F. Digital Twin and Big Data Towards Smart Manufacturing and Industry 4.0: 360 Degree Comparison [J]. *IEEE Access* (S2169-3536), 2018, 6(6): 3585–3593.

A simplified method for calculating fundamental frequency of concrete-filled double skin steel tubular structure for onshore wind turbine tower

Shou-Zhen Li
School of Civil Engineering, Chongqing University, Chongqing, China
Key Laboratory of New Technology for Construction of Cities in Mountain Area, Chongqing, China

Xuhong Zhou
Yangjiang Branch of Guangdong Advanced Energy Science and Technology Laboratory, Guangdong, China

Yuhang Wang* & Dan Gan
School of Civil Engineering, Chongqing University, Chongqing, China
Key Laboratory of New Technology for Construction of Cities in Mountain Area, Chongqing, China

Ronghua Zhu
Yangjiang Offshore Wind Energy Laboratory, Guangdong, China

Lu-Lin Ning
China Shipbuilding Industry Corporation Haizhuang Wind Power Co. Ltd., Chongqing, China

ABSTRACT: The fundamental frequency is the key parameter in determining the dynamic response of the wind turbine tower. By theoretical derivation, this study proposed a method for calculating the fundamental frequency of the concrete-filled double-skin steel tubular (CFDST) onshore wind turbine tower. Based on the theory of the Euler-Bernoulli beam, the transverse free vibration equation of the wind turbine tower was established, which was a partial differential equation with variable coefficients. The method of energy equivalence was employed to solve the mentioned equation analytically. The frequency equation was solved by the method of series expansion. The effectiveness of the proposed method was validated by comparing it with the numerical simulation. A parametric analysis was conducted to investigate the effect of design parameters on the fundamental frequency. It concludes that the proposed method is effective and accurate in calculating the fundamental frequency of the wind turbine tower in the preliminary design.

1 INTRODUCTION

With the advantages of low-cost and high efficiency, wind power energy has been widely applied across continents. As the support structure of the wind turbine, the structural reliability and construction cost of the wind turbine tower are directly related to the economic benefits of the power generation. Due to the increasing hub height and turbine sizes, it is necessary for the support structure, the wind turbine tower, to be strengthened to resist the increasing loads. The conventional tubular steel tower is facing challenges in supporting the wind turbine with greater generation capacity. The concrete-filled double-skin steel tubular (CFDST) wind turbine tower is a newly developed structural configuration. Relevant

---

*Corresponding Author: wangyuhang@cqu.edu.cn

researches indicate that this structure has wide application prospects for its advantages of rational load-bearing capacity, convenient transportation, and high construction efficiency (Han 2014; Wang 2021).

In the preliminary design of the wind turbine tower, it is necessary to evaluate the fundamental frequency and compare that with f_{1p} and f_{3p}. Herein, f_{1p} is the rotation frequency of the rotor, and f_{3p} refers to the blade passing frequency for a three-blade turbine. The design criterion is to keep the fundamental frequency far away from the periodic excitation (Harte 2007).

Currently, the main approaches to determine the fundamental frequency of the wind turbine tower are field measurement [Oliveira 2018] and finite element analysis (FEA) (Fitzgerald 2016). When it comes to theoretical analysis, relative theoretical research is quite rare. From the perspective of solving the transverse free vibration equation, this paper offers a method for analyzing the fundamental frequency of the CFDST onshore wind turbine tower. The numerical simulation conducted by FEA is used to compare with the proposed method to prove its effectiveness. Additionally, a parametric analysis was conducted to investigate the correlation between the fundamental frequency and different design parameters.

2 THEORETICAL DERIVATION

Due to the relatively large length-diameter ratio [Bernuzzi 2021], the shear effect of the wind turbine tower under the action of the transverse load is not obvious, the wind turbine tower can be regarded as an Euler-Bernoulli beam from the perspective of mechanical analysis.

The simplified configuration is shown in Figure 1, where D_{ob} and D_{ib} respectively denote the outer diameter of the outer and inner steel tube at the bottom cross-section; t_o and t_i respectively denote the thickness of the outer and inner steel tube; h denotes the height of the tower; θ denotes the tapered angle.

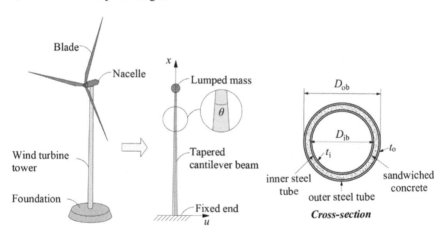

Figure 1. Simplified configuration of the onshore wind turbine tower.

The transverse free vibration equation is

$$M(x)\frac{\partial^2 u(x,t)}{\partial t^2} + \frac{\partial^2}{\partial x^2}\left[K(x)\frac{\partial^2 u(x,t)}{\partial x^2}\right] = 0 \qquad (1)$$

where $K(x)$ and $M(x)$ respectively denote the bending stiffness and the mass per unit length at any cross-section, $u(x,t)$ represents the transverse displacement of the wind turbine tower, which is the function that varies continuously with coordinate x and time variable t.

For onshore wind turbine towers, the foundation presents a huge reinforced concrete plinth with the adjunction of piles, resulting in the soil-structures interaction (SSI) is rarely considered in the design [Bernuzzi 2021]. Therefore, the tower can be regarded as being fixed at the bottom end. The boundary conditions at the bottom end are

$$u(0, t) = 0 \tag{2}$$

$$\frac{\partial u}{\partial x}(0, t) = 0 \tag{3}$$

The boundary conditions at the top end are

$$K(h)\frac{\partial^3 u}{\partial x^3}(h, t) = -\omega^2 u(h, t) m_{\text{wt}} \tag{4}$$

$$K(h)\frac{\partial^2 u}{\partial x^2}(h, t) = -\omega^2 \frac{\partial u}{\partial t}(h, t) J_{wt} \tag{5}$$

where m_{wt} denotes the mass of the wind turbine, and J_{wt} denotes the moment of inertia of the wind turbine about the x-axis or y-axis.

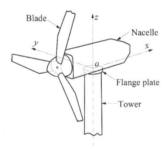

Figure 2. The rectangular coordinate system of the wind turbine.

Because the wind turbine tower usually has a variable cross-section, resulting in its transverse vibration equation, Eq. (1), becoming a partial differential equation with variable coefficients, which is hard to be solved analytically [Wang 2008]. By the method of energy equivalence, the wind turbine tower with a variable cross-section can be transformed into the one with a constant cross-section, resulting in the transverse vibration equation becoming a partial differential equation with constant coefficients.

Based on the principle that the bending strain energy and the kinetic energy are certain after the equivalence process, it has

$$K_e = \frac{\int_0^h K(x) \left[\frac{\partial \phi^2(x,t)}{\partial x^2}\right]^2 dx}{\int_0^h \left[\frac{\partial \phi^2(x,t)}{\partial x^2}\right]^2 dx} \tag{6}$$

$$M_e = \frac{\int_0^h M(x) \left[\frac{\partial \phi(x,t)}{\partial t}\right]^2 dx}{\int_0^h \left[\frac{\partial \phi(x,t)}{\partial t}\right]^2 dx} \tag{7}$$

where K_e and M_e denote the equivalent section bending stiffness and mass per unit length, respectively, and $\phi(x, t)$ denotes the deflection curve.

Replacing terms $K(x)$ and $M(x)$ in Eq. (1) with K_e and M_e, the transverse free vibration equation becomes

$$M_e \frac{\partial^2 u(x,t)}{\partial t^2} + K_e \frac{\partial^4 u(x,t)}{\partial x^4} = 0 \qquad (8)$$

By using the method of separation of variables [Clough 1993], $u(x,t)$ can be written as

$$u(x,t) = y(x)q(t) \qquad (9)$$

where $y(x)$ is the vibration mode function, and $q(t)$ is the temporally fluctuating amplitude.
After substituting Eq. (9) with Eq. (8), it becomes

$$y^{(4)}(x) - a^4 y(x) = 0 \qquad (10)$$

where

$$a^4 = \frac{\omega^2 M_e}{K_e} \qquad (11)$$

The complementary solution of Eq. (10) is

$$y(x) = A \sin ax + B \cos ax + C \sinh ax + D \cosh ax \qquad (12)$$

where the constants $A \sim D$ are determined by the boundary conditions of the wind turbine tower.

After substituting Eq. (12) with the boundary conditions Eqs. (2)-(5), simultaneous linear equations of two unknowns are obtained. The necessary and sufficient condition that the simultaneous linear equations have an untrivial solution is that the coefficient determinant of the simultaneous equations is zero

$$\begin{vmatrix} D_1 & D_2 \\ D_3 & D_4 \end{vmatrix} = 0 \qquad (13)$$

where

$$D_1 = -\cos ah - \cosh ah + m'ah(\sin ah - \sinh ah) \qquad (14)$$

$$D_2 = \sin ah - \sinh ah + m'ah(\cos ah - \cosh ah) \qquad (15)$$

$$D_3 = -\sin ah - \sinh ah + j'(ah)^3(\cos al - \cosh ah) \qquad (16)$$

$$D_4 = -\cos ah - \cosh ah + j'(ah)^3(-\sin al - \sinh al) \qquad (17)$$

$$m' = \frac{m_{wt}}{M_e h} \qquad (18)$$

After expanding the determinant, Eq (15), it becomes

$$\left[1 - m'j'(ah)^4\right] + \left[1 + m'j'(ah)^4\right]\cos ah \cdot \cosh ah \\ + \cos ah \cdot \sinh ah\left[j'(ah)^3 + m'(ah)\right] + \cosh ah \cdot \sin ah\left[j'(ah)^3 - m'(ah)\right] = 0 \qquad (19)$$

Eq. (19) is the frequency equation. With Taylor's formula, the terms cos (ah), cosh (ah), sin (ah), and sinh (ah) in Eq. (19) are expanded into the power series, where the first four terms are adopted. It has

$$2 + \left(-\frac{2}{3}m' + 2j' - \frac{1}{6}\right)\lambda + \left(-\frac{1}{6}m'j' - \frac{1}{15}j' + \frac{1}{315}m' + \frac{1}{2880}\right)\lambda^2$$
$$+ \left(\frac{1}{2880}m'j' + \frac{1}{30240}j' - \frac{1}{151200}m' - \frac{1}{518400}\right)\lambda^3 \qquad (20)$$
$$+ \left(-\frac{1}{518400}m'j' - \frac{1}{1814400}j'\right)\lambda^4 = 0$$

where

$$\lambda = (ah)^4 \qquad (21)$$

It is obvious that Eq. (20) is a quartic equation, and the specific solving process can be found in Ref. [Yacoub 2012].

After Substituting Eq. (21) into Eq. (11), it yields that

$$\omega = \sqrt{\frac{\lambda K_e}{M_e h^4}} \qquad (22)$$

3 RESULTS AND DISCUSSION

3.1 *Error analysis*

The numerical example is defined as follows: $D_{ob} = 4.4$ m, $D_{ib} = 3.0$ m, $t_o = 0.03$ m, $t_i = 0.03$ m, $h = 100$ m, $\tan\theta = 1/100$. The wind turbine, which was simulated by a coupling point, was defined with m_{wt} of 228473 kg, J_{wty} of 28592521.77 kg·m², and J_{wtx} of 50279832.33 kg·m². The steel material was defined using a density of 8500 kg/m³ and a modulus of 2.06×10^5 MPa. The purpose of defining the steel density as 8500 kg/m³ rather than the typical value of 7850 kg/m³ is to consider the effect of paint and connections on the density [Jonkman 2009]. The concrete was defined using a density of 2500 kg/m³ and a modulus of 37000MPa.

Two different deflection curves that satisfy the geometric boundary conditions of the cantilever beam are used to construct $\phi(x,t)$. They are

$$\phi(x,t) = q(t)\left(x^2 - \frac{x^4}{6h^2}\right) \qquad (23)$$

$$\phi(x,t) = q(t)\left(1 - \cos\frac{\pi x}{2h}\right) \qquad (24)$$

As shown in Figure 3, the FE model of the CFDST wind turbine tower was developed by using the software ABAQUS. The mass and moments of inertia of the wind turbine were applied through the reference point with a multi-point coupling constraint. Surface-to-surface contacts were defined to simulate the interactions between the outer/inner steel tube and the sandwiched concrete.

The comparison between the calculation results of the proposed method and the numerical simulation is shown in Figure 4. The maximum relative error is only 0.85%. It implies that the proposed method coincides well with the numerical simulation.

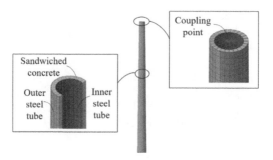

Figure 3. FE model of CFDST tower.

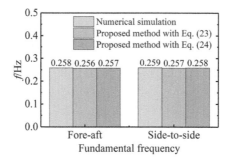

Figure 4. Comparison between the proposed method and numerical simulation.

3.2 *Parametric analysis*

Taking the numerical example introduced in section 3.1 to conduct the parametric analysis. To compare the variation of the fundamental frequency with different parameters in the same coordinate system, the following dimensionless parameters are introduced

$$D_1 = \frac{D_{ob}}{4.4 \text{ m}}, \quad D_2 = \frac{D_{ib}}{3.0 \text{ m}}, \quad H = \frac{h}{100 \text{ m}}, \quad T_1 = \frac{t_o}{0.03 \text{ m}}, \quad T_2 = \frac{t_i}{0.03 \text{ m}}, \quad K = \frac{\tan \theta_i}{1/100} \quad (25)$$

Figure 5 shows the variations of the fundamental frequency with the dimensionless parameters. It shows that the fundamental frequency increases with D_{ob}, D_{ib}, t_o, and t_i, but

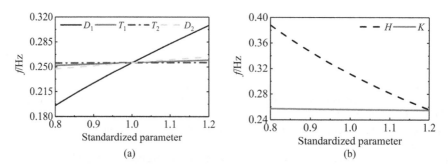

Figure 5. Variations of the fundamental frequency with the dimensionless parameters. (a) D_1, D_2, T_1, and T_2 (b) H and K.

decreases with the increase of h and $\tan\theta$. Besides, D_{ob} and h are the most sensitive parameters to the fundamental frequency.

4 CONCLUSION

In this paper, a formula for estimating the fundamental frequency of the onshore tapered wind turbine tower was established by theoretical derivation, and the following conclusions could be drawn:

(1) The comparison between the presented solution and the numerical solution shows that the proposed method is effective and accurate. Besides, compared with the numerical simulation, the proposed method is much more convenient to be conducted, which provides great convenience for analyzing the fundamental frequency of onshore wind turbine towers in preliminary design.
(2) The fundamental frequency increases with the parameters of D_{ob}, t_o, D_{ib}, and t_i. Other parameters including h and $\tan\theta$ have a negative correlation with the fundamental frequency. Among those parameters, D_{ob} and h are the most sensitive parameters to the fundamental frequency.

ACKNOWLEDGMENTS

The authors gratefully acknowledge the financial support provided by the National Natural Science Foundation of China (52278144) and Fundamental Research Funds for the Central Universities (2022CDJQY-009).

REFERENCES

Bernuzzi C., Crespi P, Montuori R (2021). Resonance of Steel Wind Turbines: *Problems and Solutions. Structures*. 32, 65–75.

Clough R.W., Penzien J (1993). *Dynamics of Structures*. 2nd Ed. New York: McGraw-Hill.

Fitzgerald B., Basu B. (2016). Structural Control of Wind Turbines with Soil Structure Interaction Included. *Engineering Structures*. 111, 131–151.

Han L.H., Li W., Bjorhovde R. (2014). Developments and Advanced Applications of Concrete-filled Steel Tubular (CFST) Structures: Members. *Journal of Constructional Steel Research*. 100, 211–228.

Harte R., Gideon P.A.G. Van Zijl (2007). Structural Stability of Concrete Wind Turbines and Solar Chimney Towers Exposed to Dynamic Wind Action. *Journal of Wind Engineering & Industrial Aerodynamics*. 95(9), 1079–1096.

Jonkman J., Butterfield S., Musial W., Scott G. (2009). Definition of a 5-MW Reference Wind Turbine for Offshore System Development Technical Report No NREL/TP-500-38060. Golden (CO): *National Renewable Energy Laboratory*.

Oliveira G., Magalhães F., Cunhá A., Caetano E. (2018). Continuous Dynamic Monitoring of an Onshore Wind Turbine. *Engineering Structures*. 164, 22–39.

Wang Y.L., Chao L. (2008). Using Reproducing Kernel for Solving a Class of Partial Differential Equations with Variable Coefficients. *Applied Mathematics and Mechanics*. 29(1), 129–137.

Yacoub M.D., Fraidenraich G. (2012). A Solution to the Quartic Equation. *The Mathematical Gazette*. 96 (536), 271–275.

Wang Y.H., Lyv X, Wang S.Q, Zhou X.H., Tan J.K. (2021). Research on Buckling Behavior of Steel Plate in Concrete-filled Double Skin Steel Tube Tower for Wind Turbine, *Special Structures*. 38(4), 24–30 (in Chinese).

Numerical study on GFRP-strengthened offshore T-joints under earthquake cyclic loading

Pingping Han
College of Pipeline and Civil Engineering, China University of Petroleum (East China), Qingdao, Shandong, China

Hong Lin*
College of Pipeline and Civil Engineering, China University of Petroleum (East China), Qingdao, Shandong, China
Center for Offshore Engineering and Safety Technology (COEST), China University of Petroleum (East China), Qingdao, Shandong, China

Alexander Moiseevish Uzdin
Department of Mechanics and Strength of Materials and Structures, Emperor Alexander I St. Petersburg State Transport University, Russia

Lei Yang
College of Science, China University of Petroleum (East China), Qingdao, Shandong, China

Haochen Luan, Chang Han, Hao Xu & Shuo Zhang
College of Pipeline and Civil Engineering, China University of Petroleum (East China), Qingdao, Shandong, China

ABSTRACT: Offshore structures located in high seismic risk marine are vulnerable to being damaged by earthquake cyclic loading. The glass fiber reinforced polymer (GFRP) is widely used in offshore structure reinforcement and repairment. This study aims to analyze the GFRP reinforcement effect on tubular T-joints of offshore platforms under earthquake cyclic loads through numerical studies based on FE software ANSYS. The static ultimate bearing capacity of the strengthened T-joints was compared to that of un-strengthened. Then, the deformation response under different seismic damage and the ultimate bearing capacity of joints strengthened with different GFRP sheets under 0.7 damage level were studied. Eventually, parametric analysis was conducted. The results demonstrated that both the number of layers and orientation of the GFRP sheets have a significant effect on the reinforcement under earthquake load.

1 INTRODUCTION

Thin-walled steel tubular T-joints (Lesani et al. 2013) with lightweight, high strength, and convenient installation has been extensively applied to offshore platforms. Since local buckling often occurs on the T-joint (Zhu et al. 2014), the structural performance of the T-joint is critical to the behavior of the whole structure. Particularly, if the T-joint was subjected to earthquake loads, the chord around the intersection of the joint will easily be damaged or even failed, and then the damage will reduce the ultimate bearing capacity of the

---

*Corresponding Author: linhong@upc.edu.cn

T-joints. Therefore, some measures have been taken to improve the ultimate bearing capacity of the T-joints, one of which is to reinforce the joint using glass fiber-reinforced polymer (GFRP).

In recent years, glass fiber-reinforced polymer (GFRP) composites attached to the surface have been widely used in strengthening and repairing various structures. As a structural reinforcement, GFRP composite (Zhou et al. 2022) is superior to a steel plate in terms of corrosion resistance, environmental durability, and stiffness-to-weight ratio. Research over the last few decades reveals (Hamzeh et al. 2020; Lesani et al. 2015; Safdar et al. 2022; Swamy & Sekar 2009; Valarinho et al. 2017; Wang et al. 2021) that the application of glass fiber composite materials for rehabilitation, retrofit and upgrade of variety of structures has proved to be of effective.

Nowadays, lots of numerical and experimental research have been done concerning GFRP in strengthening T joints. Lesani et al. (2013, 2014) performed numerical and experimental research on GFRP wrapped on tubular T/Y-joints. They reported that the ultimate load-carrying capacity of the strengthened joints increased by around 60% compared to un-strengthened joints, and the deflection was reduced by 50%. Kumar and Das (2021) conducted a detailed adhesion failure analysis of the non-planar T-joint based on ANSYS codes. Hossein et al. (2021) investigated the effects of CFRP and GFRP on the static load-bearing capacity of strengthened tubular T/Y-joints.

According to the published literature, studies on GFRP-strengthened tubular T-joints mainly focus on static loading conditions. However, the research concerning the effect of earthquake cyclic loading is rarely few. Therefore, it is of great significance to select an appropriate measure to strengthen the damaged T-joints after the earthquake.

This research aims to compare the ultimate load-bearing capacity of GFRP-strengthened and un-strengthened T-joints under earthquake cyclic loads using the finite element software ANSYS. The key parameters were studied in the numerical study, including the damage level, the number of GFRP sheets and the orientation of GFRP sheets. The effects of key parameters on the load-displacement response and Von-Mises stress response of GFRP-strengthened T-joints were investigated. It is critical to evaluate the load-bearing capacity of GFRP-strengthened tubular T-joints and their behavior under different earthquake damage levels.

2 METHODOLOGY

2.1 *Framework of ultimate bearing capacity analysis of T-joints after an earthquake*

The ultimate bearing capacity of T-joints is an important index reflecting the mechanical properties of structures. In this paper, the force-controlled quasi-static loading method was used to study the reinforcement effect of GFRP for seismic damaged T-joints by using the numerical simulation software ANSYS. Figure 1. outlines the procedures of the proposed method for calculation of ultimate bearing capacity and influence analysis of GFRP strengthened T-joints under seismic load, which includes seven steps listed as follows.

Step 1 is to establish the FE models of tubular T-joints and GFRP sheets, considering the actual working conditions of T-joints, and the layout of GFRP sheets;

Step 2 is to compare the ultimate bearing capacity of un-strengthened tubular T-joints with that of the GFRP-strengthened one without earthquake damage;

Step 3 is to simulate the actual working conditions, specifically, a compression dead load of 50% of the ultimate bearing capacity of T-joints was applied on the brace end along the axial direction;

Step 4 is to conduct a pseudo seismic simulation, according to the ultimate bearing capacity and the corresponding displacement obtained in step 2;

Step 5 is to study the deformation mode of the seismic damaged T-joints in the statics module, using the quasi-static analysis method;

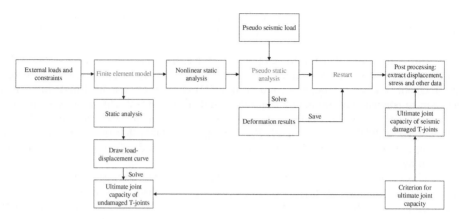

Figure 1. Flow chart.

Step 6 is to study the ultimate bearing capacity of seismically damaged T-joints. The restart analysis method was used;

Step 7 is to study the influence of different sheet parameters on the reinforcement of seismic damaged T-joints.

2.2 *Quasi-static analysis method of earthquake loading*

Here, the brace end of the T-joint was assumed subjected to pseudo-earthquake cyclic loading, and then Pushover analysis was carried out to study the damage caused by earthquake load. The damage level $\widetilde{D}$ was defined as follows (Xu et al. 2019):

$$\widetilde{D} = \frac{F_\mu - \lambda F_\mu}{F_\mu} \qquad (1)$$

where F_μ is the ultimate bearing capacity of T-joints without damage, λF_μ is the ultimate bearing capacity of T-joints considering earthquake damage. $\lambda = 1-\eta$, where η is the ultimate bearing capacity reduction coefficient considering residual deformation and stress, and η could be set between 0.2 and 0.7. The loading scheme is shown in Figure 2.

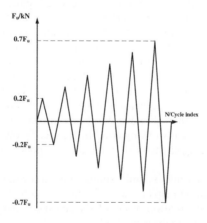

Figure 2. Loading scheme.

2.3 Definition of ultimate bearing capacity of T-joints

The ultimate bearing capacity is defined as the load-carrying capacity when the structure reaches the ultimate strength. The ultimate bearing capacity of tubular T-joint is defined (Choo et al. 2005; Vegte et al. 2005) as follows: when the load-displacement curve has a peak value before the displacement reaches 6% D (D is the outer diameter of the chord member), thus the peak value could be regarded as the ultimate bearing capacity of the T-joint; If the displacement exceeds 6% D and there is no peak yet, the load-carrying capacity at 6% D is taken as the ultimate bearing capacity of the tubular T-joint.

3 FE MODEL AND MATERIAL PROPERTIES

3.1 FE model of GFRP reinforced T-joint

In the current numerical investigations, the finite element (FE) analysis of the T-joint was conducted using the ANASY software package. The FE models of tubular T-joint and GFRP sheets were modeled using shell element 181. The welding between the chord and brace members was ignored.

Two FE models of T-joints were established including the un-strengthened specimen (T0) and GFRP-strengthened specimen (T10), respectively. The detailed geometric dimensions of the specimens are shown in Figure 3. $L = 1593$ mm and $l = 796.5$ mm are the lengths of the chord and brace, respectively; $D = 318.5$ mm and $d = 139.8$ mm are the outer diameters of the chord and brace, respectively; $T = 4.5$ mm and $t = 4.5$ mm are the wall thickness of chord and brace, respectively; and $L_C = 454$ mm and $L_B = 227$ mm are the lengths of GFRP along chord and brace, respectively.

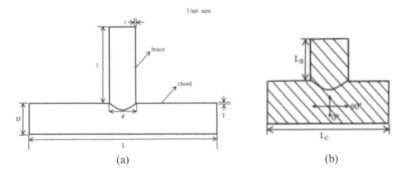

Figure 3. Geometry of the T-joint and GFRP. (a) Joint (b) GFRP.

A perfect bonded constraint was considered for the interface of GFRP and the steel substrate. The contact was simulated using Target 170 and Contact 174. Both ends of the chord were fixed. A quasi-static load was applied on the top of the vertical brace end, and the displacement in X and Z directions was constrained to prevent eccentric bending. The established FE model of T-joint and GFRP sheets were shown in Figure 4.

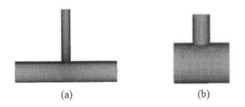

Figure 4. Finite element mesh diagram of T-joint and GFRP. (a) Joint (b) GFRP.

3.2 Material properties

In this study, the chord and brace of the T-joint were made of steel. The bilinear kinematic hardening plasticity (BKIN) model was used to define the stress-strain relation of steels in FE analysis. For steel, Young's modulus of 210 GPa with a Poisson's ratio of 0.3, and yield stress of 412 MPa and a Shear modulus of 1% of Young's modulus were used throughout the analyses.

In this study, the properties of the glass/epoxy composite used in the analyses were listed in Table 1. Here, subscripts "1" and "2" represents the longitudinal and transverse direction of the fiber, relatively.

3.3 Parametric setting

8 FE models were established to carry out the parametric study of the GFRP sheets. Here, the following parameters were considered including the number of GFRP sheets (N_{GFRP} = 0, 4, 8 and 12 layers) and the orientation of GFRP sheets (θ = 0/0/0/0, 0/90/0/90, 90/0/90/0, 90/90/90/90 and 0/45/0/45), and the detailed parameters were listed in Table 2.

Table 1. Mechanical properties of the CFRP sheets (Ganesh & Naik 1993).

E_1	38.6GPa	v_{13}	0.26
E_2	8.27GPa	G_{12}	4.14GPa
E_3	8.27GPa	G_{23}	3.10GPa
v_{12}	0.26	G_{13}	4.14GPa
v_{23}	0.33	Ply thickness	0.5mm

Table 2. Details of tubular T-joint models in the parametric study.

Specimen	N_{GFRP}	θ
T0	0	–
T1	8	[0/90/0/90] × 2
T10	4	0/90/0/90
T11	12	[0/90/0/90] × 3
T12	4	0/0/0/0
T13	4	90/0/90/0
T14	4	90/90/90/90
T15	4	0/45/0/45

4 RESULTS

4.1 Results of ultimate bearing capacity without earthquake loading

The load-displacement curves of the two specimens T0 and T10 were plotted in Figure 5. The peak load is represented by P_m, and the corresponding deflection of the chord is

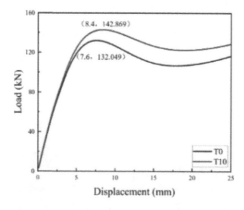

Figure 5. Load-displacement curves of T0 and T10 without seismic damage.

represented by Δ_m. As shown in Figure 5, the ultimate bearing capacity of the unstrengthened joint (T0) was 132.049 kN, while the corresponding deflection of the chord was 7.6 mm. However, the ultimate bearing capacity of the 4-layers GFRP-strengthened specimen (T10) increased to 142.869 kN, and the corresponding deflection of the chord increased to 8.4 mm. Compared to that of T0, the ultimate bearing capacity and the deflection of T10 had increased by 8.19%, and 10.53%, respectively.

The Von-Mises stress distributions at the critical state were plotted in Figure 6. The yield area of T0 was mainly located at the intersection of the chord and brace, while the yield area of T10 was significantly reduced.

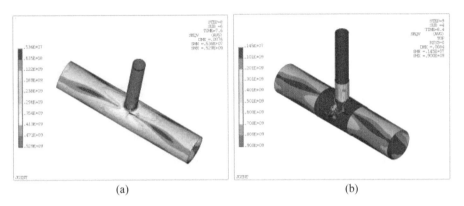

Figure 6. Critical regions of Von-Mises stress distribution. (a) T0 (b) T10.

4.2 Deformation results of T-joints under seismic damage

Seismic damage level is the key factor to evaluate the performance of tubular T-joints. Here, the top end of the brace was subjected to a compression dead load of 50% ultimate bearing capacity of T0. Then, a Pushover analysis was conducted. A pseudo earthquake load that was assumed to produce damage to T0 and T10 was applied to the brace top end of T-joints. The locations on the chord surface were made dimensionless using the effective chord shell width equal to $\sqrt{DT}$ (Lesani et al. 2013). In the longitudinal (crown line) direction, the distance was measured from the joint center toward the chord end (X). In the transverse (hoop line) direction, the distance was measured from the joint center towards the chord half on the joint shell surface (RΦ).

Due to the symmetrical geometry of the T-joint and concentric loading, the results of half of each direction were displayed. The displacement curves along the crown line and hoop line under 0.2~0.7 damage level were plotted and shown in Figure 7. The larger damage level produced a larger deformation level accordingly. The failure mode could be recognized as a local concave deformation on the chord surface around the intersection. The

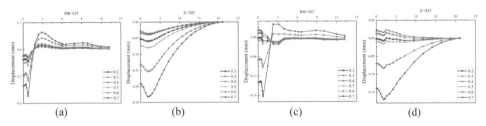

Figure 7. Chord surface deformation under different damage. (c) Along the hoop line of T0. (b) Along the crown line of T0. (c) Along the hoop line of T10. (d) Along the crown line of T10.

deformation degree was reduced dramatically after GFRP strengthening. Figure 7 denoted that the deformation along the crown line was obviously smaller than those along the hoop line, indicating that the main failure occurs at the saddle point. Figures 8 and 9 show the Von-Mises equivalent stress distribution under different damage levels. It was found that the residual stress and residual deformation occurred in the chord surface at the intersection region of the chord and brace, the plastic area further expanded with the increase of damage level. In addition, the deformation area was reduced after reinforcement. Comparing the Figures 8b and 9b, it was observed that the maximum Von-Mises stress was 411 MPa and 399 MPa respectively. From the above discussion, it can be concluded that the GFRP reinforcement is effective in reducing the stress and deformation of T-joints.

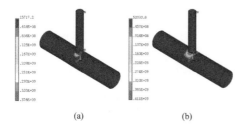

Figure 8. Von-Mises stress distribution with different damage of T0. (a) $\tilde{D} = 0.5$ (b) $\tilde{D} = 0.7$.

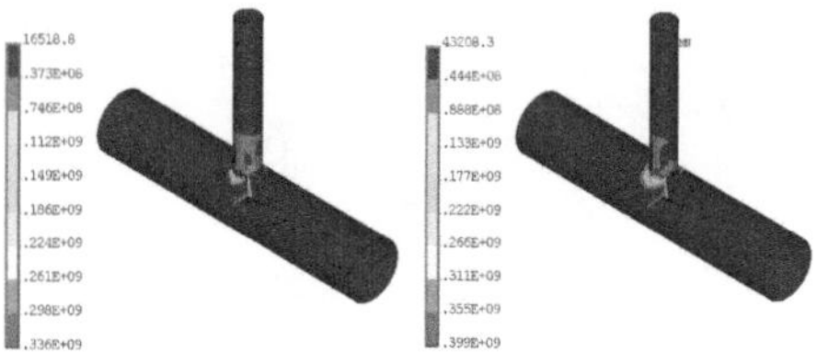

Figure 9. Von-Mises stress distribution with different damage of T10. (a) $\tilde{D} = 0.5$ (b) $\tilde{D} = 0.7$.

4.3 Ultimate bearing capacity of T-joints after an earthquake

To analyze the influence of the GFRP strengthening effect, the models of T-joints with the same damage level of 0.7 were established. Figure 10 shows the load-displacement curves of T0 and T10 under 0.7 damage level. As shown in Figure 10, the ultimate bearing capacity of T0 was 127.575kN. However, T10 had increased in ultimate bearing capacity to 142.820kN. Besides, deflection from 7.4mm up to 8.4mm has been recorded in the ultimate state. Compared to that of T0, the ultimate bearing capacity and the deflection of T10 had increased by 11.95%, and 13.51%, respectively. Compared with section 4.1, it could be seen that seismic damage has reduced the ultimate bearing capacity of joints. Moreover, the GFRP reinforcement effect under seismic damage was more obvious. The results showed that it was useful to strengthen tubular T-joints with GFRP under seismic damage.

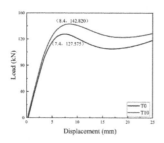 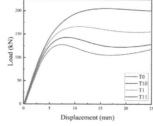 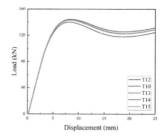

Figure 10. Load-displacement curves of T0 and T10 under 0.7 damage level.

Figure 11. Load-displacement curves of different numbers of GFRP sheets under 0.7 damage level.

Figure 12. Load-displacement curves of different orientations of GFRP sheets under 0.7 damage level.

5 DISCUSSION

5.1 Influence of the number of GFRP sheets on the ultimate bearing capacity

In this section, the influence of the number of the GFRP sheets on the ultimate bearing capacity of T-joints was studied under the condition of the same seismic damage level of 0.7. Compared with the models T0, T1, T10 and T11. The detailed results of the FE models of different numbers of the GFRP sheets were listed in Table 3. The load-displacement curves of the four investigated cases were shown in Figure 11. The results show that the ultimate bearing capacity of the specimens T0 and T11 were 127.575 kN and 204.944 kN, respectively. Compared to that of T0, the ultimate bearing capacity remarkably increased from 11.95% of the 4-layers to 60.65% of the 12-layers. The deflection of the joint increased from 8.4 mm of the 4-layers to 16.2 mm of the 12-layers, respectively. Thus, a larger bearing capacity can be achieved by increasing the number of GFRP sheets under seismic damage.

5.2 Influence of the orientation of GFRP sheets on the ultimate bearing capacity

In this section, the influence of the orientation of the GFRP sheets on the ultimate bearing capacity of T-joints was studied under the condition of the same seismic damage level of 0.7. Compared with the models T10, T12~T15. The detailed results of the FE models of different orientations of the GFRP sheets were listed in Table 4. The load-displacement curves of five investigated cases were shown in Figure 12. The results showed that case 1 was more effective for increasing the ultimate bearing capacity. Moreover, case 4 was less effective to increase the ultimate bearing capacity. Also, compared with cases 1, 2 and 5, it was observed that increasing θ (changing the orientation of the fibers on the joint from the circumferential to the longitudinal direction) reduced the effectiveness of GFRP. Thus, compared with case 2 and case 3, the first layer should select the circumferential fibers on the T-joint.

Table 3. Ultimate bearing capacity and corresponding displacement of strengthened T-joints with different N_{GFRP} under 0.7 damage level.

	P_m	Δ_m
T0	127.575 kN	7.4 mm
0/90/0/90	142.820 kN	8.4 mm
[0/90/0/90] × 2	165.964 kN	10.9 mm
[0/90/0/90] × 3	204.944 kN	16.2 mm

Table 4. Ultimate bearing capacity and corresponding displacement of strengthened T-joints with different θ under 0.7 damage level.

Case	θ	P_m	Δ_m
1	0/0/0/0	144.225 kN	8.5 mm
2	0/90/0/90	142.820 kN	8.4 mm
3	90/0/90/0	142.782 kN	8.5 mm
4	90/90/90/90	139.765 kN	8.5 mm
5	0/45/0/45	143.069 kN	8.5 mm

6 CONCLUSIONS

A numerical investigation on the strengthening of GFRP to tubular T-joint under earthquake cyclic loads was carried out in this study. Two specimens were compared, including an un-strengthened T-joint and a strengthened T-joint with four layers of GFRP. The comparative analysis was followed by parameter analysis on 8 FE models conducted by ANSYS to study the effects of the number of GFRP sheets and the orientation of GFRP sheets on the ultimate bearing capacity of the T-joints. Finally, the numerical results support the following conclusions:

(1) GFRP reinforcement reduced the local deformation of the chord surface around the intersection efficiently, i.e., the chord deformation at the intersection was restricted, and the strength of the T-joint was improved.

(2) The 4-layers GFRP-reinforced tubular T-joints can significantly increase the displacement and the ultimate bearing capacity because it produces constraints to the local deformation on the chord. Specifically, when the damage level was 0.7, the ultimate bearing capacity was improved by 11.95% and the deflection was increased by 13.51% after strengthening with 4-layers of GFRP.
(3) With more layers of GFRP sheet, the T-joint deflection increases around the intersection of brace and chord members, and subsequently the ultimate bearing capacity increases.
(4) Considering the effect of GFRP on the T-joint, circumferential wrapping ($\theta = 0°$) leads up to more increase in ultimate bearing capacity than longitudinal fiber wrapping ($\theta = 90°$).

FUNDING

This research was funded by the National Natural Science Foundation of China, China, grant No. 51879272, No. 52111530036; and the Fundamental Research Funds for the Central Universities, China, grant No. 22CX03022A.

REFERENCES

Choo, Y. S., Vegte, G. J. V. D. & Zettlemoyer, N. et al. (2005). Static Strength of T-joints Reinforced with Doubler or Collar Plates. I: Experimental Investigations. *J. Journal of Structural Engineering*. 131(1), 119–128.
Ganesh, V. K. & Naik, N. K. (1993). Some Strength Studies on FRP Laminates. *J. Composite Structures*. 24(1), 51–58.
Hamzeh, L., Hassanein, A. & Galal, K. (2020). Numerical Study on the Seismic Response of GFRP and Steel Reinforced Masonry Shear Walls with Boundary Elements. *J. Structures*. 28, 1946–1964.
Hosseini, A. S., Bahaari, M. R. & Lesani, M. et al. (2021). Static Load-bearing Capacity Formulation for Steel Tubular T/Y-joints Strengthened with GFRP and CFRP. *J. Composite Structures*. 268, 113950.
Kumar, U. & Das, R. R. (2021). Adhesion Failure Analyses of Laminated FRP Composite Made Bonded Tubular T-joints with Axially Compressed Brace. *J. Composite Structures*. 258, 113386.
Lesani, M., Bahaari, M. R. & Shokrieh, M. M. (2013). Detailed Investigation on Un-stiffened T/Y Tubular Joints Behavior Under Axial Compressive Loads. *Journal of Constructional Steel Research*. 80, 91–99.
Lesani, M., Bahaari, M. R. & Shokrieh, M. M. (2013). Numerical Investigation of FRP-strengthened Tubular T-joints Under Axial Compressive Loads. *J. Composite Structures*. 100, 71–78.
Lesani, M., Bahaari, M. R. & Shokrieh, M. M. (2014). Experimental Investigation of FRP-strengthened Tubular T-joints Under Axial Compressive Loads. *J. Construction and Building Materials*. 53, 243–252.
Lesani, M., Bahaari, M. R. & Shokrieh, M. M. (2015). FRP Wrapping for the Rehabilitation of Circular Hollow Section (CHS) Tubular Steel Connections. *J. Thin-Walled Structures*. 90, 216–234.
Safdar, M., Sheikh, M. N. & Hadi, M. N. S. (2022). Numerical Study on Shear Strength of GFRP-RC T-joints. *J. Structures*. 43, 926–943.
Swamy, B. S. K. & Sekar, M. (2009). Rehabilitation of Reinforced Concrete T-joints Using GFRP. *J. Journal of Structural Engineering (Madras)*. 36, 195–201.
Valarinho, L., Sena-Cruz, J. & Correia, J. R. et al. (2017). Numerical Simulation of the Flexural Behaviour of Composite Glass-GFRP Beams Using Smeared Crack Models. *J. Composites Part B: Engineering*. 110, 336–350.
Vegte, G. J. V. D., Choo, Y. S. & Liang, J. X. et al. (2005). Static Strength of T-joints Reinforced with Doubler or Collar Plates. II: Numerical Simulations. *Journal of Structural Engineering*. 131(1), 129–138.
Wang, X., Wang, W. W. & Cao, H. B. (2021). Analysis of GFRP-concrete Composite Bridge Deck Using a Multi-scale Modeling. *J. Jiangsu Construction*. (1), 25–28, 43.
Xu, J. X., Tong, Y. G. & Han, J. P. et al. (2019). Fire Resistance of Thin-walled Tubular T-joints with Internal Ring Stiffeners Under Post-earthquake Fire. *J. Thin-Walled Structures*. 145, 106433.
Zhou, X. Y., Qian, S. Y. & Wang, N. W. et al. (2022). A Review on Stochastic Multiscale Analysis for FRP Composite Structures. *J. Composite Structures*. 284, 115132.
Zhu, L., Zhao, Y. & Li, S. W. et al. (2014). Numerical Analysis of the Axial Strength of CHS T-joints Reinforced with External Stiffeners. *J. Thin-Walled Structures*. 85, 481–488.

Harmonic response in deep sea truss spar platform

Nan Liu*
School of Transportation Civil Engineering, Shandong Jiaotong University, China

Jisen Liu
Branch of Water Administration Supervision, Jinan, China

Wei Liu
School of Transportation Civil Engineering, Shandong Jiaotong University, China

ABSTRACT: This paper studies whether the overall structural design of spar platforms can overcome the harmful effects caused by structural resonance. Using finite element software ANSYS to take the harmonic response analysis under the premise of excluding the transient vibration at the beginning of excitation, only the steady-state forced vibration of the structure is considered. The determined response value of the displacement with the intersection in the deck and main body to the frequency, predict the continuous dynamic characteristic of the platform structure. It was verified that the resonance phenomenon occurs when the vibration frequency of the structure is close to its natural frequency, namely the resonance effect. The results show that the platform can overcome structural resonance effectively, and the analysis of the harmonic response of offshore platform structure is valuable to engineering practice.

1 INTRODUCTION

The total surface area of the earth is about 510 million square kilometers. The oceans make up about 71% of the earth's total surface area. The ocean area is so vast that there are rich oil and natural gas resources. According to statistics, the amount of marine oil resources in the world accounts for 34% of the total global oil resources, among which the proven reserves are about 38 billion tons. After hundreds of years, most parts of the world that are in the shallow sea and offshore oil and gas resources will decrease. On the other hand, due to the great charm of the deep sea, investment in the future will continue to increase, and the proportion of deep-water oil and gas will be bigger and bigger (Chen 1996) (Yan 1996). The offshore platform is the key equipment to exploit offshore oil and gas resources. This paper mainly studies the dynamic characteristics of the truss spar platform suitable for the deep sea.

Harmonic response analysis is one of the dynamic characteristics of structures, which is mainly used to determine the steady-state response of linear structures under sinusoidal or harmonic changes over time. In the analysis process, only the steady-state forced vibration of structures is calculated, and the transient vibration at the beginning of excitation is not considered. The purpose of harmonic response analysis is to calculate the frequency curve of the response value (usually displacement) of the structure at several frequencies to predict the continuous dynamic characteristics of the structure (PERA Global 1998). In this paper, the

---

*Corresponding Author: 752826457@qq.com

harmonic response analysis is mainly to verify whether the overall structure design of the truss spar platform can overcome the harmful effect caused by resonance, which has a certain reference value to the engineering practice.

2 RESEARCH METHODS

Harmonic response analysis of the deep-sea truss spar platform is mainly carried out and the finite element numerical simulation method is mainly used in the calculation process. On the premise of excluding the transient vibration at the beginning of excitation, only the steady-state forced vibration of the structure is considered. The calculation results of the displacement frequency response curves of the structure at several frequencies are used to verify whether the structure can overcome the harmful effects caused by the resonance phenomenon.

3 PARAMETER DESIGN OF TRUSS SPAR PLATFORM

The platform is located in the Gulf of Mexico, off the southeastern coast of the North American continent, in a water depth of about 1,500 m (Purath 2006). The main structure of the platform includes a deck, main body, hard cabin, soft cabin, swing plate, truss, etc. The hard cabin is located above the main body of the platform, with a cylinder structure and a total length of 68.88 m. The middle part of the platform, designed with an open frame structure, is constructed with dozens of X trusses and three hanging slabs. The lower part of the body has a ballast tank 5 meters long. In the actual construction process, considering reducing the construction difficulty, construction personnel used a common frame structure to reconstruct the main part of the platform. This idea fundamentally reduces the horizontal area of the main body, restricts the horizontal movement of the platform to a certain extent, and thus reduces the influence of external load on the platform. At the same time, to mitigate the impact of wave force on the overall structure of the platform and reduce the heave response of spar, the construction staff tried to install three heave plates (32.21 m by 32.21 m) under the main part of the platform (Chakrabarti 2007) (Haslum 1999). Specific characteristic parameters of the spar platform are shown in Table 1, and the schematic diagram of the calculation is shown in Figure 1. The spar is suitable for a water depth of 1500 meters, with the deck 15.956 m above the water level.

4 BUILDING A PLATFORM MODEL USING ANSYS

In this section, a finite element modeling of the platform is carried out using ANSYS software, which is widely used to realize multi-field and multi-field coupling analysis, among

Table 1. Basic parameters of the truss spar platform.

Components	Size (m)
Deck	40 × 40 × 2
Diameter of the main body	32.31
Soft capsule length	5.00
Frame length	96.00
Side length of the swing plate	32.31
Draft	153.924
X truss diameter	1.25
Vertical truss diameter	2.00

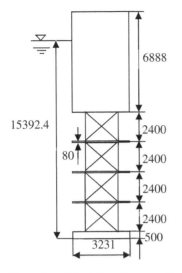
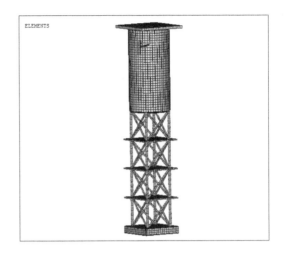

Figure 1. Calculation diagram of truss spar.

Figure 2. Finite element model of truss spar.

which PIPE59 is a unique pipe structure unit of ANSYS. It can carry out the static calculation and dynamic response analysis of the pipe structure of any offshore structure under the wave, sea current, and other loads. This function has more advantages than other CAE software. Finally, SHELL63 and PIPE59 are used as two structural units to build the platform model according to the requirements of actual harmonic response analysis. SHELL63 unit is used for the construction of the deck, upper main structure, swing plate, and bottom ballast tank model. The PIPE59 unit is used for the construction of the truss structure model, and the finite element model is shown in Figure 2.

The riser and mooring system are not considered in the calculation process. The platform adopts a steel structure as a whole. The selected material parameters are elastic modulus $EX = 2.0 \times 10^{11}$ Pa, Poisson's ratio, and density $DENS = 7850$ kg/m³.

5 HARMONIC RESPONSE ANALYSIS OF THE TRUSS SPAR PLATFORM

As a linear analysis method in ANSYS, harmonic response analysis is mainly divided into: reduction method, complete method, and modal superposition method. In the complete method, the harmonic response of the complete system matrix is calculated by a single calculation method without considering matrix reduction and mass matrix approximation (Liu 2013). The complete method is mainly adopted, and the load is applied to the water surface position of the platform by referring to the natural frequency of the spar platform during calculation (Hou 2021). The direction of loading was selected as y, x, and z-axis. After analysis and calculation, displacement-frequency response curves of four points (NODE1, NODE2, NODE3, and NODE4) on the intersecting surface between the platform deck and the main body are obtained, as shown in Figure 3. One of them, NODE2, is taken for calculation, and its displacement frequency response curves in three directions are shown in Figure 4.

Displacement frequency response curves of NODE1, NODE2, NODE3, and NODE4 under the action of unit harmonic load in the y direction are represented by UY, UY1, UY2, and UY3, respectively. Three peak values of each curve correspond to the first (0.0736Hz), third (0.2221Hz), and fifth (0.5949Hz) natural frequencies of the model, respectively, and the acting director of the unit harmonic load is Y direction. According to Hou (2021), bending

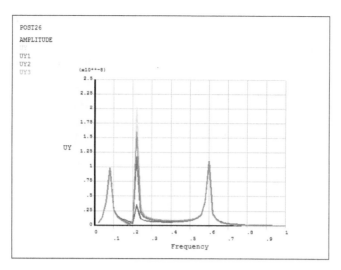

Figure 3. UY-frequency in the y direction.

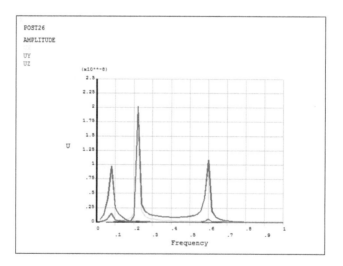

Figure 4. UX, UY, and UZ-frequency in the y direction.

deformation of the platform model in the Y direction corresponds to the first, third, and fifth modes.

Similarly, peak values of the change curve in Figure 5 correspond to the second-order (0.0736Hz) and fourth-order (0.5949Hz) natural frequencies of the model, respectively, and the action direction of unit harmonic load is x direction. According to Hou (2021), the bending deformation of the platform model in the X direction corresponds to the second and fourth modes.

The peak value of the change curve in Figure 6 corresponds to the sixth-order natural frequency of the model (0.6792Hz). The action direction of the unit harmonic load is the z-direction, and the z-axis is used as the axis of torsion. According to Hou (2021), the bending deformation of the platform model in the z-direction corresponds to the sixth-order mode exactly. It is well known that the structure itself has a natural frequency, and when the external load is applied to the structure, it will also vibrate at a certain frequency. When the

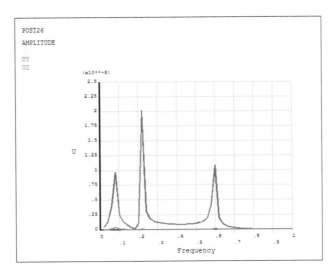

Figure 5. UX, UY, UZ-frequency in *x* direction.

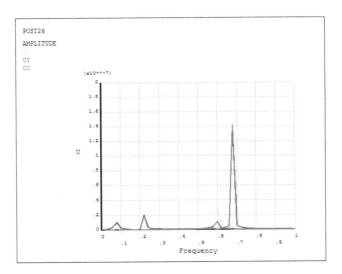

Figure 6. UX, UY, and UZ-frequency in the *z-direction*.

natural frequency of the structure is close to the vibration frequency imposed by the external load, resonance will be generated. The variation curves in Figures 3–6 above just verify the phenomenon of the resonance effect.

6 CONCLUSIONS

(1) The finite element method can accurately analyze the harmonic response of the structure.
(2) The displacement-frequency response curve at the intersection of the platform deck as well as the main body was calculated. The peak value of the curve corresponded to the natural frequency of the platform model, which verified the resonance effect.

(3) The platform can effectively overcome the structural resonance, which should be avoided as much as possible in the specific analysis and design of the structure.

ACKNOWLEDGMENTS

This work was supported by the doctoral research start-up and college fund of Shandong Jiaotong University.

REFERENCES

Chakrabarti, S. (2007) *Ocean Engineering*. **1** 621–629
Chen H. M. 1996 *China Offshore Platform*. **1** 23–24
Haslum, H. A. 1999 *Houston*, Texas, OTC-10953
Hou X. L. 2021 *Dynamics of Structure*. Xi'an Jiaotong University Press, Xi'an
Liu N. 2013 Fracture Analysis of Offshore Platform Structures, *Doctoral Dissertation of the Ocean University of China*. pp 76–102
Pera Global 1998 *Guide to ANSYS Dynamic Analysis* (Beijing: Beijing office of ANSYS Company): 28–49
Purath, B.T. 2006 *Numerical Simulation of the Truss Spar 'Horn Mountain' Using couple*. Texas: Texas A & M university
Yan S. H. 1996 *China Offshore Platform*. 1 16–19

Design and performance simulation of hydraulic system of walking launching equipment

Mingxia Shi
Suqian Transportation Industry Group Co., Ltd., Suqian Jiangsu, China

Qi Hu*
Cccc WuHan Harbour Engineering Design and Research Co. Ltd., Wuhan Hubei, China

ABSTRACT: To offer a theoretical analysis and support for similar large-scale lifting equipment to walking launching equipment, this paper mainly introduces the design, application, and performance simulation of a throttling and speed-regulating hydraulic system for a 1200t walking launching equipment. Based on the requirements of functional design and construction conditions of the walking launching hydraulic system, the overall power of the system, key valves, pipelines, and other key components are selected, calculated, and analyzed in detail. The hydraulic system is mathematically modeled, and the parameters of the hydraulic system are simulated and improved by MATLAB. It provides certain practical guiding significance and theoretical value to the technical analysis of the hydraulic system of the existing launching equipment.

1 INTRODUCTION

Launching technology is a construction method that pushes or pulls prefabricated bridge structures. In 2005, the wedge-type launching construction technology is successfully adopted in the French Millau Bridge. Since 2007, the walking launching technology and equipment system have gradually been adopted in plenty of prefabricated bridge constructions in China (Wang et al. 2021). It overcomes the key technical problems existing in the traditional methods, such as large horizontal force, adaptive control of the completed bridge alignment, and distributed monitoring of the overall launching construction. What's more, it effectively reduces the manufacturing and using costs of equipment systems, and improves the efficiency and safety of bridge construction (Jarno et al. 2009; Wang et al. 2021).

The hydraulic system is the lifeline of walking launching equipment. Its component selection, design, and system performance are directly related to the synchronous launching construction of multi-point launching equipment and the quality of bridge line information. Although the launching process is becoming more and more mature, there is still a lack of theoretical analysis and support for similar large-scale lifting equipment. The launching equipment produced by different manufacturers is also different in terms of shape, hydraulic principle, and control algorithm. Therefore, there are certain practical guiding significance and theoretical value to the technical analysis of the hydraulic system of the existing launching equipment (Jarno et al. 2009) (Chen et al. 2013).

*Corresponding Author: 13080679175@163.com

This paper mainly focuses on the hydraulic system of a 1200t launching equipment. Through the analysis of the construction requirements, the design of the corresponding hydraulic schematic diagram, the calculated parameters, selected components, and the finally established mathematical model of the hydraulic system, the system is simulated and analyzed to make a reasonable optimization (Jing et al. 2010) (Vahab Esmaeili et al. 2022).

2 PRINCIPLE OF HYDRAULIC SYSTEM

The actions performed by the launching equipment include load direction (lifting), load travel direction (launching), and load traverse direction (correction). The system is the basis for the realization of the action of the entire launching system, which determines the final running speed of the equipment, control properties, safety and reliability, and other characteristics (Fu et al. 2002).

According to the active principle, the hydraulic system is designed as three main circuits. Since the flow rate of the equipment is not large and it is generally in a discontinuous working state, the quantitative pump and a relief valve are chosen to supply oil to the system.

The Y-type proportional reversing valve is chosen for throttling and speed regulation in the system to avoid the problem of two-way conduction or leakage caused by the pressure holding of the oil circuit (Kim et al. 2002). To ensure safety and actuators' movement at a speed higher than the set speed, and to prevent the acceleration movement when the movement in the load direction is out of control, each cylinder is installed with a HAWE LHDV type balance valve, as shown in Figure 1.

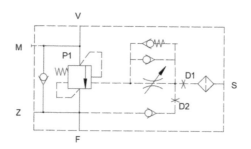

Figure 1. LHDV type balancing valve.

The externally adjustable flow damping and bypass throttling method are adopted in the LHDV-type balance valve. When using sequences valves instead of balance valves, it is easy to cause vibration at the moment of starting and stopping during the descent with load, and the hydraulic lock is more obvious. However, the LHDV-type balance valve is more adaptable under a flexible load with large vibration or swing. Its internal damping element can effectively weaken the impact when the main spool is opened, and reduce the vibration amplitude and frequency of the system.

The launching circuit and the rectifying circuit are independent circuits, each with a proportional reversing valve for speed regulation, and a hydraulic lock is installed.

3 CALCULATION OF HYDRAULIC SYSTEM PARAMETERS

Table 1 shows the specification requirements of the hydraulic cylinder. The system design pressure is 31.5 Mpa and the launching speed is designed to be 3.5 m/hs.

Table 1. System cylinder requirements.

	Pressure	Pushing force / pulling force	Stroke
Lifting cylinder	31.5 Mpa	300 t	300 mm
Launching cylinder	31.5 Mpa	Pulling force 40 t	350 mm
Correction cylinder	31.5 Mpa	40 t	60 mm

3.1 Selection of hydraulic pump

The displacement of the hydraulic pump can be preliminarily calculated according to the cylinder area, operating speed, and system pressure of the hydraulic actuator (Niu 2016). The formula is as follows:

$$Q_p = nV\eta_p \times 10^{-3} \geq 6v_{\max}A \times 10^4 + Q_Y \quad (1)$$

Where V represents the displacement of the pump, unit, cm^3/rev. n represents motor speed, unit, r/min. η_p is the volumetric efficiency of the pump and taking 0.95. $v_{\max}$ indicates the action speed of the actuator, unit, m/s. A indicates the effective action area, unit, m^2. Q_Y indicates the minimum relief flow of the relief valve, unit, L/min, which can be ignored. The known parameters can be substituted into Formula (1).

The most suitable type of PARKER F1-25-R axial piston pump is selected according to the flow rate, with a displacement of 25 ml/rev. When the motor speed is 1000 r/min, the output flow is slightly larger than the flow required by the system.

3.2 Motor selection

According to the actual flow of the selected pump, the motor power can be calculated based on Formula (2).

$$P = \frac{\Psi P_N Q_N}{60\eta_P} \quad (2)$$

Where Ψ represents the conversion coefficient of the pump, which takes 1. P_N represents the rated pressure of the system, unit, Mpa. Q_N represents the system flow, unit, L/min. η_P represents the total efficiency of the hydraulic pump (volume efficiency plus mechanical efficiency), which takes 0.95. Therefore, the Siemens three-phase asynchronous motor with a power of 15 kw and a speed of 1000 r/min can be used.

3.3 Selection of hydraulic accessories

The connection between the pump station and the equipment is a steel wire-reinforced hydraulic rubber hose. The maximum flow rate in the pipe can reach 6 m/s. The formulas for the inner diameter, flow, and flow rate of the hose are as follows:

$$A = \frac{1}{6}\frac{Q}{v} \quad (3)$$

The inner diameter of the hose $d = 9.4$ mm can be calculated according to the calculation of the diameter and system pressure, and a type 2 hose with a diameter of 10 mm according to the national standard is selected.

4 HYDRAULIC SYSTEM SIMULATION

A mathematical model of the system must be first established in system simulation. Then the relationship between system input and output can be analyzed, which is helpful to get the dynamic characteristics of the system such as response speed and vibration frequency. Thereby the modification of design parameters can be guided (Yue et al. 2010).

According to the principle of the proportional reversing valve and balance valve, the launching process can be equivalent to the circuit shown in Figure 2. The proportional reversing valve can be simplified as a throttle valve and a reversing valve. And the balance valve can be directly equivalent to a one-way valve.

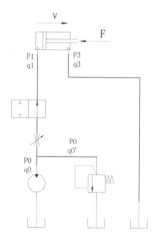

Figure 2. Equivalent oil circuit for launching.

4.1 *Proportional valve opening and cylinder speed characteristics*

The flow formula of the throttle port of the LHDV type balance is:

$$Q = kA\Delta P^m \tag{4}$$

For the throttle valve, the flow characteristic formula is as follows:

$$q_1 = C_d A_T \sqrt{\frac{2}{\rho}(P_0 - P_1)} \tag{5}$$

Where C_d is the flow coefficient. The liquid density ρ is 870 kg/m³. A_x is the valve core area, which is a parameter related to the input signal.

The oil returns pressure can be ignored for the oil cylinder. According to the balance of flow and force, it can be obtained:

$$\begin{aligned} q_1 &= A_1 v + C_1 \frac{dP_1}{dt} + \lambda P_1 \\ F + Bv + m\frac{dv}{dt} &= P_1 A_1 \end{aligned} \tag{6}$$

Where C_1 represents the liquid volume of the oil cavity and the pipeline of the oil inlet $C = V/K$. V represents the oil cavity volume. K represents the oil volume elastic modulus.

m is the mass of the moving part. B is the viscous damping coefficient. λ represents the leakage coefficient.

Considering the stable operating state of the system, the liquid volume and leakage of the oil pipe, the oil cavity, and the damping coefficient of the oil cylinder can be ignored. The relationship between the motion speed v and the throttling area A_T can be obtained by combining Formulas (4) and (5):

$$v = \sqrt{\frac{A_1^3 \rho}{2C_d^2(P_0 A_1 - F)}} A_T \tag{7}$$

It can be seen that, when the load remains unchanged, the speed of the hydraulic cylinder is proportional to the throttle valve area.

4.2 Load speed characteristics

The Laplace transform of Equation (6) can be obtained.

$$\begin{aligned} Q_1(s) &= A_1 V(s) + C_1 s P_1(s) + \lambda P_1(s) \\ F(s) + BV(s) + ms V(s) &= P_1(s) A_1 \end{aligned} \tag{8}$$

$$\frac{V(s)}{F(s)} = \frac{1}{\frac{A_1^2 W_1(s)}{1 - C_1 s W_1(s) - \lambda W_1(s)} - (B + ms)} \tag{9}$$

Among them, $W_1(s) = P_1(s)/Q_1(s)$ represents the transfer function of the joint action of the throttle valve and the oil inlet pipeline system. The influence of the pipeline and the reversing valve on the system dynamics can be ignored and only the throttle valve in the oil inlet pipeline can be considered as follows:

$$W_1(s) = -\frac{1}{C_d} \tag{10}$$

Substitute into Formula (9).

$$\frac{V(s)}{F(s)} = -\frac{C_d + \lambda}{A_1^2 + BC_d + B\lambda} \cdot \frac{\frac{C_1}{C_d + \lambda} s + 1}{\frac{mC_1}{A_1^2 + BC_d + B\lambda} s^2 + \frac{BC_1 + mC_d + m\lambda}{A_1^2 + BC_d + B\lambda} s + 1} \tag{11}$$

By ignoring the leakage of the cylinder and viscous damping, and substituting the parameters into the equation, the undamped natural frequency and damping ratio of the system are as follows:

$$\begin{aligned} w_n &= 1.76 \times \frac{10^6}{\sqrt{m}} \\ \xi &= 1.76 \sqrt{m} \end{aligned} \tag{12}$$

MATLAB is used to find the unit step response and ramp response of the transfer function.

Figure 3 is the respective response curves of the system. From the analysis of the response curve of the unit slope, it can be seen that the system is overdamped, and the problem of starting vibration is not obvious. The greater the mass of the moving object is, the longer it takes for the system to rise to a steady state. The natural frequency of the system has an inverse relationship with the mass of the moving object.

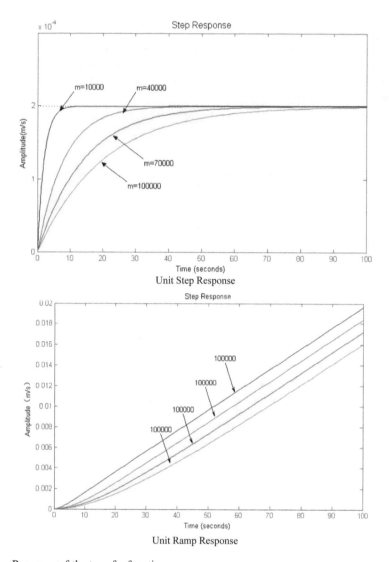

Figure 3. Response of the transfer function.

5 CONCLUSIONS

The hydraulic system of the launching equipment is taken as an example in this paper. According to the actual working conditions, the hydraulic system scheme is determined successively. The component parameter models are calculated, and finally, the dynamic response of the system load-rising process is analyzed through system simulation. In general, the main points are as follows:

1. Proportional valve and quantitative pump step-less speed regulation are selected, which are economical and efficient.
2. The LHDV-type balance valve is installed in the launching oil circuit to prevent falling.
3. The rising oil circuit of the hydraulic system belongs to the over-damping system. The larger the moving mass, the longer the time to reach a steady state.

REFERENCES

Baode Jing, Zhiyong Luo, Xilin Zhu, et al. Shift Hydraulic System Modeling and Simulation Based on Power Bond Graph. *Proceedings of 2010 International Conference on Computer, Mechatronics, Control and Electronic Engineering (CMCE 2010)*. 5 (2010).

Dexue Niu. Hydraulic Transmission and Control Technology Based on Combination Model. *Journal of Computational and Theoretical Nanoscience*. 13(12), 9431–9435 (2016).

Jarno R A, Uusisalo, Kalevi Huhtala, Matti Vilenius. Effects of Remote Control on Usability of Hydraulic Excavators. *ASME 2009 Dynamic Systems and Control Conference*. (2009).

Johnny Fu S, Liffring M, Mehdi I S. Integrated electro-hydraulic system modeling and analysis. *IEEE Aerospace & Electronics Systems Magazine*. 17(7), 4–8 (2002).

Kim D H, Park J W, Gabsang Lee, et al. Active Impact Control System Design with a Hydraulic Damper. *Journal of Sound & Vibration*. 250(3), 485–501 (2002).

Qiang Wang, Zong Liu, Guangwu Yang, et al. Research on Pipe Jacking Construction Technology Based on Upcrossing Metro and Cavle Tunnel. *IOP Conference Series Earth and Environmental Sciences*. 651(4), 042014 (2021).

Vahab Esmaeili, Ali Imanpour, Robert G Driver. Numerical Assessment of Design Procedures for Overhanging Steel Girders. *CSCE Annual Conference – Structures Specialty*. Whistler, Canada. (2022).

Yanfang Yue, Jinye Wang, Guang Yang, et al. *Study on Digital Processing Technology of the Multidimensional Graph in Mechanical Design Handbook*. (2010).

Zhou Chen, Sheng Yan, Buyu Jia, et al. Large Span Continuous Girder Bridge Jacking Steel Hoop Stress Analysis. *Advanced Materials Research*. 671–674:991–995 (2013).

Analysis of Sag effect of auxiliary cable of cable-stayed bridge and replacement of CFRP cable

Fang Pan*
Guangdong Provincial Highway Construction Co., Ltd., Guangzhou, China

Pan Wu* & Yaoyu Zhu*
China Communications Construction Company Highway Bridges National Engineering Research Center Co., Ltd., Beijing, China

Wenjin Sun*
Guangdong Provincial Highway Construction Co., Ltd., Guangzhou, China

ABSTRACT: The Huangmaohai bridge is the control project of the Huangmaohai sea-crossing channel in Guangdong Province. It will become the world's largest single-column tower three-pylon cable-stayed bridge with double cable planes. Each cable plane of the Huangmaohai bridge uses five pairs of mid-span auxiliary cables to increase the vertical stiffness. However, the influence of the sag effect cannot be ignored for the large horizontal span of auxiliary cables. Therefore, this paper compares and analyzes the influence of cable length, horizontal inclination angle, and tensile strength on the sag effect of stay cables by improving the equivalent elastic modulus method. Based on the principle of strength equivalence, stiffness equivalence, and cost equivalence, the sag effect after CFRP cable replacement is analyzed, and the appropriate CFRP cable anchor is given. The results show that: ① Reducing the cable length and increasing the horizontal inclination angle is helpful to reduce the influence of the sag effect; ② In the case of keeping the safety factor of the cable unchanged, reducing the tensile strength of the cable helps to reduce the sag effect; ③ The traditional steel cable is replaced by CFRP cable, which can effectively reduce the sag effect of the cable, and a straight tube cone bonded CFRP cable anchor with high anchoring efficiency is presented.

1 INTRODUCTION

The cable-stayed bridge has many advantages, such as large span capacity, beautiful shape, large stiffness, and so on. With a span of 400 ~ 1500 m, it has become the most common form of bridge structure. As a statically indeterminate structure system composed of three basic components of the tower, beam, and cable, the geometric nonlinear effect of the cable-stayed bridge cannot be ignored when the span is large. At present, there are three main geometric nonlinear effects of cable-stayed bridge: ① cable sag effect; ② beam-column effect; ③ Large displacement effect. The influence of the cable sag effect is more significant, so it must be considered in the design of a cable-stayed bridge system.

The stay cable material will produce elastic deformation when subjected to external loads. In addition, it will also deform under the action of gravity. This deformation is the self-

---

*Corresponding Authors: 1751115036@qq.com, 826094622@qq.com, zhuyaoy@sina.com and 826155441@qq.com

weight sag of the stay cable. According to the analysis of relevant research examples, it is shown that the displacement and internal force of the main girder change greatly when the sag effect is considered. Among them, the deflection of the main beam can be increased by 16%; the internal force of the upper chord and the lower chord of the steel truss stiffening girder will increase, and the influence of the upper chord is greater than that of the lower chord. The displacement and internal force of the bridge tower will also increase significantly, with the maximum amplitude reaching 31%. Therefore, when the horizontal projection length of the stay cable is large, the influence of the sag effect of the stay cable cannot be ignored. Carbon fiber reinforced polymer (CFRP) has the advantages of being lightweight and high strength, which can reduce the influence of the sag effect of the stay cable and has the potential to replace the ordinary stay cable.

To sum up, this paper will take the auxiliary cable of the Huangmaohai bridge as the object, analyze the influence of cable length, horizontal inclination, and tensile strength on the sag effect of stay cable by combining the equivalent elastic modulus method, and further analyze the feasibility of CFRP cable replacement.

2 ENGINEERING SITUATION AND SAG EFFECT ANALYSIS METHOD

2.1 *Engineering situation*

The Huangmaohai Bridge is a controlling project of the Huangmaohai sea-crossing passage in Guangdong Province. The span arrangement is (100 + 280 + 2 × 720 + 280 + 100) m. After completion, it will become the world's largest single-column double-cable-plane three-tower cable-stayed bridge. Each cable plane of the Huangmaohai bridge uses five pairs of mid-span auxiliary cables to increase the vertical stiffness of the three-tower cable-stayed bridge. The minimum horizontal inclination of the auxiliary cables is 23.3°, as shown in Figure 1 below.

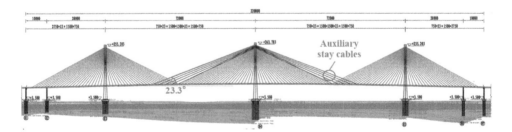

Figure 1. The Huangmaohai bridge.

2.2 *Analysis method of sag effect*

At present, the most widely used method to analyze the sag effect of stay cable is the equivalent elastic modulus method. This method is to equivalent stay cable with higher initial stress and a certain sag to a straight chord, considering the influence of stay cable weight along the vertical direction of the chord. The equivalent elastic modulus can be used to further modify the element stiffness matrix, which is very convenient to apply. The improved formula is as follows:

$$E_{eq} = \frac{E}{1 + \frac{(qL_x)^2 EA}{12T^3}} \quad (1)$$

Where: q = Unit cable length weight; L_x = Length of horizontal projection; E = Theoretical elastic modulus; A = Stay cable cross-sectional area; T = Stay cable force.

The parameters of the most unfavorable auxiliary cable of the Huangmaohai bridge are shown in Table 1 below, where the safety factor is guaranteed to be 2.5, so the maximum tension of the auxiliary cable is T = 1860MPa/2.5 × 1.30E4mm^2 = 9672kN.

Table 1. The most unfavorable auxiliary cable reference parameters.

Stay cable model	LPES7-337
Stay cable cross-sectional area A	1.30E4mm^2
Unit cable length weight q	1.057N/mm
Stay cable length L	465.5m
Maximum of stay cable tension T	9672kN
Horizontal inclination θ	23.3°
Length of horizontal projection L_x	427.5m
Tensile strength σ	1860.0MPa
Theoretical elastic modulus E	1.95E5Mpa

3 ANALYSIS OF SAG EFFECT OF STAY-CABLE

3.1 Influence of cable length

In the Huangmaohai bridge, stay cable lengths at different positions are different. When the stay cable safety factor is set to 3.2, and the ensure stay cable maximum tension is T = 1860Mpa/3.2 × 1.30E4mm^2 = 7500kN, the horizontal inclination angle θ = 23.3° and other parameters are unchanged. The influence of the sag effect of different stay cable lengths in the range of 372.4~558.6m is analyzed. The calculation results of equivalent elastic modulus under different cable lengths are as follows:

Table 2. Comparison of sag effect under different stay cable lengths.

Stay cable length L/m	Theoretical elastic modulus E/MPa	Equivalent elastic modulus E_{eq}/MPa	E_{eq}/E
372.4	1.95E + 05	1.83E + 05	93.9%
418.9	1.95E + 05	1.80E + 05	92.4%
465.5	1.95E + 05	1.77E + 05	90.7%
512.0	1.95E + 05	1.74E + 05	89.0%
558.6	1.95E + 05	1.70E + 05	87.2%

As shown in Table 2 and Figure 2, when the stay cable length increases from 372.4m to 558.6m, the ratio of the equivalent elastic modulus to the theoretical elastic modulus of the stay cable decreases from 93.9% to 87.2%, indicating that the longer stay cable length, the greater sag effect and the lower the vertical stiffness.

3.2 Influence of horizontal inclination

In the Huangmaohai bridge, the horizontal inclination of stay cables at different positions is different. When the staying cable safety factor is set to 3.2, that is, when the stay cable maximum tension is T = 7500kN, stay cable length L = 465.5m (the horizontal projection

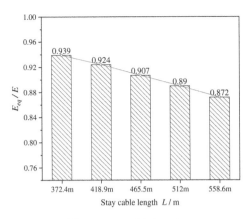

Figure 2. Variation of E_{eq}/E under different stay cable lengths.

length $L_x = 427.5m$), stay cable force and other parameters remain unchanged, and the horizontal inclination angle changes in the range of 23.3° to 43.3°. The influence on the sag effect of stay cable is calculated as follows:

Table 3. Comparison of sag effect under different stay cable horizontal inclinations.

Horizontal inclinations θ	Theoretical elastic modulus E/MPa	Equivalent elastic modulus E_{eq}/MPa	E_{eq}/E
23.3°	1.95E + 05	1.77E + 05	90.7%
28.3°	1.95E + 05	1.78E + 05	91.4%
33.3°	1.95E + 05	1.80E + 05	92.2%
38.3°	1.95E + 05	1.81E + 05	93.1%
43.3°	1.95E + 05	1.83E + 05	94.0%

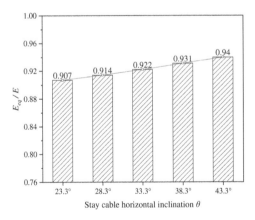

Figure 3. Variation of E_{eq}/E under different stay cable horizontal inclinations.

As shown in Table 3 and Figure 3, when the horizontal inclination of the stay cable increases from 23.3° to 43.3°, the ratio of the equivalent elastic modulus to the theoretical elastic modulus of the stay cable increases from 90.7% to 94.0%. It shows that when the stay cable length and cable forces are constant, the larger the horizontal inclination, the smaller the sag effect, and the greater the vertical stiffness of the structure.

3.3 Influence of tensile strength

When the tensile strength of the stay cable is changed, the safety factor of the stay cable is set to remain unchanged at 2.5. When the tensile strength of stay cable changes in the range of 1660MPa, 1760MPa, 1860MPa, 1960MPa, and 2060MPa, the corresponding cable tension T is 8632kN, 9152kN, 9672kN, 10192kN, and 10712kN, respectively to ensure that the stay cable length is 465.5m, the horizontal inclination angle is 23.3° and other parameters remain unchanged. The influence of different tensile strengths on the sag effect of stay cable is analyzed. The calculation results are as follows:

Table 4. Comparison of sag effect under different stay cable tensile strengths.

Tensile strength σ/MPa	Stay cable force T/kN	Theoretical elastic modulus E/Mpa	Equivalent elastic modulus E_{eq}/Mpa	E_{eq}/E
1660	8632	1.95E + 05	1.83E + 05	93.7%
1760	9152	1.95E + 05	1.85E + 05	94.7%
1860	9672	1.95E + 05	1.86E + 05	95.4%
1960	10192	1.95E + 05	1.87E + 05	96.1%
2060	10712	1.95E + 05	1.88E + 05	96.6%

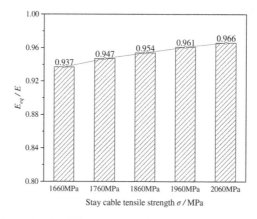

Figure 4. Variation of E_{eq}/E under different stay cable tensile strengths.

As shown in Table 4 and Figure 4, when the stay cable strength increases from 1660MPa to 2060MPa, the ratio of the equivalent elastic modulus to the theoretical elastic modulus of the stay cable increases from 93.7% to 96.6%. It shows that with the increase of the tensile strength of the stay cable, the sag effect decreases.

4 ANALYSIS OF CFRP CABLE REPLACEMENT CONSIDERING SAG EFFECT

4.1 Sag effect analysis of CFRP cable replacement

Under the same conditions, the strength, stiffness, quality, and cost of CFRP cable are about 1.32 times, 0.84 times, 0.23 times, and 2 times that of ordinary LPES7-337 stay cable, respectively. Combined with the parameters of the most unfavorable auxiliary cable and based on the design principles of strength equivalence, stiffness equivalence, and cost equivalence, that is, the area of CFRP cable is changed to 1860/2450 = 0.76 times, 1.96/1.65 = 1.19 times, and 2 times of the reference cable, respectively. The sag effect analysis of CFRP cable replacement under different criteria is as follows:

Table 5. Sag effect analysis of CFRP cable replacement under different criteria.

	A/mm^2	q/(N/mm)	σ/MPa	E/MPa	E_{eq}/MPa	E_{eq}/E
Benchmark stay cable	1.30E4	1.057	1860.0	1.95E + 05	1.77E + 05	90.7%
Strength equivalence	9.88E + 03	0.190	2450.0	1.65E + 05	1.65E + 05	99.8%
Stiffness equivalence	1.55E4	0.297	2450.0	1.65E + 05	1.64E + 05	99.2%
Cost equivalence	2.60E + 04	0.499	2450.0	1.65E + 05	1.59E + 05	96.3%

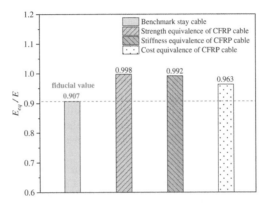

Figure 5. Variation of E_{eq}/E under different CFRP cable replacement criteria.

From Tables 5 and Figure 5, it can be seen that based on the design principles of strength equivalence and stiffness equivalence, the CFRP cables used are basically not affected by the sag effect, and the ratio of the equivalent elastic modulus of stay cable to the theoretical elastic modulus is 99.8% and 99.2% respectively. Based on the principle of cost equivalent design, the CFRP cable is less affected by the sag effect and less than the traditional steel cable. In summary, the use of CFRP cable to replace the traditional high-strength steel wire cable can effectively reduce the influence of the sag effect, and to a certain extent, it can appropriately reduce the cross-sectional area of the auxiliary cable and reduce the project cost. It is a good design option.

4.2 *Anchor of CFRP cable*

It can be seen that CFRP cable instead of ordinary stay cable can effectively reduce the stay cable sag effect, which is a good design choice, but at the same time, the performance of CFRP cable is inseparable from the efficient anchor, so it is necessary to give the appropriate CFRP cable anchor to change the CFRP cable into reality.

After comparison and analysis, a new type of straight tube cone bonded CFRP cable anchor is given in Figure 6. The anchoring mechanism is as follows: the internal structure is anchored by the friction force and extrusion force between the sleeve and the bonding medium. The bonding force between the bonding medium and the CFRP (composed of chemical adhesive force, friction force, and mechanical bite force) is used to anchor the CFRP cable. The sleeve, the bonding medium, and the CFRP cable interact and coordinate the deformation to form an anchoring system that can bear the axial tensile load. The analysis shows that the anchor efficiency of the anchor is high and it can anchor CFRP cables well.

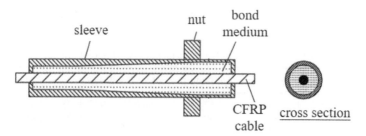

Figure 6. Straight tube cone bonded CFRP cable anchor schematic.

5 CONCLUSION

Based on the auxiliary cable of the Huangmaohai bridge, this paper analyzes the sag effect of the auxiliary stay cable and CFRP cable replacement, and draws the following conclusions:

(1) Under the condition of keeping the stay cable force unchanged, further reducing the stay cable length and increasing the horizontal inclination angle are helpful to improve the vertical stiffness of the stay cable and reduce its sag effect;
(2) Under the condition of keeping the safety factor of the stay cable unchanged, reducing the tensile strength of the stay cable helps to improve the vertical stiffness of the stay cable and reduce its sag effect;
(3) Based on the design principles of strength equivalence, stiffness equivalence, and cost equivalence, the traditional steel cable is replaced by CFRP cable, which can effectively reduce the influence of the cable sag effect and is a good design option. A straight tube cone bonded CFRP cable anchor with high anchor efficiency is presented.

REFERENCES

Hou, S. W. 2013. Experiment on the Mechanism of Bond-type Anchors for CFRP Cables. *China Journal of Highway and Transport* 26(05): 95–101.

Jia, L. J. 2019. Mechanical Performance Analysis of One Tube Composite Bonding-type Anchor of CFRP Tendons. *Journal of South China University of Technology* 47(09): 98–106.

Liu, P. 2016. Analysis of Deadweight Induced Sag of Long-span Stay Cable. *Journal of Highway and Transportation Research and Development* 33(03): 60–63.

Pan, J. Y. 1993. Study on a Nonlinear Live Load of Large Span Cable-stayed Bridge. *China Civil Engineering Journal* (01): 31–37.

Sham, S. H. R. 2016. Construction Aerodynamics of Cable-stayed Bridges for Record Spans: *Stonecutters Bridge*. *Structures* 8(1): 94–110.

Xia, G. Y. 2001. Nonlinear Analysis of Stayed Cable. *Journal of Transport Science and Engineering* (01): 47–50.

Xiao, H. Z. 2006. Study of Key Techniques for the Design of a Cable-stayed Bridge with a Span Longer Than 1000 Meters. *Bridge Construction* (S2): 1–4 + 8.

Xin, K. G. 2002. Nonlinear Static Analysis of Long-span Cable-stayed Bridges Under Dead Loads. *Journal of Tsinghua University* (06): 818–821.

Xu, K. Y. 2008. Geometric Nonlinear Static Analysis of the Long-span Cable-stayed Bridge. *Earthquake Resistant Engineering and Retrofitting* (03): 64–68.

Zhao, H. Y. 2020. Calculation Method for the Equivalent Elastic Modulus of Cable Sag Effect Based on Catenary. *Bridge Construction* 40(02): 62–668.

Numerical analysis on fully bolted beam-to-column joints in assembled steel structure

Haisheng Liang, Wei Xie & Xiaoxiang Tang*
State Grid Shanghai Economic Research Institute, Shanghai, China

Xiaoqun Luo
College of Civil Engineering, Tongji University, Shanghai, China

ABSTRACT: This article studies the numerical analysis on fully bolted joints in assembled steel structure combined with finite element (FE) method with theoretical calculation. First, the FE model of a typical column-beam joint was established in some substations. The results of the FE model and the theoretical calculations were compared to prove the effectiveness of the calculation. Parametric studies were carried out to analyze the influence of the thickness of stiffeners and end plates. The factors in the calculation as the effects of the prying action and failure modes were also put into consideration. The result showed that the setting of stiffeners and the increase of the thickness of the end plate would both improve the bearing capacity and stiffness of the joint, while the influence would become weak if the end plate is thick enough. The reasonable selection of the size of stiffeners and end plates would meet the requirement of the design value of loads and construction cost.

1 INTRODUCTION

The global mechanical performance of steel structures is mainly determined by joints due to their significant role in transmitting shear forces and bending moments. Joints connected by high-strength bolted end plates, which are also recommended to be applied in the current steel structure specifications, are widely used in prefabricated steel structures because of convenience in production, installation, and affordable cost. Joints connected by high-strength bolted end plates are generally categorized as semi-rigid connections based on the relationship between bending external moments and internal corners (Zhuo et al. 2021). In fact, these joints have sophisticated mechanical performance (Jiang et al. 2021; Mourad et al. 1996) and load transmission paths owing to the influence of the construction of nodes, material characteristics, and distribution of loading (Sajid et al. 2021; Zhuo et al. 2021). To meet the requirement of construction, the one-way bolts, an easy and effective connecting type, are used in fully bolted joints because they can be secured and linked on the outside without breaking the closed sections.

A finite element (FE) model is established to simulate a typical beam-column joint of the low-level prefabricated steel structure in a substation to analyze the mechanical performance under external loading. The results from the model are then verified and extended by theoretical calculation results. The parameters of end-plate thickness and stiffeners' thicknesses are studied in the FE mode, and the design advice for this joint is provided as well.

---

*Corresponding Author: tang_xiaoxiang@126.com

2 A BRIEF INTRODUCTION OF JOINTS

According to the plane size of the steel frame joints of the State Grid substation, the typical beam-column joints between the points of inflection are chosen for analysis. As plotted in Figure 1, the height of the column and the length of the beam are 2000 mm and 500 mm, respectively. The rectangular sectional stiffener between the flanges of the column is set at the position corresponding to the flange of the beam. To satisfy bearing capacity requirements, 10.9 class high-strength bolts with a diameter of 27 mm were arranged in 6 rows and 4 columns between the end plate and column. The distances between the vertical and horizontal bolts in the middle section are all 85 mm, the edge distance is 54 mm, and the edge distance between the bolts and the extended outer plate is 60 mm. Except for the bolts, the steel used in the beams, columns, and end plates is Q355B steel. The parameters of the specimens are shown in Table 1. The steel utilized for beams, columns, and end plates is Q355B steel, and their specimens' parameters are listed in Table 1.

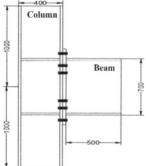

Figure 1. Size of main members.

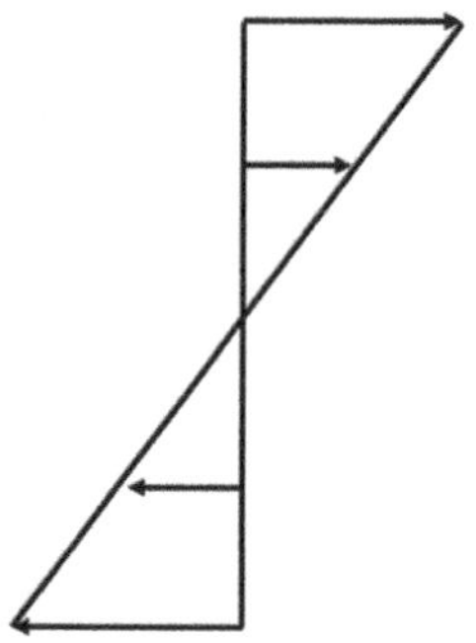

Figure 2. Force distribution of bolts.

Table 1. Parameter of members.

Section of column/mm	Length of column/ mm	Section of beam/mm	Length of beam/ mm	Section of end-plate/ mm	Section of stiffener/mm
HN400 × 400 × 14 × 24	2000	HN700 × 300 × 12 × 24	500	940 × 300	352 × 169 × 24

3 THEORETICAL ANALYSIS OF FULLY-BOLTED JOINTS

The joints connected with high-strength bolted end plates can be guaranteed to transmit the majority of the bending moments of the connected beams and columns, as the web stiffeners are arranged. However, the joints connected with bolts should be considered as a semi-rigid node for further calculations and parameter analysis, thanks to its constraint stiffness. The initial rotational stiffness and ultimate plastic flexural capacity of the jointing with a 30-mm-thick end plate are calculated theoretically in this section.

The first rotational stiffness of the joint is calculated as follows:

$$K_i = \frac{h_0^2}{\frac{1}{K_t} + \frac{1}{K_{cwc}} + \frac{1}{K_{cwv}}} \qquad (1)$$

where, h_0 is the distance between the midpoint of the upper and lower flange of the column section; K_t is the global stiffness of tension components, which is determined by the tensile stiffness of the bolt, the bending stiffness of the end plate, the bending stiffness of the column flange, and the tensile stiffness of the column web; K_{cwc} is the compression stiffness of the column web; K_{cwv} is the shear stiffness of the column web. According to the joint size information, the global rotational stiffness K_i can be calculated.

Generally, the joints may appear as three failure types under loads: bolt tensile failure, end plate failure, and column wall failure. The failure of bolts and end plates is mainly taken into account due to the column wall being restrained by the stiffener and web. The calculation process applies the assumption of friction-type high-strength bolts from Technical Regulations for Connecting High-strength Bolts of Steel Structures in China, which means the neutral axis of rotation lies in the centroid of the bolt group, and tension follows a linear distribution (Figure 2). Thus, the bearing capacity of the joint determined by the bolt failure is:

$$M_{bt} = \frac{mF_{u,b}}{h_1} \sum h_i^2 \qquad (2)$$

where m is the columns of bolts; $F_{u,b}$ is the ultimate tensile capacity of one-way bolts; h_i is the distance between the distance from the i-row bolt to the centroid of the bolt group.

This paper adopted the T-shaped connector approach of the European Standard (European Committee for Standardization 2005) to calculate the bearing capacity determined by the end plate. This standard simplifies the assembly of the end plate of each bolt row and the web of steel beams into multiple T-shaped parts; the bearing capacity can be obtained by calculating the bearing capacity of these T-shaped parts. Then, the total bearing capacity determined by the end plate is:

$$(F_{u,1} + F_{u,2})H \qquad (3)$$

where H is the height of the beam section; $F_{u,1}$ is the ultimate tensile capacity of the T-shaped flange; $F_{u,2}$ the ultimate tensile capacity of the T-shaped web. Overall, it may be said that end-plate failure carries a far heavier weight than bolt failure.

4 FINITE ELEMENT MODE ANALYSIS

This article establishes an FE model using Abaqus to further investigate the mechanical characteristics of the joints. To guarantee the precision of and the efficiency of calculation, the established model is simplified as follows:

(1) Initial geometric flaws and residual stresses are neglected
(2) The bolt rod is simplified as a cylinder, and the bolt cap and bolt rod are treated as a single unit.

Stiffeners (Q355B) composed of isosceles right triangle section stiffeners with sides of 120 mm are added at the midpoint of the beam web and the welding location between the beam flange and the end plates (as shown in Figure 3) to prevent deformation of the end plates. The above-mentioned material parameters of Q355 steel are used; the elastic modulus is 206 GPa, and the Poisson's ratio is 0.3. According to Figure 4, all constitutive relationships between the steel and bolts are based on the double-slash model. Steel has a 378.3 MPa yield strength, a 500-MPa ultimate strength, and an ultimate plastic strain of 0.06. High-strength bolts have a 960-MPa yield strength, an 1100-MPa ultimate strength, and an ultimate plastic strain of 0.008.

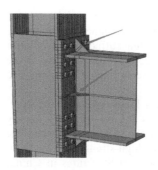 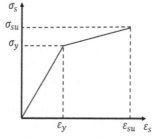 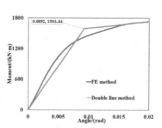

Figure 3. Result of double line method.

Figure 4. Position of stiffeners.

Figure 5. Double slash model.

5 PARAMETRIC ANALYSIS FOR JOINT

FE models of various sizes were established to study the mechanical performance of different joint models. The specific model parameters are displayed in Table 3, and all loading techniques are monotonic loading techniques.

5.1 *Stiffeners of the end plate*

This research sets up five models with different thicknesses of end-plate stiffeners under monotonic load while maintaining the thickness of the end-plate to verify and assess the impact of end-plate stiffeners on the bearing performance of the joint. The end-plate stiffener is not specified in the control model, S0.

Table 2. Comparison between the FE method and theoretical calculation.

	FE method	Theoretical calculation	Error value
Initial rotation stiffness/(N·mm/rad)	1.732×10^{11}	1.589×10^{11}	8.26%
Yield bearing capacity/(kN·m)	1593.44	1456.54	8.59%

Table 3. Parameter of models.

Model number	Thickness of stiffeners/mm	Thickness of end plates/mm	Model number	Thickness of stiffeners/mm	Thickness of end plates/mm
S0	0 (without stiffeners)	24	EP16	12	16
S6	6	24	EP20	12	20
S10	10	24	EP22	12	22
S12	12	24	EP26	12	26
S16	16	24	EP30	12	30

The results of the bending moment-corner and the ultimate bearing capacity of the joint are shown in Figure 6 and Table 4. Each moment-corner curve exhibits that the joint of a different model is in the elastic stage when the moment increases initially but is less than a certain value. As the load continues to apply, the preload of the bolt is gradually overcome

Table 4. Ultimate bearing capacity of different models.

Model number	S0	S6	S10	S12	S16
Ultimate bearing capacity/kN·m	1571.31	1702.64	1776.22	1788.56	1795.79

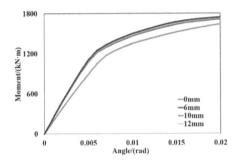

Figure 6. Models with different thicknesses of stiffeners.

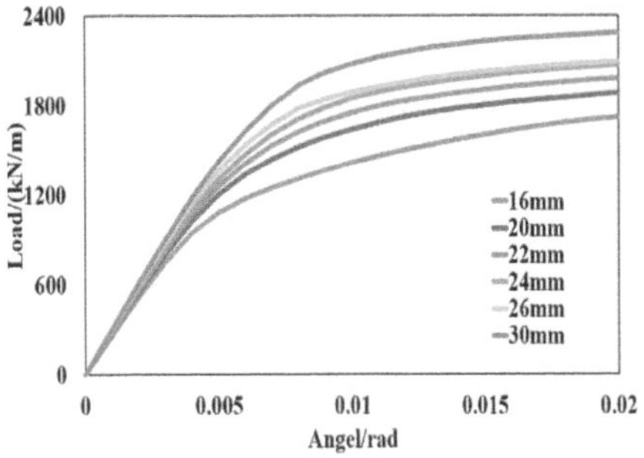

Figure 7. Models with different thicknesses of the end plate.

Table 5. Ultimate bearing capacity of different models.

Model number/mm	EP16	EP20	EP22	S12	EP26	EP30
Ultimate bearing capacity /kN·m	1366.52	1589.12	1696.17	1788.56	1926.54	2210.31

by the tension generated by the load, and the end plate near the upper beam flange is pulled away from the column wall flange. The bolt deformation increases, and the failure occurs first as the T-shaped specimen formed by the beam flange and the end plate gradually yields. The bearing capacity and the rotational stiffness of the joint are both greatly increased as the end plate stiffeners are arranged by comparing with the various results of different models. Moreover, the bearing capacity and stiffness continue to improve as the thickness of the end-plate stiffener increases. However, these two properties almost stop growing after the end-plate stiffener thickness reaches 12 mm. The fundamental reason is the stiffener exceeding 12 mm could generate a better constraint effect on the joint, and it is no longer the determining factor of the joint's mechanical performance.

5.2 Thickness of the end plate

This section studies the influence of the thickness of the end plate on the bearing capacity and rotational stiffness of the joint by comparing the results of six models with different thicknesses of the end plate while maintaining the 12-mm-thickness of end-plate stiffeners. The load is the same as in section 5.1. As shown in Figure 7 and Table 5, the end-plate thickness greatly enhances the bearing capacity and rotational stiffness of the joint. The end plate thickness and the maximum bearing capacity have a substantially linear correlation coefficient of 0.9980. The global deformation of the end plate can be reduced as the thickness of the end plate increases. As the end-plate thickness is increased to 30 mm, the lateral bending deformation of the bolt is small when the pure tensile failure occurs due to the bending moment, eventually leading to the failure of the joint. This is an appropriate joint since it allows full use of the mechanical properties of high-strength bolts.

5.3 Prying force

The above analysis of the parameters shows that the joint's bearing capacity will further increase with the restriction on the deformation of the end plate. The essence of increase is that the bolt's prying force decreases due to the displacement gap between the bolt's two sides narrowing with a decreased deformation of the end plate. When the prying force is reduced, the strength of the bolt can be fully exhibited in the tension, since the bolt is the controlling factor of the failure of the joint. Therefore, increasing the bolt's strength improves the bearing capacity of the joint significantly. As a result, the bolt's prying force can be neglected as the end plate is sufficiently thick. According to the modified formula of the bolt's bearing capacity and considering the influence of the prying force, the maximum end-plate thickness t_c is calculated as follows,

$$t_c = \sqrt{\frac{4N_{tu}^b \cdot e_2}{bf_y}} \quad (4)$$

where N_{tu}^b is the tensile capacity of a single bolt; e_2 is the horizontal distance from the bolt hole to the center of the beam web; b is half of the end-plate width. The calculation formula of prying force Q is:

$$Q = \delta \cdot \alpha \cdot \frac{e_2}{e_1} \cdot \left(\frac{t}{tc}\right)^2 \cdot N_{tu}^b \quad (5)$$

where δ is the net cross-section coefficient of the end plate; α is the ratio of the section to the plastic bending moment of the bolt. It is defined that the prying force ratio is the ratio of the prying force to the bolt's net force (the summation of the prying force and the external force). Table 6 lists the prying forces and prying force ratios of different models.

Table 6. Prying force and prying rate of different models.

Model number	S0	S6	S10	S12	S16	EP16	EP20	EP22	EP26	EP30
Prying force/kN	164.47	188.60	202.12	204.39	205.72	195.87	205.71	207.26	208.17	212.00
Prying rate%	35.68	36.99	37.62	37.72	37.78	43.17	40.69	39.31	36.41	33.70

It can be seen that the stiffeners' thickness has no significant effect on the prying force. In fact, the end-plate stiffeners are set up at the location of the welding between the beam flange and the end-plate, and they can only effectively limit the end plate's deformation and relative rotational angle of the beam sections; they, however, have the limited capacity for controlling the end-plate deformation near the bolt. As a result, the arrangement of the stiffeners cannot effectively lessen the bolt's prying force. In this instance, the increase in bearing capacity of the joint was provided by the influence of the compressed stiffeners of the end plate near the lower beam flange. The global deformation capacity of the end plate will decrease as end-plate thickness rises, resulting in an obvious decrease in the bolt's prying force ratio and a higher utilization rate of the high-strength bolt. The research shows that adding more end-plate thickness will not further increase the joints' bearing capacity as the end-plate thickness reaches 43 mm.

6 CONCLUSIONS

Based on the results of the numerical analysis of a fully bolted beam-column joint, some valuable conclusions could be drawn as follows:

(1) The bearing capacity of this type of joint determined by the plate is significantly more than that determined by the bolt, indicating the failure of the bolt typically occurs first.

Therefore, the relative strength relationship between the bolt and end plate should be carefully taken into account when designing a joint.
(2) Arrangement of stiffeners can obviously raise a joint's bearing capacity and rotational stiffness. The thickness of stiffeners, however, has little effect. Thus, the thickness of the end plate should not be excessively large in engineering applications.
(3) In this study, the joint's bearing capacity rises as the thickness of the end plate increases, and the increase's curve exhibits a linear law within a particular range. The rotational stiffness has similar properties as well. However, this increment effect on the mechanical performance appears as an upper limit. Therefore, the thickness of the end plate is determined by the specific bearing capacity of the joint.
(4) The bolt will undoubtedly be subjected to the prying force brought on by the deformation of the end plate because the end plate cannot entirely fit the flange of the column in the H-type beam-column connection. To prevent the bolt from being damaged by the joint action of bending moment and additional prying force, the design should take into account both the external tension created by the load and the prying force of the bolt.

ACKNOWLEDGMENT

This research was partially supported by State Grid Shanghai Economic Research Institute for objective projects.

REFERENCES

European Committee for Standardization (CEN). EN 1993-1-8-Eurocode 3 (2005). *Design of Steel Structures-Part 1-8: Design of Joints. S. Brussels*: CEN. (16)

Jiang K. L. & Z. F. Yang (2021). Effects of the joint surface considering asperity interaction on the bolted joint performance in the bolt tightening process. *J. Tribology International*. (9)

Liu Z. Y. & X. Y. Dong (2022). Study on the Performance of the Blind Bolted Connection Joints of the Square Steel Tube column-H-shaped Steel Beam T-stub. *J. Build Struct*. 52(03);116–123. (3)

Mourad S. & M. Korolr (1996). Design of Extended End-plate Connections for Hollow Section Columns. *J. Canadian J Civil Eng*. 23(1); 277–286. (5)

Sajid Z. & S. Karuppanan (2021). Carbon/basalt Hybrid Composite Bolted Joint for Improved Bearing Performance and Cost Efficiency. *J. Comp Struct*. 275. (11)

Zhuo D. B. & Cao H. (2021). Damage Identification of Bolt Connection in Steel Truss Structures by Using Sound Signals. *J. Struct Heal Monit*. 21(2). (7)

Estimation method of bulk material quantity of engineering cost based on BP algorithm

Shu Zong*
Jinqiao College of the Kunming University of Technology, Kunming, Yunnan, China

Bin Chen
Department of Engineering, Yunnan Institute of Technology and Information Career College, Kunming, Yunnan, China

ABSTRACT: Due to the long construction period of construction projects, it is difficult to accurately estimate the number of materials to meet the needs of the entire project. Too much material will cause waste, and too little will delay the progress of the project. It is very difficult to calculate the number of bulk materials in the bill of quantities, and the number of bulk materials varies with factors such as project scale, design habits, and construction methods, which bring difficulties to the project cost work. Therefore, this paper starts with the estimation of the number of bulk materials in the construction cost, taking a residential construction project in Yunnan as an example to establish an estimation model for the number of bulk materials based on the BP algorithm and divide the residential sample data in Yunnan into training samples and test samples. The network model training has obtained a good training effect, and the relative error between the measured value and the estimated value is within 10%, which shows the feasibility of the model proposed in this paper in the estimation of bulk material. The smaller the relative error is, the less the waste of bulk materials is, then the model can be used to reasonably control the construction cost so that the limited funds can be more reasonably used in the construction project, so as to improve the use efficiency of building materials.

1 INTRODUCTION

With the structuring and large-scale development of construction projects, engineering facilities, and construction technology becoming more and more complex, building users, managers and owners are increasingly concerned about improving the efficiency of the use of building materials. But in actual engineering, one-sided will cause a waste of talent. Therefore, rapid and accurate estimation of bulk material volume is of great significance for investment control of construction projects.

Nowadays, many scholars have proposed methods for estimating the number of bulk materials for construction costs, and have achieved good research results. For example, a scholar uses the momentum method and learning rate adaptive adjustment strategy to improve the BP NN model, according to the data of a completed project to verify the accuracy of the project cost, project index, and project volume predicted by the model, and compare and analyze based on the two results obtained without considering the experimental hardware and software environment. The verification found that the accuracy of

---

*Corresponding Author: 928288652@qq.com

considering the experimental environment is higher (Patricio et al. 2020). Some scholars use the standard BP NN to simulate the network process through MATLAB software and estimate the engineering cost and material cost of the expressway toll station building (Alao et al. 2018). A scholar used the BP neural network (NN) model to predict the unilateral cost of construction projects as well as the engineering bulk material quantities of main building materials such as steel bars and concrete. The applicability of the BP NN neural network in the estimation of engineering bulk material cost is verified (Kinderen 2018). Although many scholars use the BP algorithm to estimate the amount of engineering bulk materials, it is worth noting that the training of the BP NN model requires the collection of multiple engineering data to make the training effect better.

In this paper, the concept of the BP NN model is firstly introduced, and the BP algorithm model is proposed. The number of layers and nodes of the NN is designed when constructing the estimation model for the quantity of engineering cost bulk materials based on the BP algorithm, and in Matlab software, the training process of BP NN can be realized. Then, the model simulation training is carried out by taking the residential construction project in Yunnan as an example, which verifies the validity of the model in the estimation of bulk material quantity.

2 BP NN AND ITS ALGORITHM

The NN algorithm is to use the computer for simulation and makes predictions with the help of a large number of database resources (Cangioli et al. 2018). BP NN has a multi-layer feedforward NN, which can effectively solve the learning problem between hidden layers (HL). By repeating such a process of learning and adjustment, the error signal can finally be made close to the set target. At this time, the network stops training and the learning method of minimum mean square error is adopted in this process (Cangioli et al. 2018; Postnikova 2021). Forward propagation process.

$$S_i^k = \sum_{j=1}^{N} w_{ij} o_j^k - \theta_i \qquad (1)$$

Among them, i=(1, 2,..., k), S_i^k represents the input network, w_{ij} is the connection weight between the i-th neuron and the j-th neuron, and o_j^k is the input layer where the j-th node passes through p output network, θ_i is the threshold of the ith neuron in the HL, and N is the number of neurons in the input layer.

3 ESTIMATION MODEL OF PROJECT COST BULK MATERIAL QUANTITY BASED ON BP ALGORITHM

3.1 Model design

Although great achievements have been made in the research and application of NNs, there is still no complete set of theories to guide them. In the actual application process, it is necessary to fully understand the problem to be solved, combine it with the actual situation, and finally select a better design scheme through many improvement experiments (Sullivan 2019).

In the practical application of BP NN in engineering cost, first of all, it is necessary to clarify the requirements of the role, and at the same time grasp the type of the project. Then, select a fast and accurate model for the determination of the number of NN nodes, the number of layers, and the classification and characteristics of each neural unit. According to the input and output of the network structure selected by the node, the sample value is simulated according to the known samples in the instance, and the process of training is debugged (Melvin 2018; Samira et al. 2019). Figure 1 shows the estimation process of bulk material based on the BP algorithm model.

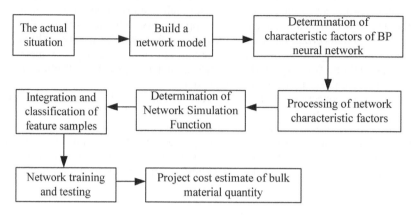

Figure 1. BP NN neural network cost and bulk material quantity estimation process.

Network layer design: For traditional NN design, the network structure is the basis of calculation, and the number of HLs also needs to be considered. For the HL, only discontinuous functions need to increase the number of HLs. When designing the number of network layers, or when an HL cannot improve the network performance, gradually increase the number of HLs. In some practical problems, if increasing the number of network nodes cannot reduce the network training error, the number of HLs should be appropriately increased. The activation functions selected in the estimation model of engineering cost and bulk material quantity established in this paper are all sigmoid functions, which are continuous functions in the definition domain, so only a single HL structure needs to be used in the network structure design (Han & Roy 2018; Tachibana & Ohtsuki 2020).

Node number design: The setting of the number of HL nodes is affected by three factors, one is the number of selected training samples, the second is the irregular content in the training samples, that is, the size of the noise, and the last one is the training The complexity of the regular content hidden in the sample (Wani & Quadri 2019).

Initial weight design: The nonlinear nature of the BP network causes the network to consider whether the learning of the network can achieve the required local minimum and whether the network can converge when selecting the initial value. When selecting the initial value, it is the most important requirement to make the initial value fall as close to the zero point as possible. Generally, the initial weights selected are randomly between (-1, 1).

3.2 *Realization of BP network on Matlab software*

MATLAB has its own toolbox, including the NN toolbox, which can realize rapid modeling and solution. When the software is continuously updated, the toolbox can also be downloaded and used through the official website according to the required toolbox (Obukhov & Krasnyanskiy 2021).

Through the NN toolbox function, people can realize the functions of BP NN model establishment, initialization, model training and learning, and error curve drawing. The establishment functions and training functions are included in the toolbox. You only need to define and assign variables when using them and do not need to consider the writing of programming language and the analysis of logical thinking. The training function can easily implement the tedious NN training process, greatly reduce the workload of the calculation, and realize an efficient solution to the problem (Hatamleh et al. 2018).

The prediction method of the BP network using Matlab software is as follows:

Open the Matlab software: directly call the NN toolbox tool in the Matlab software.

Import data: Click Import to import the input vector and output vector into the corresponding [Inputs] and [Targets] windows respectively.

Creation of BP network: Click the New Network button on the main interface, and enter various parameters in the pop-up dialog box.

After setting all parameters, click the View button to preview the schematic diagram of the network structure. Check whether the network structure settings are correct. If there is no problem, you can create the network, that is, click the Create button. After the BP network is established, the initial weight can be manually set in the Initialize option. If not set, the default weights in the program are executed.

4 ENGINEERING EXAMPLE

The case selected in this paper is a residential community in a city in Yunnan Province. The construction project was invested and developed by the Municipal Real Estate Development Co., Ltd. with a construction area of about 183,000m2. Thirty unfurnished residential buildings and commercial buildings are included, and one residential building is selected as a case for empirical analysis. The construction materials used in the project are mainly steel, concrete, cement, and wood.

When the sample database was established, there were 120 groups of Yunnan housing engineering sample data with complete information and a wide range of data, and these 120 groups of samples were divided into training samples and test samples. In order to ensure the practicability of the model, among the 120 groups of samples, a group of samples needs to be randomly selected as the test samples. At this point, there are 119 training samples and 1 testing sample. Among them, the training samples are used to train the established BP NN model. The test sample calculates the error and error percentage according to the network output vector and expected output vector obtained from the test, and observes the accuracy of the analysis model and the network performance of the model.

4.1 *Model training*

In order to intuitively analyze the performance of the BP NN algorithm, 119 groups of training samples were trained using the BP NN model, and the training results obtained are shown in Figure 2.

It can be seen from the figure that the performance gradient of the BP NN decreases rapidly when it is trained 25 times, and the error decline curve gradually becomes flat from the 25th time onward, indicating that the BP algorithm estimation model established in this paper is practical and reliable.

4.2 *Model validation*

In previous studies, most scholars are generally used to taking the total cost of construction projects or the unilateral cost index as the only output vector. It is not appropriate to measure the project cost of the proposed project by the price status of typical projects or training samples. Therefore, when establishing the estimation model, the consumption of materials should be used as the output value, which is just suitable for the estimation of the number of bulk materials. It is also necessary to consider the consumption of labor days and add the unilateral cost as the estimated output. In the process of model training and testing, two output vectors will be involved, one is the output vector of the NN, which is the predicted value of the model, and the other is the actual value of the sample, which is the so-called

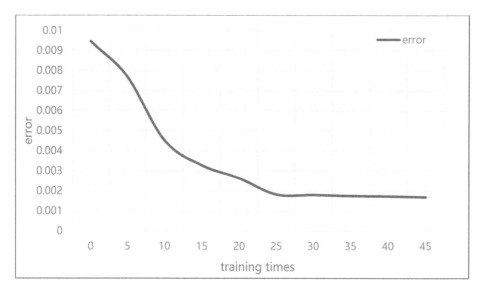

Figure 2. Training error.

expected output vector and is determined by them. The difference between them can calculate the error of the network, so it is necessary to explain the meaning of the output vector and the expected output vector respectively. The output vector is represented by Ai, and the expected output vector is represented by Bi.

Using the engineering cost estimation model obtained after training, use the sigmoid function to simulate the randomly selected test sample, and you can obtain five output values of the test sample. According to the output value and the difference between the output vector and the expected output vector, the test samples were subjected to error analysis, and the results are shown in Table 1.

Table 1. Simulation results.

Output vector	Actual value	Predictive value	Error	Relative error (%)
A1	0.1635	0.1579	0.0056	3.43
A2	0.1247	0.1285	−0.0038	3.08
A3	0.2516	0.2597	−0.0081	3.21
A4	0.0876	0.0804	0.0072	8.22

According to the output results of the test samples, the error between the unit project bulk material cost estimated by the BP NN algorithm and the actual project bulk material cost is within 10%. In general, an error within 10% of the estimated project cost is allowed within the acceptable range, which shows that the model is feasible in the estimation of bulk materials in construction cost. The error between the actual value and the predicted value is very small, indicating that the accuracy of the estimated value of bulk material is very high.

5 CONCLUSIONS

In this paper, the Yunnan residential project is simulated as a project example, and the BP NN is used to establish a project cost bulk material quantity estimation model to predict the bulk material quantity of the project, and the BP algorithm model is trained on Matlab software. The simulation results show that after 25 times of model training, the error is gradually flattened, and the estimation error of the amount of bulk material is within 10%, which shows that the BP algorithm proposed in this paper can be used to estimate the amount of bulk material, and the model training effect is very good. The estimation of the amount of bulk material can provide a reference for the project cost and increase the efficiency of engineering projects.

ACKNOWLEDGEMENT

This paper is one of the phased achievements of the scientific research fund project of the Yunnan Provincial Department of education.

REFERENCES

Alao O.O., Jagboro G.O., Opawole A. Cost and Time Implications of Abandoned Project Resuscitation: A Case Study of Educational Institutional Buildings in Nigeria[J]. *Journal of Financial Management of Property and Construction*, 2018, 23(2):185–201.

Al-Shammary A., Kouzani A.Z., Kaynak A., et al. Soil Bulk Density Estimation Methods: A Review[J]. *Pedosphere*, 2018, 28(004):581–596.

Cangioli F., Pennacchi P., Vannini G, et al. Effect of the Energy Equation in One Control-volume Bulk-flow Model for the Prediction of Labyrinth Seal Dynamic Coefficients[J]. *Mechanical Systems & Signal Processing*, 2018, 98 (Jan. 1):594–612.

Han B., Roy K. DeltaFrame-BP: An Algorithm Using Frame Difference for Deep Convolutional NNs Training and Inference on Video Data[J]. *Multi-Scale Computing Systems, IEEE Transactions on*, 2018, 4(4):624–634.

Hatamleh M., Hiyassat M., Sweis G.J., et al. Factors affecting the accuracy of cost estimate: Case of Jordan [J]. *Engineering Construction & Architectural Management*, 2018, 25(1):00–00.

Kinderen D.D. PHMSA Letter of Interpretation Bulk Shipments of Class 9 Material[J]. *The Journal of Hazmat Transportation*, 2018, 29(1):47–47.

Melvin J. Bergen-Linden Transmission Project Costs Continue to Ruffle Feathers[J]. *Platts Megawatt Daily*, 2018, 23(18):4–5.

Obukhov A.D., Krasnyanskiy M.N. Automated Organization of Interaction Between Modules of Information Systems Based on NN Data Channels[J]. *Neural Computing and Applications*, 2021, 33 (12):7249–7269.

Patricio J., Kalmykova Y., Rosado L. A Method and Databases for Estimating Detailed Industrial Waste Generation at Different Scales - With Application to Biogas Industry Development[J]. *Journal of Cleaner Production*, 2020, 246 (Feb. 10):118959.1–118959.14.

Postnikova I.V. Two-Stage Grinding as the Most Cost-Effective Option for Organizing the Process of Ultrafine Grinding of Solid Materials[J]. *Russian Journal of General Chemistry*, 2021, 91(6):1218–1223.

Samira N., Hamed M.S., Shahram A. Predicting the Project Time and Costs Using EVM Based on Gray Numbers[J]. *Engineering construction & architectural management*, 2019, 26(9):2107–2119.

Sullivan S. FERC Accepts Columbia Tariff Over Utility Protest Related to Project Costs[J]. *Inside Ferc*, 2019 (FEB.4):16–16.

Tachibana J., Ohtsuki T. Learning and Analysis of Damping Factor in Massive MIMO Detection Using BP Algorithm With Node Selection[J]. *IEEE Access*, 2020, PP (99):1–1.

Wani Z.H., Quadri S. An Improved Particle Swarm Optimization-based Functional Link Artificial NN Model for Software Cost Estimation[J]. *International Journal of Swarm Intelligence*, 2019, 4(1):38–54.

Comparative study on prediction models of EPB shield tunnelling parameters in the water-rich round-gravel formation

Jun Wang
China Railway 16th Bureau Group Beijing Rail Transit Construction Co., Ltd, Beijing, China

Jinpeng Zhao* & Zhongsheng Tan
Beijing Jiaotong University, Beijing, China

Songyan Fu
China Railway 16th Bureau Group Beijing Rail Transit Construction Co., Ltd, Beijing, China

ABSTRACT: There are many prediction models for shield tunneling parameters, and there is a lack of comparative study on the applicability of various prediction models in the same stratum. This paper combines the three models of SVR, linear regression, and BP neural network in MATLAB software to train and learn the shield tunneling parameter data of the Shuai-Nei section of Hohhot Rail Transit Line 2 and predict the tunneling speed of the shield machine in this section. The results show that the SVR regression model has the lowest prediction accuracy and is unsuitable for predicting shield tunneling speed in this stratum. After the noise reduction of the input tunneling parameters, the BP neural network model and the linear regression model can better predict the tunneling speed of the shield machine in this stratum with the accuracy of the test set being 87%. Applying the BP neural network model and linear regression model to simulate the tunneling parameters of EPB shield in the water-rich round gravel stratum is good. Among them, the accuracy rate of the training set of the BP neural network regression model is 98%, which indicates that the nonlinear mapping ability and generalization application of the BP neural network are excellent.

1 INTRODUCTION

The 21st century is the century of underground space development and utilization. With the increasing population base, urban ground traffic becomes increasingly crowded. Therefore, the construction of underground rail transit is significant. In this process, the shield construction method is gradually replacing the mining method and becoming the mainstream construction method for urban subway construction due to its fast-tunneling speed and negligible environmental impact.

In shield tunneling, the equipment automatically records dozens of tunneling parameters such as thrust, cutter head rotation speed, torque, tunneling speed, penetration, and bunker pressure. These parameters are not entirely independent but have a specific correlation. Only by reasonably selecting the tunneling parameters, especially some key parameters, can the surface settlement be effectively controlled and the safety of shield construction be ensured. However, even if the influence of machine failure or personnel operation is ignored, the tunneling parameters of the shield are still affected by many other factors, such as stratum conditions and the groundwater environment. Moreover, each factor will influence the other,

*Corresponding Author: 18115060@bjtu.edu.cn

resulting in complicated internal relations. It is a complex high-order nonlinear fuzzy system (Yao 2020). Therefore, it is not easy to accurately express the relationship between tunneling parameters with an explicit function, and it is challenging to establish a complete mathematical prediction model of tunneling parameters based on external environmental parameters.

Many scholars use different methods to simulate and predict tunneling parameters. Abroad, N. Barton (1999) considered the direction of joint structure and the point load strength of rock based on the extended Q system of rock mass classification and the average cutting force related to rock mass strength. At the same time, the tool life index and rock stress level are considered to estimate the TBM performance during tunneling. A. Bruland (2000), combined with the shield's construction parameters and indoor test data, proposed an NTNU prediction model to predict the shield's tunneling speed and other parameters. Bruines P. A (2001) proposed a TBM construction schedule prediction model using a fuzzy neural network for simulation. J. Hassanpour et al. (2011) used statistical methods to analyze the shield tunneling parameters when the Karachi water conveyance tunnel crossed the magmatic rock stratum and proposed a new prediction model, which can be used to predict the tunneling parameters such as torque and cutter head speed.

Domestic aspect, Zhang et al. (2005) studied the tunneling parameters of shield construction of Guangzhou Metro Line 3 through orthogonal experiments and proposed a prediction model of torque and tunneling speed suitable for EPB shield tunneling in the soft soil layer. Yang (2008) studied the tunneling parameters of a shield project for Guangzhou Metro by using statistical methods and combined RBF and BP neural networks to propose a torque and thrust prediction model. Li et al. (2017) used BP neural network to train and learn the shield tunneling parameters of Shenzhen Metro Line 11 and put forward a prediction model for the tunneling parameters of different types of shields under the condition of composite strata. Su (2018) studied the correlation between the main driving parameters of the Herrick mud water balance shield in the Nanjing Yangtze River tunnel project utilizing mathematical statistics and established a BP neural network model that can predict the driving speed.

Based on the shield tunnel project in the Shuai-Nei section of Hohhot Rail Transit Line 2, combined with the water-rich round gravel stratum, this paper uses the SVR regression model, linear regression model, and BP neural network regression model to predict the tunneling parameters of earth pressure balance shield, and finally compares the prediction results.

2 INTRODUCTION TO THE PREDICTION MODEL OF SHIELD TUNNELING PARAMETERS

2.1 *Principle of SVR regression model*

SVR (support vector regression) is the extension and extension of a support vector machine (SVM). The difference between SVR regression and SVM classification is that SVM aims to find a hyperplane that can make different kinds of sample points farthest from each other to classify and predict samples. SVR aims to find a hyperplane that can close all the sample points to realize the fitting between the sample and the plane. That is, specify the training set sample $D= \{(x_1, y_1), (x_2, y_2), \ldots , (x_n, y_n)\}$, $y_i \in R$, in the following form, find the regression model in the form of $f(x)=\omega^T x+b$, and make the difference between $f(x)$ and y as small as possible (Su, 2018).

SVR adopts the idea of a soft interval support vector machine to deal with regression problems.

SVR will tolerate the deviation of ε between the model output $f(\mathrm{x})$ and the model input y_i. Only when the deviation is more significant than ε will it be included in the loss function. Otherwise, the loss of the sample will be 0. The objective function (Xia 2016) of SVR is

$$\min_{\omega,b} \frac{1}{2}\|\omega\|^2 + C\sum_{i=1}^{m} l_\varepsilon(f(x_i) - y_i) \qquad (1)$$

where ω is the coefficient of the hyperplane, C is the coefficient of the common term, and l_ε is an insensitive loss function (Insensitive Loss).

$$l_\varepsilon(z) = \begin{cases} 0, if |z| < \varepsilon \\ |z| - \varepsilon, otherwise \end{cases} \tag{2}$$

In order to learn the nonlinear information in the data set and improve the fitting ability of the model for the nonlinear relationship, the inner product of the nonlinear mapping of the sample, i.e., the kernel function, is often added to $f(x)$, which is recorded as $K(x_i, x_j)$. The commonly used kernel functions are the Gaussian and sigmoid kernel functions. The expressions are

$$K(x_i, x_j) = \exp\left(-\frac{\|x_i - x_j\|^2}{2\sigma^2}\right) \tag{3}$$

$$K(x_i, x_j) = \tanh(\alpha x_i^T x_j + \theta) \tag{4}$$

2.2 Linear regression model

The linear regression model is the simplest in machine learning. The advantage of the model is that the training speed is fast and the interpretability is strong. The most significant disadvantage is the poor fitting ability of the data set with a large amount of data and complex relationships between variables. Moreover, it is easy to overfit when facing a task with too few data sets and too many features (Wang 2016). Lasso regression solves the problem of overfitting well. The regular term is introduced into the loss function to be optimized, which alleviates the problem of overfitting to a certain extent. The introduced regular term is the L1 norm of the characteristic coefficient, and the sparse solution can be obtained by optimizing the loss function according to the properties of the L1 norm (Zhou 2016). The objective function of lasso regression is

$$\min_\omega \sum_{i=1}^{m} (y_i - \omega^T x_i)^2 + \lambda \|\omega\|_1 \tag{5}$$

where ω^T is the characteristic coefficient vector, $\lambda\|\omega\|_1$ is the common term of L1 norm, and λ is a regularization parameter greater than 0. Since the available time series problems rarely have complete linear relations, and the solutions of lasso regression have partial sparse solutions, Lasso regression is mainly used for feature selection in feature engineering.

Another variant of the linear regression model, Ridge regression, uses the L2 norm in the regular term, and the L2 norm often makes the optimal solution of the loss function of Ridge regression a non-sparse solution. Therefore, Ridge regression does not have the ability of feature screening. The introduction of L1 or L2 regular terms into the loss function has also been adopted by other more complex machine learning and deep learning models to alleviate overfitting or feature screening.

In this paper, the linear regression model of linear mapping is adopted. The model maps six feature quantities to high-dimensional space and then performs nonlinear regression. The gradient descent algorithm is used for linear regression fitting. The regularization term is considered, and the parameters adopted λ Control can better prevent overfitting, so the model's generalization is poor.

2.3 BP neural network regression model

BP neural network is a multilayer feedforward neural network trained by error back-propagation. It has the characteristics of a simple structure and strong plasticity. Its neurons are arranged in layers in the network, mainly composed of input, hidden, and output layers. The algorithmic system in this network solves the learning problem of remote unit connection weights in multilayer networks. For the output layer, the output of neurons in each layer is weighted and summed with the connection weight. It is transmitted as an input value to the next layer, enhancing or suppressing the output through the weight of the connection. The output of each neuron depends on the input value of its upper layer and its activation function.

The training process of the BP neural network is as follows. Firstly, the neural network is trained with a large number of sample data from the existing training set. The output value of the finally trained neural network can approach the real value as much as possible by continuously adjusting the connection weight and the size of the bias term between the networks of each layer. After that, the trained BP neural network is tested by the test set's sample data to verify the neural network's generalization ability. If the generalization ability is good, it can provide a more accurate output prediction effect for any new input value. If the generalization ability is poor, the original neural network should be optimized by adding standard terms and other methods. The BP neural network's learning algorithm includes forward and errors backpropagation. First, the input signal is transmitted to the output layer through the input layer and the hidden layer. The learning ends if the output prediction result is the same as the actual value. Otherwise, the error is back propagated according to the original path. The weights and offsets of all connections in the original network are adjusted to reduce the error. The signal propagation mode of the BP neural network is shown in Figure 1.

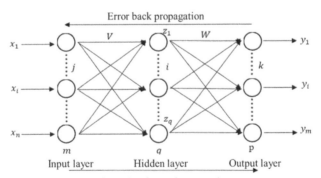

Figure 1. BP neural network structure.

3 ANALYSIS OF PREDICTION RESULTS OF SHIELD TUNNELING PARAMETERS

The data used in this paper are from the right line of the Shuai-Nei section of Hohhot Rail Transit Line 2. The driving parameter data of 400 loops (11-410) are selected as the training set, and the data of 100 loops (411-450, 511-570) are selected as the test set. The model prediction results of the training set and the test set are analyzed and compared with the actual tunneling parameters of the shield.

The accuracy rate is defined as the percentage of the deviation data between the model output value and the actual value within 5% of the total data. The deviation between the model output value and the actual value represents the ratio between the model output value and the actual value. For example, if the model output is 90 and the standard output is 96, the deviation is (96-90) / 96 = 0.0625, that is, 6.25%. If 30 of the data meet the 5% deviation range, the accuracy rate is 30 / 40 = 75%.

3.1 *Result analysis of the SVR prediction model*

The comparison between the output results of 400 training sets and 100 test sets of the SVR regression model and the actual values is shown in Figures 2 and 3.

By processing the results of the SVR prediction model, the prediction accuracy of the training set and the test set of the SVR prediction model is 61.0%, which is relatively low. It is mainly due to the intrinsic mechanism of the SVR algorithm, which has a weak nonlinear fitting ability.

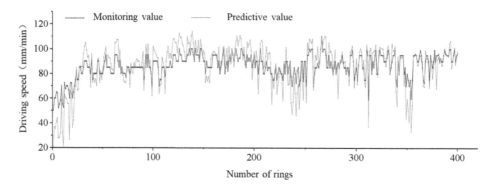

Figure 2. The training set output results of the SVR prediction model.

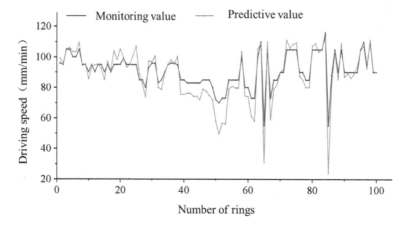

Figure 3. The validation set output results of the SVR prediction model.

3.2 *Result analysis of linear regression prediction model*

The comparison between the output results of 400 training sets and 100 test sets of the linear regression model and the actual values is shown in Figures 4 and 5.

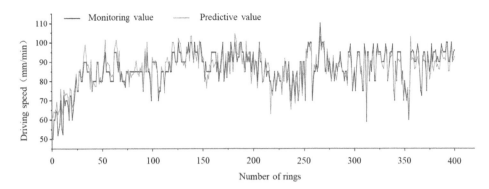

Figure 4. The training set output results of the linear regression model.

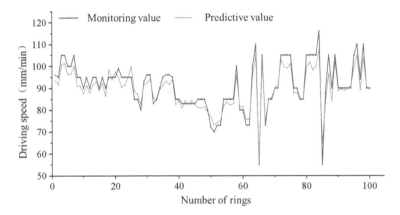

Figure 5. The validation set output results of the linear regression model.

By processing the prediction results of the linear regression prediction model, the accuracy of the training set of the linear regression model is 85.0%, and the accuracy of the test set is 87%. The accuracy is higher than that of the SVR prediction model, and the prediction effect is good. It is worth noting that noise reduction processing is performed on the data before the data training process.

3.3 *Result analysis of BP neural network prediction model*

The comparison between the output results of 400 training sets and 100 test sets of the linear regression model and the actual values is shown in Figures 6 and 7.

By processing the prediction results of the BP neural network prediction model, the accuracy of the training set of the BP neural network regression model is 98.0%, and the accuracy of the test set is 87%. Moreover, in the calculation process, the model only needs 16 steps to achieve a high degree of regression. Compared with the previous two prediction methods, the accuracy rate is the highest, which can better predict the tunneling speed of the ground shield machine. It is worth noting that noise reduction processing is performed on the data before the data training process.

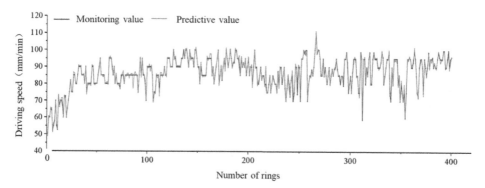

Figure 6. The training set output results of the BP neural network regression model.

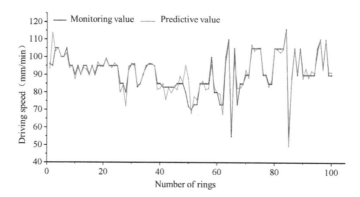

Figure 7. The validation set output results of the BP neural network regression model.

4 COMPARISON AND ANALYSIS OF PREDICTION EFFECT

The prediction results of the above three prediction models are summarized in Table 1. It can be seen from table 1 that for the prediction of shield tunneling speed in the water-rich round gravel stratum, different prediction models have different prediction results due to different internal algorithm mechanisms. The SVR prediction model has the lowest accuracy and is unsuitable for predicting shield tunneling speed in the water-rich pebble stratum. The BP neural network prediction model and the linear regression prediction model can predict the tunneling speed by inputting the tunneling parameters with an accuracy rate of 87%. The accuracy of the training set of the BP neural network regression model is 98%. It also verifies the excellent applicability of Matlab's BP neural network in the working environment of shield tunneling.

Table 1. Comparison of simulation accuracy of the regression model.

Model	SVR regression model	Linear regression model	BP neural network regression model
Training set accuracy	61	85	98
Test set accuracy	61	87	87

5 CONCLUSIONS

Based on the Shuai-Nei section shield tunnel project of Hohhot Rail Transit Line 2, this paper uses three classical regression models (SVR prediction model, linear prediction model, and BP neural network prediction model) in MATLAB to learn the tunneling parameters of EPB shield in water-rich round gravel stratum. The results show that the accuracy of the SVR regression model is low in this stratum, and it is not suitable for this stratum. The prediction accuracy of the BP neural network regression model and linear regression model for tunneling speed can reach 87%. Both models are suitable for simulating and predicting the tunneling parameters of EPB shields in the water-rich round gravel stratum. Among them, the accuracy rate of the training set of the BP neural network regression model is 98%, which indicates that BP neural network has nonlinear solid mapping ability and generalization application.

REFERENCES

Barton N. (1999). TBM Performance Estimation in Rock Using Q(TBM). *Tunnels & Tunnelling International*, 31(9):30–34.

Bruines P. A. (2001) *The Use of a Rock Mass Classification Scheme for Tunnel Boring Machines*. In: Adachi Swets, Zeitlinger ed. Modem Tunnelling Science and Technology. 583–588.

Bruland A. (2000). *Hard Rock Tunnel Boring*. Doctoral thesis. Trondheim: Norwegian University of Science and Technology.

Hassanpour J., Rostami J., Zhao J. (2011). *A New Hard Rock TBM Performance Prediction Model for Project Planning. Tunneling and Underground Space Technology incorporating Trenchless Technology Research*, 26(5).

Li, C., Li, T., Li, Z., et al. (2017). Prediction and Analysis of Shield Tunneling Parameters in Composite Stratum Based on BP Neural Network. *Journal of Civil Engineering*, 50 (S1): 145–150.

Su, X. (2018). *Data Analysis and Optimization of Shield Tunneling Parameters*. Master thesis, Shijiazhuang, Shijiazhuang Railway University.

Wang, Y. (2016). *Study on Prediction of Settlements of TBM Tunnel Construction in Guangzhou by Using SVR algorithm*. Master thesis, Jinan, Jinan University.

Xia, Y. (2021). *Research on Real-Time Prediction for Operation Parameters of TBM Based on Deep Learning*. Master thesis, Dalian, Dalian University of Technology.

Yang, Q. (2008). *Analysis and Research on Optimization of Tunneling Parameters in Shield Construction*. Master thesis, Beijing, Beijing Jiaotong University.

Yao, L. (2020). *Key Technology Of Earth Pressure Balance Shield Tunneling In Water-Rich Round Gravel Stratum*. Master thesis, Beijing, Beijing Jiaotong University.

Zhang, H., Wu, X., Zeng, W. (2005). Research on Earth Pressure Balance Shield Tunneling Test and Tunneling Mathematical Model. *Journal of Rock Mechanics and Engineering*, 5762–5766.

Zhou, Z. (2016). *Machine Learning*. Beijing, Tsinghua university press.

Study on the current situation and solutions of aquatic and biological channel connectivity in urban rivers – The Kunyu River and Shuiya Ditch in Beijing as an example

Lijuan Wang
University of Jinan, China
China Institute of Water Resources and Hydropower Research, China

Guangning Li*
China Institute of Water Resources and Hydropower Research, China

Yancheng Han
University of Jinan, Jinan, China

Songtao Liu
China Institute of Water Resources and Hydropower Research, China
China Three Gorges University, China

Shuangke Sun
China Institute of Water Resources and Hydropower Research, China

Xiuyan Wang
Shandong Province Water Transfer Project Operation and Maintenance Centre Pingdu Management Station, China

ABSTRACT: To understand the impact of an artificial intervention on rivers during rapid urban development, such as the ecological destruction of water systems and the blockage of aquatic pathways, we conducted field research on two typical rivers (the Kunyu River and the Shuiya Ditch) in Beijing. The study found that the Kunyu River is mainly a hard boundary, with physical barriers to biological exchange such as sluice gates, and that the river has a single species of aquatic plants and is rich in aquatic organisms, mainly hatchlings of loach, shrimps, minnows, crucian carp, and warbler, with few adults. Shuiya Ditch is dominated by ecological barges, and there are physical barriers such as waterfalls and barrages in the river, with fewer aquatic animals. It is thought that the water-blocking structures in urban rivers have blocked aquatic pathways and that the movement of individuals is mainly unidirectional downstream, with a lack of two-way communication, causing ecological fragmentation in many water systems and making it difficult to maintain the long-term survival of adult fish in local waters. Water system connectivity is essential for the long-term stability of river ecosystems. This paper combines the calculations of mathematical models to propose specific and feasible suggestions and measures for the restoration of aquatic pathways such as the construction of fish passages, with a view to providing a reference for the construction of urban river water ecology.

1 INTRODUCTION

With the rapid development of urbanization in China, cities have become the places where most of our residents live and reside. Urban rivers are the most densely packed with natural

---

*Corresponding Author: lgnchina@163.com

elements and the richest in natural processes in cities. They are also important ecological spaces actively used by humans and designed to provide various ecological services to them (Yu 2005). During the rapid construction and development of cities, the social benefits of river systems have become increasingly evident. Most rivers have been artificially improved and transformed, with bends cut and straightened, and buildings such as knotty rubber dams, plunges, and gates artificially constructed to connect the water system, with a relatively one-sided emphasis on engineering and economic purposes such as flood control and drainage, while neglecting the ecological functions of rivers (Wu 2015). Early planning of the urban water system landscape failed to arrange reasonable channels for the exchange of water systems and water ecosystems, resulting in the gradual isolation and degradation of local water ecology and a serious decline in long-term stability (Lin et al. 2016). Under the influence of continuous intensive human activities, how to make urban rivers healthy and sustainable is an important issue facing us today.

The issue of water connectivity in urban rivers has received attention from many scholars. In a short summary of the ecological management of urban rivers in Beijing, Yang Ling et al. (Yang et al. 2014) pointed out that water-blocking structures such as rubber dams were artificially constructed in rivers, such as the Heishui River, the Fengcun Ditch, the Chengzi Ditch, and the Xifengsi Ditch, and a large amount of hard retaining masonry was carried out at the river boundaries, leading to a reduction in biodiversity in the rivers. He Bing et al. (He et al. 2006) argue that traditional urban river regulation only considers the functions of rivers such as flood control, drainage, and sewage, but not the ecological benefits of rivers, and a large amount of hard material paving makes rivers artificial and channelized, which destroys the living environment of organisms in rivers. Yang Guang (Yang 2015) argued that the impermeability of hard revetment boundaries in the Songhua River basin hinders the connection of various parts of the water ecosystem and reduces biodiversity and the self-purification capacity of water bodies. When evaluating the effectiveness of the ecological management projects on rivers, Song Bo (Song 2017) also suggested that the bank slopes of the Guitang River with slurry blocks, concrete plastering, and block masonry led to the loss of natural features of the river ecosystem. Similarly, the artificial transformation of the Malin River led to the basic disappearance of habitat habitats such as step-deep pool structures, sandbars, and connected wetlands in the original natural river channel. Chen Gang (Chen 2015) proposed ecological restoration and conservation methods for the watershed while elaborating on the importance of water system connectivity for the survival and reproduction of aquatic organisms. All of these studies agree that maintaining the natural boundaries of rivers and the connectivity of water systems is an important condition for maintaining the healthy and sustainable development of rivers.

This paper investigates two typical rivers in Beijing, the Kunyu River (from Kunming Lake to Yuyuantan, about 10 km long) and the Shinia Ditch (which flows through Zheng Changzhuang and Liuli Bridge and joins the Lotus River, about 7.3 km long). The study included: river cross-sections, barge forms, and their respective percentages, number, form, operation of water-blocking structures, species of aquatic organisms, and water quality. Based on the results of the study, the current status of the aquatic life channel connectivity of these two rivers is evaluated and corresponding countermeasures are proposed, with a view to providing a reference for the construction of urban river water ecology.

2 STUDY AREAS

2.1 *The Kunyu River*

The Kunyu River, the waterway downstream of the Beijing-Mi Diversion Canal leading from Kunming Lake in the Summer Palace to Bayi Lake in Yuyuantan, is about 10 kilometers long and is the only waterway running through the second, third and fourth rings of Beijing, as shown in Figure 1. As the largest waterfront area in western Beijing, the Kunyu

Figure 1. Location map of the Kunyu River.

River flows through the densely built-up urban area from south to north, linking the landscapes of the Temple of the Chinese Century, Yuyuantan Park, the Central TV Tower, Jinyuan Times Shopping Centre, Linglong Tower, West Diaoyutai, Wanliu Golf and the community, including numerous cultural, entertainment, commercial and residential resources.

2.2 *Shuiya Ditch*

Shuiya Ditch is located in the territory of Lugouqiao Township, Fengtai District, and flows through Zheng Changzhuang and Liuliqiao to join the Lotus River. The main ditch is 7.3 km long and controls a watershed area of 15.5 km^2, and is a drainage channel, as shown in Figure 2. A square culvert with a bottom width of 10 meters has been built below Liuli Bridge downstream. The Shuiya Ditch treatment (Phase II) project starts from the Fourth Ring Road in the west to the southern extension of Wanshou Road in the east and will provide a landscape river channel at three locations: 302 Hospital, Xicui Road Bridge and the downstream Sports Park. From 2004 onwards, the river will be treated as the main channel of the Shiniagou, starting from the concealed exit of Yuquan Road and ending at the southern extension of Wanshou Road, with a total length of about 4.52 km. To ensure smooth traffic flow in the area, four new bridges will be built across the river, while 15 water-blocking structures will be removed. To prevent scouring at the bottom of the river, two new plunges will be built at the south extension bridge of Yongding Road and upstream of the West Fourth Ring Road respectively, with a height difference of 1.5 meters, and more than

Figure 2. Location map of the Shuiyagou.

200 meters of water storage area will be expanded below the south extension bridge of Yongding Road to form an ecological water feature.

3 RESEARCH RESULTS AND ANALYSIS

3.1 *Research results*

The study found that water-blocking structures such as plunges existed in both the Kunyu River and the Shuiya Ditch. The river was divided into several areas, the results of which are shown in Table 1. The Kunyu River is divided into three zones by the presence of a drop and a small test power station at Bayi Lake in Yuyuan Lake, a drop near Baiyunguan Street North Lane, and a barrage near the Beijing Exhibition Hall. The Kunyu River is a compound section with a hard boundary and a single aquatic plant, with a variety of small aquatic organisms such as loaches, shrimps, minnows, and carp, as shown in Figure 3. Fish and other aquatic organisms in the Kunyu River are predominantly juvenile, with few adult individuals present. There is a small drop in the Siena Ditch and two rubber dams that divide the Ditch into four control areas of 7.5 acres (Region 1), 15 acres (Region 2), 5.7 acres (Region 3), and 1.9 acres (Region 4) from upstream to downstream. The ditch is a trapezoidal section with approximately 50% ecological barges and 50% hard boundaries, as shown in Figure 4. There is a rich variety of aquatic plants in the Shui Nga Ditch, but aquatic fauna is scarce, especially fish.

Table 1. Survey results.

River name	The Kunyu River	The Shuiya Ditch
Water division	Region 1: 714 acres. From Summer Palace to Bayi Lake & From Changchun Bridge to Beijing Exhibition Hall Region 2: 189 acres. From Bayi Lake to Baiyunguan Street North Lane Region 3: 562 acres. From Beijing Exhibition Hall to Nanhai	Region 1: 7.5 acres. From Xiaotun Road to Qingta West Road Region 2: 15 acres. From West Tsingtao Road to Football Park Region 3: 5.7 acres. From Soccer Park to Wanfeng Park Region 4: 1.9 acres. From Wanfeng Park to Wanfeng Road
Section form	Duplex section	Trapezoidal section
Barge form	Approximately 100% of the hard barge	Approximately 50% eco-barge and 50% hard barge each
Water blocking structure	2 waterfalls; 1 barrage; 1 test power station	1 small waterfall; 2 rubber dams
Shoreline buffer strip	Presence	Presence
Aquatic Plant	Single species of aquatic plants and limited in number	The plant species are abundant and more numerous.
Aquatic animal	Loach, shrimp, minnow, carp, and many other species are present and abundant	Only very few fish, such as loaches, are present in small numbers

3.2 *Ecological stability of the watershed*

The size of a river is directly related to the biodiversity, complexity, and stability of its ecosystem (Yan et al. 2010). Fish are an important part of the river ecosystem and interact closely with the water environment through upstream and downstream effects. The various

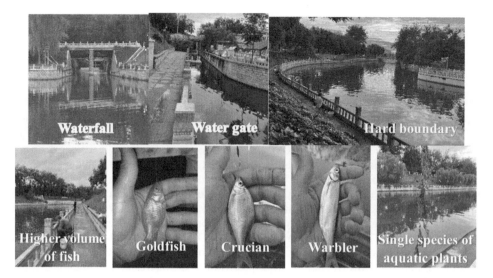

Figure 3. Field research on the Kunyu River.

Figure 4. Field research on the Shuiyagou.

water-blocking structures constructed in the course of artificial river training divide a larger body of water into several smaller bodies of water, with different indicators of the fish biomass and size that can be maintained depending on the size of the body of water. The larger the water area, the greater the diversity and complexity, and the higher the fish diversity and community stability (Gao et al. 2016). In addition, hard barges destroy the ecological community structure of the original slope barges, the use of impermeable treatment deprives them of their ability to self-purify and affects the autodynamic processes of the river, and the inability to grow plants on the barges leads to a lack of biological habitat. The widespread construction of water-blocking structures such as gates and dams in rivers has increased the local water level drop in the river, interrupting the continuity of the river and artificially causing ecological fragmentation of the water system, as well as leading to a reduction in aquatic habitat diversity. This is detrimental to the sustainable development of water ecosystems.

Studies have shown that the main fish stocking specifications vary according to the size of the water area. The three areas of the Kunyu River divided by water-blocking structures all fall within the range of small reservoirs of less than 1,000 mu in size, with small water bodies and small fish holding specifications. The Kunyu River is landscaped, with hard boundaries, a single aquatic plant, and insufficient nutrient composition to sustain a large number of adult fish. Analysis suggests that the majority of aquatic individuals within the Kunyu River have come downstream from the upper reaches and have developed gradually within the local waters. However, due to the predominantly hard boundary of the Kunyu River, the monotony of aquatic vegetation, the lack of aquatic spawning habitat, and the inadequate nutrient composition, it is difficult to sustain adult individuals.

The water body of the Shuiya Ditch is divided into four particularly small areas and the water quality is mainly medium water with poor mobility. The addition of reeds in the river channel is intended to improve the ecology, provide habitat for aquatic organisms and ensure the quality of the water ecology, but some of the reeds in the river channel are too dense and consume too much oxygen. Although the Shuiya Ditch is generally able to maintain the river's condition, there is a high risk of water quality deterioration.

Field studies and analyses of the Kunyu River and Shuiya Ditch in Beijing have revealed that there are many ecological fragmentations of water systems in urban rivers, and that local aquatic organisms have difficulty sustaining long-term survival as adults, with individual organisms moving mainly in one direction and lacking two-way communication. The use of ecological restoration measures such as ecological barges can help to improve river biodiversity too, especially in terms of plant species, but maintaining the connectivity of the river system and aquatic life pathways is the key factor to maintain the long-term stability of the water ecosystem.

4 MEASURES AND SUGGESTIONS

4.1 *Adding biological pathways such as fish passages*

The fishway is a structure that helps fish to cross over an obstacle such as a lock or dam. The addition of a fishway at the location of a structure such as drop water is an effective way to alleviate the barrier effect of the structure on biological passage and can be designed in the form of an engineered fishway (Zhang 2021) or a naturalistic fishway (Li et al. 2019). In this paper, we designed three fish passage options for waterfall buildings based on the results of the study and used Flow3D to carry out 3D numerical simulations to demonstrate the rationality of each option.

Mathematical modeling: The width of the simulated urban river is 20 m. The difference in drop height is 0.6 m, the width of the bottom is 1.2 m and the slope ratio is 1:2.

Option 1, engineered fish passage. The fish passage is rectangular in section, with a bottom slope of $J=2\%$, and the pool chamber is separated by staggered partition walls with a length of 3m and a width of 2m.

Option 2, a natural fishway. Using 3d cobbles arranged at 3m intervals with staggered seams, the overall width of the fishway is 4m.

Option 3, a side fishway. A separate channel is created on the left bank of the river, irregular in shape and approximately 4m wide, with cobbles laid at 3m intervals at the same general interval as the natural fishway.

The three options are modeled as shown in Figure 5, with conditions set to dam overflow, 1.2m water depth at the inlet boundary, and 0.8m water depth at the outlet boundary.

The results of the calculation are shown in Figure 6.

Option 1: The inlet of the fish passage is influenced by the downstream flow and the outflow of the fish passage, with the flow direction about 45° with the mouth of the fish passage, and the flow velocity of 1.2~2.5m/s. The flow velocity at the mouth of the vertical slit inside the fish passage is 0.6~1.5m/s, which is less than the common fish outbreak swimming speed. The flow velocity on both sides of the mainstream inside the pool is

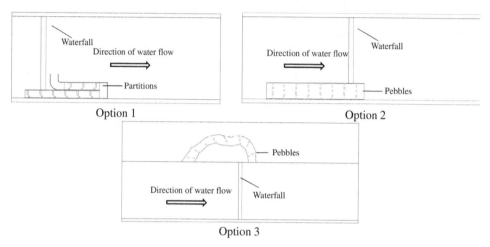

Figure 5. Mathematical models for each option.

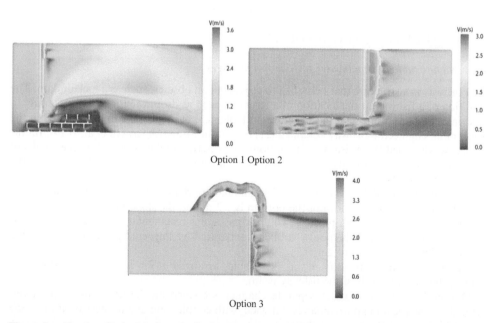

Figure 6. Graphs of calculated results for each option.

0~0.5m/s, which can become a resting area for fish to go upstream and meet the demand of fish to go upstream.

Option 2: The inlet of the fish passage is located downstream of the dam, relying on the downstream flow of the dam to attract fish, with a flow velocity of 1-2.5 m/s. The natural fish passage is rich in flow patterns, with a high flow velocity of 0.5-1.5 m/s at the wrong seam, which is less than the outbreak swimming speed of common fish, and a low flow velocity of 0-0.5 m/s on both sides of the mainstream. The fish passage is rich in flow patterns and offers the possibility for fish with different swimming abilities to move up.

Option 3: The inlet of the fish passage is located on the left bank of the river downstream of the dam, relying on the downstream flow of the dam (flow velocity 1~2.6m/s) to attract fish. The arrangement of the fishway is similar to that of a natural fishway, with the cobblestones arranged in a staggered manner to enrich the flow pattern. The flow velocity at the staggered seam is 0.5~1.5 m/s, which is less than the normal fish burst swimming speed, and the flow velocity on both sides of the mainstream is 0~0.6 m/s, which meets the fish upstream demand.

Based on the analysis of existing research experience, it is considered that all three of the above options can help fish to pass through water-blocking structures. Where possible, the design should be designed to mimic a natural fish passage, which is close to nature but does not take up additional urban space.

4.2 *Partial modification of river hard boundaries*

Using plants and natural materials to ecologically modify the hard boundaries of the river. Vegetation can have a great impact on a river by influencing its flow (Ni et al. 2006), the strength of its banks against scouring, the stability of the sediment deposited riverbed, and the morphology of the river. It is advisable to select native plants with high adaptability, low management, maintenance costs, and strong affinity as the main body, and to build a community pattern in imitation of natural plant communities. The use of wood piles, bamboo, natural stones (pebbles, marble, etc.), and other construction materials that are conducive to the restoration of the ecological functions of the river to reinforce the river bank, and the use of prefabricated concrete fish nest structures (Li et al. 2013) (fish nest bricks as in Figure 5), can be considered for the lining of the river bank below the normal water level to provide a living space for fish and other aquatic animals, which can ensure the stability of the river bank without cutting off the connection between the aquatic ecosystem in the river and the terrestrial ecosystem on the bank (Gao et al. 2009). These measures can provide good habitat and breeding conditions for the growth of aquatic plants and aquatic animals and amphibians, protect the natural environment of river biodiversity and ensure the orderly continuation of the biological chain, as well as abate surface pollution and play a role in landscape beautification.

4.3 *Other measures*

(1) Installation of natural buffer zones. The hydrological connectivity between the river and the buffer zone rocks is a key factor affecting the diversity of river species. Setting a buffer zone of a certain width on each side of the river is an important method of ecological restoration of the river, making it an ecological river with multiple ecological functions such as habitat, biological corridor, and waterfront filter zone biobank (Wang et al. 2009).

(2) Both water width and water depth can be used to reflect the relative size of a water body. However, studies have shown that river width is a significant variable in determining the structure of fish communities, while water depth has no significant effect on fish communities (Gao et al. 2016). Therefore, a reasonable allocation of aquatic organisms

Figure 7. Example image of river ecological restoration measures.

according to the volume of the river is conducive to the construction of a healthy aquatic ecosystem in terms of ecological structure, material cycling, and energy flow.

(3) Restoring the natural shape of rivers. In the future planning of urban rivers, various factors such as ecology, efficiency, and landscape are taken into account to maintain the original width and natural state of existing rivers as far as possible, to maintain the natural cross-sectional shape of rivers as far as possible, to replace artificially hardened side slopes with natural side slopes, and to create rivers with curved flow lines, "deep pools" and "shallow pools" and other natural features on which many organisms depend.

5 CONCLUSIONS

Through field research and analysis of two typical rivers in Beijing, the Kunyu River and the Shuiya Ditch, the following main conclusions were reached:

(1) The research found that both the Kunyu River and the Shuiya Ditch have water-blocking structures such as gates, dams, and plunges, and that neither has considered the connectivity of biological pathways. The rivers are divided into smaller waters, creating fragmented local ecosystems.

(2) The analysis concluded that water system connectivity is a prerequisite for aquatic life pathway connectivity and that aquatic life pathway connectivity is a necessary condition for the long-term stability of the water ecosystem. The Kunyu River has a

predominantly hard boundary, with single aquatic plant life and a lack of spawning habitat for aquatic animals, with insufficient nutrient composition to sustain adult individuals. Although the Shuiya Ditch has been designed with ecological barges, the water system is relatively closed, lacking in biological sources, with fewer aquatic animals in the river and a higher risk of water quality deterioration.
(3) Based on the typical problems identified by the findings, it is recommended that water-blocking structures be modified and biological pathways such as fish passes be added to restore the ecology of urban rivers in order to restore the connectivity of the river's aquatic pathways and provide a good habitat for aquatic life to survive.

REFERENCES

Chen Gang. Analysis of Ecological Restoration of Rivers and Increase in Aquatic Biodiversity[J]. *Water Resources & Hydropower of Northeast China*, 2015, 33(04):24–26.

Gao Guoming, Dong Jianwei. The Application and Research for the Technology of Composite (fish nest) Eco-Concrete Wall Revetment[J]. *Jilin Water Resources*, 2009(09):1–4.

Gao Xueping, Yang Rui, Zhang Chen. Study on the Aquatic Ecosystem Design Method of Artificial Lakes[J]. *Chinese Journal of Environmental Engineering*, 2016, 10(02):948–954.

He Bing, Gao Huiqiao, Xia Xudong. Study on Urban Rivers and Their Ecological Management Planning[J]. *Soil and Water Conservation in China*, 2006, (12):23–25.

Li Gang, Wang Meng, Mao Yun. *Exploring the Application of Fish Nesting Bricks in River Management in Zibo*[C]. Proceedings of the 2013 Annual Academic Conference of the Chinese Hydraulic Engineering Society—S2 Lake Governance Development and Protection, 2013:391–394.

Li Guangning, Sun Shuangke, Guo Ziqi, et al. Physical Model Test on Hydraulic Characteristics and Fish Passing Performance of Nature-Like Fishway[J]. *Transactions Of The Chinese Society Of Agricultural Engineering*, 2019, 35(09):147–154.

Lin Jiantao, Yang Dongdong, Cao Lei. Exploration and Conception of Aqueous System Protections Strategy of Qilihai Wetland[J]. *Journal Of Tianjin University Social Sciences*, 2021 (2016-5):445–450.

Ni Jinren, Liu Yuanyuan. Ecological Rehabilitation of Damaged River System[J]. *Journal Of Hydraulic Engineering*, 2006, 37(09).

Song Jie. *Evaluation of Ecological Treatment of Two Typical Damaged Rivers in Changsha Area*[D]. Changsha University of Science & Technology, 2017.

Wang Wen, Huang Suiliang, Zhang Shenghong, et al. Study on Eco-Restoration Modes for Plain Rivers in Haihe River Basin 1: Eco-Restoration Mode[J]. *Water Resources and Hydropower Engineering*, 2009, 40(04):14–19.

Wu Danzi. *Research on near-naturalization of urban channelized rivers*[D]. Beijing Forestry University, 2015.

Yan Yunzhi, Zhan Yaojun, Chu Ling, et al. Effects of Stream Size and Spatial Position on Stream-Dwelling Fish Assemblages[J]. *Acta Hydrobiologica Sinica*, 2010, 34(5):1022–1030.

Yang Guang. *Impact of the Water Environment Change of Songhua River in Harbin on the Water Ecosystem Security*[D]. Northeast Agricultural University, 2015.

Yang Ling, Deng Zhuozhi, Zhou Zhihua. A Summary of Ecological Management of Urban Rivers in Beijing and Suggestions[J]. *Water Resources Planning And Design*, 2014, (08):6–9+56.

Yu Fangzhen. A Study on Resource Security in China's Urbanization Process[J]. *Group Economy*, 2005 (2):50–51.

Zhang Haonan. *Study on the Effect of Different Baffle Plate Arrangements in Vertical Slot Fishway on the Migration Behavior of Schizothorax Prenanti*[D]. University of Jinan, 2021.DOI: 10.27166/d.cnki.gsdcc.2021.000248.

Application of 3D design with multi-platform collaboratively in the centralized control building of the Wudongde Hydropower Station

Ning Wang*, Wei Wang*, Donghai Chen* & Lijun Li*
Changjiang Survey Planning Design and Research Co., Ltd, Wuhan, China

ABSTRACT: At present, the application of 3D design and BIM technology in hydraulic engineering has changed from single specialty and single stage to the whole process and multi-specialty cooperation. It is difficult for a single software platform to meet the needs of the 3D design application in the whole process, so the multi-software collaborative design has become the main strategy at present. In the design of the K25 karst cave centralized control building of Wudongde Hydropower Station, due to the difficulties such as the complex terrain environment, building special-shaped surfaces, and complex internal structure, multi-software is used to complete the whole process of 3D design from scheme to construction drawing. In this paper, we have completed the modeling and BIM model integration of cave inwall, concrete backfill, special-shaped curtain wall, and other complex shapes through the collaboration of SketchUp, Rhino, Revit, and other 3D design software, Finally, the construction drawings are generated from BIM model. As a multi-software collaborative design technology practice, the design of the K25 karst cave centralized control building not only provides a reference for the 3D design of similar hydraulic projects, but also lays a foundation for multi-platform data fusion in the later stage of the project.

1 INTRODUCTION

With the improvement of national and industry technical standards, 3D design and BIM technology have become an important technical means for hydraulic engineering survey and design, and have been widely used in many large and medium-sized hydraulic projects. 3D design of hydraulic engineering and application of BIM technology has changed from a single stage of design and application of a single major to a full stage, the whole process involving the application of multi-specialty. The multi-specialty collaborative design has become the main trend in the development of the 3D design of hydraulic engineering at this stage.

As a typical large-scale complex hydropower project, the 3D design method of multi-platform collaboration is adopted in the design stage of the Wudongde Hydropower Station. The main body and auxiliary hydraulic structures of the dam are designed in three dimensions by CATIA, while the management house and central control buildings of the project are designed by Revit. As the central control building is located in K25 karst cave on the right abutment of the dam in the design scheme, it brings technical difficulties to the 3D design. In view of the topographical particularity and complex structure of the central control building of Wudongde Hydropower Station, this paper adopts the design method combining Revit and other software to solve the problems of cavity wall topography,

*Corresponding Authors: wangning@cjwsjy.com.cn, wangwei@cjwsjy.com.cn, chendonghai@cjwsjy.com.cn and lilijun@cjwsjy.com.cn

complex building shape, and internal layout in 3D method, and finally completes the 3D design of the central control building.

2 PROJECT INTRODUCTION

2.1 *Brief introduction on Wudongde Hydropower Station and K25 karst cave*

The dam site of the Wudongde Hydropower Station is located in the lower reaches of the Jinsha River at the junction of Huidong in Sichuan and Luquan in Yunnan. It is the first cascade hydropower station in the four cascades of the lower reaches of the Jinsha River. The main structure of the complex project is composed of water retaining structures, water release structures, and power generation structures (Hu 2014). K25 karst cave, a natural karst inclined shaft, exists close to the right abutment of the dam. Based on the stress and safety requirements of the main hydraulic structure of the dam, concrete backfilling is used at the lower part of the karst tunnel, which is flush with the dam crest at 988m elevation. After the scheme demonstration, the centralized control center of the Wudongde complex is arranged in this cave (Ni 2021). This way of arranging buildings in karst caves brings challenges to the building design.

2.2 *Project characteristics and difficulties*

(1) Unique terrain of the project
According to literature analysis and comparison, the use of natural karst caves in dam abutments to arrange central control building is the first case of a hydropower project at home and abroad (Ni 2021), so there are not many referable cases to guide the project design. Compared with conventional construction projects, the project's most striking feature is that buildings are built within the terrain (i.e. cave walls), not above the terrain. In order to accurately express the space location relationship between the inwall and the building, it is decided to adopt the 3D method to design the centralized control building.

(2) Special appearance of buildings
The appearance of the central control building is covered by a curtain with a special curved surface, which is beautiful and elegant (Ni 2021). In order to restore the intention of the scheme design, the exterior model of the central control building is precisely created by Rhino. How to accurately use 3D shape data of Rhino model to produce construction drawings in the stage of construction drawing design becomes a technical difficulty in design.

(3) Complex internal structure of the building
The scheme of the centralized control building adopts a four-story structure consisting of the first floor above ground and the third floor below ground. The interior layout includes a lobby, a central control room, a duty room, a computer room, a meeting room, a communication equipment room, and an auxiliary equipment room (such as a ventilation and smoke-exhausting room and a distribution room) (Ni 2021). The complex internal structure of the central control building and the dense layout of various pipeline systems also put forward higher requirements for engineering design methods. 3D design methods can more intuitively express this complex space layout.

2.3 *Technical features of the 3D design of the project*

(1) The whole process of 3D forward design
The architectural design of this project adopts the whole process of the 3D collaborative design method to complete the construction drawing design and submit the drawing. The

design covers the conceptual scheme design to the construction drawing design. At the concept stage, SketchUp is used to create the basic shape and layout of the building. In the preliminary design stage, the exterior curtain wall shape of the building is precisely created by Rhino and imported into Revit for integration with the architectural and structural Revit model. In the construction drawing design stage, After the Revit model is cut by a plane, necessary dimensions and text marking are added, and the formal construction drawings are finally generated.

(2) Exploration of data fusion from multi-software

The architectural design of this project is a successful exploration of multi-software collaboration and multi-data integration. From scheme design to construction drawing design, it is applied to SketchUp, Rhino, Revit, 3ds Max, and other software. Design data transfer is lossless in each software. In addition, according to the requirements of the project construction management stage, the BIM model (Revit format) is imported into our company's 3DE platform, and Revit and CATIA models are integrated to provide important data support for construction management in the later construction stage (Li 2017).

(3) 3D representation on cave inwall and concrete backfill

In order to accurately express the spatial positional relationship between the inner wall of the karst cave and the central control building, Rhino is used to create the cave inwall model, and the model is imported into Revit to integrate with the BIM model of the central control building, so as to ensure the layout of the central control building in the complex geographical environment. The concrete backfill terrain is represented by Revit mass modeling, which also plays an important reference role in drawing and calculating quantities of the centralized control building.

3 3D DESIGN OF CENTRALIZED CONTROL BUILDING IN A K25 KARST CAVE

3.1 Overall technical process

The 3D design software is used for the main design work of the project from conceptual scheme design to construction drawing. The main work steps are shown in Figure 1.

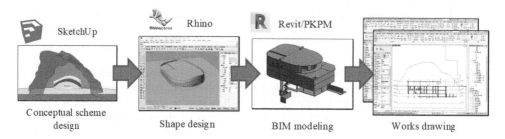

Figure 1. The overall workflow of 3D design.

3.2 Conceptual scheme design

The design inspiration of the K25 central control building is from the natural form of water flow when the dam opens and releases flood. During the scheme creation stage, the architecture forms are created through SketchUp and the design schemes are refined. After the design scheme has been determined, Rhino is used to create an accurate profile of the exterior of the building.

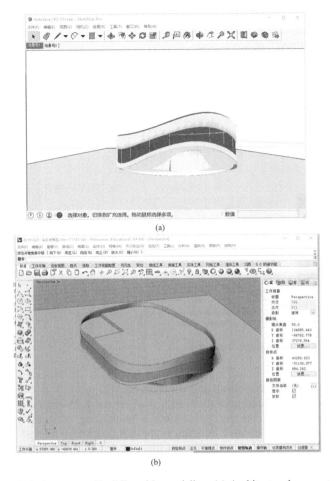

Figure 2. The technical process of building skin modeling. (a) Architectural conceptual design (SketchUp) (b) Create building accurate exterior shape (Rhino).

3.3 *Construction design*

(1) Create an architectural model

The main body of the K25 karst cave central control building is a four-story concrete frame structure building. The 3D design of this building adopts the conventional architectural design method. That is, in Revit, the BIM model of the building is created by positioning the elevation grid and using the wall, floor, column, and other components. And perform room separation and functional regionalization for each story are required. After iterative optimization and modification, the BIM model with the depth of construction drawing can be obtained and prepared for plotting.

(2) Import structural calculation model

For the 3D structural design of the project, considering the particularity of the structural design discipline (the structural design model should be associated with the structural analysis and calculation) and the limitations of the structural design of the Revit (i.e., it does not have the structural analysis and calculation function), the research group did not adopt the method of directly modeling structural components in the Revit, but first used the YJK software to conduct structural analysis and calculation. Then, the structural analysis and calculation model of YJK is imported into Revit by a plug-in, and then the structural model is modified and improved and submitted to the architect and electromechanical designer as a reference.

(3) Handle terrain (inwall of karst cave)

This centralized control building is located in the terrain (karst cave inwall), so it is obviously different from the terrain of conventional buildings. In order to accurately express this spatial positional relationship, the research group used Rhino for lofting contour terrain (DWG format) to generate accurate terrain surface, and imported it into Revit through SAT format (the process is shown in Figure 3). In this way, the terrain and buildings can be integrated into the same BIM model, so as to facilitate the later architectural design and spatial positioning.

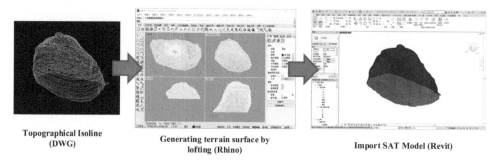

Figure 3. 3D design flow of karst cave inwall.

(4) 3D representation on cave inwall and concrete backfill

The part below the ground (elevation: 988m) where the K25 central control building is located in the project is constructed by concrete backfilling, while the space on the third floor under the central control building is the unfilled area reserved after backfilling. Since it is difficult to express such internal space layout through conventional structural components in Revit, the research group uses the Revit mass model to represent the concrete backfill area and completes the creation of the concrete backfill area model (as shown in Figure 4) through the in-place mass editing function (such as lofting, stretching and hollow shape cutting, etc.), and then integrates the backfill area model with the architectural and structural model. This can accurately reflect the spatial layout of the

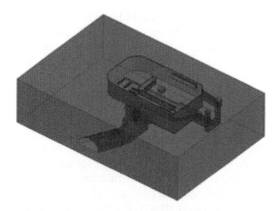

Figure 4. 3D design modeling of concrete backfill area.

underground part of the central control building, and be profitable for internal space layout optimization of the building.

(5) Integration of the BIM model

In Revit, after the modeling of the curtain, architecture, structure, foundation pit, and inwall is completed, the above model components are linked into the main model by using the model linking function (Wang 2017), and the splicing and integration of the model are completed, as shown in Figure 5. Here, it should be noted that each model component needs to use a unified positioning reference. In addition, in the process of model linking, it is also necessary to pay attention to the north of the model and the due north orientation of the project to ensure that all model components are integrated according to the same positioning and orientation.

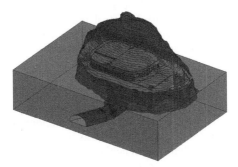

Figure 5. Model integration of centralized control building.

(6) 2D drawing from BIM model

After the detailed design of each subject is completed, these models are integrated with Revit, and the construction drawing design of the project is completed by using the drawing function of Revit (Wang 2021; Wu 2018). The steps are as follows:

Select the customized view template file, which includes predefined styles such as border, dimension style, line type, font, and fill patterns.

Based on the drawing function of Revit, add necessary dimension and text marking from the plan, elevation, and section views of the model to complete the plan, elevation, section drawings, and layout drawings.

Here, the research group uses the third-party plug-in to realize the rapid drawing of stair detail and elevation.

4 SUMMARY AND EXPECTATION

4.1 *Summary*

The 3D design of hydraulic engineering and the application of BIM technology in the whole process and life cycle has become the main trend of the development of design technology in the hydraulic design domain. In the 3D design of large-scale water conservancy and hydropower projects, there are often multi-platform collaborative designs and data integration. Taking the engineering design of the K25 karst cave centralized control building of the Wudongde Hydropower Station as an example, this paper introduces how to realize the 3D forward design and drawing through multiple 3D design software.

The main technical features of the 3D design of the project in this paper include:

(1) *3D forward design mode.* K25 central control building adopts the 3D design mode from conceptual scheme design to construction drawing design, and the main drawings are generated from the BIM model.
(2) *Modeling and drawing of special-shaped buildings.* In the scheme design stage, Rhino is used to accurately model the exterior curtain with a special-shaped surface, and it is imported into Revit to divide the curtain, thus ensuring the design intent to the maximum extent.
(3) *Modeling of the inwall of the karst cave.* As the main body of the K25 central control building is in the karst cave, Rhino lofting is used to generate the surface of the inwall. Then it is imported into Revit to provide a reference for the layout of the building scheme.
(4) *3D representation on concrete backfill.* The layout of the concrete backfill part of the underground part of the K25 central control building is complex. The project team accurately modeled the area through the Revit in-place mass model, which provides an important reference for the design of the underground part of this building.

4.2 *Expectation*

The 3D design of hydraulic engineering has developed from 3D parametric modeling in the early stage to the 3D collaborative design with BIM technology nowadays. At present, the technology of 3D design of water conservancy projects based on a single software platform is matured, and the industry has rich technical achievements for mainstream design platforms. However, due to the complexity of water conservancy projects, it is often difficult for a single platform to cover the all subject design needs of hydraulic engineering. Therefore, multi-platform collaboration has become the research hotspot of 3D design technology of hydraulic engineering. This paper discusses the application of 3D design technology with multi-platform collaboratively in hydraulic engineering around the K25 karst cave centralized control building project of the Wudongde Hydropower Station, which not only serves as a reference for the design of similar projects, but also provides a train of thought for the work collaboration and data collaboration among the multi-design platform software of hydraulic engineering.

REFERENCES

Hu Qingyi, Weng Yonghong, et al. (2014). Design and Research on the General Layout of Hydro-complex Structures of Wudongde Hydropower Station. *Yangtze River*. 45 (20), 16–20.
Li Xiaoshuai, Zhang Le (2017). Wudongde Hydropower Station BIM Design and Application. *Journal of Information Technology in Civil Engineering and Architecture*. 52 (3), 7–13.
Ni Aimin, Zhou Ziqing, Li Lijun (2021). The Innovative Architectural Design of the Centralized Control Building in K25 Karst cave of Wudongde Hydropower Station. *Yangtze River*. 52 (3), 118–122.
Wang Ning, Chen Rong, et al. (2017). Research and Implementation on the 3D Design of Hydraulic Engineering Based on BIM. *Yangtze River*. (s1):156–159.
Wang Ru, Xing Yuan, Wen Huazhou (2021). Revit-Based Construction Drawing Forward Design Application Process. *Journal of Information Technology in Civil Engineering and Architecture*. 13(1):56–62.
Wu Wenyong, Jiao Ke, et al. (2018). BIM Forward Design Method and Software Implementation of Building Structure Based on Revit. *Journal of Information Technology in Civil Engineering and Architecture*. 10(3): 39–45.

Application of residual modified gray model in dam safety monitoring

Zhaoqing Fu*

China Renewable Energy Engineering Institute Co., Ltd., Beijing, China

ABSTRACT: Due to the influence of many factors, there may exist long interrupt in the dam monitoring data, which is not suitable for the adaption of regression analysis. This paper introduced an improved gray model generalized from the GM(1, m) model to solve the problem of data interrupt in dam safety monitoring, and an improved gray model is applied to the fitting and prediction of the base displacement and stress data of a high arch dam. The results show that the residual modified non-equidistant MGM(1, m) model has high accuracy and can be applied to practical engineering.

1 INTRODUCTION

In order to master the real operation status of the dam, it is necessary to analyze the prototype observation data and establish a mathematical monitoring model to determine the safety status of the dam. Commonly used monitoring models include statistical, deterministic, and mixed models (Su 2012; Wu 2003), which are generally built based on a large sample of continuous observations. However, in practical engineering, due to various subjective or objective reasons, there may exist a long gap in the obtained observation data, which is not applicable to the establishment of a large sample data model (Su 2011). But the non-equidistant multi-variable gray model established for the data with little information and small samples can solve this problem very well (Liu 2010; Wu 2012). Non-equidistant multivariate gray model, namely MGM(1, m) model, is generalized from the GM (1,1) model (Xiong 2011), which can uniformly describe multiple variables and better reflect the internal connection of each variable, so as to achieve the prediction effect. Since the time response equation of the MGM(1, m) model is a power function, if the change of the actual data is relatively gentle, the residual value generated by the prediction will increase with the extension of the prediction time, and the increase of the residual value of different variables will be different. Therefore, to improve the prediction accuracy, this paper adapts different curve regression equations (Chen 2012) to fit the residual values of the multiple variables and corrects the calculation results of the MGM (1, m) model with those fitted residual values.

2 NON-EQUIDISTANT MGM(1, M) MODEL

For sequence $X_j^{(0)} = \{x_j^{(0)}(k_1), x_j^{(0)}(k_2), \ldots, x_j^{(0)}(k_n)\}$ ($j = 1, 2, \ldots, m$), the set distance $\Delta k_i = k_i - k_{i-1} \neq const$, $X_j^{(0)}$ will be a non-equidistant sequence.

---

*Corresponding Author: 827642951@qq.com

For sequence $X_j^{(1)} = \{x_j^{(1)}(k_1), x_j^{(1)}(k_2), \ldots, x_j^{(1)}(k_n)\}$ $(j = 1, 2, \ldots, m)$, the set $x_j^{(1)}(k_i) = \sum_{l=1}^{i} x_j^{(0)}(k_l)\Delta k_l$, $X_j^{(1)}$ will be the first-order accumulation sequence of the non-equidistant sequence $X_j^{(0)}$.

For sequence $Z_j^{(1)} = \{z_j^{(1)}(k_2), z_j^{(1)}(k_3), \ldots, z_j^{(1)}(k_n)\}(j = 1, 2, \ldots, m)$, the set $z_j^{(1)}(k_i) = 0.5(x_j^{(1)}(k_{i-1}) + x_j^{(1)}(k_i))(j = 1, 2, \ldots, m, i = 2, 3, \ldots, n)$, $Z_j^{(1)}$ will be the neighboring mean sequence of $X_j^{(1)}$.

$$X^{(0)} = [X_1^{(0)} X_2^{(0)} \ldots X_m^{(0)}]^T = \begin{bmatrix} x_1^{(0)}(k_1) & x_1^{(0)}(k_2) & \cdots & x_1^{(0)}(k_n) \\ x_2^{(0)}(k_1) & x_2^{(0)}(k_2) & \cdots & x_2^{(0)}(k_n) \\ \vdots & \vdots & \ddots & \vdots \\ x_m^{(0)}(k_1) & x_m^{(0)}(k_2) & \cdots & x_m^{(0)}(k_n) \end{bmatrix}$$ is the matrix of original

data, where $x_j^{(0)}(k_i)$ is the measured sequence of the jth variable at the timek_i.

Accumulating $X^{(0)}$ in the first order, the obtained matrix $X^{(1)}$ is:

$$X^{(1)} = [X_1^{(1)} X_2^{(1)} \ldots X_m^{(1)}]^T = \begin{bmatrix} x_1^{(1)}(k_1) & x_1^{(1)}(k_2) & \cdots & x_1^{(1)}(k_n) \\ x_2^{(1)}(k_1) & x_2^{(1)}(k_2) & \cdots & x_2^{(1)}(k_n) \\ \vdots & \vdots & \ddots & \vdots \\ x_m^{(1)}(k_1) & x_m^{(1)}(k_2) & \cdots & x_m^{(1)}(k_n) \end{bmatrix} \quad (1)$$

First-order ordinary differential equations of the MGM(1, m) model are:

$$\begin{cases} \dfrac{dx_1^{(1)}}{dt} = a_{11}x_1^{(1)} + a_{12}x_2^{(1)} + \cdots + a_{1m}x_m^{(1)} + b_1 \\ \dfrac{dx_2^{(1)}}{dt} = a_{21}x_1^{(1)} + a_{22}x_2^{(1)} + \cdots + a_{2m}x_m^{(1)} + b_2 \\ \vdots \\ \dfrac{dx_m^{(1)}}{dt} = a_{m1}x_1^{(1)} + a_{m2}x_2^{(1)} + \cdots + a_{mm}x_m^{(1)} + b_m \end{cases} \quad (2)$$

Set

$$\begin{cases} X^{(0)}(k) = (x_1^{(0)}(k), x_2^{(0)}(k), \cdots, x_n^{(0)}(k))^T \\ X^{(1)}(k) = (x_1^{(1)}(k), x_2^{(1)}(k), \cdots, x_n^{(1)}(k))^T \end{cases} \quad (3)$$

$$A = \begin{bmatrix} a_{11} & a_{12} & \cdots & a_{1m} \\ a_{21} & a_{22} & \cdots & a_{2m} \\ \vdots & \vdots & \ddots & \vdots \\ a_{m1} & a_{m2} & \cdots & a_{mm} \end{bmatrix}, B = \begin{bmatrix} b_1 \\ b_2 \\ \vdots \\ b_m \end{bmatrix}, \quad (4)$$

It can be obtained that:

$$\frac{dX^{(1)}(t)}{dt} = AX^{(1)}(t) + B \quad (5)$$

where $X^{(1)}(t) = \{x_1^{(1)}(t), x_2^{(1)}(t), \cdots, x_m^{(1)}(t)\}^T$.

Discrete MGM(1, m) model can be obtained by the difference calculation of Equation 1:

$$x_j^{(0)}(k_i) = \sum_{l=1}^{m} a_{jl} z_l^{(1)}(k_i) + b_j, \quad j = 1, 2, \cdots, m, \quad i = 2, 3, \cdots, n. \tag{6}$$

Set

$$a_j = (\hat{a}_{j1}, \hat{a}_{j1}, \cdots, \hat{a}_{j1}, b_j)^T, \quad j = 1, 2, \cdots, m \tag{7}$$

Estimation of a_j can be calculated with the least square method:

$$\hat{a}_j = (\hat{a}_{j1}, \hat{a}_{j2}, \cdots, \hat{a}_{jm}, \hat{b}_j)^T = (L^T L)^{-1} L^T Y_j \tag{8}$$

where

$$L = \begin{bmatrix} z_1^{(1)}(k_2) & z_2^{(1)}(k_2) & \cdots & z_m^{(1)}(k_2) & 1 \\ z_1^{(1)}(k_3) & z_2^{(1)}(k_3) & \cdots & z_m^{(1)}(k_3) & 1 \\ \vdots & \vdots & \ddots & \vdots \\ z_1^{(1)}(k_n) & z_2^{(1)}(k_n) & \cdots & z_m^{(1)}(k_n) & 1 \end{bmatrix} \tag{9}$$

$$Y_j = \{x_j^{(0)}(k_2), x_j^{(0)}(k_3), \cdots, x_j^{(0)}(k_n)\}^T, \quad j = 1, 2, \cdots, m. \tag{10}$$

The obtained parameter matrix $\hat{A}$ and parameter vector $\hat{B}$ are:

$$\hat{A} = (\hat{a}_{ij})_{m \times m}, \quad \hat{B} = (\hat{b}_1, \hat{b}_2, \cdots, \hat{b}_m) \tag{11}$$

The time response equation of the equidistant MGM(1, m) model is:

$$\hat{X}^{(1)}(k_i) = \{\hat{x}_1(k_i), \hat{x}_2(k_i), \cdots \hat{x}_m(k_i)\}^T = e^{\hat{A}(k_i - k_1)}(X^{(1)}(k_1) + \hat{A}^{-1} B) - \hat{A}^{-1} B \tag{12}$$

where $i = 2, 3, \cdots, n$. And

$$\hat{X}^{(0)}(k_i) = \{\hat{x}_1^{(0)}(k_i), \hat{x}_1^{(0)}(k_i), \cdots, \hat{x}_m^{(0)}(k_i)\}^T = (\hat{X}^{(1)}(k_i) - \hat{X}^{(1)}(k_{i-1}))/\Delta k_i \tag{13}$$

3 IMPROVED NON-EQUIDISTANT MGM (1, M) MODEL

3.1 *Self-adaptive residual regression correction*

Assuming there exist m influencing factors, residuals between fitted values and measured values can be described as follow:

$$\sigma_j(k_i) = \begin{bmatrix} \sigma_1(k_1) & \sigma_2(k_1) & \cdots & \sigma_m(k_1) \\ \sigma_1(k_2) & \sigma_2(k_2) & \cdots & \sigma_m(k_2) \\ \vdots & \vdots & \ddots & \vdots \\ \sigma_1(k_n) & \sigma_2(k_n) & \cdots & \sigma_m(k_n) \end{bmatrix} \tag{14}$$

where $j = 1, 2, \ldots, m$; $i = 1, 2, \ldots, n$; $\sigma_j(k_i) = x_j(k_i) - \hat{x}_j(k_i)$, $\sigma_j(k_i)$ is a residual sequence, $x_j(k_i)$ and $\hat{x}_j(k_i)$ are respectively observation sequence and time response sequence of the jth variable at the time k_i.

To avoid negative values in the fitted residuals, a constant b is added to translate the residual sequence σ_j into a non-negative sequence δ_j, i. e. $\delta_j(k_i) = \sigma_j(k_i) + b$. Then the sequence δ_j is accumulated in the first order to better reflect the changing rule of each element in δ_j.

$$\delta_j^{(1)}(k_i) = \begin{bmatrix} \delta_1^{(1)}(k_1) & \delta_2^{(1)}(k_1) & \cdots & \delta_m^{(1)}(k_1) \\ \delta_1^{(1)}(k_2) & \delta_2^{(1)}(k_2) & \cdots & \delta_m^{(1)}(k_2) \\ \vdots & \vdots & \ddots & \vdots \\ \delta_1^{(1)}(k_n) & \delta_2^{(1)}(k_n) & \cdots & \delta_m^{(1)}(k_n) \end{bmatrix} \tag{15}$$

where $\delta_j^{(1)}(k_i) = \sum_{i=1}^{k} \delta_j(k_i)$

The polynomial regression equation and power function regression equation are introduced to fit the scatter chart of $\delta_j^{(1)}(k_i)$:

$$f(x_1) = ax^3 + bx^2 + cx + d \tag{16}$$

$$f(x_2) = ax^b + c \tag{17}$$

The mean variance is used to evaluate the fitting accuracy of each regression equation, and the regression equation with the smallest variance will be chosen as the prediction equation for $\delta_j^{(1)}(k_i)$. When predicted values of $\delta_j^{(1)}(k_i)$ are calculated, inverse accumulation will be conducted, i. e. $\hat{\delta}_j^{(0)}(k_i) = \hat{\delta}_j^{(1)}(k_i) - \hat{\delta}_j^{(1)}(k_{i-1})$, and $\hat{\sigma}_j(k_i) = \hat{\delta}_j^{(0)}(k_i) - b$.

Finally, with help of the self-adaptive residual regression correction, the improved prediction result can be described as follow:

$$\hat{x}_j(k_i) = \begin{bmatrix} \hat{x}_1^{(0)}(k_1) + \hat{\sigma}_1(k_1) & \hat{x}_2^{(0)}(k_1) + \hat{\sigma}_2(k_1) & \cdots & \hat{x}_m^{(0)}(k_1) + \hat{\sigma}_m(k_1) \\ \hat{x}_1^{(0)}(k_2) + \hat{\sigma}_1(k_2) & \hat{x}_2^{(0)}(k_2) + \hat{\sigma}_2(k_2) & \cdots & \hat{x}_m^{(0)}(k_2) + \hat{\sigma}_m(k_2) \\ \vdots & \vdots & \ddots & \vdots \\ \hat{x}_1^{(0)}(k_n) + \hat{\sigma}_1(k_n) & \hat{x}_2^{(0)}(k_n) + \hat{\sigma}_2(k_n) & \cdots & \hat{x}_m^{(0)}(k_n) + \hat{\sigma}_m(k_n) \end{bmatrix} \tag{18}$$

3.2 *Model accuracy test*

There are three commonly used accuracy testing methods for the Gray model, namely residual test, posterior difference test, and correlation test. In this paper, a posterior difference test with a clear accuracy evaluation standard (Table 1 is chosen for the curacy test of the improved MGM (1, m) model (Xu 1999). The testing steps are as follows:

Step 1: Calculate the variance of σ_j.

$$S_1^2 = \frac{1}{n} \sum_{i=1}^{n} (\sigma_j(k_i) - \bar{\sigma}_j)^2 \tag{19}$$

Step 2: Calculate the variance of the original data:

$$S_2^2 = \frac{1}{n} \sum_{i=n}^{n} (x^{(0)}(k_i) - \bar{x})^2 \tag{20}$$

where $\bar{x}$ is the mean of $x^{(0)}(k_i)$.

Step 3: Calculate the posterior difference ratio C

$$C = S_1/S_2 \tag{21}$$

Table 1. Standard of prediction model chosen for any grade evaluation.

Accuracy level	The posterior difference ratio C	Probability of error P
Level 1 (Good)	<0.35	≥0.95
Level 2 (Qualified)	<0.5	≥0.8
Level 3 (Barely qualified)	<0.65	≥0.7
Level 4 (Unqualified)	≥0.65	<0.7

Step 4: Calculate the small error probability P

$$P = P\{|\sigma_j(k_i) - \bar{\sigma}_j| < 0.674 S_2^2\} \quad (22)$$

Step 5: Substitute C and P into Table 1, and judge the model accuracy level.

3.3 *Application process of the improved MGM (1, m) model*

Figure 1 shows the application process of the residual modified non-equidistant MGM (1, m) model.

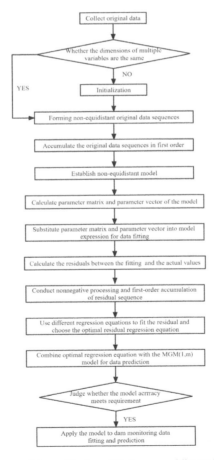

Figure 1. Application process of the residual modified non-equidistant MGM(1, m) Model.

4 EXAMPLE

This paper takes a high arch dam in Lancang River as an example to verify the validity of the residual modified non-equidistant MGM (1, m) Model. Due to monitoring instrument damage and water seepage at the dam base, the monitoring data of the arch dam base exist a long blank. The original monitoring data of the heel stress and displacement are listed in Table 2.

Table 2. Original monitoring data.

Number	Water level (m)	Compressive stress (MPa)	Displacement (mm)
1	1217.49	1.8496	3.5568
2	1219.01	1.7208	3.6243
3	1220.25	1.6543	3.6989
4	1221.26	1.5921	3.7214
5	1222.24	1.5211	3.7862
6	1223.02	1.4811	3.8383
7	1224.29	1.4057	3.9354
8	1225.08	1.3657	4.0086
9	1225.87	1.3169	4.1015
10	1226.65	1.2858	4.1282
11	1228.01	1.2237	4.1690
12	1228.74	1.1838	4.2507
13	1229.28	1.1527	4.3055
14	1230.75	1.1483	4.3464
15	1231.23	1.1217	4.5786

To better reflect the connections and changes between different variables, the water level measurement is set as the time sequence k_i. Considering the different units of compressive stress and displacement, the two variables are initialized to construct the two sets of dimensionless data $x_1^{(0)}(k_i)$ and $x_2^{(0)}(k_i)$.

Table 3. Data after initialization.

Number	Water level k_i	Compressive stress $x_1^{(0)}(k_i)$	Displacement $x_2^{(0)}(k_i)$
1	1217.49	1.0000	1.0000
2	1219.01	0.9304	1.0190
3	1220.25	0.8944	1.0400
4	1221.26	0.8608	1.0463
5	1222.24	0.8224	1.0645
6	1223.02	0.8008	1.0791
7	1224.29	0.7600	1.1064
8	1225.08	0.7384	1.1270
9	1225.87	0.7120	1.1531
10	1226.65	0.6952	1.1607
11	1228.01	0.6616	1.1721
12	1228.74	0.6400	1.1951
13	1229.28	0.6232	1.2105
14	1230.75	0.6208	1.2220
15	1231.23	0.6065	1.2873

The first 10 data sets are used as fitted samples, and the last five data sets are used as predicted samples. Substituting the data in the fitted samples into Equation 4, the parameter matrix can be obtained:

$$A = \begin{bmatrix} -0.0212 & -0.0121 \\ -0.0424 & 0.0488 \end{bmatrix} \quad (23)$$

$$B = \begin{bmatrix} 0.992 \\ 1.005 \end{bmatrix} \quad (24)$$

The MGM (1, 2) model for the two sequences is:

$$\begin{cases} dx_1^{(1)}/dt = -0.021x_1^{(1)} - 0.012x_2^{(1)} + 0.992 \\ dx_2^{(1)}/dt = -0.042x_1^{(1)} + 0.049x_2^{(1)} + 1.005 \end{cases} \quad (25)$$

The fitting and prediction results of the MGM(1, 2) model can be calculated with Equation 25. The comparison results of calculated values and monitoring data are shown in Table 4.

Table 4. Comparison of fitting value and initial data.

	Number	Water level k_i	MGM (1,2) sequence		The sequence of measured values		Residual sequence	
			Compressive stress $\hat{x}_1^{(0)}(k_i)$	Displacement $\hat{x}_2^{(0)}(k_i)$	Compressive stress $x_1^{(0)}(k_i)$	Displacement $x_2^{(0)}(k_i)$	Compressive stress $\sigma_1(k_i)$	Displacement $\sigma_2(k_i)$
Fitting	1	1217.49	1.0000	1.0000	1.0000	1.0000	0.0000	0.0000
	2	1219.01	0.9428	1.0165	0.9304	1.0190	0.0125	−0.0025
	3	1220.25	0.9107	1.0271	0.8944	1.0400	0.0163	−0.0129
	4	1221.26	0.8792	1.0395	0.8608	1.0463	0.0184	−0.0067
	5	1222.24	0.8482	1.0540	0.8224	1.0645	0.0258	−0.0105
	6	1223.02	0.8176	1.0706	0.8008	1.0791	0.0169	−0.0085
	7	1224.29	0.7874	1.0893	0.7600	1.1064	0.0274	−0.0172
	8	1225.08	0.7577	1.1102	0.7384	1.1270	0.0193	−0.0168
	9	1225.87	0.7284	1.1334	0.7120	1.1531	0.0164	−0.0197
	10	1226.65	0.6992	1.1592	0.6952	1.1607	0.0041	−0.0015
Prediction	11	1228.01	0.6705	1.1874	0.6616	1.1721	0.0089	0.0152
	12	1228.74	0.6420	1.2182	0.6400	1.1951	0.0020	0.0231
	13	1229.28	0.6137	1.2519	0.6232	1.2105	−0.0095	0.0414
	14	1230.75	0.5856	1.2884	0.6208	1.2220	−0.0353	0.0664
	15	1231.23	0.5575	1.3281	0.6065	1.2873	−0.0489	0.0408

Table 4 shows that the results calculated by the MGM (1,2) model are far away from the actual value, especially in the prediction part. Afterward, different regression equations are tried to fit the first 10 data sets, and the results show the polynomial regression equation performs better. The calculated polynomial residual regression equations of the two variables are:

$$\begin{cases} \hat{\sigma}_1(k_i) = 0.001 + 0.005(k_i - 1217.490) + 3.465 \times 10^{-4}(k_i - 1217.490)^2 - 9.3083 \times 10^{-5}(k_i - 1217.490)^3 \\ \hat{\sigma}_2(k_i) = -0.001 + 8.450 \times 10^{-4}(k_i - 1217.490) - 0.001(k_i - 1217.490)^2 + 1.163 \times 10^{-4}(k_i - 1217.490)^3 \end{cases} \quad (26)$$

The residual values of the last 5 data sets can be predicted by Equation 26, and the predicted residual values are used to modify the final predicted values, as shown in Table 5.

Table 5. Modified sequence of MGM(1, 2) model.

Number		Water level k_i	The modified MGM (1, 2) sequence		The corrected residual sequence		Relative residual sequences	
			Compressive stress $\hat{x}_1(k_i)$	Displacement $\hat{x}_2(k_i)$	Compressive stress $\sigma_1(k_i)$	Displacement $\sigma_2(k_i)$	Compressive stress $\delta_1(k_i)/\%$	Displacement $\delta_2(k_i)/\%$
Fitting	1	1217.49	1.0002	1.0013	−0.0002	−0.0013	0.0199	0.1300
	2	1219.01	0.9313	1.0189	−0.0009	0.0001	0.0996	0.0076
	3	1220.25	0.8930	1.0327	0.0014	0.0072	0.1585	0.6952
	4	1221.26	0.8585	1.0485	0.0023	−0.0022	0.2693	0.2094
	5	1222.24	0.8262	1.0659	−0.0038	−0.0014	0.4621	0.1354
	6	1223.02	0.7956	1.0842	0.0052	−0.0051	0.6452	0.4729
	7	1224.29	0.7671	1.1038	−0.0071	0.0027	0.9366	0.2428
	8	1225.08	0.7394	1.1234	−0.0011	0.0036	0.1447	0.3227
	9	1225.87	0.7128	1.1435	−0.0008	0.0097	0.1143	0.8402
	10	1226.65	0.6869	1.1640	0.0083	−0.0034	1.1932	0.2895
Prediction	11	1228.01	0.6649	1.1772	−0.0033	−0.0050	0.4944	0.4292
	12	1228.74	0.6405	1.1963	−0.0005	−0.0012	0.0731	0.0985
	13	1229.28	0.6154	1.2195	0.0078	−0.0090	1.2533	0.7397
	14	1230.75	0.5966	1.2183	0.0243	0.0037	3.9108	0.3009
	15	1231.23	0.5717	1.2426	0.0348	0.0447	5.7389	3.4710

By comparing data in Tables 4 and 5, it can be found that the improved model is closer to the measured values than the original model. Specifically, compared to the average relative predicted residual difference of 3.39% and 3.05%, the average relative residual difference of the improved model is reduced to 2.29% and 1.01%. Eliminating the data of the 15th data set and conducting the model accuracy test, the calculation result is $C_1 = 0.252$, $P_1 = 1$, $C_2 = 0.262$, $P_2 = 1$. According to indicators in Table 1, the model accuracy is judged as level 1, which means it can be applied to the data fitting and prediction of dam safety monitoring.

5 CONCLUSION

This paper establishes a non-equidistant MGM(1, m) model with the observed stress and displacement data of a high arch dam. And fitting and prediction results indicate that the model accuracy meets the requirements. The prediction length of the model is actually limited by the few observed data. From the examples of this paper, a reliable length head of 2m is predicted. Considering the case of small samples, for each additional set of known data, the data of the original model should be timely updated, and a new prediction equation should be established so as to achieve a more accurate prediction result.

REFERENCES

Chen Xijiang, Lu Tiading (2012). Residual Adaptive Regression to MGM (1, n). *Surveying and Mapping Science*, 37 (2), 36–37.
Huaizhi Su, Zhiping Wen, and Zhongru Wu (2011). Study on an Intelligent Inference Engine in the Early-warning System of Dam Health. *Water Resources Management*, 25(6), 1545–1563.
Huai-zhi Su, Zhong-ru Wu, Zhi-ping Wen (2007). Identification model for dam behavior based on wavelet network. *Computer-Aided Civil and Infrastructure Engineering*, 22(6), 438–448.
Liu Sifeng, Dang Yaoguo, Fang Zhigeng (2010). *Gray System Theory and Its Applications*. Science Press.

Su Huaizhi, Hu Jiang, Wu Zhongru (2012). A Study of Safety Evaluation and Early-warning Method for Dam Global Behavior. *Structural Health Monitoring*, 11(3), 269–279.

Su Huaizhi, Wu Zhongru, Dai Huichao (2005). The Intelligent Fusion Monitoring System for Dam Safety. *Journal of Hydropower*, 24 (1), 122–126.

Wu Bangbin, Chen LAN, Ge Cui (2012). Application of the Modified Non-equally Spaced GM (1,1) Model to Dam Sedimentation Analysis. Water Resources and Power, 30 (6), 95–97.

Wu Zhongru (2003). *Safety Monitoring Theory of Hydraulic Buildings and Its Application*. Higher Education Press.

Xiong Pingping, Dang Yaoguo (2011). Non-equally Spaced-based Multivariable MGM (1, m) model. *Control and Decision-making*, 26 (1), 45–53.

Xu Wei, Xu Hujun, Xu Haiyan (1999). Precision Test of the Applied Gray Prediction Model. *Journal of Mathematical Medicine*, 12 (2), 166–167.

Analysis and suggestions on the construction of a water-saving society in Yunnan Province

Qiu-Ju Zhu* & Sen Wang
The Pearl River Hydraulic Research Institute, Guangzhou, Guangdong, China

ABSTRACT: This paper fully enumerates the experiences and practices of some counties (districts) in the process of establishing the standard in Yunnan Province in such aspects as agricultural water-saving, industrial water-saving transformation technology, and water-saving publicity, and summarizes the achievements of industrial water-use efficiency, water-saving carrier construction and water-saving fund investment. Besides, it expounds and analyzes the existing problems in the utilization level of reclaimed water, the precise subsidy and water-saving incentive mechanism of agricultural water use, the establishment of water efficiency leaders, and water-saving benchmark units. Finally, it puts forward suggestions for the existing problems. This article was originally written by the authors.

1 INTRODUCTION

The National Water Conservation Action Plan, issued and implemented by the National Development and Reform Commission and the Ministry of Water Resources on April 15, 2019, requires that the construction of a water-saving society shall meet the standards of a water-saving society comprehensively, taking the county as the unit. By 2022, more than 30% of counties (districts) in southern China will meet the standards of a water-saving society (Ji et al. 2021; Li et al. 2021; Ma 2021; Zhang & Luo 2020).

Among the 129 county-level administrative regions in Yunnan Province, 48 counties (districts) had completed the construction of a water-saving society to meet the standards by the end of 2021, with a construction rate of about 37%, exceeding the target requirements of the National Water-saving Action Plan (Wang et al. 2021).

2 EXPERIENCE IN PRACTICE

In the creation process, the province's county (area) standard attaches great importance to the county water-saving society construction. The government as a whole organizes and leaders this work involving each department conducts and promotes water-saving society the public institutions, industrial enterprises, schools, residential quarters, and so on for the pilot demonstration of construction work, through developing and promoting water-saving society public institutions, industrial enterprises, schools, residential areas and so on for the pilot demonstration of construction work, through developing the water-saving society construction, implementing strict water management, adjusting measures to local conditions of water saving measures, and strictly strengthening the regulation of the water. A number of water-saving key projects have been completed to continuously improve the sustainable use of water resources and the carrying capacity of the water environment in the county to ensure the sustainable development of the country's economy and society, and to accumulate

---

*Corresponding Author: hnfczqj@163.com

rich experience for the comprehensive creation of a water-saving society (Fu & Peng 2021; Ling et al. 2021).

(1) Agricultural water-saving effect is very outstanding. Site review found that pipeline water delivery and drip irrigation facilities in Yunnan Province accounted for a relatively high proportion of irrigation areas. In the large control irrigation area of Malone District, Qujing City, the agricultural water association has been established to realize metering and charging, and the collection rate is relatively high. The irrigation area of Zishichong Reservoir in Shuangbai County of Chuxiong Prefecture has realized the intelligent IC card metering and billing of water consumption. A state-level modern agricultural industrial park has been established in the Mengcaig Irrigation Area of Kaiyuan City, Honghe Prefecture, implementing high-efficiency water-saving irrigation and online monitoring, metering, and charging.

(2) Key technologies are conquered for saving water and reducing emissions, and new water-saving technologies are applied in the industry to actual production. Yunnan Meisenyuan Forest Products Technology Co., LTD., Shuangbai County, Chuxiong Prefecture, has successfully solved the problems of extensive control, high water consumption, and large loss of turpentine in the existing production process by applying the key technology of water saving and emission reduction in green processing of turpentine, and has reduced the water consumption per ton of turpentine processing by nearly 70%.

(3) Water-saving publicity has various forms. All counties and districts vigorously carried out water-saving publicity activities in communities, schools and towns, widely posted water-saving publicity signs, used the network platform for water-saving publicity, and called on various sectors and the public to participate in water-saving knowledge contests. Thus, a good social water-saving atmosphere is created.

3 ACHIEVEMENTS ACHIEVED

(1) Efficiency of water use in the industry
From 2020 to 2021, the economic aggregate of Yunnan Province continued to rise, but the economic and social water use did not increase but decreased. In terms of water use efficiency, the water consumption per 10,000 yuan of GDP and the water consumption per ten thousand yuan of industrial added value in Yunnan Province showed a decreasing trend, and the effective utilization coefficient of farmland irrigation water showed an increasing trend. From 2020 to 2021, water consumption per 10,000 yuan of GDP and industrial added value per 10,000 yuan decreased by 9.8% and 24.7%, respectively, and the effective utilization coefficient of farmland irrigation water increased from 0.492 to 0.502. The comparison of industrial water use efficiency in Yunnan Province in 2020 and 2021 is shown in Table 1.

Table 1. The Changes in industrial water use efficiency during the construction of a water-saving society at the county level in Yunnan Province.

Industry water efficiency	2020 parameter values	2021 parameter values
Water consumption per 10,000 yuan of GDP (m^3/ ten thousand yuan)	64	57.7
Water consumption of ten thousand yuan of industrial value added (m^3/ten thousand yuan)	30	22.6
Coefficient of effective utilization of farmland irrigation water	0.492	0.502

In addition to Wuding County of Chuxiong Prefecture, the other 15 counties and districts of the 16 newly created water-saving society reached the standard in 2021, including water consumption per 10,000 yuan of GDP, water consumption per 10,000 yuan of industrial added value and effective utilization coefficient of farmland irrigation water, and the optimal value reached 55 cubic meters, 21.5 cubic meters and 0.577, respectively.

From 2020 to 2021, the water consumption per 10,000 yuan GDP of Fumin County, Linxiang District of Lincang City, and Kaiyuan City of Honghe Prefecture were all lower than that of Yunnan Province. In 2020, the water consumption per 10,000 yuan GDP of the above counties and districts was 14%, 8%, and 6% lower than that of Yunnan Province respectively. In 2021, the water consumption per 10,000 yuan of GDP in the above counties was 7%, 11%, and 7% lower than that in Yunnan Province respectively. The comparison of water consumption per 10,000 yuan of GDP in 16 counties and Yunnan Province is shown in Figure 1.

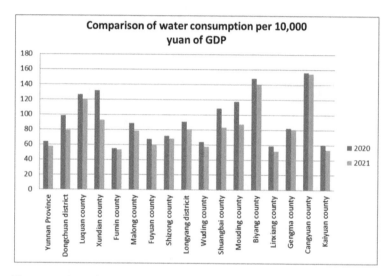

Figure 1. The comparison of water consumption per 10,000 yuan of GDP in 16 counties and Yunnan Province.

From 2020 to 2021, the water consumption per ten thousand yuan of industrial added value in Malong District, Shizong County, and Shuangbai County of Qujing City were all lower than the level of Yunnan Province. In 2020, the water consumption per ten thousand yuan of industrial added value in the above counties and districts was 26%, 31%, and 10% lower than that of Yunnan Province respectively. In 2021, the water consumption per 10,000 yuan of industrial added value in each county was 7%, 12%, and 11% lower than that in Yunnan Province respectively.

The comparison of water consumption per ten thousand yuan of industrial added value between counties and Yunnan Province is shown in Figure 2.

(2) Effect of water-saving carrier construction

By December 2021, a total of 622 water-saving units, 78 water-saving enterprises, and 117 water-saving communities were built in 16 newly created water-saving society counties (districts) in Yunnan Province in 2021. In 2021, the average construction rate of water-saving enterprises in 16 counties and districts registered in Yunnan Province was 49%, the average construction rate of water-saving units was 62%, and the average construction rate of water-saving residential communities was 27%. Among them, the

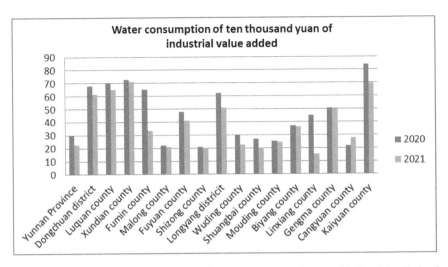

Figure 2. The comparison of water consumption per ten thousand yuan of industrial added value between counties and Yunnan Province.

number of water-saving enterprises in Wuding County of Chuxiong Prefecture is the largest, 12. The number of water-saving units in Muding County of Chuxiong Prefecture is the largest, 59. The number of water-saving residential communities built in Muding County of Chuxiong Prefecture is the largest, 21.

(3) Investment in water-saving funds has yielded results

Yunnan Province invested a total of 43.1 million yuan in direct water-saving funds and 578.1 million yuan in indirect water-saving funds from 2020 to 2021. The central funds approved by each county and district are mainly used for the construction of a water-saving society, water-saving publicity, upgrading of sewage plants, etc. The indirect funds are mainly used for the renovation of the water supply network, water-saving renovation of irrigated areas, reservoir maintenance, and high-efficiency water-saving irrigation.

Yunnan Province has given full support to water-saving agricultural irrigation, which has achieved outstanding results. Many irrigated areas have implemented high-efficiency water-saving irrigation, and metering and charging fees have been implemented. Yunnan Meisenyuan Forest Products Technology Co., LTD., with self-raised funds, has established stable scientific research cooperative relations with many universities and scientific research institutes. Besides, it has developed and applied the key technologies of water saving, emission reduction, and green processing of pine resin. It can save 4,000 tons of production water every year, as well as a lot of water and sewage treatment costs. Also, it can reduce the loss of pine resin by 390 tons, and increase the production income by more than 5.8 million yuan. After the promotion of this achievement, it can increase the revenue by more than 232 million yuan for the turpentine processing industry every year.

4 EXISTING PROBLEMS

Although there are various experiences and practices in the construction of a water-saving society at the county level in this province, there are still some problems and deficiencies.

(1) The utilization level of reclaimed water needs to be further improved. The water quality of the effluent from the reclaimed water plant in Fumin County, Yangbi County, Dali

Zhou, and Linxiang District, Lincang City does not reach the standard. Water quality monitoring reports showed that fecal coliforms exceeded the standard. This has something to do with the weak awareness of recycled water utilization.

(2) The mechanism of precise subsidies for agricultural water use and incentives for water conservation has not been effectively implemented. Among the 16 counties (districts), 12 counties (districts) had policies on precise subsidies for agricultural water use and incentives for water conservation, but no records of actual subsidies were available. This may have something to do with subsidies and incentives not being funded.

(3) The establishment of national and provincial water efficiency leaders and water-saving benchmark units needs to be strengthened. 16 counties (districts) have not successfully created a case. This may be related to the lack of effective incentive policies implemented by water-saving management departments.

5 SUGGESTIONS

Suggestions for the existing problems are as follows:

(1) It is devoted to further increasing the awareness of reclaimed water utilization, improving the quality of reclaimed water and the utilization rate of reclaimed water, realizing the sustainable utilization of water resources, and alleviating the contradiction of water shortage.

(2) They should be committed to implementing policies to provide targeted subsidies for agricultural water use and incentives for water conservation, ensuring that households in need receive subsidies and those who save water receive incentives, and implementing these subsidies and incentive policies.

(3) They should strengthen efforts to establish water-efficiency leaders and water-saving benchmarking units, and improve the water efficiency of water-saving public institutions and enterprises, which play an exemplary and leading role. The water-saving administrative department can take certain incentive measures to stimulate the initiative of the establishment unit.

6 CONCLUSION

The construction of a water-saving society is an effective carrier to save water resources and a powerful grasping hand to promote the work of water-saving in all aspects and fields. Since the implementation of the construction of a water-saving society at the county level in Yunnan Province, the total water use in the province has continued to decrease, and the water-saving effect has been remarkable, especially in agriculture. In the future, if we can maintain good practices, consolidate water-saving results, solve problems such as the utilization of reclaimed water in a planned way and step by step, and further improve the utilization efficiency of water resources based on the existing water-saving work, the construction of water-saving society will be able to go to a higher level next year and form a good trend of water-saving in the whole society (Zhou et al. 2021; Zeng 2021).

ACKNOWLEDGMENTS

This study is supported by Multi-objective Water Resources Scheduling Technology and Application in the Pearl River Basin, National Key R&D Program of China ((2017YFC0 405900). The authors would like to thank the Editors and anonymous Reviewers.

REFERENCES

Fu Qingcheng, Peng Ziwei. Discussion on the Main Practices and Achievements of the Construction of a Water-saving Society at the County Level in Henan Province [J] *China Water Resources*, 2021 (4):64–67.

Ji Haiting, Zhang xi, Pan Ru, Hua Chen, Chen Jiayu. Research on the Construction of a Water-saving society in Jintan District [J]. *Water Technology Supervision*, 2021(7):110–112,130.

Li Zhonghua, Zhang Yunfeng, Zhou Caixia. Analysis of Difficulties and Countermeasures of Building Water-saving Society in County areas of Yunnan Province [J]. *Water Technology Supervision*, 2021(3): 48–49,128.

Ling Fan, Wang Jun, Fei Jian, Wang Guangyue, Zhang Li. Construction Practice of Water-saving Society in Jianhu County [J]. *Water Resources Management*, 2021 (9):42–43.

Ma Mengchun. Reflections on the Construction of a Water-saving Society at the County Level [J]. *Farmland Water Conservancy*, 2021 (10): 102–103.

Wang Leizhi, Li Xiaotian, Zhou Yongkang, Shang Baojian, Li Lingjie, Gai Yongwei. Practice and Thinking on the Construction of a Water-saving Society Reaching the Standard in County Areas with Abundant Water [J]. *Jiangsu Water Resources*, 2021 (7): 23–26.

Zeng Y. Analysis of the Construction of a Water-saving Society in Jiading District of Shanghai [J]. *Water Resources Development and Management*, 2021 (1):74–77.

Zhang Fang, Luo Junxiao. Discussion on the Development of High-efficiency Water-saving Irrigation in Yanjiang District during the 14th Five-Year Plan Period [J]. *Water Conservancy Technology Supervision*, 2020(5): 63–64,122.

Zhou Xue, Jia Shoudong, Huang Lizhou. Discussion on the construction of a water-saving society at the county level in Shandong Province [J]. *Shandong Water Resources*, 2021 (1):23–24.

Study on the improvement of spatial effects of the old residential life circle in the context of aging suitability

Xiaoli Xu, Tong Liu*, Jing Liang, Jiahui Huang & Tengfei Wu
Northeast Petroleum University, Daqing, Heilongjiang, China

ABSTRACT: From the perspective of the interrelationship among space, people, and social activities, and based on field visits and questionnaires, the existing spatial problems and the current situation of spatial use in the old residential life circle were fully understood. At the same time, the needs of residents for space, facilities, and landscape were analyzed. Meanwhile, the spatial syntax theory and Depthmap technology were used to deeply explore the structural relationship of the outdoor activity space in the old residential life circle. Four new ideas for planning and design will improve the spatial efficiency of the old residential life circle.

1 GENERAL INSTRUCTION

China entered the aging society in 1999 (Zhou 2011), and the seventh census shows that 18.7% of the population is over 60 years old (OL2020), and it is expected that by 2040, the number of elderly people in China will account for more than 20% of the total population. As they grow older, their bodies and minds change, and they have problems such as reliance on familiar environments, frequent negative emotions, and reluctance to participate in social activities. To maintain a sense of belonging, comfort, and neighborhood, most elderly people choose to provide for the aged in the community or their native place. Whether a residential living area can provide good activity space, infrastructure and support services for the elderly has become the main indicator to measure whether a residential living area is age-friendly. In the process of rapid urbanization, many old living circles reveal a large number of problems: such as over-simplification and homogenization, aging facilities, and poor space utilization, which greatly reduce the service efficiency of aging in place and community aging, so the spatial effectiveness improvement of living circles in old living areas will become an important task to improve the quality of aging society.

This paper selects the elderly group and conducts research on the old residential life circle, and conducts a detailed investigation on the physiological function, psychological activity, activity type, and space demand of the elderly group through field visits and questionnaires. From the perspective of user needs, this paper proposes existing problems and corresponding optimization strategies for the outdoor activity space. Based on the relevant research and analysis of user needs, this paper quantitatively analyzes the community public activity space through the model of space syntax, measures the problems existing in the form and organization of the space through syntactic parameters, and proposes retrofit and optimization strategies in a reasonable range. Therefore, the spatial research object of this paper is mainly the living circle of a residential area in Daqing City. It mainly studies and analyzes four

---

*Corresponding Author: lt95112788@163.com

aspects of its road system, public space, green landscape, and supporting facilities, using its parameters to quantify the spatial efficiency of the living circle. The old residential life circle is suitable for the aging transformation to provide a certain reference.

2 OVERVIEW OF SPATIAL SYNTAX AND DEPTHMAP TECHNOLOGY

Spatial syntax theory was put forward by Billy Hillier of Bartley College, the University of London in the 1970s (Zhao 2017). Based on topology theory, it is a theory and method to study the relationship between the spatial organization and human society by quantitatively describing the residential space structure including architecture, city, and landscape (Zhang 2004). The theory explores the relationship between spatial organization and social life based on spatial grouping as the core (Qiu 2013), while the spatial grouping relationship covers a series of spatial factors and the relationship between them and their mutual influence. Therefore, for the old residential life circle, its internal spatial environment influences the social activities and social structure in the residential area through spatial grouping, so the quantitative study of the internal spatial grouping of a residential area can improve the spatial effects of the old residential life circle, enhance the vitality of the residential area, promote the neighborhood relationship, and provide a reasonable reference for the aging-friendly renovation of the old residential area.

The spatial syntactic model is mainly drawn with the help of software and its parameters are calculated by the software. Among all the syntactic software, Depthmap software (Chen 2020) can make a simple and intuitive simulation analysis, and by importing the drawn CAD drawings, the calculation of the relevant parameters and the drawing can be carried out.

3 THE CURRENT SITUATION OF AGEING IN THE LIVING CIRCLE OF A RESIDENTIAL AREA IN DAQING

With the rapid socio-economic development of China and the continuous expansion of the population base, the construction model of infinite expansion has caused a large number of new urban centers to rise on the edge of the original city, attracting government agencies, factories, enterprises, and a large number of people, forming a situation where some buildings and sites in the original urban areas are left unused, the age structure of residents is imbalanced, and a large number of building spaces are in disrepair. As an important part of the city, the problems faced by the old residential life circle are of particular concern. Through field visits and questionnaire research, the problems that exist in the spatial layout. Supporting facilities and the social creation of such living areas are mainly manifested in the following aspects.

3.1 *Lack of management of road space, the problem of mixed pedestrian and vehicle traffic, there are security risks*

Because the old residential life circle was built a long time, the automobile was not popularized during the construction period, therefore the width of the road space is designed to meet only the travel needs of non-motorized vehicles and pedestrians, and because the road is used for a long time, most of the road will be damaged road surface and other phenomena (Figure 1). In today's social environment of high car penetration, the status of such roads will cause internal traffic chaos, and walking residents' Safety is threatened, especially for the elderly, whose physical functions decline as they age, and whose mobility and responsiveness diminish, greatly reducing their chances of communicating outside (Figure 2).

Figure 1. Status of roads in old communities.

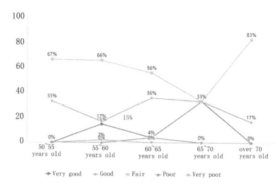

Figure 2. Analysis of age structure to road evaluation.

3.2 *The irrational layout of public space, single function, low utilization rate, and lack of vitality*

Most of the buildings in the old residential life circle adopt a row layout, forming outdoor public space with compact space and single form (Figure 3). The homogeneous outdoor public space reduces the identifiability of the space to a certain extent, and due to the development of the social economy and the improvement of the living standard of residents, people are no longer satisfied with the basic material needs, and their pursuit of multi-functionality, fun, and sharing of space is growing. The existing space does not create a high-quality communication space for residents (Figure 4), thus reducing the

Figure 3. Status of public space.

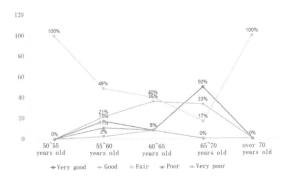

Figure 4. Analysis chart of age structure on-site evaluation.

effectiveness of the space and the overall lack of vitality, which affects the aging process of the community.

3.3 *Greening landscape form with single and low utilization rate is idle*

The old residential life circle has low standards in terms of plant configuration, landscape planning, management, and maintenance. Generally, large trees are selected on both sides of the road, and low shrubs are selected around the activity space. Other than this, there are no other regional or seasonal characteristics. The unique plant configuration makes the green space in the residential area less ornamental (Figure 5). The low shrubs do not provide good shading for the activity space and reduce the comfort of using the space (Figure 6), and

Figure 5. Status of green space.

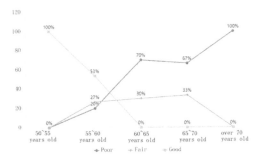

Figure 6. Analysis of age structure to landscape evaluation.

649

because the existing green landscape is encroached upon by the parking space, it reduces the spatial effects of the green landscape.

3.4 *Unreasonable arrangement of commercial fitness facilities, lack of signage and barrier-free facilities*

At the early stage of the construction of the old residential life circle, only a large number of single hard pavements and greenery were used to create outdoor space, so the lack of supporting facilities is another major problem it faces. In the process of aging-friendly community renovation, although many fitness facilities and commercial facilities were added, the lack of research on users in the early stage led to the unreasonable arrangement of facilities, which not only did not have a positive impact but also affected the normal life of residents to a certain extent (Figure 7). The lack of signage and accessibility facilities for the elderly also greatly reduces the frequency of their use of space and facilities (Figure 8).

Figure 7. Current status of supporting facilities.

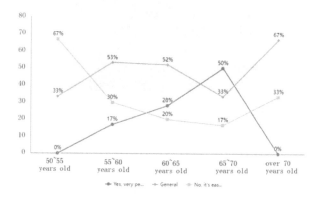

Figure 8. Analysis chart of age structure to mark evaluation.

4 AGEING-FRIENDLY RENOVATION OF A LIVING AREA IN DAQING

The research on the application of spatial syntax and Depthmap technology in aging-friendly renovation mainly includes two aspects: the establishment of the axis model and the drawing of the view field model, whose final evaluation system is reflected by the parameter indexes of the software simulation. For the spatial units inside the living circle of old residential areas, the syntactic parameters can intuitively reflect the interrelationship and influence of each spatial node, and its main technical applications and corresponding syntactic interpretations are mainly as follows.

4.1 *Assess road accessibility and improve the community transportation network.*

The axial model of the road space is constructed to quantify the accessibility, convenience, and ability of the road to attract residents to traverse through it by integration degree and connectivity degree. As shown in Figures 9 and 10, the higher the integration and connectivity of the road, the warmer its graphical color, which means that the road space has a stronger ability to carry the flow of people and vehicles and has better permeability to other spatial nodes. When enhancing spatial effectiveness, we should strengthen the quality of this type of high-quality road space, reasonably arrange the surrounding commercial facilities, and increase the integration and multifunctional public space to achieve the effect of

Figure 9. Axis integration.

Figure 10. Axis connectivity.

attracting people and enhancing vitality. For the negative road space, the degree of connectivity with other spatial nodes should be reasonably improved so that residents can reach it more conveniently.

4.2 *Quantify spatial accessibility and optimize the quality of public space.*

When the color of the axis is warmer (Figure 11), it proves that the selectivity of the spatial node is higher, and the residents have a stronger willingness to stop and interact with others at that place. The values and diagrams of axis selectivity visually express important spatial nodes and spatial units lacking in vitality and provide scientific reference for the improvement of spatial effectiveness and the goal of elderly-friendly.

Figure 11. Axis choice.

4.3 *Metric node usage rate to enhance the sense of place in the space.*

Constructing the visual field model of the living circle in the old residential area, simulating the users' perception of space and the range of sight thus measuring the efficiency of space use, is an important part of improving the effectiveness of space. In Figure 12, the warmer

Figure 12. Visual integration.

Figure 13. Visual clustering coefficient.

the color of the spatial node, the higher the integration of its visual field, indicating that the more accessible and perceptible the spatial node is, the more attractive it is to users, and the easier it is to become a generator of various social activities. In Figure 13, the cold color icon represents a public space with more openness, the opposite space with more privacy. When proposing strategies for spatial effectiveness, we should combine the accessibility of space and the degree of privacy and openness of space, create public spaces with different characteristics according to the physical and mental needs of users, and enhance the sense of place of each spatial node.

4.4 *Simulate the mobility of residents and rationalize the planning and design.*

The proxy robot simulation (Figure 14) is a human flow simulation experiment based on the visual field model, where small robots are placed to simulate the human perception of the surrounding environment and form a self-organized movement system after a period of time according to the grouping characteristics of the spatial environment. The warmer color of

Figure 14. Simulation analysis.

the spatial node represents more robots passing by, which means that more people flow is selected for that spatial node and vice versa. This method can well predict the movement of human flow in the complete spatial system, which can guide the subsequent planning and design to a certain extent, and has some reference value for improving spatial efficiency.

5 CONCLUSION

Based on space syntax and depth map technology, the space of the old residential life circle is simulated, and its advantages in the definition and quantification of space can be used to discover the existing problems of the space environment and supporting facilities, and then propose corresponding optimization strategies. This paper analyzes the spatial efficiency of the living area in the context of aging and further proposes that in the process of efficiency improvement, the spatial syntax and Depthmap technology should be used to evaluate and improve the internal road network, quantify and improve the spatial system and spatial quality, support the creation of a sense of place in space, and simulate the trend of human movement to lay a solid foundation for planning and design. The results of this study not only play a positive role in improving the spatial effectiveness of living circles in old residential areas but also provide a certain reference for the planning and design of living circles in new residential areas.

ACKNOWLEDGMENTS

This work was supported by the Fundamental Research Project Fund for Universities in Heilongjiang Province (2020QNW-01) and the Daqing City Philosophy and Social Science Planning Research Project (DSGB2022038).

REFERENCES

2020 *Seventh National Population Census Bulletin*. [OL] http://www.gov.cn/guoqing/2021-05/13/content_5606149.htm

Chen Qi, Cai Wei, Sinan, Fu Wenchu. Spatial Layout Design of Cruise Ship Deck Based on Spatial syntax[J]. *Ship Engineering*, 2020, 42(01):4–8 + 41.DOI:10.13788/j.cnki.cbgc.2020.01.02.

Qiu Lin, Sun Cheng, Jiang Hongguo. Spatial Configuration Simulation Analysis and Quantitative Evaluation of Public Library Buildings Based on Depthmap Software: The Example of the New Library of the Great Britain Library[C]//. *Computational Design and Analysis–Proceedings of the 2013 National Workshop on Teaching Digital Technology in Architecture for Architecture Faculties.*, 2013:116–120.

Sun, Quan-Sheng. Analysis of the Causes of Urbanization Problems in China and the path of Enhancement [J]. *Journal of North China Electric Power University (Social Science Edition)*, 2020(05):79–91. doi:10.14092/j.cnki.cn11-3956/c.2020.05.008.

Zhang Yu Wang Jianguo. The "Spatial Syntax" [J]. *The Architect*, 2004, (03): 33–44.

Zhao L. *Spatial Analysis of Jinzhong Fortress Settlement Based on Spatial Syntax Theory* [D]. Taiyuan University of Technology, 2017.

Zhou Yan-Min, Cheng Xiao-Qing, Lin Ju-Ying, Lin Jing-Yi. *Elderly Housing* [M]. Beijing: China Construction Industry Press, 2011.

Numerical simulation of freeze-thaw damage of root-soil complex based on discrete element method

Yuan Sun & Hui Li*
Qinghai University, Xining, China

ABSTRACT: Based on the discrete element method, this paper presents a new simulation method of freeze-thaw damage, and conducts PFC simulation of loess under different freeze-thaw cycles. Using this method, the simulation results of the unconfined compressive strength of loess under different cycles are similar to the experimental results, which proves the feasibility of this method. Then, the root model is established to simulate the UU triaxial test of the root-soil complex, and the parameters of the numerical model of the root-soil complex are calibrated, to further explore the effect of root angle on the shear strength of the root-soil composite and carry out numerical simulation research on the strength characteristics of the root-soil composite under different freeze-thaw cycles. The results show that: (1) The proposed freezing-thawing damage simulation method can better simulate the deterioration effect of loess under the freezing-thawing cycle. (2) The existence of a root system can improve the shear strength of the soil. Under the same number of freeze-thaw cycles, the shear strength of the root-soil composite is higher than that of plain soil, and the shear strength of the root-soil composite decreases first and then tends to be flat with the increase of freeze-thaw cycles. (3) When the root inclination angle is 30 degrees, it has the greatest effect on improving the strength of the soil. The inclination angle of the root system is 60 degrees, which has the smallest improvement on the strength of the soil. However, when the root inclination angle is 60 degrees, the shear strength of the root-soil composite is least affected by the freezing-thawing cycle. The research results provide a new way to explore the freeze-thaw damage mechanism of root-soil complex from the microscopic point of view, and can also provide guidance for ecological slope engineering construction in cold regions.

1 INTRODUCTION

Mingjing Jiang adopted the method of root particle expansion, carried out a root pull-out simulation experiment with the discrete element, and analyzed the influence of water content and root burial depth on the frictional mechanical characteristics of the root-soil interface (Jiang 2017). Yitao Zhang used discrete element software to simulate an unconfined compression test of microbial solidified coral sand and analyzed the crack development mode and failure mechanism of solidified coral sand (Zhang 2021). Hua Xu clarified the influence of root morphology and different hierarchical structures on the mechanical properties of the root-soil complex by establishing the ryegrass root-soil complex model, which made up for the deficiency of only emphasizing the first-level root reinforcement at present. Based on the discrete element method (Xu 2021). Di Ma endowed loess samples with adhesive strength that obeys Gaussian distribution and analyzes the influence of distribution variance on the mechanical properties of loess samples (Ma 2020). Lei Tian used a discrete element model combined with a strength reduction method to analyze the

---

*Corresponding Author: lihui@qhu.edu.cn

reinforcement effect of vegetation slope protection position on cohesive soil slope (Tian 2018). Lingzhi Huang carried out a numerical simulation of the uniaxial compression experiment of a concrete model damaged by freezing and thawing by discrete element software and analyzed the influence of the freezing and thawing effect on the mechanical properties of concrete (Huang 2021). Tantan Zhu based on DEM, according to the relationship between temperature and unfrozen water content, proposed a new method to simulate the freezing and thawing of saturated rocks (Zhu 2021). Weijie Xing established a triaxial specimen model with PFC3D software and analyzed the influence of various mesoparameters on the macroscopic mechanical properties of cohesive soil (Xing 2017). Hengxing Wang carried out direct shear tests on the root-soil complex subjected to different freeze-thaw cycles. It was found that with the increase in freeze-thaw cycles, the cohesion decreased and then stabilized, and the internal friction angle increased first and then decreased, and then stabilized (Wang 2018). Xianghua Song used the discrete element numerical simulation method to conduct a direct shear experiment simulation of an inverted triangular root system to explore the soil consolidation mechanism of the root-soil complex (Song 2021). However, at present, there are few studies on the simulation of freeze-thaw damage of loess and root-soil complex by using the discrete element method, and the proposed method can provide a new idea for the study of the microscopic damage mechanism of loess under freeze-thaw cycle. The soil consolidation effect of the root system is analyzed, and the effect of root system angle on soil is obtained.

2 EXPERIMENTAL SIMULATION OF UNCONFINED COMPRESSION OF LOESS UNDER FREEZE-THAW CYCLE

2.1 *Model establishment*

The discrete sample is a cylinder with a radius of 0.01955 m and a height of 0.08 m, which is generated by the layered Undervoltage method. According to the loess grain grading curve in reference (Jiang 2016), if the grain size is completely generated according to the real loess grain size, the number of grains is too large, and the program is difficult to run. On the basis of considering the size effect, the particle size is enlarged to generate particle size. The final number of loess particles is 6341, and the number of ice particles is 140. The minimum particle size of loess is 0.00255 m, the median particle size is 0.021 m, and the maximum particle size is 0.053 m, as shown in Figure 1. The contact model between loess particles is a parallel bonding contact model, and the contact model between ice particles and loess particles is a linear model.

Figure 1. Sample model.

2.2 *Simulation method of freeze-thaw damage*

When the temperature drops to freezing temperature, the pore water freezes and changes its phase, and the volume expansion is about 9%. Frost heave stress is generated on the surrounding soil skeleton. At the same time, due to the phase change of ice and water, the suction of the matrix increases, forming capillary cohesion (Yan 2019). In the PFC simulation freezing process, the pore ice particles are divided into ten times parts and gradually increased to 1.090513 times the original volume. In each volume expansion process, to avoid the particles flying out, a certain number of pore ice particles are randomly selected, and all the contacts of pore ice particles are traversed, which is transformed into a parallel bonding contact model to simulate the formation of ice cohesion. At the same time, the bonding strength between loess particles is improved to simulate the formation of capillary cohesion. After the sample is balanced, the pore ice particles are restored to the original volume, and a freeze-thaw damage simulation is completed.

2.3 Numerical simulation of unconfined uniaxial compression

Firstly, the walls are built at the upper and lower ends of the generated sample model, the relative velocities of the upper and lower walls are given, and the strain rate is controlled to be 0.8 mm/min. To ensure the rationality of the intergranular model parameters, it is necessary to calibrate the model samples according to the peak strength of undisturbed loess in the literature (Liu 2021) without freezing and thawing cycles. See Table 1 for the relevant model parameters. Numerical simulation experiments on the unconfined compressive strength of samples under different freeze-thaw cycles are carried out respectively. The results of the simulation experiment and the peak compressive strength of the reference are shown in Figure 2. The peak strength of the samples that have not experienced freeze-thaw

Table 1. Basic parameter of triaxial experiment model.

Parameter	Value
Linear effective modulus/(10^7 nm^{-2})	2.3e6
Linear stiffness ratio	2
Parallel bonding effective modulus/(10^7 nm^{-2})	1.6e7
Parallel bonding stiffness ratio	2
Normal bond strength of soil/(10^6 nm^{-2})	2.4e5
Tangential bond strength of soil/(10^5 nm^{-2})	2.4e5
Root-soil interface friction coefficient	0.5

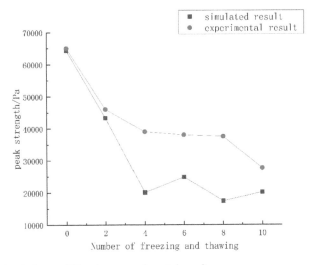

Figure 2. PFC simulation and literature experimental results.

damage and the samples that have experienced two freeze-thaw cycles are very close to the peak strength of the original text, and the unconfined compressive strength of the numerical simulation samples gradually tends to be gentle, which is consistent with the actual situation. Therefore, the feasibility of the freeze-thaw damage simulation method based on a discrete element is proved.

3 STATIC TRIAXIAL TEST AND NUMERICAL SIMULATION OF ROOT-SOIL COMPLEX

3.1 Static triaxial test of plain soil and root-soil complex

In the process of root growth, chemical substances with good cementation will be secreted, forming root-soil cohesion (Xia 2020). In order to simulate the actual interaction between root and soil as much as possible, the root-soil composite sample should be prepared immediately after the root system is cleaned. In the static triaxial test, the SLB-1 triaxial shear permeability tester controlled by stress and strain is selected.

3.2 Static triaxial numerical simulation of soil-soil composite

3.2.1 Establish a root model

According to the stress-strain curve of the straight roots of Amorpha fruticosa under axial load (Bai 2021), a single tensile test simulation was carried out. According to the experimental method of tendon tensile simulation in reference (Yang 2020), different parallel bonding normal stiffness of root particles is given according to the non-strain interval. In the stretching experiment, the leftmost particle should be fixed at first, and the uppermost particle should be given a constant rightward velocity. As shown in Figure 3, the simulation results are consistent with the measured data, indicating that the model can simulate the axial mechanical properties of the Amorpha fruticosa root system well.

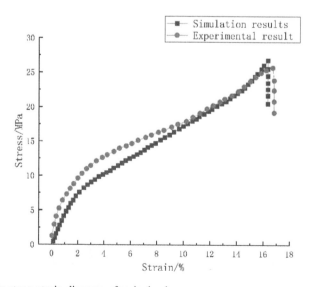

Figure 3. Tensile stress-strain diagram of a single piece.

3.2.2 *Simulated triaxial test of root-soil complex*

Five root segments are generated in the horizontal direction at 1/4, 1/2, and 3/4 heights of the sample, respectively. Considering the net pocket effect formed by the intertwined fibrous roots in the soil (Yang 2019), the bonding strength of the rest of the soil is improved. The final calibration results of relevant parameters are shown in Table 1. As shown in Figure 4, the stress-strain curve of plain soil and root-soil complex simulated by PFC is in good agreement with the experimental curve. At the same time, when the triaxial specimen is subjected to confining pressure and axial load, shear stress will be generated in the soil specimen, and the interfacial action between the root and soil will convert some shear stress of the soil into root tension, thus improving the bearing capacity of the soil. In the process of

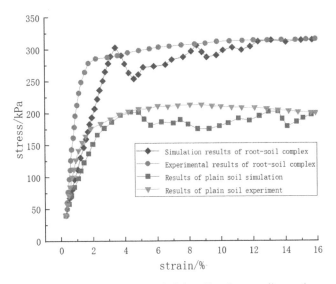

Figure 4. Simulation and experimental results of plain soil and root-soil complex.

loading, the contact between root particles shows tensile characteristics, which is consistent with reality, so the existence of a root system can effectively improve the strength characteristics of loess.

4 STUDY ON THE EFFECT OF ROOT ARRANGEMENT ANGLE AND FREEZE-THAW CYCLE ON ROOT-SOIL COMPLEX

Samples with angles of 0, 30, 45, 60, and 90 between the taproot system and the horizontal line were generated, and a numerical simulation of the static triaxial test was carried out. When the angle between the root angle and the horizontal plane is 30 degrees, the shear strength of the root-soil composite is the highest. When the angle between the root angle and the horizontal plane is 60 degrees, the shear strength is the lowest. However, the shear strength of root-soil composite without inclination angle is obviously higher than that of plain soil. Using the proposed freeze-thaw damage simulation method, the damage simulation of plain soil, root-soil complex with 30 degrees and 60 degrees inclination angle under different freeze-thaw cycles is carried out, and the triaxial shear simulation is carried out. With the increase in the number of freeze-thaw cycles, when the root inclination angle is 30 degrees, although the strength is the highest without freeze-thaw cycles, it decreases the most.

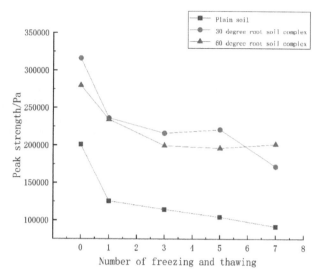

Figure 5. Peak strength of plain soil and root-soil complex under different freeze-thaw cycles.

On the contrary, when the root inclination angle is 60 degrees, the strength is the lowest without the freezing and thawing cycle, but the decline is small.

5 CONCLUSION

(1) The simulation method of freezing-thawing damage of loess based on the discrete element method proposed in this paper can simulate the freezing-thawing damage effect of loess well and provide a new way to study the meso-mechanism of freezing-thawing damage of loess.
(2) The root system of Amorpha fruticosa, a shrub, improves the ability of soil to resist freezing and thawing damage, and provides the theoretical basis for the slope protection effect of vegetation under freezing and thawing cycle in winter.
(3) When the angle between the root angle of Amorpha fruticosa and the horizontal plane is 30 degrees, the shear strength of loess increases the most. When the angle between the root angle and the horizontal plane is 60 degrees, the reinforcement effect on loess is the lowest, but with the increase of freezing and thawing cycles, the shear strength decreases the least, and the effect of freezing and thawing cycles is the least.

ACKNOWLEDGMENTS

This work was supported by the Project of the Qinghai Science and Technology Department (2020-ZJ-718).

REFERENCES

Changgen Yan, Ting Wang, Hailiang Jia. (2019). Effect of Unfrozen Water Content on Shear Strength of Unsaturated Silt During Freezing And Thawing. *Journal of Rock Mechanics and Engineering*. 38(6),1253–1260.

Di Ma, Xiaodi Guan, Huanhuan Wei. (2020) Study on Discrete Strength of Loess Particles Based on Discrete Element Method. *J. Engineering Investigation.*

Hua Xu, Haili Yuan, Xinyu Wang, et al. (2021). Effects of Root Morphology and Hierarchical Structure on Mechanical Properties of Root-soil Composites. *Journal of Geotechnical Engineering.* 44(05):926–935.

Lei Tian, Yongsen Yang, Jingjun Li. (2018). Influence of Plant Roots on Cohesive Soil Slope Stability Based on Discrete Element Method. *Journal of China Agricultural University.* 23(7),133–140.

Leqing Liu, Wuyu Zhang, Bingyin Zhang, et al. (2021). Experimental Study on Unconfined Compressive Strength and Microcosmic Law of Loess Under Freeze-Thaw Cycles. *J. Hydrogeology and Engineering Geology.* 48(4),110–115.

Lingzhi Huang, Meiwei Ke, Zheng Si, et al. (2021). Study on Meso-failure of Concrete with Freeze-Thaw Damage Under Uniaxial Compression. *Journal of Applied Mechanics.* 38(4), 1401–1407.

Lu Bai, Jing Liu, Jinghua Hu. (2021). Study on Mechanical Properties of Amorpha Fruticosa Taproot. *J. Arid Zone Research.* 38(4),1112–1118.

Michael Tobias Löbmanna, Clemens Geitnerb, Camilla Wellsteina, et al. The Influence of Herbaceous Vegetation on Slope Stability. *J. Earth-Science Reviews.* 209, 103328.

Mingjing Jiang, Haijun Hu, Tao Li. (2016). Discrete Element Analysis of Biaxial Collapsible Test of Unsaturated Structural Loess. *Journal of Underground Space and Engineering.* 12(4), 1111–1116.

Mingjing Jiang, Yungang Zhu. (2017). Discrete Element Analysis of Friction Between Remolded Loess and Plant Roots. *Journal of Water Resources and Water Engineering.* 28(2),211–215.

Weijie Xing, Xiangjuan Yu, Lei Gao, et al. (2017). Numerical Simulation of Triaxial Shear Test of Cohesive Soil Based on Particle Flow Discrete Element. *J. Science Technology and Engineering.* 17(35),12–124.

Xianghua Song, Yong Tan, Ye Lu. (2021). *Direct Shear Test and Discrete Element Numerical Simulation of Root-soil Composites of Tap-type Root System.* C.The 13th National Conference on Slope Engineering Technology. 58–67.

Xin Xia, Yuanjun Jiang, Lijun Su. (2020). Estimation Model of Ultimate Shear Strength of Rooted Soil Based on Interface Bonding. *J. Rock and Soil Mechanics.* 42(8),2174–2184.

Xingxing Wang, Lin Yang.(2018).Experimental Study on Soil Reinforcement by Herbaceous Plant Roots Under Freeze-thaw Action. *Journal of Glaciology and Geocryology.* 40(4),793–801.

Yang Yang, Chao Xu, Cheng Liang. (2020). Discrete Element Analysis of Bearing Characteristics and Failure Mechanism of Reinforced Soil. *J. Rock and Soil Mechanics.* 41(2),1–9.

Yitao Zhang, Xiangwei Fang, Fenghui Hu, et al. (2022). Unconfined Compression Discrete Element Analysis of MICP Solidified Coral Sand With Different Cementation Degrees. *Journal of Civil and Environmental Engineering.* 44(04):18–26.

Research on DEM construction in mountain areas based on airborne LiDAR data

Wei Chen*

College of Urban Construction, Wuhan University of Science and Technology, Wuhan, China

ABSTRACT: Digital elevation model (DEM) is a very important product that represents the topography digitally, which uses limited topographic elevation data to realize the digital simulation of the surface shape. To solve the difficulty of DEM production in a large range of complex mountainous areas, the method of constructing DEM using LiDAR point cloud data was proposed in this paper. Airborne LiDAR is one of the most effective and reliable methods of terrain data collection. Airborne LiDAR technology can obtain high-density point cloud data. It has the characteristics of small fieldwork and high accuracy and has unique advantages in the production of DEM products. In this paper, taking a mountainous area with dense vegetation in Gansu Province as an example, the point cloud data of this area was collected by airborne LiDAR technology to produce DEM. The accuracy of the DEM was verified by the data of the field measurement points obtained with the method of RTK and total station in the same area. The results show that this method can improve DEM production efficiency and reduce costs. At the same time, the data accuracy can meet the requirements of large-scale DEM.

1 INTRODUCTION

The digital elevation model has become an important part of spatial data infrastructure and digital Earth. DEMs play an important role in terrain-related applications. The common methods of traditional DEM production are field digital surveying and tilt photogrammetry. The traditional methods can yield high-accuracy terrain data, but they are time-consuming and labor-intensive. At the same time, for areas with dense vegetation coverage, the accuracy of these methods cannot be effectively guaranteed. With the development of surveying and mapping technology, data acquisition methods are becoming more and more diversified, not only simple and fast but also more and more accurate. Airborne lidar technology is another major technological revolution after GPS, integrating the laser scanning system, GPS, and ins. Airborne LiDAR is an active remote sensing device, which is less affected by the external environment and can work all day and all night. LiDAR data have become a major source of digital terrain information (Raber et al. 2007) and have been used in a wide of areas, such as building extraction and 3D urban modeling, hydrological modeling, glacier monitoring, landform or soil classification, riverbank and coastal management, and forest management (Liu 2008). However, terrain modeling has been the primary focus of most LiDAR collection missions (Hodgson et al. 2005). It can penetrate vegetation cover, rapidly and accurately detect geographic information data in various areas and carry out topographic mapping in difficult areas and is currently a viable technical means to determine ground elevation in densely vegetated areas. However, raw LiDAR data can contain return signals from no matter what target the laser beam happens to strike, including human-made objects (e.g., buildings, telephone poles, and power lines), vegetation, or even birds (Barber & Shortrudge 2004; Stoker et al.

*Corresponding Author: nancychw@wust.edu.cn

2006). The data collected by LiDAR has no topology blindness, and there are a large number of non-ground reflection points in the original point cloud data, so it is impossible to obtain the high-precision digital elevation model of the survey area directly. To obtain DEM, it is necessary to adopt a certain algorithm to remove non-ground points from the original point cloud data, that is, airborne point cloud filtering. Therefore, it is crucial to filter or extract bare earth points from LiDAR data. In this paper, an irregular triangular mesh iterative filtering algorithm was used to filter the airborne LiDAR point cloud data to produce DEMs.

In this paper, the complex mountainous area was selected as the experimental area, and the DEM production method of airborne LiDAR data was discussed. The accuracy of the DEM was verified by the data of the field measurement points obtained with the method of RTK and total station in the same area.

2 THE TECHNICAL PROCESS OF DEM PRODUCTION BY AIRBORNE LIDAR

The airborne LiDAR system consists of a central control unit, a POS system, a laser scanning rangefinder, and an imaging unit (digital camera). The airborne LiDAR technology uses INS to obtain the instantaneous attitude parameters during flight, GPS to obtain the coordinates of the center of the laser scanner (x, y, z), and the laser scanning system to obtain the distance from the center of the scanner to the ground point, thus calculating the spatial coordinates of the corresponding laser point on the ground (X, Y, Z), as well as obtaining information such as reflectivity and the number of laser pulse echoes. The technical process of DEM production by airborne LiDAR technology is shown in Figure 1.

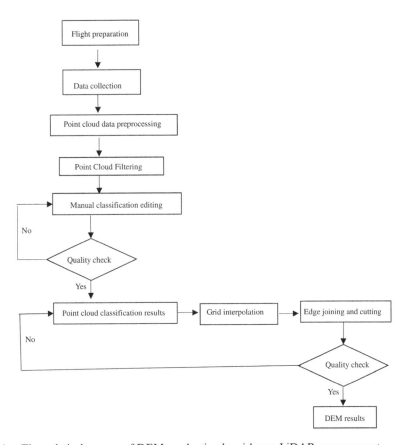

Figure 1. The technical process of DEM production by airborne LiDAR measurement.

3 EXPERIMENTAL PROCEDURES

The study area is located in Dangchang County, Longnan City, Gansu Province, with a height difference of 800 meters and lush vegetation. The vegetation coverage rate is over 90%, and the average height of trees is 8 meters to 10 meters. The scope of the survey area with a coverage of 5 km² is shown in Figure 2. The cw-30 UAV equipped with an airborne LiDAR system was used for point cloud data acquisition.

Figure 2. Study area.

3.1 *Point cloud filtering*

After the data acquisition is completed, the point cloud data needs to be pre-processed before filtering the original data, mainly including filtering out radar noise, as well as normalizing the original signal, time parameters, distance parameters, GPS and INS data, and coordinate transformation to obtain the 3D point cloud coordinate data of the ground target.

The data obtained by LiDAR contains all object points of the study area, but it needs only ground points to get DEM. Point cloud filtering is to separate non-ground points from point cloud data. The filtering algorithm used in this paper is based on the progressively encrypted irregular triangular network (PTIND) filtering algorithm proposed by Axelsson in 2000 (Axelsson 2000).

The principle of the method is as follows: first, a certain number of ground seed points are obtained to form an initial sparse irregular triangulation network (TIN), then each point is judged, and if the vertical distance and angle from the point to the triangular surface are less than a set threshold, the point is added to the set of ground points to achieve the increase of ground points. The irregular triangulation network is then recalculated using all the identified ground points, and the points in the non-ground point set are then discriminated. This iterates until no new ground points are added, or until the given conditions are met.

Non-ground points mainly include vegetation, traffic and road networks, and their ancillary facilities, buildings, and so on. The non-ground point filtering classification mainly uses the ground elevation point triangulation network model as the reference framework to design the relative height difference thresholds for low, medium, and high vegetation points, so as to separate the vegetation points. As the applicability of the algorithm will merge building points in the vegetation point set, the separated vegetation point set is processed by applying a mathematical morphology algorithm, which can separate the building points twice.

3.2 Point cloud refinement

None of the automated filter processes is 100% accurate so far (Romano 2004). Manual editing of the filtering results is still needed (Chen 2007). The data automatically filtered by the software will inevitably produce classification errors, and there will be local misclassifications and omissions, which will require a more detailed manual classification for error correction to meet production requirements. At present, there is no accurate classification method suitable for all terrain. Therefore, only manual classification can be used for error correction. Through the interactive section method, slope method, image information, and other visual environments, misclassified point clouds can be reclassified or corrected to improve the accuracy of classification.

The point cloud data of the study area after vegetation removal by filtering technology is shown in Figure 3.

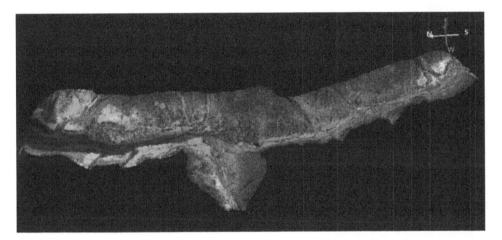

Figure 3. Point cloud data of study area.

3.3 Accuracy check

RTK and total station are used to test the elevation accuracy of the project area. The inspection points are distributed throughout the whole surveying and mapping area, which are evenly distributed and representative. The data are shown in Table 1. Through calculation, the elevation accuracy by airborne radar in complex mountainous areas is ± 0.069 m, which meets the relevant national standards for large-scale data results.

Table 1. Elevation detection data.

Point No.	Coordinates of points by field survey			Coordinates of points on the graph			Error		Mean square error m
	X/m	Y/m	Z/m	X/m	Y/m	Z/m	ΔXY	ΔZ	
1	**258.304	**173.083	**43.006	**258.331	**173.039	**42.992	0.052	0.014	±0.069
2	**226.090	**081.925	**43.024	**226.082	**081.670	**43.057	0.255	−0.033	
⋮	⋮	⋮	⋮	⋮	⋮	⋮	⋮	⋮	
100	**249.088	**123.093	**43.034	**249.076	**123.183	**42.988	0.091	0.046	

4 CONCLUSION

Advances in airborne LiDAR systems make it possible to acquire high-quality terrain data in terms of accuracy and density. Taking a complex mountainous area in Gansu Province as an example, this paper uses airborne lidar to collect data in this area and studies the DEM construction of a complex mountainous area based on airborne LiDAR data. The experimental results show that the airborne LiDAR technology can fully meet the accuracy requirements of large-scale DEM data, and the airborne LiDAR system has the characteristics of high efficiency and speed, especially in the densely vegetated areas that cannot be penetrated by oblique photogrammetry, which greatly reduces the workload of fieldwork, greatly improves the production efficiency of DEM production, and effectively reduces the human and material costs.

REFERENCES

Axelsson P (2000). DEM generation from laser scanner data using adaptive TIN models. *International Archives of Photogrammetry and Remote Sensing*, 33,110–117.

Barber, C.P. and Shortrudge, A.M. (2004). *Light Detection and Ranging (LiDAR)-derived Elevation Data for Surface Hydrology Applications.* East Lansing, MI: Institute of Water Resource, Michigan State University.

Chen, Q. (2007). Airborne LiDAR Data Processing and Information Extraction. *Photogrammetric Engineering and Remote Sensing*. 73, 109–112.

Gao Shengchao, Gao Yantao, Wang Wenjie (2022). Application of Generated High-precision DEM in Key Reservoir Region of Henan Province Based on Airborne LIDAR. *Bulletin of Surveying and Mapping*, (1):128–132.

Guo Xiaoyu, Lyu Huaquan, Huang Youju (2021). Research on High Accuracy DEM Production Based on Airborne LiDAR Data in Guangxi Difficult Areas. *Geomatics & Spatial Information Technology*, 44(7): 24–27.

Hodgson, M.E., Jensen, J., Raber, G., Tullis, J., Davis, B.A., Thompson, G., and Schuckman, K. (2005). An Evaluation of LiDAR-derived Elevation and Terrain Slope in Leaf-off Condition. *Photogrammetric Engineering and Remote Sensing* 71, 817–23.

Kasaim, Ikedam, Asahinat, et al. (2009). LiDAR-derived DEM Evaluation of Deep-seated Landslides in a Steep and Rocky Region of Japan. *Geomorphology*, 113(1):57–69.

Liu Xiaoye (2008). Airborne LiDAR for DEM Generation: Some Critical Issues. *Progress in Physical Geography*, 32(1):31–49.

Nizar Polat, Murat Uysal (2018). An Experimental Analysis of Digital Elevation Models Generated with Lidar Data and UAV Photogrammetry. *Journal of the Indian Society of Remote Sensing*.

Raber, G.T., Jensen, J.R., Hodgson, M.E., Tullis, J.A., Davis, B.A. and Berglend, J. (2007). Impact of LiDAR Nominal Post-spacing on DEM Accuracy and Flood Zone Delineation. *Photogrammetric Engineering and Remote Sensing* 73, 793–804.

Romano, M.E. (2004). Innovation in LiDAR Processing Technology. *Photogrammetric Engineering and Remote Sensing*. 70, 1202–206.

Stoker, J.M., Greenlee, S.K., Gesch, D.B. and Menig, J.C. (2006). CLICK: the New USGS Center for LiDAR Information Coordination and Knowledge. *Photogrammetric Engineering and Remote Sensing* 72, 613–16.

Wang Cong (2020). DEM Production Process Based on Airborne LiDAR Data. *Geomatics & Spatial Information Technology*,43(10):204–207.

Simulation experimental study on aggregate filling and interception in mine water inrush channel

Peili Su, Chong Li* & Feng Liu
Xi'an University of Science and Technology, Xi'an, China

ABSTRACT: The size of aggregate particle size, the way of pouring, and its accumulation form in the channel are the key factors affecting the cut-off speed and subsequent grouting effect of the mine water inrush channel. In this paper, a visual water inrush channel test platform is built to carry out different particle size aggregate perfusion simulation experiments, and the blocking effect is judged by the aggregate retention rate. The aggregate perfusion methods of 'fine first and then coarse' and 'coarse first and then fine' are compared with the 'mixed perfusion' method, and the law of aggregate accumulation morphology under different methods is summarized. Through the coupling of FLUENT and EDEM software, the test results are further improved by adding feeding ports. The results show that the difference in the aggregate retention rate of different aggregate perfusion methods can reach 18.92%, and the aggregate retention rate of 'fine first and coarse second' is the highest. The large difference in aggregate particle size range is not conducive to the rapid formation of water blocking section, the numerical simulation is consistent with the experimental conclusion. Based on the test results, this paper puts forward optimization suggestions for the closure project of a water inrush channel with a high flow rate and a small flow rate.

1 INTRODUCTION

Most of the disposal of mine water inrush channels is based on the hydrogeological conditions of the accident site, and the sealing process of grouting or first pouring aggregate to cut off the flow and then grouting is adopted. In the past treatment of water inrush accidents, Chinese scholars have conducted in-depth research and accumulated rich experience in water plugging. The water-plugging effect of 100% can be achieved by grouting aggregate, single slurry, and double slurry (Wang 2004). After determining the accurate spatial position of the collapse column, treatment measures are taken. After the water blocking section is completed by pouring aggregate, the plugging of the water roadway is completed by combining the single liquid high-pressure dynamic water grouting technology (Liu 2005). Based on the selection of a water-blocking roadway, the principle of borehole layout, aggregate perfusion, and the complete standard of the water-blocking section, the comprehensive construction technology of a water-blocking wall in water inrush roadway under dynamic water conditions was introduced (Jiang 2009). It is also a kind of treatment scheme to construct a water retaining wall by pouring aggregate, changing the pipeline flow into seepage flow, and then injecting cement slurry to effectively accumulate and condense it (Li 2010). Cement-water glass double liquid grouting method suitable for hydrostatic conditions is also proposed (Shao 2011). In view of the loose coal wall of the coal roadway, the equilibrium pressure method can try to close the drainage valve and boost the channel (Yue 2012). Engineering practice proves that large flow dynamic water interception can be achieved only by fine aggregate (Ji 2014).

---

*Corresponding Author: 1418908204@qq.com

Aiming at the problem of water inrush in the collapsed roadway, "cement single grout filling grouting technology" and "large well recharge boosting water retaining wall rapid construction technology" were adopted for treatment (Chai 2015). In view of the rapid plugging of large-flow water inrush channels, relevant rapid solidification high-strength grouting materials and new equipment were developed (Zhu 2015). At the same time, some scholars put forward the aggregate perfusion method from fine bone to coarse aggregate when normal perfusion, and coarse and fine aggregate combination when close to the top (Mou 2021).

In practical engineering, the geological environment conditions are often complex and hidden, the migration and accumulation state of aggregate after pouring aggregate is difficult to predict, and there is a lack of research on the particle size of intercepting aggregate and the selection of injection methods. Based on the above problems, the aggregate particle size and aggregate perfusion methods selected in the project often need to be adjusted in time according to the changes in water flow and the environment in the channel. This paper builds a visual water inrush channel aggregate closure simulation test platform, taking the high flow rate and small flow water inrush channel as the research object, and explores the influence of different particle size aggregates and perfusion methods on the effect of aggregate closure. Combined with numerical simulation methods, it reveals the sedimentary migration law of aggregates in dynamic water.

2 SIMULATION LABORATORY TEST

2.1 Water inrush channel aggregate interception simulation test platform

The test adopts the self-designed visual water inrush channel aggregate interception simulation test platform (Figure 1). The system includes the simulation water inrush channel device, simulation water source device, aggregate perfusion control device, and data monitoring device. In order to facilitate the construction of the test platform and meet the visualization requirements, the simulated water inrush channel device is simplified as a single constant diameter acrylic pipeline, which is composed of three pipelines and connected with an organic glass flange. The aggregate perfusion control device is set up 1 m away from the pipeline inlet. The pipeline inlet is connected to the water supply system, and the outlet is connected to the recovery device to form a complete simulation system.

1-Acrylic pipeline; 2-Pipe support; 3-aggregate perfusion system; 4-pressure sensor; 5-Paperless recorders; 6-Electromagnetic flowmeter; 7-aggregate recovery system; 8-weighing apparatus; 9-Cameras; 10-constant pressure pump; 11-Water supply barrel

Figure 1. Simulation test platform for aggregate interception of water inrush channel.

2.2 Test scheme design

According to the similarity ratio and indoor test conditions, the test adopts a 70 mm pipeline, setting the flow rate of 0.3 m/s and the flow rate of 4.15 m^3/h. Referring to the experience of relevant engineering treatment examples (Liu 2019), the aggregate size should not be greater than 1/3–1/4 of the inner diameter of the borehole. The height of the feeding device and the pouring volume of each group of test aggregate (0.04 m³) were fixed. Four groups of aggregate size range were selected: 0–2 mm, 2–5 mm, 5–8 mm, and 8–10 mm. Four kinds of aggregates with different particle sizes were paired, and the volume ratio of aggregates in different particle sizes was 1:1, a total of six combinations, See table 1 for specific test design.

The indoor simulation test can evaluate the effect of aggregate interception and water plugging according to the final residual water channel (Hui 2018). The aggregate retention rate S is obtained by calculating the ratio of the cross-section of the accumulation body to the cross-section of the pipeline. The calculation formula for the aggregate retention rate is as follows:

$$S = \left[\frac{1}{2} + \frac{1}{180°}\arcsin\frac{H-R}{R} + \frac{R(H-R)\cos\left(\arcsin\frac{H-R}{R}\right)}{\pi R^2}\right] \times 100\% \quad (1)$$

H is the final accumulation height of aggregate; R is the pipe radius.

Table 1. Test table of different perfusion schemes of aggregate.

Test number	Fine first and then coarse/(mm)	Test number	Coarse first and then fine/(mm)	Test number	Mixed perfusion/ (mm)
1-1	0-2, 2-5	2-1	2-5, 0-2	3-1	0-2, 2-5
1-2	0-2, 5-8	2-2	5-8, 0-2	3-2	0-2, 5-8
1-3	0-2, 8-10	2-3	8-10, 0-2	3-3	0-2, 8-10
1-4	2-5, 5-8	2-4	5-8, 2-5	3-4	2-5, 5-8
1-5	2-5, 8-10	2-5	8-10, 2-5	3-5	2-5, 8-10
1-6	5-8, 8-10	2-6	8-10, 5-8	3-6	5-8, 8-10

2.3 Analysis of aggregate accumulation morphology

The larger the particle size difference of aggregate combination is, the more obvious the accumulation characteristics of aggregate in the water passage are. Therefore, test numbers 1-3, 2-3, 3-3 are selected to analyze the aggregate accumulation morphology under different aggregate perfusion methods. The test results are shown in Figure 2.

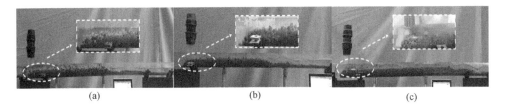

Figure 2. Aggregate stacking morphological diagram. (a)Test 1-3 aggregate heap (b)Test 2-3 aggregate heap (c)Test 3-3 aggregate heap.

In the 'fine first and then coarse' mode, the fine aggregate deposits at a certain distance from the feeding port to the bottom of the pipeline and forms a slope erosion surface. The initial accumulation position of coarse aggregate is closer to the feeding port than that of fine aggregate. After the aggregate perfusion, coarse and fine aggregates are formed in the water passage, and the coarse aggregate covers the upper layer of the fine aggregate. Due to the influence of hydrodynamic turbulence on the side facing the water surface, some surface coarse aggregates are passively carried to the back surface of the accumulation body, and the distribution of surface aggregates is concave and convex. Figure 2(a) shows a rapid decrease in water flow downstream of the accumulation.

In the 'coarse first and then fine' mode, the coarse particle aggregate rapidly deposits at the bottom of the channel and moves downstream to form the skeleton of the water-resistance section. The fine particle aggregate passively carries water to the skeleton pore, and the water resistance section is gradually dense. After aggregate perfusion, coarse and fine mixed accumulation bodies are formed in the water passage, some coarse aggregate diffuses upstream, and the diffusion distance is far smaller than that of downstream. The fine aggregate particles in the surface layer continuously scour downstream of the accumulation body by passive water, forming a gentle trailing at the foot of the backwater slope. Figure 2(b) shows that the water flow in the channel is significantly reduced.

In the 'mixed perfusion' mode, coarse aggregate, and fine aggregate slip to the bottom of the channel under the action of water thrust. Because the coarse aggregate is not easy to passive water dispersion, coarse particles gather near the grouting mouth, most fine aggregates distribute along the pipe wall in the downstream direction, and a few fine aggregates are filled in the coarse aggregate gap. After aggregate perfusion, there is no obvious boundary between coarse and fine aggregates. Figure 2(c) shows a small decrease in water flow in the channel.

2.4 *Analysis of water shutoff effect of aggregate*

The final packing height H of the aggregate at 30 cm from the feeding port was measured after the aggregate perfusion, and the aggregate retention rates were obtained by Formula (1).

It can be seen from Figure 3 that:

(1) The aggregate retention rate of 'fine first and then coarse' is the highest, with the highest retention rate of 93.7%, followed by 'coarse first and then fine', and the retention rate of 'mixed perfusion' is poor. The difference of retention rate caused by the same number of 'fine first then coarse' and 'mixed perfusion' methods were 12.13%, 13.64%, 12.23%, 13.47%, 18.92%, and 18.9%, respectively. The effect of aggregate perfusion on aggregate retention under hydrodynamic conditions was obvious.
(2) Aggregate particle size has a significant impact on the retention rate, the larger the aggregate particle size, the higher the retention rate. The test results show that the sixth group has the highest retention rate, and the first group has a lower retention rate.

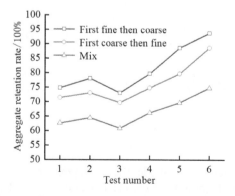

Figure 3. Effects of different perfusion schemes and particle size combinations on the water plugging rate.

The reason is analyzed. The sixth group of aggregate particle size is 5–8 mm and 8–10 mm, and the first group is 0–2 mm and 2–5 mm. The ability to move water-carrying aggregate is inversely proportional to the aggregate particle size. The coarse aggregate is difficult to carry passive water. In addition, the surface roughness of coarse aggregate is higher, the bite force between particles is strengthened, and it is easy to form deposits.

(3) The difference in aggregate particle size range is so large that the fine aggregate is lost from the coarse aggregate gap, which is not conducive to the rapid formation of water blocking section. It can be seen from Figure 3 that the retention rate under the third aggregate particle size combination is the lowest, and it is too different from the highest retention rate in the same perfusion method, which will cause time and cost waste in practical engineering.

2.5 *Numerical simulation of water blocking effect of aggregate interception*

The conclusions obtained from the study based on laboratory tests are limited, so it is necessary to use relevant software to carry out a numerical simulation to supplement the conclusions of laboratory tests. Therefore, the numerical simulation of water inrush channel closure under different aggregate perfusion methods is carried out by coupling the computational fluid dynamics software FLUENT and discrete element software EDEM.

The experiment added a No.2 perfusion hole at 1 m downstream of the original perfusion hole, making the simulation results closer to the actual working condition. Select 2–5 mm fine aggregate and 8–10 mm coarse aggregate, 'first fine and then coarse' test scheme is: after the beginning of the test, two holes are perfused with 10 s fine aggregate, when the end of the perfusion of coarse aggregate perfusion 10 s, 'first coarse and then fine' test in the order of two kinds of aggregate perfusion is opposite to 'first fine and then coarse'.

In Figure 4, the change process of aggregate accumulation shape with time was simulated according to the way of 'fine first then coarse' and 'coarse first then fine', where gray is fine aggregate and blue is coarse aggregate. It can be seen from the diagram that the aggregate

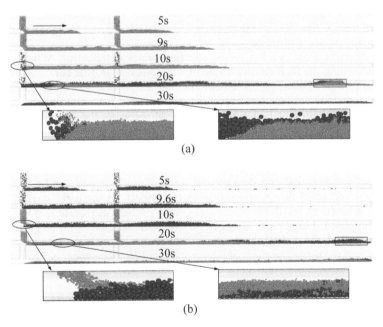

Figure 4. Influence of different aggregate perfusion sequences on interception effect. (a) Test 1–5 aggregate heap (b) Test 2–5 aggregate heap.

connection time between the two holes is basically the same when pouring coarse and fine aggregates in different orders. Among them, the connection time between the two holes of test numbers 1-5 is about 9 s, and the connection time between the two holes of test numbers 2-5 is about 9.6 s. When the simulation time is 10 s, according to the way of 'fine first and then coarse' perfusion, the accumulation body of fine aggregate itself becomes dense, and coarse aggregate is difficult to mix into it, and all of them are accumulated on the surface of fine aggregate. Some coarse particles are exposed to the flow field and move to the front end of the accumulation body under the scouring of water. According to the perfusion method of "coarse first and fine later", it can be observed from the local amplification diagram that there is a filtering effect on the coarse aggregate. The fine aggregate fills the pores of the coarse aggregate under the action of dynamic water and is easy to aggregate, reducing the porosity. When the simulation time is 20 s, the migration of the accumulation body occurs in both conditions. The reason is analyzed. The initial hydrodynamic velocity is large. Under the action of water scouring and particle arrangement, the integrity of the accumulation body is enhanced. With the continuous filling of aggregates, the residual water channel is reduced, and the flow velocity and the inlet water pressure are increased. At a certain moment, the water pressure exceeds the friction between the accumulation body and the channel, and the accumulation body appears overall migrate. When the simulation time is 30 s, it can be observed that the aggregate accumulation height of the 'fine first and then coarse' perfusion method is higher, and the height distribution of the whole accumulation body is relatively uniform.

3 CONCLUSIONS

(1) The visual water inrush channel aggregate interception simulation test platform was established, and the simulation test was carried out to obtain the aggregate retention rate and aggregate accumulation form under different perfusion methods of 'fine first then coarse', 'coarse first then fine' and 'mixed perfusion'.
(2) The effect of aggregate perfusion on water plugging is remarkable. The aggregate retention rate of 'fine first and then coarse' is up to 93.7%, and the effect of intercepting and blocking water is obvious.
(3) Aggregate particle size is a key factor affecting aggregate retention. When other conditions are certain, the final retention rate of a large-size aggregate combination is greater than that of a small-size aggregate combination; the particle size range difference is too large, aggregate retention effect in the channel is poor.
(4) Based on the coupling of FLUENT and EDEM software, the simulation of aggregate morphology and closure effect under different perfusion methods after increasing the perfusion hole is obtained, which is consistent with the results obtained in the laboratory test.

In conclusion, the 'fine first and then coarse' water shutoff method can be preferred for a high flow rate and small flow water inrush in the project. In the early stage, fine particle aggregate is selected to lay the bottom, and coarse particle aggregate is poured after the bottom is completed. Large particle aggregate should be selected as far as possible without plugging. In general, the smaller the residual channel area is, the higher the retention rate of aggregate is, and the better the effect of intercepting and blocking water is. However, there is a gap between the accumulation body of coarse aggregate, and there is still a minimum residual channel between the coarse aggregate and the inner wall of the pipeline after closing the top. At this time, fine aggregate can be replaced again to improve the skeleton stress of coarse aggregate and make the accumulation body denser until the top is successfully connected. When the flow is transformed from pipe flow to seepage, grouting reinforcement can be carried out.

REFERENCES

Chai Huichan, Li Wenping, Zheng Shitian, et al. Rapid Construction of Water Blocking Wall Technology in Collapsed Coal Roadway [J]. *Coal Mine Safety*, 2015, 46(06): 66–68.

Hui Shuang. *Simulation Test of Porous Perfusion Aggregate Plugging in Submerged Roadway in Mine* [D]. Xuzhou: China University of Mining and Technology, 2018.

Ji Zhongkui. Mine Large Flow Dynamic Water Grouting Fine Aggregate Interception Technology [J]. *Coal Engineering*, 2014, 46(07):43–45.

Ji Zhongkui. Research on Drilling Technology of Dynamic Water Interception in Mines with Extra-large Water Inrush[J]. *Coal Technology*, 2014, 33(05):12–14.

Jiang Qinming. Comprehensive Construction Technology of Water-blocking Wall in Concentrated Burst (Over) Water Roadway[J]. *Coal Mine Safety*, 2009,40(5):37–39.

Li Caihui. Interception and Plugging Technology of Mines with Large Water Inrush [J]. *Journal of Xi'an University of Science and Technology*, 2010,30(3):305–308.

Liu Jiangong, Zhao Qingbiao, Bai Zhongsheng, et al. Rapid Control of the Catastrophic Water Inrush Disaster in the Collapsed Column of Dongpang Mine [J]. *Coal Science and Technology*, 2005(05):4–7.

Mou Lin. *Research on Construction Mechanism and Key Technology of Water Blocking Wall for Roadway with Dynamic Water Conditions* [D]. Xi'an: General Research Institute of Coal Science, 2021.

Mou Lin. Development and Experimental Research on Visualization Platform for Aggregate Perfusion and Interception in Water-inrush Mine Dynamic Water Roadway[J]. *Coalfield Geology and Exploration*, 2021, 49(05):156–166.

Quansheng Liu, Lei Sun. Simulation of Coupled Hydro-mechanical Interactions During Grouting Process in Fractured Media Based on the Combined Finite-discrete Element Method[J]. *Tunnelling and Underground Space Technology incorporating Trenchless Technology Research*, 2019,84.

Shao Hongqi, Wang Wei. Rapid Construction of Water Blocking Wall to Block Water Inrush Roadway By Double-Liquid Grouting Method [J]. *Coal Mine Safety*, 2011, 42(11): 40–43.

Wang Zecai. Water Grouting and Water Blocking Technology in 8101 Working Face of Guozhuang Coal Mine[J]. *Coalfield Geology and Exploration*, 2004(04):26–28.

Yue Weizhen. Application of Balanced Pressure Method in Grouting, Interception and Water Blocking of Extremely Loose Coal Roadway[J]. *Coal Engineering*, 2012(08):40–42.

Zhu Mingcheng. Application of Drilling Controlled Grouting Technology in the Plugging of Large Water Passages[J]. *China Coal Geology*, 2015, 27 (05): 46–49.

Zhu Mingcheng. Key Technologies and Equipment for Efficient Plugging of Controlled Grouting in Water Inrush Through Boreholes in Large Hydrodynamic Channels[J]. *Coalfield Geology and Exploration*, 2015, 43(4):55–58.

A comprehensive dam safety monitoring information system for catchment/area-level hydropower station groups

Gang Cui, Qi Ling* & Lan Zhang

State Grid Electric Power Research Institute Ltd. Co./Nari Group Corporation Ltd., Nanjing, China

ABSTRACT: Based on technologies of big data analysis and network communication, the comprehensive dam safety monitoring information system for catchment/area-level hydropower station groups should be provided by adopting partitioning and shading database and microservice architecture technology. A timely and accurate knowledge and analysis related to dam safety are enabled by the remote centralized control and comprehensive information application of dam safety monitoring data. The practice serves as a technical approach to managing dam safety and evaluating its operating performance to make an upgrading shift from decentralized management to centralized control of the safety of dam groups.

1 INTRODUCTION

There is a wide variety of professional safety monitoring information systems used in hydraulic engineering currently in China (Gu 2005; Wang 2019; Yu 2006). Most of them are so technically mature that they are widely used at a large number of hydropower stations. Dams and other hydraulic structures powered by automatic monitoring all deploy information systems for collecting, storing, transmitting, and analyzing data. Original systems used by hydropower stations can be technologically upgraded to allow them to use such automatic systems for information management. However, comprehensive and professional platforms dedicated to safety monitoring for dams are rarely found. This is a special case for large power companies without mature information systems specializing in centralized safety monitoring management of catchment/area-level hydropower station groups (Zhou 2016). Some "reporting" production management approaches cannot meet the requirements for "optimizing operation" and "better and stronger performance" of catchment/area hydropower stations, preventing improvements in centralized and precise management. Therefore, it is necessary to develop a comprehensive safety monitoring information system for catchment/area-level dam groups to achieve the goal of safety monitoring, thus improving efficiency and management levels and ensuring the long-term and safe operation of dams.

From the managerial perspective, large power companies have massive subjects, extensive professional fields, and complex business types relating to safety management due to their differences in scale, structure, type, and dispatching operations. To improve management performance and efficiency, safety monitoring information platforms for catchment/area-level dam groups are required to organize, coordinate and engage in the dam safety monitoring works of various operating units (Zhang 2014). An effective dam safety monitoring mechanism is driven by dam registration, regular safety inspection, dam monitoring information, reinforcement and strengthening, emergency response plans, and other business operations, to ensure the safe operation of dams of hydropower stations.

---

*Corresponding Author: 47755393@qq.com

2 TECHNICAL PRINCIPLE

A comprehensive safety monitoring information system for catchment/area-level hydropower station groups should be designed in accordance with national and industrial standards and adopt standard communication protocols, cutting-edge computing software and hardware technologies, and network technologies. The system, based on big data analysis and network communication, can collect and transmit automatic monitoring data remotely, allowing staff to access and analyze the timely safety information and state of dam groups. As such, the system is made to manage dam safety information in a centralized manner. It is a comprehensive management platform for comprehensive dam safety information, including safety monitoring, rating and registration of integrated dams (certificate renewal), regular inspection, maintenance of dam construction, engineering strengthening, information report, technical supervision of hydraulic works, water and rain information, water inflow and outflow of dams, operations of unit and gate of flood discharging holes and tunnels and emergency response.

The comprehensive safety monitoring information system for catchment/area-level hydropower station groups is developed based on a partitioning and sharding database and microservice architecture for managing data and operation applications. It can serve diverse purposes, including calculation of monitoring amount, data monitoring, limit alert, information collation, information distribution, and management of system operations, to ensure efficient and quality hydraulic routine management on a daily basis and make an upgrading shift from decentralized management to centralized control for hydropower station groups. Besides, it provides powerful technical support for the evaluation of dam safety management and operations.

3 SYSTEM DESIGN

3.1 *Overall architecture*

The system would involve massive staff and hydropower stations, covering wide geographical areas with diverse types of data in a huge amount. Moreover, its widespread application opens a door to complicated network environments, various manufacturers of automatic dam safety information collection systems, water predicting and forecast systems and measuring systems, and different equipment. In consideration of these situations, special efforts would be made to study the overall architecture of the dam safety monitoring information system to develop an adaptable and full-featured system featuring high expandability, stability, and reliability with user-friendly operation interfaces.

Multi-tiered and distributed structures should be adopted in the whole system that comprises station-level subsystem, secondary unit subsystem, and headquarters-level main system. The system can be divided into three tiers from bottom to top:

(1) Safety subsystem for dam operations targeting the operation of hydropower stations;
(2) Safety subsystem for dam operation targeting competent secondary units;
(3) Main safety monitoring system targeting the operation of headquarters-level dam groups.

The station-level subsystem is deployed at all hydropower stations and mainly targets professional managers of hydraulic works to help them perform daily routines, such as collection and input, management and query, calculation and analysis, and monitoring and alarm of safety information, information collation, figure and report making, and file preparation and uploading.

The secondary unit subsystem is deployed at the competent units superior to all stations and mainly targets general managers of hydraulic works to help them achieve the comprehensive management of safety information on dam operations. The subsystem covers all other functions except for data collection and input in the former subsystem. Meanwhile, it includes the response coordination of emergency alarms of the former subsystem, file review and approval, and work examination and evaluation.

The main safety monitoring system for dam groups is deployed at corporations and enterprises and targets safety managers at the headquarters to help them browse through and search for monitoring information, check files and documents, inspect and urge works, review and approve important proposals, evaluate management works and perform general management.

3.2 *Network architecture*

The network architecture of the dam monitoring system is presented in Figure 1. The private intranet of power companies is used for data exchange, data collection, and data distribution, while off-site data backup and universal Web are adopted to collect real-time data of all monitoring items.

Data centers are deployed with multiple servers. Among them, the database server is primarily used to store dam monitoring service data and data at important measuring points received from all monitoring systems. Web server mainly provides Web presentation of monitoring systems. The mobile server offers access to data and information at the mobile terminal and monitors the operations of dams in real time. The application server is responsible for the storage, collation, and analysis management of monitoring data received on dams. The file server primarily works to manage and store all files, while the Web collection server aims to conduct Web general collection at the collection terminal of all monitoring systems.

The dam monitoring center of power companies can be deployed with four servers for data backup to temporarily switch to the dam center for data check (not real-time) in the case of network failure at the data center. The dam monitoring subsystems at all stations are required

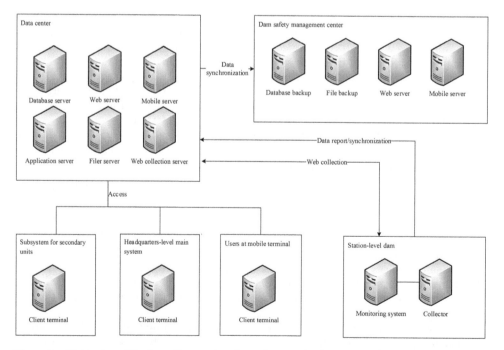

Figure 1. Network architecture of comprehensive safety monitoring systems for catchment/area-level hydropower station groups.

to report and synchronize data to the data center. All systems should comply with the Web's general collection protocols and connect with the Web's real-time collection server.

3.3 Database design

The system's database is designed based on data layering, logical independence of domain data, and data separation on the basis of data type, safety, manageability, and ease of use. According to the structural definition of data, there are structured data and unstructured data. Structured data refer to row data that are stored in the table of databases and can be logically described through two-dimensional structures, such as information data of service applications. Such data can be stored through relational databases comprising system data measuring databases and basic information databases. Unstructured data are designed to store information on technical engineering documents, monitoring system design, layout drawings, design report and drawings of construction works through all stages. To be specific, such data mean video, picture, and file stored using HDFS technology.

4 SYSTEM FUNCTION

4.1 Supervision management of hydraulic works

In accordance with national and industrial standards, the following works should be carried out, such as management of the hydraulic supervision, information input of the process and results of works relating to dam registration and regular inspection, key node control, and file search and record. A special file format should be used to define the information on dam safety registration, which allows system users to input and output such information and search for such general or detailed information by operation unit, year, and project.

4.2 Safety monitoring management

Real-time information on dam safety monitoring data should be collected to monitor the operation of the data safety monitoring system and hydraulic structures and filter the latest data for reliability verification. These steps can ensure that monitoring data go through clearly accountable reviews as required by processes for subsequent search and official use of users. These functions include information collection, system monitoring, data monitoring, data proofreading, and confirmation and treatment of abnormal data.

4.3 Information collation

The system provides commonly used functions for graphic production, such as hydrograph, contour line, distribution graph, correlogram, layout drawing, and various reporting tables like collation specification table, eigenvalue statistical table, and customized table. The system would generate tables automatically according to relevant specifications and formats or other tables based on personal customization for users, providing support for the qualitative analysis of monitoring results. Tables and graphs, once completed, can generate corresponding modules that allow users to obtain relevant monitoring results quickly for subsequent analysis by switching to other analysis periods.

4.4 Safety analysis evaluation

With existing industrial standards and mature professional theories, the system can mine and extract massive data and generalize their rules, tendency, and relationship. As such, it can judge and evaluate the safety conditions of various hydraulic structures and overall or partial

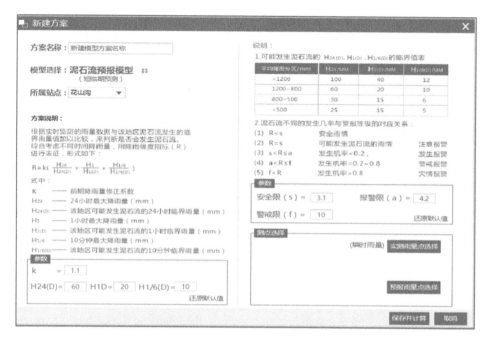

Figure 2. Typical interface design of the function of safety analysis evaluation.

side slopes, automatically identify possible safety hazards and weak sections, while locating abnormalities, analyzing root causes, and predicting risks. Moreover, it issues a warning in response to abnormalities or risks to achieve real-time safety monitoring. The system provides smart decisions on the operation, dispatching, maintenance, and management to improve the safe and science-based management level of dams.

4.5 *Management of hidden dangers and flaws*

Management of hidden dangers and flaws of dams falls into the discovery, classification, and rating, response measures (including formulating response proposals and monitoring proposals, response process track, and response acceptance), and elimination or level change of hidden dangers. The following basic information on hidden dangers should be uploaded to the system online, including name, section, date, and description, accompanied by relevant pictures and files. With registration completed, efforts should be made to examine whether such information and descriptions are real and then classify and grade the hidden danger. The system can collate and then present information on a hidden danger, including its rating, state, class, response progress, and other overall situations to help operation management units grasp the overall picture of the hidden danger and provide inspiration for formulating response plans.

4.6 *Management of flood prevention and relief*

Management of flood prevention and relief primarily provides real-time search for meteorological data, wind and cloud data transmitted by satellites and radar, and forecast analysis of meteorological data, meteorological and geological disasters and flood prevention of hydropower stations, search for information on flood prevention of hydropower stations, flood prevention organization and management of hydropower stations, earthquake alarm presentation of hydropower stations, management of flood prevention documents, coordinated management of emergence response to flood prevention, and reporting tables of supervision of flood prevention.

4.7 Emergency response management

Emergency response management mainly includes emergency response plans for the prevention of typhoons, floods, and geological disasters, and the breach of dikes and dams planned by relevant dam units beforehand. The system can remind these units of performing these emergency response plans according to the requirements for the rehearsal time or period. Furthermore, it can check the implementation of emergency response rehearsal and training conducted by hydropower stations and notify these units and departments that are inconsistent with requirements or fail to implement well, to supervise the completion of these tasks as scheduled and required. Relevant units can upload real-time video monitoring to the system to warn about all dams requiring special attention and provide the level of related emergency response plans and response measures, when safety, water quality, meteorological and hydrological conditions exceed the limit or reveal abnormalities.

4.8 Presentation of comprehensive information

The presentation of comprehensive information mainly provides basic general information, classified statistics, and dam information on hydropower stations covered in the area/catchment. Approaches to presentation include:

(1) Presentation of construction visualization: present construction sites by virtualizing real scenes through 2D/3D models, including geological model visualization, building visualization, side slop, and reservoir area visualization.
(2) Presentation of information on patrol inspection: able to plan and search for patrol routes, mark, correct, and present hidden dangers and flaws, and automatically manage information on patrol and inspection, including the presentation of operating state and on-site real scenes.

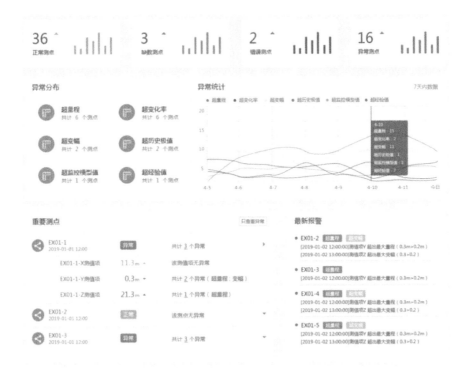

Figure 3. Interface design of presentation of comprehensive information.

(3) Presentation of information on safety monitoring: able to present installation information of measuring points in real-time, search for measuring state and safety monitoring data, and collate these data. The system allows to search, present and manage data, monitoring indicators of measuring points, design indicators, hydrograph figures, and distribution diagrams.
(4) Presentation of online monitoring and forecast: present the operating state and forecast information of hydraulic structures in 2D/3D model diagrams and further navigate to a specific engineering location, engineering area, monitoring item, or monitoring point.

5 CONCLUSION

To adapt to new economic situations and policy requirements, many large hydropower companies begin to shift their dam safety management from comprehensive management to professional management in a core belief "based on values and driven by benefits". This transition is made in response to the changes in the domestic trend of dam safety management and the requirements for workplace safety management of enterprises in recent years. It is also a necessary demand for more professional workplace safety managed through informalized approaches.

The safety monitoring system for catchment/area-level hydropower station groups is a professional platform targeting large hydropower companies to facilitate their dam safety management, covering functions of service support and information sharing. Through design, R&D, and construction, the safety monitoring information platform can cover hydropower stations across a wide area and is comprised of three-tiered systems, including a station-level subsystem, branch (subsidiary) system, and headquarters-level main system to monitor dam operations. These systems contribute to the information collection, remote transmission, management, distribution, two-dimensional and three-dimensional presentation, real-time monitoring, and feedback on the safe operation of hydropower station groups. The system would provide efficient information processing and application procedures for dam safety management to combine the comprehensive management and professional management of workplace safety. Besides, it can benefit the centralized and professional management of large power companies to improve the digitalization of dam safety operation management.

ACKNOWLEDGMENTS

The research was financially supported by the project (5246C5220003) of NARI Group Corporation Ltd, Nanjing, China.

REFERENCES

Gu Y. M., Zheng D. J., Yu P. Dam Safety Monitoring Information Management System Based on the Mixed Model of C/S and B/S [J]. *Journal of Water Resources and Architectural Engineering*, 2005, 3(2):4.

Wang P. C., Bao T. F., Zhu Q. Research on the Development of Dam Safety Monitoring Information System Based on BIM [J]. *Water Resources and Power*, 2019, (4):4.

Yu P., Yu L. M., Wu D. Design and Optimization of Dam Safety Monitoring Information System [J]. *Journal of Engineering of Heilongjiang University*, 2006, 33(001):10–13.

Zhang X. S., Li X. X. Analysis of Dam Safety Management Models for Hydropower Stations of a Catchment [J]. *Dam & Safety*, 2014, (5):4.

Zhou X. L., Wu J., Xiang N. Application and Materialization of the Comprehensive Safety Monitoring System for Tiered Power Station Groups of Qingjiang River [J]. *Water Power*, 2016, (42):98.

Human factor failure pathways in dam failures

Dandan Li* & Huiwen Wang*
Hydraulic Engineering Department, Nanjing Hydraulic Research Institute, Jiangsu, Nanjing, China

ABSTRACT: With the advancement in technology, dam failures caused by engineering quality problems are gradually decreasing. However, the proportion of dam safety management, which involves the human factor, is on the rise. This paper looks at the basic human pathway in a dam failure and divides human behavior in this context into four stages: monitoring awareness, condition diagnosis, plan making, and operation execution. The paper categorizes and analyzes the main influencing factors in each stage.

1 INTRODUCTION

Reservoir dams are important economic and social infrastructure with multiple functions such as flood control, irrigation, power generation, shipping, and ecology. They, however, also involve certain risk factors. Once a breach occurs, it not only threatens the lives and property of the people downstream but also has a critical social impact.

Dam failures in reservoirs in China have gone through several stages (Sheng 2019). In the early years, dam failure accidents were caused by a combination of factors such as low standards and poor work quality, and the lack of safety management because of the weak economic and technical base. With the advancement of technology and management, as well as the large-scale reconstruction of the diseased and dangerous reservoirs, the average annual dam failure rate has dropped significantly, and the importance of the human factor has increased. Therefore, in the future, human factor failures need to be controlled in dam safety management and dam failure prevention. Strengthening human factors reliability research is not only of great theoretical significance but also has important practical application value.

2 RESEARCH THEORIES

For the study of human cognitive-behavioral pathways, Reason proposes a model of human-caused accidents in complex systems organizations based on the defense-in-depth approach (Reason 1990). He believes that accidents are caused by a combination of factors, which can be divided into explicit and potential failures. Explicit failures such as human errors and violations can directly affect safety, while potential failures originate in the management organization, such as poor organizational culture and management errors, which are not normally visible but can threaten the reliability of the defense-in-depth strategy. When explicit failures are combined with deficiencies in the system, they lead to accidents.

Russell and Mehrabian first proposed the S-O-R (stimulus, organism, and response) model in 1974, which was modified by Jacoby in 2002. The model suggests that certain environmental factors can stimulate an individual's emotional and cognitive state leading to certain behavioral outcomes (Kamboj 2018; Laato 2020). The S-O-R model consists of three components: Stimulus, i.e., the stimulus of external conditions for the individual; Organism,

---

*Corresponding Authors: ddli@nhri.cn and 453590381@qq.com

i.e., the individual, the organism, often referred to as the consumer in sales; and Response, i.e., the response to the event, the output link. In this model, external conditions influence an individual's emotional and cognitive situation and consequently change the person's behavior, which in turn feeds back to the individual.

Rasmussen, on the other hand, divides human behavior into three categories: skill-based, rule-based, and knowledge-based, all of which correspond to different levels of performance and are in line with human information processing patterns (Jiang 2016), see Figure 1. Skill-based behavior is a direct human response in which there is a close link between the operator and the input signal, when human reliability depends only on the staff's experience and skill and is not affected by the complexity of the task, and has a low probability of error. Rule-based behavior is mainly governed by a set of rules or procedures that need to be tested in practice, if not, the behavior needs to be repeated and calibrated, making it prone to human error in complex tasks that are time-critical and cognitively difficult. Knowledge-based behavior refers to situations where there are no rules or experience to follow, such as ambiguous situations, contradictory tasks, or new situations that are completely unexplored, and operators must rely on their own knowledge and experience to analyze and make decisions, causing higher human error probability.

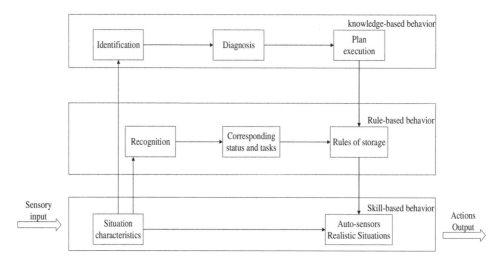

Figure 1. S-R-K three-level model cognitive map.

3 HUMAN ERROR PATH IN DAM FAILURE

The basic human-caused failure paths in dam failure accidents mean the failure paths that human behavior patterns will always follow in dam failure accidents caused by different dam types and different dam failure cause conditions. It can be judged in relation to human information processing paths and can be divided into two categories: explicit failure paths and implicit failure paths. The basic path of failure in the management of reservoir dams is caused by manifest failures, which involve the process of increasing the unreliability of the dam by unsafe actions of the operators. Potential errors, on the other hand, stem from organizational management deficiencies that act on the system's longitudinal defenses and increase the probability of human error. The discussion in this paper focuses on the explicit failure path.

Based on the various types of information processing models available, combined with other areas of research, the main operation task process is divided into four stages: monitoring awareness, status diagnosis, plan making, and operation execution, see Figure 2.

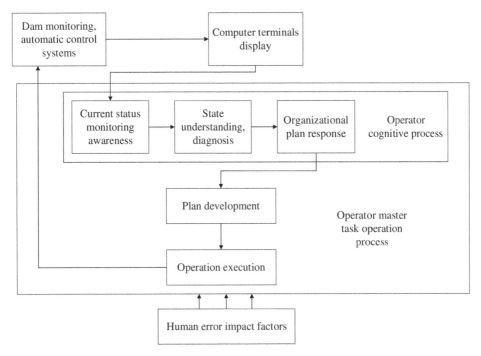

Figure 2. The operator cognitive model.

3.1 *Monitoring awareness*

Monitoring awareness is the process by which reservoir managers obtain information from the situational context. Abnormal conditions in reservoirs are mainly found during routine inspections or through monitoring equipment, alarms, and other display devices that detect parameters, where monitoring means that the operator checks the various parameters on the computer screen, and in some reservoirs, for a lack of digital displays. Monitoring refers to the inspection of the reservoir's appearance, measuring instruments, etc., by the manager. Awareness is the operator's ability to detect anomalies in the computer/measuring instruments, or a precursor of a reservoir hazard, etc. Monitoring awareness is the most frequent and important task of the operator.

There are two main types of monitoring awareness, passive data-driven monitoring and active model-driven monitoring. Data-driven monitoring is mainly influenced by the external environment, where more prominent information such as the large size and bright colors of the information on the screen is more likely to be noticed. Model-driven surveillance is operator-directed, outside of the attention of the alarm system, and is operator-led, triggered primarily by institutional requirements such as protocols or operating instructions or by the operator's subjective awareness.

Therefore, the main factors influencing human error during the surveillance awareness phase are factors in the cognitive functioning and psychological factors of the person:

1) Individual factors: Physical factors such as level of visual acuity, color weakness, color blindness, fatigue, and level of health; psychological factors such as mental state, concentration, and mood; qualitative factors such as skill level, experience, level of training, and responsibility.
2) Technical factors: Features of display devices such as human-computer system display interfaces; Alarm systems; modern technologies such as reservoir dam water and rainfall measurement, safety monitoring facilities, reservoir dam potential hazard detection technology, penetrating sensing technology, etc.

3) Organizational factors: The organization's communication and cooperation, management system, distribution of tasks, etc.
4) Environmental factors: Physical factors such as light and sound in the contextual environment; social factors such as social promotion.

3.2 *Status diagnosis*

During normal operation of the system, the operator assesses the acceptability of the various states based on protocols, skill level, personal experience, etc. In the abnormal event, the operator uses their working memory to analyze and diagnose the possible causes; for example, to determine the cause and location of a fault when there is a problem with an instrument parameter, which serves as the basis for subsequent planning and execution. This is the condition diagnosis phase.

In the state diagnosis phase, the more important influencing factor is the human quality aspect:

1) Individual factors: Psychological factors such as nervousness; quality factors such as skill level and experience.
2) Technical factors: Information, pictures, and other system display interface features, intelligent diagnostic technology.

3.3 *Plan making*

Plan making mainly refers to the daily tasks such as assigning work and planning tasks under normal system operation, as well as implementing operations based on alarm response or contingency plans during abnormal conditions. When existing protocols are not available, the plan needs to be reworked to suit the situation.

The main cognitive activities required of the operator during the plan development process are as follows:

1. Use their own state model to identify targets.
2. Select the right gauge step.
3. Assess whether the actions in the process will meet the objectives.
4. Adapt the protocols to the actual situation.

During this process, the operator may generate several candidate plans, evaluate them, and select the plan that best fits the current state model. The difficulty of developing the plans that operators need to complete varies for different incident scenarios. If there are no existing contingency plans to draw on, the operator needs to mobilize their own long-term technical knowledge and experience to make analytical decisions, which is a great stress load and makes it difficult to sustain attention resources, causing the knowledge-based behavior. At the same time, there are also rule-based and skill-based plan-making, with a decreasing probability of human error.

In this phase, human factors play an important role, in addition to human qualities, organizational factors, and task difficulty.

1) Individual factors: Psychological factors such as nervousness and stress resilience; qualitative factors such as level of knowledge and experience.
2) Technical factors: Software facilities, intelligent decision-making technology; emergency planning.
3) Organizational factors: Factors such as internal organizational climate, communication and cooperation, and safety culture.
4) Environmental factors: On the one hand, the different complexity of the occurred events and the complexity of the tasks that operators need to perform – plan making may be skill-based, rule-based, or knowledge-based, and the probability of human error rises as a

result. On the other hand, complex tasks and tight schedules can also increase the stress load on people and thus affect their reliability.

3.4 Operation execution

Operation execution refers to the completion of the action or sequence of operations identified in the previous plan-making step, mainly physical operations, such as monitoring and control, behavioral input, etc., mostly step-by-step, as required by the protocol. At the same time, the execution of certain operations may require communication and cooperation between different teams.

The execution of an operation may be a matter of an operator working alone to complete the tasks required in the plan drawn up, or it may be a matter of a coordinated team working together to complete a series of tasks. Therefore, the human factors of influence in the operational execution phase are related to both human and organizational factors.

1) Individual factors: Human physical factors; psychological factors; qualitative factors.
2) Technical factors: Hardware, such as gates and display facilities; software, such as control systems
3) Organizational factors: Communication and cooperation, coordination, level of management, and distribution of tasks in the team
4) Environmental factors: Physical, social, and engineering environment

4 CONCLUSION

In this paper, we study a large number of dam failure examples and combine them with research results from other fields to derive a cognitive behavioral model in the field of dam safety. The model of human error in the dam failure process is proposed, and the basic path of human error is divided into four nodes: monitoring awareness, status diagnosis, plan making, and operation execution. The study also analyzes the factors influencing each node, thus offering the possibility of finding human-caused failure paths.

ACKNOWLEDGMENTS

This work was supported by the National Natural Science Foundation (51909172) and Fundamental Research Funds of Nanjing Hydraulic Research Institute (Y720006 and Y721001).

REFERENCES

Jinbao Sheng, Dandan Li, Zhifei Long. (2019) *Reservoir Dam Risks and Their Assessment and Management.* M. Nanjing, Hohai University Press.
Kamboj, Shampy, Sarmah, et al. (2018) Examining Branding Co-creation in Brand Communities on Social Media: Applying the paradigm of Stimulus-Organism-Response. *J. International Journal of Information Management.*
Laato S, Islam A, Farooq A, et al. (2020) Unusual Purchasing Behavior During the Early Stages of the COVID-19 Pandemic: The Stimulus-organism-response Approach. *Journal of Retailing and Consumer Services*, 57: 102224.
Reason J T. (1990) *Human Error*[M].
Shu Jiang, Xinhui Ma, Cunyan Cui, et al. (2016) SRK-based Analysis of the Cognitive Behaviour of Aerospace Test Launch Operators. *J. Military Automation*, 35(11): 6.

Ecological park design and sustainability evaluation for metropolis based on emergy method

Junxue Zhang*
School of Civil Engineering and Architecture, Jiangsu University of Science and Technology, Zhenjiang, China

Li Huang
School of Design Art, Lanzhou University of Technology, Lanzhou, China

Yan Zhang
Nanjing Wenzhong Construction Engineering Co., LTD, Nanjing, China

Dan Xu
School of Civil Engineering and Architecture, Jiangsu University of Science and Technology, Zhenjiang, China

ABSTRACT: Ecological city has become a necessary way for urban development. Therein, The sustainable assessment is one of the core contents of an Eco-city. In this paper, an ecological park was been selected and analyzed based on emergy method. The results demonstrate that the park is in a sustainable state. From the perspective of sustainable indicators to analyze, the key sustainability indicator (ESI) is 6.28, which is within the sustainability criteria.

1 INTRODUCTION

Currently, the vast majority of residents live in cities, leading to a large number of issues, such as the urban heat island, solid pollution, water pollution, air pollution, and traffic congestion. In order to cope with these problems, the concept of sustainable development needs to be applied and practiced urgently.

When discussing the sustainability of cities, many researchers have paid continuous attention to Eco-city and produced a lot of research results, which have several angles, involving ecological security (Wang et al. 2021); ecological footprint (Wu & Huang 2022); urban symbiosis (Lu et al. 2020); ecological civilization construction (Zhang et al. 2022); ecological network (Yang et al. 2022); environment Quality (Sun et al. 2022); ecological Smart City (Xu et al. 2022); ecological Livability Assessment (Fan et al. 2021); emergy analysis (Zhang & Ma 2021); ecological literacy (Ha et al. 2022); ecological pattern (Huang et al. 2022); ecological risks (Wang et al. 2020), etc.

This paper aims to explore the sustainability of parks in the city.

---

*Corresponding Author: zjx2021@just.edu.cn

2 METHODOLOGY

2.1 *Study area*

Figure 1 displays the boundary between the city and the ecological park, which has a detailed comparison for observers. In table 2, the design details have been revealed, including green space, water bodies, flowers, buildings, corridors, etc.

Figure 1. The boundary between the city and the ecological park.

Table 1. Ecological element design in the park.

(*continued*)

Table 1. Continued

2.2 Emergy method

In this paper, the emergy approach has been applied to the assessment field. Emergy methodology was proposed by H.T. Odum firstly to evaluate the sustainability of ecological systems (Zhang & Ma 2021). The concept of emergy allows us to compare different types of energy contributions in a system with the same energy standard. In practical applications, the emergy of certain energy is measured by solar energy, so solar energy can be used as a standard to measure any kind of energy. Emergy unit is the solar joule (sej).

2.3 Sustainable indicators

(1) Environmental loading ratio (ELR): As the ratio of nonrenewable emergy and purchased emergy, EIR is defined, which can be used to explain the ecological load for the ecological park, including low values (ELR < 2), medium intensity (3 < ELR < 10), and high environmental load (ELR > 10);

Table 2. Sustainable indexes based on emergy angle.

Items	Index	Meanings
Environmental loading ratio	ELR	Environmental pressure
Emergy yield ratio	EYR	Ability to obtain emergy
Emergy Sustainability Indicator	ESI	Sustainable degree

(2) Emergy yield ratio (EYR): EYR can be calculated based on total emergy and imported emergy part, which demonstrates the ability that can produce emergy. The higher the EYR is, the better the consequence of the system is. The lower input of purchased emergy could bring about higher EYR, revealing the competitive ability of the system.
(3) Emergy sustainability index (ESI): ESI can be got from EYR and ELR, which explains the sustainability in the target system. Generally speaking, three standards can be referenced, which are $ESI < 1$ (Unsustainable), $1 < ESI < 5$ (Medium situation), and $ESI > 5$ (Sustainable) in the long term.

3 RESULTS AND DISCUSSION

3.1 Emergy calculated details

Table 3. Psychological indicators collection.

Tapes	Emergy (sej)
Green lawn	1.55E + 10
Water body	5.68E + 12
Green vegetation	9.65E + 11
Flowers	4.67E + 7
Buildings	3.42E + 12
Corridors	2.96E + 11

3.2 Results analysis

In Tables 3 and 4, the main results have been calculated and shown. In Figures 2 and 3, their trends were compared.

In view of these results, the ecological state of the park is good and is in sustainable development mode. Among them, the largest emergy is water emergy, followed by building emergy, green vegetable, corridors and green lawn. The proportion of water emergy accounts for 54.7%. The others are 32.9% of building emergy, 9.3% of green vegetation emergy, and 2.9% of corridors emergy.

From the sustainability indicators to discuss, the environmental loading ratio is 35.7, which has a relatively high value and is higher than 10. Analyzed from this index alone, a load of ecological parks is relatively heavy. But because this park is designed as an ecological park, the emergy yield ratio is very high, leading to the final sustainable index parameters being qualified.

Table 4. Sustainable indexes based on emergy angle.

Items	Index	Value
Environmental loading ratio	ELR	35.7
Emergy yield ratio	EYR	224.2
Emergy Sustainability Indicator	ESI	6.28

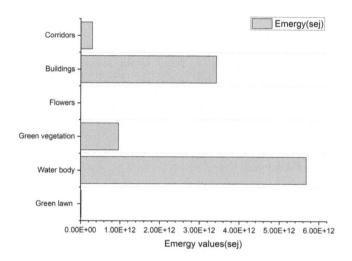

Figure 2. Emergy proportion order.

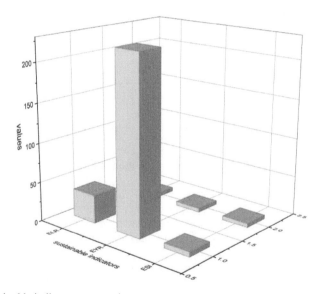

Figure 3. Sustainable indicators comparison.

ACKNOWLEDGEMENT

The work described in this paper was supported by the Open Fund of the State Key Laboratory of Silicate Materials for Architectures (Wuhan University of Technology) (SYSJJ2022-16).

REFERENCES

Changchen Ha, Guowen Huang & Jiaen Zhang & Shumin Dong. Assessing Ecological Literacy and its Application Based on Linguistic Ecology: a Case Study of Guiyang City, China. *Environmental Science and Pollution Research* (2022) 29:18741–18754.

Chunyang Lu, Shanshan Wang, Ke Wang, Yu Gao, Ruiqin Zhang. Uncovering the Benefits of Integrating Industrial Symbiosis and Urban Symbiosis Targeting a Resource-dependent City: A Case Study of Yongcheng, China. *Journal of Cleaner Production* 255 (2020) 120210.

Delu Wang, Ziyang Huang, Yadong Wang, Jinqi Mao. Ecological Security of Mineral Resource-based Cities in China: Multidimensional Measurements, Spatiotemporal Evolution, and Comparisons of Classifications. *Ecological Indicators* 132 (2021) 108269.

Dong Wang, Xiang Ji, Cheng Li and Yaxi Gong. Spatiotemporal Variations of Landscape Ecological Risks in a Resource-Based City under Transformation. *Sustainability* 2021, 13, 5297.

Feihan Sun, Chongliang Ye, Weirong Zheng and Xumei Miao. Model of Urban Marketing Strategy Based on Ecological Environment Quality. *Hindawi Journal of Environmental and Public Health* Volume 2022.

Jing Wu, Zhongke Bai. Spatial and Temporal Changes of the Ecological Footprint of China's Resource-based Cities in the Process of Urbanization. *Resources Policy* 75 (2022) 102491.

Junxue Zhang, Lin Ma. Urban Ecological Security Dynamic Analysis Based on an Innovative Emergy Ecological Footprint Method. *Environment, Development and Sustainability*, 2021.

Lei Huang, Dongrui Wang, Chunli He. Ecological Security Assessment and Ecological Pattern Optimization for Lhasa City (Tibet) Based on the Minimum Cumulative Resistance Model. *Environmental Science and Pollution Research*, 2022.

Linbo Zhang, Hao Wang, Wentao Zhang, Chao Wang, Mingtao Bao, Tian Liang, Kai-di Liu. Study on the Development Patterns of Ecological Civilization Construction in China: An Empirical Analysis of 324 Prefectural Cities. *Journal of Cleaner Production* 367 (2022) 132975.

Sunquan Xu, Yi Hou, and Li Mao. Application Analysis of the Ecological Economics Model of Parallel Accumulation Sorting and Dynamic Internet of Things in the Construction of Ecological Smart City. *Hindawi Wireless Communications and Mobile Computing* Volume 2022.

Yang Yang, Zhe Feng, Kening Wu, Qian Lin. How to Construct a Coordinated Ecological Network at Different Levels: A Case From Ningbo city, China. *Ecological Informatics* 70 (2022) 101742.

Zengzeng Fan, Yuanyang Wang and Yanchao Feng. Ecological Livability Assessment of Urban Agglomerations in Guangdong-Hong Kong-Macao Greater Bay Area. *Int. J. Environ. Res. Public Health* 2021, 18, 13349.

Study on the status and conservation of stone carvings in Lianyungang

YuYang Chen & DeHeng Zhang*
Nanjing Institute of Technology, Nanjing, China

YuQian Wu
Yangzhou University, Yangzhou, China

ABSTRACT: This paper presents the current situation of local stone carvings in Lianyungang and the protective research on stone carvings, and analyzes the impact of local climate and environment on stone carvings, discussing the mechanism of damage to stone carvings in terms of human activities, natural physical activity damage, chemical reaction effect, and biological effect. Various measures are proposed to address the current situation and problems, including the establishment of protective facilities around the area, stone carving cleaning work, and anti-weathering materials and measures.

1 INSTRUCTIONS

Currently, the tourism and cultural industry of Lianyungang, Jiangsu Province, China, benefits from the richness of local stone carvings. However, along with the rapid development of tourism and changes in the world climate, the local stone carvings are facing damage caused by human activities and changing weather. The purpose of this paper is to study the current situation of local stone carvings and propose targeted conservation measures in the face of the current stone carving conservation situation.

2 GEOGRAPHICAL AND CLIMATIC CONDITIONS

Lianyungang, a prefecture-level city in Jiangsu Province, China, is situated in the area between the subtropical and warm temperate zones, where the southern hills of Shandong Province meet the northern plains of the Huai River. It has some marine climate features because it is close to the water. The highest temperature is in July above thirty degrees, while the lowest is in January below zero degrees. The annual precipitation is 883.8 ml. Winters are chilly and dry, while summers are hot and rainy. With a yearly average wind speed of 4.3 m/s, monsoon winds predominate. Solar radiation is copious, and there are 2000 to 2600 hours of sunshine per year.

Lianyungang is a predominantly industrial city with dense industries, including chemical plants and power plants that tend to emit acid rain, with the worst acid rain rate reaching 15% in a year, with ph values as low as 4.6 (Meng 2012; Zhang 2012).

---

*Corresponding Author: 2473433272@qq.com

Table 1. Local Temperature and precipitation map of Lianyungang.

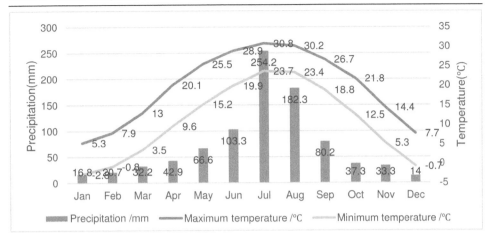

3 STONE CARVING VALUE

With 190 remaining artifacts, mostly on Kongwang Mountain and Shibang Mountain, Lianyungang is abundant in regional stone carvings. The Neolithic Age is the one with the longest history of them all. Local stone carvings are valuable because they provide a conduit for the locals' efforts to pursue production, life, and the spiritual realm. The political, cultural, economic, and military aspects of the region are all strongly tied to the stone carvings' subject matter, and all of these aspects can be artistically displayed. The durability and variety of stone carvings are intrinsic qualities, unlike other works of art. We can learn about the changes in the local environment and times from antiquity to the present by observing these stone sculptures.

The intricate indigenous stone carvings are now the tourism industry's calling card in Lianyungang, drawing thousands of visitors each year.

4 CONSERVATION STATUS

The predominant types of local rocks in Lianyungang include granite and gneiss, which are hard in texture. However, their surfaces have been exposed for a very long period, and the majority of them are constantly exposed to rain and sunlight. A few of these local rocks are also found inside caves. The yearly rainfall in Lianyungang is high, sun radiation is abundant, the temperature and humidity are high, and there are significant weathering effects, which cause the surface of the stone carvings to weather and flake, among other occurrences. The integrity and ornamental value of some stone carvings are directly impacted by tectonic fissures, level fissures, and weathering fissures, which are mostly substantial cracks that stretch farther and deeper across the surface of the carvings.

The same kinds of biological aggressions harm the exposed rock carvings, and the moss, lichens, molds, or metabolites of other plants and animals found in the nearby mountains cover the surface of the rock carvings, which has a negative impact on the cliff carvings' appearance. The integrity of the rock carvings is seriously impacted by the fact that some of the rock surfaces have been pried and broken by the slow growth of plant roots.

Some of the stone carvings have sulfides sticking to the surface, darkening the surface, which affects the decorative aspect of the stone carvings. The exposed stone sculptures are especially sensitive to deterioration by acid rain.

5 ANALYSIS OF STONE CARVING DISEASES

5.1 Human activity damage

Most of the stone carvings in the Lianyungang area have no protective fence, and visitors can easily touch the surface of stone carvings, and these actions may cause further damage to the already weathered stone carving relics. Human activities may include the following: touching the surface of the stone carvings, which loosens the surface of the stone carvings; the presence of stains and sweat on the fingers, which can have a chemical reaction with the surface of the rock; using foreign objects to carve or copy the stone carvings, which damages the surface carvings.

5.2 Natural physical activity damage

Temperature is one of the main factors for the damage to rock carvings. Most of the rock bodies are hot outside and cold inside during the day, while the rock bodies are cold outside and hot inside at night. In such a situation of uneven heat inside and outside, the volume of the rock body will repeatedly expand and contract. The rock body as a whole will be destabilized, and the rock carvings on the surface of the rock body will be weathered (He 2107).

The effect of flowing water on the rock carvings is not only the scouring effect on the surface but also the infiltration of flowing water in the rock fissures, after low-temperature freezing and volume expansion after freezing, which makes the rock fissures continue to expand. In addition, part of the flowing water into the rock fissure is mixed with salt, microorganisms, and pollutants, there is a certain unknown erosion effect on the expansion of the rock fissure.

The effect of crystalline salt is that when crystalline salt is dissolved in water, it penetrates into the rock fissures and rock carving surfaces, and under high and dry conditions, the salt will crystallize, thus generating a certain expansion pressure. With the constant change of climate, alternating wet and dry, the salt in the crevices, will keep repeating the process of dissolution and crystallization.

Most of the effects of wind and rain on the rock body are mainly abrasive, with more obvious corrosive effects such as scouring and friction on the rock surface and stone carvings, thus making the rock surface and stone carvings appear loose particles.

Table 2. Mechanism of action of physical factors.

Physical factor	Mechanism of action	Disease performance
Temperature	Repeated expansion and contraction of the volume of the rock in case of uneven heating.	Lack of rock stability
Water	Water in the fissure freezes and expands in volume	Fissure expansion
Crystallized salt	The water in the fissure evaporates and precipitate the crystalline salt, producing the expansion force	Fissure expansion
Valley wind, rain	Scrape out and wash out the stone carving surface	Rifts appear on the surface

5.3 The chemical reaction to damage the effect

The dissolution between H_2O and $CaCO_3$ is weak when the rock is only in contact with water, but it is also the first step in subjecting the rock to chemical weathering. If the combined effects of the H+ concentration in the water, the temperature of the environment in which the rock is exposed, and the presence of other impurities in the water are taken into account, they will all lead to further dissolution and dissolution of the rock, thus destroying the integrity and stability of the rock carving.

Chemical reactions will also involve the hydrolysis and hydration of the rock carvings by dissolved salts and water. Some of the weakly acidic mineral salts exist on the surface of the rock, and there is a possibility of hydrolysis reactions when water is encountered, resulting in the loss of mineral metal cations (mainly Ca^{3+}). Some of the water will enter the lattice of the rock mass minerals as crystalline water, thus forming new water-bearing minerals, resulting in an increase in volume and continued expansion of damage to the rock mass. All of these can have a corrosive effect on the rock and stone carvings.

Exposed outdoor rock carvings are also exposed to the chemical reaction of some atmospheric gases, mainly the reaction of carbon oxides to the carbonates and silicates of the rock body; the reaction of small amounts of acidic sulfides present in the air to the pigments on the surface of the rock carvings; and the acid rain formed by the mixture of pollutants produced by the local chemical industry and rainwater, mainly sulfur oxides and nitrogen oxides, which have a more serious corrosive effect on the rock carvings.

Table 3. Chemical factors mechanism.

Gaseous Fluid	Chemical Reaction	The Reaction Effect
CO_2	$CO_2+H_2O+K[(AlO_2)(SiO_2)_3] \rightarrow K_2CO_3+SiO_2+Al_2SiO_5(OH)_4$ $CaCO_3+CO_2+H_2O \rightarrow Ca(HCO_3)_2$	Weakening of stone carving surface
H_2S	$H_2S+Fe^{3+} \rightarrow FeS+2H^+$	Pigment fading
SO_2	$CaCO_3+SO_2+H_2O+O_2 \rightarrow CaSO_4 \cdot 2H_2O$	Acid rain corrosion
NO_2	$CaCO_3+NO_2+H_2O \rightarrow CO_2 \uparrow +Ca(NO_3)_2$	

5.4 Biological effects

The local rock carvings are mostly located in mountainous areas and the surrounding environment is humid, so the surface of the rock carvings is very easy to breed algae and mold, and other organisms. The growth of plant roots causes cleavage, which expands the cracks in the rock. The expansion of mold-like microorganisms covers the surface of stone carvings, making the appearance of stone carvings suffer from pollution (Sun 2021).

6 STONE CARVING PROTECTION AND RESTORATION SUGGESTIONS

For the exposed stone carvings, protective fences can be equipped around the open-air stone carvings, and this equipment can effectively organize the damage to the stone carvings from human activities and reduce the partial erosion effect of wind and rain on the rocks. However, while choosing the corresponding equipment, it is necessary to consider that the ornamental nature of the stone carvings should not be affected. And the removability of such protective equipment should also be taken into account to facilitate the restoration of stone carvings daily (Yang 2004).

For most of the carvings, it is important to clean the surface of the carvings. The focus is on cleaning the surface sludge and cleaning the surface algae and mold. To clean the surface fissures of stone carvings where inorganic salts exist, priority can be given to the use of the paper pulp adsorption method. If there are no special hard objects on the surface, you can use the vacuum cleaner to absorb dust and a soft brush to clean the surface. If there are hard materials on the surface, you can consider using solvents to soften and then clean the surface. For algae and mold surfaces, a high-temperature steam machine is used to jet steam to the stone carving surface, high temperature is applied back to soften the stone surface biological adhesions and then a soft brush is used to scrub. After overall cleaning and washing, the surface is brushed with a prepared 2A (pure water 50% + 50% acetone) solution using a stiff bristle brush to clean the scrubbing location. For the effect of water infiltration and crystallized salt on stone carvings, the hydrophobic coating can be used to cover the rock surface and the parts related to stone carvings after finishing the cleaning of the stone carving surface.

For partially damaged stone carvings to be repaired, the area should be strictly controlled. Only for open fissures or cement repair parts that have been loosened, without affecting their mechanical stability, only the part that has been loosely repaired should be removed. When removing the cracked side of the surrounding rocks cannot have new damage, chiseling should be done from the inside out, first, the middle and then the edge of the way, mortar removed cleanly after using a scalpel to remove debris such as dust inside the crack, blow dry, repair using water hard lime plus stone powder material, mix fully with pure water, blend into a repair material close to the color of the surface of the stone carving, fill in the missing area until the repair material curing.

For the weathering effect of the rock carvings, both reinforcement and protection are needed. Considering the local climate and environmental characteristics, the chemical materials for surface treatment should be resistant to acid rain and weathering. The grouting material is a colorless and transparent imported oxide resin, which ensures that the appearance of the relics is not affected while taking into account a certain degree of acid resistance. The use of the grouting method for weathering parts needs to be reinforced, which can make the material and the surface of the rock body closely fit.

Organosilicon-type materials for rock penetration reinforcement are also feasible, this material is colorless and transparent after curing, which does not affect the appearance of cultural relics, and has strong acid resistance and hydrophobia. The surface strength of the rock body is greatly improved, while the impact on the secondary restoration work of stone carvings is minimal. After painting, silicone materials need to cover the rock surface, it is recommended that after curing at a temperature above 20°C for about 7 days, remove the cover and air dry naturally for 3 days (Chen 2022; Zhu 2013). However, the penetration of silicone-type material needs to be painted several times to ensure the rock carving is saturated. Therefore, after completing the above process, the process should be repeated twice again to achieve the best results. As a representative of silicone materials, Remmers series reinforcement has been widely used. In this test, the Remmers grades 100, 300, and 500 were selected for indoor tests to compare the reinforcement effect 'Mohs hardness' with the permeability of the area of the reagent applied to the test block before and after reinforcement.

The results of the following two tests showed that after reinforcement, both Remmers 300 and 500 showed an increase in strength, and Remmers 100 was slightly less effective. In terms of infiltration performance, Remmers 500 had a slight disadvantage compared with 300 at the beginning of the test, but after a long time comparison, it was found that Remmers

Table 4. Rock reinforcement effect 'Mohs hardness results'.

Test number	*Classification*	*Mohs hardness results*
1	100	6.5
2	300	7.0
3	500	7.0
4	Original state	6.0

Table 5. Rock permeability test.

Test number	Classification	16h Amount of penetration/g	20h Amount of penetration/g	40h Amount of penetration/g	400h Amount of penetration/g
1	100	0.002	0.004	0.011	0.030
2	300	0.000	0.004	0.010	0.028
3	500	0.001	0.003	0.010	0.020
4	Original state	0.000	0.008	0.014	0.036

500 had a more outstanding performance, and the difference between 100 and 300 in terms of infiltration performance was not significant. For the actual local situation, we recommend the use of the more comprehensive performance of Remmers 500 as a reinforcing agent.

7 CONCLUSIONS

This article focuses on the local stone carvings in Lianyungang and discusses the current state of the carvings and the various effects of the local environment and climate on the carvings. This study helps us to understand the causes and conditions of local stone carvings, and this paper also proposes conservation methods and restoration recommendations based on the current status of local stone carvings, including protective fencing, stone carving cleaning, stone carving repair, and, quite importantly, stone carving reinforcement materials. This paper also compares the reinforcement effect and permeability of silicone reinforcement materials and selects a more suitable material for local stone carving reinforcement.

ACKNOWLEDGMENT

Science Innovation Training General Project for College Students in Jiangsu Province of China.(Grant Number 202211276088Y)

REFERENCES

Chen, B.X. 2022 Study on the Conservation of Cliff Carvings in Liuzhou. *Cultural Identification and Appreciation*, 230 (11): 71–74.

He, X.B. 2017 Analysis of Disease and Conservation Research of Cliff Stone Carvings in Jiuhua Mountain. *Ancient Garden Technology*, 2017 (2): 65–68,73.

Meng, J.J. & Li, H.B. 2012 Taking Rock Paintings in Lianyungang as an Example. *The Three Gorges Forum*, 2012 (04): 67–70

Yang S. & Bian, J. 2004 Investigation and Analysis of the Current Situation of the Conservation of Stone Carvings of the Six Dynasties in Nanjing. *Southeast Culture*, 2004 (2): 91–94.

Sun, S.H. & Tang, Y.J. 2021 A Review on Weathering Mechanism of Red Sandstone and Stone Carving Conservation. *Construction Technology*,50(23): 150–153.

Zhang, H.H. & He, L. 2012 *Characteristics and Causes of Climate Change in Sunshine Hours in Lianyungang*. Proceedings of the 29th Annual Meeting of the Chinese Meteorological Society, 2012: 1–10.

Zhu, Y.F. 2013 Conservation and Study of cliff Stone Carvings. *Heilongjiang History*, 2013(19): 314, 318.

Optimization design of aluminum alloy connections on PV roof against wind-uplift load

Xiaoxiang Tang, Yongfang Liu* & Haisheng Liang
State Grid Shanghai Economic Research Institute, Shanghai, China

Di Wu
Butler (Shanghai) Inc, Shanghai, China

Xiaoqun Luo
College of Civil Engineering, Tongji University, Shanghai, China

ABSTRACT: The wind-bearing capacity of PV roof systems is a part of structural safety that cannot be ignored. PV roof systems are damaged due to insufficient wind-bearing capacity. Numerical analyses were carried out on a typical type of connection for PV roofs. The bearing capacity and weak position of PV roof connections under wind-uplift load were investigated through reasonable simplification, and the main design parameters of aluminum alloy clamps in connections, such as side thickness, bottom thickness, and bottom width, were analyzed parametrically. Based on the analysis results, the influence law of each parameter on the bearing capacity of the connection was summarized. The parameter setting table for different wind-uplift load design values was obtained, the bearing capacity calculation formula of the connection was fitted, and design suggestions for the actual design were given finally.

1 INTRODUCTION

The application of PV technology in buildings is one of the important means of achieving the goal of carbon peaking and carbon neutrality in the construction field and has been widely used worldwide. To ensure the safety of PV roofs, it is important to ensure their structural safety under the combination of wind effects. At present, there are more research results on the test performance of metal roofing systems against wind-uplift load. In recent years, reasonable and controllable test methods and loading methods have been designed, and the specification of wind-uplift tests is gradually being formed (Serrette et al. 1997; Sivapathasundaram et al. 2017; Shoemaker 2009). In terms of finite element (FE) simulation analysis, scholars have put forward some methods to deal with the contact problem. To understand the wind-uplift damage mechanism more comprehensively, some theoretical studies have been conducted by numerical simulation methods based on relevant experiments, and parametric analysis and some complex model simplification methods have been proposed (Ali et al. 2003; Damatty et al. 2003; Morrison et al. 2010). However, there has yet to be a study on the wind-uplift resistance of roofing connections similar to those covered with PV modules, nor has it provided an effective theoretical formula or calculation basis for bearing capacity, which hinders the promotion of PV roofing design and application.

---

*Corresponding Author: 1056248764@qq.com

The main form of current PV roofing systems is the upright locking edge metal roof covered with PV modules. As the load-bearing capacity of the connections under the combination of wind load effects is controlled for the safety of the entire PV roof, it is necessary to conduct fine research and analysis on the wind-bearing capacity of the PV roof connected nodes.

Therefore, based on the design parameters of the new aluminum alloy fixture in the actual connection of the project, the side thickness, bottom thickness, and bottom width of the fixture are selected as the main research parameters. The wind-bearing capacity of the PV roof connection is studied theoretically by numerical analysis method.

2 NUMERICAL ANALYSIS OF THE BEARING CAPACITY OF PV ROOFING SYSTEMS

To clarify the stress state, deformation process, and damage form of PV roofing systems under wind-uplift load, this paper adopts ABAQUS nonlinear FE software to analyze the bearing capacity of PV roofing connections under wind-uplift load and the stress mechanism at the weak point. It is assumed that the PV roofing system's load-bearing capacity fails, and the corresponding wind load at this time is the damage load when the material stress reaches its ultimate tensile strength (or strain reaches ultimate strain).

2.1 Finite element model of aluminum alloy connection for PV roofing

The size and thickness of each component of the roofing system are consistent with the actual project. The width of the roof panel is 600 mm, the thickness is 0.65 mm, the length of the PV panel is 2102 mm, and the width is 1040 mm. The connection between the roof panel and the PV panel is made by aluminum alloy clamps, as shown in Figures 1(a) and 1(b). Under wind load, all components of the roofing system are subject to load. Most experts and scholars have confirmed that the connections of fixed bearings and roof panels are the weak links of the metal roofing system against wind damage through the wind damage test of large roofing systems. As a result, a relatively small, simplified model can be constructed to study the wind damage resistance of local nodes in a small area. According to the relative position of each member of the actual project, the wind load transmission path is known from the PV panel to the upper fixture, through the upper screw to the lower fixture, and then from the lower screw transition to the roof panel locking edge, and finally to the connected purlins. Accordingly, the upper and lower clamps, upper and lower screws, and the roof panel locking edge are separated to establish a connection model in a certain area, as shown in Figure 1(c).

To facilitate modeling, meshing, and connection settings between components, all components are considered in accordance with the solid. The Mises criterion is used to determine

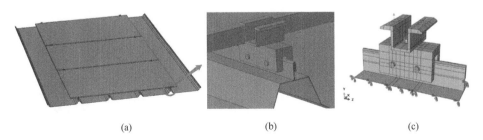

(a) (b) (c)

Figure 1. Finite element model of the PV roofing system overall and local connection schematic.
(a) Panel model (b) Connection parts (c) Connection mesh.

the yield criteria and damage for each component in the FE model. The primary materials of the connecting nodes of the intrinsic relationship model are assumed to be a bilinear model, and the material parameters are displayed in Table 1. The C3D8R eight-node hexahedral linear reduced integral solid cells are used to segment the mesh. In Figure 2, the stacking portion of the upright locking edge is established as a fully homogeneous entity. The locking edge is simplified, and it is thought that the locking edge will be less prone to damage and no relative displacement will occur. In Figure 1(c), the fixed constraint is applied to the lateral part of the locking edge.

Table 1. Material properties of components.

Components	Yield strength (MPa)	Tensile strength (MPa)	Extreme Strain
Fixture	259	313	0.1
Screws	640	800	0.008
Roof panels	345	470	0.12

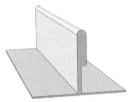

Figure 2. Simplification of locking edge.

2.2 Wind-bearing capacity of aluminum alloy connections of PV roofs

The size of the upper fixture is such that the bottom thickness is 3 mm, the bottom width is 26 mm, and the side thickness is 3 mm. Considering the geometric nonlinearity and material nonlinearity, displacement loading with incremental loading is used to simulate the wind load at the connections, and the applied position is shown in Figure 1(c).

The applied displacement load increases gradually from 0 until the connection fails. The calculated value of the ultimate load is obtained by dividing the extracted support reaction force by the corresponding wind-uplift area for conversion. Figure 3(a) shows the results of the upper fixture's load-maximum strain. It can be seen that when the wind load reaches

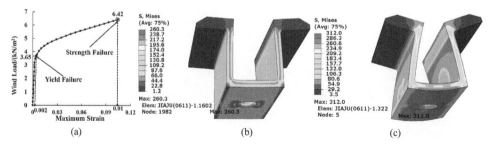

Figure 3. Results of finite element analysis of the basic model of the upper fixture. (a) Load-maximum strain curve (b) Elastic limit phase (c) Destruction phase.

3.65kN/m², the residual plastic strain generated at this time is 0.2%, and it is considered that the connection reaches the yield limit stage. The material at the stress concentration of the screw hole on the bottom side of the upper fixture is the first to enter the yielding stage. The stress cloud diagram is shown in Figure 3(b). When the wind load reaches 6.42kN/m², the maximum strain reaches the ultimate strain of 0.1, the screw hole edge of the upper fixture reaches the ultimate tensile strength, and the connection is damaged, at which time the stress cloud diagram is shown in Figure 3(c).

3 PARAMETRIC ANALYSIS OF THE WIND-UPLIFT BEARING CAPACITY OF ALUMINUM ALLOY CONNECTIONS

The aluminum alloy connection analysis model established in Section 1 is used as the basic model for parametric analysis. From the aforementioned analysis results, it is known that the wind-uplift damage location of the aluminum alloy connection may be the stress concentration at the bottom screw hole or the non-stress concentration location at the side of the upper fixture. The corresponding positions of the upper fixture dimensional parameters are shown in Figure 4, and each dimension of the upper fixture is considered to be changed for parametric analysis, as shown in Table 2. The bottom thickness of 2–4 mm (interval 0.25 mm), the bottom width of 23–29 mm (interval 1 mm), and the side thickness of 2–3.5 mm (interval 0.25 mm) were selected as the analysis parameters to carry out 23 sets of FE analysis on the load-bearing performance of the connections (two of them were repeated), and the effects of different parameters on the wind-bearing capacity of the connections were obtained.

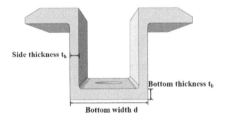

Figure 4   Corresponding parameters of the upper fixture.

Table 2.   Summary of parametric analysis conditions.

Nos.	Bottom thickness (mm)	Bottom width (mm)	Side thickness (mm)
C1-C9	2~4	26	3
C10-C16	3	23~29	3
C17-C23	3	26	2~3.5

3.1   *The effect of the bottom thickness of the fixture*

According to the actual engineering experience, the bottom thickness of the fixture was selected from 2 mm to 4 mm, with an interval of 0.25 mm, and other parameters were kept the same as the basic model. Nine models were established, and FEM was performed on the connection. The load-maximum strain results of the fixture are shown in Figure 5.

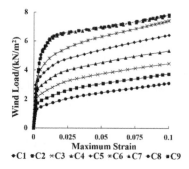
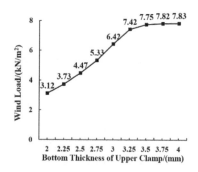

Figure 5. Load-maximum strain curves of Condition 1-9.

Figure 6. Failure load under different bottom thicknesses.

Taking the limit strain 0.1 as the maximum strain, the load of the connection was set as the limit load, and then the influence law of the bottom thickness of the fixture increases, the damage load of the aluminum alloy connection increases, and the relationship is linear within a limit.

The analysis demonstrates that increasing the bottom thickness can significantly increase the bearing capacity of the assembly when the bottom thickness is small. However, as the bottom thickness increases, the upper fixture's damage position shifts from the area surrounding the bottom screw hole to the vertical sides. Plastic damage occurs in the non-stress concentration area, and the ultimate bearing capacity is no longer controlled by the bottom thickness. Therefore, the appropriate increase in the bottom thickness of the fixture can improve the wind-bearing capacity of the connection, and the increase of the bearing capacity decreases after the bottom thickness exceeds 3.25 mm.

3.2 The effect of the bottom width of the fixture

According to the actual engineering experience, the bottom width of the fixture was selected from 23 mm to 29 mm, with an interval of 1 mm, and other parameters are kept the same as the basic model. Seven groups of analytical models are established, and FE analysis is performed on the connection. The load-maximum strain results of the fixture are shown in Figure 7, and the influence law of the bottom width on the damage load of the connection is shown in Figure 8. When the bottom width is small, increasing the bottom width will reduce the bearing capacity of the connection, but the effect of the bottom width gradually decreases with the increase in the bottom width. Therefore, under the premise of reserving

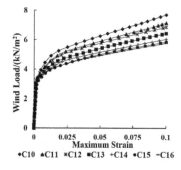
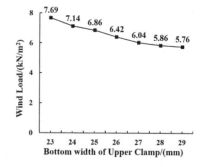

Figure 7. Load-maximum strain curves of Condition 10-16.

Figure 8. Failure load under different bottom width.

space for nut installation, reducing the bottom width of the fixture can improve the wind-bearing capacity of the connection.

3.3 Effects of the side thickness of the fixture

According to the actual engineering experience, the bottom thickness of the fixture was selected from 2 mm to 3.5 mm, with an interval of 0.25 mm, and other parameters were kept the same as the basic model. Nine models were established and FE analysis was performed on the connection. The load-maximum strain results of the fixture are shown in Figure 9, and the influence law of the bottom thickness on the damage load of the connection is shown in Figure 10. When the side thickness is small, increasing the side thickness can significantly improve the component bearing capacity, but with the increase of side thickness, the damage position of the upper fixture will shift from the vertical sides to the bottom screw hole perimeter. Plastic damage in the stress concentration area of the screw hole side also occurs, and the ultimate bearing capacity is no longer controlled by the side edge width. Therefore, the appropriate increase of the fixture's side thickness can improve the wind-bearing capacity of the connection, and the bearing capacity will no longer increase after the side thickness exceeds 2.75 mm.

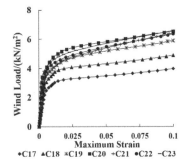
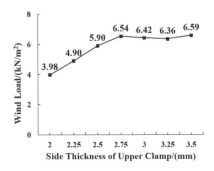

Figure 9. Load-maximum strain curves of Condition 17-23.

Figure 10. Failure load under different side width.

3.4 Summary of parameter analysis results

The above calculation results are combined to give the parameter settings of the connections of the roofing system for different design values of wind-uplift load, as shown in Table 3.

Table 3. Clamp size parameters setting table for different wind-uplift load design values.

Wind load (kPa)	Bottom thickness (mm)	Bottom width (mm)	Side thickness (mm)	Bearing capacity (kPa)
7	3.5~4	26	3	7.7~7.8
	3	23	3	7.7
	3.25	26	3	7.4
6	3	25	3	6.8
	3	26	2.75~3.5	6.4~6.6
	3	27	3	6.0

(*continued*)

Table 3. Continued

Wind load (kPa)	Bottom thickness (mm)	Bottom width (mm)	Side thickness (mm)	Bearing capacity (kPa)
5	3	26	2.5	5.9
	3	27~28	3	5.8~5.9
	2.75	26	3	5.3
4	3	26	2.25	4.9
	2.5	26	3	4.5
	3	26	2	4.0
3	2.25	26	3	3.7
	2	26	3	3.1

4 FITTING THE CALCULATING FORMULA FOR WIND-UPLIFT RESISTANCE OF ALUMINUM ALLOY CONNECTIONS

Since the wind damage position of the connection is the stress concentration at the bottom screw hole of the upper fixture or the non-stress concentration position at the side, and it is known from the previous calculation that the wind bearing capacity is roughly positively related to the bottom thickness t_b (less than 3.5 mm) and the side thickness t_h (less than 2.75 mm) and negatively related to the bottom width d within a limit, it is assumed that the wind-uplift bearing capacity of this connection has a certain proportional relationship with the dimensionless parameter, as shown in Equation 1.

$$q = \alpha \frac{t_b t_h}{d^2} + \beta \qquad (1)$$

On the basis of the parameter analysis, MATLAB is used to fit the quantitative relationship between this parameter and the wind-bearing capacity under the aforementioned 23 cases to provide the wind resistance design basis for the cases not in the table. As shown in Figure 11, the coefficients $\alpha = 530$ and $\beta = -0.696$ are obtained by fitting according to the FE calculation results, and the correlation coefficient is greater than 0.8, which is considered to be a good fit.

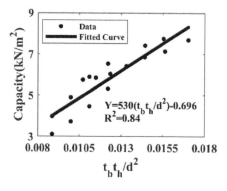

Figure 11. Fitting results of the wind-uplift capacity of aluminum alloy connections.

5 CONCLUSIONS

Through the analysis of wind bearing capacity of aluminum alloy connections of PV roofing system, the following conclusions are obtained:

(1) Under the wind-uplift load, due to the stress concentration, the screw hole of the upper fixture will leave a large residual strain after unloading, while the non-stress concentration area of the upper fixture may also be damaged by reaching the ultimate stress first.
(2) Through the parametric analysis of wind bearing capacity, it can be concluded that appropriately increasing the bottom thickness and the side thickness of the upper fixture can improve the wind resistance until the corresponding damage area is transferred to other locations. In addition, reducing the bottom width can improve the bearing capacity of the connection.
(3) Through the numerical calculation, the parameter setting table of the aluminum alloy fixture of the connection for different design values of the wind-uplift load is obtained. Using the curve fitting method, the limit load capacity calculation formula of the fixture is obtained, which provides a reference for the practical design and application of the project.

ACKNOWLEDGMENT

This research was partially supported by the State Grid Shanghai Economic Research Institute for objective projects.

REFERENCES

Ali H. M., Senseny P. E. Models for Standing Seam Roofs. *J. Wind Eng. Ind. Aerodyn.*, 2003, 91(12–15): 1689–1702.

Damatty A., Rahman M., Ragheb O. Component Testing and Finite Element Modeling of Standing Seam Roofs. *Thin-Walled Struct.*, 2003, 41(11): 1053–1072.

Morrison M. J., Kopp G. A. Analysis of Wind-induced Clip Loads on Standing Seam Metal Roofs. *J. Struct. Eng.*, 2010, 136(3): 334–337.

Serrette R., Pekoz T. Bending Strength of Standing Seam Roof Panels. *Thin-Walled Struct.*, 1997, 27(1): 55–64.

Shoemaker W. L. (2009) Design and Specification of Standing Seam Roof Panels and Systems. *ASCE.*, 16(22): 101–106.

Sivapathasundaram M., Mahendran M. Experimental Studies of Thin-walled Steel Roof Battens Subject to Pull-through Failures. *Eng. Struct.*, 2017, 143(Apr. 15): 388–406.

Author index

Bao, T. 494
Bao, Y.J. 41

Cai, Z.H. 203
Chen, B. 600
Chen, D.H. 624
Chen, H.Y. 77
Chen, J.H. 445
Chen, T.S. 77
Chen, W. 662
Chen, Y.W. 537
Chen, Y.Y. 692
Chen, Z.S. 408
Chen, Z.X. 456
Cheng, L. 90
Cheng, L.F. 77
Cong, X. 203
Cui, G. 674

Dai, Y.W. 105
Dai, Z.Y. 274
Deng, F. 235
Ding, J.H. 129
Ding, M. 209
Dou, G.Q. 408
Duan, B.F. 294

Fang, L. 221
Fang, S.F. 186
Feng, G. 171
Fu, Q.Z. 287
Fu, S.Y. 606
Fu, Z.Q. 631

Gan, D. 557
Gan, G. 41
Gao, H.J. 537
Ge, S.J. 209

Gou, X.Y. 77
Guan, Y.H. 64
Guang, H. 221
Guo, M. 310
Guo, P. 186
Guo, S.G. 547
Guo, X.P. 41
Guo, Y.C. 10

Han, B.Y. 512
Han, C. 156, 564
Han, D.Y. 348
Han, P.P. 64, 156, 564
Han, S. 203
Han, X. 41
Han, Y.C. 408, 614
Hu, G.Q. 487
Hu, H.Q. 41
Hu, J.L. 3
Hu, Q. 579
Hu, T. 235, 439
Hu, Y.C. 302
Huang, J.H. 646
Huang, L. 686
Huang, W. 10
Huang, Y. 348
Huang, Z.X. 264

Ji, R.C. 55
Ji, W. 209
Jia, X. 264
Jian, Y.B. 501
Jiang, L.C. 396
Jiang, R.A. 310
Jiang, X. 33
Jiang, Y.L. 302
Jiao, Z.K. 165

Karampour, H. 156
Kong, W. 221
Kou, W.W. 547

Lei, H.D. 243
Lei, H.G. 330
Lei, H.J. 480
Lei, L.X. 203
Lei, S.P. 165
Li, A.J. 243
Li, C. 667
Li, C.F. 111
Li, D. 424
Li, D.D. 681
Li, D.P. 501
Li, G.G. 408
Li, G.N. 614
Li, H. 655
Li, H.L. 221
Li, J. 524
Li, J.S. 10
Li, J.X. 137
Li, L.J. 624
Li, Q. 137
Li, S.-Z. 557
Li, S.H. 337
Li, S.J. 445
Li, T.S. 111
Li, W.H. 362
Li, Y. 280
Li, Z. 424
Li, Z.L. 209
Li, Z.X. 243, 537
Liang, H.S. 593, 698
Liang, J. 646
Liang, Z.Y. 530
Lin, F.G. 302
Lin, H. 64, 156, 564

Ling, Q. 674
Liu, A. 317
Liu, B.C. 408
Liu, D. 317, 337
Liu, F. 667
Liu, F.Y. 186
Liu, J.F. 472
Liu, J.S. 573
Liu, J.S. 537
Liu, K.M. 171
Liu, N. 348, 573
Liu, N.N. 439
Liu, Q. 530
Liu, Q.L. 264
Liu, S.R. 77
Liu, S.S. 171
Liu, S.T. 614
Liu, T. 646
Liu, W. 573
Liu, X.M. 33
Liu, Y.F. 698
Liu, Y.H. 243
Liu, Y.Y. 203
Luan, H.C. 564
Luan, H.C. 156
Luo, J.X. 472
Luo, X.Q. 593, 698

Ma, R. 379
Mao, G.X. 47
Miao, X.H. 501
Miao, X.W. 10
Mo, L.F. 424
Mu, G.L. 196

Nam, J.M. 396
Nam, R.C. 396
Ning, L.-L. 557

Pan, F. 586
Pan, S.B. 530
Pang, J. 494
Pang, M. 456
Pang, R.N. 186

Pang, Z.G. 111
Peng, L.H. 424

Qian, L.J. 85
Qiu, L. 171
Qiu, Q.C. 472
Qiu, Y.J. 253
Qiu, Y.Q. 137
Qu, D.X. 362

Ren, Y.W. 379
Rong, L.F. 3

Shen, F. 186
Shen, T. 99
Shen, W.Y. 501
Shen, Y.H. 227, 391
Shi, M.K. 456
Shi, M.X. 579
Shi, Q.Y. 445
Shi, Y.C. 178
Shou, L.C. 494
Shu, Q.H. 111
Su, P.L. 667
Sun, S.F. 111
Sun, S.K. 614
Sun, W.J. 586
Sun, X.B. 494
Sun, X.Y. 323
Sun, Y. 655
Sun, Y.F. 142
Sun, Z.J. 294

Tan, Z.S. 606
Tang, X. 302
Tang, X.X. 593, 698
Tian, H.T. 142
Tong, J.H. 355

Uzdin, A.M. 64, 564

Wang, B.X. 379
Wang, H. 221
Wang, H.C. 129
Wang, H.J. 203

Wang, H.W. 681
Wang, J. 606
Wang, J.W. 171
Wang, J.X. 424
Wang, K.Q. 55
Wang, L. 23
Wang, L.F. 494
Wang, L.G. 445
Wang, L.H. 111
Wang, L.J. 614
Wang, M.L. 235
Wang, N. 624
Wang, Q. 99, 137, 274
Wang, S. 640
Wang, T. 372
Wang, W. 624
Wang, X.H. 216
Wang, X.Y. 77, 614
Wang, Y. 264
Wang, Y.H. 557
Wang, Y.J. 439
Wang, Z.M. 23
Wang, Z.Y. 178
Wen, Z.H. 235
Wu, D. 698
Wu, J.J. 137
Wu, K.P. 472
Wu, P. 355, 586
Wu, T.F. 646
Wu, X.X. 280
Wu, Y.L. 23
Wu, Y.Q. 692

Xia, W. 264
Xie, H.Y. 105
Xie, J.C. 417
Xie, W. 593
Xie, W.J. 77
Xing, R. 330
Xu, D. 686
Xu, H. 64, 156, 564
Xu, L. 424
Xu, W.S. 294
Xu, X.L. 646

Xu, Y.X. 142
Xue, H.G. 424

Yan, B.B. 178
Yan, L. 77
Yang, B. 362, 445
Yang, F. 323
Yang, J. 424
Yang, J.J. 23
Yang, L. 64, 156, 564
Yang, L.P. 372
Yang, T.C. 487
Yang, W. 424
Yang, X.J. 348
Yang, Z. 537
Yao, D. 501
Ye, S.H. 547
Yin, J.F. 501
Yu, C.Y. 537
Yu, H.J. 524
Yu, J.T. 456
Yu, Z.H. 129

Yu, Z.W. 294
Yu, Z.-Y. 253
Yue, S.L. 129

Zeng, F. 142
Zeng, L. 142
Zeng, Y.W. 424
Zhang, C.H. 171, 424
Zhang, D.H. 692
Zhang, H.C. 221
Zhang, H.R. 253
Zhang, J. 150, 465
Zhang, J.J. 330
Zhang, J.K. 253
Zhang, J.X. 686
Zhang, L. 674
Zhang, M.J. 209
Zhang, P. 77
Zhang, Q.Y. 362
Zhang, S. 64, 156, 564
Zhang, T. 264
Zhang, W.R. 537

Zhang, X.N. 178
Zhang, X.W. 287
Zhang, X.Z. 99, 274
Zhang, Y. 85, 150, 235, 686
Zhang, Z.H. 47, 512
Zhao, D.K. 129
Zhao, H. 323
Zhao, J.P. 606
Zhao, M.Q. 216
Zhao, Y.C. 99, 274
Zheng, L.M. 348
Zhou, D.W. 280
Zhou, H.X. 472
Zhou, L. 120
Zhou, X.H. 557
Zhou, Z.Y. 171
Zhu, G.P. 480
Zhu, G.Q. 348
Zhu, Q.-J. 640
Zhu, R.H. 557
Zhu, Y.Y. 355, 586
Zong, S. 432, 600